The Changing World Religion Map

Stanley D. Brunn

Editor

The Changing World Religion Map

Sacred Places, Identities, Practices
and Politics

Volume V

Donna A. Gilbreath
Assistant Editor

 Springer

Editor
Stanley D. Brunn
Department of Geography
University of Kentucky
Lexington, KY, USA

Assistant Editor
Donna A. Gilbreath
UK Markey Cancer Center
Research Communications Office
Lexington, KY, USA

ISBN 978-94-017-9375-9 ISBN 978-94-017-9376-6 (eBook)
DOI 10.1007/978-94-017-9376-6

Library of Congress Control Number: 2014960060

Springer Dordrecht Heidelberg New York London

Printed on acid-free paper

Springer Science+Business Media B.V. Dordrecht is part of Springer Science+Business Media (www.springer.com)

Preface: A Continuing Journey

Religion has always been a part of my life. I am a Presbyterian PK (preacher's kid). From my father I inherited not only an interest in the histories and geographies of religions, not just Christianity, but also a strong sense of social justice, a thread that has been part of my personal and professional (teaching, research, service) life. My mother was raised as a Quaker and from her I also learned much about social justice, peace and reconciliation and being a part of an effective voice calling for ends to war, social discrimination of various types, and other injustices that seem to be a continual part of daily life on the planet. My father had churches mostly in the rural Upper Middle West. These were open country and small town congregations in Illinois, Wisconsin, Minnesota, South Dakota, Nebraska, and Missouri. The members of these congregations were Germans, Czech, Scandinavians (Norwegians, and Swedes), and English. Perhaps or probably because of these experiences, I had friendships with many young people who comprised the mosaic of the rural Middle West. Our family moved frequently when I was living at home, primarily because my father's views on social issues were often not popular with the rural farming communities. (He lost his church in northwest Missouri in 1953 because he supported the Supreme Court's decision on desegregation of schools. By the time I graduated from high school in a small town in southeastern Illinois, I had attended schools in a half-dozen states; these include one-room school house experiences as well as those in small towns.

During my childhood days my interests in religion were, of course, important in the views I had about many subjects about those of different faiths and many places on the planet. I was born in a Catholic hospital, which I always attribute to the beginning of my ecumenical experiences. The schools I attended mixes of Catholics and Protestants; I had few experiences with Native Americans, Jews and African, and Asian Americans before entering college. But that background changed, as I will explain below. My father was always interested in missionaries and foreign missions and once I considered training for a missionary work. What fascinated me most about missionaries were that they were living in distant lands, places that I just longed to know about; an atlas was always my favorite childhood book, next to a dictionary. I was always glad when missionaries visited our churches and stayed in our homes. The fascination extended to my corresponding with missionaries in Africa, Asia, and Latin America. I was curious what kind of work they did. I also found them a source for stamps, a hobby that I have pursued since primary school. Also I collected the call letters of radio stations, some which were missionary stations, especially in Latin America. (Some of these radio stations are still broadcasting.)

When I enrolled as an undergraduate student at Eastern Illinois University, a small regional university in east central Illinois, I immediately requested roommates from different countries. I very much wanted to make friends with students from outside the United States and learn about their culture. During my 3 years at EIU, I had roommates from Jordan, Samoa, Costa Rica, Ethiopia, and South Korea; these were very formative years in helping me understand cross-cultural, and especially, religious diversity. On reflection, I think that most of the Sunday services I attended were mostly Presbyterian and Methodist, not Catholic, Lutheran, or Baptist. When I entered the University of Wisconsin, Madison, for the M.A. degree, I was again exposed to some different views about religion. The Madison church that fascinated me the most was the Unitarian church, a building designed by Frank Lloyd Wright. I remember how different the services and sermons were from Protestant churches, but intellectually I felt at home. My father was not exactly pleased I found the Unitarian church a good worship experience. The UW-Madison experience also introduced me to the study of geography and religion. This was brought home especially in conversations with my longtime and good friend, Dan Gade, but also a cultural geography course I audited with Fred Simoons, whose new book on religion and food prejudices just appeared and I found fascinating. Also I had conversations with John Alexander, who eventually left the department to continue in his own ministry with the Inter Varsity Christian Fellowship. A seminar on Cultural Plant Geography co-taught with Fred Simoons, Jonathan Sauer, and Clarence Olmstead provided some opportunities to explore cultural and historical dimensions of religion, which were the major fields where geographers could study religion. The geographers I knew who were writing about religion were Pierre Deffontaines, Eric Issac, and Xavier de Phanol. That narrow focus, has, of course, changed in the past several decades, as I will discuss below.

The move to Ohio State University for my doctoral work did not have the strong religious threads that had emerged before. I attended a variety of Protestant churches, especially Presbyterian, Congregational, and Methodist. I took no formal courses in geography that dealt with religion, although I was very interested when Wilbur Zelinsky's lengthy article on church membership patterns appeared in the *Annals of the Association of American Geographers* in 1961. I felt then that this was, and would be, a landmark study in American human geography, as the many maps of denominational membership patterns plus extensive references would form the basis for future scholars interested in religion questions, apart from historical and cultural foci which were the norm at that time. My first article on religion was on religious town names; I wrote it when I was at Ohio State with another longtime friend, Jim Wheeler, who had little interest in religion. I can still remember using my knowledge of biblical place names and going through a Rand McNally atlas with Jim identifying these town names. This study appeared in *Names*, which cultural geographers acknowledge is one of the premier journals concerned with names and naming processes. Even though my dissertation on changes in the central place functions in small towns in northwest Ohio and southeast Ohio (Appalachia) did not look specifically at churches, I did tabulate the number and variety during extensive fieldwork in both areas.

My first teaching job was at the University of Florida in fall 1966. I decided once I graduated from OSU that I wanted to live in a different part of the United States where I could learn about different regional cultures and politics. I was discouraged by some former teachers about teaching in Florida, especially about the region's segregation history, recent civil rights struggles in the South and also the John Birch Society (which was also active in Columbus when I lived there). The 3 years (1966–1969) in Gainesville were also very rewarding years. These were also very formative years in developing my interests in the social geography, a new field that was just beginning to be studied in the mid-1960s. Included in the forefront of this emerging field of social geographers were Anne Buttimer, Paul Claval, Yi-Fu Tuan, Dick Morrill, Richard Peet, Bill Bunge, Wilbur Zelinsky, David Harvey, and David Smith, all who were challenging geographers to study the social geographies of race, employment, school and housing discrimination, but also poverty, environmental injustice, inequities in federal and state programs promoting human welfare, the privileges of whiteness and the minorities' participation in the voting/political process. Living in northern Florida in the late 1960s or "Wallace years" could not help but alert one to the role that religion was playing in rural and urban areas in the South. Gainesville had distinct racial landscapes. I was definitely a "northerner" and carpetbagger who was an outcast in many ways in southern culture. One vivid memory is attending a University of Florida football game (a good example of regional pride and nationalism) and being about the only person seated while the band played "Dixie." I joined a Congregational/United Church of Christ church which was attended by a small number of "northern faculty" who were supportive of initiatives to end discriminatory practices at local, university, and broader levels. At this time I also was learning about the role of the Southern Baptist Church, a bastion of segregation that was very slow to accommodate to the wishes of those seeking ends to all kinds of overt and subtle discrimination (gender, race, class) practices. The term Bible Belt was also a label that rang true; it represented, as it still does, those who adhere to a literal interpretation of the Bible, a theological position I have never felt comfortable. I soon realized that if one really wanted to make a difference in the lives of those living with discrimination, poverty, and ending racial disenfranchisement in voting, religion was a good arena to express one's feelings and work with others on coordinated efforts. Published research that emerged from my Florida experiences included studies on poverty in the United States (with Jim Wheeler), the geographies of federal outlays to states, an open housing referendum in Flint, Michigan (with Wayne Hoffman), and school levies in cities that illustrated social inequities (with Wayne Hoffman and Gerald Romsa). My Florida years also provided me the first opportunity to travel in the developing world; that was made possible with a summer grant where I visited nearly 15 different Caribbean capitals where I witnessed housing, social, and infrastructure gaps. This experience provided my first experiences with the developing world and led to a Cities of the World class I taught at Michigan State University and also co-edited several editions with the same title of a book with Jack Williams.

The 11 years at Michigan State University did not result in any major research initiatives related to religion, although it did broaden my horizons about faiths other than Christianity. I began to learn about Islam, especially from graduate students in the department from Saudi Arabia, Libya, Kuwait, and Iran. Many of these I advised on religion topics about their own cultures, especially those dealing with pilgrimages and sacred sites. Probably the main gain from living in Michigan was support for and interest in an emerging secular society. The religious "flavors" of Michigan's religious landscape ran the full gamut from those who were very traditional and conservative to those who were globally ecumenical, interfaith, and even agnostic. I continued to be active in Presbyterian and United Churches of Christ, both which were intellectually and spiritually challenging places for adult classes and singing in a choir.

When I moved to the University of Kentucky in 1980, I knew that living in the Bluegrass State would be different from Michigan in at least two respects. One is that Kentucky was considered a moderate to progressive state with many strong traditional and conservative churches, especially the Southern Baptist denomination. Zelinsky's map accurately portrayed this region as having a dominance of conservative and evangelical Protestantism. Second, I realized that for anyone interested in advancing social issues related to race and gender equality or environmental quality (especially strip mining in eastern Kentucky), there would likely be some conflicts. I also understood before coming to Kentucky that alcoholic beverage consumption was a big issue in some countries; that fact was evident in an innovative regional map Fraser Hart prepared in a small book about the South. And then there was the issue about science and religion in school curricula. With this foreknowledge, I was looking forward to living in a region where the cross-currents of religion and politics meshed, not only experiencing some of these social issues or schisms firsthand, but also having an opportunity to study them, as I did.

I realized when I moved to Lexington, it was in many ways and still is a slowly progressing socially conscious city. Southern Baptist churches, Christian churches, and Churches of Christ were dominant in the landscape and in their influences on social issues. One could not purchase alcoholic beverages on Sunday in restaurants until a couple referenda were passed in the mid-1980s that permitted sales. I think 90 of the state's 120 counties were officially dry, although everyone living in a dry county knew where to purchase liquor. One could not see the then-controversial "The Last Temptation of Christ" movie when it appeared unless one would drive three hours to Dayton. "Get Right with God" signs were prominent along rural highways. The University of Kentucky chimes in Memorial Hall on campus played religious hymns until this practice stopped sometime in the middle of the decade; I am not exactly sure why. Public schools had prayers before athletic events; some still do. Teachers in some public schools could lose their jobs if they taught evolution. Creationism was (and still is) alive and well. I was informed by university advisors that the five most "dangerous" subjects to new UK students were biology, anthropology, astronomy, geology, and physical geography. Students not used to other than literal biblical interpretations were confused and confounded by evolutional science. Betting on horses was legal, even though gambling was frowned on by some religious leaders. Cock fighting and snake handling still existed (and still do) in pockets in rural eastern Kentucky. In many ways living in Kentucky was like living "on the dark side of gray." Lexington in many ways was and still is an island or outlier. Desegregation was a slow moving process in a city with a strong southern white traditional heritage. Athletic programs were also rather slow to integrate, especially UK basketball. In short, how could one not study religion in such an atmosphere. Living in Kentucky is sort of the antipode to living in agnostic-thriving New England and Pacific Northwest. I would expect that within 100 miles of Lexington one would discover one of the most diverse religious denominational and faith belief landscapes in the United States. There are the old regular mainline denominations, new faiths that have come into the Bluegrass and also many one-of-a-kind churches, especially in rural eastern, southern, and southeastern Kentucky.

I have undertaken a number of studies related to religion in Kentucky and the South in the past three decades. Some of these have been single-authored projects, others with students and faculty at UK and elsewhere. Some were presentations at professional meetings; some resulted in publications. The topics that fascinated me were ones that I learned from my geography colleagues and those in other disciplines that were understudied. These include the history and current patterns of wet/dry counties in Kentucky, a topic that appears in local and statewide media with communities deciding whether to approve the sale of alcoholic beverages. This study I conducted with historian Tom Appleton. With regularity there were clergy of some fundamentalist denominations who decried the sale of such drinks; opposing these clergy and their supporters were often those interested in promoting tourism and attracting out-of-state traffic on interstates. Also I looked into legislation that focused on science/education interfaces in the public schools and on the types of religious books (or avoidance of such, such as dealing with Marx, Darwin, and interfaith relations) in county libraries. Craig Campbell and I published an article in *Political Geography* on Cristo Redentor (Christ of the Andes statue) as an example of differential locational harmony. At the regional level I investigated with Esther Long the mission statements of seminaries in the South, a study that led to some interesting variations not only in their statements, but course offerings and visual materials on websites. I published with Holly Barcus two articles in *Great Plains Research* about denominational changes in the Great Plains. Missionaries have also been relatively neglected in geography, so I embarked on a study with Elizabeth Leppman that looked at the contents of a leading Quaker Journal in the early part of the past century. Religions magazines, as we acknowledged in our study, were (and probably still are) a very important medium for educating the public about places and cultures, especially those where most Americans would have limited first-hand knowledge. The music/religion interface has long fascinated me, not only as a regular choir member, but for the words used to convey messages about spirituality, human welfare and justice, religious traditions and promises of peace and hope.

After 11 September 2001, I collected information from a number of churches in eastern Kentucky about how that somber event was celebrated and also what hymns they sung on the tenth anniversary. As expected, some were very somber and dignified, others had words about hope, healing, and reaching across traditional religious boundaries that separate us. I also co-authored an article (mostly photos) in *Focus* on the Shankill-Falls divided between Catholic and Protestant areas of Belfast with three students in my geography of religion class at the National University of Ireland in Maynooth. The visualization theme was integral to a paper published in *Geographica Slovenica* on ecumenical spaces and the web pages of the World Council of Churches and papers I delivered how cartoonists depicted the controversial construction of a mosque at Ground Zero. How cartoonists depicted God-Nature themes (the 2011 Haitian earthquake and Icelandic volcanic eruption) were the focus of an article in *Mitteilungen der Ősterreichsten Geographischen Gesellschaft*. I published in *Geographical Review* an article how the renaissance of religion in Russia is depicted on stamp issues since 1991. A major change in my thinking about the subject of religion in the South was the study that I worked on with Jerry Webster and Clark Archer, a study that appeared in the *Southeastern Geographer* in late 2011. We looked at the definition and concept of the Bible Belt as first discussed by Charles Heatwole (who was in my classes when I taught at Michigan State University) in 1978 in the *Journal of Geography*. We wanted to update his study and learn what has happened to the Bible Belt (or Belts) since this pioneering effort. What we learned using the Glenmary Research Center's county data on adherents for the past several decades was that the "buckle" has relocated. As our maps illustrated, the decline in those counties with denominations adhering to a literal interpretation of the Bible in western North Carolina and eastern Tennessee and a shift to the high concentration of Bible Belt counties in western Oklahoma and panhandle Texas. In this study using Glenmary data for 2000, we also looked at the demographic and political/voting characteristics of these counties. (In this volume we look at the same phenomenon using 2010 data and also discuss some of the visual features of the Bible Belt landscapes.)

What also was instrumental in my thinking about religion and geography interfaces were activities outside my own research agenda. As someone who has long standing interests in working with others at community levels on peace and justice issues, I worked with three other similarly committed adults in Lexington to organize the Central Kentucky Council for Peace and Justice. CKCPJ emerged in 1983 as an interfaith and interdenominational group committed to working on peace and justice issues within Lexington, in Central Kentucky especially, but also with national and global interests. The other three who were active in this initiative were Betsy Neale (from the Friends), Marylynne Flowers (active in a local Presbyterian church) and Ernie Yanarella (political scientist, Episcopalian, longtime friend, and also contributor to a very thoughtful essay on Weber in this volume). This organization is a key agent in peace/justice issues in the Bluegrass; it hosts meetings, fairs, conferences, and other events for people of all ages, plans annual marches on Martin Luther King Jr. holiday, and is an active voice on issues related to capital punishment, gun control, gay/lesbian issues, fair trade and employment, environmental responsibility and stewardship, and the rights of women, children, and minorities.

I also led adult classes at Maxwell Street Presbyterian Church where we discussed major theologians and religious writers, including William Spong, Marcus Borg, Joseph Campbell, Philip Jenkins, Diane Eck, Kathleen Norris, Diana Butler Bass, Francis Collins, Sam Harris, Paul Alan Laughlin, James Kugel, Dorothy Bass, and Garry Wills. We discuss issues about science, secularism, death and dying, interfaith dialogue, Christianity in the twenty-first century, images of God, missions and missionaries, and more. I also benefitted from attending church services in the many countries I have traveled, lived, and taught classes in the past three decades. These include services in elaborate, formal, and distinguished cathedrals in Europe, Russian Orthodox services in Central Asia, and services in a black township and white and interracial mainline churches in Cape Town. Often I would attend services where I understood nothing or little, but that did not diminish the opportunity to worship with youth and elders (many more) on Sunday mornings and listen to choirs sing in multiple languages. These personal experiences also became part of my religion pilgrimage.

While religion has been an important part of my personal life, it was less important as part of my teaching program. Teaching classes on the geography of religion are few and far between in the United States; I think the subject was accepted much more in the instructional and research arenas among geographers in Europe. I think that part of my reluctance to pursue a major book project on religion was that for a long time I considered the subject too narrowly focused, especially on cultural and historical geography. From my reading of the geography and religion literature, there were actually few studies done before 1970s. (See the bibliography at the end of Chap. 1). I took some renewed interest in the subject in the mid-1980s when a number of geographers began to examine religion/nature/environment issues. The pioneering works of Yi-Fu Tuan and Anne Buttimer were instrumental in steering the study of values, ethics, spirituality, and religion into some new and productive directions. These studies paved the way for a number of other studies by social geographers (a field that was not among the major fields until the 1970s and early 1980s). The steady stream of studies on geography and religion continued with the emergence of GORABS (Geography of Religion and Belief Systems) as a Specialty Group of the Association of American Geographers. The publication of more articles and special journal issues devoted to the geography of religion continued into the last decade of the twentieth century and first decade of this century. The synthetic works of Lily Kong that have regularly appeared in *Progress in Human Geography* further supported those who wanted to look at religion from human/environmental perspectives. These reviews not only introduced the study of religion within geography, but also to those in related scholarly disciplines.

As more and more research appeared in professional journals and more conferences included presentations on religion from different fields and subfields, it became increasing apparent that the time was propitious for a volume that looked at religion/geography interfaces from a number of different perspectives. From my own vantage point, the study of religion was one that could, should, might, and would benefit from those who have theoretical and conceptual training in many of the discipline's major subfields. The same applied to those who were regional specialists; there were topics meriting study from those who looked a political/religion issues in Southeast Asia or Central America as well as cultural/historical themes in southern Africa and continental Europe and symbolic/architectural features and built environments of religions landscapes in California, southeast Australia, and southwest Asia. Studying religions topics would not have to be limited to those in human geography, but could be seen as opportunities for those studying religion/natural disaster issues in East Asia and southeast United States as well as the spiritual roots of early and contemporary religious thinking in Central Asia, East Asia, Russia, and indigenous groups in South America. For those engaged in the study of gender, law, multicultural education, and media disciplines, there were also opportunities to contribute to the study of this emerging field. In short, there were literally "gold mines" of potential research topics in rural and urban areas everywhere on the continent.

About 7 years ago I decided to offer a class on the geography of religion in the Department of Geography at the University of Kentucky. The numbers were never larger (less than 15), but these were always enlightening and interesting, because of the views expressed by students. Their views about religion ran the gamut from very conservative to very liberal and also agnostic and atheist, which made, as one would expect, some very interesting exchanges. Students were strongly encouraged (not required, as I could not do this in a public university) to attend a half dozen different worship services during the semester. This did not mean attending First Baptist, Second Baptist, Third Baptist, etc., but different kinds of experiences. For some this course component was the first some had ever attended a Jewish synagogue, Catholic mass, Baptist service, an African American church, Unitarian church, or visited a mosque. Some students used this opportunity to attend Wiccan services, or visit a Buddhist and Hindu temple. Their write-ups about these experiences and the ensuing discussion were one of the high spots of the weekly class. In addition, the classes discussed chapters in various books and articles from the geography literature about the state of studying religion. And we always discussed current news items, using materials from the RNS (Religion News Service) website.

Another ingredient that stimulated my decision to edit a book on the geography of religion emerged from geography of religion conferences held in Europe in recent years. These were organized by my good friends Ceri Peach (Oxford), Reinhard Henkel (Heidelberg), and also Martin Baumann and Andreas Tunger-Zanetti (University of Lucerne). These miniconferences, held in Oxford, Lucerne, and Gottingen, usually attracted 20–40 junior and senior geographers and other religious scholars, and were a rich source of ideas for topics that might be studied. The opportunities for small group discussions, the field trips, and special events were conducive to learning about historical and contemporary changes in the religious landscapes of the European continent and beyond. A number of authors contributing to this volume presented papers at one or more of these conferences. Additional names came from those attending sessions at annual meetings of the Association of American Geographers.

Some of my initial thoughts and inspiration about a book came from the course I taught, conversations with friends who studied and did not study religion, and also the book I edited on megaengineering projects. This three-volume, 126-chapter book, *Engineering Earth: The Impacts of Megaengineering Projects*, was published in 2011 by Springer. There were only a few chapters in this book that had a religious content, one on megachurches, another on liberation theologians fighting megadevelopment projects in the Philippines and Guatemala. When I approached Evelien Bakker and Bernadette Deelen-Mans, my first geography editors at Springer, about a religion book, they were excited and supportive, as they have been since day one. They gave me the encouragement, certainly the latitude (and probably the longitude) to pursue the idea, knowing that I would identify significant cutting-edge topics about religion and culture and society in all major world regions. The prospectus I developed was for an innovative book that would include the contributions of scholars from the social sciences and humanities, those from different counties and those from different faiths. For their confidence and support, I am very grateful. The reviews they obtained of the prospectus were encouraging and acknowledging that there was a definite need for a major international, interdisciplinary, and interfaith volume. Springer also saw this book as an opportunity to emphasize its new directions in the social sciences and humanities. I also want to thank Stefan Einarson who came on board late in the project and shepherded the project to its completion with the usual Springer traits of professionalism, kindness, and commitment to the project's publication. And I wish to thank Chitra Sundarajan and her staff for helpful professionalism in preparing the final manuscript for publication.

The organization of the book, which is discussed in Chapter One, basically reflects the way I look at religion from a geographical perspective. I look at the subject as more than simply investigations into human geography's fields and subfields, including cultural and historical, but also economic, social, and political geography, but also human/environmental geography (dealing with human values, ethics, behavior, disasters, etc.). I also look at the study of religious topics and phenomena with respects to major concepts we use in geography; these include landscapes, networks, hierarchies, scales, regions, organization of space, the delivery of services, and virtual religion. I started contacting potential authors in September 2010. Since then I have sent or received over 15,000 emails related the volume.

I am deeply indebted to many friends for providing names of potential authors. I relied on my global network of geography colleagues in colleges and universities around the world, who not only recommended specific individuals, but also topics they deemed worthy of inclusion. Some were geographers, but many were not; some taught in universities, others in divinity schools and departments of religion around the world. Those I specifically want to acknowledge include: Barbara Ambrose, Martin Checa Artasu, Martin Baumann, John Benson, Gary Bouma, John Benson, Dwight Billings, Marion Bowman, John D. Brewer, David Brunn, David Butler, Ron Byars, Heidi Campbell, Caroline Creamer, Janel Curry, David Eicher, Elizabeth Ferris, Richard Gale, Don Gross, Wayne Gnatuk, Martin Haigh, Dan Hofrenning, Wil Holden, Hannah Holtschneider, Monica Ingalls, Nicole Karapanagiotis, Aharon Kellerman, Judith Kenny, Jean Kilde, Ted Levin, James

Munder, Alec Murphy, Tad Mutersbuagh, Garth Myers, Lionel Obidah, Sam Otterstrom, Francis Owusu, Maria Paradiso, Ron Pen, Ivan Petrella, Adam Possamai, Leonard Primiano, Craig Revels, Heinz Scheifinger, Anna Secor, Ira Sheskin, Doug Slaymaker, Patricia Solis, Anita Stasulne, Jill Stevenson, Robert Strauss, Tristan Sturm, Greg Stump, Karen Till, Andreas Tunger-Zenetti, Gary Vachicouras, Viera Vlčkova, Herman van der Wusten, Stanley Waterman, Mike Whine, Don Zeigler, Shangyi Zhou, and Matt Zook.

And I want to thank John Kostelnick who provided the GORABS Working Bibliography; most of the entries, except dissertations and theses, are included in Chap. 1 bibliography. Others who helped him prepare this valuable bibliography also need to be acknowledged: John Bauer, Ed Davis, Michael Ferber, Julian Holloway, Lily Kong, Elizabeth Leppman, Carolyn Prorock, Simon Potter, Thomas Rumney, Rana P.B. Singh, and Robert Stoddard. These are scholars who devoted their lifetimes to advancing research on geography and religion.

Finally I want to thank Donna Gilbreath for another splendid effort preparing all the chapters for Springer. She formatted the chapters and prepared all the tables and illustrations per the publisher's guidelines. Donna is an invaluable and skilled professional who deserves much credit for working with multiple authors and the publisher to ensure that all text materials were correct and in order. Also I am indebted to her husband, Richard Gilbreath, for helping prepare some of the maps and graphics for authors without cartographic services and making changes on others. As Director of the Gyula Pauer Center for Cartography and GIS, Dick's work is always first class. And, finally, thanks are much in order to Natalya Tyutenkova for her interest, support, patience, and endurance in the past several years working on this megaproject, thinking and believing it would never end.

The journey continues.

February 2014   Stanley D. Brunn

Contents of Volume V

Contributors

Jamaine Abidogun Department of History, Missouri State University, Springfield, MO, USA

Afe Adogame School of Divinity, University of Edinburgh, Edinburgh, Scotland, UK

Christopher A. Airriess Department of Geography, Ball State University, Muncie, IN, USA

Kaarina Aitamurto Aleksanteri Institute, University of Helsinki, Helsinki, Finland

Mikael Aktor Institute of History, Study of Religions, University of Southern Denmark, Odense, Denmark

Elizabeth Allison Department of Philosophy and Religion, California Institute of Integral Studies, San Francisco, CA, USA

Johan Andersson Department of Geography, King's College London, London, UK

Stephen W. Angell Earlham College, School of Religion, Richmond, IN, USA

J. Clark Archer Department of Geography, School of Natural Resources, University of Nebraska-Lincoln, Lincoln, NE, USA

Ian Astley Asian Studies, University of Edinburgh, Edinburgh, UK

Steven M. Avella Professor of History, Marquette University, Milwaukee, WI, USA

Yulier Avello COPEXTEL S.A., Ministry of Informatics and Communications, Havana, Cuba

Erica Baffelli School of Arts, Languages and Cultures, University of Manchester, Manchester, UK

Bakama BakamaNume Division of Social Work, Behavioral and Political Science, Prairie View A&M University, Prairie View, TX, USA

Josiah R. Baker Methodist University, Fayetteville, NC, USA

Economics and Geography, Methodist University, Fayetteville, USA

Holly R. Barcus Department of Geography, Macalester College, St. Paul, MN, USA

David Bassens Department of Geography, Free University Brussels, Brussels, Belgium

Ramon Bauer Wittgenstein Centre for Demography and Global Human Capital (IIASA, VID/ÖAW, WU), Vienna Institute of Demography/Austrian Academy of Sciences, Vienna, Austria

Whitney A. Bauman Department of Religious Studies, Florida International University, Miami, FL, USA

Gwilym Beckerlegge Department of Religious Studies, The Open University, Milton Keynes, UK

Michael Bégin Department of Global Studies, Pusan National University, Pusan, Republic of Korea

Demyan Belyaev Collegium de Lyon/Institute of Advanced Studies, Lyon, France

Alexandre Benod Research Division, Department of Japanese Studies, Université de Lyon, Lyon, France

John Benson School of Teaching and Learning, Minnesota State University, Moorhead, MN, USA

Sigurd Bergmann Department of Philosophy and Religious Studies, Norwegian University of Science and Technology, Trondheim, Norway

Rachel Berndtson Department of Geographical Sciences, University of Maryland, College Park, MD, USA

Martha Bettis Gee Compassion, Peace and Justice, Peace and Justice Ministries, Presbyterian Mission Agency, Presbyterian Church (USA), Louisville, KY, USA

Warren Bird Research Division, Leadership Network, Dallas, TX, USA

Andrew Boulton Department of Geography, University of Kentucky, Lexington, KY, USA

Humana, Inc., Louisville, KY, USA

Kathleen Braden Department of Political Science and Geography, Seattle Pacific University, Seattle, WA, USA

Namara Brede Department of Geography, Macalester College, St. Paul, MN, USA

John D. Brewer Institute for the Study of Conflict Transformation and Social Justice, Queen's University Belfast, Belfast, UK

Laurie Brinklow School of Geography and Environmental Studies, University of Tasmania, Hobart, Australia

Interim Co-ordinator, Master of Arts in Island Studies Program, University of Prince Edward Island, Charlottetown, PE Canada

Dave Brunn Language and Linguistics Department, New Tribes Missionary Training Center, Camdenton, MO, USA

Stanley D. Brunn Department of Geography, University of Kentucky, Lexington, KY, USA

David J. Butler Department of Geography, University of Ireland, Cork, Ireland

Anne Buttimer Department of Geography, University College Dublin, Dublin, Ireland

Éric Caron Malenfant Demography Division, Statistics Canada, Ottawa, Canada

Lori Carter-Edwards Gillings School of Global Public Health, Public Health Leadership Program, University of North Carolina, Chapel Hill, NC, USA

Clemens Cavallin Department of Literature, History of Ideas and Religion, University of Gothenburg, Göteborg, Sweden

Martin M. Checa-Artasu Department of Sociology, Universidad Autónoma Metropolitana, Unidad Iztapalapa, Mexico, DF, Mexico

Richard Cimino Department of Anthropology and Sociology, University of Richmond, Richmond, VA, USA

Paul Claval Department of Geography, University of Paris-Sorbonne, Paris, France

Paul Cloke Department of Geography, Exeter University, Exeter, UK

Kevin Coe Department of Communication, University of Utah, Salt Lake City, UT, USA

Noga Collins-Kreiner Department of Geography and Environmental Studies, Centre for Tourism, Pilgrimage and Recreation, University of Haifa, Haifa, Israel

Louise Connelly Institute for Academic Development, University of Edinburgh, Edinburgh, Scotland, UK

Thia Cooper Department of Religion, Gustavus Adolphus College, St. Peter, MN, USA

Catherine Cottrell Department of Geography and Earth Sciences, Aberystwyth University, Aberystwyth, UK

Thomas W. Crawford Department of Geography, East Carolina University, Greenville, NC, USA

Janel Curry Provost, Gordon College, Wenham, MA, USA

Seif Da'Na Sociology and Anthropology Department, University of Wisconsin-Parkside, Kenosha, WI, USA

Erik Davis Department of Religious Studies, Rice University, Houston, TX, USA

Jenny L. Davis Department of American Indian Studies, University of Illinois, Urbana-Champaign, Urbana, USA

Kiku Day Department of Ethnomusicology, Aarhus University, Aarhus, Denmark

Renée de la Torre Castellanos Centro de Investigaciones y Estudios Superiores en Antropologia Social-Occidente, Guadalajara, Jalisco, Mexico

Frédéric Dejean Institut de recherche sur l'intégration professionnelle des immigrants, Collège de Maisonneuve, Montréal (Québec), Canada

Veronica della Dora Department of Geography, Royal Holloway University of London, UK

Sergio DellaPergola The Avraham Harman Institute of Contemporary Jewry, The Hebrew University of Jerusalem, Mt. Scopus, Jerusalem, Israel

Antoinette E. DeNapoli Religious Studies Department, University of Wyoming, Laramie, WY, USA

Matthew A. Derrick Department of Geography, Humboldt State University, Arcata, CA, USA

C. Nathan DeWall Department of Psychology, University of Kentucky, Lexington, KY, USA

Jualynne Dodson Department of Sociology, American and African Studies Program, Michigan State University, East Lansing, MI, USA

David Domke Department of Communication, University of Washington, Seattle, WA, USA

Katherine Donohue M.A. Diplomacy and International Commerce, Patterson School of Diplomacy and International Commerce, University of Kentucky, Lexington, KY, USA

Lizanne Dowds Northern Ireland Life and Times Survey, University of Ulster, Belfast, UK

Kevin M. Dunn School of Social Sciences and Psychology, University of Western Sydney, Penrith, NSW, Australia

Claire Dwyer Department of Geography, University College London, London, UK

Patricia Ehrkamp Department of Geography, University of Kentucky, Lexington, KY, USA

Paul Emerson Teusner School of Media and Communication, RMIT University, Melbourne, VIC, Australia

Chad F. Emmett Department of Geography, Brigham Young University, Provo, UT, USA

Ghazi-Walid Falah Department of Public Administration and Urban Studies, University of Akron, Akron, OH, USA

Yasser Farrés Department of Philosophy, University of Zaragoza, Pedro Cerbuna, Zaragoza, Spain

Timothy Joseph Fargo Department of City Planning, City of Los Angeles, Los Angeles, CA, USA

Michael P. Ferber Department of Geography, The King's University College, Edmonton, AB, Canada

Tatiana V. Filosofova Department of World Languages, Literatures, and Cultures, University of North Texas, Denton, TX, USA

John T. Fitzgerald Department of Theology, University of Notre Dame, Notre Dame, IN, USA

Colin Flint Department of Political Science, Utah State University, Logan, UT, USA

Daniel W. Gade Department of Geography, University of Vermont, Burlington, VT, USA

Armando Garcia Chiang Department of Sociology, Universidad Autónoma Metropolitana Iztapalapa, Iztapalapa, Mexico

Jeff Garmany King's Brazil Institute, King's College London, London, UK

Martha Geores Department of Geographical Sciences, University of Maryland, College Park, MD, USA

Hannes Gerhardt Department of Geosciences, University of West Georgia, Carrolton, GA, USA

Christina Ghanbarpour History Department, Saddleback College, Mission Viejo, CA, USA

Danilo Giambra Department of Theology and Religion, University of Otago-Te Whare Wānanga o Otāgo, Dunedin, New Zealand/Aotearoa

Banu Gökarıksel Department of Geography, University of North Carolina, Chapel Hill, NC, USA

Margaret M. Gold London Guildhall Faculty of Business and Law, London Metropolitan University, London, UK

Anton Gosar Faculty of Tourism Studies, University of Primorska, Portorož, Slovenia

Anne Goujon Wittgenstein Centre for Demography and Global Human Capital (IIASA, VID/ÖAW, WU), International Institute for Applied Systems Analysis (IIASA), Laxenburg, Austria

Vienna Institute of Demography/Austrian Academy of Sciences, Vienna, Austria

Alyson L. Greiner Department of Geography, Oklahoma State University, Stillwater, OK, USA

Daniel Jay Grimminger Faith Lutheran Church, Kent State University, Millersburg, OH, USA

School of Music, Kent State University, Kent, OH, USA

Zeynep B. Gürtin Department of Sociology, University of Cambridge, Cambridge, UK

Cristina Gutiérrez Zúñiga Centro Universitario de Ciencias Sociales y Humanidades, El Colegio de Jalisco, Zapopan, Jalisco, Mexico

Martin J. Haigh Department of Social Sciences, Oxford Brookes University, Oxford, UK

Anna Halafoff Centre for Citizenship and Globalisation, Deakin University, Burwood, VIC, Australia

Airen Hall Department of Theology, Georgetown University, Washington, DC, USA

Randolph Haluza-DeLay Department of Sociology, The Kings University, Edmonton, AB, Canada

Tomáš Havlíček Faculty of Science, Department of Social Geography and Regional Development, Charles University, Prague 2, Czechia

C. Michael Hawn Sacred Music Program, Perkins School of Theology, Southern Methodist University, Dallas, TX, USA

Bernadette C. Hayes Department of Sociology, University of Aberdeen, Aberdeen, Scotland, UK

Peter J. Hemming School of Social Sciences, Cardiff University, Cardiff, Wales, UK

William Holden Department of Geography, University of Calgary, Calgary, AB, Canada

Edward C. Holland Havighurst Center for Russian and Post-Soviet Studies, Miami University, Oxford, OH, USA

Beverly A. Howard School of Music, California Baptist University, Riverside, CA, USA

Martina Hupková Faculty of Science, Department of Social Geography and Regional Development, Charles University, Prague 2, Czechia

Tim Hutchings Post Doc, St. John's College, Durham University, Durham, UK

Ronald Inglehart Institute of Social Research, University of Michigan, Ann Arbor, MI, USA

World Values Survey Association, Madrid, Spain

Marcia C. Inhorn Anthropology and International Affairs, Yale University, New Haven, CT, USA

Adrian Ivakhiv Environmental Program, University of Vermont, Burlington, VT, USA

Maria Cristina Ivaldi Dipartimento di Scienze Politiche "Jean Monnet", Seconda Università degli Studi di Napoli, Caserta, Italy

Thomas Jablonsky Professor of History, Marquette University, Milwaukee, WI, USA

Maria Jaschok International Gender Studies Centre, Lady Margaret Hall, Oxford University, Norham Gardens, UK

Philip Jenkins Institute for the Study of Religion, Baylor University, Waco, TX, USA

Wesley Jetton Student, University of Kentucky, Lexington, KY, USA

Shui Jingjun Henan Academy of Social Sciences, Zhengzhou, Henan Province, China

Mark D. Johns Department of Communication, Luther College, Decorah, IA, USA

James H. Johnson Jr. Kenan-Flagler Business School and Urban Investment Strategies Center, University of North Carolina, Chapel Hill, NC, USA

Lucas F. Johnston Department of Religion and Environmental Studies, Wake Forest University, Winston-Salem, NC, USA

Peter Jordan Austrian Academy of Sciences, Institute of Urban and Regional Research, Wien, Austria

Yakubu Joseph Geographisches Institut, University of Tübingen, Tübingen, Germany

Deborah Justice Yale Institute of Sacred Music, Yale University, New Haven, CT, USA

Akel Ismail Kahera College of Architecture, Art and Humanities, Clemson University, Clemson, SC, USA

P.P. Karan Department of Geography, University of Kentucky, Lexington, KY, USA

Sya Buryn Kedzior Department of Geography and Environmental Planning, Towson State University, Towson, MD, USA

Kevin D. Kehrberg Department of Music, Warren Wilson College, Asheville, NC, USA

Laura J. Khoury Department of Sociology, Birzeit University, West Bank, Palestine

Hans Knippenberg Department of Geography, Planning and International Development Studies, University of Amsterdam, Velserbroek, The Netherlands

Katherine Knutson Department of Political Science, Gustavus Adolphus College, St. Peter, MN, USA

Miha Koderman Science and Research Centre of Koper, University of Primorska, Koper-Capodistria, Slovenia

Lily Kong Department of Geography, National University of Singapore, Singapore, Singapore

Igor Kotin Museum of Anthropology and Ethnography, Russian Academy of Sciences, St. Petersburg, Russia

Katharina Kunter Faculty of Theology, University of Bochum, Bochum, Germany

Lisa La George International Studies, The Master's College, Santa Clarita, CA, USA

Shirley Lal Wijesinghe Faculty of Humanities, University of Kelaniya, Kelaniya, Sri Lanka

Ibrahim Badamasi Lambu Department of Geography, Faculty of Earth and Environmental Sciences, Bayero University Kano, Kano, Nigeria

Michelle Gezentsvey Lamy Comparative Education Research Unit, Ministry of Education, Wellington, New Zealand

Justin Lawson Health, Nature and Sustainability Research Group, School of Health and Social Development, Deakin University, Burwood, VIC, Australia

Deborah Lee Department of Geography, National University of Singapore, Singapore, Singapore

Karsten Lehmann Senior Lecturer, Science des Religions, Bayreuth University, Fribourg, Switzerland

Reina Lewis London College of Fashion, University of the Arts, London, UK

Micah Liben Judaic Studies, Kellman Brown Academy, Voorhees, NJ, USA

Edmund B. Lingan Department of Theater, University of Toledo, Toledo, OH, USA

Rubén C. Lois-González Departamento de Xeografía, Universidade de Santiago de Compostela, Galiza, Spain

Naomi Ludeman Smith Learning and Women's Initiatives, St. Paul, MN, USA

Katrín Anna Lund Department of Geography and Tourism, Faculty of Life and Environmental Sciences, University of Iceland, Reykjavik, Iceland

Avril Maddrell Department of Geography and Environmental Sciences, University of West England, Bristol, UK

Juraj Majo Department of Human Geography and Demography, Faculty of Sciences, Comenius University in Bratislava, Bratislava, Slovak Republic

Virginie Mamadouh Department of Geography, Planning and International Development Studies, University of Amsterdam, Amsterdam, The Netherlands

Mariana Mastagar Department of Theology, Trinity College, University of Toronto, Toronto, Canada

Alberto Matarán Department of Urban and Spatial Planning, University of Granada, Granada, Spain

René Matlovič Department of Geography and Applied Geoinformatics, Faculty of Humanities and Natural Sciences, University of Prešov, Prešov, Slovakia

Kvetoslava Matlovičová Department of Geography and Applied Geoinformatics, Faculty of Humanities and Natural Sciences, University of Prešov, Prešov, Slovakia

Hannah Mayne Department of Anthropology, University of Florida, Gainesville, FL, USA

Shampa Mazumdar Department of Sociology, University of California, Irvine, CA, USA

Sanjoy Mazumdar Department of Planning, Policy and Design, University of California, Irvine, CA, USA

Andrew M. McCoy Center for Ministry Studies, Hope College, Holland, MI, USA

Daniel McGowin Department of Geology and Geography, Auburn University, Auburn, AL, USA

James F. McGrath Department of Philosophy and Religion, Butler University, Indianapolis, IN, USA

Nick Megoran Department of Geography, University of Newcastle-upon-Tyne, Newcastle, UK

Amy Messer Department of Sociology, University of Kentucky, Lexington, KY, USA

Sarah Ann Deardorff Miller Researcher, Refugee Studies Centre, Oxford, UK

Kelly Miller Centre for Integrative Ecology, School of Life and Environmental Sciences, Deakin University, Burwood, VIC, Australia

Nathan A. Mosurinjohn Center for Social Research, Calvin College, Grand Rapids, MI, USA

Sven Müller Institute for Transport Economics, University of Hamburg, Hamburg, Germany

Erik Munder Institut für Vergleichende Kulturforschung - Kultur- u. Sozialanthropologie und Religionswissenschaft, Universität Marburg, Marburg, Germany

David W. Music School of Music, Baylor University, Waco, TX, USA

Kathleen Nadeau Department of Anthropology, California State University, San Bernadino, CA, USA

Caroline Nagel Department of Geography, University of South Carolina, Columbia, SC, USA

Pippa Norris John F. Kennedy School of Government, Harvard University, Cambridge, MA, USA

Government and International Relations, University of Sydney, Sydney, Australia

Orville Nyblade Makumira University College, Usa River, Tanzania

Lionel Obadia Department of Anthropology, Université de Lyon, Lyon, France

Daniel H. Olsen Department of Geography, Brigham Young University, Provo, UT, USA

Samuel M. Otterstrom Department of Geography, Brigham Young University, Provo, UT, USA

Barbara Palmquist Department of Geography, University of Kentucky, Lexington, KY, USA

Grigorios D. Papathomas Faculty of Theology, University of Athens, Athens, Greece

Nikos Pappas Musicology, University of Alabama, Tuscaloosa, AL, USA

Mohammad Aslam Parvaiz Islamic Foundation for Science and Environment (IFSE), New Delhi, India

Valerià Paül Departamento de Xeografía, Universidade de Santiago de Compostela, Galiza, Spain

Miguel Pazos-Otón Departamento de Xeografía, Universidade de Santiago de Compostela, Galiza, Spain

David Pereyra Toronto School of Theology, University of Toronto, Toronto, Canada

Bruce Phillips Loucheim School of Judaic Studies at the University of Southern California, Hebrew Union College-Jewish Institute of Religion, Los Angeles, CA, USA

Awais Piracha School of Social Sciences and Psychology, University of Western Sydney, Penrith, NSW, Australia

Linda Pittman Department of Geography, Richard Bland College of the College of William and Mary, Petersburg, VA, USA

Richard S. Pond Department of Psychology, University of North Carolina, Wilmington, NC, USA

Carolyn V. Prorok Independent Scholar, Slippery Rock, PA, USA

Steven M. Radil Department of Geography, University of Idaho, Moscow, ID, USA

Esther Long Ratajeski Independent Scholar, Lexington, KY, USA

Daniel Reeves Faculty of Science, Department of Social Geography and Regional Development, Charles University, Prague 2, Czechia

Arthur Remillard Department of Religious Studies, St. Francis University, Loretto, PA, USA

Claire M. Renzetti Department of Sociology, University of Kentucky, Lexington, KY, USA

Friedlind Riedel Department of Musicology, Georg-August-University of Göttingen, Göttingen, Germany

Sandra Milena Rios Oyola Department of Sociology and the Compromise after Conflict Research Programme, University of Aberdeen, Aberdeen, Scotland, UK

C.K. Robertson Presiding Bishop, The Episcopal Church, New York, NY, USA

Arsenio Rodrigues School of Architecture, Prairie View A&M University, Prairie View, TX, USA

Andrea Rota Institute for the Study of Religion, University of Bern, Bern, Switzerland

Rainer Rothfuss Geographisches Institut, University of Tübingen, Tübingen, Germany

Jeanmarie Rouhier-Willoughby Department of Modern and Classical Languages, Literatures and Cultures, University of Kentucky, Lexington, KY, USA

Rex J. Rowley Department of Geography-Geology, Illinois State University, Normal, IL, USA

Bradley C. Rundquist Department of Geography, University of North Dakota, Grand Forks, ND, USA

Simon Runkel Department of Geography, University of Bonn, Bonn, Germany

Joanna Sadgrove Research Staff, United Society, London, UK

Michael Samers Department of Geography, University of Kentucky, Lexington, KY, USA

Åke Sander Department of Literature, History of Ideas and Religion, University of Gothenburg, Göteborg, Sweden

Xosé M. Santos Departamento de Xeografía, Universidade de Santiago de Compostela, Galiza, Spain

Alessandro Scafi Medieval and Renaissance Cultural History, The Warburg Institute, University of London, London, UK

Anthony Schmidt Department of Communication Studies, Edmonds Community College, Edmonds, WA, USA

Mallory Schneuwly Purdie Institut de sciences sociales des religions contemporaines, Observatoire des religions en Suisse, Université de Lausanne – Anthropole, Lausanne, Switzerland

Anna J. Secor Department of Geography, University of Kentucky, Lexington, KY, USA

Hafid Setiadi Department of Geography, University of Indonesia, Depok, West Java, Indonesia

Fred M. Shelley Department of Geography and Environmental Sustainability, University of Oklahoma, Norman, OK, USA

Ira M. Sheskin Department of Geography and Regional Studies, University of Miami, Coral Gables, FL, USA

Lia Dong Shimada Conflict Mediator, Methodist Church in Britain, London, UK

Caleb Kwang-Eun Shin ABD, Korea Baptist Theological Seminary, Daejeon, Republic of Korea

J. Matthew Shumway Department of Geography, Brigham Young University, Provo, UT, USA

Dmitrii Sidorov Department of Geography, California State University, Long Beach, Long Beach, CA, USA

Caleb Simmons Religious Studies Program, University of Arizona, Tucson, AZ, USA

Devinder Singh Department of Geography, University of Jammu, Jammu, Jammu and Kashmir, India

Rana P.B. Singh Department of Geography, Banaras Hindu University, Varanasi, UP, India

Nkosinathi Sithole Department of English, University of Zululand, KwaZulu-Natal, South Africa

Vegard Skirbekk Wittgenstein Centre for Demography and Global Human Capital (IIASA, VID/ÖAW, WU), International Institute for Applied Systems Analysis, Laxenburg, Austria

Alexander Thomas T. Smith Department of Sociology, University of Warwick, Coventry, UK

Christopher Smith Independent Scholar, Tecumseh, OK, USA

Ryan D. Smith Compassion, Peace and Justice Ministries, Presbyterian Ministry at the U.N., Presbyterian Mission Agency, Presbyterian Church (USA), New York, NY, USA

Sara Smith Department of Geography, University of North Carolina, Chapel Hill, NC, USA

Leslie E. Sponsel Department of Anthropology, University of Hawaii, Honolulu, HI, USA

Chloë Starr Asian Christianity and Theology, Yale Divinity School, New Haven, CT, USA

Jeffrey Steller Public History, Northern Kentucky University, Highland Heights, KY, USA

Christopher Stephens Southlands College, University of Roehampton, London, UK

Jill Stevenson Department of Theater Arts, Marymount Manhattan College, New York, NY, USA

Anna Rose Stewart Department of Religious Studies, University of Kent, Canterbury, UK

Nancy Palmer Stockwell Senior Contract Administrator, Enerfin Resources, Houston, TX, USA

Robert Strauss President and CEO, Worldview Resource Group, Colorado Springs, CO, USA

Tristan Sturm School of Geography, Archaeology and Palaeoecology, Queen's University Belfast, Belfast, UK

Edward Swenson Department of Anthropology, University of Toronto, Toronto, ON, Canada

Anna Swynford Duke Divinity School, Duke University, Durham, NC, USA

Jonathan Taylor Department of Geography, California State University, Fullerton, CA, USA

Francis Teeney Institute for the Study of Conflict Transformation and Social Justice, Queen's University Belfast, Belfast, UK

Mary C. Tehan Stirling College, University of Divinity, Melbourne, Australia

Andrew R.H. Thompson The School of Theology, The University of the South, Sewanee, TN, USA

Scott L. Thumma Professor, Department of Sociology, Hartford Seminary, Hartford, CT, USA

Meagan Todd Department of Geography, University of Colorado, Boulder, CO, USA

Soraya Tremayne Fertility and Reproduction Studies Group, Institute of Social and Cultural Anthropology, University of Oxford, Oxford, UK

Gill Valentine Faculty of Social Sciences, University of Sheffield, Sheffield, UK

Inge van der Welle Department of Geography, Planning and International Development Studies, University of Amsterdam, Amsterdam, The Netherlands

Herman van der Wusten Department of Geography, Planning and International Development Studies, University of Amsterdam, Amsterdam, The Netherlands

Robert M. Vanderbeck Department of Geography, University of Leeds, West Yorkshire, UK

Jason E. VanHorn Department of Geography, Calvin College, Grand Rapids, MI, USA

Viera Vlčková Department of Public Administration and Regional Development, Faculty of National Economy, University of Economics in Bratislava, Bratislava, Slovakia

Geoffrey Wall Department of Geography and Environmental Management, University of Waterloo, Waterloo, ON, Canada

Robert H. Wall Counsel, Spilman Thomas & Battle, Winston-Salem, NC, USA

Kevin Ward School of Theology and Religious Studies, University of Leeds, West Yorkshire, UK

Barney Warf Department of Geography, University of Kansas, Lawrence, KS, USA

Stanley Waterman Department of Geography and Environmental Studies, University of Haifa, Haifa, Israel

Robert H. Watrel Department of Geography, South Dakota State University, Brookings, SD, USA

Gerald R. Webster Department of Geography, University of Wyoming, Laramie, WY, USA

Paul G. Weller Research, Innovation and Academic Enterprise, University of Derby, Derby, UK

Oxford Centre for Christianity and Culture, University of Oxford, Oxford, UK

Cynthia Werner Department of Anthropology, Texas A&M University, College Station, TX, USA

Geoff Wescott Centre for Integrative Ecology, School of Life and Environmental Sciences, Deakin University, Burwood, VIC, Australia

Carroll West Center for Historic Preservation, Middle Tennessee State University, Murfreesboro, TN, USA

Gerald West School of Religion, Philosophy and Classics, University of KwaZulu-Natal, Scottsville, South Africa

Mark Whitaker Department of Anthropology, University of Kentucky, Lexington, KY, USA

Thomas A. Wikle Department of Geography, Oklahoma State University, Stillwater, OK, USA

Justin Wilford Department of Geography, University of California, Los Angeles, CA, USA

Joseph Witt Department of Philosophy and Religion, Mississippi State University, Mississippi State, MS, USA

John D. Witvliet Calvin Institute of Christian Worship, Calvin College and Calvin Theological Seminary, Grand Rapids, MI, USA

Teresa Wright Department of Political Science, California State University, Long Beach, CA, USA

Ernest J. Yanarella Department of Political Science, University of Kentucky, Lexington, KY, USA

Yukio Yotsumoto College of Asia Pacific Studies, Ritsumeikan Asia Pacific University, Beppu, Oita, Japan

Samuel Zalanga Department of Anthropology, Sociology and Reconciliation Studies, Bethel University, St. Paul, MN, USA

Donald J. Zeigler Department of Geography, Old Dominion University, Virginia Beach, VA, USA

Shangyi Zhou School of Geography, Beijing Normal University, Beijing, China

Teresa Zimmerman-Liu Departments of Asian/Asian-American Studies and Sociology, California State University, Long Beach, CA, USA

Chapter 165
Islam and Assisted Reproduction in the Middle East: Comparing the Sunni Arab World, Shia Iran and Secular Turkey

Zeynep B. Gürtin, Marcia C. Inhorn, and Soraya Tremayne

165.1 Introduction

The birth of Louise Brown, the world's first *in vitro* fertilization (IVF) baby, in 1978 in the UK heralded a new dawn in the treatment of infertility. What made this baby so remarkable was the method of her conception, which occurred with techno-medical assistance outside of the female body, literally *in glass*. However, also remarkable has been this technology's rapid global spread and proliferation. In the years hence, an estimated five million "miracle babies" have been born using assisted reproductive technologies (ARTs) in many different countries. ARTs, despite the range of (financial, practical, emotional, psychological, as well as religious and moral) difficulties they may pose, have ultimately been embraced by many cultures because they address the fundamental desires of men and women to become parents. However, reproductive technologies have not been transferred into cultural voids: local considerations, be they cultural, economic, or political, have shaped and sometimes curtailed the way these Western-generated technologies are both offered to and received by non-Western subjects (Inhorn 2003a). In this chapter, we outline the global problem of infertility and the technological possibilities afforded by ARTs to remedy them. We then turn to a discussion of how Islam has responded to the novel

Z.B. Gürtin (✉)
Department of Sociology, University of Cambridge, Cambridge CB223PO, UK
e-mail: zbg20@cam.ac.uk

M.C. Inhorn
Anthropology and International Affairs, Yale University, New Haven, CT 06520, USA
e-mail: marcia.inhorn@yale.edu

S. Tremayne
Fertility and Reproduction Studies Group, Institute of Social and Cultural Anthropology, University of Oxford, Oxford OX2 6PF, UK
e-mail: soraya.tremayne@anthro.ox.ac.uk

© Springer Science+Business Media Dordrecht 2015
S.D. Brunn (ed.), *The Changing World Religion Map*,
DOI 10.1007/978-94-017-9376-6_165

ethical and social dilemmas created by assisted reproduction, and examine three different Muslim Middle Eastern contexts: the Sunni Arab world, Shi'a Iran, and secular Turkey. The comparative analysis of these contexts demonstrates that, while there are similarities there are also important differences, and thus an expectation of a monolithic "Islamic approach" to ARTs is misguided and over-simplistic. Our comparative approach of how Islam has influenced and impacted ART practice in three specific settings illuminates not only the extent to which religion must be constantly interpreted when faced with novel dilemmas, but also the inter-relationship between local and global considerations at the nexus of medicine, commerce, law, and family-making.

165.2 Infertility and Assisted Reproductive Technologies

Infertility is defined as the inability for a reproductive-aged couple to conceive a child after a year (or longer) of regular unprotected intercourse (Zegers-Hochschild et al. 2009). It is a universally occurring health problem that affects more than 80 million people worldwide, although its prevalence is subject to great global variation from less than 5 % in some countries to more than 30 % in others (Vayena et al. 2002). On average, one in ten couples will experience either primary or secondary infertility during their lifetime. The social science scholarship in this area has described stigmatization, social isolation, and gendered repercussions as central tenets of the infertility and involuntary childlessness experience; however, since the social meanings of infertility are always the discursive product of a hegemonic cultural system, the particulars of men's and women's experiences will change according to time and place, region and religion (Jenkins and Inhorn 2003).

The causes and etiology of infertility also present great global variation, from growing public health concerns surrounding delayed childbearing and age-related fertility problems in some developed countries, to the staggering frequency of iatrogenic infections in some developing nations. Indeed, a couple's involuntary childlessness can be caused by a range of physiologically male or female factors, ranging from low sperm count to azoospermia in men, and premature ovarian failure to blocked fallopian tubes in women. These underlying causes, however, may or may not receive social acknowledgment. For example, despite the widespread existence of male-factor infertility, accounting for around 50 % of all cases, globally the major blame and burden of infertility falls mostly on women's shoulders (Inhorn and Van Balen 2002). Infertile women, and increasingly infertile men, are massive users of biomedical health services. In fact, particularly in developing countries where health infrastructures are less comprehensive and resorted to less often, infertility is a leading cause of healthcare seeking behavior (Inhorn 2003b), although biomedical treatments may be accompanied or supplemented by a range of "ethnogynecological" practices (Inhorn 1994; Inhorn and Birenbaum-Carmeli 2008).

Different causes of infertility fluctuate not only in their prevalence in different regions, but also with regards to how they may be treated and how well they may

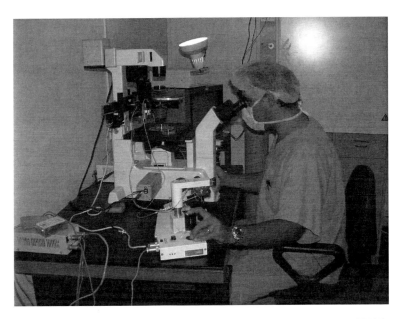

Fig. 165.1 Intracytoplasmic sperm injection (ICSI) being performed in a Lebanaese IVF laboratory (Photo by Marcia Inhorn)

respond to treatment. IVF, a technology that was originally developed to address infertility caused by blockages of the fallopian tubes - by fertilizing eggs outside the body and then transferring them directly into the uterus - is now widely used in the treatment of a range of female factor and "unexplained" fertility problems. Intra-cytoplasmic sperm injection (ICSI), a variation of IVF developed in 1991 in Belgium, involves the injection of a single sperm into the egg under a high-powered microscope, and has revolutionized the treatment of male infertility problems (Inhorn 2011a) (Figs. 165.1 and 165.2). Both of these technologies can be used in conjunction with the freezing, storage, and subsequent thawing of sperm, eggs and embryos, enabling reproductive cells to be preserved over time and to travel across geographical distances. In sum, these techniques have helped millions of involuntarily childless men and women conceive their much longed-for children.

However, IVF or ICSI with the gametes of the intending parents is not always possible; in intractable cases of male and female infertility, where viable gametes are unavailable or pregnancy to term proves impossible, couples may opt for "third-party reproductive assistance" using donor sperm, donor eggs or embryos, or com missioning a surrogate to gestate the fetus. Indications for the use of donor sperm include, for example, cases of male sterility, severe sperm abnormalities and genetic disorders. Egg donation may be indicated by genetic conditions, poor ovarian function, and advanced maternal age. The use of donor sperm is also growing as a means of family creation by lesbian couples and single women, and donor eggs and surrogacy are utilized by some gay and single men wishing to parent. In sum, the

Fig. 165.2 Clinic board of IVF baby pictures from Beirut, Lebanon (Photo by Marcia Inhorn)

technological possibilities offered by ARTs have been used in myriad ways, to enable procreation in ever expanding scenarios, including, for example, posthumous conception, pregnancy by women in their 60s and 70s, and the birth of high-order multiples. While these potential "treatments" are increasingly regarded as an acceptable option in some cultures, in others they remain prohibited, heavily stigmatized, or absolutely unacceptable.

Shifting the boundaries of our understanding of the creation of life and family formation, it is not surprising that ARTs present epistemological and ethical challenges and create new dilemmas for bioethicists, regulators, and religious leaders, as well as for men and women facing fertility problems. Although different countries have adopted various approaches to the regulation of ARTs - from legislation to professional guidance to a free market model (see Jones et al. 2007) - religious authorities have often expressed their views and sought to influence the (bio)ethical reasoning of both individuals and collective decision-makers (such as governments or regulators). Although Schenker rightly cautions that "it is often difficult to dissociate the influence of distinctly religious factors from other cultural conditions" (2005: 310), he argues that there are at least three factors that determine the influence of religious viewpoints on the practice of assisted reproduction: the size of the community; the authority of religious views within the population; and the unanimity or diversity of opinion present. As will become clear in the comparative discussion of the three different contexts in this chapter, depending on the wider socio-cultural conditions, religious rulings on ARTs can be both deterministic and surprisingly flexible.

165.3   Islam and Assisted Reproduction Technologies

Islam as a religion not only accepts, but positively endorses and encourages the seeking of biomedical treatment for infertility. In general, biomedicine and science are valorized, and in the case of infertility treatment in particular, Muslim couples are encouraged to create their own biological children through IVF. IVF is seen as a beneficial medical treatment and presents no significant ethical problems, as long as the gametes used for fertilization belong to a heterosexual married couple intending to parent (Clarke 2006a; Inhorn 2002, 2003a, b, 2012). This is particularly important since adoption of children is explicitly forbidden in the scriptures. (Infertile Muslim couples may foster an orphaned child, but permanent adoption is disallowed in family law in the vast majority of Muslim countries). In the absence of child adoption as a solution to infertility, IVF and related technologies are seen as the only options by many infertile Muslim couples. While there is broad agreement throughout Sunni Islam regarding which aspects of ARTs are and are not acceptable, and specifically on the prohibition of all forms of third-party reproductive assistance, there has been greater heterogeneity in Shia Muslim responses (Clarke 2007, 2008; Inhorn 2006a, b, c, d, 2012; Inhorn et al. 2010). This has led to a variety of ART practice in the Muslim Middle East, which includes the "Shia Crescent" within a majority Sunni Muslim region. The case studies below explore in greater depth the Islamic responses to assisted reproduction – comparing the Sunni Arab world with Shi Iran and secular Turkey – demonstrating the similarities and differences that exist between these three different Muslim contexts and revealing the over-simplicity of assuming a monolithic "Islamic response" to ARTs (Table 165.1).

165.3.1   Sunni Arab World

In 1980, only 2 years after Louise Brown's birth, the Grand Shaikh of Egypt's Al-Azhar University had issued the first *fatwa* permitting IVF to be practiced by Muslims. By 1986, the first IVF center had opened in Egypt, with the first Egyptian IVF baby, Hebbatallah Mohamed, born in 1987. By 1990, Egypt's first experiment in state subsidization of IVF for the poor came to fruition with the birth of a full-fledged IVF clinic in a public maternity hospital in Alexandria (Inhorn 1994). Then, soon after its invention in 1991, ICSI spread across the Mediterranean to Egypt, where it was introduced in an IVF clinic in Cairo in 1994 (Inhorn 2003a). By 1996, Egypt already hosted ten private IVF clinics in major cities. By the year 2003, the Egyptian IVF industry had truly blossomed, with approximately 50 clinics, 5 of them at least partially state subsidized (Inhorn 2010). In 2003, Al-Azhar University itself, through its Department of Obstetrics and Gynecology and International Islamic Center for Population Studies and Research, had opened a state-subsidized IVF clinic to serve the Cairene poor and to provide training for physicians and embryologists. Similar stories of diffusion and expansion were found throughout

Table 165.1 Assisted reproduction in the Muslim Middle East: comparing Sunni Arab World, Shia Iran, and secular Turkey

	Sunni Islam	Shia Iran	Secular Turkey
Fertility treatment for married heterosexual couples	Yes	Yes	Yes
Fertility treatment for unmarried couples	No	No	No
Fertility treatment for same-sex couples	No	No	No
Fertility treatment for single women	No	No	No
Artificial insemination with donor sperm	No	Yes	No
IVF	Yes	Yes	Yes
ICSI	Yes	Yes	Yes
The cryopreservation and storage of sperm	Yes (for medical necessity)	Yes	Yes (for medical necessity)
The cryopreservation and storage of eggs	Yes (for medical necessity)	Yes	Yes (for medical necessity)
The cryopreservation and storage of embryos	Yes (for use in a future frozen cycle)	Yes	Yes
Use of donor sperm	No	Yes	No
Use of donor eggs	No	Yes	No
Use of donor embryos	No	Yes	No
Use of surrogacy	No	Yes	No
PGD	Yes (for genetic screening)	Yes	Yes
Saviour siblings	Yes	Yes	Yes
Pre-implantation sex selection	No	Yes	No
Multi-fetal pregnancy reduction	Yes	Yes	Yes
Posthumous conception	No		No
Post divorce conception	No		No
Post menopausal conception	Possibly, only with the use of own eggs	Yes	Possibly, only with the use of own eggs
Embryo research	Yes (up to 14 days post-fertilization)	Yes	
Human cloning	No		No
Cross-border reproductive care	Yes, but not if third-party reproductive assistance is not used	Yes	Yes, but not if third-party reproductive assistance is not used

Source: Authors

Sunni Muslim countries during this period. In 1997, a global survey of ART clinics in 62 countries was published; 8 Middle Eastern Muslim countries (Egypt, Iran, Kuwait, Jordan, Lebanon, Morocco, Qatar, and Turkey) and 3 South and Southeast Asian Muslim countries (Indonesia, Malaysia, and Pakistan) were represented. All of these Muslim countries were practicing IVF and ICSI, yet importantly, none of them practiced donor insemination or any other form of third-party reproductive assistance. As noted by the study authors, the use of third-party reproductive assistance "is considered adultery and leads to confusion regarding the lines of genealogy, whose purity is of prime importance in Islam" (Meirow and Schenker 1997: 134). The ban on sperm donation and all other forms of third-party assistance has been clearly spelled out multiple times in *fatwas* and bioethical decrees issued in the Sunni Muslim countries.

Following the issuance in 1980 of the original Al-Azhar *fatwa*, the Islamic Fiqh Council issued a nearly identical *fatwa* banning all forms of third-party assistance in its seventh meeting held in Mecca in 1984. Subsequently, *fatwas* supporting ARTs but banning third-party assistance have been issued in Kuwait, Qatar, and the United Arab Emirates (Serour 2008). In 1997, at the ninth Islamic law and medicine conference, held under the auspices of the Kuwait-based Islamic Organization for Medical Sciences (IOMS) in Casablanca, a landmark five-point bioethical declaration included recommendations to prevent human cloning and to prohibit all situations in which a third party invades a marital relationship through donation of reproductive material (Moosa 2003). As noted by Islamic legal scholar Ebrahim Moosa (2003: 23):

> In terms of ethics, Muslim authorities consider the transmission of reproductive material between persons who are not legally married to be a major violation of Islamic law. This sensitivity stems from the fact that Islamic law has a strict taboo on sexual relations outside wedlock (*zina*). The taboo is designed to protect paternity (i.e., family), which is designated as one of the five goals of Islamic law, the others being the protection of religion, life, property, and reason.

Such a ban on third-party reproductive assistance of all kinds is effectively in place in the Sunni Muslim world, which represents approximately 80–90 % of the world's more than 1.5 billion Muslims (Inhorn 2003a; Meirow and Schenker 1997; Serour 1996; Serour and Dickens 2001). In Sunni Egypt, as well as the Sunni-dominant Arab nations of North Africa (Algeria, Libya, Morocco, Tunisia), the Arab Gulf (Kuwait, Oman, Qatar, Saudi Arabia, United Arab Emirates, Yemen), and the Levant (Jordan, Palestine, Syria), third-party assisted reproduction is not practiced—at least knowingly—in IVF clinics. In the Sunni countries, this ban on donors and surrogacy has been instantiated through antidonation bioethical codes, antidonation professional codes for obstetricians and gynecologists, and antidonation laws that specify the punishments that will ensue if an IVF practitioner wrongfully undertakes any form of third-party assisted conception. Such punishments range from permanent clinic closing to confiscation of all profits derived from donation to physician imprisonment and even the death penalty (although this has never happened and is not bound by legislation).

Yet, the ban in the Sunni world seems to derive less from the threat of legal punishment than from the force of Islamic morality. Namely, the majority of Sunni Muslims—both physicians and their patients—ardently support the Sunni ban on third-party donation, for three important reasons: (1) the moral implications of third-party donation for marriage; (2) the potential for incest; and (3) the moral implications of donation for kinship and family life. With regard to marriage, Islam is a religion that can be said to privilege—even mandate—heterosexual marital relations.

As is made clear in the original Al-Azhar *fatwa*, reproduction outside of marriage is considered *zina* (adultery), which is strictly forbidden in Islam. Although third-party donation does not involve the sexual "body contact" of adulterous relations, nor presumably the desire to engage in an extramarital affair, it is nonetheless considered by Sunni Muslim religious scholars to be a form of adultery, by virtue of introducing a third party into the sacred dyad of husband and wife. It is the very fact that another man's sperm or another woman's eggs enter a place where they do not belong that makes donation of any kind inherently wrong—or *haram* (religiously forbidden)—and hence threatening to the marital bond. The second aspect of third-party donation that troubles marriage is the potential for incest among the offspring of unknown donors. Moral concerns have been raised about the potential for a single anonymous donor's offspring meeting and marrying each other, thereby undertaking an incestuous union of half-siblings. In a small country such as Lebanon, with only four million inhabitants, such unwitting incest of the children of an anonymous donor is a real possibility, a moral concern that has also been raised in neighboring Israel (Kahn 2000). The final moral concern voiced by Sunni Muslims, including clerics, IVF physicians, and patients themselves, is that third-party donation confuses issues of kinship, descent, and inheritance. As with marriage, Islam is a religion that can be said to privilege—even mandate—biological inheritance. Preserving the biological "origins" of each child—meaning its relationship to a known biological mother and father—is considered not only an ideal in Islam, but a moral imperative. The problem with third-party donation, therefore, is that it destroys a child's *nasab* (lineage or genealogy), which is immoral in addition to being psychologically devastating to the donor child.

It is important to emphasize that these moral concerns are taken very seriously. To our knowledge, not one single IVF clinic in a Sunni-dominant Muslim country practices third-party assisted conception. Although physicians are sometimes asked about gamete donation and surrogacy by IVF patients who cannot conceive a child in any other way, they tell them that it is "against the religion," and therefore, not performed. Those patients who are committed to pursuing third-party assisted reproduction are told that they must travel "outside" to Europe, North America, or Asia. Such cases of Sunni Muslim "reproductive tourism" are certainly beginning to occur (Gürtin 2011; Inhorn 2011b), however, the vast majority of infertile Sunni Muslim couples abide by the religious ban on donation and surrogacy, agreeing with the moral justifications for it. For example, in ethnographic interviews undertaken by Inhorn with nearly 600 infertile individuals and couples in Egypt (1988–89, 1996), Lebanon (2003), United Arab Emirates (2007), and "Arab Detroit" (2003–5, 2007–8), only a handful of Sunni Muslim couples (<10) were willing to contemplate any form of third-party donation. Of the few men and women who "approved" of the practice, their approval was most often a "last resort" when no other ART option could be expected to solve the infertility problem. Furthermore, only egg donation was approved of, because it allowed the infertile wife to experience a pregnancy and could be compared to the *halal* (religiously permitted) practice of polygyny. Sperm donation, on the other hand, was not; it was said to confuse patrilineal descent and constitute a form of *zina,* or a wife's "extramarital" acceptance of another man's sperm. Most importantly, men argued that a donor child "won't be my son" (Inhorn 2006b, 2012): in their view, sperm donation would be like "raising another man's child."

165.3.2   Shia Iran

In opposition to the Sunni rulings on assisted reproduction, leading clerics in Iran have legitimized third party donation in all its forms: sperm, egg, and embryo donation, and surrogacy. However, there is great heterogeneity within Shia Islamic responses to assisted reproduction. Unlike Sunni scholars, Shia scholars in Iran are reluctant to engage in formal collective deliberations on issues of global importance and prefer instead to rely on individualistic independent reasoning. It is this individualistic practice of *ijtihad* that has paved the way for the Shia to engage dynamically with most forms of biotechnology (Inhorn and Tremayne 2012), and has in this case resulted in the development of a wide diversity of opinions among Shia *marja's* (sources of emulation). These opinions can sometimes disagree and take opposing views on the interpretation of the Qur'an, which has historically led to senior *marja's* forming their own groups of followers. However, this diversity has also led to considerable "flexibility" for the Shia in the regulation and practice of ARTs.

Although initially both Sunni and Shia religious authorities restricted the use of ARTs to married couples using their own gametes (Mahmoud 2012), by the beginning of the new millennium, the Iranian Shia had found solutions within the religious rules that allowed the use of all forms of ARTs, including most importantly third-party donation. To be able to practice third-party donation within religiously sanctioned parameters, the Shia in Iran extended the definition of marriage to include *mutʿa* marriages, a form of "temporary marriage" that is only practiced by Shia Muslims (Haeri 1989). They ruled that if a donor became a legitimate – albeit temporary - spouse, then eggs or sperm could be donated within the confines of legal marriage, but without any sexual contact taking place between the donor and the recipient (Inhorn 2003a, b; Clarke 2006b; Tremayne 2009). Several Iranian religious leaders engaged in further debates, legally approving of embryo donation and surrogacy on the same grounds. Such approval has most recently been extended to allow stem cell research in Iran, and has been applied to other forms of biotechnological advances, including organ donation and transgender surgery. Indeed, it could be argued that the Shia have gone further in embracing all forms of third-party donation than most Western Christian countries (Inhorn et al. 2012).

To understand the reasons for and the speed by which such "liberal" (Clarke 2009) decisions have been made and accepted into practice in Iran, it is essential to realize the constitution of the legislative councils which have made them. These councils are themselves a part of Iran's theocratic regime, made up of political as well as religious leaders, and the decisions they pass become "official." However, those who do not wish to abide by these decisions can turn instead to their own *marja'* without the worry of breaking any rules. Furthermore, in deciding on the legitimacy of various ARTs, religious leaders do not necessarily act alone; instead, they engage with experts from various disciplines including specialists in Islamic law, medicine, and psychology, to explore the legal and bioethical ramifications of these biotechnologies on society, the family, marriage, and potential children. The approval of third-party donation, for example, was the result of many years of intensive interdisciplinary debate among several *marja's*. Having said this, it must be noted that it was the endorsement of the supreme religious leader, Ayatollah Khamene'i, that gave third-party donation "official" legitimacy in 1999. However, approval or disapproval by Ayatollah Khamene'i does not mean that all Shia leaders are in agreement with him; Shia religious leaders are deeply divided among themselves on this and other divine matters. To date, no definitive, universal conclusion has been reached among the Shia jurisprudents on ARTs. Indeed, there may never be a consensus among Shia leaders about whether or not third-party donation should be permitted and herein lies the potential for great flexibility in ART practice: Shia Muslims are free to adhere to the views of different *marja'* as they see fit, regardless of the "official" decision.

As a result of the diversity of *marja'* opinions, potential Shia Muslim users of third-party donation (that is, doctors and patients) have been able to exercise a great degree of agency and control over actual clinical practices, reinforcing independent understandings of what constitutes kinship and relatedness (Tremayne 2009). Such practices have led to some surprising and counter-intuitive outcomes. For example, siblings of both sexes may donate and receive gametes from each other to make embryos (Garmaroudi Naef 2012). Although this would, in theory, be a breach of kinship rules, in particular the prohibition on sibling incest, the prescription that donation is allowed "as long as no bodily contact or touch and gaze takes place between the parties involved" (Ayatollah Khamenei 1999) is extended to include any two parties acting as a gamete donor and recipient. The ensuing ethical and legal problems in these and other similar cases are abundant. In discussing the validity of the ethical decisions made in Iranian fertility clinics, Tappan (2012) raises serious questions about Shia "flexibility" and bioethics. He argues that while Islamic law, presented as *fatwas*, or legal opinions of Islamic scholars, plays a key role in Islamic bioethics, the assertion that "Islamic bioethics" is synonymous with *fatwas* does not bear out in Iranian fertility clinics. There, clinicians and ethical committees consider a wide range of sources, including civil law, Western bioethical notions, and *ijtihad* and do *not* limit themselves to *fatwas*. These efforts, in Tappan's view, are part of the wider articulation of Islamic bioethics that includes, but goes beyond, mere reference to Islamic law. Yet, Tappan argues that Iranian clinicians and jurists have failed to unfold deeper, more foundational grounds for Islamic bioethics, and for the application of important theological, ethical, and legal principles. For example, the rights of the future child, to be born from third-party donation, are rarely invoked in clinical discussions.

The justification for allowing the use of third-party reproductive assistance in Iran has been to ensure the stability and happiness of the family through the birth of children, and thereby to reduce the suffering of infertile couples. Indeed, the focus throughout these Shia jurisprudential debates has remained on the family, which is considered the foundation of society. Nonetheless, the dynamic array of ART practices that are allowed in Iran has opened the way for myriad bioethical, legal, and personal dilemmas. Research on the impact of third party gamete donation has shown that lawmakers, physicians, and patients - each with their own concerns and agendas—are not always equipped to deal with the complex ethical and interpersonal problems that may be generated (Tremayne 2012). Religious texts and authorities cannot always solve these contemporary dilemmas; particularly with regards to the use of sperm donation, for example, religious permmissibility may not translate into cultural acceptance or social approval. Thus, Iranian couples who resort to sperm donation may make great efforts to do so "secretly," and continue to struggle personally and emotionally with their actions, leading to unfavorable outcomes that may have drastically negative implications for the family, for the women, and for the donor-conceived children (Tremayne 2012).

165.3.3 Secular Turkey

Although Turkey is a country committed to secularism in medical ethics (Arda, 2007), its pattern of ART regulation can be thought of as distinctly "Sunni Muslim in character" (Inhorn et al. 2010). As has been described for the Sunni Arab world earlier in this chapter, in Turkey too, the use of assisted reproduction occurs within the strict parameters of heterosexual marriage. The Assisted Reproduction Treatment Centers Directorate, under the Ministry of Health, regulates the practice of ARTs in accordance with a comprehensive piece of legislation which provides definitions, outlines prohibitions, and details all the necessary requirements (including building and physical environment specifications, equipment, materials, and personnel) for clinics to obtain an ART practice license. This Statute on Assisted Reproduction Treatment Centers was first introduced in 1987 (Official Gazette 19551, 21 August 1987), as a pre-emptive framework for ART practice in Turkey, and has subsequently been updated five times, with the latest changes taking place in March 2010 (Official Gazette no. 27613, 6 March 2010). This legislation makes it clear that ARTs are confined to the treatment of married heterosexual couples using their own gametes, and all forms of third-party reproductive assistance are forbidden. In fact, in 2010 Turkey also banned its citizens from seeking treatment with donor gametes in other jurisdictions, thereby becoming the first country to regulate against reproductive tourism (Gürtin 2010, 2011).

The extent of the influence of Islam on Turkey's ART regulation is a controversial topic without an easy answer (see debates from the Ethics, Law and Moral Philosophy of Reproductive Biomedicine conference in 2006 reported in Gürtin 2012). While on the one hand Turkey is a secular country, on the other it has a predominantly (Sunni) Muslim population whose views on such fundamental and ethical matters are undoubtedly shaped by their religious affiliation and by the broader aspects of a Muslim culture. Indeed, some commentators have argued that the ART law "was created by a committee established by the ruling government without the necessary consideration for, or consultation with, couples requiring third-party reproduction. Two fears were integral to the decision made by the committee: the fear produced by the incomprehension of the new technology and the religious issues of a country with a predominantly Muslim culture" (Isikoglu et al. 2006: 321). Turkey's highest religious authority, The Presidency of Religious Affairs, lends support to the regulation of ARTs, explaining that IVF "is no longer permissible if a foreign element is included, meaning if the sperm, eggs or womb belong to a person outside of the husband-wife couple; because according to the general principles of the religion of Islam, there is an imperative for a legitimate child to belong, whether by sperm or egg or womb, to a wedded husband-wife couple" (Presidency of Religious Affairs, 2006, author's translation).

It would, however, be over-simplistic to stipulate a homogenous position against third-party assisted reproduction among Turkish men and women informed solely by Muslim morality and culture, since individuals do not always care about, follow, nor even have an accurate knowledge of the teachings of their religion. While it is generally true that third-party reproductive assistance (particularly the use of donor sperm) is a stigmatized taboo in Turkey, there have been reports for years of Turkish couples surreptitiously crossing borders to nearby Cyprus and Greece in order to access ARTs with donor gametes, particularly donor eggs. Indeed, according to Irfan Şencan, the director of the Ministry of Health's Treatment Services department, the ban on reproductive travel was introduced in 2010 as a response to the growth of this phenomenon in recent years. Although it is difficult to quantify the occurrence of such activities or even to gain an accurate picture of public opinions on assisted reproduction and third party reproductive assistance (since there are only a small number of studies on this subject, with some methodological limitations, see Baykal et al. 2008; Isikoglu et al. 2006; and Kilic et al. 2009), the available evidence points to a diversity of opinions among Turkish people.

According to the first available data from Turkey regarding public opinion towards egg donation, there are high rates of approval for this form of fertility treatment, with only 15 % of respondents showing "complete objection" (Isikoglu et al. 2006). Moreover, more than half of the women and two-thirds of the men (wrongly) thought that their religion (i.e. Islam) would allow egg donation if they needed it, and more than half stated that they would prefer egg donation to adoption. Acceptance of egg donation was also shown to vary according to different scenarios, with medical conditions receiving highest rates of approval (from 81.03 % of women and 79.76 % of men); followed by age-related infertility (68.10 % and 60.12 %); inherited medical or mental problems in the family (45.26 % and 54.76 %); and finally the existence of a previous child with disability (49.57 % and 42.86 %). However, Baykal et al. (2008) and Kilic et al. (2009) both found significantly lower levels of acceptance for egg donation than Isikoglu et al. (2006) in their studies of infertile Turkish women, at 23.3 % and 26 % respectively, which suggests more clarification is required.

Baykal et al. (2008) investigated the attitudes of 368 women who had applied for infertility treatment using self-completion questionnaires consisting of 38 items determining socio-demographic status, previous history with infertility, and opinions. They found acceptance rates of 23.3 % for egg donation, 15.1 % for surrogacy and only 3.4 % for sperm donation. These highly differential rates reflect the different statuses traditionally attributed to the role of men and women in procreation in Turkey, but also more generally in Muslim cultures (Inhorn 2006b), and have also been shown to exist to some extent among Turkish communities living in Europe (Gürtin-Broadbent 2009). However, it is difficult to know how to interpret what is meant by "acceptance" in this study, since they also report that when asked what they would do if their IVF treatment failed, 59.7 % of respondents answered that they would "do nothing," 38.3 % would pursue adoption, and only 2 % would consider gamete donation. Similarly, Kilic et al. (2009) presented a sample of 250 women who had applied for infertility treatment with a questionnaire containing 7 socio-demographic items and 5 questions on attitudes towards surrogacy and egg donation, for which they found acceptance rates of 24 % and 26 % respectively. Although the acceptance rates for egg donation among infertile women in the studies by Baykal et al. (2008) and Kilic et al. (2009) are consistent, there are also a range of confusing divergences. For example, in the study by Kilic et al. (2009) patients' (anticipated) responses to IVF failure suggested adoption to be the preferred solution (59.6 %), followed by accepting egg donation (26 %), doing nothing (25.6 %), and accepting a surrogate mother (24 %). Interestingly, Turkey is one of the very few Muslim countries—along with "secular" Tunisia and Shia Iran—to allow child adoption, which is otherwise widely prohibited across the Sunni Arab world.

Overall, these studies, despite their limitations, clearly suggest heterogeneity in the views of Turkish infertile men and women and the general public with regards to third-party assisted reproduction. These views cannot simply be equated with a "Sunni" or "Shia" position, but reveal that, particularly in this secular context, individuals must deliberate according to their own local moral worlds to reach decisions which may or may not be aligned with the official position of their religion.

165.4   Concluding Remarks

In this chapter we have examined the globalization of ARTs with a specific focus on Islamic perspectives to these technologies. The comparisons provided here – between the Sunni Arab world, Shi Iran, and secular Turkey – demonstrate clearly that there is not a single monolithic "Islamic" response to ARTs. Rather, Islam has provided a multiplicity of responses to the practice and regulation of assisted reproduction, engaging with the novel ethical, social, and relational dilemmas of these technologies, while simultaneously taking account of religious scriptures and contemporary contexts. Indeed, although there are similarities, there is also variation at the regulatory, practice and attitudinal levels between different "Islamic" contexts. Most importantly, there is a divergence between Sunni and Shia authorities: while the former have unanimously prohibited all forms of third-party reproductive assistance as a crucial caveat to the use of fertility treatments, the latter have provided a diversity of opinions enabling the use of donor eggs, donor sperm, donor embryos and surrogacy. This difference of opinion between Sunni and Shia Islam, as well as between the *fatwas* of different Shia clerics, has resulted in fascinating social choreographies, whereby desperate infertile men and women switch their allegiances to more permissive clerics, non-Muslim practitioners display *fatwas* absolving their practices on the walls of their clinics (Clarke 2008), and Sunni couples from surrounding Middle Eastern countries engage in surreptitious trips for cross-border reproduction using third-party assistance (Inhorn 2009, 2012; Inhorn et al. 2010). Interestingly, in secular Turkey, despite a legal ban on third-party reproductive assistance, predominantly Sunni Muslim men and women have displayed a variety of opinions and personal attitudes towards gamete donation.

Comparisons, although relatively infrequent in the scholarly literature on ARTs, are an interesting and useful way to get a deeper insight into many of the central questions surrounding the expansion and spread of ARTs. Such comparisons – based on law, religion, culture, politics, practice, and attitudes – demonstrate the rate and pace of globalization of ARTs, help to delineate the similarities and differences of ART practice in different global locations; and promote an understanding of the underlying reasons (moral, legal, economic, etc.) that lead to the heterogeneity of ART practice around the world (Inhorn et al. 2010). In this chapter, we hope to have shown that the similarities and differences between the Sunni Arab world, Shia Iran and secular Turkey demonstrate that while religion is an extremely important factor in generating personal and cultural responses to new technologies and their social applications, these responses are not over-determined by religion. Thus, to imagine that all Muslims or all Islamic authorities will respond in an identical manner to emerging dilemmas would be to misunderstand how religion informs ethical, moral and social deliberations at both a collective and an individual level, and to miss the crucial nuances in how religion and culture interact.

Our work, and the work of colleagues in the field of medical anthropology and beyond, seeks to ethnographically explore some of these nuances with respect to assisted reproduction (see Inhorn and Birenbaum-Carmeli 2008 for a survey of this scholarship): Inhorn's research spans Egypt, Lebanon, the United Arab Emirates and Arab America; Tremayne focuses on Iran; and Gürtin on Turkey and on Turkish migrants in the UK. Some of the questions we have pursued include the stigma and disruption of involuntary childlessness (for example, Inhorn 1994); attitudes towards assisted reproduction (for example, Inhorn 2003a); changing gender identities and relations in these regions as a result of ARTs (for example, Inhorn 2012); the reproductive experiences and opinions of diasporic or migrant populations from these regions (Gürtin-Broadbent 2009; Inhorn and Fakih 2006; Tremayne 2012); and practices of cross-border reproductive care (e.g. Gürtin 2010, 2011; Inhorn 2009, 2010; 2011a, b). However, significant gaps remain regarding academic research into the reproductive experiences of Muslim men and women, particularly in Europe, South and South East Asia and North Africa. We encourage new researchers to address these gaps, to explore the impact of emerging technological and social possibilities, and to contribute to a vibrant, fast-developing field of scholarship.

References

Baykal, B., Korkmaz, C., Ceyhan, S. T., Goktolga, U., & Baser, I. (2008). Opinions of infertile Turkish women on gamete donation and gestational surrogacy. *Fertility and Sterility, 89*, 817–822.

Clarke, M. (2006a). Islam, kinship and new reproductive technology. *Anthropology Today, 22*, 17–20.

Clarke, M. (2006b). Shiite perspectives on kinship and new reproductive technologies. *ISIM Review, 17*, 26–27.

Clarke, M. (2007). Closeness in the age of mechanical reproduction: Debating kinship and bio-medicine in Lebanon and the Middle East. *Anthropology Quarterly, 80*, 379–402.

Clarke, M. (2008). New kinship, Islam, and the liberal tradition: Sexual morality and new reproductive technology in Lebanon. *Journal of the Royal Anthropological Institute, 14*, 143–169.

Clarke, M. (2009). *Islam and new kinship: Reproductive technology, anthropology and the Shari'ah in Lebanon*. New York: Berghahn.

Garmaroudi Naef, S. (2012). Gestational surrogacy in Iran: Uterine kinship in Shia thought and practice. In M. C. Inhorn & S. Tremayne (Eds.), *Islam and assisted reproductive technologies: Sunni and Shia perspectives* (forthcoming). New York: Berghahn.

Gürtin, Z. B. (2010). *Problems with legislating against "reproductive tourism"*. Commentary for BioNews, 550. London: Progress Educational Trust.

Gürtin, Z. B. (2011). Banning reproductive travel? Turkey's ART legislation and third-party assisted reproduction. *Reproductive Biomedicine Online, 23*, 555–565.

Gürtin, Z. B. (2012). Assisted reproduction in secular Turkey: Regulation, rhetoric, and the role of religion. In M. C. Inhorn & S. Tremayne (Eds.), *Islam and assisted reproductive technologies: Sunni and Shia perspectives* (forthcoming). New York: Berghahn.

Gürtin-Broadbent, Z. (2009). "Anything to become a mother:" Migrant Turkish women's experiences of involuntary childlessness and Assisted Reproductive Technologies in London. In L. Culley, N. Hudson, & F. van Rooij (Eds.), *Marginalised reproduction: Ethnicity infertility and reproductive technologies* (pp. 117–134). London: Earthscan.

Haeri, S. (1989). *The law of desire: Temporary marriage in Iran*. London: IB Tauris.

Inhorn, M. C. (1994). *Quest for conception: Gender, infertility, and Egyptian medical traditions*. Philadelphia: University of Pennsylvania Press.

Inhorn, M. C. (2002). The "local" confronts the "global": Infertile bodies and new reproductive technologies in Egypt. In M. C. Inhorn & F. van Balen (Eds.), *Infertility around the globe: New thinking on childlessness, gender, and reproductive technologies* (pp. 263–283). Berkeley: University of California Press.

Inhorn, M. C. (2003a). *Local babies, global science: Gender, religion, and in vitro fertilization in Egypt*. New York: Routledge.

Inhorn, M. C. (2003b). Global infertility and the globalization of new reproductive technologies: Illustrations from Egypt. *Social Science and Medicine, 56*, 1837–1851.

Inhorn, M. C. (2006a). Making Muslim babies: IVF and gamete donation in Sunni versus Shi'a Islam. *Culture, Medicine & Psychiatry, 30*, 427–450.

Inhorn, M. C. (2006b). "He won't be my son." Middle Eastern Muslim men's discourses of adoption and gamete donation. *Medical Anthropology Quarterly, 20*, 94–120.

Inhorn, M. C. (2006c). *Fatwa*s and ARTS: IVF and gamete donation in Sunni v. Shi'a Islam. *Journal of Gender, Race & Justice, 9*, 291–317.

Inhorn, M. C. (2006d). Islam, IVF, and everyday life in the Middle East: The making of Sunni versus Shi'ite test-tube babies. *Anthropology of the Middle East, 1*, 37–45.

Inhorn, M. C. (2009). Globalization and reproductive tourism in the Muslim Middle East: IVF, Islam, and the Middle Eastern state. In C. H. Browner & C. F. Sargent (Eds.), *Reproduction, globalization, and the state* (pp. 126–138). Durham: Duke University Press.

Inhorn, M. C. (2010). Assisted motherhood in global Dubai: Reproductive tourists and their help-ers. In W. Chavkin & J. Maher (Eds.), *The globalization of motherhood: Deconstructions and reconstructions of biology and care* (pp. 180–202). New York: Routledge.

Inhorn, M. C. (2011a). Male infertility and intracytoplasmic sperm injection (ICSI) in the Middle East. *Middle East Institute Viewpoints: Public Health in the Middle East, 1*, 26–29.

Inhorn, M. C. (2011b). Globalization and gametes: Reproductive "tourism", Islamic bioethics, and Middle Eastern modernity. *Anthropology and Medicine, 18*(1), 87–103.

Inhorn, M. C. (2012). *The new Arab man: Emergent masculinities, technologies, and Islam in the Middle East*. Princeton: Princeton University Press.

Inhorn, M. C., & Birenbaum-Carmeli, D. (2008). Assisted reproductive technologies and culture change. *Annual Review of Anthropology, 37*, 177–196.

Inhorn, M. C., & Fakih, M. H. (2006). Arab Americans, African Americans, and infertility: Barriers to reproduction and medical care. Special issue on "Health Disparities in Infertility". *Fertility and Sterility, 85*(4), 844–852.

Inhorn, M. C., & Tremayne, S. (Eds.). (2012). *Islam and assisted reproductive technologies: Sunni and Shia perspectives*. New York: Berghahn.

Inhorn, M. C., & van Balen, F. (Eds.). (2002). *Infertility around the globe: New thinking on child-lessness, gender, and reproductive technologies*. Berkeley: University of California Press.

Inhorn, M. C., Patrizio, P., & Serour, G. I. (2010). Third-party reproductive assistance around the Mediterranean: Comparing Sunni Egypt, Catholic Italy and multisectarian Lebanon. *Reproductive Biomedicine Online, 21*(7), 848–853.

Isikoglu, M., Senol, Y., Berkkanoglu, M., Ozgur, K., Donmez, L., & Stones-Abbasi, A. (2006). Public opinion regarding oocyte donation in Turkey: First data from a secular population among the Islamic world. *Human Reproduction, 21*, 318–323.

Jenkins, G. L., & Inhorn, M. C. (2003). Reproduction gone awry: Medical anthropological per-spectives. *Social Science and Medicine, 56*, 1831–1836.

Jones, H. W., Cohen, J., Cooke, I., & Kempers, R. (2007). IFFS surveillance 07. *Fertility and Sterility, 87*(4), S1–S67, Supplement 1.

Kahn, S. M. (2000). *Reproducing Jews: A cultural account of assisted conception in Israel*. Durham: Duke University Press.

Khamenei, Haj Seyyed Ali. (1999). *Ojoubeh al-Esteftaat* (Questions No s. 1271 to 1277). Tehran; Nashre Amir Kabir.

Kilic, S., Ucar, M., Yaren, H., Gulec, M., Atac, A., Demirel, F., Karabulut, C., & Demirel, O. (2009). Determination of the attitudes of Turkish infertile women towards surrogacy and oocyte donation. *Pakistan Journal of Medical Science, 25*, 36–40.

Mahmoud, F. (2012). Controversies in Islamic evaluation of assisted reproductive technologies. In M. C. Inhorn & S. Tremayne (Eds.), *Islam and assisted reproductive technologies: Sunni and Shia perspectives* (forthcoming). New York: Berghahn.

Meirow, D., & Schenker, J. G. (1997). The current status of sperm donation in assisted reproduc-tion technology: Ethical and legal considerations. *Journal of Assisted Reproduction and Genetics, 14*, 133–138.

Moosa, E. (2003). Human cloning in Muslim ethics. *Voices Across Boundaries* (Fall), 23–26.

Schenker, J. G. (2005). Assisted reproductive practice: Religious perspectives. *Reproductive BioMedicine Online, 3*, 310–319.

Serour, G. I. (1996). Bioethics in reproductive health: A Muslim's perspective. *Middle East Fertility Society Journal, 1*, 30–35.

Serour, G. I. (2008). Islamic perspectives in human reproduction. *Reproductive BioMedicine Online, 17*(suppl. 3), 34–38.

Serour, G. I., & Dickens, B. M. (2001). Assisted reproduction developments in the Islamic world. *International Journal of Gynecology & Obstetrics, 74*, 187–193.

Tappan, R. (2012). Moore than *Fatwas*: Ethical decision making in Iranian fertility clinics. In M. C. Inhorn & S. Tremayne (Eds.), *Islam and assisted reproductive technologies: Sunni and Shia perspectives* (forthcoming). New York: Berghahn.

Tremayne, S. (2009). Law, ethics and donor technologies in Shia Iran. In D. Birenbaum-Carmeli & M. C. Inhorn (Eds.), *Assisting reproduction, testing genes; global encounters with new bio-technologies* (pp. 144–164). New York: Berghahn.

Tremayne, S. (2012). The "down side" of third party donation: The "happy family rhetoric" in Iran. In M. C. Inhorn & S. Tremayne (Eds.) *Islam and assisted reproductive technologies: Sunni and Shia Perspectives* (forthcoming). New York: Berghahn.

Vayena, E., Rowe, P. J., et al. (2002). Current practices and controversies in assisted reproduction: Report of a WHO meeting. Geneva: World Health Organization.

Zegers-Hochschild, F., Adamson, G. D., de Mouzon, J., Isihara, O., Mansour, R., Nygren, K., Sullivan, E., van der Poel, S., on behalf of ICMART and WHO. (2009). The international Committee for Monitoring Assisted Reproductive Technology (ICMART) and the World Health Organization (WHO) Revised Glossary on ART Terminology, 2009. *Human Reproduction, 24*(11), 2683–2687.

Chapter 166
The Perpetration of Abuse in Intimate Relationships: Does Religion Make a Difference?

Claire M. Renzetti, Amy Messer, C. Nathan DeWall, and Richard S. Pond

166.1 Introduction: Religion and Intimate Partner Violence

There is widespread consensus that intimate partner violence (IPV) is a serious public health problem that affects a sizable number of Americans. Although estimates of IPV rates vary depending on the sample studied and how questions are worded, commonly cited estimates are that from 7.6 to 11.5 % of men and 12 to 25 % of women are physically and/or sexually assaulted by an intimate partner each year (Tjaden and Thoennes 2006; Black et al. 2011). Although some researchers have argued that women are as likely as men to assault their intimate partners (see, for example, Archer 2000; Dutton 2006; Straus 2007; Dutton et al. 2009), the vast majority of studies show that there are important differences in IPV perpetrated by women and men. For example, studies show that women's and men's motivations for using violence against an intimate partner differ. Men are more likely to use violence when they perceive themselves losing control of the relationship or when they interpret their partners' words or behavior as challenges to their authority. In contrast, women are more likely to use violence, especially severe physical violence, in self-defense, when they believe they are in imminent danger of being attacked, or in retaliation for being attacked (Barnett et al. 1997; Dobash et al. 1998; Miller 2001; Rajan and McCloskey 2007). Moreover, men's violence is typically

C.M. Renzetti (✉) • A. Messer
Department of Sociology, University of Kentucky, Lexington, KY 40506, USA
e-mail: claire.renzetti@uky.edu; amy.messer@uky.edu

C.N. DeWall
Department of Psychology, University of Kentucky, Lexington, KY 40506, USA
e-mail: cnathandewall@gmail.edu

R.S. Pond
Department of Psychology, University of North Carolina, Wilmington, NC 28403, USA
e-mail: pondr@uncw.edu

© Springer Science+Business Media Dordrecht 2015
S.D. Brunn (ed.), *The Changing World Religion Map*,
DOI 10.1007/978-94-017-9376-6_166

more severe; women are more likely to be injured when they are assaulted by their intimate partners, and their injuries are more serious and more likely to require medical treatment or hospitalization (Archer 2000; Tjaden and Thoennes 2000; Menard et al. 2009). And men are more likely than women to kill their intimate partners (Fox and Zawitz 2007).

In light of the data on the incidence and seriousness of male-perpetrated intimate partner violence, it is not surprising that a great deal of research has been undertaken to identify factors that may increase or decrease men's likelihood of abusing their female partners. Researchers report, for instance, that social class and employment status are related to IPV perpetration, with lower socioeconomic status and poverty, as well as unemployment, being associated with an increased risk of IPV perpetration by men (Renzetti 2009; Renzetti and Larkin 2009; Brush 2011). Researchers have also found that younger men (aged 18–30), men with substance abuse problems, and those who endorse male dominance in intimate relationships are also at greater risk for IPV perpetration (Luthra and Gidycz 2006; Bennett and Bland 2008; DeKeseredy and Dragiewicz 2009; Basile and Black 2011).

Interestingly, however, few researchers until relatively recently have examined how religion may be related to men's IPV perpetration. This is surprising because there are several good reasons to hypothesize that religion may influence the likelihood of IPV perpetration. On one hand, many religious traditions have supported and reinforced patriarchal ideology, such as the notion of male dominance and the importance of men as authority figures and instrumental leaders of their households. One might speculate, therefore, that men who are strongly religious and adhere to these patriarchal ideals might be more inclined to use violence against an intimate partner whom they judge as having committed a "sin" or who usurped male authority in some way (Nason-Clark 2000). On the other hand, given that religious values and teachings also encourage pro-social behavior, including empathy and a concern for the welfare of others, one might hypothesize that adherence to these kinds of religious tenets would reduce men's likelihood of using violence against an intimate partner (see, for example, Johnson 2011). In this case, religion might motivate intimate partners to be more accommodating to one another and to compromise with one another when there is disagreement, so that conflicts are more likely to be defused rather than to escalate to violence.

But the study of religion and intimate partner violence is complicated by a number of factors. For one thing, religion and religious expression are highly diverse. Dollahite et al. (2004), for instance, note that the United States is one of the most religious and religiously diverse countries in the world today in terms of the population's voluntary participation in religious activities and institutions. They cite research that identifies more than 2,600 distinct faith communities in the U.S. and Canada, with 216 major Christian denominations alone. As Dollahite et al. (2004: 412) also point out, "there is as much diversity within major faith groups as between them." Social scientists often assign these diverse faith communities to rather broad categories, such as "Liberal, Moderate, Conservative," but critics of this approach maintain that, in practice, these labels may have little to do with specific markers of religiosity or with individuals' identification with religious movements, and they tend to "conflate religious, economic, social and political ideas into one monolithic measure" (Woodberry et al. 2012: 65). One's affiliation with a particular faith group may be less important with regard to behavioral outcomes than *how* one is religious.

In this chapter, we discuss research studies, including our own recent work, that have examined the effects of religiosity on male perpetration of intimate partner violence. And in doing so, we will identify some of the major gaps in this body of research and suggest areas and topics for future study that would help us better understand the potential of religiosity as a protective factor or a risk factor for IPV perpetration.

166.2   Dimensions of Religiosity and Their Relationship to IPV Perpetration

Dollahite et al. (2004: 413) remark that religiosity is not only complex, it is also multifaceted. They identify three major dimensions of the construct: (1) *religious practices* (for example, outward, observable expressions of faith), (2) *religious beliefs* (for example, personal beliefs, framings, meanings), and (3) *religious communities* (support, involvement and relationships grounded in a congregation or other religious group). Although each of these dimensions of religiosity may be distinguished from one another, they are undoubtedly interrelated. Yet, researchers examining the effects of religiosity on IPV perpetration have typically measured religiosity narrowly in terms of just one of these dimensions or even with a single indicator of one dimension.

166.2.1   *Religious Practices and IPV Perpetration*

Most research on the relationship between religiosity and IPV perpetration has used rather narrow measures of religious practice. Ellison and colleagues (Ellison and Anderson 2001; Ellison et al. 2007), for example, measured religiosity in terms of frequency and consistency of attendance at religious services. Their research consistently supports the hypothesis that religiosity reduces likelihood of IPV perpetration. Ellison and Anderson (2001) found that among both men and women, those who attended religious services more often and more consistently were less likely to perpetrate IPV, even after controlling for religiosity's effects on increasing social integration and social support, and decreasing substance abuse and psychological problems. Moreover, the protective effects of religiosity with regard to IPV perpetration remained significant regardless of whether data measuring IPV perpetration were from self-reports or from partner reports – an important finding given the tendency of IPV perpetrators to underestimate the frequency and severity of their violence. In a more recent analysis, Ellison et al. (2007) found that the protective effect of religiosity (that is, church attendance) varies by race/ethnicity: It strongest for African American men and women and also for Hispanic men, groups that are at especially high risk for IPV perpetration.

Cunradi et al. (2002) also found that attendance at religious services is associated with reduced IPV perpetration, but they report that alcohol consumption appears to be significantly more important than religiosity in predicting men's likelihood of perpetrating IPV. In a study of a large, nationally representative sample of married adults, DeWall (2010) measured religiosity in terms of frequency of engaging in religious services as well as participants' self-assessed importance of religious beliefs in their daily lives. He found that although alcohol abuse was associated with greater IPV perpetration, this was the case only among participants low in religiosity. In his study, high religiosity served as a protective factor against IPV perpetration and also appeared to break the commonly observed link between alcohol use and IPV.

166.2.2 Multidimensional Measures of Religiosity

Researchers critical of using attendance at religious services as the sole measure of religiosity have instead used scales composed of multiple items that may subsequently be summed to yield a religiosity "score." For instance, in our recent study of religiosity and IPV perpetration (Renzetti et al. 2012), we used the Religious Commitment Inventory (RCI-10; Worthington et al. 2003) to capture types of religious practice (apart from attendance at religious services) as well as the importance of religious beliefs. The RCI-10 is composed of 10 items with a 5-point Likert response scale ranging from "not at all true of me" (1) to "totally true of me" (5). Items cover such aspects of religious practice as reading books or magazines about one's faith, making financial contributions to one's religious organization, and spending time in private religious thought or meditation. Items measuring the importance of religious beliefs include, "My religious beliefs lie behind my whole approach to life," and "Religious beliefs influence all my dealings in life."

A national sample of 251 adult men, ranging in age from 18 to 85 (mean = 43.3 years) who had been in a committed intimate relationship for at least 5 years completed a web-based survey that included the RCI-10 (to measure the independent variable, religiosity) as well as four subscales from the revised Conflict Tactics Scales (CTS2; Straus et al. 1996) – minor physical assault, major physical assault, minor psychological aggression, major psychological aggression (to measure the dependent variable, IPV) – using an 8-point Likert scale (0 = never happened, 8 = happened more than 20 times) . Perpetration of these acts was measured for the past year.[1]

---

[1] Men were recruited to the study in the fall of 2011 by the national online sampling and survey administration service, Zoomerang (which has since been acquired by Survey Monkey). There is

Scores on the RCI-10 could range from 10 to 50; the mean religiosity score for this sample was 31.43 (SD = 3.76). The sample was then divided into two groups – low religiosity (scores ranging from 10 to 29) and high religiosity (scores ranging from 30 to 50) – for bivariate analysis.

Scores on the scale of minor physical violence ranged from 0 to 30 (Mean = 3.68, SD = 8.08), and 26.1 % of the men reported perpetrating at least one act of minor physical violence in the past year. In addition, 19.9 % reported perpetrating as least one act of major physical violence in the past year (range = 0–24, Mean = 4.56, SD = 10.64). Not surprisingly, perpetration of psychological aggression was more common: 66.7 % of the men reported perpetrating at least one act of minor psychological aggression during the past year (range = 0–24, Mean = 4.44, SD = 5.04); 24.1 % reported perpetrating at least one act of major psychological aggression in the past year (range = 0–24, Mean = 3.49, SD = 7.49). For bivariate analysis, results for each type of violence were collapsed into two groups: men who had perpetrated at least one act of that type of violence in the past year, and men who had not perpetrated that type of violence in the past year.

Chi-square statistics were calculated to test the relationship between religiosity and IPV perpetration. In all analyses, we controlled for race/ethnicity and income, since previous studies have shown that both race/ethnicity and income have independent effects on the likelihood of IPV perpetration. None of the results of these tests were statistically significant, although they all approached significance. Men who scored high in religiosity on the MCI-10 perpetrated fewer acts of both minor ($p = .07$) and major psychological aggression ($p = .07$) than those men who scored low in religiosity. But to our surprise, men who scored high in religiosity on the MCI-10 also perpetrated more acts of both minor ($p = .10$) and major ($p = .06$) physical violence than men who scored low in religiosity.

Although these findings are at first glance puzzling – why would religiosity be associated with a reduction in psychological IPV, but an increase in physical IPV? – they remind us, as noted previously, that religiosity in and of itself may not be meaningfully related to likelihood of perpetrating IPV without considering *how* a person is religious.

---

no way to determine, therefore, the representativeness of the sample. Nearly 81 % of the men were White, 6.1 % were Black, 5 % were Asian American/Pacific Islander, 1.1 % were American Indian/ Alaskan Native, and 3.8 % were multi-racial; 3.2 % chose not to report their race. In addition, regardless of their racial identification, 7.3 % of the men identified as Hispanic. About 15 % of the men reported an annual income less than $25,000, while 17.6 % reported annual incomes between $25,001 and $40,000; 28.4 % between $40,001 and $70,000; 28.4 % between $70,001 and $90,000; and 20.7 % over $90,000. More than one third of the men (38.7 %) had been in their current intimate relationship from 5–7 years; 12.3 % , 7.1–10 years; 9.6 %, 10.1–13 years; 5.7 %, 13.1–15 years; 5.7 %, 15.1–20 years; and more than a quarter (25.7 %) for more than 20 years. About 3 % of the men, though, did not report the length of their current intimate relationship.

166.2.3 Kinds of Religiousness and IPV Perpetration

Pargament (2002) points out that the influence of religiosity on specific behavioral outcomes depends to some extent on the degree to which one's religion is integrated into one's everyday life. As social psychological research on religiosity has shown, in order for religious values to influence attitudes and behavior, they must be internalized (Blais et al. 1990). When religious values are internalized, the individual "takes on" the values as his or her own. Drawing on self-determination theory, Richard Ryan and his colleagues (Ryan et al. 1993) identified two styles of religious internalization that affect how people regulate their own behavior. They label these introjected religious self-regulation and identified religious self-regulation.

According to self-determination theory, *introjection* is a self-regulating style in which an individual is motivated to adopt a value or rule (or a religious teaching or practice) because he or she wishes to gain or wishes not to lose the approval of others; they want to avoid guilt and shame. In contrast, *identification* is a self-regulating style in which an individual is motivated to adopt a value or rule (or a religious teaching or practice) because it gives him or her enjoyment, it is intrinsically satisfying to him or her, and it corresponds to his or her personal values and beliefs. Ryan et al.'s (1993) research and that of others (for example, Blais et al. 1990) show very different emotional and behavioral outcomes for each of these types of religious self-regulation. Those high in introjected regulation have higher rates of depression, lower self-esteem, lower perceived self-efficacy, and tend to use less adaptive coping strategies. In contrast, those high in identified regulation are more likely to express greater life satisfaction, higher self-esteem, a greater sense of self-efficacy, and tend to use more positive and adaptive coping strategies (Ryan et al. 1993). Consequently, in our recent study of the relationship between religiosity and IPV perpetration, we also examined whether *how* men are religious, in terms of their style of religious self-regulation, affects their likelihood of perpetrating IPV.

The 251 men in our sample, completed the Religious Self-Regulation Questionnaire (RSQ; Ryan et al. 1993), which is composed of 12 items – 6 that measure introjected religious self-regulation, and 6 that measure identified religious self-regulation – using a 7-point Likert scale (1 = not at all true, 7 = very true). Items that measure introjected religious self-regulation include: "One reason I think it's important to actively share my faith with others is because I would feel bad about myself if I didn't," and "A reason I think praying by myself is important is because if I don't, God will disapprove of me." Items that measure identified religious self-regulation include: "When I turn to God, I most often do it because I enjoy spending time with Him," and "An important reason why I attend church is by going to church I learn new things."

We found that men who scored high in introjected religious self-regulation perpetrated significantly more acts of minor physical violence ($r=.312$, $p<.001$), major physical violence ($r=.317$, $p<.001$), minor psychological aggression ($r=.221$, $p<.001$), and major psychological aggression ($r=.315$, $p<.001$). With regard to the relationship between identified religious self-regulation and IPV perpetration, however, all of the correlations were weak and none was statistically significant, although they were in the expected direction. That is, men who scored high on identified religious self-regulation perpetrated fewer acts of minor physical violence ($r=-.062$, ns), major physical violence ($r=-.062$, ns), minor psychological aggression ($r=-.062$, ns), and major psychological aggression ($r=-.064$, ns).

It is worth reiterating that introjected religious self-regulation has been found to be associated with several negative psychological outcomes, including depression, low self-esteem, low perceived self-efficacy, and use of less adaptive coping strategies. Importantly, research with IPV perpetrators has found that they typically have low self-esteem and low perceived self-efficacy, and they tend to use negative coping strategies (Edleson and Tolman 1992; Holtzworth-Munroe and Mehan 2004). In contrast, men (and women) who are high in identified religious self-regulation have been found to express greater life satisfaction, high self-esteem, high perceived self-efficacy, and tend to use positive, adaptive coping strategies. Although the correlations between identified religious self-regulation and IPV perpetration in our study were weak and not statistically significant, they were in the expected direction. Thus, religiosity per se, whether high or low, may be less functionally important than *how* adherents to a religion are religious. This point is further underlined when one examines the intersection of beliefs and behavior.

166.2.4   Religious Beliefs, Gender Attitudes, and IPV Perpetration

Researchers have found no relationship between religious denomination (for example, Cunradi et al. 2002) or denominational homogamy (for example, Ellison et al. 1999) and likelihood of perpetrating intimate partner violence. Nevertheless, some denominations are associated with more conservative values and beliefs – with regard to gender roles and intimate relationships as well as other social issues – than other denominations, and within a specific denomination, congregations often vary in their level of traditionalism. Given that researchers report that traditional, as opposed to egalitarian, attitudes about gender roles – that is, beliefs that support stereotypical traits and behavior for women and men – are one of the most consistent predictors of attitudes that support violence against women, one might expect that adherents to religious beliefs, values, and teachings that promote traditional gender roles would be more likely to perpetrate, or at least to legitimate, violence against an intimate partner, particularly if that partner violates strongly held norms and beliefs (Berkel et al. 2004). The research that has tested this hypothesis has produced equivocal findings at best. For example, while Berkel et al. (2004) report that the college students in their sample who held traditional gender role attitudes were more likely than those who held egalitarian gender role attitudes to express more support for the use of violence against women, Cunradi et al. (2002) found that rates of IPV perpetration were higher among men in religious groups classified as liberal than those in religious groups classified as fundamentalist. Recall that in our study (Renzetti et al. 2012) men who scored high in religiosity as measured by the RCI-10 were significantly less likely to perpetrate both minor and major psychological IPV. Interestingly, we (Renzetti et al. 2011) also found that this relationship was even stronger for men who endorsed items on the Old Fashioned Sexism Scale $(p<.001)$.[2]

Cunradi et al. (2002: 149) speculate that such findings reflect the centrality of the patriarchal family unit among fundamentalists, which may lower their IPV risk, whereas couples in more liberal religious groups "may be less socially constrained by religiously ordained norms and rules governing discourse within family life." As Pargament (2002) points out, while psychologists have been critical of religious fundamentalism for its association with rigid thinking and authoritarianism, it holds some clear advantages for adherents, including greater marital happiness, satisfaction and commitment, reduced marital conflict, and more effective marital problem-solving strategies. Additional research shows that highly religious couples report that their religious beliefs help them cultivate a shared vision and purpose, enhance their relational virtues, reduce stress levels in their marriage, facilitate relational reconciliation following a conflict, and enhance partners' willingness to forgive one another (Lambert and Dollahite 2006; see also Wilcox and Wolfinger 2007).

Pargament (2002: 176) cautions that the helpfulness of religiousness may vary across context and situation and, moreover, that "the efficacy of religion may have less to do with specific religious beliefs and practices and more to do with the degree to which religion is well integrated into individuals' lives." Both of these hypotheses are certainly worthy of further empirical investigation with regard to how religiosity influences men's likelihood of perpetrating intimate partner violence.

166.3 Directions for Future Research

As we noted in the introduction, research on the relationship between religiosity and intimate partner violence perpetration is relatively undeveloped. Much of the research to date has been limited by narrow measures of religiosity that are unidimensional and fail to capture the complexity of this construct. We sought to overcome this limitation by using the RCI-10 and the RSQ. Another promising measure is the Brief Multidimensional Measure of Religiousness/Spirituality (BMMRS; Fetzer Institute 1999), which is designed to tap 12 different dimensions of religiosity, including daily spiritual experiences, religious and spiritual coping, and religious support. This last element is noteworthy, since most studies have neglected the "religious communities" dimension of religiosity identified by Dollahite et al. (2004). Pargament (2002: 174) also stresses the need for religious measures "linked to particular contexts and situations" because these are "stronger predictors of

---

[2] The Old-Fashioned Sexism Scale (Swim and Cohen 1997) is composed of five items such as "Women generally are not as smart as men," and "I would be more comfortable having a man as a boss than a woman," for which respondents use a 5-point Likert scale (1=strongly agree, 5=strongly disagree) to respond. See also Swim et al. 2005.

well-being than global distal religious measures." In addition, he reminds us that some of the influence of religion and religiosity on behavior may be direct, but some may also be mediated "by more finely delineated religious dimensions" (178). Future studies of IPV perpetration, therefore, should develop explanatory models that include both direct and mediated effects of religion/religiosity.

A related concern stems from the measures of intimate partner violence that are typically used in these studies. In secondary analyses of large, national survey data-sets (for example, Ellison et al. 2007), IPV perpetration is usually measured with one or two items, such as "During the past year, how many fights with your [partner] resulted in you hitting, shoving or throwing things at her?" Methodological research has shown that such narrow measures result in gross underestimations of IPV frequency and severity (Jaquier et al. 2011). The most commonly used instruments, though, are the Conflict Tactics Scales (CTS) and Revised Conflict Tactics Scales (CTS-2). Both of these have been extensively criticized, largely for their inability to capture the context, meaning, and motivations underlying the violent event (see DeKeseredy and Schwartz 1998).

In addition to more nuanced measures of both religiosity and IPV, future studies should include more diverse samples. Although there are some notable exceptions (see, for example, Landau et al. 2002; Nojomi et al. 2007; Vakili et al. 2010), much of the research to date has sampled largely Christian groups, many of which reside in the U.S. or other Western industrialized nations. Future research should strive to include members of non-Christian religions and spiritual groups, including New Age, "congregation free" spirituality, and other religious minorities (Dollahite et al. 2004). Comparative research conducted in societies in which non-Christian groups dominate and in which rates of IPV have been found to vary significantly from rates in the U.S and Western industrialized countries would also enrich our understanding of the direct and mediated effects of religion/religiosity on intimate partner violence perpetration. And given that levels and types of religiosity as well as rates of IPV perpetration vary by place of residence (e.g., rural, suburban, urban communities; southern United States relative to other regions of the country), researchers need to attend to regional variations in their studies of the relationship between religion/religiosity on intimate partner violence perpetration.[3]

---

[3] Many researchers, for instance, have documented high levels of religiosity in the southern United States compared with other regions of the country, although some have argued that religiosity in the South is likely to erode as greater interregional migration occurs. Interestingly, Smith et al. (1998) found that, at least in terms of church attendance and importance of faith, religiosity actually increases among those who move to a region where religious commitment is already high. Therefore, migration to the South increases religiosity among those who move there, but religiosity tends to decrease among those who migrate to regions with lower religious commitment. Smith et al.'s research also highlights the need to study not only migration – interregional as well as international – with regard to changes in religiosity, but also with respect to perpetration of intimate partner violence. There is a growing body of research exploring this relationship; see, for example, Raj and Silverman 2002; Morash et al. 2007; and Grzywacz et al. 2009. For a discussion of regional variations in IPV perpetration, see Websdale 1998; and DeKeseredy and Schwartz 2009. Many studies indicate that rates of all forms of violence are higher in the southern United States as well as in rural regions, and researchers generally theorize that this is due to cultural

Nearly all of the research on religiosity and IPV has been correlational and cross-sectional. Such studies do not allow us to draw conclusions about causality. Although we assume that specific religious beliefs and practices promote or inhibit certain behaviors, such as IPV perpetration, it may be the case that aggressive, violent, or hostile people, particularly those who are hostile to women, are attracted to or seek out religions that support their attitudes. Longitudinal studies would likely give us greater confidence in our causal inferences about the effects of religiosity on IPV perpetration, especially given that religious beliefs and commitment typically change over time for many people (Pargament 2002; Dollahite et al. 2004).

Finally, we join several other scholars who call on researchers to be more open to and less negatively biased toward individuals, couples, and families "who take their religion very seriously" (Stark and Finke 2000: 14; see also Dollahite et al. 2004; Johnson 2011). The empirical literature indicates that most religions, even in their fundamentalist forms, may be both beneficial and detrimental to adherents (Pargament 2002). However, we are unlikely to identify the benefits as well as the detriments of religion in relation to intimate partner violence perpetration if our research is colored by what Johnson (2011: xii) calls "the last acceptable prejudice" in an age of political correctness. As Pargament (2002: 168) point out, "Questions about the general efficacy of religion should give way to the more difficult but appropriate question: how helpful or harmful are particular forms of religious expression for particular people dealing with particular situations in particular social contexts according to particular criteria of helpfulness or harmfulness?" To discover the complex but accurate answers to this question, though, requires us, as social scientists, to remove the blinders of our prejudice against highly religious people.

References

Archer, J. (2000). Sex differences in aggression between heterosexual partners: A meta-analytic review. *Psychological Bulletin, 126*(5), 651–680.

Barnett, O. W., Lee, C. Y., & Thelan, R. (1997). Gender differences in attributions of self-defense and control in interpartner aggression. *Violence Against Women, 3*(5), 462–481.

Basile, K. C., & Black, M. C. (2011). Intimate partner violence against women. In C. M. Renzetti, J. L. Edleson, & R. K. Bergen (Eds.), *Sourcebook on violence against women* (2nd ed., pp. 111–130). Thousand Oaks: Sage.

Bennett, L., & Bland, P. (2008). *Substance abuse and intimate partner violence*. Harrisburg: VAWnet. Retrieved February 23, 2011, from www.vawnet.org

---

variations in acceptance of attitudes supporting violence as a way to solve conflicts, including conflicts between intimate partners. However, Brownridge (2002), who has conducted research on regional variations of violence against women in Canada, argues that one must specifically examine adherence to norms of patriarchal domination to understand such regional differences. As we noted previously, attention to such gender norms must also be included in studies of the effects of religiosity on IPV perpetration.

Berkel, L. A., Vandiver, B. J., & Bahner, A. D. (2004). Gender role attitudes, religion, and spirituality as predictors of domestic violence attitudes in white college students. *Journal of College Student Development, 45*(2), 119–133.

Black, M. C., Basile, K. C., Breiding, M. J., Smith, S. G., Walters, M. L., Merrick, M. T., Chen, J., & Stevens, M. B. (2011). *The National Intimate Partner and Sexual Violence Survey: 2010 summary report.* Atlanta: Centers for Disease Control and Prevention.

Blais, M. R., Sabourin, S., Boucher, C., & Vallerand, R. J. (1990). Toward a motivational model of couple happiness. *Journal of Personality and Social Psychology, 59*(5), 1021–1031.

Brownridge, D. A. (2002). Cultural variation in male partner violence against women: A comparison of Québec with the rest of Canada. *Violence Against Women, 8*(1), 87–115.

Brush, L. D. (2011). *Poverty, battered women and work in U.S. public policy.* New York: Oxford University Press.

Cunradi, C. R., Caetano, R., & Schafer, J. (2002). Religious affiliation, denominational homogamy, and intimate partner violence among U.S. couples. *Journal for the Scientific Study of Religion, 41*(1), 139–151.

DeKeseredy, W. S., & Dragiewicz, M. (2009). *Shifting public policy directions: Gender-focused versus bidirectional intimate partner violence.* Toronto: Ontario Women's Directorate. Report prepared for the Ontario Women's Directorate.

DeKeseredy, W. S., & Schwartz, M. D. (1998). *Measuring the extent of woman abuse in intimate heterosexual relationships: A critique of the Conflict Tactics Scales.* Harrisburg: VAWnet, the National Online Resource Center on Violence against Women. Retrieved February 13, 2012, from www.vawnet.org

DeKeseredy, W. S., & Schwartz, M. D. (2009). *Dangerous exits: Escaping abusive relationships in rural America.* New Brunswick: Rutgers University Press.

DeWall, C. N. (2010, October). *God give me self-control strength: Unlocking the mystery between religiosity and self-control.* Paper presented at the Centre for Research on Self and Identity, University of Southampton, Southampton, UK.

Dobash, R. P., Dobash, R. E., Cavanagh, K., & Lewis, R. (1998). Separate and intersecting realities: A comparison of men's and women's accounts of violence against women. *Violence Against Women, 4*(4), 382–414.

Dollahite, D. C., Marks, L. D., & Goodman, M. A. (2004). Families and religious beliefs, practices, and communities: Linkages in a diverse and dynamic cultural context. In M. Coleman & L. H. Ganong (Eds.), *Handbook of contemporary families: Considering the past, contemplating the future* (pp. 411–431). Thousand Oaks: Sage.

Dutton, D. G. (2006). *Rethinking domestic violence.* Vancouver: University of British Columbia Press.

Dutton, D. G., Corvo, K. N., & Hamel, J. (2009). The gender paradigm in domestic violence research and practice part II: The information website of the American Bar Association. *Aggression and Violent Behavior, 14*(1), 30–38.

Edleson, J. L., & Tolman, R. M. (1992). *Interventions for men who batter: An ecological approach.* Thousand Oaks: Sage.

Ellison, C. G., & Anderson, K. L. (2001). Religious involvement and domestic violence among U.S. couples. *Journal for the Scientific Study of Religion, 40*(2), 269–286.

Ellison, C. G., Bartkowski, J. P., & Anderson, K. L. (1999). Are there religious variations in domestic violence? *Journal of Family Issues, 20*(1), 87–113.

Ellison, C. G., Trinitapoli, J. A., Anderson, K. L., & Johnson, B. R. (2007). Race/ethnicity, religious involvement, and domestic violence. *Violence Against Women, 13*(11), 1094–1112.

Fetzer Institute. (1999). *Multidimensional measurement of religiousness/spirituality for use in health research.* Kalamazoo: Fetzer Institute.

Fox, J. A., & Zawitz, M. W. (2007). *Homicide trends in the United States, 2007.* Washington, DC: Bureau of Justice Statistics.

Grzywacz, J. G., Rao, P., Gentry, A., Marin, A., & Arcury, T. A. (2009). Acculturation and conflict in Mexican immigrants' intimate partnerships: The role of women's labor force participation. *Violence Against Women, 15*(10), 1194–1212.

Holtzworth-Munroe, A., & Mehan, J. C. (2004). Typologies of men who are maritally violent. *Journal of Interpersonal Violence, 19*(12), 1369–1389.

Jaquier, V., Johnson, H., & Fisher, B. S. (2011). Research methods, measures, and ethics. In C. M. Renzetti, J. L. Edleson, & R. K. Bergen (Eds.), *Sourcebook on violence against women* (2nd ed., pp. 23–45). Thousand Oaks: Sage.

Johnson, B. R. (2011). *More God, less crime: Why faith matters and how it could matter more.* West Conshohocken: Templeton Press.

Lambert, N. M., & Dollahite, D. C. (2006). How religiosity helps couples prevent, resolve, and overcome marital conflict. *Family Relations, 55*(4), 439–449.

Landau, S. F., Björkqvist, K., Lagerspetz, K. M. J., Österman, K., & Gideon, L. (2002). The effect of religiosity and ethnic origin on direct and indirect aggression among males and females: Some Israeli findings. *Aggressive Behavior, 28*, 281–298.

Luthra, R., & Gidycz, C. A. (2006). Dating violence among college men and women: Evaluation of a theoretical model. *Journal of Interpersonal Violence, 21*(6), 713–721.

Menard, K. S., Anderson, A. L., & Godboldt, S. M. (2009). Gender differences in intimate partner recidivism: A 5-year follow-up. *Criminal Justice and Behavior, 36*(1), 61–76.

Miller, S. L. (2001). The paradox of women arrested for domestic violence: Criminal justice professionals and service providers respond. *Violence Against Women, 7*(12), 1339–1376.

Morash, M., Bui, H., Zhang, Y., & Holtfreter, K. (2007). Risk factors for abusive relationships: A study of Vietnamese American women. *Violence Against Women, 13*(7), 653–675.

Nason-Clark, N. (2000). Making the sacred safe: Women and abuse and communities of faith. *Sociology of Religion, 61*(4), 349–368.

Nojomi, M., Agaee, S., & Eslami, S. (2007). Domestic violence against women attending gynecologic outpatient clinics. *Archives of Iranian Medicine, 10*(3), 309–315.

Pargament, K. I. (2002). The bitter and the sweet: An evaluation of the costs and benefits of religiousness. *Psychological Inquiry, 13*(3), 168–181.

Raj, A., & Silverman, J. (2002). Violence against immigrant women: The roles of culture, context, and legal immigrant status on intimate partner violence. *Violence Against Women, 8*(3), 367–398.

Rajan, M., & McCloskey, K. A. (2007). Victims of intimate partner violence: Arrest rates across recent studies. *Journal of Aggression, Maltreatment & Trauma, 15*(3/4), 27–52.

Renzetti, C. M. (2009). Intimate partner violence and economic disadvantage. In E. Stark & E. Buzawa (Eds.), *Violence against women in families and relationships: Victimization and community response* (pp. 73–92). Santa Barbara: Praeger/Greenwood.

Renzetti, C. M., & Larkin, V. M. (2009). *Economic stress and domestic violence.* Harrisburg: VAWnet. Retrieved October 4, 2012, from www.vawnet.org

Renzetti, C. M., DeWall, C. N., & Pond, R. (2011, November). *What's religion got to do with it? Religiosity and perpetration of violence against an intimate partner.* Paper presented at A New Frontier for Gender-based Violence: A Community Dialogue, Xavier University, Cincinnati, OH.

Renzetti, C. M., Messer, A., DeWall, C. N., & Pond, R. (2012, November). *By the grace of God: Religiosity, religious self-regulation, and perpetration of intimate partner violence.* Paper presented at the annual meeting of the Society of the Scientific Study of Religion, Phoenix.

Ryan, R. M., Rigby, S., & King, K. (1993). Two types of religious internalization and their relations to religious orientation and mental health. *Journal of Personality and Social Psychology, 65*(3), 586–596.

Smith, C., Sikkink, D., & Bailey, J. (1998). Devotion in Dixie and beyond: A test of the "Shibley Thesis" on the effects of regional origin and migration on individual religiosity. *Journal for the Scientific Study of Religion, 37*(3), 494–506.

Stark, R., & Finke, R. (2000). *Acts of faith.* Berkeley: University of California Press.

Straus, M. A. (2007). Processes explaining the concealment and distortion of evidence on gender symmetry in partner violence. *European Journal on Criminal Policy and Research, 13*(3–4), 227–232.

Straus, M. A., Hamby, S. L., Boney-McCoy, S., & Sugarman, D. B. (1996). The Revised Conflict Tactics Scales (CTS2): Preliminary psychometric data. *Journal of Family Issues, 17*(3), 283–316.

Swim, J. K., & Cohen, L. L. (1997). Overt, covert, and subtle sexism: A comparison between the attitudes toward women and modern sexism scales. *Psychology of Women Quarterly, 21*(1), 103–118.

Swim, J. K., Mallett, R., Russo-Devosa, Y., & Stangor, C. (2005). Judgments of sexism: A comparison of the subtlety of sexism measures and sources of variability in judgments of sexism. *Psychology of Women Quarterly, 29*, 406–411.

Tjaden, P., & Thoennes, N. (2000). Prevalence and consequences of male-to-female and female-to-male intimate partner violence as measured by the National Violence Against Women Survey. *Violence Against Women, 6*(2), 142–161.

Tjaden, P., & Thoennes, N. (2006). *Extent, nature, and consequences of rape victimization: Findings from the National Violence Against Women Survey.* Washington, DC: U.S. Department of Justice, National Institute of Justice.

Vakili, M., Nadrian, H., Fathipoor, M., Boniadi, F., & Morowatisharifabad, M. A. (2010). Prevalence and determinants of intimate partner violence against women in Kazeroon, Islamic Republic of Iran. *Violence and Victims, 25*(1), 116–127.

Websdale, N. (1998). *Rural woman battering and the justice system: An ethnography.* Thousand Oaks: Sage.

Wilcox, W. B., & Wolfinger, N. H. (2007). Living and loving "decent": Religion and relationship quality among urban parents. *Social Science Research, 37*(11), 828–843.

Woodberry, R. D., Park, J. Z., Kellstedt, L. A., Regnerus, M. D., & Steensland, B. (2012). The measure of American religious traditions: Theoretical and measurement considerations. *Social Forces, 91*(1), 65–73.

Worthington, E. L., Jr., Wade, N. G., Hight, T. L., Ripley, J. S., McCullough, M. E., Berry, J. W., Schmitt, M. M., Berry, J. T., Bursley, K. H., & O'Conner, L. (2003). The Religious Commitment Inventory-10: Development, refinement, and validation of a brief scale for research and counseling. *Journal of Counseling Psychology, 50*(1), 84–96.

Chapter 167
Chinese Hui Muslim Pilgrims—Back Home from Mecca: Negotiating Identity and Gender, Status and Afterlife

Maria Jaschok and Shui Jingjun

167.1 Introduction

Since the early 1990s, when pilgrimages by Chinese Muslims to Mecca were taken over by the Pilgrim Travel Bureau of the State-controlled Chinese Islamic Association, the number of *Hajjis* (male pilgrims, also used generically) and *Hajjas* (female pilgrims) (Figs. 167.1 and 167.2) has sharply increased. So great is the demand for pilgrimage, however, that the limited quota of official pilgrimage groups is insufficient, and desperate would-be pilgrims are prepared to pay exorbitant fees to independent travel agencies in order to fulfill one of the most important Muslim duties. Against this background of growing numbers of state-organized and private pilgrims, including female and male Hajjis (with women frequently outnumbering men), we argue that emerging issues related to the growing impact of pilgrimages on both individual Hajjis themselves and on their home communities deserve closer investigation.

167.1.1 From "Periphery" to "Center"

In her introduction to an unprecedented exhibition on *Hajj, journey to the heart of Islam,* held at the British Museum in London in 2012, Karin Armstrong ponders the enduring and universal urge of believers among all major religions to create their

M. Jaschok (✉)
International Gender Studies Centre, Lady Margaret Hall, Oxford University,
Norham Gardens OX2 6QA, UK
e-mail: maria.jaschok@lmh.ox.ac.uk

S. Jingjun
Henan Academy of Social Sciences, Zhengzhou, Henan Province, China
e-mail: shji7979@aliyun.com

© Springer Science+Business Media Dordrecht 2015
S.D. Brunn (ed.), *The Changing World Religion Map,*
DOI 10.1007/978-94-017-9376-6_167

部分朝觐的男哈吉

Fig. 167.1 Male Hajjis from Sangpo, central China (Photo by Shui Jingjun)

部分朝觐的女哈吉

Fig. 167.2 Female Hajjis from Sangpo, central China (Photo by Shui Jingjun)

"sacred geography" as representations of enduring mythical realms of belief (Armstrong 2012; Bianchi 2004; Hammoudi 2005; Porter 2012; Shariati 2005; Trojanow 2007). Exploring the reasons for embarking on pilgrimages that were often dangerous, a strain on energy and financial resources and posing unpredictable challenges to life and limb, Armstrong suggests as an important motivation the pilgrim's search for "a therapy of distance." That is, dissatisfied with life in their home countries, pilgrims "make a symbolic break with profane existence … and leave the spiritually peripheral for the "centre"" (Armstrong 2012: 20). While Armstrong and fellow-contributors to an exhibition companion volume focus on the richness of ritual and pilgrimage practices in engendering Muslim believers' rupture from everyday life and their journey into faith, we are particularly interested in the reflections by Hajjis upon the return into their local community on the minutiae of ordinary social interaction that is part of the crowdedness of *Hajj* (pilgrimage) everyday life. What has lingered in their memories? What impressions have Hajjis brought back with them and how have comparisons with other Islamic practices observed during the Hajj translated into changes at home? In what way does the reinvigorated "sacred geography," by which Muslims anywhere and far from their spiritual centre may know the direction of prayer, connect home community with the source of Islamic faith and Muslim practice? What changes are affected in their identity as Chinese citizens? Using theories of social interaction and social identity, Clingingsmith et al. (2008) conducted a comparative study of successful and unsuccessful Pakistani Muslim applicants for the Hajj. They found that the collective experience of Hajj, a salient feature of pilgrimage travel did not, as the theories would suggest, lead to feelings of hostility towards Muslims of countless other pilgrimage groups encountered in the course of Hajj or culminate in a rejection of non-Muslims. Their findings suggest that, on the contrary, those who had taken part in the Hajj are inclined to greater tolerance in the face of greatly diverse Islamic practices, and that Hajjis tend to develop a preference for multicultural harmony and unity. And that these qualities in turn made for greater identification with the *Ummah* (the world community of Muslims) as well as the intensification of commitment to Muslim practice. However, the authors also pointed out that whereas opportunities during the Hajj for observations of more liberal gender practices among certain pilgrim groups contributed towards a more positive appreciation of women on the part of conservative Pakistani Muslims, ultimately the adjustments were less than radical. They noted that Pakistani Hajjis' "altered views on women reflect a movement away from local prejudices against women and towards fairer treatment within Islam, rather than a more general trend towards feminism" (Clingingsmith et al 2008: 5). The gender of the pilgrim, on the other hand, was notable when it came to satisfaction with the Hajj. Less satisfaction was observed among Hajjas who were generally less well pleased with their experience of pilgrimage than were their male counterparts. The authors attribute the gender difference to physical exertions of a pilgrimage but also considered that perhaps the opportunities Pakistani women had themselves to observe women from other more liberal Muslim contexts may have played a part in creating certain displeasure.

While there are similarities with the Pakistani study, the Chinese study shows up certain dissimilarities, whether in the degree of openness to other kinds of Muslimness, the impact of Hajj on local Muslim traditions and identity or indeed in relation to gender. As will be discussed below, the question of gender is certainly notable in evaluating the impact of pilgrimage on Chinese men and women, but Chinese women are arriving at very different conclusions compared to the Pakistani women in Clingingsmith's study (2008). Moreover, effective government control of the pilgrimage industry in China and strong avowals of nationalistic sentiment by Chinese Muslims, members of a vulnerable minority religion in their home country, lead us to suggest that "Muslim China" in all its diversity is likely to continue carrying a distinct separate identity within the wider "sacred geography" of Islam (Ah 1995; Liang 2005; Wang 1994; Xie 2009). We explore this question at length in our conversations with Chinese returned Hajjis. While undoubtedly they have become more concerned about living up to the ideal of a Hajji as a role model for all, making their mark on education and female dress code in particular, few are actively engaged in shaping and changing decision-making in local politics. Whereas these are critical of certain weaknesses in Muslim conduct at home, compared to what is perceived as superior religious piety on the part of other Muslim nationalities, a perception shared by nearly all Hajjis, none of the returned pilgrims saw a need for radical intervention, let alone for application locally of the Sharī'ah (Islamic law). It is our finding that Hajjis' identity as Chinese citizens ultimately overrides the wish to advocate the establishment of Sharī'ah and of a Muslim state. "China is not Saudi Arabia," Chinese Muslim informants are often heard to say. Conversations with Sangpo Muslims reveal the importance of the distinction retained between preservation of the universalism of Islamic core pillars of faith and what is considered the unsuitable and improper transplanting of aspects of Arab-Muslim culture into local practices. Sangpo Hajjis are representative of Chinese Hajjis elsewhere in their strong sense of a Chinese national identity that makes their communities and sites of worship both an integral part of the "sacred geography" of Islam and yet retain distinct Chinese characteristics.

The rural community of Sangpo Hui[1] Muslim village in central China is a useful case study of negotiations of multiple identities. Abdel Haleem suggests in relation to Muslim Egyptian culture, that "Hajj is more keenly felt and has more impact on people in Muslim villages than in the cities" (Abdel Haleem 2012: 58). This observation holds true of Sangpo village, an ideal case study of the intersections of growing global Islamic influences with strong localized Muslim traditions embedded in tight, comprehensive mechanisms of government control at a time when Chinese mainstream society pushes for a model of secular modernity.

---

[1] Estimates of the Muslim population in China vary between 18 million to 25 million, or more. The majority belong to the Sunni tradition and its various branches. The Hui nationality is the most populous nationality among ten ethnic minorities which together constitute the Muslim population in China, and with whom this study is concerned.

167.1.2   China Inside the "Sacred Geography"

We ask how Chinese pilgrims are preparing themselves for their journey to Mecca, and with whose support? What changes in their conduct and in social interaction may be noted after their return from the pilgrimage? What is the perceived impact of Hajjis on the religious and secular affairs of their local society? How does gender play into the role that returned Hajjis are asked to play in their community? How have opportunities for interaction with Muslims from other countries translated into personal and communal examinations of national and ethnic identities? The pilgrimage involves the entire social network in which the Hajjis are embedded, including family members, friends, community leaders, religious and secular organizations at all levels of the pilgrimage industry. The deepening local impact of returned pilgrims, their role as interpreters of "the outside world" and of global Islam in particular, and their influence on societal affairs beyond the religious sphere, are framed by the domestic context of China's repressive religious policies and the volatility of Middle East politics.

167.2   China's Muslim Pilgrimage Industry and Culture

Ever since the early 1990s, the number of Chinese Muslims making their mandatory pilgrimage to Mecca has been gradually increasing. According to statistics gathered by the Villagers' Committee in Sangpo, Henan Province, from 1993 to the spring of 2008, the number of Hajjis to Mecca, organized by the Pilgrim Travel Bureau of the Chinese Islamic Association, had risen to 158. Males accounted for 44 % and Hajjas accounted for 56 % of the total number of pilgrims. Most pilgrims joined the pilgrimage tours under the auspices of the Chinese Islamic Association, however, the number of Hajjis completing their pilgrimages by private means is ever increasing.[2]

Counting both State-organized and private pilgrims, the total number of pilgrims in 2010 was around 200 or so, about 4 % of the total population of Sangpo Village. A thriving pilgrimage industry has brought changes far beyond the religious sphere.

---

[2] In May 2008, WEMC researchers arrived at Sangpo, a large and affluent Hui Muslim village in central China, and conducted 33 in-depth interviews. Several categories of people were interviewed: male and female Hajjis, registered pilgrims, villagers not ready to make a pilgrimage, leading members administering national religion affairs and administrators of local mosques. Interviews focused on the place of Islamic faith in village life and the impact of returned Hajji on social and gender change. Sangpo Village has 11 mosques, five male and six female mosques.

167.2.1   The History of Pilgrimage and Modern Hajjis

Historically, few residents in Sangpo had the capability or opportunity to make a pilgrimage. Only two local men ever went to Mecca. For example, Pang Shiqian was a native son of Sangpo Village and famous for his extensive Islamic scholarship; he made a pilgrimage in 1939 when he led a congregation of pilgrims made up of Chinese students then studying in Egypt.

A milestone for Sangpo Muslims came in 1993, when the first pilgrim from their village joined a pilgrimage tour organized by the Chinese Islamic Association. After 2000, the official quota for pilgrims grew more generous, giving financially solvent believers a chance to fulfill one of the core duties of a pious Muslim.

167.2.2   Preparing for Hajj

The Chinese Islamic Association, present at all levels of administration of Islamic affairs in China, takes general charge of organizing the pilgrimage, including making detailed practical and logistical arrangements (Ah 2005; Ma 2003; Xie 2009). Owing to the economic disparity between widely dispersed and diverse Muslim communities in China, the Association ensures that surplus quotas for pilgrimage places are transferred from poverty-stricken areas, where few Muslims are able to afford pilgrimages, to the wealthier areas where available quotas do not match the high level of demand for participation. Such a reallocation of quotas has contributed greatly to the easing of frustrations of earlier years when the selection of pilgrims under conditions of unequal wealth engendered much anger over lack of organizational efficiency.

An important aspect of the preparatory work of the Islamic Association is the proper conduct of pilgrims in their encounter with Muslims from other countries. A 60-year old Hajja talked about her participation in meetings organized by officials from the Chinese Islamic Association.

> During the two days of study in Zhengzhou [provincial capital of Henan], we were told about the situation in Saudi Arabia and what worship we would be doing. We were also told not to mind how other Muslims worship. We should just worship in our own way. We were told not to interfere with other [non-Chinese] Muslims nor bring bad customs and habits back home. That is what we were taught.

Such guidelines would be repeated and reinforced at countless other meetings, elaborating in greater detail on appropriate conduct abroad. Ignorant of customs and language, Chinese Muslims were instructed not to wander about Mecca, and not to venture forth alone unless in a group of at least four persons.

All Hajjis and non-Hajjis agree on the criteria of what constitutes a good pilgrim. Financially, the expenses of pilgrimage must pose no additional burden on the normal life of family members. Also, sound health is a prerequisite so as to complete the entire course of the challenging pilgrimage. The third criterion relates specifically to women; they must not go unaccompanied. But as a villager explained:

> In Sangpo, we are all related, and it is easy to fulfill the condition stipulated by the Qur'an as the women are all in the company of their family members.

Many villagers are not able to join the Hajj, thus participation in the traditional customs of ceremonial send-off and festive welcome is particularly poignant, turning each act of pilgrimage into a collective celebration of both shared religious piety and spiritual blessing.

An elderly Hajji, who made a pilgrimage in 1996, describes what happened when he returned home.

> After we alighted from the bus, we walked around the village. At the door [of every house], there were tables, which we call tables for guests, offering a pot of tea in readiness for the returning Hajjis. This is the etiquette … there were also fruits, savories and some other beverage prepared for the returning Hajjis …. The villagers went to meet us when they heard that we had returned home. Everybody came to meet us.

167.3 Pilgrimage: A Journey into Islamic Faith

For almost all Hajjis, their journey begins with study and introspection. An elderly Hajja recounts how she felt before the pilgrimage.

> I didn't become anxious until I was ready to go on the pilgrimage. Then I feared I couldn't complete my required courses of prayer for lack of relevant knowledge. During that time, I would go to the mosque to learn some rudimentary knowledge whenever I had free time. The Ahong [Imam] there explained to us what we must read.

She studied every day before she went on pilgrimage.

The pilgrims talked of a sense of awe in anticipation of the journey which they had for so long desired. But they also spoke of the pressure of expectations of their home community towards Hajjis. All of these factors were felt to bring about changes in their behavior as they prepared for the pilgrimage. An old Hajja talked about her psychological state:

> I had complicated feelings at that time. On the one hand, I was very excited and happy about what was to come as my dream of many years would finally become true. I was delighted, of course. On the other hand, I was scared and worried. My feelings were very complicated …. and what's more, I feared that I might not complete my pilgrimage as the days of prayer can be tiring. You might not be able to complete the course of prayer even if you failed in a trifling thing. I was quite tense and worried because my legs are weak. I was afraid I couldn't overcome my poor health.

But she carefully prepared for her pilgrimage.

> I started to do things more carefully than ever in case I would say or do something wrong. I became more cautious than I used to be.

Her neighbor had similar experiences. She added:

> I was scared because those old Hajjis told us that the place [Mecca] was packed with lots of people. People died from being jostled to death every year. With all the Muslims from

around the world gathering there, I just couldn't imagine how many people there would be. I feared having an accident. What worried me most was the failure to complete my entire course of duties. It's hard to tell you about my complicated feelings at that time. I became quite different, more cautious about things before engaging in them.

The men were similarly preoccupied. One Hajji said:

First you must get mentally prepared when going on a pilgrimage. All those evil things [I committed] and all my guilt, how should I make changes? Some preparations are necessary. Some young people said they were afraid to go on a pilgrimage because they would be prevented from smoking after they returned home. I told them that they could smoke cigarettes but just fewer. Someone said, he liked telling lies and couldn't do that again once he went there. Or someone didn't dare to go because he had no idea what to do if he couldn't get rid of his bad habit of swearing. I told them to pay attention to those bad habits and get rid of them after they returned from a pilgrimage. It's a good thing. It is important to prepare for everything involved in a pilgrimage. You must go there to change yourself and not concern yourself with what is irrelevant.

167.3.1   In the Holy Land

For most of the Hajjis, the pilgrimage was the first time they had been abroad. Once they stepped on the soil of Saudi Arabia, their personal impressions were intense. They made comparisons and formed opinions, reflecting their personalities and individual circumstances. But they had one experience in common, the strong surge of emotions upon arrival in the Holy Land. Male and female Hajjis talked about their emotions when they saw the Ka'ba.

Ah, how emotional I was! I just couldn't help shedding tears. It's hard to tell you about my feelings at that time. I just could not control myself …When I was ending my pilgrimage, I walked around the Ka'ba. I didn't want to leave at all. I just walked several steps and turned round again. Every time I turned round, I couldn't hold back my tears. I was unwilling to go away in my inner heart, but I had to. How sad I was.

His fellow Hajji said:

After arriving there, my heart started beating so fast. I shed tears at the mere sight of the Al-Masjid al-Ḥarām [holy sanctuary of Mecca]. I just couldn't help it…. I cried so hard when I saw the Al-Masjid al-Ḥarām, I cried when I stayed in Jabal Arafat [Plain of Arafat], and I cried when I was leaving the Ka'ba. Those were the moments that impressed me most deeply…when I was ending my pilgrimage, I prayed in my heart to Allah for one more chance of pilgrimage. If only I could have one more pilgrimage.

The women were even more emotional.

Overwhelmed, just overwhelmed. I couldn't hold back my tears when I saw the Ka'ba from a distance. All of us were shedding tears with an emotion beyond description. We all couldn't help crying… Such complicated feelings that I cannot put them into words.

The men added:

That was a really precious place, Allah's holy palace, and you were right near Him. We all wished we could stay longer. We had inexpressible feelings. We were all determined to be true and worthy Hajjis after returning home. Because we had been so close to Allah, confessing our weaknesses to Allah, we would not dare to risk committing any more sinful acts.

167.3.2 "Returning Home" to Mecca

The feeling they had when arriving in Mecca what that of 'returning home.' A Hajja said:

> When I arrived in Mecca, I felt that I returned back to my real home. I began to cry at the mere sight of the Ka'ba. I just couldn't help doing so with all that emotion I felt... Mecca was like our own old home and utterly familiar to us all. We never felt strange when we stayed there. If only we could have stayed on forever.

A 70-year old Hajji also had such feelings.

> I was so emotional the moment I arrived there. It seemed like I was back home.

The sense of familiarity extended to encounters, however fleeting, with their foreign fellow Hajjis. That sense of kinship with a hugely diverse community of pilgrims lingered as an abiding and deep impression.

> It seemed as if we were family when we were in Saudi Arabia. We just got along with each other in perfect harmony. No one quarreled or fought each other. If two cars bumped into each other, the drivers just got out of their cars, wished each other well by shaking hands and then drove away. No damages were claimed. We felt as if we were a big family; we felt entirely at home …. Everybody did their course of prayer, with all their heart and soul, putting all other things aside.

An elderly Hajja reminisced about the difficulty of communication with fellow pilgrims whose language, appearance and extravert conduct made them rather strange to a woman more used to the circumspect ways of communicating in a provincial milieu. She recalled the embrace from unknown women, not a custom back home; on the other hand, such extraordinary conduct, it was felt, suited the heightened emotion of a pilgrimage and formed part of a unique experience.

An older Hajji, of a scholarly inclination, was deeply impressed by the mutual tolerance shown by adherents of different Islamic organizations. He said,

> With the pilgrimage to the Ka'ba, Muslims from all over the world got together. No matter you were white, black or yellow, you were treated equally. We didn't interfere with others' affairs and they didn't interfere with ours. We did things in our own ways. Even if you intended to interfere, there were just too many different traditions …. You just did your own course of prayer without meddling in other's affairs. The doctrine of Islam is sacred, but its application by different people means we have different views and understanding of how to make Islam part of our lives. You do as you think it should be, and I do as I think it should be. So people there get along in perfect harmony without fight or quarrel.

He added:

> Making a pilgrimage is to learn from each other..., we all follow the central Islamic doctrine, and whatever [Islamic] tradition we belong to, we are in basic harmony with each other. This is a more sound harmony than the current so-called "harmonious society" [his reference was to Chinese Government policy of *Hexie*, the officially sanctioned ideal of harmony that binds together disparate peoples and cultures].

167.3.3 "Muslim Identity" at a Distance

From a religious minority background, Chinese Hajjis reflect back on their ambivalence as they entered their first experience of the Ummah when millions of Hajjis gather in Mecca in an over-powering kinship of faith. On the one hand, they felt their position in Mecca shift from the periphery, their assigned place in China, to the centre, using Armstrong's terminology (see above). But, on the other hand, the Ummah they encountered is multi-national, multi-ethnic and multi-cultural. What stood out in their reflections was not their religious identity, a dominant marker in their home country, but their identity as Chinese nationals. Against the maelstrom of fellow-pilgrims in Mecca, an intensely felt nationalism played into complex emotions. Hajjis could be heard to say:

> We promote the image of Chinese Muslims [as pilgrims] and we mustn't bring shame on the Chinese .... When we got there [Mecca], they called us "China's Muslims," our garments featured large letters saying "China's Muslims." Others then know that we are Chinese Muslims as soon as they see us.

A female *Ahong* (Imam) said:

> I was in another country after all. I felt strongly I was Chinese, a Chinese Muslim. I represented the image of the Chinese Muslim. That's what I felt. Many Hajjis felt very proud when they talked of the popularity Chinese Muslims enjoy in Muslim countries.

A male pilgrim said:

> We Chinese pilgrims enjoy a good reputation. We just did what we were supposed to do, not affronting anyone. For example, after completing our mandatory course of prayers, we went shopping in a most orderly manner.... The foreigners respected us very much when they saw us, for we have a good reputation.

The guidelines given during preparation for Hajj are seen as having been of crucial assistance, confirming their pride in representing a country that has often treated them, and still all too often treats them, as marginal peoples.

An elderly Hajja said with an air of complacency:

> The foreigners just put their thumbs up when they saw the Chinese Hajjis. We Chinese people are used to standing still while worshipping. So no matter how crowded it was, how much we were jostled or trampled upon, we just stood still. The foreigners couldn't do it; some of them were touching their heads or scratching their itchy parts. There were many petty actions. On one occasion, a foreigner saw us worshipping. He put his thumbs up and said, "The Chinese people are great."

167.3.4   Affirming Local Muslim Practices

One of the most important concerns was the manner of worship and differences in the way diverse nationalities conducted themselves in prayer. Before going abroad, the Hajjis had been told to expect "foreign customs" to be practiced by "foreign Muslims" in ways that might appear "strange" to them. They had been taught not to interfere with others, and to remain true to their own conduct.

A Hajja proudly recalled how "foreign Muslims" complimented the Chinese way of worship.

> We Chinese people were well-behaved during the worship while the foreigners had a more casual way of worship. They indulged in petty gestures such as constantly touching their face or scratching their head. The Chinese Hajjis stood stock-still while worshipping. I certainly identify with our own way. So do many other foreigners. The Chinese Hajjis received a warm welcome in Saudi Arabia. Many foreigners put their thumbs up when they saw us during the worship. But we didn't say anything bad about their way of worship. We shouldn't carp on about what are trifling details.

Her neighbor added:

> They [foreign Muslims] were somewhat casual. I think I could accept this because they are foreigners. No one knew where they studied or what Islamic school they belonged to. We just did our own thing, neither providing objections nor expressing approval. In my view, no matter how they [non-Chinese Muslims] worshipped, after all, all of us are worshipping Allah.

More opportunities to compare Chinese and "foreign Muslims" were provided by dress codes as observed by Muslims from different nationalities; in particular the female religious dress became a source of curiosity, comparison and judgment.

167.3.5   Comparing Female Islamic Dress Code

Women took greatest interest in the many forms of religious dress worn by fellow-Hajjas. A Hajja who had made her Hajj in 2007 had her own view of the garments worn by Chinese female pilgrims.

> Our head scarves are different from Indonesian ones but quite similar to Syrian scarves. We have a great diversity. Yet, on the whole, we Chinese were easy to recognize in Mecca. All of us, male and female, wore the robes and waistcoats we were given by the Chinese Islamic Association. Our headdresses were of different color and material. Pilgrims from Xi'an wore white-colored scarves while those from our village wore floral prints, but also black and gauze materials. These were truly motley kinds of garments. Looking back, I think that we were not dressed in a properly tidy manner. For example, Turkish women wore clothing much more neat and tidy compared to us. They wore real pilgrim garments – as did pilgrims from Indonesia and Malaysia – all dressed entirely in white. We were such a mess, far too colorful.

Similar self-criticism came from a recently returned Hajja. She said:

> We each were handed a gown with "Chinese Muslim" printed on it. It made us look neat. All the Chinese wore the same style of garment. It would prevent us getting lost. If you got

lost and went to the police for help, they would soon recognize you as Chinese because of your garment and then send you back to the correct address. But the problem was that our colors varied from area to area. Green-colored gowns represented Henan and pilgrims from Gansu wore beige colors.

Many Hajjas debated over appropriateness of wearing Hijab to conceal more of their body. The majority of women interviewed appreciated such demonstration of female modesty but did not wish this practice for themselves.

We are dressed neatly enough, wearing clean shoes and socks. Others may wear long robes to cover themselves, but then they show bare feet. Is this proper?

Considered immodest in China's Muslim practice, the sight of bare feet deeply impressed Chinese informants. Again, justification of such greatly divergent conduct could only be made in relation to difference in culture and race.

The foreigners wore shoes with their feet quite bare. So the top of their feet looks black and the bottom of their feet appear yellow .... The foreign Muslims really just look too alike in certain ways. On the other hand, they came in all skin colors and facial features. There were all sorts.

The pilgrimage allowed these pilgrims from a remote rural interior part of China to experience the global nature of the global Muslim community. The process, however, by which these pilgrims located themselves within the Ummah was shaped by familiar categories of "Us" and "Them" that allowed acceptance of Otherness as expressed in appearance, norms and conduct without undue influence on their own local traditions. Reflections by returned Hajjis on the impact of the Hajj, and on their awareness of moral and religious deficiencies in need of redressing as well as pilgrims' affirmation of a unique Chinese Muslim identity forms the final section of the chapter.

167.4 After the Pilgrimage – Hajjis and Local Communities

The rapid increase in the numbers of pilgrims in Sangpo, the formation of a distinct, local Hajji group, numerous accounts of other Muslim practices (practices of 'foreign Muslims as Sangpo Muslims were used to refer to non-Chinese Muslims), the nature of impact on the community and expectations placed on the Hajjis, all of these factors have made pilgrimages and Hajjis much talked-about in Sangpo. Underlying all conversations are issues of identity and allegiance. Have pilgrimages created tighter bonds with Muslims abroad to supersede identification with the Chinese nation? What is already apparent is that increasing opportunities for interaction with Muslims of other nationalities have led to further affirm and crystallize the local characteristics which shape Chinese local Muslim cultures. The opportunity for pilgrimage has made many Hajjis realize more clearly their unique situation. Through comparison with other Muslim contexts, they acknowledge the traditional Muslim culture which Chinese Muslims have inherited (and developed) over hundreds of years, importantly so, the institutions of female madrassas and women's mosques, led by female Ahong.

Pragmatism and considerations of their political home environment, where Muslims have ethnic and religious minority status, make Hajjis reluctant to aspire to ways of life practiced by Muslims outside of China. Yet a realistic appraisal of the limitations under which they practice their faith are coupled with an unbroken spirit of pride and confidence in their achievements.

> Compared with the foreign Islamic traditions, our [practice of] Chinese Islam is not that good…. Nowadays, China's Islam is experiencing the problem of aging. The old often go to worship in mosques, but most young people don't go there. China is a Han Chinese majority country. The minor nationalities have to make a living. So, too few young people go to worship in the mosque.

Pondering further, the elderly man observes that China is, however, superior in certain respects.

> The manner of worship in China is better than elsewhere. According to the Qur'an and al-Hadith, you can't move more than three times during the worship. After raising your hand, you can't be looking at the front. It is more than enough if you move three times during the worship. We are quite serious about it…I feel Chinese Hajjis are quite good. They strictly abide by the Islamic doctrine.

The Hajjas have their particular interpretation of the Chinese Muslim situation.

> They [Saudi Arabian women] worship at home, obey strict rules; they are not permitted to leave their homes. But in the present-day world, you have to make money for a living. Otherwise, how can we survive without making a living? As Muslims say, there are two life-times, a worldly life and Houshi (afterlife), for each person. You have to keep yourself alive and live your worldly life also …. After retirement I shall take more account of my afterlife. When I was young, I cared only for worldly affairs and my children. But I've changed since reaching old age.

The women's mosque is the important place for middle-aged women to make preparations for their afterlife. As it is the fate of all to grow old, both young and old Muslim women are concerned to continue the tradition of female Ahong-led women's mosques. The pilgrimage has not changed the male and female Hajjis' views on the indispensable role that the women's mosque plays in China, but has made them recognize how religious traditions are influenced by contingencies of history, kinship systems, culture and material circumstances.

167.4.1 To Become a "New" Muslim – Conduct and Action

Being a Hajji means, so the Hajjis say, to be strict with oneself and set a good example to others. Doing good deeds for the community is to demonstrate a higher spiritual state.

> In my lifetime, I really want to thank Allah for guiding me to the way of Muslim practice and offering me a chance of being a Muslim. After I returned home from pilgrimage, I never idled away my time. I am thinking about worship every day. I think I shall be a "new" Muslim … As a new Muslim, I shall carefully do my course of prayers and do more good deeds for others.

Referring to changes in her family, a Hajji says:

[With my children] I've become stricter than ever. In Saudi Arabia, the young worship and read the Qur'an. That's so good. When I returned home, I told my children to get more knowledge of Islam. Although making money is important, I have to take account of my afterlife as well as my worldly life. Under my influence, my daughter has begun to wear a Gaitou [headscarf which covers hair and neck]. We run a factory, so it's impossible for me to worship five times in the mosque. But we will pray whenever there is time.

Her male neighbor experienced similar changes.

[Before the pilgrimage] I would stay in the mosque if and when I'm at leisure. Since the pilgrimage, I care more about substance …. I will go to worship when it's time to go. The first worship in the morning is the most significant. In the past, I was too lazy. But now I will wake up at the right time without an alarm clock, for I'm thinking about it all the time. So I just wake up when the time arrives. … We can't act the way we used to. If we keep to old ways, what's the use of a pilgrimage at the cost of several thousand yuan [equivalent of many months of wages]?

Other Hajjis see the pilgrimage as an opportunity to contribute to their country.

As a Hajji, you have to make contributions to the country. As a Chinese Hajji, you should contribute to China's religious culture … . The economy in our village enjoys such good fortune, but [religious] education can't catch up with standards required…. Although I'm not the Ahong, I still bear responsibility for our backward education. But the backward religious education also has something to do with the Imam. It is the leader's responsibility, for it's the leader who takes charge of educational affairs. As a Hajji, I can't say I don't have the responsibility. As a Hajji, I should take responsibility for every legal violation, whether this might be of the laws of Islam or of our country … . As a Hajji, I should sacrifice everything I possess for both our nation and our religion. That is what I have in mind. Allah has bestowed a sound body on me, so I will give away my wealth in my life time.

Serving Islam has become conflated with serving the nation; to be a good Muslim is inseparable from being a good Chinese citizen. Imperceptibly, political contingencies have been internalized and turned into a paradigm of the exemplary Muslim Chinese.

One of the most distinguishing features of Chinese Islam is the presence of women's mosques as equal in law and entitlements to men's mosques, even if this institution is contested in certain Muslim communities in China (Jaschok and Shui 2000). Debates over continuation or abolition of a practice unique in global Islam show up the limits to which Chinese Muslims absorb "foreign" influences into their habitual religious conduct, whatever the concern among conservative male Ahong over "aberrant" practices.

167.4.2 Women's Mosque and Local Muslim Traditions

Although some Hajjis expressed their conviction that home remains the proper place for women's worship, there was a general consensus among men and women that the uses of women's mosques to Chinese women's spiritual lives were undeniable, even crucial, and that, moreover, such a venerable tradition must be recognized to have made a unique historical contribution to Muslim women's rights, nationally and globally.

Hajjis stressed the need to understand how different national situations result in different traditions. A female Ahong explained:

> I think it [the construction of women's mosques] has something to do with our national situation. Saudi Arabia is a Muslim country and everyone receives comprehensive education in the Islamic doctrine at an early age. The women in Saudi Arabia do not leave their homes. But when it comes to us women in China, well, where can we go? There is a general shortage of religious knowledge in our village. Without the women's mosque, ordinary women would have even less knowledge. If they worship at home, there is a possibility that they make mistakes when praying. If you go to a women's mosque, you can study and communicate with one another. We must continue to have women's mosques.

Listening to the Ahong, an elderly Hajji added:

> Equality between men and women is advocated in China. There should be mosques for both men and women. In the past, when the economy was not good, we could not do our ablutions if we had a power-cut or ran out of water. It is really convenient to live close to a mosque …. Under such good conditions, how can we not worship? Both men's mosques and women's mosques provide us with a lot of convenience, encouraging all of us to worship.

More explanations were forthcoming from a fellow Hajja:

> I feel the women's mosque is good because women can get together to study. It gives me the feeling that we are truly equal with men … . We saw no women's mosques on our Hajj, but then women don't leave the house to worship. Actually, we didn't see local women when we were doing our Hajj. For them, a women's mosque is unnecessary. Our country is different from that country [Saudi Arabia]… [In China] the women's mosque provides a place for studying, learning about Islam and about how to raise a good Muslim family. How could we learn all of this by ourselves? Do you think I can make it at my age? So I think it necessary to build the women's mosque in our place. Actually, we are making a kind of contribution to Islam [worldwide].

Another Hajja related her experience of building a women's. She said:

> I built the women's mosque. It had once been a men's mosque built by a man called Zhang. It was demolished during the Cultural Revolution …. I suggested to the village leadership that a women's mosque should be built for the women. And so it happened.

167.4.3   The Changes to Muslim Women's Dress

While many Hajjis emphasize their pride in the local customs Chinese Muslims have developed over time, considering these appropriate to the Chinese context, an exception is the dress code of women. Here obvious adjustments have occurred and are considered laudable demonstrations of religious modesty. Influenced by the Hajjis returned from pilgrimage, more and more Muslim women in Sangpo have begun to wear head-scarves which cover hair and neck, leaving the face free.

Representative of the views of the majority of returned Hajjis, a Hajja gave her views on changes to women's dress in Sangpo.

> Thanks to Allah, more and more young people in Sangpo are beginning to wear scarves. My daughter wore her head-scarf for the first time this year and never took it off. She has also become fussy about dress requirements and no longer wears skimpy dresses. The same happened to my daughter-in-law. Allah's guidance and her own disappointment with worldly life have much to do with this. At first, when Hajjas came back and wore scarves, the others didn't understand what is going on … You know, in the past, after the economic reform and opening-up of China [in the early 1980s], people wore simple white caps. It's been no more than five years since the scarves emerged. At first some people in the village poured scorn on those long dresses and scarves. But now, even those who have never made the Hajj, they also wear modest dresses and cover their hair. This is a good development. This applies especially to those younger people and women aged 30 years or so. Someone told me that some local women wouldn't leave the house without putting on their head-scarves.

The new custom of wearing scarves was introduced and promoted by several older female Hajjis. With the increasing adoption of modest dresses also by young women, Muslim business entrepreneurs have started to offer dresses and scarves in varied designs and colors, using and adapting local motifs to create desirable fashion and inspire consumption of modest dresses reflecting local taste and catering to aspirations for a modern Muslim look. A Hajja observed:

> Several old Hajjas were wearing it [the long head-scarf] after returning from Hajj. Everyone was surprised. Now the styles are becoming ever better and thus more beautiful to wear, therefore, more and more women love wearing Muslim dress.

167.4.4 Religious Education

Among the most troubling and recurring concerns expressed by all Hajjis, when they compared observations made during the Hajj with their situation at home, were perceived grave deficiencies in religious education given to Chinese Muslims. Their recurrent complaint was how poorly they compare to Muslims elsewhere when it comes to knowledge of the scriptures.

> During those days of travel, I found Chinese Muslims had too little knowledge of religion. Why is that? No matter we were in Al-Masjid al-Ḥarām or near the Ka'ba, we would see the foreigners reading the Qur'an. Even young people were reading the scriptures. But we Chinese people just chanted from memory; we couldn't understand the Qur'an at all except for a little knowledge some of us had had learnt at the mosque. Among the Chinese, only the Ahong (Imams) and Hailifan (students of the Qur'an) could read the scriptures.

Sangpo Hajjis see it as their important duty to support religious education and nurture a pious religious culture in general. Hajjis have been known to offer subsidized study places to young women and men from poor families at local mosques, enabling young Muslims to get a religious education. They also offer their services privately, seeking to popularize, as they say, the Islamic idea of peace and harmony for the good of the entire society.

167.5 Concluding Reflections

The long journey which Chinese Hajjis must make, starts out in their local community, not infrequently a village distant from the next transport hub, moves on to provincial capital cities, such as Zhengzhou, Lanzhou and Urumchi (some Hajjis joining in the group of pilgrims would also pass through Hong Kong or other territories) before leaving the country for Saudi Arabia, often passing through other unfamiliar countries before returning home. These cross-province, cross-region, cross-country and cross-cultural journeys, propelled by religious beliefs, have brought about personal changes in the Hajjis and also, to a certain degree, in the community in which the Hajjis live. These changes are complex and the outcomes of active exchanges of new external influences and deep-rooted familiar convictions. The "therapy of distance," to use Armstrong's language, the intensity by which old habits are ruptured and renounced, has allowed for new insights and ideas to enter Hajjis' mind-sets and personal conduct. But very quickly such religious harvest would be tempered by local realities and by the pragmatism of Muslims evinced by Hajjis used to compromise in a non-Muslim country. The changes, some embryonic, some tentative, some more influential, that have taken place during the years of thriving pilgrimage activities in Sangpo, may be summarized as follows:

167.5.1 Emergence of a Distinct Social Group

A new social group has come into being. Although everyone we interviewed expressed different views on the role played by Hajjis in the community, they agree that the mostly elderly Hajjis have become a widely accepted, distinct group in the village. From reasonably well-to-do backgrounds, Hajjis are acknowledged to deserve respect for their religious accomplishment, but their personal and social conduct receives critical attention. If the conduct is not what might be expected, criticism will refer to ideals of moral self-discipline, to the need to demonstrate exemplary personal morality and religious observance. Their function as paradigmatic Muslim role models exerts moral pressure on Hajjis and shapes the conduct of the community.

167.5.2 Negotiating Chinese Fe/male Muslim Identity

The experience of Hajj has enabled observations and comparisons with Muslim norms and practices elsewhere. China's opening of her borders has brought the outside world much closer to all Chinese, including Chinese Muslims. But it is another "West" to which Chinese Muslims refer – the Muslim countries of the Middle East – that has brought questions of identity and belonging for Muslims into sharp focus. Confronted with the diversity of the Ummah, male and female Hajjis are offered stark contrasts and varied opportunities for comparison, leading to self-reflections and questions over their place in the global community of Muslims. Most Hajjis have discovered what they consider painful deficiencies when it comes to literacy in scriptural language and religious education, identifying gaps in life-style and general piety of the Muslim population in China. However, crucially, they are all equally emphatic when it comes to recognizing the value and worth of their own traditions and customs that have been developed over time to suit local circum-stances. The strict observance of the protocol of worship and prayer is an example on which Chinese Hajjis comment with pride. Most divergent from other Muslim traditions, the institution of female Ahong-led independent women's mosques in Chinese Muslim history, however contested, is defended by most Hajjis as a proud and distinguished contribution to global Islam. Understanding better the diversity of the Ummah has led to greater acceptance of divergent traditions, in turn intensifying the proud identification with their own unique Muslim history.

167.5.3 Local Institutional Changes

The increasing number of people making a pilgrimage has created new institutions and communal facilities and given rise to a class of officials in charge of the effi-cient management of ever increasing numbers of Hajjis. This development is prom-ising to change the local dynamics of power, adding to the local branch of the Chinese Islamic Association and the village leadership a further stratum of author-ity. It remains to be seen how members of the new Village Committee, specifically established to organize the Hajj, will turn their important (and powerful) positions to use. How opportunities granted by the annual Hajj with its enormous religious, cultural and commercial implications, might turn mediators between Hajjis and non-Hajjis, between the village and the central Government, into gate-keepers of a changing local power dynamic. Whose authority will come to dominate the borders of the community travelled with such regularity by both Muslims pilgrims and those in the business of managing, and controlling, their pilgrimage?

167.5.4 Women, Hajj and Gender Change

The conversations with the Hajjis tell us how men and women are sensitive to different situations and encounters, interpreting them differently and drawing dif-ferent lessons. For instance, the Hajjas paid more attention to women's dress and conduct, acutely conscious of the gendered nature of public spaces and of codes of female modesty alien to their own experience. Their emphatic and often reiterated appreciation of the role of women's mosques in their home country suggests that distinctions made between "foreign" Muslims and Chinese Muslims are as much an affirmation of women's rights as part of all Chinese women's rights and entitlements, here harmoniously overlapping with the Chinese State's rejection of all foreign intervention in all religions, as they are acknowledgement of diversity within Islam.

Both men and women agree on the need for reform and change after many years of closed borders, on the one hand, and on the other hand, there is a genuine pride in representing China, in being Chinese Hajjis. If, as Armstrong maintains, pilgrims engage in a "therapy of distance" to leave behind them the frustrations of worldly and 'peripheral' preoccupations for the spiritual "Center," thus distancing themselves from local belonging, the experiences, reflections and reactions of Sangpo Hajjis also intimate the strength of familiar ideals, social ties and nationalist sentiment which are shaping narrations of return journeys (Armstrong 2012). Could thus the "peripheral" become invested with the significance of a newly reconstituted 'centre' of a collective patriotic Muslim identity?

Acknowledgements This chapter uses materials from a report produced for the "Women's Empowerment in Muslim Contexts" research program consortium (WEMC). The research and writing was done by members of the China Research Component, led by Jaschok in collaboration with Shui Jingjun. The WEMC consortium was funded by UK aid from the UK Department for International Development (DFID, 2006–2010) for the benefit of developing countries.

References

Abdel Haleem, M. A. S. (2012). The importance of hajj: Spirit and rituals. In V. Porter (Ed.), *Hajj, journey to the heart of Islam* (pp. 27–66). London: The British Museum Press.

Ah, Y. M. (1995). Issues of improving the work and strengthening views on the Muslim pilgrimage. *Zhongguo Musilin (Chinese Muslim), 1*, 19–21.

Armstrong, K. (2012). Pilgrimage: Why do they do it? In V. Porter (Ed.), *Hajjj, journey to the heart of Islam* (pp. 19–26). London: British Museum Press.

Bianchi, R. R. (2004). *Guests of God: Pilgrimage and politics in the Islamic world*. Oxford: Oxford University Press.

Clingingsmith, D., Asim Ijaz Khwaja, A. I., & Kremer, M. (2008). *Estimating the impact of the hajjj: Religion and tolerance in Islam's global gathering* (Faculty Research Working Papers Series 08-022). Cambridge: Harvard University, John F. Kennedy School of Government.

Hammoudi, A. (2005). *A season in Mecca. Narrative of a pilgrimage*. Cambridge: Polity Press.

Jaschok, M., & Shui, J. J. (2000). *The history of women's mosques in Chinese Islam*. Richmond: Curzon Press.

Liang, H. H. J. (2005). Hajj sentiment. *Zhongguo Musilin (Chinese Muslim), 3*, 43–45.

Ma, T. (2003). Development and changes concerning the China Islamic Association Hajjj organization. *Zhongguo Musilin (Chinese Muslim), 1*, 5–6.

Porter, V. (Ed.). (2012). *Hajj journey to the heart of Islam*. London: The British Museum Press.

Shariati, A. (2005). *Hajjj: Reflection on its rituals*. Teaneck, NJ, USA: Islamic Publications International.

Trojanow, I. (2007). *Mumbai to Mecca: A pilgrimage to the holy sites of Islam*. London: Haus Publishing.

Wang, X. (1994). Documentary of Saudi Arabia in Muslim pilgrimages from Henan. *Zhongzhou jin gu, 6*, 31–34.

Xie, J. (2009). From China Muslim pilgrimage to three splendid decades of reform and opening up. *Zhongguo Musilin (Chinese Muslim), 1*, 21–23.

Chapter 168
The Freedom of Wandering, the Protection of Settling in Place: Gendered Symbolizations of Space in the Practices of Hindu Renouncers in Rajasthan

Antoinette E. DeNapoli

> *Religion is the quest, within the bounds of the human, historical condition, for the power to manipulate and negotiate one's 'situation' so as to have a 'space' in which to meaningfully dwell. Jonathan Z. Smith, Map Is Not Territory.* (1978: 291)
>
> *Wandering is the search for meaning. It's research. I can do research wherever I sit. Sad Giri Maharaj.* (July 7, 2011)
>
> *If home is where a person is at peace with himself, where he can honestly say there is nowhere else he would rather be, then the desert had become my home. Michael Jackson, At Home in the World.* (1995: 51)

168.1 Introduction

The landlocked, north Indian state of Rajasthan (lit., "the land of kings") is typically known for its arid climate, rolling sand dunes of the great Thar Desert which consists of over 2,000 sq km (772 sq mi) and stretches over the western part of the region, the Aravalli Mountains, and its high-caste, royal Rajput culture.[1] But Rajasthan is equally a place pulsating with a vibrant and varied ascetic (*sādhu*) culture. *Sādhus* are mobile people, moving here, there, anywhere. The freedom of movement that their religious position makes possible also enables the transmission of religious ideas. Historically, *sādhus* have played a crucial role in spreading

---

[1] Rajputs belong to the larger caste community of Kšhatriyas, the second highest-caste following the Brahmins. In her book Religion and Rajput Women, Lindsey Harlan identifies three types of Rajput caste communities in Rajasthan, namely the royal, noble, and village Rajputs. See Harlan (1992), Religion and Rajput Women: The Ethic of Protection in Contemporary Narratives (Berkeley: University of California Press). Royal Rajputs have a significant place in many Rajasthanis' idea of Rajasthan as a place, region, and culture.

A.E. DeNapoli (✉)
Religious Studies Department, University of Wyoming, Laramie, WY 82071, USA
e-mail: adenapol@uwyo.edu

© Springer Science+Business Media Dordrecht 2015
S.D. Brunn (ed.), *The Changing World Religion Map*,
DOI 10.1007/978-94-017-9376-6_168

and shaping the dominant religion of the state and, more globally, of the country, namely, Hinduism. *Sādhus* are just as much emplaced people, situating themselves in dwellings here, there, anywhere conducive to the formation of a spiritual life. Ashrams, which symbolize the dominant "moods and motivations" associated with Indian asceticism (Geertz 1973), and which constitute a signal feature of the religious landscape of Rajasthan from rural villages to urban cities, tell a compelling story of the rootedness of religion, and of the *sādhus* who transmit it, in the land.

This chapter examines the lives and practices of the female *sādhus* with whom I worked for 3 years in the region of Mewar, a former princely state in southwest Rajasthan, in the districts of Udaipur (17,200 sq km) and Rajnagar (4,500 sq km), a surface area covering 21,700 sq km. The female *sādhus* have taken initiation into one of the two pan-Indian ascetic traditions, namely the Daśanāmi and Nāth orders. Although the scholarship on female ascetics tends to use the grammatically feminine form of *sādhu*, which is "sādhvī," to speak about them, the *sādhus* I met never use this term in their self-descriptions. Rather, to these *sādhus*, 'sādhvī' describes married women who become possessed by the goddess in order to heal and treat individuals afflicted with (spiritually treatable) diseases. To distinguish between themselves as a class of religious practitioners and women possessed by the goddess, the *sādhus* employ the term "sādhu." In referring to women *sādhus*, the *sādhus* use a kinship form of address, such as 'māī' "mātā-jī," terms that translate as "mother" or they simply say "lady sādhu." This chapter shows that notions of place and space, which are underrepresented categories in the scholarship on asceticism, figure prominently in what I call the *sādhus*' "rhetoric of renunciation," their practices of singing, storytelling, and textual recitation. From these they generate their constructions of identity and asceticism. I argue that the *sādhus*' ideas of place "perform" or create gendered symbolizations of space conceived and expressed through the competing idioms of "wandering" and "settling" and, thus, ground asceticism in a more, but not exclusively, "locative" notion of place (Smith 1978: 101). Their practices unsettle the dominant view of asceticism in the scholarly discourse as illustrative of a "utopian" model that transcends place (Smith 1978: 101).

The word "sādhu" describes the generic term for an ascetic in South Asia and ideally characterizes people who have left behind, or renounced, normative social-cultural expectations like work and financial responsibilities, kinship and marital obligations, householding and family, sexual activity, former social identity, including caste (birth group) and class, and religious-ritual practices tied to the (re)production of such institutions and structures. Indologist Patrick Olivelle explains that renunciation (*sannyās*) represents an anti-nomian mode of life, an anti-structure, that repudiates the values of life in-the-world. Its practitioners, who are thought to have symbolically "died" to the world at the time of their ritual initiations as *sādhus*, seek to escape from worldly life (*sansār*), a seminal concept discussed in the Sanskrit texts as the endless cycle of death and rebirth (Olivelle 1992). *Sādhus* exist as the living dead on the periphery of the world. Both the classical and medieval literature on *sannyās* across Indian religions, according to

Olivelle, employ the shared religious grammar of "going forth [*parivrajati*]" to represent the shift from home into homelessness (Olivelle 2007: 177). The tension between these opposing ideals identifies two recurring conceptual dichotomies in renunciant theology, in which home signifies what *sādhus* hope to escape: the world, death, attachment, finiteness, limitation, constriction, entrapment, weakness, illusion and impermanence (Olivelle 1992; Wilson 1996; Dumont 1960). In contrast, homelessness symbolizes all that *sādhus* intend to gain: ultimate reality, transcendence, detachment, infinity, expansion, immortality, liberation (*mokṣ*), autonomy and power. Significantly, this movement from home to homelessness is not only symbolic, but physical as well. By virtue of their world renunciation *sādhus* become "homeless wanderers," free agents, who wander the earth alone "without asking for directions ... like a worm along the path pointed out by the sun," in search of truth (*brahman*) (Olivelle 2007: 177). *Sādhus* remain forever without home and without any place in the world.

The notion of statutory wandering demonstrates a pronounced theme in most of the textual and scholarly discourse on renunciation, as well as in renunciant rhetoric in contemporary South Asia. But the *sādhu* who wanders eternally without place, without a stable residence, represents an ideal and, as Olivelle aptly observes, a theological *fiction* (Olivelle 2007: 177). It is derived from the dominant Sanskritic model of Brahmanical *sannyās*, in which wandering, as Olivelle claims on the basis of his reading of the Brahmanical texts with which he works, communicates the cultural experiences of danger, difficulty, and gain through pain (Olivelle 2007: 185). These experiences highlight gendered masculine values that identify an androcentric renunciant ethos of wandering as "conquest" (Olivelle 2007: 186). In Brahmanical *sannyās*, homeless wandering indexes a root metaphor for liberation from *sansār*, the sphere of death, impermanence, and illusion, whereas home signals *sansār*. But the interpretative issue at stake hardly ends there. The symbolic currency behind the male renouncer objective of homelessness also concerns the gendering of home and, by extension, *sansār* as feminine. Liz Wilson writes,

> ... the plight of the would-be renouncer at home is akin to the constriction of the fetus in the womb, a space that is conventionally described in Buddhist as well as Hindu texts as extremely tight and disgustingly impure. The comforts of hearth and home cause the renouncer great discomfort. The safety and privileges of the home can make one a prisoner, for the life of domesticity impedes movement, imposes myriad obligations, and makes brahmacarya [celibacy] difficult .... disengagement from the social world through the semi-nomadic life of a celibate renouncer is described in expansive terms suggestive of both physical freedom and moral autonomy. If the life of the renouncer is...lived in the somewhat utopian space of a society of renouncers, the life of a householder is lived in the dystopian space of the home, a space that is as cramped and oppressive as the womb. (Wilson 1996: 28)

In this framework, space itself is gendered. The sphere of the home/domicile represents "female" space and everything else beyond the home represents "male" space. Thus, Brahmanical *sannyās* constitutes, as anthropologist Meena Khandelwal has observed, "a tradition that was created by and for elite men" and, more precisely, a male religious space (2004: 5). Its expectation of homeless wandering all but guarantees that *sannyās* will appeal to men, and not to women, whose worlds are deeply emplaced in the home. Moreover, Brahmanical *sannyās* illustrates, to use the language of religious studies scholar Jonathan Z. Smith, a utopian vision of worldly transcendence. Here, *sādhus* have "no place to stand" (Gil 1998: 283). Most religions, Smith observes, map their worlds in terms of two competing ideas of place, namely the *locative* and the *utopian* (Smith 1978: 289–309, 1987, 2004). These opposing ideal types of human mapping strategies represent "the power to manipulate and negotiate one's "situation," so as to have "space" in which to meaningfully dwell" (Smith 1978: 291). Unlike the utopian model already discussed, the locative example, as we might expect, orients human beings in their worlds and helps them to make sense of the potential threats of chaos, danger, and mean-inglessness. "A locative map," Smith says, "guarantees meaning and value through structures of congruity and conformity" (Smith: 292) Utopian models, however, are less persuaded to deny the discrepancy between the world that people construct and the world as it is. Utopian models recognize the incongruity between the imagined and the real; they depict the patterns of transcendence, rebellion, and paradox, which highlight *sannyās* ideals *par excellence*.

But if the more dominant and utopian understanding of wandering as an experi-ence of danger, difficulty, and gain through pain constitutes a gendered masculine theological fiction maintained by the elite orthodox Brahmanical model, it behooves scholars to ask the ways in which *sādhus* in contemporary South Asia negotiate the discrepancy between the (utopian) textual ideal and their everyday (locative) reali-ties. As Smith says, citing philosopher of religion Paul Ricoeur, "… I do believe that religion is, among other things, an intellectual activity—and…it is the perception of incongruity that gives rise to thought (Smith 1978: 294)."

Examining female *sādhus*' practices offers a compelling case study of the ways that humans map and navigate their worlds, because female *sādhus* have an ambigu-ous and contested place in Indian society and *sannyās*, a religious institution traditionally seen as male-dominated (Khandelwal 2004: 23–46) (Fig. 168.1). Anthropologist Judith Okely has argued that paying attention to the lives of "atypical" people helps scholars to recognize not only the "cracks of resistance" to cultural norms, but also the methods that such individuals use to interpret those discrepan-cies and, in effect, locate themselves in the world (Okely 1991). What distinguishes *sādhus* as a class of women from householders is the freedom of movement that comes by virtue of their religious position as ascetics. Religion scholar Lindsey Harlan, who specializes in the study of married Rajput women's practices in Rajasthan, discusses the settled and enclosed lives of householders:

> For many women in India, the domicile is not simply a place of residence or grounding, but also a location that they leave or enter as brides and that designates some degree of confine-ment. For women in different parts of India and from different social locations in the same part of India, there are varying degrees and modes of restriction that bind women to their natal or conjugal homes and discourage or prohibit their participation in public life beyond their domicile. (Harlan 2007: 66)

In Rajasthan, and elsewhere in India, female *sādhus* are respected and suspected for their independence and unusual life choices (Khandelwal et al. 2006; Khandelwal 2004). As atypical people, female *sādhus*' worlds are fraught with tensions.

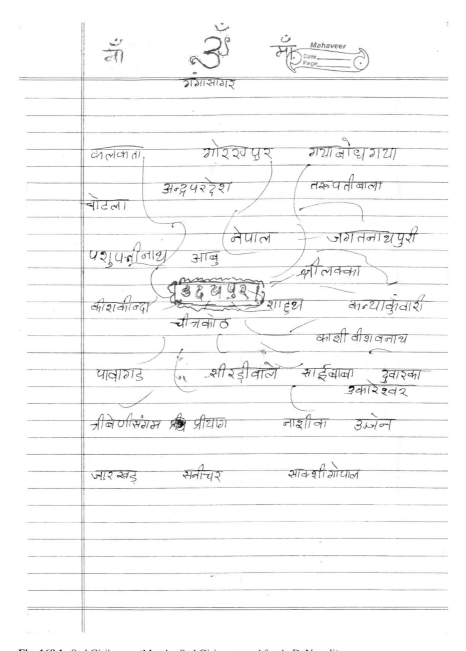

Fig. 168.1 Sad Giri's map (Map by Sad Giri, prepared for A. DeNapoli)

Because female renunciation is generally seen as a trespassing into male space, female *sādhus* are often viewed as transgressing normative gender roles (Khandelwal 2004: 21). My research with Rajasthani *sādhus*, as I have discussed elsewhere, questions this dominant perception of transgression perpetuated in the scholarly and popular literature on *sannyās* (DeNapoli 2009, 2014). The women I worked with suggest that they are not joining the "boy's club" in their becoming *sādhus*. Rather, they are pursuing an unusual religious way of life for Indian women through the authority of a divine command, and in doing so, carving out their own female space in *sannyās* (that is, they are creating their version of *sannyās*) that validates their concerns and experiences as women.

The female *sādhus'* rhetoric of renunciation performed through/as their songs, stories, and sacred texts makes explicit the tension between home and homelessness, wandering and settling. These *sādhus*, too, struggle to make this well-known opposition of ideal types of place meaningful through their applications or, as I will show, reworkings of the dominant ideals to their lives. This tension, however, is never resolved in Indian asceticism, and we should not expect the *sādhus'* asceticism to resolve it either. Their negotiations of this tension, however, reveal an alternative and gendered understanding about their experiences of wandering and settling, which enable them to generate particular ideas about *sādhu* identity and an alternative vision of *sannyās*, what I term on the basis of my analysis of their rhetorical performances as "devotional asceticism." In the practices of the *sādhus* notions of place receive as much "place" as those of wandering. The view of place vocalized in their asceticism recalls the wise words of Archytus: "All existing things are either in place or not without place" (cited from Casey 1996: 14). And, the fact that the *sādhus'* performances are emplaced in community-based devotional (*bhakti*) contexts known as *satsang* ("gathering of truth"), practices that predominantly occur in their ashrams, which they identify as places of "peace," "friendship," and "love," all which suggest that their constructions of identity and asceticism are linked to the power of their idea of such places. Thus, the *sādhus'* performing their rhetoric of renunciation from within the settled spaces of their ashrams, places that are rooted in the geography of the land, creates and enacts a "reterritorialization" or claiming as real territory/space of what is perceived by people who readily accept the dominant the male-centered definitions of asceticism as a "place-less place."

Unlike the androcentric ethos communicated in the standard Brahmanical model, the *sādhus'* practices reject such radically idealistic notions of space and place and articulate a more situated ethos of renunciant wandering as a "quest" or "search," and the values emphasized in their representations of this experience have to do with freedom, friendship, and (trans)formation. While several of these values are, indeed, found in the Brahmanical model, the *sādhus* recast their meanings to underscore specifically female experiences of wandering. Similarly, the notion of place, of settling, for the *sādhus* is connected to the values of protection, peace and love and friendship, and moral character. In what follows I will focus on the stories and songs that I documented from Sad Giri, a 30-year old (2011) Daśanāmi *sādhu*, because her understandings of wandering and settling represent and express a shared view of spatiality amongst the female *sādhus*. But just as significant, Sad Giri's background as an "unlettered" (she has never attended school), low-caste woman from the butcher (*kasāī*) community, one of the lowest communities within the hierarchical caste system; her creating a "*satsang* business" with the help of her family in order to raise funds for the building of her own ashram in her natal village;

and her continuing to live with her natal family while keeping an ashram distinguishes Sad Giri as atypical amongst female *sādhus* in Rajasthan. Sad Giri's life and practices are suggestive of the emerging developments happening in *sannyās* in South Asia. Also, paying attention to her life and practices shows Sad Giri's negotiating the reality of a rapidly changing (and globalizing) Indian culture with the ideals of the ancient institution of *sannyās*.

168.2 "I Cannot Stay in One Place:" The Wandering (and Dwelling) of Sad Giri

Sad Giri is unique for a female *sādhu*. She moves *and* settles in a manner that I have never encountered in my work with female *sādhus* in Rajasthan. She stands out as unusually independent, especially for a woman of her age and caste. Less than a week after our first meeting in 2011 at a public ritual ceremony in Gogunda village (Udaipur district), in which the memorial (*samādhi*) shrine of Maya Nath, a recently deceased female Nath *sādhu* was established at her ashram (Maya Nath passed away 40 days before my arrival in India that summer), and in which there were over 100 *sādhus* in attendance (11 of which were women), I met Sad Giri at her natal village located near the holy city of Nathdwara. When Sad Giri invited me to come to her village for the day, I jumped at the opportunity, as I was curious to learn more about a woman who had taken her *sādhu* vows only 2 years before in 2009, though, as she describes in her personal narratives, she had been wearing the "white cloth" of a celibate student (*bramacārinī*) for 17 years before she formally renounced 2 years ago.

Sad Giri was standing at the intersection of a dirt road leading to the street of her parent's house, waiting for me and my field assistant, Manvendra Singh. A married, Rajput male in his 30s Manvendra, though educated in the best English medium Catholic schools of Udaipur, speaks the standard Hindi that many American scholars crave to hear. His father, Ranchor Singh, worked as a research associate to many foreign scholars until his death in 1999, when Manvendra was only 19 years old. I met Manvendra through a colleague while studying Hindi in Udaipur 10 years ago; he has been a loyal co-pilgrim in my research wanderings ever since.

As Manvendra's car, a black Maruti Hyundai hatchback, turned on the corner of the street where Sad Giri was standing, I noticed that she was talking on a mobile phone. She signaled with her empty hand in the direction of her parent's house and Manvendra parked the car in the empty lot across the street from the house. Upon getting out of the vehicle, Manvendra said respectfully to Sad Giri, "Maharaj,[2] we would have met you inside the house. You didn't have to wait outside for us." Sad Giri's response struck me as a defining motif of her life: "I cannot stay at one place. I don't enjoy it. I *have* to move, I *have* to get up and go." Much of our talk on that

---

[2] Maharaj is a formal title of respect used for Hindu sādhus. It translates as "great king."

day, and in the days that followed, was marked by the idioms of independence and wandering. Sad Giri emphasizes these themes in the stories she tells about her life: "I like to wander [*bhramaṇ karnā*]. I'll just take my vehicle and go. I'll go to Haldigathi, Nathdwara, Eklingji.[3] This or that thought will come into my head and I'll move. I keep the vehicle with me. Wherever I want to go, I just go. When I stop [at some place], I park the vehicle. But otherwise I get up and go [wherever I want]." The vehicle to which Sad Giri's refers concerns her scooter.[4] I remember seeing it at Maya Nath's *samādhi* ceremony, sitting with all the other vehicles. I also recall men (*sādhus* and householders) operating them, whether they were coming to, or leaving from the ceremony. The scooter Sad Giri keeps with her symbolizes her unusual freedom of movement, to come and go as she wishes. It provides her the mobility she would not otherwise have without it.

Significantly, the scooter indexes the financial independence that Sad Giri has been able to achieve through her "*satsang*" performances. Sad Giri's engaging in *satsang*, an event in which she sings *bhajans*, tells stories, and recites sacred texts from memory, hardly distinguishes her as unusual among the *sādhus* with whom I worked. What does, however, separate her from the other female *sādhus* as a class of unusual religious practitioners is the fact that Sad Giri has created a professional business from her *satsangs* and earns money doing it. In many ways, Sad Giri's life turns upside-down the dominant definitional parameters of *sannyās*. As people ideally located at the world's periphery, *sādhus* traditionally cannot (and should not) touch money, as it contradicts the radical ideals of their way of life. Similarly, *sādhus* historically do not earn money. They make their living, so to speak, from the donations they receive from their householder-devotees. Hence, *sādhus* depend on others for their own survival in the world.

But not Sad Giri. Her life also turns topsy turvy the typical distinctions between *sādhus* and householders. She financially provides for her family. Though *kasāīs*, Sad Giri's family does not work in their community's craft of animal butchering. Instead, her father and brother Ramesh manage a small bicycle repair shop in the village, from which they earn very little money. Much of the financial responsibility for the family, therefore, has fallen on Sad Giri's shoulders.[5] But with her earnings she has purchased mobile phones, digital cameras, and clothes for her family (which were shown to me on that first day), and the Honda motorcycle that her brothers, Ramesh and Kamlesh, share. Sad Giri's financial independence has given her the

---

[3] Nathadwara and Eklingji, are religiously significant and, in the case of Haldigathi, culturally significant places. Nathadwara, the name of a town and a temple, and Eklingji, a temple, are both regionally and nationally known places of pilgrimage. The Nathadwara temple houses one of the most famous images of Krishna (Sri Nath), and the Eklingi temple houses a regionally popular form of the deity Shiva. For this reason, the town of Eklingi is appropriately called Kailash Puri (named after Shiva's residence in the Himalayas) Haldigathi, though, is believed to be the place where it is widely believed that the famous military hero Rana Pratap Singh fought the Mughal forces in April 1959, and where the Maharana's beloved horse, Chetak, died.

[4] It's a scooty, a type of scooter made for women.

[5] I was repeatedly struck by Sad Giri's statements that, despite being a *sādhu*, she enacts the role of the eldest brother in her family, because of all the responsibilities she has in taking care of them.

physical freedom to wander beyond the normative sphere of marriage and householding, which is not only atypical for Indian women but also for women of her caste, and settle permanently in the *sādhu* world. A *sādhu* who moves between her natal home and ashram regularly and, given the close proximity between both locations, relatively easily, Sad Giri is certainly not bound to her natal place in the way that unmarried women of her age and caste are. According to her father, who seemed proud of his daughter, Sad Giri is the first (and only) woman to take *sannyās* in her community and her family. In carving a space for herself in a world dominated by high-caste male *sādhus*, Sad Giri depends on herself. She often says, "I did all this myself." Everything from her mobile phone, to her scooter, to her business shouts of Sad Giri's unusual independence as a female *sādhu*.

Sad Giri's life and practices illustrate three spatial categories of wandering, which I characterize as local, extra-local, and, in light of her business, work wandering. In this framework, *local wandering* describes Sad Giri's moving between (a) her parent's house and ashram, both of which are located in her natal village, and (b) her natal village and local towns (for example, Nathdwara, Haldigathi, and Kailash Puri) and/or, more broadly, the city (Udaipur). *Extra-local wandering* represents two sub-categories of wandering, namely (a) *regional wandering* and (b) *translocal/pan-Indian wandering*. For example, Sad Giri's moving from her natal village in Mewar region to a different region in Rajasthan, such as the place of Pokaran, which is situated between the cities of Jaisalmer and Jodhpur in northwestern Rajasthan, where she makes the Ramdevra pilgrimage (a pilgrimage made in honor of the local folk hero-deity, Ramdev) every year; this travel constitutes an extra-local/regional wandering. Sad Giri speaks often her wanderings to Pokaran for the annual Ramdevra pilgrimage. While sitting in the main room of her family's house 1 day, Sad Giri showed me a framed photo of her and Ramesh, which was taken after they had completed their first Ramdevra pilgrimage while she was in still living in the state of "white *sannyās* [*brahmacārya*]." Here is a transcription of that conversation:

[A]ntoinette: Is this you and Ramesh?

[S]ad [G]iri: Yes. We walked [*chaltā-chalā*] all the way to Ramdevra.

[A]: You walked all the way to Ramdevra?

[SG]: I walked all the way to Ramdevra. I journeyed by foot from here [in her village].

[A]: You walked very far. Ramdevra is close to Jaisalmer, isn't it?

[Sad Giri's father]: It's close to Jaisalmer, that's correct.

[SG]: I walked all the way. That's when I was still in [the state of] white sannyās.

[A]: Was there any particular reason for your trip?

[S]: I go every year. People in my community go every year to see [darśan] Baba Ramdev [the folk hero whose image is established in the temple in Pokaran]. I made two journeys on foot. The others I made by train. I have traveled the four corners of the earth. I have traveled [*bhramaṇ karnā*] to Calcutta [in Bengal], Gorakpur [a city on the border between Uttar Pradesh in India and Nepal], Pashupati Nath [in Gorakpur, UP]; Jagganath Puri [in Orissa]; Jharkhand [a new Indian state east of Bihar], Hariyana, and Punjab.

[M]anvendra Singh: Have you traveled to Girnar [in Gujarat]?

[SG]: I have not gone there. But I have been to Gujarat. I go there a lot. I have wandered [*bhramaṇ karnā*] all sides of the earth. I have been everywhere. I like to wander [smiles].

Sad Giri speaks at length here about the type of wandering that I have identified in my research as extra-local/translocal wandering, to describe the movement that happens specifically outside of Rajasthan. In narrating her idea of the concept of wandering more globally, though, Sad Giri uses the terms *bhramaṇ* and *yātrā*—the former translates as "wandering," "travelling," or "journeying," and the latter generally translates as "pilgrimage." Sad Giri employs these terms interchangeably, suggesting that "wandering" and "pilgrimage" constitute similar (or perhaps the same) activities to her. She distinguishes between the method of wandering (for example, walking on foot or moving by train or some other form of transportation) and the actual activity of wandering (*bhramaṇ karnā/yātrā karnā*). Mostly, though, Sad Giri underscores the *prima facie* religious significance of wandering for her unusual way of life and, thus, performs her idea of wandering from place to place, and the freedom it provides, as an important characteristic of *sādhu* identity.

Furthermore, she constructs local, regional, *and* pan-Indian movements all as types of religious wanderings. This is significant for two reasons. First, in Sad Giri's perspective, local/regional wanderings are as religiously (trans)formative as transregional ones. Hence, she challenges the dominant model of pilgrimage in most of the current academic discourse as a primarily utopian experience of moving to a place "out there."[6] Olivelle voices this view in his claim that "[i]n pilgrimage and asceticism … [p]ilgrims and ascetics leave their social space and either wander around or travel to a distant "center out there" where their former social identities are merged into a new sacred identity shared by fellow ascetics or pilgrims" (2007: 183).

Second, by associating her local movements, either by vehicle or on foot, in the village, town, and/or city with "wandering," Sad Giri challenges the more dominant model of what "counts" as wandering in *sannyās*. The Brahmanical literature typically considers only the wandering that happens outside of the village, outside of the local, as a legitimate form of renunciant journeying. Olivelle explains in conjunction with the ancient Brahmanical practice of ritualized walking, a term that we may substitute with "wandering:" "[T]he most significant ritual walks involves [sic] a movement away—either really or in ritualized form—from social space, whether it be home or village. It is such walking that is valourised, not the walking that takes place *within* the social space of home or village" (2007: 185, italics in original). This representation of wandering, however, depicts a gendered masculine idea of moving in space, as a "conquering" of space, in which *sādhus* always move through it, without ever settling in it. The female *sādhus* with whom I worked, as Sad Giri's narrative shows, experience space differently than what the textual model suggests, and I believe it is important to include in discussions of wandering their

---

[6] Ann Gold's research on the idea of pilgrimage amongst Rajasthani pilgrims as documented in her monograph, Fruitful Journeys: The Ways of Rajasthani Pilgrims (1988), describes that Rajasthani pilgrims distinguish between regional and pan-Indian/translocal pilgrimage through their use of the terms jātrā and yātrā, respectively. Gold, moreover, distinguishes between types of jātrā pilgrimage, such as pilgrimages known as "sinking flowers," submerging the ashes of deceased loved ones in the Rajasthani holy town of Pushkar, etc. See Gold (1988).

understandings of local/regional journeying (without having to "conquer" space) as (valid) renunciant wandering.

Perhaps the most radical way that Sad Giri redefines the concept of renunciant journeying is through the movement that happens in the context of her business. To Sad Giri her business travelling is as much a renunciant practice as that of journeying to Ramdevra, or Jagganath Puri. This third category of movement, which I have characterized as *work wandering*, has enabled Sad Giri to move in social spaces that she would not typically wander and meet people with whom she would not typically interact as a low-caste *kasāī* woman. While Sad Giri's religious position grants her the freedom (and the authority) to meet people from all kinds of social locations (e.g., gender, caste, class), including those outside of the caste system; and while her *sādhu* status, which signifies her religious purity and, therefore, (in theory) neutralizes her caste impurity, makes intercaste interactions not only possible but also desirable, her business explains why Sad Giri can move as easily as she does into unfamiliar territories for a woman of her age and social group. Thus, her business serves as a type of "vehicle" just as much as a train or going on foot. The fact that people pay Sad Giri money (she earns between 5–10 thousand rupees—approx. $100–$200 USD—per *satsang*, a fee contingent on the its location and duration) to perform in their personal/professional spaces more than simply enables, but requires her to cross traditionally unchartered thresholds and navigate this newly emerging social-religious space of renunciation. Her patrons are Hindus from a variety of castes, but most are wealthy Marwari transplants (that is, Rajasthanis who are originally from the Marwar region) from the *jāṭ* (merchant/farming) community and have earned their wealth and high social status by purchasing quarries in the thriving local marble and stone industry. Sad Giri's business wanderings, though, keep her firmly planted in Mewar. She refuses to go to other regions due to the high costs of this kind of wandering. In my 10 years of fieldwork in Rajasthan I have never met a *sādhu* (female or male) like Sad Giri. Her life and practices shatter the conventional model of *sannyās*. Mewar's (Udaipur's) growing marble industry and tourism and rise of a middle-class with cash to spend are creating economic and social opportunities that Sad Giri seems to have seized in her attempt to carve a valid place for herself in *sannyās*. And with the money earned from her business wanderings, Sad Giri has purchased (and built) her own permanent place to settle—her ashram (Fig. 168.2).

The ashram is a modest building consisting of concrete walls and ceilings. The outside walls show what is now, after 12 years, a fading pink color. A saffron-colored flag, which the *sādhus* popularly characterize as the color of *sannyās* (that is, the color of power and truth), and which is the color of the clothing worn by Daśanami and Nath *sadhus*, placed at the front of the building, indicates its status as a "working" ashram, a place for people to come, take the blessings of the *sādhu*(s) living there, and relax and enjoy themselves. Sad Giri describes her ashram as "pakkā," a finished/permanent structure, differentiating it from "kachchā," an unfinished/ impermanent structure. It has a small shaded courtyard where the *dhunī* (fire pit), a feature of most ashrams, burns continually. The *dhunī* constitutes a symbol for the radical life of *sannyās*; it expresses *sannyās*' dominant ideals, particularly

Fig. 168.2 Sad Giri's Satsang Business Card (Photo by A. DeNapoli)

those concerning detachment and the impermanence and illusory nature of *sansār*. The *dhunī* reminds residents and visitors of any ashram that *sansār* represents a temporary place; that humans' "real place" is, as the *sādhus* say, "in God's home." To that extent, the ashram signifies a place on earth that evokes through means of the people and practices localized in that space a sense of God's place in heaven (*vaikuṇṭh*). The ashram as symbol and structure juxtaposes the tensions between home and homelessness. And yet, even as the ashram serves as a sacred center within *sansār*, which many of the sādhus associate with a state of dreaming, its world-denying ideals situate it at *sansār*'s periphery (Fig. 168.3).

The courtyard leads into the ashram. Here lies the womb of the entire structure, that creative space where ideas, dreams, and hopes are generated, incubated, and sprout into form. The room stretches 16 ft (4.8 m) wide by 80 ft (24 m) long. The whitewashed walls hold posters of deities. There is a kitchen at the southern end of the building, with aluminum pots, cups, and cutlery placed on the two shelves on the wall. A calendar that reads, "Kelwa Marble," keeps track of time. Sad Giri, Manvendra, and I seat ourselves on tattered burlap sacks made from used rice bags. Sad Giri keeps the east and west doors of the ashram open to let the wind circulate. The eastern door opens onto green scenery with cacti and trees sharing the same space. The sounds of women's voices in the fields nearby reverberate in the air. In this peaceful place Sad Giri shares the story about twin sisters who used to be berry trees. By telling this story to her audience from the situated location of her ashram, Sad Giri performs her understanding of (the act of) settling in a place as an equally important characteristic of *sādhu* identity. Here is that story.

Fig. 168.3 Sad Giri and her mother sitting in her ashram (Photo by A. DeNapoli)

168.3 "What Your Heart Says:" The Tale of the Twin Sisters Who Were Berry Trees

[S]ad [G]iri: A Brahmin had twin daughters.

[A]ntoinette: Uh-huh.

[SG]: They did something wrong [in a previous birth]. So, there were born as berry trees.

[A]: Berry trees, Maharaj? The twins were born as berry trees?

[SG]: In a previous birth, they were berry trees. A sāhu came while wandering [ghumtā-ghumtā]. He said to himself, "I need to rest." One berry tree was fixed by his feet, and the other tree was fixed by his head. His head touched one tree, and his feet touched the other tree. Because of his penance [tapasyā], the berries of those trees fell on the ground. His penance was so strong; it was unbreakable [aṭūṭ]. It was a very good penance.

[A]: Uh-huh.

[M]anvendra Singh: Both berries fell from the trees because of his penance?

[SG]: Yes. Then both of them went to God. Both of the souls [ātmā] went to God and stood in front of him. Their souls stood [before God] since their bodies could not. God said, "Tell me, where do you want to take your next birth?" Both said, "Anywhere but a poor man's house. Any house is good, Lord [bhagvān], but don't put us in a poor man's house." So, the twins were born in a Brahmin's house. They both took birth over there. Twin sisters were born in a Brahmin's house. Both had been berry trees earlier.

[A]: Uh-huh.

[SG]: They took birth and grew up together. They reached an age suitable for marriage. Their father said [to their mother], "Both girls have reached an age suitable for marriage. It's good if we get them married. You cannot keep an unmarried girl in the home." Their mother said, "Let's ask the twins and then we'll start the search [for their husbands]." The father said, "Girls! Come here." They came. "We want both of you to get married." The twins said, "But father, we don't want to get married." "Why not, daughters?" "Father, we have only one wish, and only after it is fulfilled we will get married." "Ok," tell me what your wish is?" They said, "We want to travel [*bhraman karnā*]. We want to travel the entire earth and only after that we will get married." The father said, "Alright. That's a good thing. Let's call a horse cart." There were no automobiles earlier. The parents put all the food and water, however much the twins needed, into the cart. The father sent for a charioteer and told him to take the twins traveling [*bhraman karānā*]. He ordered, "Take my daughters travelling and earn the merit of this activity."

The twins traveled together. They met a sādhu. They thought to themselves, "In our last birth we were berry trees." That's why they took birth, so they could meet that sādhu who released them from their last births…. They traveled the earth to meet that sādhu. They wanted his blessings. "Only after we meet that sādhu will we marry. Until the time that we don't meet him, we will not marry," they promised each other. "We will not marry until we meet that sādhu. Even if we have to be born again." That's why they left home. While wandering they came to Kāśi Viśvanāth.[7] That sādhu was doing penance over there. His face was shining like anything, because of his devotion [bhakti]. The sādhu received both of them, and they each bowed to him. They said, "Oh, sādhu-jī! Do you recognize us?" "I am a sādhu. You're girls. Why would I recognize you? What do we have to do with each other? You've come for blessings, so that's good. Otherwise, how would I know you?" "Sādhu-jī," said the twins, "we know you." "How do you know me? Tell me!" "Sādhu-jī, in our last births, we both were berry trees. You were a sādhu just as you are now a sādhu. Because of your grace, we went to heaven. We praised you. Then god asked us, 'Where do you want to be born again?' We were born at a Brahmin's house. We were born as twins. It was our wish to travel. That's why we're not married yet. We wished to meet you and receive your blessings. That's why we didn't marry."

[Sad Giri comments]: If a woman is married would her husband let her travel?

[MS]: He won't let her travel.

[SG]: A woman travelling alone? Would you [to Manvendra Singh] let your woman travel alone?

[MS]: No.

[SG]: [speaking in the voice of a husband]: "You stupid! You're a woman. Why do you have to travel?" This is what men say to their wives. They won't just automatically let [their women] go out the door? If someone looks at her, he will ask, "Who is that man, and why is he looking at you?" Men will ask such questions, right? [Manvendra Singh shakes his head in agreement] You will ask such questions because she is your wife. "Who's that man you're talking to? What do you have to do with him?"

[A]: O.k.

[SG]: [returns to the story; speaking as the twins]: "That's why, father, we will not marry." They finally met the sādhu they were searching for. The sādhu said, "O.k., daughters. Now you can go. Do whatever your heart desires." The twins said, "Sādhu-jī, we met you. Is there anything greater left to do in this world?" "Get married," he said. "What's the use of marrying? Here, we get the benefit [lābh] of bhakti. We no longer wish to be married." The twins refused to marry. "Sādhu-jī, we no longer want to return home [ghar],

---

[7] Kāśi Viśvanāth refers to one of the most popular (and oldest) temples in the pilgrimage town of Kāśi (or Varanasi), located in the north Indian city of Benares ("city of lights"), that patronizes the deity, Shiva (also known as Lord Viśvanāth, Lord of the Universe).

nor do we want to marry. We'll settle [þhaharnā] here with you. We'll serve you. We'll do bhakti. Whatever happens, our lives are in your hands now. We'll only do bhakti and our souls will get peace. But if we get married, we'll quarrel everyday with our husbands." The twins settled with that sādhu. "Do what your heart wishes," he said.

Sad Giri's "breakthrough into performance" happens in the company of an audience consisting of me, Manvendra Singh, Sad Giri's mother, and her brother Kamlesh (Hymes 1975). This story arises from a conversational context in which Sad Giri describes the limitations, real and imagined, on women's freedom to do "what their hearts say." In her view, travel constitutes the most powerful form of female freedom, and marriage its worst enemy. Sad Giri says,

> If your parents tell you to get married, will you do what they want? Or, will you do what your heart wants? Don't listen to anybody. Just do what your heart wants to do. "Hey girl, will you do as your heart tells you?" "Or, will you do as we tell you?" "Hey girl, what's wrong? You don't have a place in your heart for a husband?" If a girl has decided not to marry, her parents cannot force her to marry. But her father will ask, "Why daughter?" "Because I want to travel first!" "I want to travel first! I don't want to marry!" (Fig. 168.4)

The twins echo these exact words to their parents. The tale that Sad Giri performs about the berry tree twins is loosely based on a narrative from a popular text known as the *Gītā Sār* ("essence of the *Gītā*"). The story appears in the fourth chapter of the *Gītā Sār*. The question that frames the story's narration is, "Can a life form experience release [from *sansār*] just by touching someone who reads daily the *Gītā Sār*?" Both the *sādhus* with whom I worked, and the householders with whom I lived while conducting field research, have told me that the *Gītā Sār* takes the

Fig. 168.4 Sad Giri showing a book she received during her wanderings (Photo by A. DeNapoli)

essential teachings of the *Bhagavad Gītā* (henceforth, *Gītā*) and makes them intelligible to the lay person in the form of stories, like the berry tree twins tale. Even though it is not in the *Gītā*, reading this story and, by extension, reading the *Gītā Sār* is akin to reading the *Gītā* itself.

Sad Giri's narrative performance looks different from the text-based narrative because both tales are woven from a different set of priorities on the part of their "storytellers." Thus, my analysis focuses on Sad Giri's performed narrative, since her priorities in telling this tale are what help to shape how her audience understands what wandering and settling mean to her.

In this performance Sad Giri emphasizes one of several reasons for the twins' desire to wander. They want to see the world before their parents arrange their marriages. Notice that Sad Giri's telling suggestively constructs marriage as a prison that incarcerates women, removing their freedom to travel. In this framework, marriage turns women into caged birds, cages to which men hold the key. Men decide to release "their women" from the cage or, as Sad Giri's stinging commentary implies, to keep them trapped in it. The twins seem to recognize the power of marriage and its representatives, men, to stymie the independence of women and decide to marry with the stipulation that they will wander first. Sad Giri's telling challenges the Brahmanical representation of entrapment in *sansār* "as a male dilemma" and its "gendering … as feminine, as a prison in which women are the agents of incarceration" (Wilson 1996: 4). Their setting forth from home into homelessness enacts a release from the possibility of suffocating marriages. It is not an escape from "the extremely tight and disgustingly impure" womb that the home stands for in the textual literature (Wilson 1996: 28). Instead, home indexes a place of safety and protection, where the twins are nurtured with love, affection, and support from their parents (and, in this idealized natal home unmarried females, i.e., daughters and sisters, enjoy more freedom of movement than in their conjugal/affinal homes after marriage).

Wandering, in this tale, functions as more than a trope for personal autonomy; it also communicates the human search/quest for meaning, truth, and knowledge and, more precisely, a *sādhu* whom the twins believe manifests the salvific wisdom they seek. The word Sad Giri uses for wandering in this storytelling event happens to be the term she uses in her personal narratives of wandering—"bhramaṇ." At the same time, *bhramaṇ* suggests the idea of wandering as a type of research on the human condition. In the commentary that unfolds hours after Sad Giri shares her tale, she elaborates on her understanding of wandering as a form of research. She explains,

> **[SG]:** Sister [speaking to Antoinette], wandering [bhramaṇ] is all about doing research [bhramaṇ karnā]. You have come to India from America. You are travelling [bhramaṇ karnā] a lot. You are learning a lot from your wandering. This is your research, right? [Antoinette nods in agreement]. Your wandering is your research [bhramaṇ]. You have gone to a lot of places. I, too, have traveled to many places…Wherever I go, I learn something [new]. You will teach what you have learned from your wanderings. I, too, am also doing research. I have wandered the four corners of the earth [for research]. I can do research wherever I am, whether I am sitting or moving. I don't need to wander. I am prepared to do research wherever I sit. I am always ready to do research. All this is by the grace of my guru. I'm sitting here [with you] and the research is happening.

In speaking about her wandering as research, Sad Giri also spends time describing the people whom she meets and befriends:

> When I traveled to Gujarat [two years ago], I met a Māī [lady sādhu]. She taught me to sing bhajans in Gujarati. The Gujarati bhajans I know I learned from her. She also taught me the Sati Toral story [a popular tale about the virtuous woman Toral][8] that I just told you. I am illiterate [anpaḍh]. I only remember what I hear. This Māī also gave me a book [Sad Giri asks Kamlesh to retrieve the book from the family house]. It contains the Sati Toral story. Wherever I go I meet people who teach me. They teach me and give me books [she smiles]. So much I learned by listening to the people I met during my travels.

Sad Giri's narrative situates the significance of wandering as research in the context of the people she meets and the friendships she makes along the way. Friendship, to Sad Giri, represents as much a result as a reason for wandering (and settling). The motif of friendship is significant, as it suggests the idea that the knowledge a traveler gains while wandering is made meaningful in connection with the people who transmit it in particular places. Friendships act as a scaffolding for knowledge. Traveling signifies the ongoing web of relationships of the traveler, and those connections contextualize her knowledge in a transformative way. The friendship Sad Giri made with the Gujarati *sādhu* has played a transformative role in her life, particularly in her learning (and teaching) new *bhajans* and stories and, of course, getting books.

Therefore, the image that Sad Giri crafts of wandering through her practices illustrates an interactive, relational, and reciprocal experience, rather than the solitary and antisocial experience emphasized in the Brahmanical texts. Sad Giri's tale of the twins expresses an ethos of what I term "gain through love." In her telling the twins travel with the intention of meeting the *sādhu* who released them from their last births as berry trees. And, they wander together, accompanied by their charioteer. Apart from "earning the merit" of causing them to travel, the charioteer fulfills the implied role of protecting the twins while they are "on the road." There seems to be little, if any, physical or mental suffering involved in their wanderings. By contrast, in the textual model, the traveler gains truth and wisdom *through pain* (Olivelle 2007). In Brahmanical *sannyās*, ascetic torture of the body is tied to the experience of *mokṣ*. The suffering involved in ascetic wandering is thought to be spiritually profitable because "striving," meaning ritual-religious exertion, effort, and/or toiling, leads to self-transformation (Olivelle 1993: 7–34).

Sad Giri's story, however, challenges the Brahmanical view of gain through pain for self-transformation by underscoring the idea of spiritual profit [*lābh*] in the context of devotional practice. *Bhakti* creates transformation. The performance context itself helps to define what *bhakti* is—singing *bhajans*, telling stories, and reciting (or remembering) texts, like *Gītā Sār*. Thus, even as Sad Giri performs her rhetoric of renunciation within this particular *satsang* event she "earns" the spiritual rewards of her *bhakti*. Just as significant is that the story's accentuating *bhakti* suggests another reason for the twins' travels: the freedom to practice *bhakti* to *bhagvān* on their own terms. Wandering functions as a means for expressing *bhakti* to the divine. But notice, though, that the *sādhu* initially tells the twins to get married.

---

[8] The Sati Toral story is widely popular in the north Indian state of Gujarat.

Sādhus traditionally encourage their constituents to follow their prescribed social duties (*dharm*), which is determined on the basis of caste, gender, and life stage. Recall that the twins were born in "a Brahmin's house." As the highest group in the caste hierarchy, Brahmins, in theory, are the most orthodox in following the normative institutions of marriage and householding. So, what else can the *sādhu* say to two unmarried Brahmin girls wandering alone? But they refuse. They break their promise to their parents, who let them travel on the condition that the search for their daughters' husbands would begin upon the twin's return home. But in their words: "What's the use of marrying? ... if we get married, we'll quarrel everyday with our husbands." The *sādhu* accepts their decision not to marry, and as if by default, settle with him in his place. He says, "Do what your heart wishes."

The twins feel compelled to wander and meet the *sādhu* who released them from their last births. They *have* to travel the earth. In wandering, the twins not only move to the message of their hearts, but also keep a promise they made to each other in their previous and current births. That promise has to do with finding the *sādhu* from their last incarnation in order to praise him for enabling the twins to receive human births in their present incarnations. Through the power of the "unbreakable penance" of the *sādhu*, the twins became liberated from their limiting berry forms and eligible for birth in a human form. The significance of their human birth has to do with the widely recognized Hindu understanding that liberation from *sansār* happens only in a human body. Thus, the twins' meeting the *sādhu* in their last births lives has given them the precious opportunity to escape *sansār* once and for all in their current births. What is more, by meeting that *sādhu* in this life, and by serving him as a form of God on earth with love and devotion, the twins take their destinies into their own hands and ensure that they will meet God in God's home at the end of their present lifetimes. Recall from Sad Giri's narrative the twins' apt words that their settling in the place of the *sādhu* will allow them to receive the profit of their *bhakti*. The profit of which they speak concerns the reward of eternal communion with God. The twins' wandering in search of the *sādhu* helps them to avoid profitless wandering in *sansār*.

The concept "sansār" literally translates as "wandering from life to life." Another term used in colloquial Hindi is 'bhataknā,' which means to "wander aimlessly" in *sansār*. Up until the moment that the twins received their human births they had, indeed, been wandering aimlessly in *sansār*. On the day Sad Giri narrates the berry twins' tale, she sings a *bhajan* about the unprofitable movement of wandering in *sansār*. The song is entitled "Wake Up, Traveler."

Traveler, you've been asleep for so long. Wake up!
In the beginning you slept in your mother's womb comfortably.
When you emerged from your mother's womb, you made a promise.
But you have forgotten it.
Traveler, you've been asleep for many days. Wake up!
The second sleep was in the lap of your mother.
Your sister, aunty, and mother loved you.
They loved you, and everyone was happy.
You've been asleep for so long, traveler.
Now wake up! You must wake up!

The third sleep was when you lied down with [your] woman.
You slept in the bed with her.
From this, the family was made.
Wake up, traveler. Wake up!
You've been asleep for so long, traveler.
Now wake up! You must wake up!
The fourth sleep was in the cremation ground.
You're stretching your legs [there].
You've been asleep for so long, traveler.
Wake up, traveler! Wake up!

The *bhajan*, which Sad Giri attributes to the renowned, medieval Indian *bhakti* poet-saint Kabir, suggests that all beings live as travelers wandering aimlessly and, metaphorically, asleep in *sansār*. In her commentary Sad Giri explains that as long as the soul sleeps it "has to wander in the *caurāsī*," meaning, it must take birth in the 84 million life forms within the sphere of *sansār*. The twins, who have been asleep for so long, have finally awakened in the *sādhu*'s place. There, they abandon all potential plans for marriage and settle with him. The twins' journeying in *sansār* has finally come full circle. Their wandering has finally brought them *home*. What is more, their wandering has all along been a quest for "home place" (Smith 1987: 28–29; Jackson 1995). By settling with the *sādhu* the twins make his ashram in Kāi their home. In his research with nomadic Walpiri aboriginals of the Tanami Desert in Central Australia, anthropologist Michael Jackson sets out on a journey to understand the ways that the Walpiri, who do not build or dwell in houses, and who are constantly on the move, construct and experience notions of home. Jackson distinguishes between ideas of "house" and "home." He writes,

> Home was a central place to which you or your thoughts constantly return. But home was also a group of people without whom your life would cease to have meaning. I had to remind myself not to get carried away by the poetic resonances of Walpiri metaphors of country and ignore the field of kinship and affinity—the sites of individual lives—that give those metaphors vitality and value. (Jackson 1995: 66; italics in original)

Sad Giri's berry twin tale similarly distinguishes between notions of house and home. In living with their parents the twins are certainly housed, but they are not "at home in the world" (Jackson 1995). Why not? After all, wasn't their parent's house a "field of kinship and affinity"? Despite the love and comforts of their parent's house, it lacks peace. The peace the twins seek exists at the *sādhu*'s ashram. "We'll settle [*thaharnā*] here with you. We'll serve you. We'll do *bhakti*. Whatever happens, our lives are in your hands now. We'll only do *bhakti* and our souls will get peace," the twins implore the *sādhu*. Moreover, as their words express, peace comes from the practice of *bhakti*. If the twins were to return and remain at their parent's house, they would get love, but they would never get peace. Their parents would expect them to marry and have a family. To leave their natal house and settle in their conjugal (husbands' family's) houses. In the words of the *sādhu* Ganga Giri, "what *bhakti* can be done in householding [*gṛhasth*]?"

Sad Giri's construction of her ashram in her natal village has been a quest for peace. The emotional support she receives from her family indicates that, like the berry twins, Sad Giri, too, receives love, affection, and comfort in their house.

In most of my meetings with Sad Giri, her mother and brother Kamlesh settle by her side. They do whatever she asks them to do. Having her family nearby seems to afford Sad Giri some peace-of-mind. But their house is not Sad Giri's home. She can never have the kind of peace she has at her ashram at her parent's place. There, three generations of family make their home in a space consisting of only two rooms. Children move here and there; adults come and go. This universe bustles with life. Yet the priorities and values of this world make abundantly clear to Sad Giri how far from home she is. While sitting in her parent's house, Sad Giri emphasizes to me and Manvendra Singh that we should move to her ashram where "it is peaceful." "The heart speaks freely when it feels at peace," she says.

Sad Giri feels at home at her ashram because she finds peace there. Away from the expectations of householding and marriage that surround (and suffocate) her as long as she lives in her parent's house, Sad Giri spends her time in her ashram focusing on her religious practices. The ashram provides the space for Sad Giri to develop and transform herself spiritually. But, unlike the twins, Sad Giri has yet to settle permanently in her home. Sad Giri lives her life "betwixt and between" two places, her parent's dwelling and her ashram. During the day Sad Giri settles at her ashram. At night, however, she returns to her parent's house. The majority of the *sādhus* I worked with refuse to accept Sad Giri's *sādhu*-status, despite her having taken initiation as a Daśanāmi, because she wanders between what is seen as two opposing worlds. Until Sad Giri settles permanently in her ashram (or, as I was told, any ashram), the other *sādhus* will see her only as a householder, and not as a *sādhu*. Although most of those *sādhus* spend time with householders, many of whom are the *sādhus*' biological kin, by settling permanently in their ashrams (and not with their families) they distinguish themselves as "real" *sādhus*. In this way, place creates as well as differentiates *sādhus*. Where a *sādhu* settles says a lot about who she is.

Having an ashram she owns, though, offers Sad Giri a form of protection she would not otherwise have if it belonged to someone else. The sense of safety Sad Giri feels in her ashram makes it a home for her. "No one can kick me out," she says. She implies here that she can kick people out of her ashram, particularly those whom she thinks have questionable characters. Not surprisingly, the people whom Sad Giri suspects the most, and whom she keeps from settling permanently in her ashram, concerns those *sādhus* without any home place. She explains,

> This ashram is "safe." This is my room. No one has the right to be here. I will not stay here with a [male] sādhu. But if you [Antoinette and Manvendra] stay, I can stay here. I can make food; we'll have satsang. Householders can stay here, but not sādhus. This is how I feel. I am a sādhu. What if that sādhu [who wants to stay in the ashram] is bad? Male sādhus today aren't good. I'm speaking openly. They might wear the clothes of sannyās, but the way they look at women is very bad. I won't allow a sādhu to stay here with me. If he has to stay here, I will return to my room in my parent's house. If you stay, I can stay. But if I'm alone, I won't stay with sādhus. I'm speaking from my heart. Householders can stay because they won't look at me in the wrong way. I believe this 100 %…Male sādhus won't see me as their sister. They don't have anything to do with anyone. They only rape women and run away. If I will say something to [a sādhu], he will leave this village and go to another [village]. Where can we catch him? He'll leave everything and go. But where would a householder with home, land, children, and a wife go? If a householder does this, his life is over. Sādhus can run away. They can go anywhere. How can you catch a sādhu?

The association Sad Giri makes between home place and the moral character of *sādhus* is apparent. *Sādhus* without any home place lack moral fiber. They cannot be trusted. They have no sense of accountability to others because they have nothing to do with anyone. Disconnected from any place, and without any sense of home, these *sādhus* have neither peace nor love in their hearts; they wander the earth like dangerous animals, predators always on the move. In much of the Brahmanical literature, *sādhus* who act mad or like animals are seen as embodying the ideals of *sannyās*. In this framework, wandering signifies and brings about the loss of a sense of a person's humanity, of being human in the world. Both the Brahmanical texts and the *sādhus'* practices indicate that people's, and in the context of this chapter, *sādhus'* ideas of humanity—and morality—are emplaced, constituted, and nourished in their experiences of home. Whereas the classic model renounces that experience, the lived asceticism of the Rajasthani *sādhus* embraces it. For them, home defines who a *sādhu* is. The idea of homelessness that the Brahmanical model values has no place in Sad Giri's view of what "counts" as *sannyās*.

What I also find significant about Sad Giri's narrative concerns the unusual way she describes her ashram as her own "personal space." Our conversations are punctuated with her emphasizing, "this is my own ashram," "this is my own room," or "this ashram belongs to me." Traditionally, ashrams constitute public spaces (and refuges) where anyone can wander and settle. Sad Giri, though, suggests that her ashram represents not only an extension of (personal) domestic space but also a private *and* female space, the entry into which she controls.[9] She explains,

> I built this ashram ten years ago. It's been ten years. I bought a plot of land for 13,000 rupees [approx. $260 USD]. I didn't depend on anyone but myself. I made [this ashram]. I am responsible for its craftsmanship [kārīgar]. I built it myself. I will have my samādhi [memorial for burying a sādhu] here. No one can stop me from doing this. It's my land and ashram. I built my own personal [nizī] ashram. It doesn't belong to anyone else. No one else paid for it. I paid for it myself. I bought the land with my own money. I did all this myself.

168.4   Conclusions: Sādhus' Singing Themselves at Home in the World

Sad Giri's practices illustrate several fascinating, and in my experience, recently emerging paradigmatic shifts happening in the ways that female *sādhus* in Rajasthan, and perhaps elsewhere in India, think about and experience classic (and predominantly gendered masculine) renunciant concepts like wandering,

---

[9] I say "female" and "private" space here to prevent any immediate conflation of these concepts of space, that is, of female as private, and of private as female. Sad Giri is creating what is traditionally a public and male space into a private and female space. In making this distinction between female and private, I am paying heed to Lindsey Harlan's caution in conjunction with women's practices in the domicile: "...there are public activities in which women engage within the domicile, and...this fact should prevent any facile contradistinction between domestic and public, or between "female" and "male" realms" (Harlan 2007: 66).

settling, ashram, and *sannyās*. More specifically, her practices suggest that female *sādhus* move in space and create a sense of place in distinctly female ways; that *sādhus*, regardless of their gender, can be at home in the world; and that the ashram functions as their home place. Unlike the Brahmanical depiction of *sādhus* as perpetually homeless wanderers, who belong nowhere, Sad Giri's stories and songs present a more locative vision of the everyday lives of *sādhus* and their nego- tiating ideals with reality. Settling in an ashram, whether it belongs to them or not, constitutes an experience of home for the *sādhus*. Sad Giri defines her idea of an ashram by explaining what happens there. "*Satsang* happens in the ashram," she says. "*Prem-bhakti* [love-devotion] happens." Religion scholar Mary MacDonald observes that understanding the ways that religious people value places requires scholars to focus on "the ritual events enacted there" (2003: 9). Sad Giri's concept of ashram pushes back at the standard Brahmanical depiction of ashram as a place of psychosomatic suffering, penance, and striving. By contrast, in Sad Giri's ash- ram, and in the ashrams of most of the *sādhus*, people sing *bhajans*, tell stories, recite and remember sacred texts, laugh and cry together, serve each other as forms of God, love and care for one another, and share their lives, hopes, and dreams together. While activities other than those illustrative of *satsang* take place in the ashrams of the *sādhus*, they accentuate these activities as most influential in shaping their identities and asceticism.

In the *sādhus'* asceticism, the idea of ashram as home place resonates with Jackson's representation of the concept in the context of the Walpiri world as "a universe…made of skeins of relationships" (Jackson 1995: 64). The people in their ashrams and the practices enacted there create and sustain an experience of home for the *sādhus*. Stories like the berry twin tale, and songs like "Traveler, Wake Up!," render the ashram as a powerful and meaningful home place, but just as important help to create the participants' understandings and experiences of that place as home. Ideas of home, place, and space, as scholars have claimed, are culturally crafted in specific locations (Feld and Basso 1996; Jackson 1995; Smith 1987, 1978). Jackson writes, "My experiences in Central Asia had led me to explore the notion of home, not as something given but as something made, not as a bounded entity but as a mode of activity" (1995: 149). Thus, the *sādhus'* practices, unlike the Brahmanical ideal that de-territorializes and displaces the sense of *sannyās* from geography, reterritorialize it in the geography of Rajasthan and wherever else they may wander or settle in their quests for freedom, friendship, and transformation.

Whenever Sad Giri performs her rhetoric of renunciation alone and/or in the company of others in her ashram, she makes herself at home in the world. But even if Sad Giri remains far from her ashram, she creates a sense of the ashram as home place and evokes the feelings of peace and love, friendships, and memories emplaced there wherever she moves in the world by means of *satsang*. Hence, Sad Giri's routes of wandering constitute types of "place," where, simply because she is there, the ashram/home/place has been created. In the ashram, Sad Giri feels safe, protected, and at peace; she can imaginatively enter into the worlds of her songs, stories, and texts and wander, even as she is settled in place, in search of God's home, which, for the *sādhus*, is the only real place in the world. She can meet God from the very spot where she sits (and dwells) in her ashram. Here, Sad Giri has a place to stand. But whether she settles or wanders here, there, anywhere, Sad Giri maps her world by following the tracks of her heart.

Acknowledgements   I thank Stan Brunn for giving me the opportunity to contribute this essay in this volume and for inspiriting me to wander into a new and fascinating theoretical field of research. I also thank Tyler Roberts, who came to the University of Wyoming in February 2012, to give a public lecture on research from his forthcoming book, *Encountering Religion: Responsibility and Criticism after Secularism*. In his presentation entitled, "Placing Religion: The Social Sciences and the Humanities," Dr. Roberts discussed Jonathan Z. Smith's notions of "locative" and "utopian" models in a way that helped me to conceptualize clearly the ideas discussed in this chapter. Finally, I thank my colleagues at the University of Wyoming Paul V.M. Flesher and Michael Carl Brose for reading and commenting on drafts of this chapter. Their prescient comments helped me to expand and development the arguments in this chapter. The research on which a large portion of this chapter is based, particularly the research that I conducted in Rajasthan in the summer of 2011, was funded by a University of Wyoming *International Travel Grant* and a *Basic Research Grant*. The research I conducted between 2004 and 2006 was funded by an American Institute of Indian Studies (AIIS) *Junior Scholar Dissertation Fellowship* and an Emory University *Internationalization Grant*. The AIIS also awarded me an advanced *Hindi Language Fellowship* for the academic year 2002–2003.

References

Casey, E. S. (1996). How to get from space to place in a fairly short stretch of time. In S. Feld & K. Basso (Eds.), *Senses of place* (pp. 13–52). Santa Fe/New Mexico: School of American Research Press.

DeNapoli, A. E. (2009). Duty, destiny, and devotion in the oral life of female *Sādhus* in Rajasthan. *Asian Ethnology, 68*(1), 81–109.

DeNapoli, A. (2014). *Real Sadhus sing to God: Gender, asceticism, and vernacular religion in Rajasthan*. Oxford: Oxford University Press.

Dumont, L. (1960). World renunciation in Indian religions. *Contributions to Indian Sociology*. The Hague: Mouton and Company.

Feld, S., & Basso, K. (Eds.). (1996). *Senses of place*. Santa Fe: School of American Research Press.

Geertz, C. (1973). *The interpretation of cultures: Selected essays*. New York: Basic Books.

Gil, S. (1998). No place to stand: Jonathan Z. Smith as *Homo Ludens*, The academic study of religion *Sub Specie Ludi. Journal of the American Academy of Religion, 66*(2), 283–312.

Gold, A. (1988). *Fruitful journeys. The ways of Rajasthani pilgrims*. Berkeley: University of California Press.

Harlan, L. (1992). *Religion and Rajput women: The ethic of protection in contemporary narratives*. Berkeley: University of California Press.

Harlan, L. (2007). Words that breech walls: Women's rituals in Rajasthan. In T. Pintchman (Ed.), *Women's lives, women's rituals in the Hindu tradition* (pp. 65–84). Oxford: Oxford University Press.

Hymes, D. (1975). Breakthrough into performance. In D. Ben Amos (Ed.), *Folklore: Performance and communication* (pp. 11–17). The Hague: Mouton.

Jackson, M. (1995). *At home in the world*. Durham/London: Duke University Press.

Khandelwal, M. (2004). *Women in ochre robes: Gendering Hindu renunciation*. Albany: SUNY Press.

Khandelwal, M., Hausner, S. L., & Gold, A. (Eds.). (2006). *Women's renunciation in South Asia: Nuns, yoginis, saints and singers*. New York: Palgrave Macmillan.

Macdonald, M. (2003). Introduction. In M. Macdonald (Ed.), *Experiences of place* (pp. 1–20). Cambridge: Harvard University Press.

Okely, J. (1991). Defiant moments: Gender, resistance, and individuals. *Man, 26*(1), 3–22.

Olivelle, P. (Ed.). (1992). *Samnyāsa Upanišads: Hindu scriptures on asceticism and renunciation*. Oxford: Oxford University Press.

Olivelle, P. (1993). *The Āœrama system: The history and hermeneutics of a religious institution*. Oxford: Oxford University Press.

Olivelle, P. (2007). On the road. The religious significance of walking. In *Theatrum Mirabiliorum Indiae Orientalis: A volume to celebrate the 70th Birthday of professor Maria Krzysztof Byrski, Rocznik Orientalistyczny* (pp. 173–187). Warszawa: Dom Wydawniczy Elipsa.

Smith, J. (1978). *Map is not territory: Studies in the history of religion*. Chicago/London: University of Chicago Press.

Smith, J. (1987). *To take place: Toward theory in ritual*. Chicago/London: The University of Chicago Press.

Smith, J. (2004). *Relating religion: Essays in the study of religion*. Chicago/London: University of Chicago Press.

Wilson, L. (1996). *Charming cadavers: Horrific figurations of the feminine in Indian Buddhist hagiographic literature*. Chicago/London: University of Chicago Press.

Chapter 169
Religious Identity and Gender on the Edges of the Nation: The Leh District of India's Jammu and Kashmir State

Sara Smith

169.1   Introduction

In a 2004 interview in the Leh District of India's Jammu and Kashmir State (J&K), the conversation turned to Buddhist-Muslim intermarriage, often at the heart of inter-religious tensions in ways that this chapter will elucidate. Akbar, a middle-class Muslim professional in his 40s, steered our discussion to youth and discipline, and brought up the Ladakh Buddhist Association (LBA), an influential organization in this Buddhist-majority district. Speaking about LBA work to prevent Buddhist-Muslim intermarriage, he said, "I think this is their main agenda to control these things." Akbar surprised me by suggesting how they could be more effective.

> But they are not doing any homework. They don't have a code of conduct; they don't look at what the Buddhist girls are wearing, what society they are keeping, and in what activities they are involved. This is the main thing, to control the girls right from the house to the school and the college. And if they can deny a girl to marry a Muslim, and if they can threaten, or have an agitation on this point, why can't they make a code of conduct? A dress code?

As Akbar's remarks highlight, at the intersection of politics and religion, gender performance can become a point of tension and concern. While a romanticized past of Muslim-Buddhist harmony is idealized, today's discourse suggests the need for firm boundaries between those of different religions. This chapter explores the complex and contradictory ways that gender becomes a site for the politicization of religious identity in Leh District. As global and national understandings of what it means to be Buddhist or Muslim have gained circulation, and religion has crystalized as a political force in post-partition South Asia, the relationship between Leh's Buddhist majority and the Muslim minority has deteriorated. Conflict has been

S. Smith (✉)
Department of Geography, University of North Carolina, Chapel Hill, NC 27599, USA
e-mail: shsmith1@email.unc.edu

© Springer Science+Business Media Dordrecht 2015
S.D. Brunn (ed.), *The Changing World Religion Map*,
DOI 10.1007/978-94-017-9376-6_169

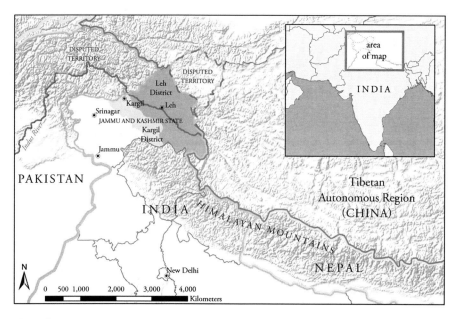

Fig. 169.1 Location of Leh District in India's Jammu and Kashmir region (Map by Timothy Stallmann)

articulated in part through women's bodies, and new interpretations of religious identity are emerging, particularly since the late 1980s.

Leh District, the focus of this research, is part of India's Ladakh region, a high altitude mountainous desert that borders Pakistan and the Tibetan Autonomous Region (Fig. 169.1). Ladakh became part of J&K State during the nineteenth century, when conquest and treaty brought together independent kingdoms into a Muslim-majority princely state with nominal independence from the British Empire. Upon partition in 1947, both India and Pakistan claimed J&K. Maharaja Hari Singh opted for India in October of 1947, but the region has been troubled since this time with claims made by Pakistan and a Kashmiri movement for independence beginning in the 1980s. Ladakh, populated by Buddhists and (predominantly Shia) Muslims, is separated from the rest of the state by mountain ranges, cultural practices, religion, and language. The Muslim majority in J&K is Sunni, speaks Kashmiri, and has little in common with Ladakhi Muslims. A poor mountain region with a population comprised of religious, linguistic, and ethnic minorities, Ladakh has been neglected by J&K State and for the most part ignored in negotiations over the fate of the state.

In 1979, the region was split into two districts: Kargil, with a Shia Muslim majority and Buddhist minority, and Leh, with a Buddhist majority and Shia Muslim minority (there are also Ladakhi Sunni Muslims and Christians, as well as Hindus and Sikhs who are not ethnically Ladakhi). Since the mid-twentieth century, political movements centered in Leh have repeatedly called for greater autonomy for the district, asking to break away from J&K. Echoing national and regional politics, this movement took on a communalist tone in the 1980s, when the LBA called for a social boycott of Ladakhi Muslims. The politicization of religious identity has had lasting effects on interactions between Buddhists and Muslims in ways that are profoundly gendered and embodied.

Buddhists and Muslims are developing an embodied religious boundary by forcibly preventing inter-religious marriages, and new interpretations of religious doctrine now discourage women from using family planning. This embodied religious boundary echoes the territorial delineation of India and Pakistan along Hindu and Muslim lines, though it has thankfully not resulted in the kinds of embodied and gendered violence that occurred in the wake of partition. Emergent practices of differentiation attend and sometimes conflict with expressions of religious identity from dress to vegetarianism. This chapter engages with the ways that Buddhist and Muslim women contend with the politicization of religion and life as religious minorities on the contested borders of the Indian nation. I draw on ethnographic work in Leh town and surrounding villages: 28 interviews conducted in 2004, 65 interviews conducted in 2007-09, a survey with 192 Ladakhi women in 2008, and youth oral history and photography projects in 2008 and 2010. This is part of a larger project, in which I have worked to mediate reified notions of the geopolitical as a macro process (the India-Pakistan conflict) by exploring how individuals reflect on, produce, and resist micro-geopolitics, for instance, by affirming or negating political-religious boundaries through marriage or child-bearing (S. Smith 2012). When clerks at the courthouse will not register the marriage licenses of Buddhist-Muslim couples, what are the repercussions? Will women eschew family planning to manage the demographic future of their religious group? When neighbors' religious identity takes on new meaning, what markers of religious difference become salient?

169.2   Feminist and Embodied Geographies of Religion

Agnew (2006: 183) has suggested that religion is today, "the emerging political language." The last decade has seen a rise in scholarship on geographies of religion, as noted by Kong (2010), Proctor (2006), and others (for example, Yorgason and della Dora 2009). This spate of recent scholarship may be driven by political events, but also by a turn to the more than representational (Yorgason and della Dora 2009). Reviews (Kong 2001, 2010) and special issues (Proctor 2006; Yorgason and della Dora 2009; Agnew 2006; Holloway and Valins 2002) have called for and observed a trend toward greater attention to intersectionalities of religion, class, gender, and sexuality (Dwyer 2010; Hopkins 2007), to sites beyond the officially sacred (Gökariksel 2009), and to affective experiences of religion (Holloway 2006). Crosscutting these new approaches has been engagement with religion through the body (Kong 2010; Holloway 2006; Bailey et al. 2007; Gökariksel 2009), and promising work has explored this through gender (Gökariksel 2009; Halvorson 2005; Hopkins 2007, 2009; Kong 2010; Dwyer 2010). Hopkins (2009) calls for a feminist approach to geographies of religion, and in particular, the ways that gender and religious identity intersect.

As gender is an iterative, bodily performance (Butler 1990), so, too, religious identity is produced through the performative (Kong 2010). This means that the bodily is called into religious practice, but also that religious identity is made through nuanced performances and experiences that intersect with other aspects of life experience and identity such as gender and class. In Leh, what it means to be Buddhist or Muslim is a changing, flexible, iterative process. Bodies are sites upon which religious identity is inscribed through dress, ritual, and practice, but are also the sites through which subjectivities are made and managed (Gökariksel 2009; Holloway 2006; Smith 2011; Bailey et al. 2007). Bailey et al. (2007) have demonstrated the importance of paying attention to embodied religiosity and performance through their exploration of Methodist practice in nineteenth century Cornwall, observing that the production of religious subjects exceeded sacred space and played a role in the formation of young subjectivities. In Cornwall, temperance movements involving the observation and regulation of young people's lives and bodies rendered those bodies both a site of inscription of Methodist values, as well as a means for protecting the future of the church; by attending to this embodied religiosity and its effects, Bailey et al. (2007) suggest a new spatiality for the study of religion. In a nuanced treatment of veiling practice, Gökariksel (2009) describes how the self is transformed not only through veiling as an embodied practice, but also through the body's movement through different spaces. Working from the body outward, Holloway (2003) demonstrates that bodily performance produces sacred space. I seek to build on these trends by exploring the ways that the performance of gender and religion is brought into conversations about politics and the geopolitical future of Leh.

Kong (2010) points to the importance of balancing our attention to microgeographies of religious practice and macro-geopolitical effects, and in particular to think through the ways that microgeographic practices contribute to global understandings of religion and politics. By addressing how intimate practices such as family planning and marriage are interpreted as religious and territorial statements, and the ways that this interpretation spills out into other assessments of daily life, I hope to take up Kong's suggestion that we approach the microgeographic as constitutive of national and global political trends. As she argues (Kong 2010: 769), "the ways in which religion is experienced and negotiated are also multifaceted and multiscaled, from the body to the neighborhood, city, nation, and across nations."

169.3 Embodying Politics and Religion

Leh District is the single Buddhist-majority district in J&K, and one of only six such districts in India. In a political framework in which claims for rights and sovereignty are often made on the basis of difference, religious identity has become increasingly politicized, particularly given Leh's geopolitical vulnerability as the site of disputed borders with Pakistan and China and the implicit post-partition assumption that sovereignty can be determined by majority religious identity.

The politicization of religion reached a crescendo in the late 1980s. Buddhist politicians called for Union Territory status, which would separate Leh's administration from the rest of the state. Pointing to discrimination and neglect by the J&K government, Buddhist activists drew attention to their cause (greater autonomy) by "boycotting" local Muslims; that is, enforcing sanctions on Buddhist-Muslim interaction of any kind. Although the boycott ended in 1992, political discussion continues to be framed in terms of religious identity. There has been excellent work on the genealogy of these tensions (Aggarwal 2004; van Beek 2000, 2001; Bertelsen 1996; Gutschow 2006; Srinivas 1998); we have yet to develop, however, an account that traces out the ways these tensions are normalized in every day practice, particularly with respect to embodied and bodily forms of intimacy and self-expression. My interviews indicate that this is a primary way people experience and engage with the politics of religious identity.

In an electoral democracy in which politicians often rely on religion to mobilize blocs of voters, numerical advantage is crucial. In a 2007 interview with a senior member of the Ladakh Buddhist Association (LBA), I was told their top priority is keeping Leh District Buddhist – that is, promoting population growth by discouraging Buddhist women from using family planning. In 2007, the LBA sent letters to Sonam Norbu Memorial Hospital in Leh (the only place tubal ligation services are available in the district), demanding they cease providing family planning services. Buddhist religious leaders began condemning family planning. Rumors spread that family planning was no longer available, and Buddhist women from Leh town traveled to rural areas to convince their rural counterparts – Buddhist women with the least education and least economic resources – to have more children in order to keep up with perceived Muslim population growth. Those who once enthusiastically embraced family planning are now discouraging their daughters and daughters-in-law from using it, though apparently without success. Buddhist women of child-bearing age are being told that Buddhists will "die off," and Buddhist leaders threaten that family planning will have dire consequences for the cycle of reincarnation. While Buddhists have a solid majority in Leh District, they perceive Ladakh (that is, Leh and Kargil Districts) to be vulnerable losing its Buddhist majority. While Muslim women do not reiterate demographic fears, their religious leaders have also told them that family planning is a sin.

In the sections that follow, I explore the ways that gendered and territorial language becomes a vector for politicized religious identity through concerns around marriage and family planning. I then go on to discuss the embodied and gendered ways that religious difference is taken up in conversation.

169.3.1   Territorial Marriage and Reproduction

The relationship between Buddhists and Muslims is described in gendered, territorial, and generational terms, and as a temporal shift from the past. In a 2004 interview, Jafar, a middle class Shia man in the education sector, described conflict as a result of youth and education: "It is the younger and educated people who disturbed the communal harmony here. Otherwise the old generation, they were living like brothers." Jafar went on to say, "As a Muslim, I always used to talk in conferences and in meetings, that we should not marry with the ladies of other communities. Because it is the main cause for communal disharmony." When I asked him why this might be the case, he explained, "Now, because of education, people are angrier…Like Buddhists, Muslims are also angrier. Muslims think that our women should not marry with the Buddhists." At the end of the 1989 social boycott, when a truce was negotiated between Buddhists and Muslims by religious and political leaders, part of the agreement was an informal ban on intermarriage (which goes against the Indian constitution).

Karima, a Muslim women in her 30s, told me that today there were hardly any intermarriages: "Maybe there might be one or two, who went way over the mountains and stayed there for years before coming back… Otherwise they will just start a demonstration immediately." When I asked why, Karima deployed a common language of risk and extinction: "They say that the Buddhist girls are going to run out. And the Muslim girls. If the Buddhists convert to Islam and Muslims convert to Buddhism…then that will be the end. That's what they say. This started with the disturbance, with the agitation." On both sides, there are suspicions that marriages between Buddhists and Muslims are part of a larger conspiracy of conversion, as articulated by Yangzes, a Buddhist, "…some will go because they fall in love, other girls get deceived, and some are coerced after being given money or something like that." Ahmed, likewise, suspected intentionality among Buddhists seeking to marry Muslim women: "Among Buddhists, despite the fact that there are so many women, they just pick one of our women. I think some people do it just to create conflict."

The gendered language used by Yangzes and Ahmed echoes a territoriality linked to the assumption that in intermarriage, the woman will convert. Yangchan told me that she was against intermarriage because the woman would not be able to sincerely practice her religion. Her own sister had married a Muslim, and Yangchan said, "There might be in 100, there might be one, who with a lot of difficulty, can do that. It's better if Muslims stay among Muslims and practice Islam and Buddhists stay among Buddhists and practice Buddhism." She went on to say, "The man never converts. And our Buddhists, they would never convert to Islam. It's our women, they might do it." There are of course exceptions, and I did hear from time to time of a Muslim man who had converted in the past and married into a Buddhist family.

In discussions about intermarriage, young men and women are described as sides of an equation. Akbar explained it as a trade balance: "We discourage our boys from marrying Buddhist girls, or marrying outsiders…. Because if there is one import from other community outside, you have one surplus girl in our community. So where will we send that girl?" Again, for Akbar, the solution to these problems is discipline, and he mentioned announcements in the mosque on Friday discouraging intermarriage.

> But the Buddhists think that if a Muslim boy goes with a girl, out of love, they start protesting. They create problems. …And a few Muslim girls also marry to outsiders. We have not protested. Why? They are supposed to make the law right from the beginning, in your own house. You educate them in your own line; you educate them in your own ways.

I was unable to find a couple that had intermarried successfully after 1992. Every person I spoke with was aware of the intermarriage ban brokered at the end of the social boycott, and most supported it. While 83 % of the Buddhists and Muslims I spoke with had relatives across the religious line (i.e., Buddhists had Muslim relatives, Muslims had Buddhist relatives), a similar percentage stated that the practice should be banned today. There was a sense of risk, danger, and inevitability around inter-religious romance, as summed up by Angmo, a Buddhist in her 60s: "but can they block it? They'll just runaway, the girls being bad." I turn now to the ways that gendered and territorial language plays out in the case of family planning.

Contraception, widely adopted in the late 1980s and 1990s when it was first available in Ladakh, has now become controversial and politicized. While Buddhist and Muslim women continue to use contraception, and desired family sizes are low across religious identification (most women desire two to three children), there is an outpouring of political and religious rhetoric around its use, particularly around the use of tubal ligation to limit family size. Some Buddhists fear that contraception will lead to a die-off of Buddhists in the future, or to demographic decline with political consequences: in the words of one interviewee, the district, "being run by Shia Muslims." This fear is driven at least in part by electoral politics and the assumption that vote blocs are managed along religious lines. Thus, the demographics of the district become a determining part of its political and territorial future.

Concerns around contraception are articulated in terms that blur lines between politics and religion. Buddhist (and some Muslim) arguments against contraception draw on the idea that children come with their own luck or fate, and that trying to manage this through small family size (and thus more resources for each child) is futile. Nilza, for instance, says:

> In the past, we really didn't have family planning. As many children as a woman could have, she would have that many. Some could have eight, nine, ten, twelve, those who only had a few would have five. It was like that … all of those [children], thanks to Konjok [the Buddhist triple gem, which includes the Buddha, his teachings, and the Buddhist community], they wouldn't have to beg, they would all get taken care of.

Nilza describes women facing a difficult choice between following religious ideas about contraception and the compulsions of contemporary economic life: "women tell one another, you need to do this, you need to do that. [Having many children] is not good for you, it's not good for your body, and you won't be able to afford the education, or good food." Nilza has heard that Muslims do not use family planning: "Those who are very successful, and have good educations, they do use it … if they are rich, or if they have good education, they do use it. But for those who are ordinary, they don't do it …. That's what people are saying."

My survey suggested perceptions of Muslim avoidance of contraception are unfounded, and that most women in Leh today, regardless of religion, prefer to have two to three children. It is not only through references to fertility and marriage, however, that Buddhist-Muslim difference is made. In the section that follows, I trace subtle forms of marking and interpreting difference.

169.3.2   Subtle and Embodied Religious Difference

> Author: Do you think there's a difference between Buddhist and Muslim women's lives?
> Saeda: Yes, there is. Buddhists can go in any field and participate. Muslims have some restrictions. For instance, in the festival. You've seen that there aren't Muslim women participating in the festival.
> Author: Oh, do you mean dancing and performance?
> Saeda: Yes, yes. There is a little bit less freedom, when compared to Buddhists.
> Author: But for instance, what about in fields like teaching, medical, and so on?
> Saeda: Those are no problem.

Beyond the territorializing language deployed to discuss family planning and the risk of intermarriage, religious difference emerges in a range of more subtle cues attributed to religious identity. While religious identity is defended in part through the prevention of intermarriage and in discourse about reproductive practices, what it means to be Buddhist or Muslim is also marked in the commonplace assumptions about masculinity, femininity, and daily comportment and dress. Saeda, a young Muslim woman, suggests that Muslim women have "a little bit less freedom." Spalzes, a Buddhist woman in her 70s describes Buddhist boys as lazy: "the Muslim boys are active and *hushar* and successful. Our Buddhist boys are wilted. They don't take responsibility and they live off their parents and don't advance. Those that are educated don't try to find things to do." These gendered markings of difference are tied implicitly and explicitly to political events: Dorje, for example, suggested the communal tension of the 1980s fueled Muslim ambition.

> I think that Muslims had gone for a little more ambition after that … more hardworking. That was the Muslim feeling. At that time there used to be a lot of Muslim boys idling in the market. Today you will find none. They used to be very, very uninterested in school.

Intermarriage is viewed through a lens that cross-tabulates religious identity and particular performances of gender. My research assistant, Hasina, and I, had a long conversation with a Shia woman in her 40s, about a recent scandal in which a Muslim man with three children had run away to be with a Buddhist woman. As the discussion developed, both Hasina and Laila expressed surprise that the Buddhist woman in question was "a simple type" rather than, "a fashion-*chokan*." Both expressions use an English word ("simple," "fashion"), to describe two tropes often counterposed in gossip: the "simple," girl (here, meaning humble, un-prepossessing), in contrast to the girl who "does fashion," portrayed as more likely to engage in illicit affairs. In these counterposed tropes, the simple girl would dress nicely and neatly, but perhaps in a more modest fashion and in less flashy colors and patterns, while the girl who "does fashion," would wear that year's cuts and patterns of salwar kameez or wear jeans, as well as spend more time styling her hair or perhaps wear makeup. In addition to the outward expression of fashion, a girl who "does fashion," might be assumed to be more self-indulgent and willful than a "simple girl."

This start to the conversation then proceeded through a discussion of the Muslim understanding that while Buddhists protest when a Buddhist woman marries a Muslim and converts, Muslims do not do the same thing: Laila said, "We Muslims, we totally don't do protests. Among them, they always protest. If one girl goes, then they just sweep together everyone from the whole *yul* to do a protest." When I asked Laila if Muslim girls often ran away with Buddhist boys, she returned to the tropes of discipline raised by Akbar in the opening to this chapter, and added assumptions about piety: "Muslim girls, they stay in their own religion, they've understood themselves. They'll understand and appreciate their own religion." Laila's rendering of Buddhist-Muslim difference was ambiguous and nuanced. Asked if there was a difference between Buddhist and Muslim women's lives, she downplayed such difference: "Aside from religion, there's nothing. The religion is different, but the lifestyle is just the same. It's what Allah gives us, that's all there is. The way of saying things is different. For instance, Buddhists say *Konjok* and Muslims say *Khuda* [God], otherwise it's the same." Although Laila embraces the idea that there is not much difference between Buddhist and Muslim women, nevertheless tropes appear in her interview: Buddhist woman who does fashion (as an imaginary that shapes her interpretation of the gossip she relays), and the pious Muslim woman who would not convert.

Sakina, an energetic woman who had converted from Buddhism to Sunni Islam upon her wedding to a Muslim man, insisted there were differences between Buddhist and Muslim women's lives:

> [Buddhist women] do get more freedom to go where they want to go. They can go anywhere. Muslim women, they are a little stricter. They can't go around in pants and a coat. They can't show their hair. They can't be as "modern," as Buddhists ... They can be with boys and wear pants and shirts and so on [rather than salwar kameez].

I pushed Sakina on this point, observing that her depiction went against some of my own observations, and she clarified that, "Here there's not a huge difference, but in dress and so on it is different." She gave the example of Buddhist women joining the police force, and went on to say, "Muslim don't do that...they do the work where you go to the office and come back in the evening. Not like the police. They can become a teacher, those kinds of things."

Speaking to the kind of narrative described by Saeda and Sakina, that Buddhist women are "more free," Laila was dismissive:

> If you stay in your own religion you'll be happy. That's all there is to it. For Muslims, you're supposed to cover your hair and not show all this. That's difference, that's nice. What's happy about going around uncovered? There's nothing happy in that. For women, it's happier. A woman, she's as valuable as gold, that's what we have in our religion.

At times, it is difficult to distinguish differences due to religious practice and differences attributed to religious identity. As religion takes on new significance, narratives about what it means to be Buddhist and Muslim begin to crystalize with the other religion as a constitutive other – a foil against which meaning is made of religious difference. Thus, experiences and practices such as styles of dress or education are reinterpreted as being part of a cluster of variables constituting religious identity. This clustering highlights difference, and sometimes leads to the overlooking of commonality. Buddhist and Muslim women of previous generations were less frequently sent to school, a common narrative in oral histories with women in their 50s and older. Diskit, a Buddhist woman nearing her 70s, told me about education: "They would say, "a girl studying, where is she going to get to?" Really, it was shameful. For girls, if you could do work inside the house, then when you got married, that would be useful." Some Muslim women, however, would interpret their own similar experiences as a mark of religious difference. Faeda and Haadiya, Sunni Muslims in their 60s and 40s, described a lack of education in the past as specifically Muslim: "For us, at that time, for Muslim girls, they said that too much studying wasn't good." Haadiya contrasted her education and life with that of her mother and grandmother, again, interpreting their lack of education to Muslim difference even though her story echoes Diskit's: "My mother, my grandmother, staying in purdah, not getting out, praying…my mother studied inside the house, with a tutor. Studying is a difference [in my life]. In the past, Muslims mostly wouldn't let girls study."

Narratives around religious difference are described in relation to an evolutionary sense of time and a gradual acquisition of religious knowledge tied to modernity and a shedding of inappropriate or ill-informed behaviors that blurred the boundaries between those of different religions. Similarities fell away sometimes through intentionality and other times through economics and the gradual recomposition of families. This sometimes came up in dress, as talked about by Faeda below, and often came up in discussions about weddings and other rituals. Faeda marks change with a description of *perak*, the cobra-shaped turquoise headdress still worn by Buddhist women at dances and weddings: "In the past, the fathers, they were all Muslim, but the mothers, since they were from Buddhists, then they had *perak*. From their grandmothers' side, they had *perak*. My mother, she had her own *perak*." In many Muslim families the valuable *perak* inherited from Buddhist relatives (given from mother to eldest daughter upon marriage) were gradually sold off. In Faeda's family, they were sold since the women no longer had occasions to wear them. In the same conversation, she also mentions participating in Buddhist fasting when she was young.

Like Faeda, Mohammad, a Sunni man in his late 70s, also portrayed the past as a time of more similarities, to the surprise of the young Muslim woman present at the interview.

> In the past, Buddhist and Mussalman weddings were the same…Now that there are many educated people and there are many Moulvis, we don't do that anymore. … in the past Mussalman would also dance and sing. Mussalman and Boto would mix and have picnics and archery festivals and there they would dance together. We used to have a great archery festival here. At that time all the boys and girls would dance. At that time, Buddhists would not get a dance. The Mussalman girls would dance so much they wouldn't get a chance. It was such a happy time. Nowadays, they say not to do that… All that, those traditions are over now.
>
> Mohammad's wife: Now the moulvis say not to dance, and not to have picnics. They even say not to go to shows.
>
> Young woman: and they say not to sing
>
> Mohammad's wife: and not to sing
>
> Young woman: and not to watch TV
>
> Author: Really?
>
> Mohammad: And not to watch TV.

Author: Do people do that, do they obey?

Mohammad: About the TV, they don't obey it…The moulvis all say, "Listen to the news, but don't watch the singing and dancing. If they are talking about farming or something, you can watch that, but not the singing and dancing."

Interpretations of religious distinction, such as an exaggeration of Muslim difference in the past by the women who associate a lack of education for women with Islam, or the young woman who believes that Muslims have never danced at weddings, suggest that while there has been a change in religious differentiation and the interpretation of religious difference, that change can be invisible for younger generations, who may assume that current practices stretch back into the past, when in fact religious differentiation may be rather more pronounced today.

169.4   Conclusion: Disciplining the Boundaries of Religion and Gender

As suggested by Akbar's remarks early in this chapter, political and geopolitical tensions drive desires for firm boundaries and their maintenance. This is often described in terms of self-control and community preservation, and evokes a desire for discipline. The firming up of these boundaries speaks to Kong's (2010) observation of the ways that micro-practices both mirror and produce the macro-scale geopolitical significance of religion.

Discussing efforts to raise awareness about religion, Akbar's descriptions of religious difference are imbued with gendered language:

… We are fortunate that most of our boys are not involved in liquor and other things. But Buddhists, they are a free society. They use liquor on every occasion. With the result that their boys are involved in such things. Which spoils the whole society…. And you can see the dress code of the girls: they hardly wear anything on their body. But we, being a small community, we control it. And we try to indoctrinate the girls right from the beginning.

Akbar's comments reflect broader concerns about youth morality and religiosity in Ladakh today. There are crosscutting narratives at work here. While some describe modern youth as discarding religion, a contrary trend is also observed, as by Nawang, a Buddhist in his 30s:

Personally, I think that when the Buddhists got strict with their religion, I think that the Muslims also got strict. When Buddhists realized they have to practice Buddhism seriously, and see themselves as Buddhists seriously, I think the Muslims also did the same, or felt the same. I guess it increased since 1989…I think that right now in our neighborhood all those kids go and study Koran and stuff, they have to

Describing a post-1989 resurgence in interest in religion, Nawang said: "They hardly used to have any teachings for students at temple, and especially after that time, they felt like, Buddhist young people are losing their Buddhism, and they don't know anything about Buddhism, and so they have organized all these talks for students." Aaliya, a Sunni Muslim, said that people are more religious today: "They study religion more today. Both Buddhists and Muslims. In the past everyone was illiterate so they didn't understand much. It's Boto and Muslim. They understand the advice more today." I was often told that there was greater religious literacy among today's generations and that there were more religious teachers and events for laypeople than in the past.

As argued by Yorgason and Della Dora (2009: 635, 630), without attention to the role of religion, we "[omit] fundamental features of social space," and the ways that it "complicate[s] categories and experience." Expressions of religious difference and conflict blur with stories about intergenerational change. I asked Faeda if inter-marriage in the past had built strong relationships between the religious communi-ties. She responded, "But in the past, all the people, they were a little upright." And went on to say, "Nowadays, where will you get people like that? Nowadays, every-one is looking after his or her own interests."

Individuals may subscribe to geopolitical narratives about marriage, but that may not prevent them from falling in love with the wrong person. They may participate in pro-natal geopolitical talk, but fears about their own bodily integrity may discour-age them from having babies for a political project. As these tensions play out, they dislodge distinctions between micro and macro scales as the ways that religious difference is performed, discussed, and managed, is tied to larger political ques-tions. For boundaries to have meaning, bodies must be marked with difference. If individuals refuse that difference, the map becomes meaningless. This research has sought to demonstrate that geopolitical territorialization – the very production and defense of territory – is enacted and refused in the most intimate of practices, and is tied in subtle and overt ways to religious identity and changing understandings of what that identity means. These changing understandings are enmeshed in attitudes toward intergenerational transformation and gendered performance: as religious identity takes on new political meanings, collective and iterative performances of gendered religious identity, and the interpretation of these performances, become a subject of concern.

What can we take from the Leh case study? Building on recent developments in geographies of religion, I echo calls to attend to the political use of religion (Agnew 2006; Proctor 2006; Yorgason and della Dora 2009), but also propose that moving forward it is crucial to examine how new political interpretations of religion inter-sect with class, gender, and sexuality. Bailey et al. (2007), Gökariksel (2009), and Holloway (2006) have already indicated the value of attention to embodied religiosity; the Leh case study suggests the promise of tracing how embodied religi-osity and markers of religious identity are shaped by and shape political narratives. We might ask, when, how and why is the embodied performance of religion under-stood to be political? Finally, I suggest that we examine the subtle and embodied shifts in how religion is not only marked and performed, but how religious differ-ence is interpreted and projected onto daily life. Some of these shifts may be quite subtle, but may mark the maintenance of or emergence of deep divisions. Furthermore, the ways that daily practice becomes interpreted as being due to reli-gious difference may help us understand the changing role of religion in political and social life.

Acknowledgements This chapter has benefited tremendously from helpful suggestions from Stanzin Tonyot. None of this research would have been possible without the incredible generosity of the people who not only gave their time to answer my uncomfortable questions, but also did so with grace, hospitality, and insight. Thanks to Timothy Stallmann for producing the map. The fieldwork for this research was supported by the Fulbright-Hays Doctoral Dissertation Research Abroad Program, the Society of Women Geographers, the International Dissertation Research Fellowship Program of the Social Science Research Council with funds provided by the Andrew W. Mellon Foundation.

References

Aggarwal, R. (2004). *Beyond lines of control: Performance and politics on the disputed borders of Ladakh, India*. Durham: Duke University.

Agnew, J. (2006). Religion and geopolitics. *Geopolitics, 11*, 183–191.

Bailey, A. R., Harvey, D. C., & Brace, C. (2007). Disciplining youthful Methodist bodies in nineteenth-century Cornwall. *Annals of the Association of American Geographers, 97*, 142–157.

Bertelsen, K. B. (1996). *Our communalized future: Sustainable development, social identification, and the politics of representation in Ladakh*. PhD dissertation, Department of Anthropology, Aarhus University, Denmark.

Butler, J. (1990). *Gender trouble: Feminism and the subversion of identity*. New York: Routledge.

Dwyer, C. (2010). Veiled meanings: Young British Muslim women and the negotiation of differences. *Gender, Place and Culture, 6*(1), 5–26.

Gökariksel, B. (2009). Beyond the officially sacred: Religion, secularism, and the body in the production of subjectivity. *Social & Cultural Geography, 10*, 657–674.

Gutschow, K. (2006). The politics of being Buddhist in Zangskar: Partition and today. *India Review, 5*, 470–498.

Halvorson, S. (2005). Growing up in Gilgit: Exploring the nature of girlhood in Northern Pakistan. In G. Falah & C. Nagel (Eds.), *Geographies of Muslim women: Gender, religion, and space* (pp. 19–43). New York: Guilford Press.

Holloway, J. (2003). Make-believe: Spiritual practice, embodiment, and sacred space. *Environment and Planning A, 35*, 1961–1974.

Holloway, J. (2006). Enchanted spaces: The séance, affect, and geographies of religion. *Annals of the Association of American Geographers, 96*, 182–187.

Holloway, J., & Valins, O. (2002). Editorial: Placing religion and spirituality in geography. *Social & Cultural Geography, 3*, 5–9.

Hopkins, P. E. (2007). Young people, masculinities, religion and race: New social geographies. *Progress in Human Geography, 3*(2), 163–177.

Hopkins, P. E. (2009). Women, men, positionalities and emotion: Doing feminist geographies of religion. *ACME: An International E-Journal for Critical Geographies, 8*, 1–17.

Kong, L. (2001). Mapping "new" geographies of religion: Politics and poetics in modernity. *Progress in Human Geography, 25*, 211–233.

Kong, L. (2010). Global shifts, theoretical shifts: Changing geographies of religion. *Progress in Human Geography, 34*(6), 755–776.

Proctor, J. (2006). Introduction: Theorizing and studying religion. *Annals of the Association of American Geographers, 96*, 165–168.

Smith, S. (2011). She says herself, 'I have no future': Love, fate, and territory in Leh District, India. *Gender, Place and Culture, 18*(4), 455–476.

Smith, S. (2012). Intimate Geopolitics: Religion, marriage, and reproductive bodies in Leh, Ladakh. *Annals of the Association of American Geographers, 102*(6), 1511–1528.

Srinivas, S. (1998). *The mouths of people, the voice of God: Buddhists and Muslims in a frontier community of Ladakh*. New York City: Oxford University Press.

van Beek, M. (2000). Beyond identity fetishism: "Communal" conflict in Ladakh and the limits of autonomy. *Cultural Anthropology, 15*, 525–569.

van Beek, M. (2001). Public secrets, conscious amnesia, and the celebration of autonomy for Ladakh. In T. B. Hansen & F. Stepputat (Eds.), *States of imagination: Ethnographic explorations of the postcolonial state* (pp. 365–390). Durham: Duke University Press.

Yorgason, E., & della Dora, V. (2009). Geography, religion, and emerging paradigms: Problematizing the dialogue. *Social & Cultural Geography, 10*(6), 629–637.

Chapter 170
Zooming-In on Terms and Spaces: Women's Perspectives and Cognitive Mapping in a West Bank Settlement

Hannah Mayne

170.1 Introduction

"Dangerous," "violent," "armed and furious," and "charismatic" are common adjectives. "Fervent believers," "religious radicals," and "eccentric fundamentalists" are frequently quoted descriptions. Bearded men with large skullcaps fill the pictures.[1] Over the past few decades, these terms and images, repeated in major news media and academic literature, are those used to characterize Jewish settlers in the occupied territories of the West Bank. Indeed, there are those in the outposts and other radical pockets whose characters fit these descriptions and, undeniably, any participation in the settlement project is necessarily implicated in the Israeli occupation of the region; however, they who can be described with such potent language are a small minority within the large social grouping classified as "West Bank settlers."[2] Such categorical words signify conjured constructions of weighted ideas, valences, moods, and relationships. Invoking them all too often forecloses further conversation and investigation. Their repeated enunciation across time, space, a diversity of media forms, and a multiplicity of discursive realms reproduces the heavy contents of the words' baggage of associations.

---

[1] Many of these words and pictures were found by searching through recent articles about the West Bank in major news websites, including The New York Times and the BBC. Others can be found in The Fundamentalist Project (discussed below) and similar academic publications. I also conducted a limited study on popular opinion concerning West Bank settlers, and such terms emerged in most responses (Personal correspondence, August 2011).

[2] For example, among the first ten articles that come-up when searching the phrase "West Bank settlers" on The New York Times website at the time of writing, eight concern the topic of settlers in small, unauthorized outposts.

H. Mayne (✉)
Department of Anthropology, University of Florida, Gainesville, FL 32601, USA
e-mail: hannahmayne@gmail.com; h.mayne@mail.utoronto.ca

© Springer Science+Business Media Dordrecht 2015
S.D. Brunn (ed.), *The Changing World Religion Map*,
DOI 10.1007/978-94-017-9376-6_170

In this chapter, I attempt to interrogate the term "settler" in the context of the West Bank, to dig beneath common compositions of meaning by reflecting more closely on those whom this word identifies.[3] I do this by focusing on settlers in a relatively large, established community, in contrast with those who live in unauthorized outposts or the small number of especially volatile locales.[4] In addition, I attend to the voices of female settlers, faintly heard and muffled by the loud blare of news flashes reporting on men's ideas and behaviors. Working within the anthropological discipline and with the method of cognitive mapping, I am interested in moving beyond the thin surface of stereotypical notions, beyond the image of black and white religiously fanatical characters, to encounter settlers directly and personally, to uncover underlying rhythms and textures, to catch sight of inner trajectories of motivations, thoughts, and action. Many people don't want to listen to these voices because they issue from a site of perceived homogeneity and religious fundamentalism, because they emanate from the bodies of those engaged in the occupation of the Palestinian people, in a region of contentious politics, wide-ranging discrimination, pain, suffering, and desire. Yet, I think it is worthwhile to be attentive, in the project of understanding the inner workings of a conflict of global significance.[5]

Contemporary anthropologist, Kathleen Stewart, sees ethnographic analysis as the investigation of how disparate images, words, associations, gestures, and processes throw themselves together and form something (2008: 73), whether it be a constructed compilation of fear, shooting, hate, and violence; or, a composition of safety, home, comfort, and peacefulness. Stewart warns against pinning down singular social meanings, and suggests, rather, to trace the disparate processes of assemblage. Such a poetic sense of the social "literally can't be seen as a simple repository of systemic effects imposed on an innocent world [or neutral social and geographic landscape] but has to be traced through the generative modalities of impulses, daydreams, ways of relating, distractions, strategies, failures, encounters, and worldings of all kinds" (73). It is this poetic attunement to imaginations

---

[3] More than two decades ago, anthropologist Susan Harding (1991b) pointed out the broad and vague nature of the term "fundamentalist," showing its hazy shape in modern imaginations. She called for a problematization of such words, and nuanced, local readings to understand these people who are so often pushed to society's fringe. More specifically, anthropologist Joyce Dalsheim (2011) contends that hegemonic discourse surrounding the West Bank conflict marginalizes subtle differences, presents a polarized perspective, and creates a situation of mutual "othering." "Settlers" are thus maintained as out there, crazy, and extreme. Opinions that do not fit comfortably into neat, pre-existing categories are most often dismissed (138), and Dalsheim therefore calls for scholars to pay heed to marginalized voices, especially "those 'others' we have grown accustomed to hating" (149).

[4] I distinguish between authorized and unauthorized settlements in the West Bank. The former are authorized by the State of Israel, while the latter are not. According to international law, in the opinion of most experts, all authorized and unauthorized settlements are illegal. I maintain the distinction, nonetheless, because I am doing research in a milieu where such a difference is in fact, I argue, socio-culturally significant.

[5] I too struggle with my own involvement in such a contested landscape and the ways in which my sojourns in this region support the continuation of the occupation. However, I believe that discomfort can also be a productive location from which to engage in investigation.

and representations that I attempt to introduce into the context of West Bank settlements.[6]

My research involves several months of fieldwork in the fall of 2011 and spring of 2012, mostly in Nofharim, a pseudonym for a Jewish settlement north of Jerusalem. I focus largely on numerous, lengthy conversations with fifteen adult women in this community. I also hung around in Nofharim, attended events, traveled often in and out of the community, and talked about such places with people within and outside. By carefully attending to the women's words, gestures, and even silences, as well as surrounding experiences, new representational compositions emerged about life in this area. More specifically, I ask two questions: By sensitively considering the voices of women, what image emerges of West Bank settlements, their inhabitants, and the broader conflict? And, secondly, what happens when we move beyond the dramatic and ostentatious narratives of the fanatically ideological, demographically-minor, unauthorized "hilltop" or "outpost" dwellers, and investigate an established settlement populated by modern religious Jewish families? In other words, what can these very particular perspectives demonstrate regarding settler participation in the project of occupation?

I would like to suggest that such research, among religious Jewish women in an established settlement, reveals a social, cultural, and even political situation that is more complex than the simplified and polarized depictions of mainly abstract male ideologues that all too commonly fill popular and academic imaginations.[7] The female settlers in my study cannot be painted with rifles and hatred on one side of a political canvas, aggressively guarding the land beneath their feet against Palestinian[8] rivals at opposing ends. Their experiential residence on this territory, rather, reflects other thoughts, desires, and emotions. For many women with whom I spoke, pragmatic

---

[6] Over the past few decades, anthropologists have struggled with questions of voice and representation – whose voices are and are not reflected in ethnographic publications, and how subjects are interpreted and represented (Appadurai 1988; Clifford and Marcus 1986; Marcus and Fischer 1986). Feminist scholars, specifically, have called for deeper sensitivity and careful attention in the way power plays out in socio-cultural domains, highlighting the presence of those who are overlooked or marginalized in academic inquiry (Abu-Lughod 1993; Behar and Gordon 1995; Wolf 1996). How do we render the depth of another's experiences? How do we avoid essentializing or simplifying the true complexity and heterogeneity of those we study? Who are we, as outsiders, to speak for them, to paraphrase or explain their language? How do we maintain analytic distance, while, at the same time, getting intimately close in knowledge, understanding, and experience? I see Stewart's approach as one method which responds to and tackles these concerns. Through attunement and careful representation, through tracing subtlety and detail, we can ideally get closer to an illustration that portrays the depth and ambiguity inherent in human experience, while capturing the shadows of our subjects' own words and views, and also leaving room for intersecting scholarly analysis.

[7] Dalsheim and Harel (2009) provide a comprehensive listing and critique of such academic representations. As mentioned above, a brief search of major online news portals also brings up numerous articles and images of radical, male settlers and their extremist activities.

[8] I use the term Palestinians throughout to identify Palestinian Arabs living in the West Bank specifically. The situation of Palestinian Arab citizens of Israel, living outside the West Bank and within the pre-1967 borders of Israel, is substantially different and beyond the scope of this chapter.

reasons (economic, educational, and environmental), in addition to the religiously ideological, underlie their reasons to move to a settlement. The primacy of these practical factors, in turn, lead to a normalization of the settlement's military circumstances, allowing everyday routines to carry on without obvious stress or discomfort. Such normalization processes, I want to suggest, consequently induce an overlooking of the conflict itself, including intellectual distance from the Palestinian population. Feelings towards Palestinians thus do not involve rage or violence, but rather sadness as well as inner anxieties that emerge from lack of knowledge and close proximity to unknown others.

To trace the resonances of social experience, Stewart looks at events. She does not seek the dramatic public episodes and conflicts of Max Gluckman and Victor Turner,[9] however, but rather momentary interactions. She looks at the subtleties and nuances of minute situations, digging for underlying, sometimes unuttered messages and the potentialities of new themes and directions in future conversations, attunements, and perceptions. Anthropologist Lila Abu-Lughod comparably recommends writing "ethnographies of the particular," which are wary of generalizations and focus on particular individuals in order to "subvert the most problematic connotations of culture: homogeneity, coherence, and timelessness" (1991: 154). After I proposed an early outline of this chapter, the editor of this volume suggested that I try to obtain cognitive maps from some of the women in my study. As I followed his advice, I realized how the exercise of cognitive mapping (a growing method in cultural geography) can be seen as a form of both Stewart's quiet yet richly evocative event, and Abu-Lughod's concentration on the particular, and it thus presents itself as a potentially productive method in ethnographic research, and narrative ethnography particularly. With careful attention to both a specific subject's process, and his or her drawn creation, much can be learned about larger compositions of meanings and associations that lie beneath a place's name and geographic spot. Stan Brunn specifically argues for cognitive mapping's constructive possibilities in locations of conflict and tension. I have therefore brought the exercise to the West Bank settlement where I work, and I do think that the maps, paired with intensive fieldwork, valuably illustrate and bring to the fore deeper trajectories and forces pulsing across this heated landscape.

170.2   Research Contexts

Nofharim is located in the occupied territories of the West Bank. The 1947 UN Partition Plan designated this area for a future Palestinian state. In 1948 it was captured by Jordan. In 1967, the area was taken under Israeli control, but not annexed. Palestinians living there today suffer from an ambiguous relationship with the Israeli state – they live under Israeli occupation, and they have neither citizenship rights and privileges, nor political independence.

---

[9] Classic examples include Gluckman (1940), Turner (1957), and Turner (1974).

For many Israelis, great import is attributed to maintaining control of this mountainous area because of reasons of strategic defense. Certain religious Jewish factions put additional pressure on the state to exercise sovereignty over this region because the West Bank area (referred to by them as Judea and Samaria) encompasses the geographic topography of utmost importance in the Bible. (According to Jewish tradition, it is here that Abraham walked, Jacob dreamed, the tabernacle stood, and prophets spoke. Ironically, the densely populated Tel Aviv metropolitan area is not mentioned in ancient religious texts.) At the same time, Israeli governments have been afraid to give West Bank Palestinians citizenship because this could threaten the country's Jewish majority. Israel's population is gauged to be nearly 7.6 million, and approximately 20 % are Israeli Arabs.[10] The estimated Palestinian Arab population in the West Bank (not included in the statistics for Israel) is 2.6 million.[11] (Jewish Israeli settlers in the West Bank, excluding those living in East Jerusalem, number approximately 300,000.) Since the Oslo Accords and the establishment of the Palestinian Authority in 1993–1994, efforts have been made to give Palestinians in the West Bank increasing autonomy over their towns and cities. Conversely, however, this has entailed the development of progressively stronger and more opaque divisions between Palestinian and Jewish populations in the region, and more vigorous and violent methods of surveillance and control over Palestinians.

What is most interesting about Nofharim is its remarkable proximity to a large Palestinian city. Before the Oslo Accords, Nofharim residents traveled through this city when they drove from Jerusalem, and even shopped there. Today, it is illegal for inhabitants on either side to cross to the other. However, they can see each other. Gazing from a balcony on the Western edge of Nofharim, one can see the figure of a Palestinian man or woman walking along the street. During the past three decades, Nofharim residents have watched houses, malls, and hotels filling the previously green and brown slopes of this Palesitnian city's suburban sprawl. Until a few years ago, Palestinian residents have watched construction develop in Nofharim. From the perspective of an outsider, the geographic proximity on one hand, and the extreme lack of accessibility on the other, are odd and unsettling.[12]

Culturally and religiously, the residents of Nofharim identify with the broad modern orthodox Zionist movement in Israel. As such, they struggle to balance

---

[10] https://www.cia.gov/library/publications/the-world-factbook/geos/is.html and www1.cbs.gov.il/reader/ Accessed June 2012.

[11] https://www.cia.gov/library/publications/the-world-factbook/geos/we.html Accessed June 2012.

[12] They may drive past each other along the roads, and their cars may sit together in traffic jams, but with the exception of the major supermarket, Jewish settlers and Palestinians from the West Bank rarely occupy the same geographic spaces and they therefore hardly ever interact. This was not always entirely so. Until the Oslo Accords, Nofharim residents bought their food in the nearby Palestinian town, had their cars fixed there, and even went shopping there for shoes and clothes. Yet, according to the first residents who came to Nofharim, developing personal relationships with Palestinian residents was not something that happened often. In this sense, Jerusalem presents a particularly contrastive example (See Romann and Weingrod (1991) for an illustration of the complicated ways Jewish Israelis and Arab Palestinians negotiate life together in the very different setting of Jerusalem – a somewhat geographically unified urban center.)

modernity and tradition: in the public and secular domain, women are educated and many have professional careers, while in the religious sphere, women have separate roles and realms. The domestic household, interestingly, has become a middle ground between these two, with a general increase in participation of fathers in childcare responsibilities, especially among younger couples.

Nofharim, like many established and state-authorized settlements, is mainly populated by middle-class families. Because of its location, many adults commute daily to their places of work in Jerusalem, which is a driving distance of twenty-five minutes. The total population of Nofharim is approximately 1,700.[13]

Nofharim is a closed community; all newcomers are screened before they are allowed to rent or purchase a home. I was able to stay for temporary visits because of personal contacts in this settlement.

In addition to ethnographic fieldwork, photography and attunement to visual dimensions of this space were also significant parts of my research. Supplementing descriptions and analysis, I photographed the physical "normalness" of the streets, buildings, and homes, and, at the same time, also tried to represent the remarkable spatial proximity of Nofharim to neighboring Palestinian suburbs. To the eye, Nofharim appears calm, clean, and middle class. This peaceful residential and psychological sphere is so far and yet so close to what lies beyond its boundaries (Fig. 170.1).

Fig. 170.1 One of the main streets at the top of the hill in Nofharim, with street signs and middle class homes like residential areas across Israel (Photo by Hannah Mayne)

---

[13] www.cbs.gov.il/ishuvim/ishuv2010/bycode.xls Accessed October 2011.

170.3   Comparative Research

Much of the ethnographic literature on Jewish settlements in the West Bank portrays an image of fundamentalist groups, led by charismatic male leaders, and their unusual, extremist lifestyles. Until recently, most of this literature was represented in *The Fundamentalism Project*, published in a series of five volumes in 1991.[14] Each of the articles in these volumes, which aim to cover the topic of fundamentalism in Jewish society, discusses as a main example, Gush Emunim, a political movement in the 1970s that aspired to establish settlements in the West Bank, Gaza, and the Golan Heights. Most of these examinations of Gush Emunim portray the members as radical religious extremists pitted against secular modernism, and with the assumption that the organization represents all settlers. Recent representations – in academia as well as public discourse – continue to reproduce this view, and only a minimal number of scholars have begun to interrogate these stereotypes.[15] (From my own ethnographic research, the socio-political group of Gush Emunim has fallen apart as an organized body, and the majority of those who live in the settlements today do not affiliate with any such organization nor feel represented by one.)

Two decades ago, feminist philosopher, Sandra Harding, argued for the de-centering of men's experiences in studies of social worlds (1991a: 13). Following Harding, I would like to contend that an examination of women and their experiences generates a very different set of stories, and an undocumented landscape of thoughts and emotions towards this territory, and towards the Palestinian population.[16] In contrast with ethnographic publications that concentrate on the fundamentalist ideologies of male settlers, my conversations with women in an established community highlight practical benefits for habitation in this area, the advantages of living in an intimate, cooperative community, and also sensitive

---

[14] For specific examples, see entries by Aran, Don-Yehiya, Heilman, Liebman, and Sprinzak, among others, in Marty and Appleby's The Fundamentalism Project (1991).

[15] In terms of ethnographic publications, Feige (2009), despite comprehensive breadth, attributes the ideologies of contemporary settlers to those of Gush Emunim, painting this group as an expansive homogeneous group of religious warriors, fighting against modern secularism. Fischer (2007), though providing valuable critique of earlier publications and the binary structures of their theoretical foundations, maintains the spotlight on the confined ideological world of the West Bank yeshivot and their male teachers and students. In contrast, and responding to earlier publications, Dalsheim and Harel (2009) show how Gush Emunim ironically began to lose strength right at the time when scholars began to write about it. "Gush Emunim survived discursively as a trope that subsumes all religiously motivated settlers" (225), they write, arguing that these outdated representations fail to take account of the diversity within the religious settler population, as well as the similarities between Jewish settlement on both sides of the Green Line. Dalsheim (2011) offers an exception that departs from previous trends, exploring the uneven, surprising contours of settler belief and action, and demonstrating the multi-dimensional heterogeneity of political and religious ideologies within this population.

[16] A small handful of academic studies have been written on female settlers. However, they all focus on small groups of particularly ideological women in especially radical locations (El-Or and Aran 1995; Feige 2009; Neuman 2004). In contrast, I attend to women in large established settlements, where land claims and political activism are engaged less aggressively.

approaches to the Palestinian "Other." Would I hear a generally different scope of responses if I spoke with men? Is it possible that women stress the practical aspects of life in such a location, while men emphasize the political struggle? Are mothers more interested in a safe, nurturing, and productive environment for their children? Are some women more sympathetic to the suffering of others? I do not have definitive answers to these questions; however, I would like to suggest that, perhaps due to the themes and approaches taught and reproduced through often separate social spaces, or perhaps owing to their more prominent role in childbearing and raising, women in this milieu express overall more practical and compassionate attitudes. If this is true, it is even more reason to hear the voices of women in the context of political conflict. Though they represent half of those who populate the settlements, their input seems to be droned out by a discourse of zealous and militant language mostly uttered (or imagined to be uttered) by men. Of course, a much wider and broader study of this kind also begs an ethnography of Palestinians living on the other side of the barbed wire fence, and the ways in which the normalization of the settlements has affected their everyday lives.

170.4   Cognitive Mapping

I asked three mothers in Nofharim to draw maps of their community. Though I was initially anxious about how they would respond to such an odd or conceivably childish request, they in fact were intrigued and ultimately seemed to enjoy the opportunity to engage in such a creative activity. I specifically engaged one woman who is outgoing in her personality and also very involved in the social life of the community, one woman who is partially involved, and one woman who is relatively shy and uninvolved in the community. One of these women came to the community when it was first established, and the other two came less than ten years ago. Their ages range from early thirties to mid-forties.

All three women speak basic English, though I spoke Hebrew with them as well. As they drew each place, I labeled it in English on their map. Shira and Naamah preferred to label some of the places themselves, but in many cases, forgot to write in English, and so I leaned over and wrote the translation below their words. Their images are Figs. 170.2, 170.3, 170.4, and 170.5.

170.4.1   Shoshana

Shoshana is the first woman I asked to engage in this exercise. She came to Nofharim as a child thirty years ago, the daughter of one of the founding families. She has lived here, pretty much, ever since. Even throughout national service and university studies in Jerusalem, her home base was at her parent's house in Nofharim. When she married in the early nineties, her husband came to live here too, and it is here that

Fig. 170.2 Shoshana's map. The *dark arrows* point to spaces that she would ideally like to change or improve. In addition, Shoshana pointed out "the fence that we don't see" and along its side, the edge of the community where her children don't feel safe (Map by Shoshana for Hannah Mayne)

they are raising their seven children. Shoshana is a part-time teacher for special education in a nearby settlement.

Perhaps it is because she has lived in Nofharim for so long, or maybe it is just her relaxed personality – but either way, Shoshana is remarkably nonchalant about the contestation surrounding the location of her home community. "What's the big deal?" she kept asking, "What is there to worry about?" Shoshana is the type of person who looks at the glass half full, who doesn't dwell on anxieties and worries. As far as she is concerned, Nofharim offers her family a positive, safe, comfortable place to live. She is distantly aware of the Palestinian suburb in full view from her

Fig. 170.3 Shira's first map. Shira sketched the interiority of the community. When I asked her, she added a fence, and drew an *arrow* to represent her deep-lying anxiety about terrorists entering. Though she has faith in the fence, she knows it is not full-proof. Swirls indicate the places in Nofharim where Shira would like to develop family parks (Map by Shira for Hannah Mayne)

back balcony, but as long as there isn't any trouble coming from that direction, she isn't terribly troubled, emotionally or intellectually.

What was also interesting about speaking to Shoshana was that her husband, Dror, seemed to represent the opposite perspectives. Dror, when he joined our conversation one afternoon, emphasized a definitively ideological attitude towards the project of settlement inhabitance. For him, living in Nofharim is an active political move. For Shoshana, it seems more like a natural, and convenient choice, living a couple blocks from her parents, and providing her children with the same religious and suburban atmosphere which she had enjoyed as a child. Shoshana and Dror's differences are obviously the result of a combination of factors, but I think that gender perhaps plays a role. Mothers, like Shoshana, express especially pragmatic benefits of living in Nofharim, including a quiet lifestyle, good air, intimate and caring community, social services, and good schools, all at an arm's reach.

Shoshana is tall, and her posture is quite relaxed and gentle, reflecting her disposition. She dresses as most national religious women, and like the majority of women I met in Nofharim, with an elegant hat or kerchief covering most of her short brown hair, and modest, simple clothes – her shirt covers her elbows, and her skirt reaches almost to the floor.

After explaining the assignment, I asked Shoshana to draw a map of Nofharim. I invited her to draw the places that are most important to her, to identify what she

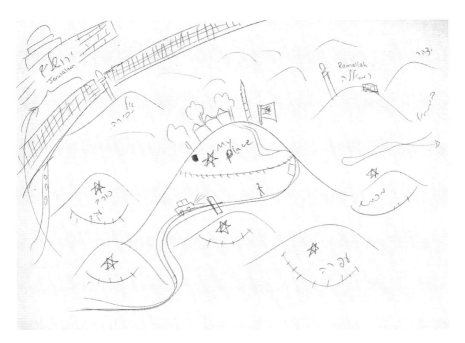

Fig. 170.4 Shira's second map. After rejecting an initial second map that depicted the wall between Nofharim and a neighboring Palestinian city, Shira re-started and sketched this broad-view picture. Here, the "primitive" wall that Shira so hates separates the rest of Israel from Jewish settlements and Palestinian towns alike (represented by *Jewish stars* or *mosque towers*). Shira added fences around all the Jewish communities, and also pointed out two locations where Jewish female drivers had been recently stopped and forced out of their cars, which were then kidnapped (Map by Shira for Hannah Mayne)

most likes, as well as areas that she feels could be improved or changed. I also asked her where she feels safe, and where she is scared.

The important spaces, those Shoshana drew first, included the synagogue, her father's house, the view of Jerusalem, Palestinians in the nearby city (marked by lots of little houses, "and big houses too" she told me), houses in Nofharim, paths between the houses, children, and people generally. Later, she laughed as she reflected on the absence of her own home on her map. Since they're renting, she and her husband perceive their apartment as a temporary space of dwelling. Her father's house, in contrast, is one of the earlier structures in the community and also the space where she spent the second half of her childhood. It thus features prominently on her map.

What Shoshana likes most, she told me, is the eastern side of the community with its open views towards rugged, rocky hills and trees. She then identified four areas of community life that she would ideally change.

First, the roads and paths could be improved, and other exterior areas in the community could be made to be more aesthetically pleasing.

The synagogue and surrounding social relations comprise a second matter that, for Shoshana, needs attention. The synagogue is a meeting place, in addition to

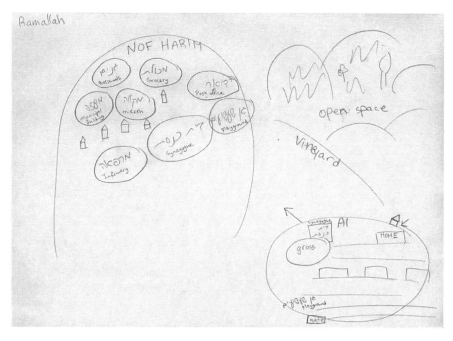

Fig. 170.5 Naamah's map. Naamah first drew the area where she lives at the *bottom* of the hill, and then the *upper part* with all its services and resources. With *arrows* she indicated the places where she wants change (Map by Naamah for Hannah Mayne)

being a house of prayer, she explained. Social friction comes to the surface and manifests performatively in this public space, despite the fact that the issues of tension are unrelated to the synagogue itself. Conflicts between Eastern European and North African Jews may take the appearance of differences in custom, but actually symbolize deeper anxieties and strife; individuals who are snubbed in the social sphere will not be given ritual honors during prayer services. "But again, these issues are not connected to the synagogue itself and what is done there, it's just that they come about there," Shoshana repeated. This is a social matter that she would like to be changed.

Thirdly, Shoshana would like to be able to build on the land that lies just beyond her father's house. However, the residents of Nofharim cannot construct further buildings because the land is owned by Palestinians who have not been located. This is hard for Shoshana because it means that the community is unable to expand, and also, more personally, because she and her husband want to buy a proper home in Nofharim and the land is there and theoretically available. It's frustrating for her, and others.

The fourth problem that Shoshana identified is the separateness of the Russian immigrants who came to Nofharim in the 1990s. They live in a corner of the community, and stay apart and mainly uninvolved. Shoshana wishes there to be more integration. This last comment fed into a wider, more general expression of concern about the overall closed-ness across the community. "People don't go and visit, and just drink tea at friends' homes. In that sense, Nofharim is more like a city," Shoshana sighed.

In terms of security, Shoshana feels safe everywhere. She told me this while shrugging her shoulders and laughing a bit, acknowledging that affective textures of safety imply a sense of familiarity. Thus, those who are less familiar with an area may not feel safe there, while those who have intimate knowledge of a space will by nature feel at home. Safety, in this respect, is a subjective matter and is disconnected from rational, technical particulars.

170.4.2 Shira

Shira is the second woman I asked to engage in cognitive mapping. Working part-time, she is the community director of events and social activities. In many ways, this vivacious and endlessly enthusiastic woman has her thumb on the pulse of Nofharim. I assumed she would willingly engage with the cognitive map project, and indeed, her jam-packed drawings reflect her energy.

Shira began by drawing her home, many other houses, trees, and an assortment of community buildings. The aspects that she would improve or change mainly include the expansion of social spaces, such as parks (theoretical locations of which are marked by spiraling circles), and the development of a communal storage for re-usable and recyclable goods and materials.

In terms of feeling secure, Shira has faith in the fence. "I don't feel like I even live within a fence because Nofharim is big. It protects me."

"And beyond it?"

"I see the Arab Palestinians, but I have no problem with that. They can stay there. That's totally fine. Before the fence, we used to go there even, and get our cars fixed and buy our bread for the Sabbath. But now we can't."

Shira doesn't feel safe, however, walking at night near the less inhabited areas, or the archeological site. The fence, she recognizes, isn't one-hundred percent secure. "Bad things can happen," she said, drawing a dark arrow to represent the vector of potential violence. "I lived somewhere where something did happen. But usually, I don't take it into mind. I'm used to it. We lock the house at night." She sketched a prominently sized lock beside the image of her house. "We even have Arab workers come and work here."

She turned to me directly, "This person said to me the other day, 'I don't buy from Arabs.' But how can they make a living?! Some are terrorists, and some are *not*. I don't get it." She shook her head, "Some say all of them are bad."

Shira also explained how she doesn't feel safe on the roads, but that she just has to go. With the pen, she showed me places where there were recent issues. In the past few months, cars have been kidnapped by young Palestinian men right on the road, and apparently now there is a group trying to kidnap women as well. When driving in front of a Jewish Israeli woman, the perpetrators will suddenly stop their car, blocking the road. The woman will be trapped, forced out of her car, and left on the side of the highway without anything.

"But inside Nofharim, I feel safe. The kids can go out whenever they want, wherever," Shira added.

Shira's first map is a cluttered close-up of Nofharim itself. She took a another piece of paper and made a second, very different map. Whereas the first one was intimate, full, and marked by familiarity, the second was more open and simple. Jewish stars identify hills inhabited by Jewish settlers, tall mosque towers distinguish the Palestinian areas. In this second map of the landscape, Shira drew a wall between Nofharim and the neighboring Palestinian areas.

"Is there a wall there?" I asked.

She turned to me and suddenly laughed. "You are right! I can't believe that I did that!" She took another piece of paper and re-started. "I forgot that I am living *within* the wall! That's so funny! Add that to your research!"

On this third map, she outlined the wall diagonally, separating Jerusalem on one side, and the Palestinian villages and Jewish settlements in the West Bank on the other. Shira marked Jerusalem by drawing a mosque. I expressed my surprise.

"That's the image that symbolizes it, no? Ok, here, I'm drawing a Western Wall for you… and the Israeli museum icon," she replied humorously.

"Why do you think you initially drew a wall between Nofharim and the neighboring Palestinian city?"

"Because it's another world, a place I'll never be. I'll never know the people. It's also hard for me to understand that there's a wall between me and the rest of Israel. How does the wall protect Jews?! It's ridiculous! Cars go, buses go… there are tunnels and rockets. It's like the wall in China – it's primitive, in so many ways! It's so ugly. Both sides hate it. It divides families, land, even animals! Imagine a goat on one side, and his water source is on the other… now he can't go."

Shira sat back and sighed deeply. "But it's not my responsibility. I can just complain."

"You can't do anything, you think?"

"I feel like I have no control."

We discussed politics, how leaders don't do what they say they will do. "All we know is that God is there, above us. That's all we can be sure of, and all we can trust."

170.4.3   Naamah

Naamah is the third woman with whom I met, regarding the activity of cognitive mapping. Naamah is more shy than Shoshana and Shira, and she cautiously warned me about her lack of artistic abilities. Nevertheless, before I knew it, she pulled out a small table and set to work.

Below the main part of the community, which spills over the top of the hill, lies a small cluster of mobile homes (sometimes referred to as caravans or house trailers).

Though one may expect them to be cold and cramped, the interiors are surprisingly warm and cozy. Naamah and her husband, their two daughters, and their son live in one of these white rectangular blocks. Naamah began by drawing this small residential section, at the bottom of her paper. She sketched a rectangle to symbolize her mobile home, then drew some additional strokes to reflect the other mobile homes in her area. She immediately drew the small synagogue and a narrow children's park. These are the central areas in her life in Nofharim. Naamah then moved beyond, up the hill, and to the larger community. She drew a bigger, central hill. Behind it she immediately wrote the name of an adjacent Palestinian city, but otherwise left that space blank and uncharted. Returning to the larger hill, she filled it in with multiple spheres representing the spaces she encounters most often: the store, the nursery schools, the *mikvah*, the municipal building, the synagogue. She made especial note of the infirmary, which, she chuckled, she uses every week. She also marked out the post office with distinct energy. "Nofharim is special, it's not the city," she explained. When she takes her girls to school in the morning, they go into the post office themselves to drop off the daily mail, and then continue on to their school. "Only in Nofharim," Naamah smiled, "That is something that could never happen in the city."

Only after marking these main points did Naamah suddenly remember to add houses. Naamah's perspective is particularly focused on the routines of childcare and work. For both personal and practical reasons, social life and leisure time with others stands second to her usual flood of responsibilities. The theme of houses, however, reminded Naamah about something that she would like to change: She dreams of building a house. Not necessarily at the top of the hill, but in fact, Naamah would love a house right behind the location where her mobile home currently sits. "And of course, there would be houses all around," she added. She drew an arrow at the imagined spot. (Houses cannot be built in the mobile home area because the area is technically owned by yet unidentified Palestinians. According to the state, the area can only be used for temporary dwellings.) Naamah also would like a shortcut made for the route up the hill, and this too is represented by an arrow.

"Beyond Nofharim… what do you think about the landscape?" I asked.

"I don't know. This is my world. I'm so busy and focused on my home, my family," Naamah responded, laughing a bit apologetically. "I'm in my own little world. Other women, like Shira, are out there and in touch with things happening all over. But I'm not like that."

Naamah considered her map. She marked out the open spaces of the eastward views, indicated by hills and trees. She added the vineyard.

I asked about safe spaces, but Naamah shrugged, "I feel safe everywhere inside Nofharim."

"Even walking up the path towards the main part of the community?"

"Yes. I mean, I don't like the hike up the hill, but yes, I certainly feel safe there."

Naamah and I sat over her map for a couple more minutes, but she had nothing else to add. "This is my world here," she told me, sheepishly apologizing for its elementary and simple character.

170.5 Analysis of the Maps

What can the maps tell us? Within the borders of the community, women build and stress the existence of normal lives. They mainly feel safe. They perform everyday routines, similar to those one might see in any communally oriented residential suburb. Religion is a salient and underlying base across time and space, symbolized by a centrally located synagogue that is one of the first items to be sketched on the women's maps. Many modern religious Jews, including the inhabitants of Nofharim, want to raise their children in a religious community, with a religious school that teaches their values, with like-minded neighbors, and with community activities that reflect their interests. For mothers especially, accessible resources and dependable social networks make raising multiple children much easier. The location is also beautiful. The maps thus show women's intellectual focus on their settlement community. What lies immediately beyond the fence, in contrast, is another story. In their words and drawings, Palestinian areas emerge as unknown, "out there," and imaginatively vague [17] (Fig. 170.6).

Fig. 170.6 The view eastward from the upper part of Nofharim, looking out onto mobile homes, the Judean hills, further villages, and the Jordanian mountain range in the distance. This is the vista that the women described in their list of valuable and appreciated features

---

[17]What lies directly outside the settlement's fence is psychologically off the radar. However, Jewish settlements in the region, as well as the nearby metropolitan centers of Jerusalem and Tel Aviv, are intricately linked. Everyday, many Nofharim residents exit the community gates in order to drive to work, bus to school, shop at the major supermarket along the highway, or visit friends and family.

Many of the women with whom I spoke in Nofharim recounted primary practical, economic, and social reasons which motivated them to move to Nofharim. They all believe in settling this most biblically valued land with Jewish inhabitance, but pragmatic rationales seem to have propelled them to this place. Shoshana returned as an adult because her parents were already there. Religious and political ideologies did not emerge significantly in our conversations; instead, she stresses the positive, technical and aesthetic benefits. Shira also wanted to live in this sacred territory, more religiously significant than the rest of the country, but again, she was drawn by the opportunity to raise her kids in a semi-rural environment, like she had also enjoyed as a child in the north. Naamah and her husband moved to Nofharim because it was geographically halfway between her and her husband's places of work, and much more affordable than living in Jerusalem. For them, it is an autho-rized, established community, with good services, schools, and not in the spotlight of extremist views and performed illegalities. Though Naamah and her husband were only able to obtain a mobile home, it was in a location that was recognized by the state. Naamah, more even than the other two women, shrugged her shoulders about religiously ideological motivations. With her disabled son, she has directly pressing issues on her mind. Just as it was for Shoshana and Shira, Nofharim offered Naamah and her husband a financially viable place to call home, with a religiously comfortable community, not too far from a major metropolitan area.

These economic and pragmatic motivating forces are reflected in the objects which Shoshana, Shira, and Naamah drew first on their maps and the items which they identify as the positive aspects: Naamah immediately pointed out the proximity of the children's playground and synagogue, the resources she utilizes in the upper, main part of the community, and the overall accessibility of all these locations of which she makes much use. Shira stressed the busy and active social buzz because it is this aspect which she most values. Shoshana made sure to draw her father's house, open spaces, trees, and wide views. All three women create a picture of life that is full, with all that they need right nearby. In certain cases, they may want improvements, but these are imagined as future developments, and ultimately, they all seem relatively satisfied with what they have. Even Shoshana, with her list of desired changes, only provided such an inventory when I provoked the question.

Hadas Weiss (2011), an anthropologist who has worked in established West Bank settlements, also points to the practical draws that motivated her immigrant informants to move to this contested region. She argues that through the processes of normalization, immigrants and other socially aspiring individuals become impli-cated in the settler project, an enterprise that was in fact not necessarily their goal. Tracing the genealogy of capital value in the settlements, Weiss writes:

> The West Bank became a popular residential frontier for a growing metropolitan workforce, offering private homes in reasonable commute from Jerusalem and Tel-Aviv… Settlements were advertised in the same way as outlying towns in Israel proper; individuals relocated to them in pursuit of upward mobility, much like suburbanites anywhere. (Weiss 2011: 114–115)

To a large degree, normalization of the settlements reflects one side of the state's ambiguous relationship towards the West Bank. Political interests and religious motivations, reproduced through daily discourse and imagination, push a narrative of peaceful family life, construction, and development, affordable to the Israeli middle class. From my experience, this hegemonic discourse is so powerful that it can obscure the magnitude of the steel and barbed wire security fences and the heavy military protection necessary to sustain a relatively safe Jewish inhabitance. My numerous conversations with fifteen women in Nofharim, where none save one mentioned the structures of security that lie beneath their windows, emphasize this point. Once inside the patrolled borders of the community, it is all too easy to fall into a rhetoric of normalcy.[18]

Such a focus on pragmatic benefits encourages a normalization of this lifestyle, an institutionalized process of rendering the extraordinary as ordinary. Aspects of settlement life that are out of the ordinary fall to the side of the intellectual radar. In other words, the stressed "normalness" of the inner community overshadows the "abnormalness" at its borders and beyond. Surveillance and security measures are seen as necessary precautions, but they are pushed to the margins of concern. Dwelling on them would only cause fear and anxiety. Not Shoshana, Shira, nor Naamah drew the fence that surrounds Nofharim, the guarded gates at the entrances, or any other form of security apparatus, until I provoked the topic. Rather, their drawings illustrate a picture of "normal life" and safety within the community's boundaries. Surrounding circumstances and military presence are absent or abstracted.

Such processes of normalization, I want to suggest, lead to an overlooking of the conflict, and a distancing of the Palestinian population. The women in Nofharim expressed feelings towards Palestinians that do not involve rage or violence. They do not identify anything particularly negative about the Palestinian spaces beside their settlement. Even when I prodded, the women just shrugged. For Naamah, a nearby Palestinian city is simply represented by its name. Shira and Shoshana drew little houses and symbolic mosques, but otherwise, none of the women drew any other forms of social life. No people, no details of community centers, schools, etc. as they had drawn in their own community. Palestinians are "over there" in another world that these women do not know or understand, despite the remarkable proximity.

When I asked about areas to improve, the women did not make any remarks about the territory immediately outside Nofharim either. I prodded this question, too, but again, it is completely beyond their main concerns. The political conflict is imagined to be sad and distant. Shira reflects this in her frustration, and ultimate passivity with regards to the wall. She hates it, as do many women with whom I spoke. It only darkens and deepens dividing lines, it only pushes sides further away from each other. But what can they do, they ask themselves. They feel hopeless, and since there is

---

[18] Feige (2009) writes about normalcy, as well, in his ethnographic treatment of the settlement of Ofra. Feige, however, sees normalcy as a statement actively shouted out by settlers in order to conceal the ugly shadows of occupation and to distract from underlying fundamentalist convictions (193–195). In contrast, I look at this process less as linear, and more as occurring cyclically and manifesting passively, as Weiss (2011) suggests. I am interested in how pragmatic desires and the perspective of settlements-as-normal, motivate participation in the settlement project, and how these circumstances encourage normalization of severe militaristic and security measures and further normalization of settlement life.

much to get done in the house and in the community, they relegate the issue some-where else, often to the domain of state responsibility.

Flags barely appear on the women's maps (the only case is Shira's second illustra-tion where she uses Jewish stars of David to signify Jewish settlements, and the Isreali flag to designate Nofharim). These women are busy with their daily lives; their focus is not on political protesting, driving stakes into the ground, and aggressively claiming the land. Likewise, they do not draw soldiers, or Palestinian terrorists. There isn't a hateful, necessarily hostile imagination of the Palestinian "Other" as an inherent enemy. Rather, I want to argue, there is an underlying blank slate.

Ignorance is, of course, not a-political. Its shape is concealed, and among its consequences are anxiety, mistrust, and ever-increasing distance and tension. It is buried, deep within psyches, not addressed because it is so often unrecognized as an issue of concern. Normalization processes do this. Uncovering the undertows of ignorance and distance thus reveals a much more complicated picture. On the per-sonal and societal levels, and across numerous geographic nodes where discord lies, how does entrenched intellectual ignorance feed feelings of threat and anxiety, and later defensive segregation and discrimination? When does long-term ignorance slip into moral exclusion? Where do ethics and legality fit in? Approaching any kind of answer, should, I think, include an understanding of underlying processes and knowledge of individual experiences.

170.6   Conclusions

In this chapter, I have attempted to interrogate common conceptual images of Jewish settlers in the occupied territories by focusing closely on women's lives, to encounter individuals face to face – to listen to how they think about their own lives as well as the broader circumstances of their contested lifestyles. I challenge common assumptions about religious and political fundamentalists, and, as such, disturb the binary between religious and secular. Indeed, as these women show through their words and maps, pragmatic, non-religious incentives underlie many of their motiva-tions to move to this region. They are certainly committed to the project of Jewish settlement in this Biblical territory, but their ultimate choice to move symbolizes their ambiguous consideration for both religious and modern convictions.

Thus, for the women with whom I spoke in Nofharim, the experience of being a settler involves a composition of positive feelings and practical advantages, very different from common resonances that echo from the term "settler." Analysis of the women's perspectives uncovers economic forces, social values, and normalization processes that conceal what is unusual and block-out what lies outside geographic, social, and cultural boundaries. As seen through the personal specificity of the wom-en's maps and words, the situation is intensely complex.

Just as it is in film-making, both broad view and close-up perspectives are essential in social research. Here, I have attempted to zoom-in on intimate compositions, because in this geographic and political context, the big pictures tend to dominate. When the particular and the poetic are examined carefully, especially in places of conflict and strife, what is revealed? How can assumptions, established categories, and binaries be challenged, to consider what emerges from underneath? In what ways can stereotypical compositions of meaning be deconstructed, to encounter people face to face, to understand the complex trajectories and resonances that lie beneath social and political conflict? As I have demonstrated in this chapter, a close-up view of women living in an established West Bank settlement reveals a socio-cultural landscape that is more complicated, and challenging than common depictions of this region.

References

Abu-Lughod, L. (1991). Writing against culture. In R. G. Fox (Ed.), *Recapturing anthropology: Working in the present. School of American research advanced seminar series* (pp. 137–162). Santa Fe: School of American Research Press.

Abu-Lughod, L. (1993). *Writing women's worlds: Bedouin stories.* Berkeley: University of California Press.

Appadurai, A. (1988). Introduction: Place and voice in anthropological theory. *Cultural Anthropology, 3*(1), 16–20.

Behar, R., & Gordon, D. (1995). *Women writing culture.* Berkeley: University of California Press.

Central Bureau of Statistics, Government of Israel. T'chunot Demographiyot [Statistical profiles]. http://www.cbs.gov.il/ishuvim/ishuv2010/bycode.xls. Accessed Oct 2010.

Central Intelligence Agency, United States of America. (2012). The world factbook (Entries on Israel and the West Bank). https://www.cia.gov/library/publications/the-world-factbook/. Accessed June 2012.

Clifford, J., & Marcus, G. E. (1986). *Writing culture: The poetics and politics of ethnography.* Berkeley: University of California Press.

Dalsheim, J. (2011). *Unsettling Gaza: Secular liberalism, radical religion, and the Israeli settlement project.* Oxford/New York: Oxford University Press.

Dalsheim, J., & Harrel, A. (2009). Representing settlers. *Review of Middle East Studies, 43*(2), 219–238.

El-Or, T., & Aran, G. (1995). Giving birth to a settlement: Maternal thinking and political action of Jewish women on the West Bank. *Gender and Society, 9*(1), 60–78.

Feige, M. (2009). *Settling in the hearts: Jewish fundamentalism in the occupied territories.* Detroit: Wayne State University Press.

Fischer, S. (2007). *Self-expression and democracy in radical religious Zionist ideology.* Ph.D. dissertation, Department of Anthropology and Sociology, Hebrew University, Jerusalem.

Gluckman, M. (1940). Analysis of a social situation in modern Zululand. *Bantu Studies, 14,* 1–30.

Harding, S. (1991a). *Whose science? Whose knowledge?: Thinking from women's lives.* Ithaca: Cornell University Press.

Harding, S. (1991b). Representing fundamentalism: The problem of the repugnant cultural other. *Social Research, 58*(2), 373–393.

Marcus, G. E., & Fischer, M. M. J. (1986). *Anthropology as cultural critique: An experimental moment in the human sciences.* Chicago: University of Chicago Press.

Marty, M. E., Appleby, R. S., & American Academy of Arts and Sciences. (1991). *The fundamentalism project* (Vol. 5). Chicago: University of Chicago Press.

Neuman, T. (2004). Maternal 'anti-politics' in the formation of Hebron's Jewish enclave. *Journal of Palestine Studies, 33*(2), 51–70.

Romann, M., & Weingrod, A. (1991). *Living together separately: Arabs and Jews in contemporary Jerusalem*. Princeton: Princeton University Press.

Stewart, K. (2008). Weak theory in an unfinished world. *Journal of Folklore Research, 45*(1), 71–82.

Turner, V. (1957). *Schism and continuity in an African society: A study of Ndembu village life*. Manchester: Manchester University Press.

Turner, V. (1974). *Dramas, fields and metaphors: Symbolic action in human society*. Ithaca: Cornell University Press.

Weiss, H. (2011). Immigration and West Bank settlement normalization. *Political and Legal Anthropology Review, 34*(1), 112–130.

Wolf, D. (1996). Situating feminist dilemmas in fieldwork. In D. Wolf (Ed.), *Feminist dilemmas in fieldwork*. Boulder: Westview Press.

Chapter 171
The Geography of Jewish Intermarriage in Five U.S. Urban Areas

Bruce Phillips

171.1 Introduction

Mixed marriage is arguably the most important social force shaping American Jewish life in the twenty-first century. The sociological literature on mixed marriage has focused on two themes: identifying the antecedent predictors of mixed marriage and assessing the Jewish connections of the mixed-married couple. These connections include formal affiliations (especially synagogue membership), informal connections, and connections to the Jewish future with regard to how children are raised. The one connection that has never been examined is geographic: where do mixed-married couples live relative to other Jews, and does it make a difference for other Jewish connections?

The literature on "spatial assimilation" argues that ethnic groups assimilate as they leave urban ethnic enclaves (Massey 1985) for the superior amenities offered by non-ethnic suburbs such as quality of education, access to employment, exposure to crime and social prestige (Massey and Denton 1985 p. 94):

Jews are now the most affluent and educated ethnic and religious group in the United States (Smith 2005), but this has not universally been translated into spatial assimilation. As significant numbers of Jews have made individual suburbs into Jewish residential enclaves (Phillips 2007a, b), Jews can choose to live in affluent suburbs with an identifiable Jewish character and significant Jewish population concentration. For Jews suburbanization is not necessarily concomitant with spatial assimilation.

Mixed-married couples constitute a group of Jews to which spatial assimilation might well apply. Intermarriage in general has long been linked to assimilation

B. Phillips (✉)
Loucheim School of Judaic Studies at the University of Southern California,
Hebrew Union College-Jewish Institute of Religion, Los Angeles,
Los Angeles, CA 90007, USA
e-mail: baphucusc@gmail.com

© Springer Science+Business Media Dordrecht 2015   3249
S.D. Brunn (ed.), *The Changing World Religion Map*,
DOI 10.1007/978-94-017-9376-6_171

(Gordon 1965), and this is also true of Jewish mixed-marriage (Cohen 2006). If marriage to a non-Jew is associated with assimilation, then it should be expressed as spatial assimilation via the propensity to live among non-Jews. This study addresses two questions with regard to the geography of mixed marriage in the context of spatial assimilation theory:

1. To what extent do intermarried couples live near other Jews? Do they live in "core" Jewish areas or do they live far from Jewish population concentrations?
2. Does living in a "Core" Jewish area have outcomes for intermarried couples? Are they more likely to be engaged in Jewish life as a consequence of living in a Jewish area?

171.2 Methods

This research compares the residential patterns of in-married and mixed-married couples (defined as a Jew currently married to a non-Jew)[1] in five different metropolitan areas: Los Angeles, Chicago, Baltimore, Atlanta, and Phoenix. All five Jewish communities have conducted at least two population surveys so that the residential patterns can be examined over time. Los Angeles, Chicago, and Baltimore are stable in terms of Jewish population growth. Atlanta and Phoenix are Sunbelt metro areas that have experienced rapid population growth (Jewish and general). Including these two Jewish boomtowns brings in two related questions: (1) do rapid growth communities have "core" areas and (2) are residential patterns of mixed marriage affected by the rapid and recent Jewish population growth?

In each of the five metro areas I identified one or more "core" Jewish areas. A "core" Jewish area is defined in four ways. The first is share of the Jewish population. An area with a high proportion of Jewish households (relative to the metro area as a whole) is considered a *core* area. The second definition is Jewish population concentration. Areas of higher Jewish density are by definition "core" because living there implies more interaction with other Jews. Jewish density was computed based on zip code. The five studies used here had household weights that produce the estimated number of Jewish households that each case represents. To obtain Jewish density the aggregated number of Jewish households in each zip code was divided by the *total number of occupied housing units*. The studies used for the density analysis were all conducted within 5 years of the 2000 census because of the availability of zip code estimates and to facilitate comparisons across communities. *Zip code density* is measured in quartiles relative to each community because density varied considerably among the five metro areas studied. The third dimension is *historical*. Some metro areas have long had established Jewish population centers that have come to be identified as Jewish. This longevity has led to the establishment of identifiable Jewish businesses and synagogues. The fourth definition is *spatial*. Areas far from the "Jewish center of gravity" in a given metro area are considered "marginal."

---

[1] Conversionary marriages are treated as in-marriages.

171.3   Chicago

171.3.1   The Jewish Geography of Chicago

The Jewish population of Chicago decreased steadily from 272,000 in 1951 to 248,000 in 1982 (Friedman and Weinberg 1983). Between 1982 and 2010 Chicago saw an upswing in Jewish population, growing to 292,000 in 2010. During this period the Jewish population in Chicago continued to move to the suburbs. The 1982, 1991, 2001, and 2010 Chicago Jewish population surveys reveal a steady process of suburbanization. In 1982, half of all Jewish households in the Chicago metropolitan area lived in the City of Chicago, declining to 34 % in 2010. I use the seven planning areas designated by the Jewish Federation of Chicago, five of which are suburban (Fig. 171.1).

Chicago has three core Jewish areas, three marginal Jewish areas, and one area that could be considered "near core." "City North" is a core area in historical terms. The "City North" area is comprised of the traditional Jewish neighborhoods of the "North Side" that often appear in the novels of Saul Bellow. "City North" can also be considered core based on its share of Jewish households. The share of Jewish households in this area declined by half between 1982 and 2010, but this area still had the largest single share of Jewish households of any of the seven areas. Jewish density in City North (13 %) is in the mid-range for the metro area. It is also geographically "core" by virtue of being next to the other two core areas.

The "Near North" suburbs include such heavily Jewish inner-ring suburbs as Skokie and Evanston. Skokie has the highest concentration of Holocaust survivors in America (Strum 1999). The Near North had the second highest Jewish density of all seven areas (21 %). Between 1982 and 2010 the share of Jewish households in the Near North has consistently been between 18 and 20 %. The affluent North Shore Suburbs have also retained a steady share of Jewish households at between 16 and 17 %. This area also had the highest Jewish density in 2001 at 34 %. The three core areas are contiguous and in 2010 accounted for just over half (54 %) of all Jewish households.

The "Rest of Chicago" and the Western and Southern suburbs constitute the three marginal Jewish areas of Chicago. None of these areas has ever held more than 10 % of all Jewish households. The three marginal areas were also characterized by low Jewish density in 2001 at between 4 and 7 % Jewish. Located in the south and west of the metro area, they are far from the Jewish core areas situated in the northeast section of the metro area.

The Northwest suburbs fall somewhere in between core and marginal. An area of recent Jewish population growth, the share of Jewish households in the Norwest suburbs more than doubled from only 7 % in 1983 to 18 % in 2010. Based on household share the Northwest suburbs would be considered "core." Because Jewish density in the Northwest (17 %) is only half that in the North Shore (34 %), it is considered as "near-core."

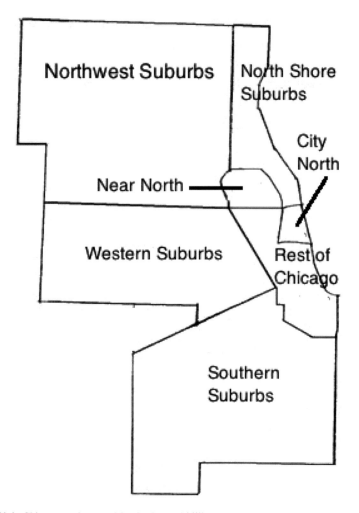

Fig. 171.1 Chicago study area (Map by Bruce Phillips)

171.3.2 The Geographic Distribution of Mixed-Marriage in Chicago, 1982–2010

The relative spatial distributions of in-married and mixed-married couples have remained remarkably stable over the 30-year period from 1982 to 2010 (Table 171.1). In-married couples were consistently more likely than mixed-married couples to live in the three "core" areas: City North, North Shore and Near North Suburbs. Conversely, mixed-married couples have been consistently more likely than in-married couples to live in the two of the three marginal areas: the "Rest of Chicago"

Table 171.1 Distribution of in-married and mixed-married couples: Chicago, by area, 1982, 1991, 2001, 2010

Area	1982		1991		2001		2010	
	In-married	Mixed-married	In-married	Mixed-married	In-married	Mixed-married	In-married	Mixed-married
City North	35	46	24	15	19	21	23	15
Rest of Chicago			3	19	5	10	3	11
Near North Suburbs	24	14	28	13	24	13	25	13
North Shore Suburbs	24	14	26	8	27	12	25	11
Northwest Suburbs	7	10	12	16	17	24	18	30
Western Suburbs	11	15	3	19	3.	16	5	20
Southern Suburbs			5	9	5	5	3	1
Total	100 %	100 %	100 %	100 %	100 %	100 %	100 %	100 %

The Northwest Suburbs contained the largest share of mixed-married couples in both 2001 and 2010. This is an artifact of size. A separate analysis of the Northwest suburbs revealed that in-married couples were concentrated in the highest Jewish density zip codes while mixed-married couples were more likely to live in the lower density Jewish zip codes within the Northwest Suburbs and the Western Suburbs. The biggest gap between mixed-married and in-married couples is found in two places: the "Rest of Chicago" and the Western Suburbs. In 2010 mixed-married couples were four times as likely as in-married couples to live in the "Rest of Chicago" and 4.5 times as likely to live in the Western Suburbs. As the share of Jewish households in the Western Suburbs has increased, the gap between mixed-married and in-married couples living there has diminished. This suggests that in-married couples have followed mixed-married couples into the Western Suburbs in recent years. Perhaps as the number of Jewish households in the Western suburbs increased, it became more "Jewish" and thus more attractive to in-married couples. Mixed-married couples were twice as likely as in-married couples to live in the Southern suburbs only in 1991. This may be an artifact of small sample size, however, because the Southern suburbs cover such a large expanse with relatively few Jews.

171.3.3 Jewish Density and Mixed-Marriage in Chicago

Table 171.2 compares the residential patterns of in-married and mixed-married couples by Jewish density of zip code, categorized in quartiles. Mixed-married couples were concentrated in zip codes of lowest Jewish density. Nearly half (49 %) of the mixed-married couples resided in zip codes in the lowest Jewish density category, and another 24 % lived in the second lowest density quartile for a total of 73 %. Conversely, the plurality (42 %) of in-married Jews lived in the top quartile of Jewish density with another 21 % in the second highest quartile so that fully 63 % of in-married couples chose to live in zip codes that were at least double the overall Jewish density for the Chicago area.

171.3.4 The Impact of Area and Jewish Density on Jewish Behaviors

Does living either in a core area or in a higher density zip code have an impact on the Jewish connections of mixed-married couples? Three kinds of Jewish connections are considered for Chicago: synagogue membership, Jewish friendship network, and how children were raised.

Synagogue Membership For mixed-married couples in Chicago synagogue membership was not affected either by Jewish density or by the particular area of residence.

Table 171.2 Distribution of in-married and mixed-married couples: Chicago, by Jewish density, 2001

Jewish density	In-married	Mixed-married
Lowest quartile	9	49
Second quartile	28	24
Third quartile	21	13
Highest quartile	42	14
Total	100 %	100 %

Jewish Friendship Networks   The 2001 Chicago study asked "Among the people you consider your closest friends, would you say none, a few, some, most, or almost all are Jewish?" In-married respondents overall were much more likely than-mixed-married respondents to report that all or most of their close friends were Jews: 68 % vs. 20 %. For both in-married and mixed-married couples the proximity of Jewish neighbors increased the number of close Jewish friends. Mixed-married respondents living in the zip codes of lowest Jewish density were more likely to have few or no Jewish friends than those in the top category (62 % vs. 31 %); conversely, mixed-married respondents in the top Jewish density category were far more likely to have all or mostly Jewish friends than those in the bottom category (49 % vs. 6 %).

How Children Were Raised in Mixed-married Families   The impact of geography on how children are raised in mixed marriages relates to parallel research on mixed race children. Holloway et al. (2009) studied the impact of neighborhood on the racial identification of mixed race children. Using 1990 Census data at the neighborhood (that is, i.e. census tract) level they examined how racially mixed Asian-White, Latino-White, and Black-White couples listed their children on census forms and found that racially mixed children living in white neighborhoods were more likely to be reported as white, while racially mixed children in more diverse neighborhoods were more likely to be reported as black, Asian, or Latino. "…as the proportion white increased, parents were more likely to report their children as white and less likely to report them as having the race of the minority parent" (p. 536). Holloway et al. theorized that the influence of neighborhood went in two directions. Mixed race couples might choose the neighborhood in which to raise their children, depending in part on how they wish them to identify. The neighborhood can also influence the parents' choice of racial designation in line with what is normative in the particular neighborhood. Does this also apply to marriages between Jews and non-Jews?

The pattern documented for mixed race couples by Holloway et al. (2009) was mirrored by mixed-married couples in Chicago. Almost all (95 %) of the children with two Jewish parents were being raised as Jews as compared with just over one-third (36 %) of children in mixed-married families. Like the parents of mixed-race children, mixed-married Jewish respondents in Jewish zip codes of highest Jewish density were more likely to report their children being raised in Judaism than those living in the lowest density zip codes (68 % vs. 22 %).

171.4   Los Angeles

171.4.1   The Jewish Geography of Los Angeles

Los Angeles is the second largest Jewish community in the United States with a Jewish population that has remained relatively stable. There were 503,000 Jews living in Los Angeles in 1979 and 519,000 in 20 years later (Herman 1998). The Jewish geography of Los Angeles has been conceptualized as seven socio-ecologies based on physical ecology, built environment, and socioeconomic status (Phillips 2007a).

Based on household share alone there is no identifiable core area. With the exception of the Santa Clarita and Simi Valleys in 1997 and the Southeast County in 1979, Jewish households were evenly distributed by area. The Santa Clarita and Simi Valleys in 1997 and the Southeast County in 1979 were both marginal areas based on household share. In 1979 the Southeast County area of Los Angeles consisted of a collection of working class communities (Nicolaides 2002) that were several miles away from the rest of Jewish Los Angeles. Only 3 % of Jewish households were found here in 1979. By the time of the 1997 study there were only a handful of Jews left in the Southeast County, but a new marginal area had emerged. In 1997 the Simi Valley and the Santa Clarita Valley accounted for only 5 % of Jewish households. These two valleys are physically isolated from the San Fernando Valley, separated from it by 5 mi (8 km) or more of open land. The Newhall Pass separates the Santa Clarita Valley, from the San Fernando Valley and the Santa Susana Pass separates the Simi Valley from the San Fernando Valley. When the commuting time on crowded freeways (the only access to either valley) is taken into consideration, these two valleys are more than just physically isolated.

Using Jewish density as a criterion, two core areas and one "secondary core" area can be identified. The Valley Hills (sometimes referred to as the "Hebrew Hills") and West Los Angeles (including Beverly Hills) are core areas with 48 and 26 % Jewish density respectively (Fig. 171.2). Although the overall Jewish density in the Urban Core was only 7 %, the combined the high visibility Jewish neighbor-

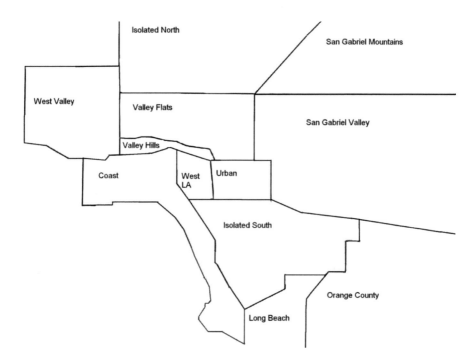

Fig. 171.2 Los Angeles study area (Map by Bruce Phillips)

hoods in the Pico-Robertson/Fairfax area was 39 % Jewish. Jewish density in the marginal Santa Clarita and Simi Valley area, by contrast, was only 3 % in 1997, half the overall Jewish density of 6 % for the Federation study area.

171.4.2   The Geographic Distribution of Mixed-Marriage in Los Angeles, 1979 and 1997

Los Angeles is a very different city from Chicago (Dear and Dishman 2001; Abu-Lughod 1999), but the respective patterns of Jewish residential dispersion are remarkably similar when it comes to mixed-marriage. Table 171.3 compares the geographic distribution of in-married and mixed-married couples by socio-ecology for 1979 and 1997. Physical isolation has long been associated with higher rates of intermarriage in Los Angeles. The last row of the table is labeled "Isolated Area," because this category refers to different geographies in 1979 and 1997. In 1979 the "isolated area" consisted of the blue-collar communities of Southeast Los Angeles County. In 1997 the Santa Clarita and Simi Valleys constituted the "isolated area."

In both 1979 and 1997 the greatest difference between the distribution of in-married and mixed-married couples was in the "isolated" areas. In 1979 mixed-married couples were 6 times as likely as in-married couples to live in the physically isolated southeast county area and in 1997 they were 3.3 times as likely to live in the physically isolated Santa Clarita and Simi Valleys. The share of Jewish households in the Coastline communities grew by 22 % between 1979 and 1997. Like the Western Suburbs in Chicago, the gap between mixed-married and in-married stributions diminished as the Coastline socio-ecology became more Jewish. In 1979 mixed-married couples were 1.8 times as likely as in-married couples to live there, but by 1997 they were only 1.3 times as likely. Something similar happened in the West Valley, which grew by 18 %. In 1979 mixed-married couples were 1.3 times as likely as in-married couples to live in the West Valley, but by 1997 in-married

Table 171.3 Distribution of in-married and mixed-married couples: Los Angeles, by area, 1979 and 1997

Socio-ecology	1979		1997	
	In-married (%)	Mixed married (%)	In-married (%)	Mixed married (%)
Urban	18	11	13	13
West LA	20	10	21	10
Coastline	13	24	16	21
Valley Flats	20	23	15	20
Valley Hills	11	6	15	7
Valley West	15	20	18	16
Isolated Area	1	6	4	13
Total	100	100	100	100

Table 171.4 Distribution of in-married and mixed-married couples: Los Angeles by Jewish density, 1997

Jewish density category	In-married (%)	Mixed married (%)
Bottom (0–10 %)	17	43
Second (10–22 %)	24	23
Third (23–36 %)	25	22
Top (37+ %)	34	13
Total	100	100

couples were slightly more likely to live there. Mixed-married couples were only half as likely as in-married couples to live in West Los Angeles and the Valley Hills, the two core areas in both 1979 and 1997.

As in Chicago, mixed marriage and Jewish density in Los Angeles were inversely related. In both Los Angeles and Chicago mixed-married couples were far more likely to live in zip codes of low Jewish density: fully 43 % of the mixed-married couples lived in the lowest Jewish density quartile while only 13 % lived in the highest quartile (Table 171.4).

171.4.3 The Relative Impact of Area and Jewish Density on Jewish Behaviors

Synagogue Membership As in Chicago, in-married couples in Los Angeles were far more likely than mixed-married couples to belong to a synagogue. Just over half (52 %) of in-married couples in Los Angeles reported belonging to a synagogue versus just 11 % of mixed-married couples (data not shown). As in Chicago, synagogue membership was not related to Jewish density for mixed-married couples.

Jewish Friendship Networks As in Chicago, Jewish density was associated with Jewish friendship networks in Los Angeles; in-married couples in Los Angeles had more Jewish friends than did mixed-married couples. Three-quarters of in-married respondents reported that all or most of their close friends were Jewish, as compared with only a quarter (27 %) of the mixed-married respondents. For both in-married and mixed-married respondents in Los Angeles, the proximity of other Jews fostered Jewish friendships. In-married couples in the zip codes of highest Jewish density were 1.5 times as likely to have all or mostly Jewish friends than those in the lowest. Mixed-married couples in the highest Jewish density category were 1.9 times as likely those in the bottom category to report having all or mostly Jewish friends. The obverse is true for having few or no close Jewish friends: in-married and mixed-married respondents who live in the lowest Jewish density zip codes are respectively 9 and 10 times more likely to report having few or no Jewish friends than their counterparts in the highest density Jewish zip codes.

For mixed-married couples friendship patterns varied little by area with one important exception: mixed-married couples in the isolated Simi and Santa Clarita Valleys had many fewer Jewish friends than mixed-married couples elsewhere. This difference makes a powerful point: non-Jewish Anglos were six times more likely than Jews to live in the Simi and Santa Clarita Valleys and this socio-ecology was the also the "whitest" in terms of its racial composition (Phillips 2007a). The mixed-married couples who chose the Simi and Santa Clarita Valleys were following the overall residential pattern of non-Jewish whites. Their close friendships were consistent with their choice of where to live. For mixed-married couples being geographically isolated from other Jews and living overwhelmingly with non-Jews in the Simi and Santa Clarita Valleys conforms to the model of spatial assimilation.

How Children Were Raised in Mixed-married Families   As in Chicago, Jewish density was strongly associated with raising Jewish children in Los Angeles. All of the children in mixed-married families in the highest density zip codes were being raised exclusively in Judaism. Conversely, two-thirds of children in mixed-married homes in the lowest density zip codes were being raised as Christians.

171.5   Baltimore

171.5.1   The Jewish Geography of Baltimore

The earliest demographic study of Baltimore, conducted in 1968, estimated the Jewish population of that community to be 106,000. The Baltimore Jewish Museum estimates the Jewish population declining to 94,000 by 1975.[2] It had further declined to 87,000 by 1985 (Tobin 1985). To what extent the decline in Jewish population was simply part of the larger pattern of migration out of the eastern Seaboard and to what extent a result of the 1968 riots is not known. After 1985 the Jewish population remained largely stable, growing to 91,400 in 1999 and 93,400 in 2010 (Ukeles and Associates 2010).

For the past four decades Baltimore Jewry has remained concentrated in the Northwest Baltimore metro area, although the Jewish center of gravity has shifted within that geographic sector (Fig. 171.3). Park Heights was the core of Jewish Baltimore in 1968, containing half (49 %) of Jewish households with a Jewish density of 50 %. The "Upper" section of Park Heights was estimated to be almost entirely (83 %) Jewish. At the time of the 1968 study, Lower Park Heights was in the process of turning Black (Mcdougall 1993). The report of the 1968 study (Associated Jewish Charities 1968) noted, "The movement of Jews has been from once heavily populated concentrations in Forest Park and Lower Park Heights to Upper Park Heights, Reisterstown Corridor and Liberty." Randallstown/Liberty Road was a secondary core area in 1968, accounting for another quarter (24 %) of all Jewish households with a Jewish density of 47 %. Park Heights had become a diminished Jewish core by 1985

---

[2] http://www.jhsm.org/timeline-1950-present

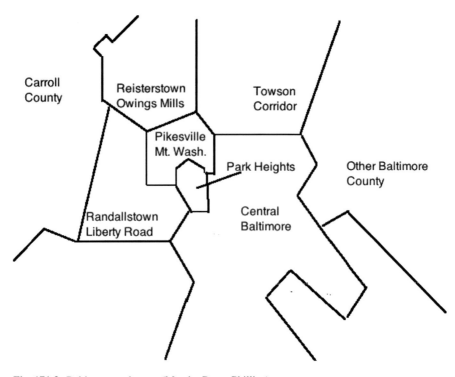

Fig. 171.3 Baltimore study area (Map by Bruce Phillips)

with still the largest share of Jewish households (34 %), but a smaller share than in 1968 (Jewish density was not available for the 1985 study). Pikesville had emerged as a new Jewish core area with a quarter (24 %) of Jewish households. In the meantime Randallstown/Liberty had declined to only a 12 % share of Jewish households. By the turn of the twenty-first century Pikesville had become the Jewish core of Baltimore with a 28 % share of Jewish households and 75 % Jewish density. The neighboring areas of Owings Mills and Mt. Washington formed a secondary layer around the Pikeville core with a 14 % and 15 % share of Jewish households respectively in 1999. Jewish density in these two areas was also substantial with 31 % Jewish density in Owings Mills and 47 % in Mt. Washington. By 2010 the share of Jewish households in Pikesville had increased to 31 % and Owings Mills retained its secondary core position with a 13 % share of Jewish households (down from 15 % in 1999). It was no longer clear that Mt. Washington could be considered part of the Baltimore Jewish core because its share of Jewish households had decreased by half between 1999 and 2010.

Even as Baltimore has retained a Jewish core (with a shift from Park Heights to Pikesville) the Jewish population also dispersed. In 1968 just 6 % of Jewish households lived in the non-specified "outlying" areas. In 1985, only 8 % of Jewish households resided in the three outlying areas of Towson, "Other [that is, southeast] Baltimore County," and Carroll County. The proportion of Jewish households in these three areas increased steadily to 13 % in 1999 and 18 % in 2010. Jewish density was also lower in the outlying areas. In 1968 it was estimated to be only 1 %. In 1999 Carroll County was only 4 % Jewish, and the Towson Corridor was only 7 % Jewish. "Other Baltimore County" was more Jewish at 17 %. These are all over-estimates since they are based on the zip codes in which Jewish households were interviewed. Suburban areas such as Carroll County include many zip codes in which no Jewish households were interviewed. Jews were interviewed in only 3 of the 15 zip codes in Carroll County in 1999.

The share of Jewish households in "Central Baltimore" increased slowly from 9 % in 1968 to 12 % in 1985 and 1999 to 15 % in 2010. This probably reflects the gentrification of such areas as the "Inner Harbor." Because Central Baltimore covers a large area it had only a 5 % Jewish density in 1999, and thus does not qualify as a core area despite its 15 % share: that share is spread out over too large an area.

171.5.2   The Geographic Distribution of Mixed-Marriage in Baltimore, 1985–2010

Intermarriage increased in Baltimore from 5 % of all married couples in 1968 to 15 % in 1985. Intermarriage remained stable over the next 14 years, rising to just 16 % in 1999. By 2010 the couple rate of intermarriage had risen to 21 %, still the lowest rate of the five studies considered here and among the lowest in the United States. Even with a lower overall rate of mixed-marriage, in-married couples in Baltimore, like those in Chicago and Los Angeles, were still more likely to live in low Jewish density zip codes.

Because of the continued residential concentration combined with dispersion into the outer suburban ring, Baltimore provides an excellent case study for comparing the residential choices of in-married and mixed-married couples. Baltimore Jews have four kinds of communities to choose from:

1. Live in a heavily Jewish area in which Jews predominate (Pikesville and Mt. Washington).
2. Live in an established Jewish area of intermediate (but still high) Jewish density (Owings Mills).
3. Live in a low density Jewish zip code (Carroll County, Other Baltimore County, Central Baltimore).
4. Live in a formerly heavily Jewish area with currently low Jewish density (Park Heights, Randallstown/Liberty).

The 1968 study did not report mixed-marriage by area, but it did find that Jews in Central Baltimore and the "outlying areas" were more open to intermarriage. That study reported "Jews living in the Lower Park Heights area for example seem to be most adamant regarding Jewish young people dating and marrying Jews only. This feeling declines somewhat in the other areas. In fact, in the Inner City and Outlying areas there are the most substantial support for mixed dating and intermarriage" (Associated Jewish Charities 1968: 95).

Table 171.5 presents the comparative distribution of in-married and mixed-married Jewish households by area for 1985, 1999, and 2010. The preference shown by in-married couples for heavily Jewish areas is consistent over time. In-married couples are more likely than mixed-married couples to reside in a core area, even as the core areas have changed. In 1985 68 % of in-married couples were concentrated in the two core areas of Park Heights and Pikesville as compared with just 18 % of the mixed-married couples. Conversely, over half (56 %) of the mixed-married couples resided in the outlying areas (Towson, Central Baltimore, Other Baltimore County, and Carroll County) versus just 8 % of the in-married couples. In 1999 in-married couples were more dispersed, but still concentrated in the core and secondary core areas: 38 % of the in-married couples lived in the two core areas of Park Heights and Pikesville as compared with just 8 % of mixed-married couples. The proportion of in-married couples living in the outlying areas had doubled since 1985 (from 8 to 15 %), but this was still substantially smaller than the 64 % of mixed-married couples found there. By 2010 the in-married couples were still concentrated in the two core areas (52 %) and the secondary core areas (22 %), and the mixed-married couples were still only half as likely to live in the core and secondary core (22 % and 13 %) respectively. Conversely, mixed-married couples were still far more likely than in-married couples to live in the outlying suburbs (58 % vs. 16 %).

Randallstown/Liberty and Reisterstown did not consistently fit the overall pattern, but they are exceptions that prove the rule. Randallstown/Liberty Road is "Jewish" in historic terms rather than by virtue of its Jewish population concentration. No intermarriage data were available in the 1968 report, but in-married and mixed-married couples were almost equally likely to live in Randallstown/Liberty in 1985. After 1985 in-married couples were more attracted to this predominantly African-American area than were mixed-married couples. In-married couples were twice as likely as mixed-married couples to live there in 1999 and four times as likely in 2010. The pattern for Reisterstown was inconsistent. Like Randallstown/Liberty it was heavily Jewish (40 %) in 1968, declining to just 7 % in 1999. In 1985 mixed-married couples were more likely to live there, in 1999 in-married couples were more likely to live there, and in 2010 in-married and mixed-married couples were equally likely to live there.

Despite population shifts, dispersion into the outlying suburbs, and the emergence of a new core area in Pikesville, the overall pattern persisted (with the minor exception of Reisterstown) over a quarter century: in-married couples remained concentrated in the core and secondary core areas and mixed-married couples gravitated to the outlying suburbs and Central Baltimore.

171.5.3 Mixed-Marriage and Jewish Density in Baltimore

To be consistent with other studies, Jewish density was divided into quartiles (Table 171.6). This was more problematic in Baltimore than Los Angeles and Chicago because Baltimore Jewish households are concentrated in relatively few

Table 171.5 Distribution of in-married and mixed-married couples: Baltimore, by area, 1985, 1999, 2010

Named area	1985		1999		2010	
	In-married (%)	Mixed-married (%)	In-married (%)	Mixed-married (%)	In-married (%)	Mixed-married (%)
Pikesville	32	9	31	6	39	21
Park Heights	34	9	7	2	13	1
Randallstown/Liberty	15	13	5	3	4	1
Mt. Washington	0	0	16	4	8	3
Owings Mill	9	8	19	17	14	10
Reisterstown	2	6	8	4	7	6
Towson Corridor	2	16	3	23	5	10
Central Baltimore	5	27	8	19	6	19
Other Bal. County	1	9	2	13	3	16
Carroll County	0	4	2	9	2	13
Total	100	100	100	100	100	100

Table 171.6 Distribution of in-married and mixed-married couples: Baltimore, by Jewish density, 1999

Jewish density	In-married	Mixed-married
Bottom quartile	16	53
Second quartile	18	17
Third quartile	35	24
Top quartile	31	6
Total	100 %	100 %

zip codes where Jewish density is quite high. In 1999 the top quartile was 75 % or more Jewish. In other words, one quarter of all the Jewish households in Baltimore lived in a zip code in which at least three out of four households was Jewish. Zip codes in the third quartile were between 30 and 70 % Jewish, still very high compared with the other communities analyzed here. Almost two-thirds (68 %) of in-married couples were concentrated in the two top quartiles of Jewish density. More than half (53 %) of the mixed-married couples in 1999 resided in the lowest quartile zip codes (less than 7 % Jewish) and only 6 % resided in the top quartile zip codes. Mixed-married couples were less likely than in-married couples to reside in the third quartile zip codes, but one-in-four mixed-married couples did live there.

171.5.4 The Relative Impact of Area and Density on Jewish Behaviors

Synagogue Membership Because of the overlap between area and Jewish density was so much greater in Baltimore than in Chicago and Los Angeles, and because Jews were so much more concentrated in Baltimore, a special typology of density and area was created to understand the impact of geography on Jewish behavior. The typology combined the zip codes and areas into a dichotomy:

1. *Core Zip Codes:* Pikesville, Mt. Washington, Owings Mill, Reisterstown, Park Heights, and Randallstown/Liberty *excluding* the zip codes in those areas in the bottom quartile of Jewish density.
2. *Non-Core Zip Codes:* Towson Corridor, Central Baltimore, other Baltimore County, and Carroll County *plus* zip codes in the lowest Jewish density quartile in the core zip code areas.

In 1999 almost all (89 %) of the in-married couples resided in one of the Core Zip Codes and the mixed-married couples were almost evenly split between the Core and Non-Core Zip Codes (48 % and 52 % in 1999).

Synagogue membership for in-married and mixed-married couples varied by the area-density typology for 1999. The combination of density and area influenced synagogue membership for both in-married and mixed-married couples. In-married and mixed-married respondents living in the "Core Zip Codes" were more likely than their counterparts in "Non-Core Zip Codes" to report paying dues to a synagogue. Mixed-married respondents residing in core zip codes were six times more likely than those in non-core zip codes to report paying dues to a synagogue: 41 % vs. 7 %. Living in a core zip code also influenced synagogue membership among in-married couples: 71 % of those in core zip codes reported paying dues to a synagogue vs. 47 % of those living in a non-core zip code.

In Chicago and Los Angeles neither area nor density had much if any impact on formal affiliation. In Baltimore, by contrast, the combination of density and area had a strong association with synagogue membership for both in-married and mixed-married couples. One important difference between Baltimore and the other two communities is that Jewish densities are so much higher in Baltimore. There are two opposite but not mutually exclusive reasons why higher Jewish residential concentrations in Baltimore reinforce synagogue membership: (1) Jews in the core zip code areas are known to each other and thus may feel it is expected of them to support a synagogue and (2) Jews who want to be affiliated move to areas where Jewish institutions are most accessible.

How Children Were Raised in Mixed Marriages   The area-density typology worked better than either density or area alone for examining how children were being raised in mixed-marriages because there were so few cases of children with mixed-married parents in core areas such as Pikesville and in the highest density category. In Baltimore, as in Los Angeles and Chicago, living with other Jews was associated with how children were being raised in mixed-marriages: 78 % of the children of mixed-marriages in Core zip codes were being raised in Judaism as compared with 55 % in Non-Core zip codes.

171.6   Atlanta

The two metro areas that experienced the most rapid growth, Atlanta and Phoenix, were the most residentially dispersed and also had the highest rates of mixed-marriage 49 % in Atlanta and 40 % in Phoenix. These rates of mixed marriage were considerably higher than in Los Angeles (27 %), Chicago (33 %[3]) and Baltimore (21 %).

171.6.1   The Jewish Geography of Atlanta

Atlanta and Phoenix are representative of the rapid growth Jewish communities of the Sunbelt. By the mid-1950s Los Angeles and Chicago were already well established as the second and third largest Jewish communities in the United States. Atlanta, by contrast, had a relatively tiny Jewish population in the 1950s. A study

---

[3] This is the couple rate for the 2000 study. The couple rate for 2010 is 37 %.

conducted in 1946 estimated 10,200 Jews living in the Atlanta metro area, mostly in the city of Atlanta itself (National Jewish Welfare Board 1947). Twenty years later the 1984 study estimated that the Jewish population had grown six-fold to 59,000. By the time of the next study in 1996 the Jewish population had increased yet again to 77,000. At the time of the most recent study in 2006, the Atlanta Jewish population had increased to 120,000.

Although the report of the 1983 study includes numerous tables on demography and Jewish affiliations, it did not contain any information about intermarriage. The 1983 data could not be re-analyzed because the North American Jewish Data Bank concluded that the dataset was not usable.[4] The 1996 study showed a couple inter-marriage rate of 38 %, increasing to 50 % in 2006.

With Jewish populations of approximately the same size, Atlanta is the opposite of Baltimore in three important ways. First, Baltimore is a stable community; 53 % of the survey respondents were born in Baltimore as compared with less than 20 % in Atlanta. Only 10 % of the Baltimore respondents had migrated there within the decade prior to the study as compared with 31 % in Atlanta. Second, intermarriage in Atlanta in 2006 was more than double the rate in Baltimore in 2010. The third difference is residential dispersion. Where Jews remained concentrated in the Northwest section of metro Baltimore, Atlanta Jews have continued to disperse throughout the metro area (discussed in the next section).

Because Jewish population growth and residential dispersion occurred at the same time and on a large scale, Atlanta never developed the kinds of "core" Jewish areas identified in Chicago, Los Angeles, and Baltimore. One indication of the lack of a clearly identifiable Jewish core this is the lack of consistency among the geo-graphic designations used in the different study reports. The 1983 study reported Jewish population using six areas based on county (Cohen 1983), and the 1996 study used a completely different geography consisting of five large geographic areas. The 2006 study introduced a third Jewish geography of Atlanta, the one used in this study. As the report explains, "Jewish Atlanta is as much a collection of diverse geographic communities as it is one cohesive community. These areas vary not only by the number of Jewish persons, but also in their Jewish character, extent to which they attract newcomers, age structure and income" (Ukeles and Miller 2006). The five areas (Fig. 171.4) used in 2006 (and in this analysis) were: In-town, North Metro Atlanta, East Cobb Expanded, Sandy-Springs-Dunwoddy, Gwinnett+East Perimeter, North and West Perimeter, and South Area. "In-town" refers to the city itself, where almost all Atlanta Jews lived at the time of the 1946 study. Even before launching a full population study in 2006, the Atlanta Federation was aware of Jewish growth in the "North Metro Atlanta" area and it commissioned a special survey there in 2004 (Ukeles and Associates 2004). That study found North Metro Atlanta to be an area of young families who had recently moved to Atlanta. Sandy Springs-Dunwoody is a small area located just to the north of "In-town." East

---

[4]The dataset submitted to the North American Jewish Data Bank had been corrupted (e.g. cases missing, variables mislabeled, etc.) and did not include sampling weights.

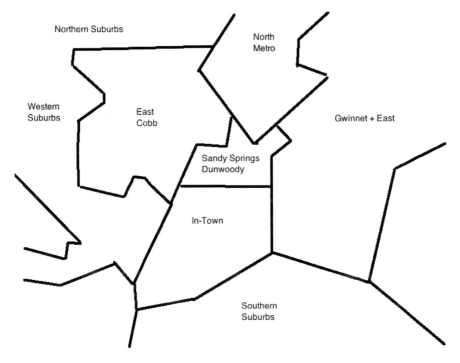

Fig. 171.4 Atlanta study area (Map by Bruce Phillips)

Cobb Expanded is a larger area to the northwest of Sandy Springs-Dunwoody. Gwinnet + East of Perimeter is a large expanse to the east and northeast of "In town." The final two designations were residual areas of extremely low Jewish density, as the 2006 report explains, "Data on North & West Perimeter and the South are not included, given the relative low number of interviews in each of these sub-areas, which were included for Jewish household and population estimates only" (Ukeles and Miller 2006). Despite the relatively low number of completed interviews, I reanalyzed the 2006 study including these areas precisely to see whether mixed-married couples were found to disproportionately live in these outlying areas.

171.6.2   Atlanta's Changing Residential Distribution

Based on share alone, "In-town" was the core area in 1996 with a 46 % share of Jewish households, followed by Sandy Springs/Dunwoody with a 19 % share of Jewish households. In the decade between the 1996 and 2006 surveys the Atlanta Jewish population dispersed significantly throughout the metro area. The total number of Jewish households in Atlanta increased by 61 %, but the 1996 "core" areas, In-town and Sandy Springs/Dunwoody, lost Jewish households. As a result their respective shares of Jewish households decreased from 45 to 27 % and 19 to 11 % respectively. The East Cobb Expanded area gained Jewish households, but grew at less than half the overall rate so that its share of Jewish households also decreased slightly from 16 to 14 %. Jewish growth was greatest in the outlying suburbs where the number of Jewish grew by at least 200 %: North Metro (199 %), Gwinnet+East (204 %), North and West Perimeter (560 %), Southern Area (1,710 %). The combined share of Jewish households in these outlying areas increased from under a quarter in 1996 (20 %) to almost half (49 %) in 2006. Growth had been so rapid in Atlanta that in 2006 that 2 % of the respondents were not sure of their zip code.[5]

The rapid growth and subsequent extensive dispersion of Jewish households in Atlanta complicates the identification of a reference "core" area in 2006. Based on share of Jewish households, Sandy Springs/Dunwoody could no longer be considered a core area with only 11 % of Jewish households in 2006. "In-town" still had the largest share at 27 %, but this had fallen sharply from 45 % in 1996, and was only 7 % larger than the share in the outlying area of North Metro (20 %). Based on Jewish density, In-town and Sandy Springs would be core areas at 12 % and 14 % respectively. These are high densities compared to the rest of Atlanta but are low compared with Chicago, Los Angeles, and Baltimore.

Because of the difficulty of identifying core areas in Atlanta based on population share and density, two attitudinal items from the 2006 study were used to help identify a possible Jewish core area in Atlanta:

- "Right now, how important is it to you to be part of the Jewish community in Greater Atlanta?"
- "To what extent do you feel like you are part of the Jewish community of Greater Atlanta?"

Based on these two questions Sandy Springs/Dunwoody was the most Jewish area of Atlanta, with 52 % of the respondents agreeing that being a part of the Jewish community is very important and 39 % saying that they feel "a lot" that they are part of the Jewish community. By contrast, only 10 % of respondents in Gwinnet+East Perimeter agreed that "being part of the Jewish community is very important" and only 3 % saying that the felt "a lot" connected to the Jewish community. Based on these two questions In-Town is the second most Jewish area, a finding that might be explained by the presence of an Orthodox enclave in the Toco Hills section of Atlanta. East Cobb and North Metro fall in the middle. The sample sizes for the North and West Suburbs and South Area were small (48 and 38 cases respectively) and the responses to the attitudinal questions for these two areas were not included in the published report. To confirm that two areas were indeed Jewishly marginal the two attitude items discussed above were analyzed for the North and West Suburbs and South Area. The results confirmed that these areas resemble Gwinnet+East in terms of weak orientation toward the Jewish community.

---

[5] These cases are excluded from tabular data presented.

171.6.3   The Geographic Distribution of Mixed-Marriage in Atlanta, 1996–2006

New migrants to a Jewish community have more opportunities to disperse than do established residents. Established residents need a compelling reason to relocate elsewhere in the metro area, whereas new migrants by virtue of their move must choose where to live upon arrival. A sizeable proportion of newcomers to Atlanta were found among both in-married and mixed-married couples: 36 % of in-married and 42 % of mixed-married couples moved to Atlanta between 1996 and 2006. (There was also sizeable movement within Atlanta). In-married couples had lived at their current residence for an average of 9.3 years, as compared with 5.7 years for mixed-married couples. Thus, a significant proportion of all married couples had moved to and/or moved within Atlanta since the 1996 study; the residential choices of more than a third of all couples were made since the 1996 study, with mixed-married couples (who were more likely to have been both recent in-migrants to Atlanta and recent movers within Atlanta) making even more residential choices.

As in the three metro areas already discussed, mixed-married couples in Atlanta were more likely than in-married couples to live in the outlying areas in both 1996 and 2006 (Table 171.7). Moreover, the difference between the residential distribution of mixed and in-married couples grew considerably between 1996 and 2006. Mixed-married couples were twice as likely as in-married couples to live in the Northern and Western Suburbs in both 1996 and 2006. They were twice as likely to live in the South Area in 1996 and four times as likely to live there in 2006. Mixed-married couples were 1.6 times as likely as in-married couples to live in the Gwinnet+East area in 1996 and 2.3 times as likely to live there a decade later. In 1996 in-married couples were 1.7 times as likely as mixed-married couples to live in the most Jewish area of Sandy-Springs-Dunwoody, but almost ten times more likely to do so in 2006. The patterns changed for

Table 171.7 Distribution of in-married and mixed-married couples: Atlanta, by area, 1996 and 2006

| Area | Year of study | | | |
| | 1996 | | 2006 | |
	In-married (%)	Mixed married (%)	In-married (%)	Mixed married (%)
In-town	38	44	28	19
Sandy Springs-Dunwoody	24	14	19	2
East Cobb Expanded	18	14	13	16
North Metro	12	13	23	18
Gwinnett & East	7	11	9	21
Northern and Western Suburbs	1	2	7	16
South Area	1	2	2	8
Total	100	100	100	100

Table 171.8 Distribution of in-married and mixed-married couples: Atlanta, by Jewish density, 2006

Jewish density	In-married	Mixed-married
Bottom quartile	18	37
Second quartile	17	32
Third quartile	24	19
Top quartile	41	11
Total	100 %	100 %

East Cobb and In-town. In 1996 mixed married couples were more likely to live in the City of Atlanta, but in 2006, with the continued expansion of the metro area, it was in-married couples that were more likely to live in the city. Since the "In-Town" area is adjacent to Sandy Springs/Dunwoody this could be interpreted as a preference for a Jewish area (in relative terms at least). Further evidence for this interpretation is that "In-town" respondents were the second most communally oriented after those in Sandy Springs/Dunwoody. The relative propensity of in-married and mixed-married couples to live in East Cobb stayed pretty much the same.

Given that Atlanta has a less readily identified Jewish core, another way to think about the relationship between intermarriage and geography is the couple intermarriage rate for each area. Sandy Springs-Dunwoody had by far the lowest rate of intermarriage in both studies. The far outlying areas of Gwinnett & East, the Northern and Western Suburbs, and the South Area had the highest rates of intermarriage in both 1996 and 2006. Jewish density has a whole different meaning in Atlanta than in Baltimore. As a result of in-migration, Jews were more dispersed throughout the Atlanta metro area than in Chicago, Los Angeles, and Baltimore (all stable communities). As a result, Jewish density was lowest in Atlanta. The most Jewish zip codes in Baltimore were more than 50 % Jewish in both 1999 and 2010. In Atlanta the most Jewish zip code was only 24 % Jewish. The median Jewish zip code density in Atlanta was only 6 % as compared with 22 % in Los Angeles, 21 % in Chicago, and 20 % in Baltimore.

Table 171.8 shows the distribution of in-married and mixed-married Jews by Jewish density in 2006. Even though Jews were the most widely dispersed in Atlanta, in-married couples nonetheless gravitated to the zip codes of greatest Jewish density. In-married Jews were most likely to live in the top quartile of Jewish density (41 %), followed by the second densest quartile (24 %). Conversely, mixed-married couples were most likely to live in zip codes in the lowest density quartile (37 %) followed by the second to lowest (32 %). Mixed-married couples were twice as likely as in-married couples to live in the Jewish zip codes of lowest Jewish density. In-married couples were 3.7 times as likely as mixed-married couples to live in the most Jewish zip codes.

171.6.4 The Relative Impact of Area and Density on Jewish Behaviors in Atlanta

Synagogue Membership and other Affiliations As in the other three Jewish communities considered so far, in-married couples in Atlanta were far more likely to belong to a synagogue than mixed-married couples: 63 % vs. 8 %. In-married couples were also more likely than mixed-married couples to belong to a Jewish community center (19 % vs. 3 %). The Atlanta study included a question about Jewish cultural engagement that is examined in this analysis because (unlike synagogue membership) it does not require a substantial outlay of money: "In the past year, have you or anyone else in the household visited a Jewish Museum in Atlanta, or attended a Jewish or attended a Jewish cultural event in Atlanta (such as a holiday festival, a play, a Jewish film, musical performance, Jewish art exhibit)." As compared with the 8 % of mixed-married couples that belonged to a synagogue, and the 3 % that belonged to a JCC, an impressive 23 % had visited the Jewish museum or attended a cultural event.

Because Jewish density was so low in Atlanta the top two and bottom two categories were collapsed for this analysis. Mixed-married couples living in higher density zip codes were twice as likely as those living in lower density zip codes to belong to a synagogue and to the JCC. The biggest difference was for cultural engagement: 53 % of mixed-married respondents in the higher density zip codes reported visiting the Atlanta Jewish museum or attending a Jewish cultural event as opposed to only 9 % of those in the lowest density zip codes. Cultural engagement is particularly good test of the impact of Jewish density because mixed-married respondents might well have learned about such events through other Jews. Even in the zip codes of highest Jewish density only 12 % of mixed-married couples belonged to a synagogue so it is unlikely that they would have learned about such events via a synagogue bulletin or email announcement.

Area was also associated with synagogue membership for mixed-married couples in Atlanta. Even though there were only eight interviews with mixed-married couples in Sandy Springs-Dunwoody, the results were nonetheless statistically significant. Mixed-married respondents in Sandy Springs-Dunwoody were virtually the only ones to report a membership in the Marcus Jewish Community Center, which is located in Dunwoody. On the other hand there are synagogues located throughout Atlanta, and but it was mixed-married respondents in Sandy Springs-Dunwoody who were by far the most likely to report a synagogue membership. They were also the most likely to report visiting the Jewish museum or attending a cultural event. More than half (53 %) of the mixed respondents who lived "In-Town" reported visiting the Jewish museum or attending a cultural event. This might be because the Jewish museum is located in downtown Atlanta as are several performance venues that might have hosted a Jewish cultural event. Their cultural engagement is a product of geography. Mixed-married couples living "In-Town" were less likely than other mixed-married respondents to answer "very" or "somewhat important" to the question, "Right now, how important is it for you to be part of the Jewish Community in Greater Atlanta?" Even though they did not place much value on being part of the Atlanta Jewish community they nonetheless partook of Jewish cultural offerings nearby.

How Children Were Raised in Mixed-marriages Virtually every Atlanta child with two Jewish parents (97 %) was being raised exclusively in Judaism, but only a third (32 %) of the children of mixed-married parents were being raised this way. Almost as many were being raised in a religion other than Judaism as were being raised in

Judaism. The percentage of children of mixed-married couples being raised in Judaism differed by area of Atlanta. Although the sub-sample sizes were small, the differences were nonetheless statistically significant. Because there were only 9 children of mixed-married parents in the South Area sub-sample, this was combined with other outlying suburbs in the Northern and Western suburbs area. The impact of living in a "core area" is seen in the 85 % of children with mixed-married parents in the Sandy Springs-Dunwoody area being raised exclusively as Jews. Conversely, in the outlying Northern, Western, and Southern areas only 17 % of the mixed-married children were being raised exclusively Jewish and 57 % were being raised in a religion other Judaism (most likely as Christians).

The Jewish density of the zip code where the mixed-married couple lived was also associated with how they were raising their children: 53 % of children with mixed-married parents who lived in the highest Jewish density quartile were being raised exclusively Jewish as compared with 20 % in the bottom quartile and between 36 and 39 % in the second and third quartiles. Even with relatively low overall Jewish density in Atlanta, living with other Jews was associated with mixed-married parents raising their children in Judaism.

171.7 Phoenix

171.7.1 The Jewish Geography of Phoenix

Like Atlanta, Phoenix is a rapid growth Sunbelt community. The Phoenix Jewish population doubled from 38,000 Jews in 1983 to 83,000 in 2002. The Atlanta Jewish population also doubled in size over the same period from 59,000 in 1985 to 120,000 in 2006. In both Atlanta and Phoenix 42 % of the respondents were newcomers to their respective metro, having arrived in the community within the 10 years prior to the study. Mixed-marriage increased considerably over the 19 years between the two Phoenix Jewish population studies, from 24 % in 1983 to 40 % in 2002. As in Atlanta, identifying a "core" Jewish area was not straightforward. The Jewish geography of Phoenix was largely determined by a topography of valleys (Fig. 171.5): Central Phoenix, the Northeast Valley (including Scottsdale and Paradise Valley), the Northwest Valley (including Glendale, Peoria, Sun City) and Tri-Cities (Chandler, Gilbert, Mesa, Tempe, and Awahtukee).

Based on share alone Central Phoenix was the core area in 1983 with 54 % of Jewish households. It was replaced as the core area in 2002 by the Northeast Valley (Paradise Valley/Scottsdale), which contained 41 % of all Jewish households. The number of Jewish households in Central Phoenix increased by 31 %, much less than the overall growth rate for Jewish Phoenix of 138 %. The number of Jewish households in the Northeast Valley, by contrast, quadrupled from less than 4,000 in 1983 to more than 18,000 in 2002. Even though lagging behind the Northeast Valley, Central Phoenix still had a substantial share of Jewish households in 2002, accounting for almost one-in-three (30 %). Together these two areas accounted for

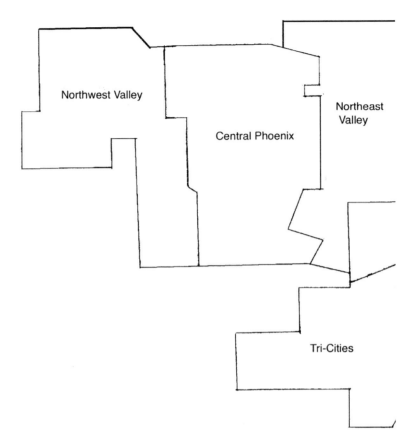

Fig. 171.5 Phoenix study area (Map by Bruce Phillips)

the vast majority of all Jewish households in both 1983 and 2002 (75 % in 1983 and 71 % in 2002).

With a median Jewish zip code density of 8 %, Phoenix resembles the other Sunbelt community, Atlanta (overall Jewish density of 6 %). Based on Jewish density, the Northeast Valley (Paradise Valley/Scottsdale) stands out as the only core area with 14 % Jewish density as compared with 6 % in Central Phoenix.

Of the two remaining areas, "Tri-Cities" and Northwest Valley, the former stands out as the least dense Jewish area (4 %), with a decreasing share of Jewish households (falling from 19 % in 1983 to 16 % in 2002). The share of households in the Northwest Valley doubled from 6 to 12 % as the area changed from a Jewish retirement enclave to more of a family area. In the 1983 study most of the Jewish households interviewed in the Northwest Valley lived in the retirement community of Sun City. By 2002 the Northwest Valley had diversified: 22 % of all Northwest Valley Jewish households were married couples with children, the

Table 171.9 Distribution of in-married and mixed-married couples: Phoenix, by area, 1983 and 2002

| Area | Year of study | | | |
| | 1983 | | 2002 | |
	In-married (%)	Mixed married (%)	In-married (%)	Mixed married (%)
Central Phoenix	54	53	23	27
Northeast Valley	24	17	46	29
Northwest Valley	8	4	17	12
Tri-Cities	14	25	14	32
Total	100	100	100	100

same as the entire study area. The retirement legacy continued to be evident, as this was still a relatively older area with few households headed by young singles and young couples.

171.7.2   The Geographic Distribution of Mixed-Marriage in Phoenix, 1983–2002

The relative preferences of in-married and mixed-married couples for particular areas have remained remarkably consistent over the two decades separating the two studies (Table 171.9). This finding is striking because mixed-married couples were more likely than in-married couples to have been recent in-migrants. In both studies in-married couples were more likely to live in the Northeast and Northwest Valleys, while mixed-married couples were more likely to live in the Tri-Cities area. The preference of mixed-married couples for the Tri-Cities strengthened over time. In 1983, 25 % of mixed married couples lived there as against 32 % in 2002. In-married and mixed-married couples were equally likely to live in Central Phoenix in 1983, but in 2002 mixed-married couples were slightly more likely to live there (27 % vs. 23 %).

171.7.3   Intermarriage and Jewish Density

The relative distribution of in-married and mixed-married couples was only weakly associated with Jewish density in Phoenix (Table 171.10). In-married couples were 1.3 times as likely as mixed-married couples to live in the highest Jewish density quartile, but this is a much smaller gap than found in the other communities. In Atlanta with a low overall Jewish density similar to Phoenix, in-married couples were 3.7 times as likely to live in the highest density zip codes. Conversely, mixed-married couples were 1.3 times as likely to live in the lowest quartile of Jewish density, but this difference is much smaller than the ratios of 2.1, 2.5, 3.3, and 5.4 found in Atlanta, Los Angeles, Baltimore, and Chicago respectively.

Table 171.10 Distribution of in-married and mixed-married couples: Phoenix, by Jewish density, 2002

Jewish density	In-married (%)	Mixed married (%)
Bottom quartile	21	28
Second quartile	19	14
Third quartile	28	34
Top quartile	32	24
Total	100	100

171.7.4   The Relative Impact of Area and Density on Jewish Behaviors

Synagogue Membership As in the other communities, in-married respondents in Phoenix were far more likely than mixed-married respondents to report a synagogue membership: 48 % vs. 10 %. Mixed-married respondents in the core Northeast Valley were the most likely to report paying dues do a synagogue (22 %), followed by those in Central Phoenix (11 %). Synagogue membership among mixed-married couples in the Northwest Valley and Tri-Cities was almost non-existent (3 and 2 %). Among mixed-married couples density was associated with synagogue membership only in the top quartile where it was 23 %. Unlike the other four metros, synagogue membership remained the same as density decreased in the other three quartiles.

How Children Were Raised As expected (and consistent with synagogue membership), the children of mixed-married parents in the Northeast Valley were the most likely to be raised exclusively Jewish, although there were only 16 such children in this area. Although only 2 % of the mixed-married couples in the Tri-Cities area belonged to a synagogue, 40 % of their children were being raised exclusively in Judaism. This figure was significantly higher than the 10 % in Central Phoenix and 20 % in the Northwest Valley. It is not readily apparent as to why this should be so. Also consistent with low rates of synagogue membership in the Northwest Valley, the children of mixed-married parents in this area were the least likely to be raised exclusively Jewish. The children of mixed-married parents in Tri-Cities and Central Phoenix did not fit the expected pattern. Contrary to expectations a large percentage of the children of mixed-married parents in the Tri-Cities (40 %) and a small minority of those in Central Phoenix (10 %) were being raised exclusively Jewish. Central Phoenix is closer to being a core area than Tri-Cities in terms of Jewish population concentration, but mixed-married parents in the presumably "marginal" Tri-Cities area were four times more likely to raise their children exclusively in Judaism.

Density was a better predictor than area of how children were being raised, but only when the two bottom and two top categories were combined: 36 % of children in the two highest density categories (8 % +) were being raised exclusively as Jews as against 19 % of those in the two lowest categories. The reason density was a better predictor than area is that 98 % all such children in Central Phoenix were living in the two lowest density categories while 96 % of those in Tri-Cities were living in zip codes of 8 % or greater Jewish density. Raising Jewish children did not, however, translate into joining a synagogue in the Tri-Cities area.

Although Phoenix generally conformed to the patterns found in the other four metro areas, the patterns were the weakest in Phoenix. Whether this is because the West overall differs significantly from the rest of American Jewry (Sheskin 2004) or whether this because Phoenix itself is somehow peculiar is beyond the scope of this analysis. Even though the patterns observed in the other four communities were less pronounced in Phoenix, they nonetheless were evident there.

171.7.5   Summary and Conclusion

Five different metropolitan areas were examined here (Table 171.11). Chicago and Los Angeles are the second and third largest Jewish communities in the United States, both with stable populations. Baltimore, Atlanta, and Phoenix all have comparable Jewish population sizes (around 100,000). Baltimore had a stable Jewish population while Atlanta and Phoenix are rapid growth Sunbelt Jewish communities. At least two studies for each metro area were used to confirm that differences in the residential patterns of in-married and mixed-married couples have been consistent over time. Having established that in-married and mixed-married couples had differential residential patterns in each metro area the impact of living in a "Jewish" area on Jewish behaviors was considered. The summary discussion is divided into two parts: how distributions differed by metro area and how those distributions affected the Jewish behaviors of mixed-married couples.

171.7.6   Comparative Distribution of In-Married and Mixed-Married Couples

Although the patterns varied by community, in-married and mixed-married couples tended to live apart from each other. In each metro area mixed-married couples were concentrated in the outlying suburbs furthest from the "core" Jewish area. In some of the outlying suburbs virtually all the couples were mixed-married. Conversely, in each metro area, with the exception of Los Angeles, in-married Jews were concentrated in the core (and sometimes the secondary core) area. Core areas are most evident in the metros with stable Jewish populations.

In each of the five metro areas mixed-married couples were far more likely than in-married couples to live in the lowest Jewish density (bottom quartile) zip codes. Conversely, in-married couples were far more likely than mixed-married couples to live in the highest Jewish density (top quartile) zip codes. The association between Jewish density and mixed marriage was weakest in the two Western communities, particularly in Phoenix.

Table 171.11 Summary of findings

Metro area	Geographic distribution	Density	Synagogue	Other	How children are raised
Chicago	In-married couples concentrated in 3 core areas, mixed-married couples in two marginal areas. Northwest suburbs, the anomaly, was explained by density	Mixed-married couples concentrated in the two lowest density quartiles, in-married in the two top quartiles	No relation between synagogue membership and either area or density for mixed-married couples	Mixed-married couples in core areas and higher density zip codes had more Jewish friends	Related only to density
Los Angeles	Mixed-married couples concentrated in isolated areas. Top and bottom density quartiles related to distribution but not intermediate quartiles	Mixed-married couples concentrated in lowest density quartile, in-married couples in highest. No difference in the two intermediate categories	Weakly associated with area and not with density	Density related to friendship, but not area with the exception of the isolated area	Related only to density
Baltimore	In-married more likely to live in core areas, mixed-married in outlying suburbs	Half of mixed-married couples lived in the lowest quartile. Two-thirds of in-married couples in the two top quartiles	Area and density combined into dichotomous typology. Synagogue membership highest for mixed-married couples in the "core" zip codes	n/a	Related to area-density typology
Atlanta	Only one small core area., Sandy Springs. Mixed marriage rates 70 %+ in the outlying suburbs	In-married couples concentrated in the top two quartiles and mixed-married in the bottom two quartiles	Mixed-marrieds in top two density quartiles more likely to belong to synagogue. Area related only for Sandy Springs	Mixed-married couples in top two density quartiles more likely to report cultural Jewish engagement	Strongly related to density and to area in Sandy Springs
Phoenix	In-married couples concentrated in core area of Northeast, mixed-married couples in tri-cities	Relationship between density and mixed marriage weakest in Phoenix	Highest in core area. Density related	n/a	Not related to density. Highest in core area, otherwise area not related

171.7.7   Impact on the Jewish Connections
of Mixed-Married Couples

Given the disparity in residential patterns between in-married and mixed-married couples, was living among other Jews associated with the Jewish connections of mixed-married couples? Three kinds of Jewish connections were investigated: synagogue membership, non-formal Jewish connections, and how children were raised.

Synagogue Membership   The association between living in a core area and living in a higher density zip code and synagogue membership among mixed-married couples varied considerably by community.

In the two largest communities, Los Angeles and Chicago, there was either no association (Chicago) or a weak association (Los Angeles) observed. In Baltimore, with its high concentration of Jewish population in the northwest corridor, living in a core Jewish area *or* living in a high-density zip code increased synagogue membership rates for both in-married and mixed-married couples.

In Atlanta, where the Jewish population grew substantially and became increasingly dispersed, mixed-married couples were concentrated in the outlying suburbs in the north, east, west, and south of metro Atlanta. Outside of the single core area of Sandy Spring, area was not associated with synagogue membership in Atlanta, but Jewish density was. Living in the top two density zip codes doubled synagogue membership among mixed-married couples. In Phoenix, by contrast, density was not associated with the proclivity of either in-married or mixed-married couples to join a synagogue with one exception: mixed-married couples living in the highest density zip codes were between three and six times more likely to belong to a synagogue than those in residing in the lower three quartiles of Jewish density. Even though inconsistent, the association between living in a Jewish area and joining a synagogue was impressive, given the substantial cost of synagogue membership.

Non-Formal Connections   Only three of the five studies asked questions that would get at non-formal Jewish connections. Los Angeles and Chicago asked about friendship networks and Atlanta included a question about Jewish cultural connections. The relationship between zip code Jewish density and having Jewish friends or consuming Jewish culture was very strong. When the Jewish connection did not entail a substantial outlay of cash, living among other Jews strengthened the Jewish connections of mixed-married couples.

How Children Were Raised   The most dramatic impact of the geography of mixed marriage was on how children were raised. The research on children of inter-racial couples found that how parents listed their inter-racial children was influenced by racial make-up of their census tract. The same was true of every metro-area but Phoenix. The higher the density of the zip code, the greater the proportion of children being raised exclusively in Judaism. Although zip code Jewish density was not associated with how children were raised in Phoenix, living in the core "Northeast Valley" (including Scottsdale and Paradise Valley) was. Whether Phoenix is typical of other mid-sized Western Jewish communities is beyond the scope of this research.

171.7.8   Implications of the Association Between Place and Jewish Engagement for Mixed-Married Couples

The association between living in a core Jewish area and Jewish engagement among mixed-married couples is theorized as bi-directional. A mixed-married couple might choose to live in a Jewish area so as to have access to other Jews and to Jewish resources. Conversely, a mixed-married couple might end up in a predominantly Jewish area for reasons other than being near other Jews (commute to work, quality of life, reputation of schools, proximity to Jewish relatives, etc.), but ends up being influenced by the presence of other Jews. It is just as likely that both factors are at work, with mixed-married couples that locate in areas of higher Jewish density both seeking out a more Jewish area and subsequently being influenced by living in such an area.

Perhaps the most interesting finding has not been mentioned. At the beginning of the twenty-first century Jews, unlike other non-Hispanic whites, live within metropolitan areas (Katz and Lang 2003). Within these metro areas there are distinct Jewish residential enclaves. In-married couples are more likely than mixed-married couples to live in those enclaves. Mixed-married couples more than in-married couples have voted with their cars and moved away from city centers and established Jewish areas. Those mixed-married couples that live near and with other Jews are more likely to participate in Jewish life and pass on a Jewish identity to their children. As with mixed-race couples, mixed-married Jewish couples might have chosen a more Jewish area to begin with or might have been influenced by living in one, or both. Jewish organizations and institutions making "outreach" to mixed-married families would do well to start close to home.

171.7.9   Implications for Future Research

The analysis of these five communities suggests that the analysis of local Jewish geography needs to be expanded. Core Jewish areas were the least defined in the two rapid growth communities (Atlanta and Phoenix). This suggests that core areas have important historical roots and that newer Jewish communities may not develop them. Both Atlanta and Phoenix are Sunbelt communities, and Los Angeles also had less clearly defined core areas, so region maybe a factor along with growth. The analysis of Jewish residential distribution should be expanded to include additional old industrial cities such as Detroit, Cleveland, and Pittsburgh (also in the North American Jewish Data Bank) as well as more rapid growth Sunbelt communities such as San Francisco, Las Vegas, and San Diego.

Informal Jewish connections were found to be important for mixed-married couples living in higher density Jewish zip codes. Unfortunately, only three of the five studies included such questions. Given that mixed-marriage will continue to increase, local Jewish population studies would do well to include more such questions. These would provide useful information for organizations that do outreach programming as well as providing scholars with a better understanding of importance place for the Jewish engagement of mixed-married couples.

References

Abu-Lughod, J. L. (1999). *New York, Chicago, Los Angeles: America's global cities*. Minneapolis: University of Minnesota Press.

Associated Jewish Charities of Greater Baltimore. (1968). *The Jewish community of Baltimore: A population study*. Baltimore: Associated Jewish Charities of Greater Baltimore.

Cohen, G. (1983). *The 1983 metropolitan Atlanta Jewish population study. Final report*. Atlanta: Atlanta Jewish Federation.

Cohen, S. M. (2006). *A tale of two Jewries: The "inconvenient truth" for American Jews*. New York: Jewish Life Network/Steinhardt Foundation. Report.

Dear, M. J., & Dishman, J. D. (Eds.). (2001). *From Chicago to LA: Making sense of urban theory*. Los Angeles: Sage.

Freidman, P., & Weinberg, E. (1983). *Chicago Jewish population study, 1982*. http://www.jewish-databank.org/Archive/C-IL-Chicago-1982-Main_Report.pdf

Gordon, M. (1965). *Assimilation in American life*. New York: Oxford University Press.

Herman, P. (1998). *Los Angeles Jewish population survey '97*. Los Angeles: Jewish Federation of Greater Los Angeles. Printed report.

Holloway, S. R., Wright, R., Ellis, M., & East, M. (2009). Place, scale and the racial claims made by white-minority parents for their multiracial children in the 1990 census. *Ethnic & Racial Studies, 32*(3), 522–547.

Katz, B. J., & Lang, R. E. (2003). *Redefining urban and suburban America: Evidence from census 2000* (Brookings metro series). Washington, DC: Brookings Institution Press.

Massey, D. (1985). Ethnic residential segregation: A theoretical synthesis and empirical review. *Sociology and Social Research, 69*, 315–350.

Massey, D. S., & Denton, N. A. (1985). Spatial assimilation as a socioeconomic outcome. *American Sociological Review, 50*, 94–106.

Mcdougall, H. (1993). *Black Baltimore: A new theory of community*. Philadelphia: Temple University Press.

National Jewish Welfare Board. (1947). *A study of the Jewish population of Atlanta, Georgia, 1947*. Atlanta: National Jewish Welfare Board.

Nicolaides, B. M. (2002). *My blue heaven: Life and politics in the working-class suburbs of Los Angeles, 1920–1965*. Chicago: University of Chicago Press.

Phillips, B. A. (2007a). Faultlines: The seven socio-ecologies of Jewish Los Angeles. In B. Zuckerman & J. Schoenberg (Eds.), *The Jewish role in American life: An annual review* (Vol. 5, p. 5). Lafayette: Purdue University Press.

Phillips, B. A. (2007b). *Lakeville revisited: Spatial assimilation among Chicago Jews*. Paper presented at the Association for Jewish Studies. Available at: http://www.jewishdatabank.org/Archive/C-IL-Chicago-2001-Spatial_Assimilation_Article.pdf

Sheskin, I. (2004). *Geographic differences among American Jews*. United Jewish Communities report series on the National Jewish population survey 2000–01 report #8. http://www.jewish-databank.org/Archive/NJPS2000_Geographic_Differences_Among_American_Jews.pdf

Smith, T. W. (2005). *Jewish distinctiveness in America, a statistical portrait*. New York: American Jewish Committee.

Strum, P. (1999). *Freedom for speech we hate*. Lawrence: University Press of Kansas.

Tobin, G. A. (1985). *Jewish population study of Greater Baltimore, 1985*. Baltimore: Associated Jewish Charities of Great Baltimore.

Ukeles Associates Inc. (2004). *The Jewish community study of north metro Atlanta: 2004*. Atlanta: The Jewish Federation of Greater Atlanta in Partnership with the Marcus Jewish Community Center of Atlanta.

Ukeles Associates Inc. (2010). *The 2010 Greater Baltimore Jewish community study*. Baltimore: Associated Jewish Charities of Baltimore.

Ukeles, J., & Miller, R. (2006). *The Jewish community centennial study of Greater Atlanta: 2006*. Atlanta: The Jewish Federation of Greater Atlanta.

Chapter 172
The Transnational Debate Over Homosexuality in the Anglican Communion

Robert M. Vanderbeck, Joanna Sadgrove, Gill Valentine, Johan Andersson, and Kevin Ward

172.1 Introduction

For several decades, the Anglican Communion has been consumed by debates regarding homosexuality. The issues involved are multi-faceted, but include the morality of same-sex relationships, the blessing of same-sex unions, the ordination of gay clergy, and the consecration of gay bishops (Clatworthy 2008; Hassett 2007; Vanderbeck et al. 2010, 2011). This chapter examines how debates over homosexuality are serving to re-shape the international Anglican Communion, a globally significant network of Christian churches historically linked to the Church of England, producing new geographies of connection and disconnection between Anglicans globally. The discussion is theoretically situated in, and extends, the growing body of research on transnational

R.M. Vanderbeck (✉)
Department of Geography, University of Leeds, West Yorkshire LS2 9JT, UK
e-mail: r.vanderbeck@leeds.ac.uk

J. Sadgrove
Research Staff, United Society, Harling House, 47-51 Great Suffolk Street, London SE1 OBS, UK
e-mail: jos@weareus.org.uk

G. Valentine
Faculty of Social Sciences, University of Sheffield,
ICOSS, 219 Portobello, Sheffield S1 4DP, UK
e-mail: g.valentine@sheffield.ac.uk

J. Andersson
Department of Geography, King's College London,
K7 42 Strand Campus, London WC2R 2LS, UK
e-mail: johan.andersson@kcl.ac.uk

K. Ward
School of Theology and Religious Studies, University of Leeds, West Yorkshire LS2 9JT, UK
e-mail: k.ward@leeds.ac.uk

© Springer Science+Business Media Dordrecht 2015
S.D. Brunn (ed.), *The Changing World Religion Map*,
DOI 10.1007/978-94-017-9376-6_172

religion. The concept of transnationalism—broadly refering to complex processes of interconnectivity and movement that transcend national boundaries—has become increasingly important in the study of religion. While the contemporary context of globalization has given impetus to this turn towards transnationalism, it should be stressed that transnational dynamics are far from new to religious communities, and, indeed, religious communities have often been at the forefront of developing forms of transnational practice and exchange. As Levitt (2004: 1) argues, "religious life has long been global," and transnational dynamics have been evident among some religious communities for centuries (Kong 2001). Nevertheless, wider literatures on transnationalism have often given scant attention to religious belief and practice, or problematically conflated religion with national identity or ethnicity.

Research on transnational religion has tended to emphasize the lives, experiences, beliefs, and practices of migrant communities (Levitt 2004; Sheringham 2010) and the influence of these migrants on host societies. Our emphasis is not on transmigrants as such, but rather more broadly on how transnational religious debates are transforming relations within the Anglican Communion, even when the great majority of members have relatively little direct connection to, or experience of, the often spatially distant others with whom they are in (or, increasingly, out of) communion. While traditional scalar categories such as parish, diocese, and province are important for understanding the shape of the sexuality debates and their implications for the Communion, new networks and assemblages are emerging that both interact with and challenge the organization of the Anglican polity.

We draw on research from a larger project on the sexuality debates within Anglicanism called 'Sexuality and Global Faith Networks: A Social Topography.' The project involved interviews with a range of key actors across the Communion; observation at several important Communion events; and a series of parish-level case studies in England, U.S. and South Africa. Additional fieldwork was also carried out in Uganda and Lesotho. Data collection for the project took place between 2008 and 2010. The chapter proceeds as follows. We begin by outlining in greater detail some of the key parameters of the sexuality debates in the Anglican Communion. We then turn to an exploration of new networks and alliances that have formed across national/provincial borders, and how these networks have responded to recent developments in the Communion related to the sexuality debates. Drawing on select examples from parish case studies, we next explore how local ideas about the Communion are being challenged and shaped by the wider transnational debate over homosexuality. In the final section, we raise questions about the future of the Anglican Communion and Anglican identity more broadly.

172.2   The Anglican Communion and the Sexuality Debates

The Anglican Communion claims approximately 80 million baptized members worldwide. Geographically, the Communion is composed of a network of 44 churches, comprised of 34 autonomous provinces (constituted on a national basis or as supranational regions), four united churches (autonomous churches that consist of historical unions between Anglicans and other traditions, including congregational churches), and six extra-provincial churches (five of which are under the jurisdiction of the Archbishop of Canterbury). The Church of England (CoE) nominally contains the largest number of baptized Anglicans, but the Church of Nigeria now claims the largest number of regularly practicing Anglicans, with the Church of Uganda also likely exceeding the CoE in this respect. The Archbishop of Canterbury officially acts as the Communion's "focus of unity," but neither he nor any Communion body can make binding decisions for the provinces, which have their own structures of governance. The churches share histories tied to missionary activity and/or British colonization (Ward 2006), and they are voluntarily linked to the so-called Instruments of Communion (the Archbishop of Canterbury, the Lambeth Conferences, the Primates Meeting, and the Anglican Consultative Council). Across the Communion, both within and between provinces, there is a wide diversity of styles of worship, with Anglo-Catholic, evangelical, and other traditions and approaches.

Although it is now common to refer to the "worldwide" or "global" Anglican Communion, it is important to recognise that this notion is a comparatively recent one (see Ward 2006 for a fuller history). As anthropologist Miranda Hassett (2007: 253) notes, "Many speak of current controversies [over human sexuality] as if they were tearing up an established, settled way of being a global Communion," when in fact the Communion has been very much an evolving project. The first of the decennial Lambeth Conferences of Anglican bishops was convened in 1867; however, only since about the 1960s – a period that corresponds with African decolonization – have there emerged concerted efforts to produce a fellowship that exhibits a "high degree of interconnectedness and mutual awareness" (Hassett 2007: 253). Many commentators have wondered whether the various local manifestations of Anglicanism can maintain and nurture a sense of unity across sometimes significant differences. Many are also concerned about the issue of unequal power relations between the provinces, given vastly different levels of wealth and histories of colonization: can unity be produced in ways that do not privilege the richer provinces of the traditional Anglican core (for example, Douglas and Pui-Lan 2001)? Given that the fastest growing churches in the Communion are in the global South (particularly Sub-Saharan Africa), can a postcolonial Anglican identity remain centered on Canterbury? Indeed, this very question has been publically asked by the Primates of several Anglican provinces, including Archbishop Henry Luke Orombi, Primate of Uganda, who argues:

> The peculiar thing is that this one man [the Archbishop of Canterbury], who is at the center of the communion's structures, is not even elected by his peers. Even the Pope is elected by his peers, but what Anglicans have is a man appointed by a secular government. Over the past five years, we have come to see this as a remnant of British colonialism, and it is not serving us well. The spiritual leadership of a global communion of independent and autonomous provinces should not be reduced to one man appointed by a secular government. (Orombi 2008)

Orombi's remarks were made specifically in relation to a perceived failure of the current institutional structures of Anglicanism to halt developments related to understandings of human sexuality in some of the provinces which he and many other self-identified "orthodox" Anglicans view as inconsistent with scripture and church tradition.

172.3   New Transnational Alliances

To date, the most divisive debates between provinces have centered on trends within The Episcopal Church (TEC) in the U.S. and elsewhere towards the widened acceptance of homosexual relationships. Opponents of these trends often argue that the increased acceptance of homosexuality represents the incursion of secular agendas into the life of the church, with church leaders in some provinces increasingly allowing themselves to be swayed by transformations in the wider "cultural context" (Davies 2008), most notably by the legislative and rhetorical frameworks of equality and human rights. In contrast, many supporters of widened gay inclusion strongly reply that their motivations for pursuing this are very much religiously inspired and consistent with (even demanded by) Christian ethics of love, care, and respect for family (Clatworthy 2008; Vanderbeck et al. 2010).

Increasingly outnumbered within their own national contexts, conservatives in TEC and, to a lesser extent, other provinces in the global North, have sought numerical support from outside their provinces. American conservatives have had particular success forming alliances in Sub-Saharan Africa, a region in which many countries have been advancing their own public rhetorics regarding the risks that "Western" secular culture pose to national moral values. Forming alliances with leaders of provinces in the global South, it was hoped, would provide a kind of transnational moral authority to American conservatives, a minority in TEC but, they argued, part of a strong global Anglican majority. Anglican churches in Africa were often represented by conservatives in the North as "young," "vigorous," and "fervent" compared to TEC and the Anglican Church of Canada, which were experiencing declining membership and influence, like mainline denominations across the global North (see Jenkins 2002; Hassett 2007; Sadgrove et al. 2010).

The emerging network is one in which many of the most publicly prominent figures have included the Primates of Nigeria (Archbishop Peter Akinola, recently retired), Uganda (Orombi), Kenya (Archbishop Benjamin Nzimbi, recently retired), and Rwanda (Archbishop Emmanuel Kolini). Representing some of the largest provinces, they have vigorously queried the nature of power and authority in the Communion, including (as articulated by Orombi, above) the centrality of the Church of England and the influence of TEC. Although the Primates and their supporters frequently charge that liberal Anglicans in the North continue to approach their interactions with African churches with neo-colonial mindsets, their critics have often replied that African Anglican leaders approach questions of homosexuality with mentalities that are still effectively colonized, adopting what critics views as literalistic interpretations of scripture brought to them (and often since largely abandoned or moderated) by their former colonizers (Hassett 2007; Hoad 2007; Jenkins 2000, 2004; Rubenstein 2004, 2008; Valentine et al. 2013a, b). Questions have been raised about the financing of these transnational networks. A much noted report by Naughton (2007), for example, argues that emerging networks between U.S. Anglican conservatives and African church leaders have been heavily financed by wealthy right wing donors, part of a larger political project of both fracturing mainline Protestant churches in the U.S. (weakening their ability to oppose far right politics) while promoting anti-gay policies in Africa (Kaoma 2009; see also Sadgrove et al. 2012).

The emerging alliances had a significant and lasting impact at the Lambeth Conference of 1998. Since 1867, bishops from around the Communion have held the decennial Lambeth Conference under the presidency of the Archbishop of Canterbury. The Lambeth Conference is not a legislative body, but nevertheless its resolutions are deemed to be important articulations of the consensus of the Anglican Episcopate. In the past, the Lambeth Conferences have been sites for enacting debates over other matters related to human sexuality and gender, including contraception, divorce, polygamy, and female ordination. The 1998 Lambeth Conference is most remembered for its discussions of human sexuality, which culminated in the passage of Resolution 1.10, which "reject[s] homosexual practice as incompatible with Scripture." While promising to "listen to the experiences of homosexual people," the bishops resolved that it could not "advise the legitimising or blessing of same-sex unions nor ordaining of those involved in same-gender unions." This resolution passed by a large majority: 526 votes in favour, 70 opposed and 45 abstentions. Although the vote partially reflected the strength of the emerging alliances between conservative bishops in different provinces, a number of more 'liberal' bishops apparently voted for the resolution out of fear that something with even more condemnatory language, and no commitment to listening to the experiences of gay and lesbian Christians, would emerge if they opposed it (Hassett 2007; Hoad 2007).

In the decade following Lambeth 1998, a number of developments served to escalate tensions. Within the Anglican Church of Canada, the diocese of New Westminster created a liturgy for blessing same-sex unions in 2002. It was also widely known that some CoE and TEC clergy were offering blessings to same-sex couples. By far the most notable event, however, was the 2003 consecration of Gene Robinson as Bishop of New Hampshire by TEC, a move which attracted extensive international media attention. Robinson (who was in a longterm relationship with a male partner) was the first openly gay man to be consecrated in the Communion, and has since become a very significant figure in struggles over the status of non heterosexuals in Christian churches and wider society (Robinson, for example, delivered the invocation at the launch event of the inaugural weekend for Barack Obama in January 2009). The responses to the consecration included a statement from Archbishop Akinola of Nigeria, on behalf of what is known as the working committee of the Primates of the Global South (it is important to recognize how this constructs a problematically unified vision of the "global South," with some conservative primates purporting to represent the "Southern" position when, in reality, churches in South Africa and Brazil, among others, have explicitly distanced themselves from much of their rhetoric). Akinola stated that the consecration "clearly demonstrates that authorities within the Episcopal Church consider that their cultural-based agenda is of far greater importance than obedience to the Word of God." (Global South Anglican 2003). He went on to declare that "a state of impaired communion now exists both within a significant part of [TEC] and between [TEC] and most of the provinces within the Communion" (Global South Anglican 2003).

Within the U.S. and Canada, a number of parishes and several dioceses broke away from their provinces but, wishing to remain within the broader Communion, sought affiliations with provinces in the global South. What emerged was a complex process often referred to as "Anglican realignment." Primates of several provinces (which have included Nigeria, Rwanda, Uganda, Kenya, Southeast Asia, and the Southern Cone of South America) offered episcopal oversight and/or consecrated new American bishops who were then sent back to the U.S. to minister to the parishes who have left TEC, as part of a missionary outreach from the global South. As a result of these processes of "border crossing" or "cross-border intervention," situations arose in which two competing bishops (one under the authority of TEC, the other a missionary bishop under the authority of another province) were covering a single geographic area. However, only those bishops recognized by Canterbury (those still part of TEC) were invited to the 2008 Lambeth Conference, a situation which further angered a number of conservatives, many of whom felt that bishops from TEC who supported Robinson's consecration should be sanctioned or banned from Lambeth.

172.4 Transnational Responses

In reaction to dissatisfaction with the emerging plans for the Lambeth Conference 2008, a range of existing coalitions of conservatives co-ordinated a separate international meeting to convene just prior to Lambeth, with many conservative bishops also announcing their plans to boycott Lambeth. This coalition included the "Primates of the Global South," traditionalists within TEC, UK-based Anglican Mainstream, and the Diocese of Sydney (which is strongly evangelical in its orientation). This meeting constituted the largest public expression to date of the growing transnational orthodox Anglican movement, linking theological conservatives from disparate corners of the Communion. The Global Anglican Future Conference (GAFCON) convened in Jerusalem in June 2008 with approximately 1300 delegates. This included laity, clergy, and the primates of six provinces—Nigeria, Kenya, Uganda, Rwanda, West Africa, and the Southern Cone. The majority of delegates came from Sub-Saharan Africa, specifically Nigeria and Uganda – because the organisers of GAFCON reportedly wished to more closely reflect the numerical constitution of the wider Communion in contrast to Lambeth Conferences (where, for example, the number of TEC bishops is disproportionately large compared to the size of TEC, given the large number of geographical diocese in the US, some with quite small populations).

Although opposition to the acceptance of homosexual practice was acknowledged as the "presenting issue," GAFCON attempted to publicly foreground a positive, missional identity for the group, with the stated goal of renewing the Anglican Communion. At GAFCON, movement leaders sought to portray the group to the global media as simultaneously united in spirit and purpose while at the same time needing to work through certain areas of cultural, theological, and other difference between its members (for example, in relation to female ordination; approaches to scriptural understanding). Despite attempts by many meeting organizers to portray the meeting as about more than opposition to homosexuality—in part showing awareness that the movement was increasingly being satirized by opponents as "sex-obsessed" (for example, Coward 2009), the acceptance of homosexual relationships was central to the ensuing conversation around GAFCON's so-called "renewed vision of Anglican orthodoxy" (Global Anglican Future Conference 2008: 41). Even in relation to the "presenting issue," however, a range of differences were evident at GAFCON related to conceptualizations of homosexuality, the treatment of lesbian and gay members of society, and acceptable forms of discourse. For example, some British delegates expressed discomfort with rhetoric espoused by Archbishop Akinola of Nigeria, who gave at best ambiguous answers to questions from the media about the violence experienced by gays and lesbians in African countries and elsewhere (Akinola argued, "Every society has its own standard of life. In African societies, we have what we call taboos—the 'don'ts'—and if you break the taboos, there are consequences"). (See Sadgrove, 2010 for a fuller discussion of how these differences were negotiated at GAFCON). Nevertheless, delegates were publicly united in a view, enshrined in the final Jerusalem Declaration, that the only appropriate context for sexual activity was heterosexual marriage. In contrast, despite the delegates representing provinces with sometimes substantially different positions on female ordination (for example, the Church of Nigeria opposes this while the practice is permitted in Uganda), the issue was explicitly defined as secondary, thus allowing movement participants to remain within the boundaries of the emerging orthodoxy without having to hold a particular stance on women's role in the church.

Religious orthodoxies, while often making claims to unproblematic historical foundations, are in fact fluid and dynamic constructions that seek to define ideological boundaries in relation to particular issues that come to be labelled as heresies (Henderson 1998; Berlinerblau 2001). Leaders of the transnational orthodox Anglican movement have indeed often labelled supporters of gay inclusion and equality as heretics (see Sadgrove et al. 2010) The final Jerusalem Declaration provides a formalized boundary structure for this emergent orthodoxy within Anglicanism. The declaration is at the core of the Fellowship of Confessing Anglicans (FCA), a group formed after the Jerusalem meeting to carry forward the agenda of GAFCON, and all members of FCA must explicitly assent to the declaration. While welcomed

R.M. Vanderbeck et al.

by many (but far from all) conservative leaders within the Communion, many Anglicans have also argued that a binding statement of confession is at its core fundamentally "un-Anglican," given how the CoE and Communion as a whole traditionally encompasses a disparate range of practices and traditions (the CoE itself was engineered as a *via media*—or middle road—between Protestantism and Catholicism to appeal to a broad range of sixteenth century religious opinion). From this perspective, the Jerusalem Declaration, rather than being an unproblematic representation of historical "orthodoxy," represents a significant departure from the historical self-understanding of the Communion in which the terms of being "in communion" were fairly flexible, such as the relatively broad theological principles of the Chicago-Lambeth Quadrilateral of the 1880s (see Ward 2006).

In the aftermath of GAFCON, a number of disaffected conservatives in the U.S. and Canada formed the Anglican Church in North America (ACNA), which was designed to provide an alternative home for those 'orthodox' Episcopalians who wanted to leave (or had already left) TEC or the Anglican Church of Canada. ACNA, which held its inaugural assembly in June 2009, currently claims about 100,000 members across 21 dioceses (nearly 1000 congregations). One notable characteristic of ACNA is that an ACNA-affiliated diocese can simultaneously seek to remain dually affiliated with an Anglican province (although this affiliation is not recognized by the Anglican Communion office). For example, the Diocese of Fort Worth simultaneously claims affiliation to ACNA and the Anglican Province of the Southern Cone, and, therefore, the Diocese can attempt to represent itself as inside the Communion without being part of the province that Canterbury views as having jurisdiction in that territory. Both the Church of Nigeria and the Church of Uganda, which consider themselves to be in states of "impaired communion" with TEC, have declared themselves to be in full communion with ACNA. A recent report by the Archbishops of Canterbury and York recommends "an open-ended engagement" (Williams and Sentamu 2011: 3) between the CoE and ACNA, but no definitive decisions have been made regarding whether the CoE will ultimately recognize ACNA in addition to (or in replacement of) TEC and the Anglican Church of Canada as legitimate expressions of global Anglicanism (Table 172.1).

The formation of new transnational networks has not, however, simply been a characteristic of the anti-gay wing of the Communion. Although the phenomenon has been less researched, pro-gay groups have also begun to strengthen alliances in disparate part of the Communion. For example, some of these linkages were evident at the 2008 Lambeth Conference. Although the official business of the Conference is open only to the invited bishops and a select few other visitors, the three-week event attracts a large cross-section of Anglican/Episcopal pressure groups, charities, and other organizations seeking to disseminate information and promote their positions in various ways. Integrity USA (the leading pro-gay group of American Episcopalians) and Changing Attitude UK (a similar British-based group) supported the attendance of several African activists at Lambeth 2008. Although on a much smaller scale that the GAFCON movement, these linkages were considered important not just to provide support to pro-gay Anglicans in provinces where they are often tiny minorities living under oppressive circumstances, but also to challenge

Table 172.1 Churches of the Anglican Communion, 2012 (Data source: Anglican Communion Office 2013)

The Anglican Church in Aotearoa, New Zealand & Polynesia
The Anglican Church of Australia
The Church of Bangladesh (United)
The Anglican Church of Brazil
The Anglican Church of Burundi
The Anglican Church of Canada
The Church of the Province of Central Africa
Anglican Church in the Central American Region
Anglican Province of the Congo
The Church of England
Hong Kong Sheng Kung Hui
The Church of the Province of the Indian Ocean
The Church of Ireland
The Nippon Sei Ko Kai (The Anglican Communion in Japan)
The Episcopal Church in Jerusalem & The Middle East
The Anglican Church of Kenya
The Anglican Church of Korea
The Church of the Province of Melanesia
The Anglican Church of Mexico
The Church of the Province of Myanmar (Burma)
The Church of Nigeria (Anglican Communion)
The Church of North India (United)
The Church of Pakistan (United)
The Anglican Church of Papua New Guinea
The Episcopal Church in the Philippines
Province de L'Eglise Anglicane au Rwanda
The Scottish Episcopal Church
Church of the Province of South East Asia
The Church of South India (United)
Anglican Church of Southern Africa
Anglican Church of the Southern Cone of America
The Episcopal Church of the Sudan
The Anglican Church of Tanzania
The Church of the Province of Uganda
The Episcopal Church
The Church in Wales
The Church of the Province of West Africa
The Church in the Province of the West Indies
The Church of Ceylon (Extra-provincial to the Archbishop of Canterbury)
The Episcopal Church of Cuba (Extra-provincial-metropolitan council)
Bermuda (Extra-provincial to Canterbury)
The Lusitanian Church (Extra-provincial to the Archbishop of Canterbury)
The Reformed Episcopal Church of Spain (Extra-provincial to the Archbishop of Canterbury)
Falkland Islands (Extra-provincial to the Archbishop of Canterbury)

forms of anti-gay rhetoric (see Vanderbeck et al. 2010 for a fuller analysis of pro-gay collective action at Lambeth). In particular, the participation of African Anglicans allowed for direct challenges to rhetoric that there was anything inherently "un-African" about homosexuality, as is often asserted by African primates and their supporters (cf. Hassett 2007; Hoad 2007). It also created space to challenge accusations that Northern gay groups were essentially trying to colonize African countries and churches with Euro-American ideas about sexual identity. As Davis Mac-Iyalla, representing Changing Attitude Nigeria, told the global news media at Lambeth, "Homosexuality does exist in Africa. It's not a Western thing, as our African bishops would want people to believe… The bishops need to know that what is coming out of their mouths … is putting us in a very difficult position" (Landau 2008). Their discourse, for example, is believed to have contributed to the climate of persecution that has developed as both Nigeria and Uganda have moved to tighten legal controls on homosexuality (see Sadgrove et al. 2012 on the anti-homosexuality bill in the Ugandan parliament).

172.5   An Anglican Covenant?

Levels of polarization have seemingly increased dramatically in recent years. Fearing an irreparable fracturing of the Communion in the wake of Gene Robinson's consecration, the Archbishop of Canterbury appointed Robin Eames, the Anglican Primate of All Ireland, to lead a commission to study the state of relations in the Communion. The commission published the Windsor Report in 2004, which made several recommendations (Lambeth Commission on Communion 2004). Among these was a recommendation that the churches of the Communion voluntarily enter into an "Anglican Covenant," through which churches obliged themselves to consult the rest of the Communion before making important decisions (the consecration of Robinson was taken to be emblematic of a decision about which the wider Communion should have been consulted). The report also called for a moratorium on further consecrations of openly gay individuals to the episcopate. Many supporters of widened gay inclusion in the Communion were deeply critical of the report for seemingly taking for granted that Robinson's consecration was a theological error, a point which is far from obvious to Christians who support gay inclusion (for example, Clatworthy and Taylor 2005).

In late 2009, a document called The Anglican Communion Covenant was circulated to the churches for discussion and (it was hoped) approval through each church's governance structure (Covenant Design Group 2009). The proposed Covenant sought to impose "relational consequences" (point 4.27) for those churches which took action which could not be accepted by other covenanted churches, including potential exclusion from participation in the Instruments of Communion. In essence, this would create a two-tiered Communion of provinces, an inner circle signed up to and abiding by the Covenant, and a loosely attached outer circle who either chose not to agree to the Covenant or who were deemed to be in violation of it. The Archbishop of Canterbury has described the document, in pushing for provinces its affirmation, as a means "to help us…intensify our fellowship and our trust" (Williams 2009). In contrast, religious historian and supporter of gay inclusion Diarmaid MacCulloch (2012) has described the Covenant as:

> an effort to increase the power of centralizing bureaucracy throughout the worldwide Anglican communion. However much the promoters denied it, the principal aim was to discipline Anglican churches in the United States and Canada, which had the gall to think for themselves and, after much prayer and discussion, to treat gay people just like anybody else.

To date, the covenant has been widely popular with neither churches that support aspects of gay inclusion nor those that oppose them. At the time of writing, eight churches have agreed to enter into the Covenant (Anglican Communion News Service 2012a), including the provinces of Ireland, the West Indies, the Southern Cone, Papua New Guinea, and Southern Africa. However, despite the Covenant often being represented as a measure to appease the African Provinces, Nigeria and Uganda, amongst others, have yet to accept the document. Among reasons cited for this are a reluctance to enter into a sacred covenant with churches with whom they already considered themselves in states of "impaired Communion," and a belief that the drafting of the document had been dominated by the wealthier provinces of the North. As articulated by Bishop John Akao (2011: 24), now Primate of the Church of Nigeria.

> Unfortunately, the original idea of covenant to bring back erring members who have embarrassed the Communion and torn its fabrics apart, was adopted by the Anglican Establishment, by fashioning a covenant which in motive, content and thrust deviate from the original objective of healing and unifying the communion. The present covenant to the African Anglicans, is crafted to persuade orthodox Anglicans to accept and commit to fellowshipping with revisionist groups who have perpetrated aberrations but who unrepentantly defy various moves and resolutions to bring them back on course [….] As long as there is no cohesion, the idea of a covenant will remain impracticable.

The Covenant has also been voted down by the Scottish Episcopal Church, and TEC declined to take any official position on the Covenant at its General Convention in July 2012 (Davies 2012). Perhaps most notably, the covenant—despite the lobbying of the Archbishop of Canterbury—has been rejected by the CoE. In May 2012, the majority of dioceses in the CoE voted to reject the Covenant, with roughly 80 % of bishops supporting the covenant but roughly half of the clergy and laity opposing it (MacCulloch 2012). This poses particularly interesting questions for the future of the Communion: Communion membership has traditionally been based on communion with Canterbury, yet can this be sustained if the CoE is outside the Covenant?

172.6 Life in a Transnational Religious Field: Perspectives at the Parish Level

The defeat of the Covenant within the CoE is suggestive of how the views and experiences of ordinary parishioners can differ substantially from bishops and others in positions of church leadership. Despite the entreaties of their most senior figure and many other leaders, ordinary parishioners in England have not felt compelled to approve the Covenant despite it being presented as necessary to foster transnational bonds of religious fellowship and preserve the Communion. In this section, we explore some key dimensions of the transnational debate over homosexuality draw-ing on field research (interviews, focus groups, and participant observation) with local parishes in England, U.S. and South Africa (the dominant country in the Anglican Church of Southern Africa). The case studies that we chose were a mix of parishes with different stances and mixes of opinion on issues of homosexuality. The fieldwork took place between 2009 and 2010. These case studies were not meant to provide "representative" pictures of opinions within their respective prov-inces or even dioceses, given the wide diversity of Anglican parishes both within and between provinces and dioceses. Rather, the examples we have selected are intended to illustrate the complex ways in which this transnational debate is experi-enced and interpreted by ordinary people (see Vanderbeck et al. 2010 for a fuller discussion of methodological issues and challenges associated with the research).

As illustrated in the previous sections, bishops and other leaders in the Communion often discussed Communion-level affairs with a sense of urgency, indi-cating a need to rejuvenate the Communion and (certainly for supporters of the Covenant) intensify bonds of accountability between churches. Our research in local parishes revealed a very mixed picture in terms of feeling of commitment to an international communion. Parishioners had vastly differing levels of knowledge about, and interest in, the churches with whom they were considered in communion. Indeed, we encountered many parishioners in all three provincial contexts who, while very aware that a debate was raging over issues of homosexuality (in part trig-gered by the consecration of Gene Robinson, or simply "the gay bishop" to some respondents), expressed little knowledge of the Communion itself or its institutional frameworks. Some respondents reported that their first real awareness that they were actively part of a global Communion corresponded with the intensification of international debate about homosexuality.

Although many respondents reported feeling favorably that they were part of a "global" church, their feelings about debates in the Anglican Communion were influ-enced by their ambivalences about the wider subject of Christian factionalism. In all of the case studies, a proportion of parishioners were converts from other Christian or non-religious backgrounds, and even many respondents who were born into the Anglican church expressed a willingness to consider abandoning the tradition if they developed too great a sense of discomfort with developments in their provinces or the wider Communion (whether related to homosexuality or otherwise).

Ambivalences about Anglican identity were evident in both theologically "liberal" and "conservative" parishes, as well as those with more mixed or moderate approaches. Two of the parishes in which we conducted fieldwork were actively gay-affirming, with relatively high proportions of gay and lesbian members: one in the U.S. (TEC-1) and one in England (COE-1). Many members of TEC-1 had taken a relatively active interest in the affairs of the national church. In particular, at the time of the interviews, TEC-1 was actively working to oppose what was known as BO33: a resolution that had been adopted at the 2006 General Convention of the Episcopal Church. The resolution called upon "Standing Committees and bishops with jurisdiction to exercise restraint by not consenting to the consecration of any candidate to the episcopate whose manner of life presents a challenge to the wider church and will lead to further strains on communion." In short, the resolution was understood to be a ban on the consecration of any additional gay-identified bishops, for the purpose of appeasing anti-gay forces in the Anglican Communion. The congregation actively worked to build opposition to BO33, and lesbian and gay members of the congregation talked about the "hurt" that had been caused by BO33. Some parishioners compared the effect of this resolution to being excluded from a "table" or a "family gathering," as articulated by the respondents below:

It's hurtful to those of us that are gay and lesbian and have had that happen repeatedly in our lives and have struggled with that in our families of origin and that's why we've come to the church, a welcoming accepting church to say we are behind you, we care about you, we welcome you but then when push comes to shove and when there's friction from another member of the family and one can equate this to a family function, a family gathering, to have the church say "well you know we won't bring you along this time or we'll hide you for this moment or we won't let you be as part of the wedding party on the day" (white male, middle aged)

The pain of being a second-class citizen is so deep that any concession to bigotry is kind of like saying, "Oh, actually we are second-class citizens" If there is still space, then you can get a place at the table but you're not one of the first people to get seated at the table. (white male, middle age).

Given the strength of the anti-gay sentiment emerging from some foreign churches (particularly the Church of Nigeria), a number of respondents at TEC-1 had conflicted feelings about their involvement in the international Anglican Communion. One man suggested that TEC should cut off all mission funding to foreign churches with anti-gay views, although others challenged this view and hoped that there could be a way for the Communion to hold together. However, it was repeatedly emphasised by most TEC-1 parishioners that it would be wrong to compromise its gay-affirmative position for the sake of relationships with distant churches (BO33 was effectively overturned by the Episcopal Church at its General Convention in summer 2009, in part due to organizing by pro-LGBT parishes such as TEC-1).

A similar stance was taken by many parishioners at the gay affirmative English parish COE-1. Some respondents worried that their relationships with churches in other parts of the world were damaging the faith of LGBT Christians in their own parishes:

The thing I feel about the Anglican Communion is that … a diversity of views is good as long as those views don't actually become destructive, and I think we're at a stage now where … they're beginning to destroy, and not just destroy the church as a Communion but to destroy individuals …. What these people [anti-gay bishops in African provinces] are saying is actually damaging people here's relationship with God. (white male, young adult, COE-1)

From this perspective, the transnational debate over homosexuality was proving damaging to lesbians and gays locally. This particular young male, who identified as gay, in fact often concealed his CoE identity from his non-religious gay friends because he had encountered questions and teasing in the past regarding why he attended a church that included leaders like Peter Akinola. At the same time, many respondents expressed appreciation for the diversity of views in the Anglican Communion, and hoped to be able to maintain a situation where there was a "breadth of theology and experience and churchmanship …. Otherwise, you know, if we are content to let one wing go, it diminishes all of us and we are left with a narrower church" (white male, middle aged, COE-1). However, few expressed any willingness to compromise the gay-affirmative position of their parish for the sake of relations with foreign provinces. In fact, several expressed a willingness to leave the CoE altogether if necessary.

We found both interesting parallel and contrasts in field research in South Africa, with rspondents giving responses to the subject of the Anglican Communion that were fraught with ambivalences. South Africa is by far the most sizable member of the Anglican Church of Southern Africa (ACSA). Anglican account for a relatively small proportion of South Africans (4 %), although many key figures, such as now retired Archbishop Desmond Tutu, have been Anglicans. South Africa is renowned for having the world's first constitution prohibiting discrimination based on sexual orientation. Nevertheless, opinion polls suggest that a substantial majority of people hold negative views regarding homosexuality (for example, Roberts and Reddy 2008). The leadership of ACSA, although in some ways espousing supportive views about homosexuality, have been cautious over issues of gay marriage. One of our South African case studies (ACSA-1) was a township parish with a predominantly Zulu population in the Kwa-Zulu Natal region. Within ACSA-1, denominational identity was a complex matter, involving struggles within the parish over what it meant to be Anglican. A major source of concern was how the parish should be positioned in relation to other churches in the township, particularly Methodist and Pentecostal churches (which were viewed as growing faster). Some respondents expressed frustration that scripture was not, in their view, taken seriously enough within the Anglican Church when compared with more charismatic "Bible believing" churches in the township:

The Anglican Church seems to be allowing certain things to, we seem to be very accommodating and we don't want to hurt … by speaking the truth to say what the Bible says and now we begin to wonder, when do we decide the Bible is the word of God? And what makes us decide "no, not this; we can change this?"(Zulu female, retired)

These anxieties over how scripture was used in the Anglican Church reflect wider concerns over their parish's relationship with both their own province and foreign churches. TEC was deemed by some respondents to be too "accommodating" of lesbians and gays, and ACSA's unwillingness to take a strong stance against homosexuality was felt to threaten Anglicanism's position in relation to other denominations locally:

> I think America is too accommodating. The word of God is the word of God when it suits a certain trend of whatever. When it does not it doesn't matter. It doesn't matter….. [The Anglican church] says "everybody is the same" but it's not helping this church to gain more members. You find that in those other churches the charismatic churches where they take Bible as the word of God as it is, many people are going there. They're joining the churches in droves. (Zulu female, middle aged)

Several respondents suggested that ACSA-1 had sometimes been ridiculed or teased by members of other "bible believing" churches in the township because of ACSA-1's links to 'the gay bishop' in America.

At ACSA-1, Relationships with other black Christians in the township were seemingly prioritized over those with predominantly white Anglican churches in their own diocese. The history of racial and linguistic segregation had contributed to a sense of strong attachment between the township's many Zulu speaking congregations. For example, as one respondent explained, when attending a funeral for someone in the township, people sang the same "African" choruses as their Methodist and Pentecostal neighbors and friends:

> We are in an African Church. We are in African communities whereby you even sing hymns that when you go to the Methodist Church you know you're related to them; you sing the same choruses and everything. That's the culture of the black people in church you know. Where you're not very much different. (middle aged female, ACSA-1)

Several respondents suggested that they worshipped in an Anglican church because that had been the tradition in their families, but felt little commitment to the idea of denominations. One respondent discussed how, for him, being Anglican was merely a "surname;" that is, it comes after the primary identity of "Christian," which is of far greater significance. As another respondent emphasized:

> Being an Anglican doesn't mean anything to me…. I'm no longer influenced by being an Anglican. I'm influenced by the word of God…. If your spirituality has taken a step up, then you know that God is above Anglicanism. (Zulu female, retired, ACSA-1)

Despite group interview respondents expressing a clear desire for ACSA to take a more "biblical" view, there was at least one self-identified lesbian who participated in the congregation and felt relatively comfortable there. One respondent explicitly noted that there was a need to take mind of New Testament messages regarding being non-judgemental rather than simply Old Testament injunctions regarding sexuality, and another stressed that any parents could give birth to a gay child and this would present moral dilemmas. Thus, there was a certain degree of negotiation and contestation about how, practically speaking, gays and lesbians should be treated. Nevertheless, given the daily problems related to poverty and health care confronted by the township, these concerns were deemed more important as a focus of parish life than debates over homosexuality which seemed to preoccupy many Communion leaders.

172.7 The Future of the Communion?

Understanding the changing geographies of Anglicanism requires a recognition of how religious identities, beliefs, and practices are shaped across multiple scales and through complex transnational networks. The intensification of transnational flows of discourses, people, and money—rather than serving as a force connecting people worldwide who share an Anglican identity—has often seemingly created senses of disconnection to the wider Communion, accentuating perceived differences at least as much as it has served to foster senses of connection and belonging. The threat of permanent schism within the Communion is widely discussed, with some suggesting that it has in essence already arrived, so damaged are relations between provinces (for example, MacCulloch 2012) and, in some cases, within them. It is uncertain whether the Communion will continue to exist for long in a form resembling its present structure, or, if not, what will ultimately replace it.

This is not to suggest that efforts to preserve the Communion have been abandoned, even as the Covenant process seems to be disintegrating. The Anglican Communion has created an official ministry called Continuing Indaba which attempts to facilitate dialogue between parishes across the Communion in hopes of producing "conversation across difference" (Anglican Communion News Service 2012b). Indaba is a Zulu concept interpreted within the context of the Communion as "a community process for discernment on matters of significance" in which each participant has equal voice (2012b). The 2008 Lambeth conference, rather than operating on a more traditional model of voting on resolutions, used an adapted model of Indaba based on conversations between small and medium-sized groups of bishops to try to find common ground and a common story from which to proceed. Continuing Indaba has been designed to develop and extend this model, but at the level of parishes rather than simply of bishops, and this approach to ongoing dialogue at various levels is hoped to stem the prospects of a permanent fracturing of the Communion. A number of different provinces have agreed to participate in cross-cultural encounters and facilitated conversations to date, based on meetings between triads of dioceses (Continuing Indaba 2012). Pilot conversations have taken place between: (1) New York (U.S.-Derby (England)-Mumbai (India); (2) Toronto (Canada)-Jamaica-Hong Kong; (3) Western Tanganyika (Tanzania)-El Camino Real (California, USA)-Gloucester (England); and 4) Saldanha Bay (South Africa)-Ho (Ghana)-Mbeere (Kenya).

However, parishes from Nigeria, Uganda, and many other provinces most closely associated with the GAFCON movement have not been involved to date. At least in the view of the current Primate of Nigeria, the appetite for ongoing conversation is limited, with no plans to involve their parishes or dioceses:

> The present covenant distracts the orthodox Anglican voices from the major issues currently in contention in the Communion. It seeks to surreptitiously engender perpetual talking and dissipation of valuable time, energy and human as well as material resources in endless meetings which so far have led nowhere, whereas the erroneous teaching and practices are being consolidated. (Akao 2011: 24)

Jenny Te Paa (2009), the dean of Te Rau Kahikatea (College of St. John the Evangelist) of Aotearoa New Zealand, has differentiated between visions of communion that involve becoming "one with" the other from those that demand becoming "one of" the other. It remains an open question whether those involved in the complex transnational networks of the Anglican Communion will develop ways to be "one with" each other, given the unlikelihood of Communion-wide agreement on issues of human sexuality in the foreseeable future.

References

Akao, J. (2011, March 18). Church of Nigeria and the proposed Anglican Covenant. *Church Times*, p. 24. Retrieved June 2, 2012, from www.churchtimes.co.uk/uploads/documents/Anglican%20 Covenant_18%20March.pdf

Anglican Communion News Service. (2012a, May 30). *2012 standing committee bulletin Day 1*. Retrieved June 2, 2012, from www.anglicancommunion.org/acns/news.cfm/2012/5/30/ACNS5112

Anglican Communion News Service. (2012b). *Continuing Indaba is celebrated as a 'wonderful gift' to the Communion*. Retrieved June 2, 2012, from www.anglicancommunion.org/acns/news.cfm/2012/5/29/ACNS5111

Anglican Communion Office. (2013). Provincial directory. http://www.anglicancommunion.org/tour/index.cfm

Berlinerblau, J. (2001). Toward a sociology of heresy, orthodoxy and doxa. *History of Religions, 40*, 327–351.

Clatworthy, J. (2008). *Liberal faith in a divided church*. Hants: O Books.

Clatworthy, J., & Taylor, D. (Eds.). (2005). *The Windsor report: A liberal response*. Hants: O Books.

Continuing Indaba. (2012). *The pilot conversations*. Retrieved June 2, 2012, from http://continuingindaba.com/report/conversations/

Covenant Design Group. (2009). *The Anglican Covenant*. Retrieved June 2, 2012, from www.anglicancommunion.org/commission/covenant/docs/The_Anglican_Covenant.pdf

Coward, C. (2009, April 24). Sex and the city: Anglican Mainstream's homosexual obsession. *Changing Attitude Blog*. Retrieved June 2, 2012, from http://changingattitude-england.blogspot.co.uk/2009/04/sex-and-city-anglican-mainstreams.html

Davies, M. (2008). *The gospel and the U.K.'s cultural context*. From: http://www.gafcon.org/resources/the_gospel_an.d_the_uks_cultural_context/

Davies, M. (2012, July 12). Convention wrap-up: Re-envisioning church for the 21st century. *Episcopal News Service*. Retrieved July 13, 2012 from http://episcopaldigitalnetwork.com/ens/2012/07/12/convention-wrap-up-re-envisioning-church-for-the-21st-century/

Douglas, I. T., & Pui-Lan, K. (Eds.). (2001). *Beyond colonial Anglicanism: The Anglican Communion in the twenty-first century*. New York: Church Publishing.

Global Anglican Future Conference. (2008). *The way, the truth, and the life: Theological resources for a pilgrimage to a global Anglican future*. Prepared by the Theological Resources Team. Retrieved June 2, 2012, from www.gafcon.org/images/uploads/gafcon_way_truth_life.pdf

Global South Anglican. (2003) *Statement of the primates of the global South in the Anglican Communion*. Retrieved June 2, 2012 from: www.globalsouthanglican.org/index.php/comments/statement_of_the_primates_of_the_global_south_2_Nov_2003

Hassett, M. (2007). *Anglican Communion in crisis: How Episcopal dissidents and their African allies are reshaping Anglicanism*. Princeton: Princeton University Press.

Henderson, J. B. (1998). *The construction of orthodoxy and heresy: Neo-Confucian, Islamic, Jewish, and early Christian patterns*. Albany: State University of New York Press.

Hoad, N. (2007). *African intimacies: Race, homosexuality, and globalization*. Minneapolis: University of Minnesota Press.

Jenkins, W. (2000). Ethnohomophobia? *Anglican Theological Review, 82*, 551–563.

Jenkins, P. (2002). *The next Christendom: The coming of global Christianity*. Oxford: Oxford University Press.

Jenkins, W. (2004). Episcopalians, homosexuality, and world mission. *Anglican Theological Review, 86*, 293–316.

Kaoma, K. (2009). *Globalizing the culture wars: U.S. churches, American conservatives, and homophobia*. Political Research Associates. Retrieved June 2, 2012, from www.publiceye.org/publications/globalizing-the-culture-wars/

Kong, L. (2001). Mapping 'new' geographies of religion: Politics and poetics in modernity. *Progress in Human Geography, 25*, 211–233.

Lambeth Commission on Communion. (2004). *The Windsor report*. Retrieved June 2, 2012, from www.anglicancommunion.org/windsor2004/

Landau, G. (2008, July 31). Gay Nigerian tells of death threats. *BBC News* Online. Retrieved June 2, 2012, from http://news.bbc.co.uk/1/hi/world/africa/7535533.stm

Levitt, P. (2004). Redefining the boundaries of belonging: The institutional character of transnational religious life. *Sociology of Religion, 65*, 1–18.

MacCulloch, D. (2012). The Anglican Church can start afresh. The *Guardian*. Available online at: www.guardian.co.uk/commentisfree/2012/mar/25/anglican-covenant-bishops-division

Naughton, J. (2007). *Following the money: Donors and activists on the Anglican right*. Retrieved June 2, 2012, from www.canticlecommunications.com/bios.aspx

Orombi, H. L. (2008, August 1). The church cannot heal this crisis of betrayal. The *Times*, p. 24.

Roberts, B., & Reddy, V. (2008). Pride and prejudice: Public attitudes toward homosexuality. *Human Sciences Research Council Review, 6*(4). Retrieved June 2, 2012, from www.hsrc.ac.za/HSRC_Review_Article-121.phtml

Rubenstein, M. J. (2004). An Anglican crisis of comparison: Intersections of race, gender, and religious authority, with particular reference to the Church of Nigeria. *Journal of the American Academy of Religion, 72*, 341–365.

Rubenstein, M. J. (2008). Anglicans in the postcolony: On sex and the limits of Communion. *Telos, 143*, 133–160.

Sadgrove, J., Vanderbeck, R. M., Ward, K., Valentine, G., & Andersson, J. (2010). Constructing the boundaries of Anglican orthodoxy: An analysis of the Global Anglican Future Conference (GAFCON). *Religion, 40*(3), 193–206.

Sadgrove, J., Vanderbeck, R. M., Andersson, J., Valentine, G., & Ward, K. (2012). Morality plays and money matters: Towards a situated understanding of the politics of homosexuality in Uganda. *Journal of Modern African Studies, 50*(1), 103–129.

Sheringham, O. (2010). Creating 'alternative' geographies: Religion, transnationalism and everyday life. *Geography Compass, 4*, 1678–1694.

Te Paa, J. (2009, July 12). Jenny Te Paa addresses the House of Deputies, *Episcopal Life Online*. Retrieved June 2, 2012, from www.episcopalchurch.org/107152_112408_ENG-HTM.htm

Valentine, G., Vanderbeck, R. M., Sadgrove, J., Andersson, J., & Ward, K. (2013a). Transnational religious networks: Sexuality and the changing power geometries of the Anglican Communion. *Transactions of the Institute of British Geographers, 38*(1), 50–64.

Valentine, G., Vanderbeck, R. M., Sadgrove, J., Andersson, J., & Ward, K. (2013b). Producing moral geographies: The transnational dynamics of homophobia within a global faith network. *Geographical Journal, 179*(2), 165–176.

Vanderbeck, R. M., Valentine, G., Ward, K., Sadgrove, J., & Andersson, J. (2010). The meanings of communion: Anglican identities, the sexuality debates, and Christian relationality. *Sociological Research Online, 15*(2). Retrieved June 2, 2012, from www.socresonline.org.uk/15/2/3.html

Vanderbeck, R. M., Andersson, J., Valentine, G., Sadgrove, J., & Ward, K. (2011). Sexuality, activism, and witness in the Anglican Communion: The 2008 Lambeth conference of Anglican Bishops. *Annals of the Association of American Geographers, 101*, 670–689.

Ward, K. (2006). *A history of global Anglicanism*. Cambridge: Cambridge University Press.
Williams, R. (2009). *A message from the Archbishop of Canterbury on the Anglican Communion Covenant*. Retrieved June 2, 2012, from www.archbishopofcanterbury.org/articles.php/1504/
Williams, R., & Sentamu, J. (2011). *The Church of England and the Anglican Church in North America*. Available at: http://churchofengland.org/media/1389262/gs%20misc%201011%20 -%20acna.pdf

Chapter 173
Religion and State in Marriage, Cohabitation and Civil Partnership: Examples, Typologies and Contestations from the United Kingdom

Paul G. Weller

173.1 Introduction

This chapter explores a number of "case studies" in the relationships between the state, marriages, cohabitation and civil partnership as *points of intersection* between religious tradition, personal belief and commitment, familial and social belonging, the conditions of modernity, and the structures and laws of modern bureaucratic states.

The examples are presented with special reference to the United Kingdom, but also as having resonance with, and potential significance for, similar debates in other parts of the world. In terms of its focus, the chapter seeks especially to illuminate the variety of *religious* perspectives on these matters – not because other perspectives do not matter, but because in more secularizing societies, the religious perspectives are least well understood and are most likely to be characterised by simplistic stereotypes. The chapter is therefore written from a Religious Studies perspective, but drawing in an inter-disciplinary way also on the contributions of the disciplines of theology, history, law and aspects of the social sciences.

173.2 Terminologies

The intersections that are explored in this chapter cluster around key terminologies such as "religion," "state," "marriage," "cohabitation" and "civil partnership." All of these words – and also the social realities to which they refer – come loaded with

P.G. Weller (✉)
Research, Innovation and Academic Enterprise, University of Derby, Derby DE22 1GB, UK

Oxford Centre for Christianity and Culture, University of Oxford, Oxford, UK
e-mail: p.g.weller@derby.ac.uk

© Springer Science+Business Media Dordrecht 2015
S.D. Brunn (ed.), *The Changing World Religion Map*,
DOI 10.1007/978-94-017-9376-6_173

inherited meanings that are located in, and formed by, specific social, historical, political, legal and religious contexts, all of which, of course, are also changing.

Although a word that is often used popularly in an unproblematised way, both as a word and also in terms of the social realities to which it points, "religion" is itself a far from self-evident concept. But as a "working definition," one which the present author (Weller 2003: 66) has proposed elsewhere is that "religion" is perhaps best understood as:

> ...a way of living in which some form of identification (either in a weaker and more general sense, or in a stronger and more specific sense of alignment with particular movements, communities and/or organisational forms) is often (though not always or necessarily) to be found in conjunction with different forms of 'believing' (in various combinations of certain values, ideals and doctrines) and can be expressed through 'practice' (that is related to shared symbols, rituals, observances and ethical orientations).

As used in this chapter, "state" is intended to refer to a particular formation of social and political life that is typical of the condition of modernity. Even where a state may have a strong individual in a leading position, because of its bureaucratic structure it is fundamentally different from the earlier social and political formations of empires or kingdoms which are characterized by more personalised systems of allegiance.

"Marriage" is a terminology used for the publically, and often legally, contracted and regulated form of personal, social and economic relationships between individuals, in which either religion and/or the state become involved in its wider recognition, while "cohabitation" is more technical terminology that is generally used in contemporary English to signify the relational personal, social and economic arrangements of those who, in more popular parlance are described as "living together" – in other words who have set up home together not on the basis of the institution of marriage, but in relation to which, nevertheless, some form of de facto legal recognition can often accrue.

"Civil Partnership" is a phrase that is increasingly used in contemporary English to signify the relatively newer forms of public and legal recognition that pertain to a relationship between men or women of the same sex, and which in all public and legal aspects is parallel to marriage in terms of the legal rights, obligations and responsibilities that accrue to it, but where the terminology of marriage remains reserved to the publically contracted and regulated form of relationship between a man and a woman.

The changing context for these terminologies and the social realities to which they refer is that of a social and legal evolution in which the wider *Gesellschaften* of modern civic societies have emerged out of the *Gemeinschaften* of more organically formed pre-modern communities, in which religion often played a socially constitutive role. As a result of this, what we are dealing with are the notions and practices of "marriage," "cohabitation" and "civil partnership" as *points of intersection* between religious tradition, personal belief and commitment, familial and social belonging, the conditions of modernity, and the structures and laws of modern bureaucratic states (see Berman 1974; Mitchell 1967; Edge 2006).

In a context in which perhaps the majority of the public media and political discussion of "marriage," "cohabitation" and the "civil partnership" takes place primarily from a legal, social and political perspective, it is the argument of this chapter that it is also important to consider the perspectives of a range of key world religious traditions, their teachings and perspectives on marriage, human sexuality and related matters (see Parrinder 1980; Browning et al. 2009) in a deeper and properly rounded – and not simply a popular or superficial – understanding of these religious perspectives.

173.3   Broad Historical Perspectives on Marriage

One of the failings in contemporary discussion of marriage is an assumption that modern and particular nation-state forms of this have been the norm throughout time and space when clearly they have not. Apart from the broad social, cultural and religious differences around marriage being based on monogamy, polygamy or polyandry, even within each of these basic forms of "marriage," there have been substantial differences when considered throughout time and space (see Dowell 1990).

Thus, in England and Wales, it was only following the 1836 *Marriage Act* that civil marriages were recognized as a legal alternative to church marriages. Until then, marriage in England and Wales had been a matter for ecclesiastical law and practice which, due to the particular form of close relationship between the state and the Church of England, known as "establishment," was also a recognised part of the law of England (see Hastings 1991; Morris 2009). In other words, until then the only legally recognized "marriage" was what might be called "religious marriage," the records of which we kept as part of church parish registers.

At the same time, much earlier in English history, before parish records became established and prior to there being priests located in a settled and comprehensive parish system, religious marriage was often a later recognition through the celebration of Christian rites conducted by a visiting cleric, of what had earlier been a de facto union of people, sometimes constituted on the basis of earlier pre-Christian traditions.

173.4   Of Gretna Green, Plural Jurisdictions and Shared Polities

In focusing on the context of the country from which the author comes – namely the country that is officially known as the United Kingdom of Great Britain and Northern Ireland – it should be noted that in seeking to understand this context in relation to the topic under consideration, the key word is "united" (even if often more in aspiration rather than reality!) rather than "unitary." In other words, while many people outside of the United Kingdom (UK for short) tend to think of it as a single social, legal and political entity it is, in fact, *not* that – or at least not without some degree of qualification. Because of the diverse English, Welsh, Scottish and Irish national traditions that are a part of its history, and even more so because of recent developments in political devolution, it is not a unitary "nation state" but rather might better be characterized as what the author has elsewhere (Weller 2005) called a "four-nations-state."

To take one example related to the theme of this book: within the UK, in the Scottish area of Dumfries and Galloway, is located the Scottish border village of Gretna Green. This was a village that became famous because it became associated with the phenomenon of what came to be known as a "Gretna Green marriage." This was because it was to the village of Gretna Green, within the jurisdiction of Scots law that, in the English language of times past, young couples from England "eloped" in order to get married according to the possibilities presented by Scottish law.

This happened because in England, following the introduction of Lord Hardwicke's *Marriage Act* of 1753, parental consent was required in order for people under 21 years of age legally to marry. By contrast, in Scotland at the time, it was possible, without parental consent, for boys to marry at the age of fourteen and girls at the age of twelve. Scots law also provided for what were called "irregular marriages." In these, if a declaration was made before two witnesses, almost anybody could conduct a ceremony of marriage. Thus the blacksmiths of Greta Green became involved in officiating at marriage ceremonies that were described as having been conducted "over the anvil," a role which popularly became known as that of "anvil priests." Therefore, in order to circumvent the constraints that English law placed on the passions of youth, many young people took advantage of the legal pluralism of the United Kingdom and "eloped" to get married in Scotland, with Gretna Green being the first village that many encountered over the Scottish border. And, in fact, although Gretna Green is the most famous instance of this, in fact a number of other Scottish border villages also served in this way including Coldstream Bridge, Lamberton, Mordington and Paxton Toll.

In time, the contrast between English and Scottish jurisdictions became less extreme, and from 1856 onwards, Scots law was changed to require 21 days of residence before marriage could take place under its provisions. Today, in England Wales, the legal age for marriage without parental consent is 18, but with parental consent, it can be 16. Under the *Age of Marriage Act, 1929*, marriage between persons either one of whom was under sixteen years old was made void. But in contrast to England, in Scotland parental permission has not subsequently been required for those marrying between the ages of sixteen and eighteen.

Gretna Green remained one of the world's favorite locations for marriages. Furthermore, the terminology of a "Gretna Green marriage" became widespread to describe a marriage that takes place in a different jurisdiction to that from which the parties to the marriage come, and in order for them to avoid the restrictions that exist in their home jurisdiction(s). This tradition, of course, can also be found in the USA in the famous example of Las Vegas in Nevada, which is now known as the "Wedding capital of the world" and has no required waiting period before which a marriage can be contracted.

At first sight, all of this might be seen as of purely curiosity value to our main theme except that it was to Scotland that a member of the British Royal Family, Princess Anne, the Princess Royal, went in 1992 (to Crathie Kirk, near Balmoral Castle) in order to be able to be married in church to Timothy Laurence. This was because she had previously been divorced from her first husband Captain Mark Phillips, and a second marriage in church was not possible according to the rites of the "established" Church of England. However, it was possible in the different (albeit also to some extent "established," but in the case of Scotland, Presbyterian, rather than Anglican) Church of Scotland!

These opening examples about the relationships between the state, religion and marriage in are in some ways highly particular to the UK, consequent upon the particular constitutional arrangements that ensued from the 1603 union of the English and Scottish Crowns and the 1707 union of the Parliaments. But perhaps these particularities could also act as a wider reminder that, in approaching the topic of "state and religion: marriage, cohabitation and civil partnership" it is not, in fact, *necessary* – as may be often assumed – for an individual polity to have a singular and exclusive operative model.

173.5   Archbishop Rowan Williams and the Shariah Controversy

This chapter now moves on from history to the contemporary context of what the author (Weller 2005) has elsewhere called a "three dimensional" society. That is to say to a contemporary society that I argue is characterized by its long *Christian inheritance*; an inheritance that is modified by the *secularity* of the Enlightenment and accelerating in the nineteenth and twentieth centuries; and which, in the latter part of the twentieth and early part of the twenty-first centuries, is also now also changed by its increasing *religious plurality* that is largely consequent upon migration of substantial numbers of Muslims, Hindus, Sikhs and others, as well as patterns of adoption by indigenous people of these and other world religious traditions, especially Buddhism (see Weller 2007; Weller 2008).

Within this "three dimensional" context for religion, state and society, a specific "cameo" will be explored around the issues under consideration in this paper. This relates to a controversy that, in February 2008, blew up around Dr. Rowan Williams, the then (Anglican) Archbishop of Canterbury, and which related to a lecture (Williams 2008a) and associated radio interview (Williams 2008b) that he gave on the topic of "Civil and Religious Law in England – a Religious Perspective."

The lecture was given to around 1,000 people at the Royal Courts of Justice, as one of a series of lectures given by senior Muslims, lawyers and Christian theologians on the broad topic of "Islam in English Law." The controversy that ensued was concerned with the wisdom and practicality of at least what was being reported to have been the Archbishop's approach to the issue of so-called "supplementary" or "overlapping" jurisdictions rooted in religious perspectives, customs and law – and very specifically in relation to incorporation into civil law of aspects of the Islamic Shari'ah of Muslims.

While the specifics of this particular controversy belong to the UK, and more particularly to England, some of the contours of the controversy might usefully serve as a mirror within which related questions and issues in other societies can more clearly be reflected by means of comparison and contrast. In Williams' lecture, he explained its purpose as being "… to tease out some of the broader issues around the rights of religious groups within a secular state, with a few thoughts about what might be entailed in a creating a just and constructive relationship between Islamic law and the statutory law of the United Kingdom."

Within these terms of reference, it is significant that the controversy that was generated from this focussed almost exclusively on the actually very tentatively phrased *questions* that Rowan Williams raised about what *possibilities* there might be for recognizing aspects of Shari'ah law as being something which Muslims might properly elect to have as their agreed basis for dispute resolution in a limited number of personal matters, including marriage. In exploring such possibilities, Williams drew upon the work of the Jewish legal theorist Ayelet Shachar and her book on *Multicultural Jurisdictions: Cultural Differences and Women's Rights* who argued that we need to "work to overcome the ultimatum of 'either your culture or your rights'" (Shachar 2001: 114). In exploring the theme, Williams (2008a) referred to examples of so-called "supplementary jurisdictions" or "overlapping jurisdictions." Of these examples he said that:

> …both jurisdictional stakeholders may need to examine the way in which they operate: a communal/religious nomos, to borrow Shachar's terminology, has to think through the risks of alienating its people by inflexible or over-restrictive applications of traditional law, and a universalist Enlightenment system has to weigh the possible consequences of ghettoising and effectively disenfranchising a minority, at real cost to overall social cohesion and creativity. Hence 'transformative accommodation:' both jurisdictional partners may be changed by their encounter over time, and we avoid the sterility of mutually exclusive monopolies.

The fact that the controversy that developed around Williams' lecture centred upon the Islamic-related aspects of what was declared to be a lecture about broader matters is not accident and needs to be seen against the background of the way in which – due to the forms of its claimed implementation in a number of countries – what is the historically rich and complex notion of Islamic Shari'ah tends to be viewed in the West. As Tariq Ramadan (2004: 31) put it in his book on *Western Muslims and the Future of Islam*: "… the idea of Shariah calls up all the darkest images of Islam. It has reached the extent that many Muslim intellectuals do not dare even to refer to the concept for fear of frightening people or arousing suspicion of all their work by the mere mention of the word."

As an academic theologian as well as a religious leader, the Archbishop's thinking had been set out as a characteristically dense and in many places highly qualified paper. But in the world of "twenty-four hour news" and "media sound bites" what occurred was a "media event" in the Baudrilliardian sense. So in looking at the controversy that followed his lecture, and a BBC radio interview about it, it is important to distinguish between what the Archbishop *actually* said; his declared *intentions*; and then the way in which the controversy generated a life of its own that was more or less strongly linked with either the original text of his lecture or the radio interview based upon it.

The nature and extent of the reactions that were evoked can be seen in the kind headlines given to the story by the national British newspapers (see Paulli 2008). These ranged from the rather more measured but nevertheless wildly inaccurate *Guardian Unlimited* on-line edition headline of 5th February 2008 of, "Let shariah rule, says Archbishop" to the "What a burkha!" exclamation from *The Sun* newspaper. As the on-line *Guardian Unlimited* newspaper article by Paulli went on to point out and to illustrate by quotations, while their way of expressing it varied, the press was more or less uniformly opposed to the Archbishop's suggestion, in a BBC interview on his lecture, that the adoption of some aspects of Islamic Shariah law in Britain was likely to be "unavoidable." They were also opposed to his entertaining the possibility of some measure of "plural jurisdiction" in which Muslims could decide that some matters of potential dispute – including aspects relating to marriage – might be resolved in either secular or Shari'ah courts. Thus *The Sun* editorialised that the Archbishop was not just a "batty old goat," but also that he was a "dangerous threat to the nation." To emphasize its point, it juxtaposed a picture of the London bus destroyed in the 7/7 bombings with its report calling Dr. Williams' comments "a victory for terrorism." Another popular newspaper, *The Express*, described the Archbishop as "a mediocrity and, as a self-confessed 'bearded Leftie'," while *The Daily Mail* suggested that he stick to tending his own flock.

Some of the broadsheet newspapers appeared to give more consideration to the content of what the Archbishop had said, rather than trading in insults. Thus the *Daily Telegraph* acknowledged that, as pointed out by Rowan Williams, Shari'ah Councils already exist in the UK and that Orthodox Jews have had limited recourse to the Jewish Beth Din, or religious courts, on matters relating to marriage and divorce. But *The Telegraph* went on to charge the Archbishop with "political ineptitude," and concluded that, "On this most inflammatory of subjects the archbishop would have best kept silent." *The Guardian* saw the Archbishop's comments as an illustrative point of collision between secular and religious perspectives. While accepting that the Archbishop was aiming for was some tolerance of the role of religion in public life, *The Guardian* argued that he had succeeded only in highlighting why it might be better for religion to be altogether excluded. *The Independent* wondered whether the Archbishop really meant to say what the papers were reporting. But it concluded that, "News has little room for the subtleties of academic gavottes around delicate subjects. A canny religious leader – or at any rate his press office – ought to know that."

What was very interesting in all the debate that went on is that there was very little reference to what might be seen as a secularized version of "supplementary jurisdiction" and which has increasingly emerged in relation to marital matters in the U.S. and has also spread from there more widely. This is the growth of "pre-nuptial agreements" by which marital parties set down in advance how their financial and other matters are to be dealt with in the event of the dissolution of their marriage. Of course, "pre-nuptial agreements" are also the subject of some controversy within and between different legal systems. But where they are at least partially countenanced by public and political opinion, and by legal practice, the question needs to be asked about how different these are from what might also be seen as a pre-nuptial agreement between individuals to have their marital affairs governed by the norms and values of their religious tradition.

Of course, in *both* instances there is the question of *how freely* the individuals involved were able to enter into such agreements and what effect power relations may have in this context. This issue is especially relevant in contexts where patriarchy may have resulted in substantial gender imbalance that can be reinforced by traditional religious practice (see Saghal and Yuval-Davis 1992). And of course, in some instances – such as the existence of domestic violence or the threat of it related to so-called honour-based systems – there is a real issue which needs to be dealt with as is underlined by the personal experience, practical work and research on this carried out by, among others, Jasvinder Sanghera (2007).

But today, generally speaking, domestic violence is clearly stated by the majority of religious leaders to be contrary to the principles of Islam and indeed of other religious traditions. It is also far from being the preserve of cultural or religious minority groups. Therefore the question has to be asked about how far those dangers differ fundamentally from the wider imbalance in power and social relations relating to gender that are embedded in our wider societies as the by-product of our capitalist and patriarchal social relations. This is not to justify the problem in either context. But one senses that it is possible that when religion (and especially Islam) is involved, a different kind of critique can emerge and about which the question should at least be asked to whether such critique might not partake of at least some aspects of Islamophobia (Runnymede Trust 1997; Runnymede Trust 2004).

173.6 Marriage in World Religious Traditions

Having briefly opened up some of the issues involved at the intersection between "religion, tradition and modernity" by reference to the Williams controversy, before looking at the typological models for the kind of issues involved it is important to provide at least a "flavor" of the principal meanings of marriage(s) in a range of world religious traditions.

Note that the title of Rowan Williams' lecture was "Civil and Religious Law in England – a *Religious* Perspective" (emphasis mine). This is important balancing corrective because the majority of social, legal and political discussion of "marriage," "cohabitation" and "civil partnership" – even where it is rejected – is conducted in relation to its bourgeois institutional embodiment. But as the nineteenth century Danish philosopher and theologian Søren Kierkegaard (1843) reminded us, there can be a serious tension (if not often an outright contradiction) between bourgeois views of morality and social institutions and full-blooded religious perspectives.

In UK (and other European and North American societies), the principal historical relationship in these matters has been forged between the Christian religion and the state. That said, in itself, it is not a simple thing, since the various traditions of Christianity themselves embody different understandings of marriage, and the mode of relationship to these varied understandings also varies according to the history of the societies concerned. Indeed, even when under the control of the Church and of ecclesiastical law and practice, the clergy's role in record-keeping of marriages held in churches across Europe played an important part in the emergence of the modern bureaucratic state.

However, while the bourgeois institution of marriage became elided with aspects of its religious meanings, to engage adequately with the problematics involved, it is important to take into account understandings of marriage that are informed by scriptural, traditional and theological imperatives and values, and not only those understandings that arise from the place that marriage has in the policies and practices of the modern bureaucratic state.

Within Christianity, for many centuries marriage was held in less regard than celibacy, until in the Reformation when it was argued that the marriage was an honorable "estate" through which service could be rendered to God as much as through priestly or monastic celibacy. But the kind of marriage then sanctioned was not necessarily that of the "romantic love match" variety that has become the norm in western societies. Rather, among the more privileged and powerful strata of societies, marriages were often arranged as an embodiment or social, political and economic considerations – elements of which remain to this day among the Royal Families of Europe.

Nevertheless, in broad terms some outlines can be delineated. Thus, there is the Catholic view of "marriage" as one of the sacraments of the Church – in other words as one means through which grace is conferred to believers. It is this "sacramental" nature of marriage which is at the root of it being understood as "indissoluble" because a "sacrament" is not only concerned with an agreement or contract between two people, but necessarily involves the divine. At the same time Catholic tradition and ecclesiastical law has always recognized the existence of legitimate grounds for the "annulment" of marriage – for example where no sexual consummation has taken place. The possibilities and limits for such "annulment," and power involved in the granting or withholding of it, of course lay at the root of the events around King Henry VIII of England's marital status and dynastic aspirations that ultimately resulted in the creation of what we know today as the Church of England (Furlong 2001).

By contrast, in broad terms the general Protestant Christian view has had a more "practical" perspective on marriage as linked with the channeling of sexual expression into the good ordering of society, coupled with a view of that relationship at once more "companionate" and "contractual." At the same time, this is not a "contractual" approach seen purely in terms of human agreement, as in civil contract law. Rather, it might perhaps more appropriately be thought of as a "covenantal" approach, in which human agreement is important, but is also seen as being informed and undergirded by the divine. In time, most Protestant Christian traditions came to recognize the possibility of divorce, and in many Protestant Churches it is possible for divorced people to remarry in church, while for the Catholic Church that still remains impossible.

These broad fundamental "Catholic" and "Protestant" orientations have, by direct influence *upon* UK, European and North American societies and their legal systems, as well as by indirect reactivity *against* the religious traditions concerned, profoundly shaped the approach to marital matters of these different societies and civil law systems. Thus in the inheritance of the Revolutionary and Republican tradition of France – which emerged reactively to what was seen as a despotic power of a Catholic religious tradition embodied in ecclesiastical structures, personnel and laws – "marriage" was established as something always to be contracted in a civil ceremony.

In such a system, religious believers may, of course, also elect to take part in a religious ceremony. And from the perspective of the believers concerned such a service may constitute the *real* substance of the marriage, in relation to which the civil proceedings are only a necessary legal form in order to give social effect to aspects of the wider recognition of what believers view as the core aspects of their marriage. But in this model, a clear "separation" between the religious and the civil is aspired to. And this is on the basis that such "separation" respects the differently constituted spheres of the "private" (seen as relating to religion) and the "public" (seen in terms of the obligations of citizenship, from which considerations of religion are to be excluded). This approach – which is also found in many other European societies and legal systems – argues that such "separation" is ultimately for the benefit both of the religions and of the societies as a whole.

However, alternative approaches exist, one of which is found in English law. In England legally recognized marriages can – and increasingly are – conducted by a local Registrar of Births, Marriages and Deaths who is a civil appointment and leads a civil ceremony to that end. But they can also be conducted – without need of an additional civil ceremony – by recognized religious functionaries as long as within the forms of service that these functionaries use, they also use the common minimum of legal words which imply consent.

In accordance with the special constitutional role in England of the Church of England as an established form of religion, priests of the Church of England are, by virtue of their ordained office, legally empowered to preside at marriage ceremonies, the legal effect of which is recognized by the civil law. Religious functionaries of other Christian traditions – and indeed representatives of other religious, and also humanist groups – can become "recognized persons" for the solemnisation of marriage. And in fact, a marriage binding in civil law can be conducted by a person, including a religious leader, who is not legally recognized, so long as it is carried out in the presence of a recognized person and in a place that is legally recognized for the purpose.

Such an approach as this represents a more intersectional approach to marriage, religion, society and law than that which is found in the French Republican tradition. The intention of this approach is to connect the civil law with the customary practice of religious groups, while also tying that customary practice into the requirements of basic legal minima for it to function in a way that can be recognized by the wider society.

In other European Union countries, there are other models of how the interplay between religion and tradition, state and modernity has played out in relation to marriage, cohabitation and civil partnership. While considerable complexity is already entailed in the Christian heritages of the UK and other European Union countries, these societies and legal systems are now having to wrestle with challenges arising from a far wider range of cultural and religious plurality than was the case in their histories until the twentieth century, and especially the last quarter of that century, including in relation to marriage (Prickett 1985).

Thus, as noted earlier in this chapter, and particularly arising from the labour migrations and refugee movements of population especially from the Two Thirds world, the religious landscape of most – although not yet all – European societies has been significantly transformed. Of course, that tension remains important in societies where Christianity is still the predominant religious tradition. But the rise of the secular tradition and the introduction of a wider degree of religious and cultural plurality alters the social and religious equation in important ways, and the social actors and movements that exist in the "three dimensional" configuration of Christianity, secularity and religious plurality are themselves both subjects within, and subject to, ongoing change with regard to the balance between these elements.

For example, because the secular tradition of Europe was in many ways reactive to the Christendom tradition, the civil law has had relatively little practical experience of addressing the more widely and diversely religious (with the important exception of in colonial societies). Of course, the partial exception to this in both UK and some other European Union societies relates to the relatively longer presence of Jewish minorities. This means that – notwithstanding the appalling history of anti-Semitism – ways have been developed of dealing with the different perspectives and approaches brought by Jewish tradition. And while the popular invocation of a "Judeo-Christian" tradition is usually inadequate to the specificity of each religion and tends to subsume the distinctively Jewish, there are some broad assumptions that are, at least in some measure, shared between Christian and Jewish perspectives. Because of this, the binary relationship of the "two-dimensional" inheritance of the nineteenth and early twentieth centuries was not so fundamentally challenged until societies become informed by an even wider religious and cultural diversity.

For example, while polygamy was practiced in Jewish history, for many centuries it has not been the form of Jewish "religious marriage." With the historical exception of the Church of Jesus Christ of Latter-day Saints (Davies 2003), and among some contemporary Mormon and other smaller groups, polygamy has also generally not been recognized as a Christian form of "religious marriage." At the same time, among some Christians in Africa, there has been a practical and pragmatic recognition of pre-existing polgymous relationships as being a "lesser" evil than the wives being disowned in a context where they and their children have been economically dependent upon their husband.

But in African indigenous religious traditions, polygamy has been widely practiced, including in modern times, as also among Muslims where it is permitted, rather than recommended, and the requirements associated with it are set very highly for the husbands involved. There are also, of course, societies such as the predominantly Hindu Nepal, in which polyandry has been and still is practiced. Against this background it is interesting to note that, with one important exception, even at its most reactive to the Jewish and Christian heritage (as in the self-consciously anti-religious traditions of Marxism-Leninism) the secular tradition of Europe has reflected a general assumption about marriage as being a monogamous relationship between a male and a female, albeit that the relative ease of divorce has created a situation which has sometimes been described as one of "serial monogamy."

173.7 Same Sex Civil Partnerships and Towards Same Sex Marriage

A more recent exception to the general assumption about monogamous relationships between a man and a woman has been the debates and legal developments that have come to the fore in European societies, in which ways have been sought legally to recognize relationships between people of homosexual and lesbian sexual orientation. Thus the question of what are often popularly called "gay marriages" poses in sharp form the problematic relationship between "religious marriages" and civil contracts by which individuals ally their lives and property.

In the first instance, although the term "gay marriages" is often popularly used without distinction between the various possible forms of legal recognition for homosexual partnerships, in the United Kingdom, at the time of writing, the legal recognition of "marriage" between persons of the same sex is not yet possible, though legal recognition of their partnership is.[1] Thus the *Civil Partnership Act*, 2004, gave to same-sex couples a range of rights and responsibilities that were comparable to those that adhere to heterosexual couples in civil marriage.

Under this Act, those who are "civil partners" are therefore entitled to the same property rights; the same exemptions on inheritance tax, social security

---

[1] Since the original writing of this chapter, same-sex marriage has become possible, but as with historic law on marriage, differentially so within the various jurisdictions of the United Kingdom. For England and Wales, under the *Marriage (Same Sex Couples) Act, 2013* and which came into force on 13 March 2014, and the first same-sex marriages took place on 29th March 2014. In Scotland, in February 2014 the Scottish Parliament passed the *Marriage and Civil Partnership (Scotland) Act, 2014* and the first same-sex marriages are expected there by the end of 2014. The Northern Ireland Executive has stated that it does not intend to make same-sex marriages possible there, and in Northern Ireland, same-sex marriages contacted in other jurisdictions are treated as civil partnerships.

and pension benefits; and also the same ability to get parental responsibility for a partner's children, as are same-sex couples in civil marriage. They also have responsibility for reasonable maintenance of their partner and their children and they have next-of-kin rights in hospitals. For the formal dissolution of "civil partnerships" there is a process that is similar to that of the possibility of divorce for married same-sex couples.

Since the *Equality Act*, 2010, in England and Wales same-sex couples have been able to enter into a "civil partnership" in the context of a religious building and ceremony if those responsible for the building are willing to conduct a ceremony of this kind. However, at the time of writing the debate has moved on to whether legally recognized "marriage" should be made available to same sex couples who wish it and the UK Government has been conducting a formal consultation (Government Equalities Office 2012) on proposals to enable this. Such marriage is, at the time of writing, already available in the following jurisdictions: elsewhere in the world, namely: Argentina, Belgium, Canada, Iceland, Mexico (Mexico City only but recognized throughout Mexico), Netherlands, Norway, Portugal, South Africa, Spain, Sweden, United States of America (some states only).

The UK Government proposals include those of enabling same-sex couples to have a civil marriage (in other words only civil ceremonies in a register office or approved premises, such as a hotel); to make no changes to religious marriages in the sense that no religious organisation would be forced to conduct same-sex religious marriages as a result of the proposals; that civil partnership registrations on religious premises would continue on the basis that they are currently possible – in other words on a voluntary basis for faith groups; that civil partnerships for same-sex couples would be retained as well couples already in a civil partnership being allowed to convert that into a marriage; and finally that, for the first time, individuals would be able legally to change their gender without having to end their marriage.

Although the proposals do not require any religious body to conduct either a civil partnership or a marriage between persons of the same sex, they have caused considerable controversy among religious bodies. In particular, bishops in the Roman Catholic Church have taken a strong line in arguing, also in relation to the wider society beyond the specific sacramental practice of the Church, that marriage by its nature is necessarily between a man and a woman and so cannot take place between persons of the same sex. In Scotland, this went so far as to lead to Cardinal O'Brien, the leader of Scotland's Catholics, suspending direct communication on the matter with the Scottish Government (BBC News Scotland 2012).

In relation to the debates around either marriage or civil partnerships between men or between women, it is important to note that some religious people who may morally be opposed to homosexual lifestyles can nevertheless be accepting or even supportive of the state recognizing "civil partnerships" for gay and lesbian people. However, at the same time from the perspective of their religious beliefs they find problematic an equivalence of language in which "gay marriage" rather than "civil partnership" is referred to. In fact, within religious communities themselves, the issues involved in relation to "gay partnerships" have proved to be a lightening rod in relation to tensions and conflicts involving religion(s), tradition and modernity.

These issues involved in these debates can be found within the history (for Christianity, see Boswell 1980) of scriptural hermeneutics and contemporary practice in all religions, but have been played out particularly publically within Christianity and especially within the established Church of England where the issues involved have become so contentious that they have threatened to bring about a split in the Church (see Bates 2005), as also in the wider global Anglican Communion.

Within the wider society, a unified approach to human rights law, policy and practice such as that which is embodied in the UK's single *Equality Act*, 2010 and is under girded by the work of the Equality and Human Rights Commission, on occasion brings rights to religious freedom, belief and conscience into contention and sometimes conflict with rights relating to sexual orientation and practice. For example, even prior to the most recent *Equality Act*, 2010, under the *Equality Act (Sexual Orientation) Regulations*, 2007, there had been a big public, political and religious debate around the previous practice of some religiously-based adoption agencies (particularly those with a Roman Catholic Christian foundation) which had previously had a policy of not placing adoptees with same sex couples. At the same time, not all religiously-related agencies had that policy. This resulted in nearly all such agencies that did not close on the basis that they did not want to comply with a law that they felt impinged upon their religious freedom not to implement same-sex adoption, severing their formal ties with the Catholic Church in order to continue their work.

An exception to the general pattern in this regard has been the Catholic Care agency which serves the Catholic Dioceses of Leeds, Middlesbrough and Hallam. This particular agency has continued a legal battle with the UK Charity Commission in relation to interpretation of the Regulations and the Act, and the provision within them made for limited discrimination to be allowed when in pursuit of specific charitable objectives. The debates involved have been seen by many as a litmus test case in relation – according to one's perspective – to either the importance of religious freedom or the unfair privileging of religious exemptions relative to generally enforceable equalities law.

173.8 "Marriage, Cohabitation and Civil Partnership" as Religion/State Typologies

By reference to a range of different states and societies, the question of how "marriage," "cohabitation" and "civil partnership" is handled in relation to "religion, tradition and modernity" could be considered as both illustrative and reflective of a number of broader "typologies" in the relationship between religion, modernity, law, state and society. In his book on *The Limits of the Law*, the legal academic Antony Allott (1980) helpfully identified a range of such models (*italics* mine for emphasis) as being of:

1. *fusion*, where law and religion are seen as part of the same cultural context, and there is no structural separation between them; such is true of what one may call the "integrated societies;"

2. *infusion*, where – without there being such structural fusion between the law and religion – a ruler or other promulgator of laws seeks to make them conform to his own religious perceptions;
3. *co-ordination*, where the legal system and the religious system exist on equal and parallel lines;
4. *subordination*, where religion is subordinated to law and controlled by it – one might call this the politicization of religion;
5. *toleration*, where the law tolerates, without being either inspired by or seeking to reinforce, the dictates of religion;
6. *suppression*, where the law is inimical to religion and purports to outlaw it.

There are, of course, other possible typologies and variants on these and, in concluding, it needs to be emphasized that the nature of models that they *are* models and so they never entirely reflect empirical realities. Arguably, most social, legal and state systems include elements of more than one model. Because of this, and well as because of the variety of perspectives that will be espoused by individuals and groups, states and societies, it is unlikely that what emerges from an analysis of will be *answers* of "Yes" or "no" to particular models. Rather, what might be possible could be to reach at least some agreement about *questions* around "*how far*" it may be appropriate to apply particular models, and about to what extent, and in what ways, such models might need *modification*?

Such questions involve those about how far the state can and/or should impose unitary legal and social policy frameworks without exception; about how far exceptions to basic approaches can be accepted and implemented for particular cases without complete fragmentation of approach; and about the scope, in modern societies, that there could be for the co-existence of plural approaches within single polities?

173.9   "Religious Literacy," "How Far?" Questions, and the Way Forward

If such "*how far*" questions are likely to constitute the most balanced and appropriate approach to the issues involved in the relationship between "religion, modernity and tradition", then an important corollary for public officials and systems (including those concerned with the development and administration of the civil law) is what, in the UK, we are increasingly referring to as the need for "religious literacy."

Note that what is referred to here is "religious *literacy*" and not "*expertise.*" This is because what is needed from public officials is a *sufficiency* of religious literacy relative to the practical and policy needs of their responsibilities. At the same time, it should be noted that "religious literacy" is not something that is concerned purely with "*information,*" or only with knowledge *about* religions. While not implying the necessity of religious belief or commitment, "religious literacy" does imply some degree of "*sensitivity*" to religions and a "*feel*" for them in a way that goes beyond the mere accumulation of factual knowledge. In other words, "religious literacy" is concerned with the development of a proper *understanding* of how religions function for individuals and for the groups who seek to live by them.

The purpose of advocating such "religious literacy" is not in order to argue that social policy and legal developments should necessarily *endorse* religious perspectives. But it is to argue that developments in law and social policy should, as a minimum, establish a *connection with* the realities by which individuals and groups live. Law cannot and does not exist in a form abstracted from the concrete realities of the societies in which it operates. And for the effective operation of the law in a democratic society – where both some degree of both consensus and consent is needed – there is a need for an at least reasonable degree of understanding of the values, mores, guidelines and imperatives that shape the ways individuals and groups inspire, order and regulate their own lives either alongside of apart from the state and its legal systems.

Thus public servants and legal systems should not pretend that these alternative realities do *not* exist. In addition, while understanding *facts* about them, they should not fail to appreciate the *significance* of these religious perspectives for those who live by them, since while religious believers who are citizens are likely respect the civil law, they do not understand systems of public law as their ultimate points of reference. Rather, that ultimacy is reserved for that to which their religious traditions point.

In conclusion, when considering these difficult questions, it is important that "how far" questions remain just that in order that a dethroned absolutism of religion is not replaced with an absolutism of law. This is because the specific social and historical conditions of the evolution of law are always likely to be culturally specific, limited, and therefore potentially exclusive of minority cultural approaches within a pluralistic society. By contrast, a "how far" approach leaves the evolution of law and of social policy open to democratic engagement and provisional outcomes by means of which individuals and groups contract to live together within the intersection between religion, tradition and modernity in the context a framework of reasonable stability and *for the time being*. Such an approach is, of course, a work in progress and not a solution of perfection. But then the empirical history of attempts at perfection, whether of a religious or a secular character, is one in which such attempts have at best also been found wanting and, at worst, have been detrimental to human rights.

Acknowledgements By kind permission of the conference organisers, this chapter draws upon work presented "Religion and State in Marriage, Cohabitation and Civil Partnership: Examples, Typologies and Contestations from the United Kingdom" at the Conference on: "Family as a Value: In Religion, Tradition And Modernity," 26th–27th November 2010 organized by the Journalists and Writers Foundation, Turkey at the Rixos Tekirova Hotel, Antalya, Turkey and presented to the conference as a conference paper on: "Religion and State in Marriage, Cohabitation and Civil Partnership: Examples, Typologies and Contestations from the United Kingdom" in a pre-conference volume (2010) on *International Conference on Family as a Value in Terms of Religion, Tradition and Modernity, 26–27th November 2010, Rixos Tekirova-Antalya*, pp. 397–412. That paper, in turn, drew on work originally undertaken in preparing for a paper on "The State and Religious Marriages in European and North American Societies: Examples, Comparisons, Typologies and Implications" that was presented at the colloquium on: "The State and Religious Marriages: Confrontation and Cohabitation," 12–13th February 2009, for the Research Group Société, Droit et Religions de L'Université de Sherbrooke (SoDRUS), Sherbrooke, Quebec, Canada, participation in which was made possible by the award of a British Academy Overseas Conference Grant.

References

Allott, A. (1980). *The limits of the law*. London: Butterworths.

Bates, S. (2005). *A church at war: Anglicans and homosexuality*. London: Hodder and Stoughton.

BBC News Scotland on-line. (2012, August 19). *Cardinal Keith O'Brien snubs gay marriage talks with Scottish government*. www.bbc.co.uk/news/uk-scotland-19305232. Accessed 28 Oct 2012.

Berman, H. (1974). *The interaction of law and religion*. London: SCM Press.

Boswell, J. (1980). *Christianity, social tolerance and homosexuality: Gay people in western Europe from the beginning of the Christian era to the fourteenth century*. Chicago: University of Chicago Press.

Browning, D., Christian Green, M., & Whittle, J. (2009). *Sex, marriage, and family in world religions*. New York: Columbia University Press.

Commission on British Muslims and Islamophobia. (1997). *Islamophobia: A challenge for us all*. London: Runnymede Trust.

Commission on British Muslims and Islamophobia. (2004). *Islamophobia: Issues, challenges and action*. Stoke-on-Trent: Trentham Books.

Davies, D. (2003). *An introduction to Mormonism*. Cambridge: Cambridge University Press.

Dowell, S. (1990). *They two shall be one: Monogamy in history and tradition*. London: Collins.

Edge, P. (2006). *Religion and law: An introduction*. Aldershot: Ashgate.

Furlong, M. (2001). *C of E: The state its in*. London: Hodder and Stoughton.

Government Equalities Office. (2012, March). *Equal civil marriage: A consultation*. www. homeoffice.gov.uk/publications/about-us/consultations/equal-civil-marriage/consultation-document?view=Binary. Accessed 28 Oct 2012.

Hastings, A. (1991). *Church and state: The English experience*. Exeter: University of Exeter Press.

Kierkegaard, S. (1843, this translation 1992). *Either/or: A fragment of life*. London: Penguin.

Mitchell, B. (1967). *Law, morality and religion in a secular society*. Oxford: Oxford University Press.

Morris, R. M. (2009). *Church and state in 21st century Britain: The future of church establishment*. Basingstoke: Palgrave Macmillan.

Parrinder, G. (1980). *Sex in the world's religions*. Oxford: Oxford University Press.

Paulli, M. (2008, February 8). *Let Shariah rule, says archbishop, "Guardian Unlimited"*. www. guardian.co.uk/news/2008/feb/08/wrap.michellepauli. Accessed 28 Oct 2012.

Prickett, J. (Ed.). (1985). *Marriage and the family: Living faiths series*. Cambridge: Lutterworth Press.

Ramadan, T. (2004). *Western Muslims and the future of Islam*. Oxford/New York: Oxford University Press.

Saghal, G., & Yuval-Davis, N. (1992). *Refusing holy orders: Women and fundamentalism in Britain*. London: Virago.

Sanghera, J. (2007). *Shame*. London: Hodder and Stoughton.

Shachar, A. (2001). *Multicultural jurisdictions: Cultural differences and women's rights*. Cambridge: Cambridge University Press.

Weller, P. (2003). The dimensions and dynamics of religious discrimination: Findings and analysis from the UK. In N. Ghanea (Ed.), *The challenge of religious discrimination at the dawn of the new millennium* (pp. 57–81). Leiden: Martinus Nijhoff Publishers.

Weller, P. (2005). *Time for a change: Reconfiguring religion, sate and society*. London/New York: T & T Clark.

Weller, P. (Ed.). (2007). *Religions in the UK: Directory, 2007–10*. Derby: Multi-Faith Centre at the University of Derby and University of Derby.

Weller, P. (2008). *Religious diversity in the UK: Contours and issues*. London/New York: Continuum.

Williams, R. (2008a, February 7). *Civil and religions law in England – A religious perspective*. www.archbishopofcanterbury.org/articles.php/1137/archbishops-lecture-civil-and-religious-law-in-england-a-religious-perspective. Accessed 28 Oct 2012.

Williams, R. (2008b, February 7). *Interview with Christopher Landau of the BBC Radio 4 "World at One" programme, on the Temple lecture "Civil and religious law in England: A religious perspective"*. www.archbishopofcanterbury.org/articles.php/707/archbishop-on-radio-4-world-at-one-uk-law-needs-to-find-accommodation-with-religious-law-codes. Accessed 28 Oct 2012.

Chapter 174
Religion and Attitudes Towards Gay Rights in Northern Ireland: The God Gap Revisited

Bernadette C. Hayes and Lizanne Dowds

174.1 Introduction

Disputes over gay rights policies have occupied a central place on both national and international political agendas over the last three decades. Political institutions ranging from the U.S. Congress to the British House of Commons to the European Parliament and, more recently to governments in a number of African states such as Uganda and Malawi, have all been drawn into the fray. In fact, some commentators go so far as to suggest that public disputes and related court battles over homosexual rights have now replaced abortion as the most significant and divisive moral issue within many contemporary western industrialized societies (see Wilcox 2007; Rayside 2007; Badgett 2009). This is particularly the case in the U.S. where the question of gay rights continues to dominate the political landscape (see Wilcox et al. 2006; Rimmerman and Wilcox 2007; Whitehead 2010). Recent legislative activity suggests that not only will the issue of gay rights remain on the public agenda for some time to come, as pro- and anti-gay lobbyists seek to advance their various positions, but this appears to be particularly the case when the question of gay marriage is considered.[1]

---

[1] For example, subsequent to the passage of the Defence of Marriage Act (DOMA) in 1996 – it defines marriage as "a legal union between one man and one woman" – no less than 30 states introduced ballots to ban same-sex marriages (see McVeigh and Diaz 2009). However, in November 2012, three states – Maine, Maryland and Washington – approved of same-sex marriage via a popular vote. Previously, six other states – New York, Connecticut, Iowa, Massachusetts, New Hampshire, Vermont – and the District of Columbia had passed legislation to allow same-sex marriage, although all were enacted either by legislators or court rulings.

B.C. Hayes
Department of Sociology, University of Aberdeen, Aberdeen AB24 3QY, Scotland, UK
e-mail: b.hayes@abdn.ac.uk

L. Dowds (✉)
Northern Ireland Life and Times Survey, University of Ulster, Belfast, UK
e-mail: lizanne.dowds@ntlworld.com

© Springer Science+Business Media Dordrecht 2015
S.D. Brunn (ed.), *The Changing World Religion Map*,
DOI 10.1007/978-94-017-9376-6_174

While the issue of gay rights has not generated as much legislative activity across Europe, few would dispute that important legal changes have occurred in this area in recent years, particularly in relation to gay marriage and civil partnerships (see Takacs and Szalma 2011). For example, currently, six European nations – the Netherlands (2001), Belgium (2003), Spain (2005), Norway (2008), Sweden (2009), Portugal (2010) and Iceland (2010) – legally recognize same-sex marriages, and seventeen countries, now allow registered partnerships as a legal option for same-sex couples, one of the most recent being Ireland which legalized civil unions among gay couples in 2011.[2] Moreover, among those countries that have legalized same-sex marriages or civil partnerships, ten also permit same-sex adoptions, albeit with varying restrictions.

As in many other European nations, same-sex sexual relations for males and females aged 16 years or over is legal in both Great Britain and Northern Ireland and, since the 2004 Civil Partnership Act, gay couples have been legally entitled to engage in civil partnerships in the UK. In fact, the first civil partnerships in the United Kingdom occurred in Northern Ireland in December 2005. Under the Civil Partnership Act, these civil partners now enjoy almost the same benefits and legal rights of marriage, ranging from tax exemptions and joint property rights, to next-of-kin status and shared parenting responsibilities. However, while gay couples are legally permitted to adopt children in Great Britain, the position in Northern Ireland remains unclear where adoption legislation is currently under review. Although gay and lesbian couples are able to foster children in Northern Ireland, under the Adoption Order 1987, they are not legally permitted to adopt them. Finally, at least as far as the UK is considered, the introduction of the 2010 Equality Act suggests that inequalities related to the lives of lesbian, gay and bisexual (LGB) people will remain firmly on the political agenda.[3] In fact, the parties in the current coalition government – Conservative/Liberal Democrat – have pledged to introduce a law to legalize same-sex marriages in England and Wales before the next general election in 2015. The Scottish government has promised to make same-sex marriage legally available by then. It remains to be seen, however, whether this legislation will also apply to Northern Ireland, given its current devolved government and recent unsuccessful attempt to legalize same-sex marriage.[4]

---

[2] The seventeen countries include: Denmark (1989), Norway (1993), Sweden (1994), Iceland (1996), the Netherlands (1998), France (1999), Belgium (2000), Germany (2001), Finland (2002), Switzerland (2004), United Kingdom (2005), Slovenia (2005), Czech Republic (2006), Hungary (2009), Austria (2010), Ireland (2011) and Liechtenstein (2011).

[3] In fact, as recently as June 2011, lawyers advised the Church of England that because of the Equality Act, they can no longer take into account the sexual orientation of applicants when deciding their eligibility for Episcopal office, although they also advised that they cannot put forward clergy in active same-sex relationships. Currently, Church of England clergy may enter a civil partnership on the condition that they remain celibate.

[4] In October 2012, the Northern Ireland Assembly narrowly rejected – 49 against as compared to 45 in support – a motion proposing that same-sex couples should have the right to marry. The motion was overwhelmingly rejected by politicians who represent the Unionist/Protestant community (all DUP and all but three UUP members rejected the motion), but received universal support by politicians within the Nationalist/Catholic tradition (Sinn Fein/SDLP).

This is not to suggest, however, that these changes have been without their opponents. In fact, throughout the process, religious authorities of all denominations both in the U.S., Britain and elsewhere have been notably oppositional in their views. This is particularly so in the U.S. where the Christian Right has led a concerted campaign to mobilize opposition to these various legislative changes, most notably in relation to gay marriages (see Cahill 2007). Furthermore, there is some evidence to suggest that opposition to gay rights is not just restricted to the Evangelical denominations but has also found favor among religious traditionalists across all denominations. This is not to understate, however, important differences between the religious traditions in relation to this issue, but to suggest that division concerning gay rights may be greater within rather than between religious traditions. As Campbell and Robinson (2007: 150) put it: "Your level of traditionalism within your church matters more than which church you go to."

Religious authorities in Britain have also voiced publicly their opposition to these legislative changes. For example, not only have both Anglican and Catholic religious authorities spoken out strongly against the adoption rights of gay and lesbian couples but they also sought exemptions to the 2009 Equality Bill, which became law in April 2010, such as the prohibition of gay or lesbian individuals working in religious schools. In fact, so great was the opposition of religious authorities to these legislative changes that many religious leaders, including a former head of the Anglican Church, Lord Carey, publicly accused the British government of discriminating against members of the Christian faith and undermining their beliefs in relation to marriage, conscience and worship (Carey 2010).[5] This is not to discount, however, some ongoing and notable divisions within the Anglican Church in relation to this issue, particularly when the question of gay bishops in civil partnerships are considered,[6] as well as the increasingly vocal opposition, particularly by the Catholic Church, in relation to the proposed bill to allow both civil and religious same-sex marriage ceremonies in Scotland.[7]

Despite the opposition of religious authorities to gay rights, to date, the relationship between religion and attitudes towards gay rights among the mass public has not been comprehensively investigated (see Olson et al. 2006 for a detailed discussion of this issue). This is particularly the case when the question of

---

[5] In an open letter to the Times, he appealed to Ministers to block plans to allow civil partnership ceremonies to take place in churches, arguing that not only could this result in clergy being prosecuted if they refused to carry out the ceremonies but he also accused the government of 'blurring' the clear distinction between homosexual partnerships and heterosexual marriage.

[6] Notable examples in the Anglican church include the marked schism that occurred between liberals and mainstream Anglicans versus evangelical hardliners over the appointment of Gene Robinson as an American bishop in 2003 as well as the ongoing row concerning the appointment of Dr Jeffrey John – who entered a civil partnership with his lifelong male companion in 2006 – as a bishop in Britain.

[7] Arguing that the only legitimate definition of marriage is the "life-long union of a man and a woman," the leader of the Catholic Church in Scotland, Cardinal Keith O'Brien, described gay marriage as 'a grotesque subversion of a universally accepted human right.' (BBC News Scotland, 26th August 2012).

same-sex marriages and the adoption of children by gay couples is considered. Furthermore, this relationship remains irrespective of whether the U.S., the UK or many other European nations are considered. Our arguments in support of this position are threefold.

First, as Olson et al. (2006) note, while a good deal of research has been conducted about the relationship between religion and public opinion about attitudes towards homosexuality, much of the analysis is based on small surveys and has been restricted either by geography or by the types of people included (often college students). Second, even when nationally representative samples are used, a detailed analysis of this issue has not been undertaken because of the extremely limited nature of the information available on religion. Usually only two to three measures of religion are common, such as religious affiliation, church attendance rates and a belief in God. Third, even when some additional information on religion is available, previous studies have limited their investigation to attitudes about homosexuality generally rather than public perceptions concerning the adoption and marriage rights of gay couples. Furthermore, even when attitudes toward marriage and adoption rights are considered, to date, not only have the vast majority of studies been restricted to the United States but they have tended to investigate these rights in terms of a binary position, comparing those who hold a negative opinion in relation to these issues versus those who do not.

It is with these omissions in mind that this chapter focuses on the impact of religion on attitudes towards same-sex sexual relations and the adoption and marriage rights of gay couples. Using nationally representative data from Northern Ireland, the investigation proceeds in two stages. First, attitudes towards gay rights are presented briefly. More specifically, this analysis focuses not only on attitudes towards same-sex sexual relations, but also on those towards gay marriages as well as the adoption rights of gay couples. Second, building on this investigation, we then examine the impact of a range of religious measures in relation to these views. These religious measures include: religious practices, one's belief in and about God, and subjective religious views. By including such a diversity of measures, this investigation not only marks the first systematic analysis of religion and public opinion towards gay rights in this society but also allows us to investigate the net influence of each of these various religion measures on public opinion about gay marriages and adoption rights – two issues that are central to current debates about gay rights within contemporary industrialized societies. Finally, contrary to previous research, attitudes towards same-sex sexual relations and gay rights are compared in terms of three categories of opinion: those who hold a negative view of this issue; those who don't know or hold a neutral position; and those who adopt a positive or supportive stance in terms of their opinion.

The use of Northern Ireland as a case study to investigate this issue may be considered appropriate for the following three reasons. First, unlike many other western post-industrialized nations, Northern Ireland remains a deeply religious society. Currently, just under 90 % of the adult population consider themselves either Roman Catholic (36 %), Church of Ireland/Anglican (17 %), Presbyterian (21 %) or some other Protestant denomination (11 %),[8] and just 1 % identify themselves as "non-Christian." This is not to deny, however, the small minority, or 13 % in 2008, who claim no religious affiliation, but whose religious origins are predominantly from within the Protestant tradition. It is important to note, however, that this religious affiliation is more than merely nominal: Northern Ireland has traditionally manifested one of the highest levels of religious observance found in Europe, second only to that of Poland and Malta (Fahey et al. 2006).

Second, despite the passage of legislation, the issue of gay rights remains a controversial issue within this society. In fact, as recently as 2008, Iris Robison, who was then a member of both the Northern Ireland Assembly and the House of Commons and who describes herself as a "born again" Christian, caused much controversy when she said that homosexuality was an "abomination" that made her feel "sick" (see Duggan 2012). Surveys of public attitudes towards the gay community suggest that she may not be alone in her views. For example, a recent survey conducted by the Equality Commission for Northern Ireland – the 2011 Equality Awareness Survey – found that just over a quarter of respondents would object to having a gay or bisexual neighbor and four in ten would object to a close relative forming a relationship with a gay or bisexual person. Moreover, there was evidence to suggest a hardening in attitudes, or increasing opposition in relation to this issue, in recent years. For example, in the three time periods in which the survey was conducted, the proportion of respondents who objected to a close relative forming a relationship with a gay or bisexual person rose from 29 % in 2005 to 35 % in 2008, and further increased to 42 % in 2011 (see Equality Commission for Northern Ireland 2011: 21).

Finally, unlike many other European societies where the introduction of legislation in support of gay rights has been the result of widespread internal public pressure from its citizens within their national borders, much of the impetus for the introduction of legalisation concerning same-sex issues has come from outside Northern Ireland society. In fact, to the objection of many political and religious representatives within Northern Ireland, most notably Ian Paisley and members of the Democratic Unionist Party (DUP), the legalisation of same-sex civil partnerships was imposed upon Northern Ireland by the British government at Westminster.[9]

---

[8] These include just under 4 % of Methodists and 2 % of Baptists, and under 1 % of Free Presbyterians as well as number of additional and even smaller denominations that can be classified as Protestant and which, in many cases, have a membership of less than a hundred people (see Mitchell 2006).

[9] This was not the first time, Ian Paisley, founder of the fundamentalist Free Presbyterian church in Northern Ireland, had objected to gay rights legislation within this society. In the early 1980s, he spearheaded a campaign against the decriminalisation of homosexuality, "Save Ulster from Sodomy," when the British government sought to legalise the age of consent at 21 for same-sex sexual relations in Northern Ireland, and thus bring it in line with the rest of Britain. Despite his objections the British government extended the British legislation to Northern Ireland in 1982 and, as a result of further changes in Britain, again in 2008 so that the age of consent throughout the United Kingdom is now 16 years of age. All changes were strongly opposed by both Ian Paisley and other church leaders – both Protestant and Catholic – in Northern Ireland.

More recent public pronouncements, particularly from among some of the larger Protestant churches, concerning the legalisation of same-sex marriage, suggests that they are by no means alone in their views. For example, in September 2011, the leader of the Presbyterian Church – the largest Protestant denomination in Northern Ireland – wrote to all members of the Northern Ireland Assembly stating its opposition to a proposed motion to legalise same-sex marriage.[10] Even the Methodist Church, which advocates understanding and tolerance towards gay people, caused widespread outrage and disquiet in September 2011, when it rejected a male volunteer and long-standing member of the church to work on a youth project in light of his sexual preferences.[11] This is not to deny, however, some growing and stark divisions, particularly between political representatives of the two communities – Protestant/unionist and Catholic/nationalist – in relation to this issue.

174.2   Previous Research on Religion and Attitudes Towards Homosexuality and Gay Rights

Over the last four decades, a great deal of research has focused on the relationship between religion and attitudes towards homosexuality and gay rights in both the U.S., Britain, and elsewhere. Although the relationship between religion and attitudes towards homosexuality is complex, one of the most consistent findings to emerge from this research is that religion, as measured by individuals' religious affiliations, behaviours, and beliefs has an important influence on attitudes towards homosexuality, although the net influence of each of these various measures on attitudes towards gay rights has yet to be assessed empirically.

Earlier research on the relationship been religion and gay rights has focused on the role of religious practices in maintaining the moral order. Drawing on the theoretical insights of Durkheim ([1912] 1995), particularly in relation to the suggested overriding importance of rituals, this research has investigated the degree to which religious practices, such as church membership or belonging to a religious institution and attending religious services, has an important effect on moral beliefs, including attitudes towards homosexuality. For example, there is now a substantial body of work to suggest that church membership or religious affiliation has an especially strong and negative influence on opinions about homosexuality, including the rights of lesbians and gay men, and this is also the case when other religious practices, such as prayer but particularly church attendance, are considered (see Fisher et al. 1994; Hayes 1995;

---

[10] In a strongly worded letter, the Presbyterian Church stated that not only was the issue of gay marriage "a significant one for the whole of society" and not just a matter of conscience for Christian people and churches, but that it would "effectively demolish generations and centuries of social norms established on Judeo-Christian values."

[11] In pointing to his unsuitability for the position and justifying its decision to reject his offer, the Methodist Church issued the following public statement: "A loving marriage relationship is seen as the only appropriate relationship within which sexual intercourse can take place."

Finlay and Walther 2003; Meerendonk and Scheepers 2004; Burdette et al. 2005; Olson et al. 2006; Finke and Adamczyk 2008). Furthermore, for all religious denominations, but particularly among those within the evangelical Protestant tradition, there is some evidence to suggest that their high rates of church attendance and its derivative congregational activities and social and friendship networks are an important contributing factor in explaining their anti-gay views (Olson et al. 2006).

This is not to deny, however, some important divisions both between and within these religious traditions in relation to this issue. For example, studies have also shown that Jews, liberal Protestants and people who are religiously non-affiliated have the most liberal attitudes in relation to this issue while Catholics and moderate Protestants tend to espouse moderate but generally tolerant attitudes. By contrast, evangelical Protestants, or members of Conservative Protestant denominations, have the most homophobic views, although, at least as far as the United States is concerned, there is some evidence to suggest that they have become more tolerant of lesbians and gay men over the last two decades and this seems to be particularly the case among the young (Greeley and Hout 2006). This is not to discount, however, some important differences within these traditions in relation to this issue, with some studies suggesting that it is the religiously devout across the various traditions who hold the most negative views (Campbell and Robinson 2007; Adamczyk and Pitt 2009).

More recent research, however, calls into question this interpretation, arguing that the relationship between religion and moral beliefs depends not so much upon religious practices, such as church membership and attendance rates, but upon one's relationship with God. Pointing to the decline in both church membership and attendance rates, according to this perspective, it is both one's belief in and about God, and not religious acts, which are the key factors in accounting for moral attitudes. While earlier studies highlighted the important influence of a belief in God, more recent research, however, stresses not so much a belief in God *per se* but rather an individual's perceptions of God as the key predictor of attitudes towards moral issues (see Greeley 1995; Froese and Bader 2008; Whitehead 2010). Greeley (1995), one of the first sociologists to propose this view, suggests that not only does one's belief about God determine an individual's religious beliefs and attitudes but it also extends to more secular issues, such as attitudes towards the death penalty and other socio-political views. For example, individuals with more gracious or maternal images of God will adopt a more liberal stance in relation to a range of social and moral issues than those who hold a more authoritarian or paternal view. In other words, according to this perspective, it is the nature of one's perceptions or beliefs about God, and not simply a belief in God, that are the primary predictors of attitudes in this instance.

Recent national and international research lends considerable support to this view (see Froese et al. 2008; Froese and Bader 2008; Whitehead 2010). In fact, there is now a growing body of evidence to suggest that it is one's beliefs about God, and not religious acts, such as church attendance or even Biblical literalism, which are the decisive factors in determining a range of moral outlooks, including attitudes towards gay rights. For example, Froese and Bader (2008), in a study of eight nations, including the United States, Australia, New Zealand and five European countries, found that variation in images of God had a significant effect on sexual morality attitudes, including attitudes towards homosexuality, and that this relationship remained even when a range of variables, including religious affiliation and church attendance, were included in the analysis. Furthermore, in all but one nation – Australia – belief in a personally engaged God was a much stronger predictor of absolutist sexual morality attitudes than church attendance, with individuals holding such views being significantly more likely to condemn certain sexual acts, including homosexuality, than those who did not. As Froese and Bader (2008: 700), in explaining their findings, put it: "In the end, going to church is not what makes many Americans politically conservative on these issues, but instead it is the type of God in which they believe."

Previous research also points to the salience of religion in predicting moral attitudes. For example, Wilcox et al. (2006) in a study of attitudes towards gay marriage in the United States found that it was subjective religious salience, and not church attendance, which emerged as the key predictor of attitudes in this instance. More specifically, individuals who stressed the importance of religion in their lives were significantly more likely to oppose same-sex marriage than those who did not. More recent, cross-national research, lends further support to this view. For example, Adamczyk and Pitt (2009) in a study of 33 western and non-western nations, found that not only were subjective views concerning the importance of religion a key predictor of attitudes towards homosexuality, but it also had greater effect in countries with a strong self-expressive cultural orientation, such as the United States. In fact, some analysts go so far as to suggest that it is this factor – the importance of religion in one's daily life – which sets the United States apart, especially in relation to citizens in most European countries, and is the single most important barrier to the recognition of same-sex relationships in the U.S. (see Rayside 2007).

In summary, the relationship between religion and attitudes towards homosexuality remains complex. While a number of studies point to the importance of religious practices, such as church membership and attendance rates, in determining attitudes towards same-sex sexual relations and gay rights, others stress the overriding influence of one's belief in and perceptions of God, while still others stress the importance of religious salience or subjective religious views. It is with these three competing interpretations in mind – religious practices, beliefs in and about God, and subjective religious views – that this study provides the first comprehensive and systematic test of the role of religion in predicting public attitudes towards same-sex sexual relations as well as the marriage and adoption rights of lesbians and gay men within Northern Ireland society.

174.3   Data and Measures

The data come from the 2008 Northern Ireland Life and Times Survey (NILT), which is publicly available and can be downloaded, free of charge, from the Northern Ireland Access Research Knowledge (ARK) website: www.ark.ac.uk. Initiated in 1998, the NILT is a nationally representative survey of the adult population, aged 18 years and older, that is conducted annually. The 2008 NILT survey was based on 1,216 face-to-face interviews and had a response rate of 60 %.

Attitudes toward gay rights, the dependent variable, are measured in terms of the three following issues: same-sex sexual relations, legally treating same-sex partnerships as marriage and the rights of same-sex couples to adopt children. The independent variables, or the range of religion measures, included a variety of religious practices, such as affiliation and church attendance rates, indicators of one's beliefs in and about God, as well as two measures of subjective religious views (Table 174.1 for specific questions and response categories). With the exception of affiliation, which was included as a series of dummy variables in the analysis, all religion measures are recoded to range from 0 (the least religious position) to 1 (the most religious position).[12]

In addition, we included a number of demographic control variables such as gender, education, class, work status, age, marital status and political orientation. Previous studies in the United States, Britain and elsewhere in Europe document the fact that women, the more socio-economically privileged, the young, the unmarried and those with a left-wing political orientation are notably more tolerant in their views towards gay rights than men, the lesser-privileged, older, married and right leaning individuals (see Hayes 1997; Wilcox et al. 2007; McVeigh and Diaz 2009). With the exception of age (coded in terms of years), all control variables were included as a series of dummy variables (coded 0 and 1) in the analyses. For the purposes of the multivariate investigation, the three dependent variables have been recoded in terms of the following three categories: (a) not wrong/approve; (b) neither-don't know; and (c) wrong/disapprove, with the wrong/disapprove category being used as the basis for the contrast. Based on a multinomial logistic regression analysis, the figures in each equation are the parameter estimates, while the standard error for each estimation is shown in parentheses.

174.4   Results

Table 174.2 begins the investigation by focusing on attitudes towards same-sex issues, such as same-sex sexual relations as well as the marriage and adoption rights of gay people. The results suggest that at least as far as an expressed opinion on each of these three same-sex issues are concerned, the Northern Ireland public is notably divided in terms of their views. For example, whereas a notable minority of respondents – just over two-fifths in this instance – thought that same-sex sexual relations was "always wrong," less than a fifth and exactly a quarter of individuals adopted an identical position – "strongly disapproved" – when the marriage and adoption rights of gay couples were considered. Differences in opinion were even more marked among those who expressed the most positive stance in relation to these three issues, although in this instance it was an endorsement of same-sex sexual relations which commanded

---

[12]Although additional bivariate analysis demonstrates that some of these religious measures are significantly associated with each other, in no instance are the zero-order correlations greater than 0.70, suggesting that multicollinearity is not a problem in this instance. The variance inflation factor (VIF) for each of the various analyses – less than 3.43 in all instances – lends some further support to this finding.

Table 174.1 Measures

	Variable	Question	Response categories
Attitudes towards gay rights	Same-sex sexual relations	And what about sexual relations between two adults of the same sex, is it?	Always wrong, almost always wrong, wrong only sometimes, not wrong at all, can't choose
	Civil partnerships	Some people live with partners of the same sex. Do you approve or disapprove of laws that treat these partnerships somewhat like marriage?	Strongly approve, approve, neither approve nor disapprove, disapprove, strongly disapprove, can't choose
	Adopt children	Do you approve or disapprove of allowing same-sex couples to adopt children?	Strongly approve, approve, neither approve nor disapprove, disapprove, strongly disapprove, can't choose
Religion practices	Affiliation	Do you regard yourself as belonging to any particular religion? If yes, which?	
	Church attendance	Apart from special occasions such as weddings, funerals, baptisms and so on, how often nowadays do you attend services or meetings connected with your religion?	Nine point scale ranging from 1 for never to 9 for several times a week
	Other church activities	How often do you take part in the activities or organisations of a church or place of worship other than attending services?	Nine point scale ranging from 1 for never to 9 for several times a week
Beliefs in and about God	Belief in God	Please tick one box below to show which statement comes closest to expressing what you believe about God	I don't believe in God, I don't know whether there is a God and I don't believe that there is a way to find out, I don't believe in a personal God, but I do believe in a Higher Power of some kind, I find myself believing in God some of the time, but not at others, while I have doubts, I do feel that I do believe in God, I know God really exists and I have no doubts about it
	Personal relationship with God	I have my own way of connecting with God without churches or religious services	Strongly agree, agree, neither agree nor disagree, disagree, strongly disagree, can't choose
	Active God	Do you think that God is directly involved in your affairs?	Yes definitely, yes probably, no probably not, no definitely not, don't believe in God, can't choose
	Angry God	Do you think that God is angered by human sin?	Yes definitely, yes probably, no probably not, no definitely not, don't believe in God, can't choose

(continued)

Table 174.1   (continued)

	Variable	Question	Response categories
Subjective religious views	Self-religious	Would you describe yourself as...	Extremely religious, very religious, somewhat religious, neither religious nor non-religious, somewhat non-religious, very non-religious, extremely non-religious, can't choose
	Salience of religion	How important is religion in your daily life?	Very important, somewhat important, not very important, not at all important, can't choose

Data source: Northern Ireland Life and Times Survey, 2008

Table 174.2   Attitudes toward same-sex issues

Issue	Attitude	Percent (N = 1,093)
Same-sex sexual relations	Always wrong	43.5
	Almost always wrong	7.8
	Can't choose/don't know	17.5
	Sometimes wrong	6.7
	Not wrong at all	24.4
Marriage rights	Strongly disapprove	17.3
	Disapprove	21.8
	Neither/don't know	38.1
	Approve	16.1
	Strongly approve	6.6
Adoption rights	Strongly disapprove	25.0
	Disapprove	29.2
	Neither/don't know	29.2
	Approve	12.3
	Strongly approve	4.5

Data source: Northern Ireland Life and Times Survey, 2008

the most support. For example, while around a quarter of respondents expressed the view that same-sex sexual relations was "not wrong at all," only a tiny minority of individuals – less than 8 % in both cases – strongly approved of legally treating same-sex partnerships like marriage or allowing same-sex couples to adopt children.

This is not to deny, however, the notable number of individuals who were undecided in terms of their views, although it is indecision in relation to the right to treat gay partnerships like marriage which emerged as the key issue in this instance. For example, whereas just under a fifth of respondents expressed a neutral opinion – "can't choose/don't know" – in relation to same-sex sexual relations, the equivalent proportion who adopted an identical stance – "neither/don't know" – in relation to gay marriage was nearly double this amount at 38 %. Thus, at least as far as Northern Ireland is concerned the results are clear. While the vast majority of adults are divided in terms of their acceptance, or lack thereof, of same-sex sexual relations, either disapproval or indecision characterize their views about gay marriage and, albeit to a somewhat lesser extent, their attitudes towards the adoption rights of gay couples.

Turning now to the main focus of our investigation, or the net impact of religion on attitudes towards gay rights, the following three tables report the results of three logistic regression analyses focusing on attitudes towards same-sex sexual relations as well as the marriage and adoption rights of gay couples. In all three cases, it is religious practices – affiliation and church attendance – which emerge as the key predictors of attitudes in this instance. This is not to deny, however, the notable influence of one's beliefs in and about God, particularly when attitudes towards same-sex sexual relations and those concerning the adoption rights of gay couples are considered. By contrast, subjective religious views failed to emerge as a significant net predictor of attitudes toward same-sex issues and, with one notable exception, this lack of relationship remained regardless of whether attitudes towards same-sex sexual relations or the marriage and adoption rights of gay couples were considered.

Focusing initially on same-sex sexual relations, the results in Table 174.3 are clear. Religious identification has a strong and consistent net effect on attitudes towards same-sex sexual relations. As a group, it is members of the Protestant faith who emerge as the most negative in their views. For example, in comparison to their Catholic counterparts (the omitted category or reference), not only are members of the Protestant community notably less likely to view same-sex sexual relations as "not wrong" than "wrong" but this is also the case among individuals who adopt a neutral stance – "don't know" – in relation to this issue. This is not to deny, however, the significant and somewhat unexpected negative effect in relation to the non-affiliated, albeit only between those who hold a bipolar – "wrong" versus "not wrong" – view, a phenomenon that we suggest may be explained by their predominantly ex-Protestant origins (Hayes and McAllister 1995).

A similar, albeit somewhat less consistent, pattern emerges when church attendance rates are considered. As the results demonstrate clearly, church attendance is also a significant negative predictor of attitudes towards same-sex sexual relations. However, in this instance, it was only in relation to those who expressed either a negative or positive view – "not wrong" versus "wrong" – that these differences in attendance rates emerge as a statistically significant finding. Individuals who attend church regularly are notably less likely to view same-sex sexual relations as 'not wrong' than 'wrong' as compared to either their more irregular or non-attending counterparts. Thus, at least as far as the net effect of church attendance on attitudes towards same-sex sexual rates are considered, it is only among those who hold a bi-polar view that religious practices emerge as a significant net predictor of attitudes.

This is not to suggest, however, that practices are the only religious measures predicting attitudes towards same-sex sexual relations. Similar to previous research in this area (Froese and Bader 2008), other influences include the positive effect of having a personal relationship with God as well as the negative effect of believing in a wrathful God. However, it was again only in relation to those who expressed a bi-polar view – "not wrong" versus "wrong" – that differences in perceptions concerning one's relationship to God and beliefs about God achieve statistical significance. For example, while individuals who believed in an angry God were

Table 174.3 Religion and attitudes toward same-sex sexual relations

| | | Logistic regression coefficient | | | |
| | | Not wrong versus wrong | | Don't know versus wrong | |
		Estimate	(SE)	Estimate	(SE)
Control variables	Male	−.96**	(.19)	−1.01**	(.21)
	Married	.12	(.19)	−.15	(.20)
	Age	−.04**	(.01)	−.03**	(.01)
	Education:				
	Tertiary[a]	−	−	−	−
	Secondary	−.44*	(.21)	−.57*	(.25)
	No qualification	−.99**	(.27)	−.69*	(.29)
	Middle class	.35	(.20)	−.17	(.22)
	Labour active	−.11	(.22)	−.91*	(.24)
	Left ideological position	−.70	(.51)	−.53	(.58)
Religious practices	Affiliation:				
	Catholic[a]	−	−	−	−
	Anglican	−1.44**	(.26)	−1.55**	(.33)
	Presbyterian	−1.61**	(.25)	−1.05**	(.27)
	Other protestant	−2.54**	(.37)	−1.81**	(.36)
	Non-affiliated	−.95**	(.34)	−.37	(.39)
	Church attendance (attend)	−1.22**	(.39)	.13	(.44)
	Other church activities (attend)	−.15	(.31)	−.13	(.33)
Beliefs in/ about God	Believe	−.50	(.37)	−.35	(.42)
	Personal relationship	.68*	(.34)	.42	(.36)
	Active God	−.43	(.44)	.25	(.51)
	Angry God	−1.06**	(.37)	−.47	(.42)
Subjective religious views	Self-religious	−.28	(.46)	−.42	(.52)
	Salience of religion	−.16	(.42)	−.35	(.48)
	Constant	4.89**	(.49)	3.45**	(.65)
	Nagelkerke R-squared	.391			
	(N)	(1,012)			

Data source: Northern Ireland Life and Times Survey, 2008
[a]Omitted category of comparison; **$p \leq 0.01$; *$p \leq 0.05$

significantly less likely to view same-sex sexual relations as 'not wrong' than "wrong" no differences emerged between those who were undecided on the issue versus those who held a negative view.

Social background characteristics are also important. As the results also indicate clearly, both gender and age emerge as consistent, albeit negative, determinants of attitudes towards same-sex sexual relations. Both males and older individuals are notably less supportive of same-sex sexual relations than either females or the young. Furthermore, this relationship remains irrespective of whether those who hold a negative versus a positive or undecided view are considered. Overall, the model is reasonably well predicted by this analysis, producing a Nagelkerke R^2 of 0.39, or 39 % of the variance explained.

Similar results emerge when attitudes towards the marriage rights of gay couples are considered (Table 174.4). Religious identification is again a strong and consistent net predictor of attitudes, with members of the Protestant faith being the most nega-

Table 174.4 Religion and attitudes toward marriage rights of gay couples

		Logistic regression coefficients			
		Approve versus disapprove		Don't know versus disapprove	
		Estimate	(SE)	Estimate	(SE)
Control variables	Male	−.88**	(.19)	−.74**	(.17)
	Married	.28	(.21)	.03	(.17)
	Age	−.03**	(.01)	−.02**	(.01)
	Education:				
	Tertiary[a]	–	–	–	–
	Secondary	−.20	(.23)	−.08	(.20)
	No qualification	−.57*	(.28)	−.31	(.24)
	Middle class	.35	(.21)	.11	(.18)
	Labour active	−.14	(.24)	−.17	(.20)
	Left ideological position	−.42	(.54)	.53	(.46)
Religious practices	Affiliation:				
	Catholic[a]	–	–	–	–
	Anglican	−1.60**	(.29)	−.91**	(.24)
	Presbyterian	−1.54**	(.27)	−1.17**	(.23)
	Other protestant	−2.39**	(.39)	−1.53**	(.27)
	Non-affiliated	-.88*	(.37)	−.53	(.34)
	Church attendance (attend)	−1.09**	(.42)	.02	(.36)
	Other church activities (attend)	−.07	(.34)	−.56*	(.27)
Beliefs in/about God	Believe	−.52	(.41)	−1.04**	(.36)
	Personal relationship	.61	(.36)	.32	(.30)
	Active God	−.31	(.49)	−.21	(.42)
	Angry God	−.62	(.40)	−.27	(.35)
Subjective religious views	Self-religious	−.76	(.50)	−.07	(.43)
	Salience of Religion	.18	(.45)	.04	(.39)
	Constant	3.91**	(.61)	3.18**	(.53)
	Nagelkerke R-squared	.269			
	(N)	(1,012)			

Data source: Northern Ireland Life and Times Survey, 2008
[a]Omitted category of comparison; **$p \leq 0.01$; *$p \leq 0.05$

tive in their views. Furthermore, this relationship again holds regardless of whether those who hold a negative versus either a positive or undecided view are considered. In both cases, Protestants are consistently more likely to express a negative stance in relation to this issue than Catholics. Church attendance is also a significant negative predictor of attitudes towards the marriage rights of gay couples, although it is again only among those who were bi-polar – "approve" versus "disapprove" – in their views. Contrary to our previous analysis, however, in all but one instance, beliefs in and about God fail to emerge as a significant net determinant of attitudes in relation to this issue. Social background characteristics are again important with both gender and age emerging as a consistently negative net predictor of attitudes towards the marriage rights of gay couples. Overall, the model is reasonably well predicted by this analysis, producing a Nagelkerke R^2 of 0.27, albeit much less so that in our previous investigation of attitudes towards same-sex sexual relations.

Similar results are echoed when attitudes towards the adoption rights of gay couples are considered (Table 174.5). Both religious identification and church attendance are again strong and consistent net predictors of attitudes, with members of the Protestant faith and regular church attenders again expressing the most negative views. As in our earlier analysis on attitudes towards same-sex sexual relations, claiming a personal relationship with God and believing in a wrathful God also emerge as significant net predictors of attitudes in this instance. For example, individuals who believe in a wrathful God are significantly more likely to adopt a negative stance in relation to this issue. Furthermore, this relationship holds regardless of whether those who hold a negative versus either a positive or undecided view are considered. Social background characteristics are also significant with both men and older individuals being the most disapproving in their views. Overall, the model is again reasonably well predicted by this analysis – Nagelkerke R^2 is 0.25 – albeit again much less so than in our earlier investigation of attitudes towards same-sex sexual relations. This suggests that, at least as far as Northern Ireland is concerned, these religion measures are a much better predictor of attitudes towards same-sex sexual relations than public views concerning gay marriages and the right of gay couples to adopt children.

In summary, the results are clear. In all three cases, it is religious practices – affiliation and church attendance – which emerge as the key predictors of attitudes in this instance. This is not to deny, however, the notable influence of one's beliefs in and about God, particularly when attitudes towards same-sex sexual relations and those concerning the adoption rights of gay couples are considered. Furthermore, there is some evidence to suggest that at least as far as the effect of religious practices on attitudes towards same-sex sexual relations and the adoption rights of gay couples are concerned, it is religious affiliation, and not church attendance, which is the key predictor of attitudes in this instance. As the interaction effects between denomination and church attendance in Table 174.6 demonstrate, even among regular church attenders, it is Protestants who are emerge as the most negative in their views. Irrespective of whether Anglicans, Presbyterians or Other Protestants are considered, members of the Protestant faith who attend church on a regular basis are

Table 174.5 Religion and attitudes toward adoption rights of gay couples

		Logistic regression coefficients			
		Approve versus disapprove		Don't know versus disapprove	
		Estimate	(SE)	Estimate	(SE)
Control variables	Male	−.88**	(.21)	−.51**	(.17)
	Married	.19	(.21)	.09	(.17)
	Age	−.04**	(.01)	−.02**	(.01)
	Education:				
	Tertiary[a]	−	−	−	−
	Secondary	−.37	(.24)	−.36	(.20)
	No qualification	−.16	(.29)	−.39	(.24)
	Middle class	−.03	(.22)	−.26	(.18)
	Labour active	−.35	(.25)	−.42*	(.20)
	Left ideological position	−.39	(.57)	−.29	(.46)
Religious practices	Affiliation:				
	Catholic[a]	−	−	−	−
	Anglican	−1.36**	(.31)	−.72**	(.24)
	Presbyterian	−1.30**	(.28)	−1.05**	(.23)
	Other protestant	−1.87**	(.42)	−1.32**	(.29)
	Non-affiliated	−.91*	(.35)	−.25	(.31)
	Church attendance (attend)	−1.23**	(.43)	−.25	(.36)
	Other church activities (attend)	.11	(.36)	.03	(.28)
Beliefs in/about God	Believe	.33	(.42)	−.05	(.34)
	Personal relationship	1.02**	(.39)	.23	(.30)
	Active God	−.56	(.49)	−.42	(.41)
	Angry God	−.85*	(.40)	−.73*	(.33)
Subjective religious views	Self-religious	−1.31*	(.51)	−.59	(.43)
	Salience of religion	.35	(.47)	.14	(.39)
	Constant	3.34**	(.63)	2.93**	(.52)
	Nagelkerke R-squared	.251			
	(N)	(1,012)			

Data source: Northern Ireland Life and Times Survey, 2008
[a]Omitted category of comparison; **$p \leq 0.01$; *$p \leq 0.05$

significantly more likely than their regularly attending Catholic counter-parts (omitted category of comparison) to view same-sex sexual relations as wrong. A similar result emerges when opinions about the adoption rights of gay couples are considered, although in this instance it is Presbyterians and Other Protestants who emerge as the most "disapproving" or negative in their views. Thus, at least as far as the religiously affiliated population in Northern Ireland is concerned, there is some evidence to suggest that denomination is more important than the strength of a person's ties to religion in determining views.

Table 174.6 Interaction between denomination and church attendance among the religious affiliated (n = 885)

	Logistic regression coefficient					
	Same-sex relations (wrong vs other)		Marriage rights (disapprove vs other)		Adoption rights (disapprove vs other)	
	Estimate	(SE)	Estimate	(SE)	Estimate	(SE)
Denomination:						
Catholic[a]	–	–	–	–	–	–
Anglican	.48	(.38)	1.06**	(.39)	.64	(.35)
Presbyterian	.43	(.39)	.74	(.40)	.28	(.37)
Other Protestant	1.00*	(.49)	1.10*	(.49)	.66	(.46)
Church attendance (attend)	−.33	(.42)	.07	(.44)	−.05	(.39)
Interactions:						
Catholic × Attends[a]	–	–	–	–	–	–
Anglican × Attends	2.01**	(.72)	−.05	(.65)	.38	(.64)
Presbyterian × Attends	1.72**	(.65)	1.06	(.62)	1.65**	(.64)
Other Protestant × Attends	2.16**	(.79)	1.17	(.71)	1.45*	(.72)

Data source: Northern Ireland Life and Times Survey, 2008
[a]Omitted category of comparison; ** $p \leq 0.01$; * $p \leq 0.05$. The analysis controls for all the other variables in the model, as outlined in the previous tables

174.5   Conclusion

Over the last four decades, a great deal of research has focused on the relationship between religion and attitudes towards homosexuality in the United States, Britain and elsewhere. This is particularly the case when the influence of religious practices on attitudes towards homosexuality is considered. For example, there is now a considerable body of research to suggest that religious practices, most notably both religious affiliation and church attendance rates, have a strong and consistent influence on opinions about homosexuality, including attitudes towards gay rights. More recent research, however, calls into question this interpretation. Pointing to the centrality of God to one's religious and derivative moral views, proponents of this perspective stress the importance of one's belief in and perceptions of God, and not religious acts, as the primary factor in distinguishing moral values. Still other research, however, emphasize the influence of subjective religious views, namely the importance of religion in one's daily life, as the primary factor in determining attitudes towards gay rights and this is particularly the case when the United States is considered.

   The results of this analysis lend only partial support to these competing explanations. Our arguments in support of this interpretation are threefold. First, religion, as measured by religious practice, is a significant factor in determining public opinion about same-sex issues. Irrespective of whether attitudes towards same-sex sexual relations or the marriage and adoption rights of gay couples are considered, both religious affiliation and church attendance stand out as significant net predictors of attitudes. Second, the influence of religion, however, cannot be reduced simply to religious practices. The findings also highlight the importance of one's beliefs about God in determining views concerning same-sex issues, most notably when attitudes towards same-sex sexual relations and the adoption rights of gay couples are considered. Finally, our results also provide some further, albeit indirect, confirmation that the impact of religion on same-sex issues may vary by national religious context. Contrary to previous U.S. findings, subjective religious views, namely the importance of religion in one's daily life, were not significantly related to attitudes in this instance.

More broadly our results suggest that, unlike popular opinion in the U.S., the current bitter and deep-seated divisions about homosexuality among religious authorities in the UK are not representative of the views of the general public at large. As the results also show, the Northern Ireland public, although not supportive, is also not uniformly opposed to same-sex marriage and/or the adoption rights of gay couples. Furthermore, although religious context has a notable effect on citizen opinion in relation to same-sex sexual relations this is much less the case when attitudes towards the marriage and adoption rights of gay couples are considered. Moreover, the fact that the demographic measures perform much more consistently than the religious variables in the various models clearly show the importance of other factors, besides religion, in shaping attitudes towards same-sex issues. This is not to understate the effect of religion – both religious practices and beliefs about God – on attitudes towards gay issues, but rather to suggest that attitudes towards same-sex issues are not solely determined by a religious framework.

Finally, while the present findings are important and indicate a need for further exploration of the role of religion in determining attitudes towards same-sex issues, some limitations of the analysis merit notation. First, because of the absence of suitable data, individual attributions of the cause of homosexuality could not be included in the investigation. Previous research has demonstrated that not only are individuals who believe homosexuality to be the result of choice much less likely to support gay rights than those who view it as biologically determined but this is particularly the case among members of the evangelical Protestant tradition (Haider-Markel and Joslyn 2008). In addition, the possible ameliorating effects of gay friendships or interpersonal contacts with gay individuals should also be considered. There is now a growing body of research to suggest that, irrespective of religious background, individuals who have family members or close acquaintances who are gay or lesbian are notably more accepting of gay rights issues than those who do not (Wood and Bartkowski 2004). And, while future research should include an investigation of the relationship between these additional issues and our various religion measures, one thing remains certain, however. Both religious practices and one's belief in and about God are important differential predictors of same-sex issues, the dynamics and consequences of which deserve future research attention.

References

Adamczyk, A., & Pitt, C. (2009). Shaping attitudes about homosexuality: The role of religion and cultural context. *Social Science Research, 38*(2), 338–351.

Badgett, M. V. L. (2009). *When gay people get married: What happens when societies legalize same-sex-marriage*. New York: New York University Press.

Burdette, A. M., Ellison, C. G., & Hill, T. D. (2005). Conservative Protestantism and tolerance toward homosexuals: An examination of potential mechanisms. *Sociological Inquiry, 75*(2), 177–196.

Cahill, S. (2007). The anti-gay marriage movement. In C. A. Rimmerman & C. Wilcox (Eds.), *The politics of same-sex marriage* (pp. 155–191). London: University of Chicago Press.

Campbell, D. C., & Robinson, C. (2007). Religious coalitions for and against gay marriages: The culture war rages on. In C. A. Rimmerman & C. Wilcox (Eds.), *The politics of same-sex marriage* (pp. 131–154). London: University of Chicago Press.

Carey, G. L. (2010, March 23). Open letter to the Times. *The Times*. p. 11.

Duggan, M. (2012). *Queering conflict: Examining lesbian and gay experiences in Northern Ireland*. Surrey: Ashgate.

Durkheim, E. [1912] (1995). *The elementary forms of religious life*. New York: Free Press.

Equality Commission for Northern Ireland. (2011). *Do you mean me?* Belfast: Equality Commission.

Fahey, T., Hayes, B. C., & Sinnott, R. (2006). *Conflict and consensus: A study of values and attitudes in the Republic of Ireland and Northern Ireland*. Leiden: Brill.

Finke, R., & Adamczyk, A. (2008). Cross-national moral beliefs: The influence of national context. *The Sociological Quarterly, 49*(4), 617–652.

Finlay, B., & Walther, C. S. (2003). The relation to religious affiliation, service attendance, and other factors to homophobic attitudes among university students. *Review of Religious Research, 44*(4), 370–393.

Fisher, R. D., Derison, D., Polley, C. F., III, Cadman, J., & Johnston, D. (1994). Religiousness, religious orientation, and attitudes towards gays and lesbians. *Journal of Applied Social Psychology, 24*(7), 614–630.

Froese, P., & Bader, C. (2008). Unraveling religious worldviews: The relationship between images of God and political ideology in a cross-cultural analysis. *The Sociological Quarterly, 49*(4), 689–718.

Froese, P., Bader, C., & Smith, B. (2008). Political tolerance and God's wrath in the United States. *Sociology of Religion, 69*(1), 29–44.

Greeley, A. M. (1995). *Religion as poetry*. New Brunswick: Transaction Publishers.

Greeley, A., & Hout, M. (2006). *The truth about conservative Christians: What they think and what they believe*. Chicago: University of Chicago Press.

Haider-Markel, D., & Joslyn, M. R. (2008). Beliefs about the origins of homosexuality and support for gay rights: An empirical test of attribution theory. *Public Opinion Quarterly, 77*(2), 291–310.

Hayes, B. C. (1995). Religious identification and moral attitudes: The British case. *British Journal of Sociology, 46*(3), 457–474.

Hayes, B. C. (1997). The influence of gender on public attitudes toward homosexual rights in Britain. *International Journal of Public Opinion Research, 9*(4), 361–385.

Hayes, B. C., & McAllister, I. (1995). Religious independents in Northern Ireland: Origins, attitudes and significance. *Review of Religious Research, 37*(1), 65–83.

McVeigh, R., & Diaz, M. E. D. (2009). Voting to ban same-sex marriage: Interests, values, and communities. *American Sociological Review, 74*(6), 891–915.

Meerendonk, B. V., & Scheepers, P. (2004). Denial of equal civil rights for lesbians and gay men in the Netherlands, 1980–1993. *Journal of Homosexuality, 47*(2), 63–80.

Mitchell, C. (2006). *Religion, identity and politics in Northern Ireland*. Aldershot: Ashgate.

Olson, L., Cadge, W., & Harrison, J. T. (2006). Religion and public opinion about same-sex marriage. *Social Science Quarterly, 87*(2), 340–359.

Rayside, D. (2007). The United States in comparative perspective. In C. Rimmerman & C. Wilcox (Eds.), *The politics of same-sex marriage* (pp. 341–364). London: University of Chicago Press.

Rimmerman, C. A., & Wilcox, C. (Eds.). (2007). *The politics of same-sex marriage*. London: University of Chicago Press.

Takacs, J., & Szalma, I. (2011). Homophobia and same-sex partnership in Europe. *Equality, Diversity and Inclusion: An International Journal, 30*(5), 356–378.

Whitehead, A. L. (2010). Sacred rites and civil rights: Religion's effect on attitude toward same-sex unions and the perceived cause of homosexuality. *Social Science Quarterly, 91*(1), 63–79.

Wilcox, C. (2007). Preface. In C. A. Rimmerman & C. Wilcox (Eds.), *The politics of same-sex marriage* (pp. ix–xv). London: University of Chicago Press.

Wilcox, C., Brewer, P. R., Shames, S., & Lake C. (2007). If I bend this far I will break? Public opinion about same-sex marriage. In C. A. Rimmerman & C. Wilcox (Eds.), *The politics of same-sex marriage* (pp. 215–242). London: University of Chicago Press.

Wilcox, C., Merolla, L. M., & Beer, D. (2006). Saving marriage by banning marriage: The Christian right finds a new issue in 2004. In J. C. Green, M. J. Rozell, & C. Wilcox (Eds.), *The values campaign? The Christian right in the 2004 elections* (pp. 56–75). Washington, DC: Georgetown University Press.

Wood, P. B., & Bartkowski, J. P. (2004). Attribution style and public policy attitudes toward gay rights. *Social Science Quarterly, 85*(1), 58–74.

Chapter 175
Moral Hazard, Governing Culture and Localized Christian Right Gay Panic in Indiana

Christopher A. Airriess

175.1 Introduction

Christian conservatives possess a long historical engagement and indeed privileged position in effecting the character of national culture and identity in the U.S. Only since the 1970s, however, have religious institutions become institutionally embedded in the political process (Diamond 1998); no longer occupying the marginalized political spaces of American life, a politicized Christian conservative community emerged to become what we commonly refer to today as the 'religious or Christian Right' (Fetner 2008: 1–22).[1] The political activism of the religious right is anchored specifically in the "hot button" issues of feminism, abortion, and gay rights that are perceived to threaten the centrality of "traditional values" in an "exceptionalist" American civilization (Adam 2003; Herman 2000). The success of the Christian Right to directly embed conservative Christian values into mainstream politics rests on the ability to efficiently construct and deploy grassroots institutionalized networks at both national and local scales (Diamond 1995). While the ability of the Christian Right to legislatively change social policy in some states was only marginally successful during the 1970s and 1980s, their power to significantly affect legislative change with reference to gay rights was transformed in the 1990s and the first decade of the twenty-first century as this social movement became embedded in the national Republican Party, and especially in Republican majority

---

[1] The use of the term Christian Right in this chapter refers to Evangelical Protestants despite socially conservative Roman Catholics, Jews, and Muslims also possessing aggressive anti-gay sentiments. Nevertheless, in the Roman Catholic Church, theological arguments against homosexuality are not based on biblical scripture, but through teachings on natural law; practicing homosexual activity is a moral sin, but not being a homosexual. This seemingly contradictory distinction is based on the belief that all non-procreative sex is sinful (Fejes 2008).

C.A. Airriess (✉)
Department of Geography, Ball State University, Muncie, IN 47306, USA
e-mail: cairries@bsu.edu

© Springer Science+Business Media Dordrecht 2015
S.D. Brunn (ed.), *The Changing World Religion Map*,
DOI 10.1007/978-94-017-9376-6_175

state governments (Green et al. 1998; Oldfield 1996). As a result, the anti-gay agenda of the religious right has made significant strides through political opportunism in many state legislatures.

The purpose of this chapter is to conceptually contextualize a moment in conservative Christian political action producing the revocation of the Indiana Youth Group (IYG) specialty auto license plate in Indiana by a group of religiously conservative Republican senators only 10 weeks after the plate was first issued on January 1, 2012. The IYG is a support organization for lesbian, gay, bisexual, and transgender (LGBT) teenagers. To better understand this anti-gay political action this chapter is situated in the two seemingly unconnected, but conceptually fruitful and complimentary moral hazard and governing cultures literatures. The concept of *moral hazard* possesses explanatory value because the specialty license plate is perceived by anti-gay agents as a quotidian or commonplace object to not only promote a sinful gay lifestyle as culturally acceptable, but more importantly as a recruiting tool to increase the ranks of gays in Indiana. The *governing cultures* literature is especially rewarding because it investigates how governments institutionally shape cultural identities through both state and non-state actors.

This chapter first provides a brief introduction to the concepts of moral hazard and governing cultures followed by how these concepts respectively inform our understanding of the multiple and contradictory anti-gay ideological foundations of the Christian right. This is then followed by a description of why political opportunities for religiously-led gay rights battles are more successful at local geographical scales using the state of Indiana as an example. The empirical section of this chapter then harnesses the moral hazard and governing culture lenses to frame the role of religious conservatives as both non-state and state actors in the eventual revocation of the IYG specialty license plate. While this chapter provides just one local scale example of the political opportunism of anti-gay religious organizations, we are able to gain a fuller appreciation of the Christian Right's deep influence to govern a pluralistic culture in a time moral panic.

175.2   Conceptualizing Moral Hazard and Governing Culture

Despite being anchored in the discipline of economics, the concept of moral hazard is useful to explain the social world of because of the complex interaction of the economic with individual and institutional cultural behavior (Rose and Miller 2008). The concept of moral hazard is ideologically linked to conditions of risk and conservative neo-classical economics, and its evolution and application has changed based on large scale economic and political structural transformations (Beck 1995). First used by the fire insurance industry in the Victorian period and thus overly concerned with morals, the concept of moral hazard is centered on the "interrelated dynamics of character and temptation" (Baker 2000: 563) with reference to risk. In other words, moral hazard as perceived from the insurer's perspective was centered on the calculation of the moral character of the insured to reduce the risk of fraudulent behavior. The unsettling sense of individual temptation as a moral hazard is, of course, the doubt that it insinuates (Ewald 1999) because "moral hazard works through the individual psyche" (Stone 2002: 53). Simultaneously, the insuring institution contract could also produce moral hazard if it entices an otherwise moral person to engage in immoral behavior because of being over insured. By the mid-twentieth century, however, this earlier paradigm of responsibility was replaced by a paradigm of solidarity whereby various federal government programs of the 'welfare state' reduced the individual risk of unforeseen personal events by making individuals part of a larger national community of citizens anchored in the redistribution of risk, fairness, and civil rights. This risk paradigm of social solidarity based on citizenship began to crumble in the 1970s with the neoliberal retreat of state institutions in managing individual risk (Ericson et al. 2000) and the establishment of a paradigm of precaution that essentially marks a retreat to the Victorian era paradigm of individual responsibility anchored in moral character and temptation.

The governing cultures lens redefines culture as "the institutions, symbol systems, and forms of regulation and training responsible for forming, maintaining and/or changing the mental and behavioral attributes of the population" (Bennett 1989: 10, 1999: 372). It is especially useful in understanding culture change at the local scale because it combines culture with statist approaches and thus entertains both agency and structure in the analysis of culture change in the modern era (Marston 2004). The governing culture approach is also rewarding because it is connected with the growing distrust of large government institutions. Unlike highly hierarchical state-based governance, a localized de-centered governance structure or "governance beyond the state" is based on horizontal networks and "interactive relations between independent actors who share a high degree of trust within inclusive participatory institutional or organizational associations" (Swyngedouw 2005: 1995). Perceived as grassroots in nature, individuals and non-government institutions become political stakeholders rather than just political citizens and thus develop greater self actualized agency as a result of their stakeholder status embedded in network matrices. These new identities of empowerment involve not only how to govern, but that these stakeholders possess the right to govern (Swyngedouw 2005; Dean 2003). The concept of governing culture is especially useful to explain how institutions reduce the temptations associated with moral hazard. For example, the imbrication of the two concepts would be: if government provides welfare benefits for the poor this would incentivize social behaviors that discourage the poor from seeking gainful employment. This is why moral hazard is often harnessed to govern culture because it is widely held cultural logic is used to undermine sympathies for marginalized populations (Smart 2008).

175.3   Moral Hazard, Governing Culture and Christian Anti-gay Ideology

Much of the Christian Right's anti-gay public narrative today does not include references to God or Bible passages for fear of alienating public support, but their most essential theological beliefs, nevertheless, inform their more public practices to eliminate moral hazard. Much of the Christian Right's struggle with gays is theologically anchored in biblical literalism and the foundational narrative, at least for many its leaders, is the apocalyptic Book of Revelation eschatology, or the belief in the end of times for the world especially in its pre-millennial form (Boyer 1992). Essentially, Christ's second coming will not take place until God's kingdom is established worldwide for 1,000 years and this will only transpire if all forms of evil are destroyed. This foundational belief leads the Christian Right to interpret a wide variety of social, political, and economic evils challenging the end of times scenario. For example, because the nuclear family with a submissive wife is central to leading a conservative Christian life, gays, by virtue of their inability to biologically reproduce offspring, construct alternative social spaces that are barriers to their eschatological vision of God's kingdom (Fetner 2008: 6–8). The more extreme anti-gay groups or individuals perceive pro-gay groups as representing a dangerous force of darkness or even the anti-Christ (Burack 2008; Herman 2000) as eschatologically imagined.

How does the concept of moral hazard inform the modern Christian Right's anti-gay narrative? First, the fundamental anti-gay narrative is that gays are not genetically predisposed to being gay or possess a "gay gene," but being gay is a "learned pathology" supported by so-called neutral scientific research (Robinson and Spivey 2007; Harrub et al. 2004; Hicks 2003). Thus as individuals are able to choose to be Christian or not, gays are able to choose to be "normal" and eliminate risky behavior caused by the moral hazard of temptation. This in part explains the establishment of the 'ex-gay' movements such as Exodus International or People Can Change that provide gays with conversion therapy to reorient their homosexual temptations; saving homosexuals means saving America (Burack 2008). The situational moral hazard of temptation is perceived to increase with gay couples and child adoption because the anti-gay movement perceives same-sex marriage as sin, not conforming to the notions of the "normal family," and because gays as overly sexualized and promiscuous, gay couple adoption allows for situational moral hazard of child abuse and pedophilia. At the very least, a confused sexual orientation on the part of the adopted child would be the result (Hicks 2006).

Because it is the family that functions as the first line of defense for children against the moral hazard temptations of the secular world, it is not surprising that the names of anti-gay organizations who refer to themselves as pro-family often possess the words children or family in their names. The first anti-gay organization, for example, was Anita Bryant's Save Our Children Inc. in 1977 and later changed to Protect America's Children. A well regarded anti-gay book *Children at Risk: What You Need to Know to Protect Your Children* (1990) is authored by anti-gay activist James Dobson who is the founder of the pro-family organization Focus on the Family (1988). Two other high profile and national anti-gay organizations are the Family Research Institute (1987) and the American Family Association (1988).

How might the practice of governing culture by the Christian right eliminate the moral hazard of homosexuality? Believers in an apocalyptic eschatology easily construct conspiracy scenarios that promote anti-Christian evils such as homosexuality. Principal among these threats are a variety of large scale institutions whether public or private, associated with modernity and secular humanism. The perceived monolithic power of large governments should only be wielded by God who ideally only works through individuals and churches. Large governments are perceived to be immoral and thus conspiratorially promote a pro-gay, anti-family, and an anti-God agenda in public schools, the military, the legal and welfare system as well as government funded institutions such as National Public Radio and the Public Broadcasting System (Burack 2008; Herman 2000: 147). A more extreme view is that these government institutions indirectly assist gays in recruiting heterosexuals into the gay lifestyle or at least as institutions, promote the "gay cause" (Green 2000: 125).

Ideologically anchored in American exceptionalism, Christian right leaders believe that America is pre-ordained as a Christian God-centered state and that the government should exercise the will of God by promoting Christian values. While leaders of the Christian right believe that the individual is responsible for his/her own morality, the state should legislate morality, particularly to combat the perceived enemies of Christianity comprising the forces of modernity and secular humanism that threaten their moral and utopian vision for the country (Herman 2000: 147–148). There is, of course, an obvious contradiction to this ideological view because the Christian Right is simply replacing one type of big government for another. Indeed, actualizing conservative Christian theology as an apocalyptic imaginary at the national scale elevates religious belief to "religion as ideology" (Williams 1996: 371–372) or what is referred to as "religious politics" or "moral politics" that is simply a form of "neo-fascism" (Berlet 2008).

Concurrent with reducing or eliminating altogether this perceived monolithic and secular government, is a parallel process of geographically scaling down power to smaller scale agents, whether it be state or local governments, and ultimately families and individuals. For the Christian Right, "the state does not simply need to be rolled back; rather, ownership must be transferred" (Herman 2000: 150). Indeed, during times of perceived political, economic and social crisis, new conceptions or configurations of power emerge in response to this moral panic (Barkun 1985). During past "Great Awakenings" in U.S. history that were characterized by "a search for a new meaning, order, and direction in a society which finds that rapid change and unexpected intrusions have disrupted the order of life" (McLoughlin 1983: 108), conservative Protestants challenged the power of national political elites. In the present period of perceived moral crisis, elites continued to be challenged, but scaled down governance structures have emerged that allow for a transition from political citizenship to "stakeholder" governance as well, one of which is a "rightholder or an individual that possesses the right or entitlement to engage the political process" (Schimitter 2000 in Swyngedouw 2005). As individual members of a larger and networked community with shared cultural or religious identities, evangelical Protestants feel that they too have the right to govern (Dean 2003), theologically anchored in Christian "dominionism" (Goldberg 2007).

175.4 Christian Political Opportunities at the National and Local State Scales

For much of U.S. history, fundamentalist Protestants have traditionally isolated themselves from a rapidly modernizing, pluralistic, secularizing, and increasingly ethnically diverse country. By the 1940s, however, fundamentalist religious institutions experienced further fragmentation. While these new institutions continued to practice a faith anchored in biblical literalism and the desire for spatial isolation from political and secular influences, they importantly established networks between like minded churches and engaged in aggressive evangelism. These quasi-religious networks which are distinctive to non-denominational modern Evangelical Protestantism included the establishment of liberal arts colleges, bible institutes, youth bible camps, radio programming, and global missionary activity in an effort to mold an evangelical Protestant identity (Fetner 2008). The National Association of Evangelicals was established in 1942 thus giving birth to the modern Evangelical Protestant religious movement that became particularly strong in the South and rural Midwest (Smith et al. 1998).

175.4.1 National Scale Political Opportunities

This process of institutional networking among evangelical Protestants was critically important to govern the perceived moral hazards of the pro-gay movement that emerged in the 1960s. During the 1950s and 1960s, there existed little in way of a gay moral hazard as this sub-culture was geographically and socially marginalized in a handful of American cities. The social marginalization of gay culture was also enabled by sensationalized accounts in the national media that conflated the "common sense" knowledge of homosexuality with child pornography and molestation (Fejes 2008). It was only with the perceived social and moral panic accompanying the civil rights movement in the 1960s, plus the increased normalization of gay culture and gay rights organizations in the national media beginning in the 1970s, did Christian Right organizations embark on their national scale mission of governing culture through political institutions. Indeed, it was during the 1970s and 1980s period that a variety of non-denominational and national scale Christian Right organizations emerged such as Focus on the Family (1977), the American Family Association (1977), the Moral Majority (1979), American Values Coalition (1980), and the Christian Coalition (1989). While these were multiple issue based groups, homosexuality was a common project, and it was children that became a common anti-gay theme. In her successful 1977 crusade against a lesbian and gay anti-discrimination ordinance in Miami, Anita Bryant, for example, imbricated moral hazard and governing culture narratives by claiming that such a law would "violate my rights and all the rights all the decent and morally upstanding citizens…to provide their children with a morally healthy environment" (Fejes 2008: 2).

While these Christian Right organizations were very successful in mobilizing a conservative Christian base and fund raising, their governing culture victories in the 1970s through morality politics were less than completely successful despite possessing a larger base, financial resources, and a privileged position in American culture when compared to pro-gay activists (Burack 2008). Since the 1980s, however, Christian Right anti-gay activists have dramatically increased their governing culture power through moral hazard narratives positively impacting national scale Republican Party support. Unlike in most other Western industrialized countries, federal government legal protection to accord gays cultural and not just political citizenship status has "proceeding at a glacial pace" (Adam 2003: 261). For example, the 1993–1995 103rd Congress with both Democratic House of Representative and Senate majorities and a Bill Clinton White House, was marked by the failure to repeal the ban on gays in the military as well as the passing Defense of Marriage Act (DOMA) which defines marriage as between one man and one woman. In addition, inclusion of gays under the provision of federal civil rights laws has always floundered, unlike in Canada where gays are legally protected and the argument has always been couched as a human rights issue (Smith 2008).

It is within the context of the civil rights issue that both anti-gay Christian Right and pro-gay organizations have fused both moral hazard and governing culture narratives. The anti-gay moral hazard narrative is that public institutions could also produce conditions for individual temptation if institutional laws or policy invite seduction of otherwise moral individuals. Indeed, because evangelical Protestant belief is anchored in legislating morals, their actions are able to reduce moral temptation by being oppositionally proactive (Green 2000). These actions are necessary because the Christian Right strongly believe that pro-gay organizations are "ruthlessly aggressive in pursuing their goals" and are "the most well organized and most disciplined pressure group[s] in the country today" (Herman 2000: 143). The goal then is to deprive gays of basic civil and human rights and thus question their cultural citizenship status in a democracy. As a result, benefits of cultural citizenship accorded to citizens in heterosexual families such as retirement, survivor, disability, and health care benefits to name just a few do not apply to same sex marriages (Donovan 1997).

Gay rights organizations promoting the same sex marriage cause do so on federal civil rights grounds because as a minority community such a basic right is self-evident, and is thus a profoundly moral question. The federal government then possesses a "moral opportunity" to address an issue of fairness and equality so central to democracy (Stone 2002). Nonetheless, anti-gay groups strategically harness a civil rights argument against the inclusion of gay rights as a democratic right claiming that unlike other ethnic, racial or disability communities that deserve to be covered by civil rights laws, being homosexual is a choice, thus making them a "special interest" group (Hardisty 1999). In doing so, this antigay narrative of choice is deployed to convince the American public that the inclusion of gays in civil rights laws would simply increase the moral hazard and thus risk of a multitude of other questionable special interest groups to be tempted to seek civil rights protection (Burack 2008: 73–76).

175.4.2 Local Scale Political Opportunities

While it is true that the Christian Right has been able to greatly influence social policies of the Republican Party at the national scale (Hacker and Pierson 2005), this is not solely because the Evangelical electorate has grown substantially[2] or because of the growth of a more socially conservative electorate, but because of the distinctive nature of American political institutions that favor the success of social movements at more localized geographical scales (Smith 2008).

With both groups being dissatisfied by the absence of federal government efficacy to promote their respective goals, they have turned to more local scale political institutions. This is not surprising because in many respects, and certainly with respect to the issue of gay rights, it is more local, political institutions rather than national that possess this legal efficacy (Smith 2008). The tradition of federalism that gives greater power to state political institutions under the rubric of state's rights in an effort to promote more local or grassroots democracy has reduced the power of the federal government because states possess their own constitutions that can be amended, and independent state courts allow for political opportunities for social movement actors (Smith 2008). Indeed, between 1995 and 2012, constitutional amendments or state statutes to ban same-sex marriage were passed in 42 states. Only six states (five in New England and Iowa), the District of Columbia, and two Native American Indian tribes permit same sex marriage. Almost half the states possess constitutional bans; all the South and Great Plains states are characterized by this most strict legal sanction. The balance of states possess a combination of civil unions with statutes or constitutional bans on same sex marriage or no law on same sex marriage.

---

[2] While some scholars have identified a modern religious awakening period beginning in the 1960s (Fogel 2000), others claim that no such revival exists because despite much media attention given to the rise of evangelical Protestantism, religious television programming, and the proliferation of megachurches, church attendance has not increased in the past three decades, and in fact, church attendance peaked in 1958 (Wuthnow 2007). Similarly, the growth of evangelical Protestants has not been unusually great either. For example, between 1982 and 2006, the percentage of respondents to who believed in biblical literalism decreased from 38 to 33 %. In addition, evangelical Protestant affiliation only increased from 17–20 % to 25–28 % between the early 1970s to early 2000s and much of that growth is through natural increase or attracting former Roman Catholics, especially Latinos (Wuthnow 2007).

Whether Democratic or Republican parties in the state control legislative power, Indiana possesses a socially conservative and individualistic culture, certainly in comparison to other Midwestern states. This conservative culture is borne out by the absence of gay friendly legislation. For example, in 2004 the state banned same sex marriage with plans to amend the state constitution. Nor does the state possess anti-discrimination laws based on sexual orientation and gender identity for non-public employees or laws prohibiting housing discrimination based on sexual orientation and gender identity. Nor does the large and culturally diverse city of Indianapolis possess domestic partner benefits for public employees. Lastly, the state does not have a hate crimes statute.

Much like other states, Indiana's existing political conservatism has become increasingly more conservative and Republican. For example, every year between 1981 and 2011, the Indiana Senate has been majority Republican and that since 2001, Democrats have never comprised more than 36 % of Senate members. The 2010 election witnessed dramatic gains in Republican majorities; in 2009, Republicans comprised 66 % of the Senate membership, but increased their share to 74 % of membership. The same is true for the House as well; after slim Democratic majorities in seven of the ten years throughout the 2000s, the Republican share of House members was 48 % in 2009, but after the 2010 elections Republicans became the majority with 60 %. This coupled with a two term Republican governor plus a greater mainstreaming of evangelical Protestantism, provided an ideal political opportunity to inject a more socially conservative agenda to govern the state's culture. This electoral 'perfect storm' was heavily influenced by the Tea Party and their libertarian and anti-government sentiment. Importantly, the Tea Party has been gradually populated by an evangelical Protestant sub-culture producing imbrications of economic libertarianism, anti-government, anti-gay, and pro-life narratives that possess whiffs of Christian eschatology, dominionism and moral panic (Berlet 2012; Clement and Green 2011). Indeed, a 2012 popular media book by David Brody, the chief political correspondent for the Christian Broadcasting Network titled *Teavangelicals* describes the intertwined nature of these two institutional ideologies.

The ability to deploy greater Christian Right governing culture power, however, is enabled by non-state actors comprising networks of like-minded cultural institutions. The most powerful is Advance Indiana or what is referred to as a civil society cultural governor "pursu[ing] a new politics and public management, advocating, facilitating, guiding and moderating the new forms of democratic civil society" (Bang 2003: 20). Founded in 1980, Advance America describes itself as "the state's largest pro-family, pro-church, pro-private and home school, and pro-tax reform organization (www.advanceamerica.com/about.html). NUVO magazine, an alternative weekly publication in Indianapolis describes the multi-issue politics of Advance America as "a blend of patriotism and religion meant to shape government by giving a voice to a variety of fears and frustrations that resonate with evangelicals" (McPhee 2008). Indeed, the institutional process giving shape to government is overtly displayed on the organization's home page: "Advance America's staff read and review every bill filed in the Indiana General Assembly (more than 1,500 in some years). They evaluate each bill, offer testimony before legislative committees, talk directly with legislators, draft amendments and bills, sound the alarm when individuals need to call their legislators, and train people how to lobby" www. advanceamerica.com/about.html. Sounding the alarm is only possible by its Rapid Action Network comprising 45,000 families, 1,500 businesses, and 3,700 churches statewide. This Rapid Action Network was instrumental in lobbying the state legislature in 2006 to produce the 'In God We Trust' standard automobile license plate (Airriess et al. 2012) and aggressively lobbied for the revocation of the IYG specialty license based on a moral hazard narrative.

175.5   The IYG Specialty License Issue

Founded in 1987, The mission statement on the organization's homepage is "Indiana Youth Group (IYG) provides safe places and confidential environments where self-identified lesbian, gay, bisexual, transgender, and questioning youth are empowered through programs, support services, social and leadership opportunities and community service" (www.indianayouthgroup.org/about-us). With three full time and six part time staff, IYG also provides LGBT sensitivity training to faith-based organizations, public schools, mental health organizations and corporations. IYG has also been instrumental to establishing gay/straight alliances in public schools. As with most other non-profit organizations in the state, one of the fund raising opportunities is through the purchase of specialty auto license plates of the sponsoring non-profit organization. In May 2012, there existed over 100 specialty plate designs with the most numerous representing 33 colleges and universities and 53 "special interest" organizations. These "special interest" organizations range from the National Rifle Association, the Boy Scouts, the Autism Society, and Habitat for Humanity. These specialty plates carry an extra fee and part of that fee is donated to the sponsoring organization.

    IYG's negative experience to obtain their own specialty plate was a harbinger of things to come.[3] The IYG first applied to the Bureau of Motor Vehicles (BMV) in early 2008, but was told that no organization received permission for specialty plates that year and that the BMV was revising its specialty plate requirements for 2009. In that year, the only requirement that changed was that organizations needed 500 signatures; IYG met that requirement and reapplied, but was again turn down without being informed by the BMV which requirement was not met. In response, the IYG with the assistance of the American Civil Liberties Union of Indiana sued the BMV in 2010 requesting that the decision making process be more public and transparent. The BMV then asked the IYG to drop the suit and in return the BMV suggested that their specialty plate request would "most likely" be granted if the

---

[3] Unless cited, much of this information describing IYG's experience with the BMV as well as the license plate revocation process in the state legislature is from internal IYG documents and numerous newspaper sources.

IYG reapplied in 2011. The plate was approved, and became available on January 1, 2012; with 737 plates issued, it was the most popular specialty license plate among the ten new 2012 specialty plates until its revocation 10 weeks later. The popularity of the plate was greatest in urban counties with approximately half of the counties in the state with no sales. The two urban places with the greatest number and per capita sales were Indianapolis and Bloomington; characterized by a more educated and youthful population because it is home to Indiana University, per capita sales were slightly higher in Bloomington.

175.5.1   Governing Moral Hazard: Non-state Actors

Within two weeks of the plate's availability, the local Indianapolis Fox News affiliate posted a story about the plate on its website claiming that Indiana was the first state to possess such a LGBT-themed plate (Maryland was actually the first and South Carolina now possesses one) and this was when the Christian Right began its battle against the "pro-homosexual" license plate.

In response to the news story, the first serious attempt to subvert the IYG specialty license plate and thus govern culture at the local or state scale was pursued by Advance Indiana. In the effort to influence those that govern culture, Advance Indiana distributed an electronic flyer through their Rapid Action Network to apply grassroots pressure on state legislators to eliminate the IYG plate (Fig. 175.1). As expected, there is a complete absence of a religious narrative, for fear of alienating its audience; instead, the flyer is anchored in the imbricated moral hazard and governing cultures narratives. In terms of moral hazard, the license plate is represented as an instrumental technique to recruit or 'target' innocent children as young as 12 years old based of course on the belief that being gay is a choice in that children have yet to determine their sexual orientation. Simultaneously, Advance Indiana's statement "it is not ok for the State of Indiana to put its stamp of approval on a pro-homosexual group" harnesses the logic that the state is in effect institutionally producing a moral hazard temptation or incentive on the part of "predatory" adult gays. Micah Clark, the executive director of the American Family Association of Indiana,[4] and who works closely with Advance America on a wide variety of issues, stated much the same without any reference to religion when he asked "[s]hould the state really be part of funding groups in which 12 through 20-year olds are shown how to

---

[4]The American Family Association of Indiana is the state affiliate of the Christian-based national organization based in Tupelo, Mississippi. Its original mission in the late 1970s was to combat "indecency" in the popular media, but in the 1980s began a long tradition of targeting gay culture. An example of its virulent and conspiratorial anti-gay discourse comes from Bryan Fischer, AFA's Director of Issue Analysis for Government and Public Policy, when he claimed on his radio program in 2010 that "[h]omosexuality gave us Adolph Hitler, and homosexuals in the military gave us the Brown Shirts, the Nazi war machine and six million dead Jews" (www.splcenter.org/get-informed/intelligence-files/groups/american-family-association). The Southern Policy Law Center has labeled the AFA a hate group.

URGENT LEGISLATIVE ALERT

Warning: Discretion should be utilized before viewing the Indiana Youth Group website and no minor should be permitted to access the site! We did not list some of the items from their website because we are a pro-family group!

BMV Approves Pro-Homosexual License Plate!

INDIANA AA9999
INDIANA YOUTH GROUP

Help Protect Children – Revoke License Plate Now!
This Should Not Happen In Indiana!

A **pro-homosexual group**, the Indiana Youth Group, that <u>targets children as young as 12 years of age</u> was given a special license plate by the Bureau of Motor Vehicles recently!

Other Groups Have License Plates That Support Worthy Causes!

Other groups have a special license plate that has been approved by the legislature or the Bureau of Motor Vehicles. These include worthy causes such as: Riley Children's Hospital, military veterans, and Indiana colleges and universities. That's ok!

But it is <u>not</u> ok for the State of Indiana to put its seal of approval on a **pro-homosexual group** **that targets children as young as 12 years of age!**

It's not ok for the State of Indiana to approve a special license plate to be used by a <u>pro-homosexual group that targets children as young as 12 years of age</u> to help them raise money for their efforts!

You Won't Believe What They Do!

Here is some information from the website for the Indiana Youth Group, www.IndianaYouthGroup.org, (this is not a church youth group!)

Meetings:
- Girls Who Like Girls Group
- Sex, Dating and Relationships
- Guys Who Like Guys Group
- Why Worry? Free HIV/AIDS, STD/STI Testing

Remember, this pro-homosexual group targets children as young as 12 years of age!

Help Revoke The License Plate Now!
Urgent Action Required!

Please contact your State Senator and State Representative with the following message:

"Please immediately revoke the special license plate for the Indiana Youth Group. This is a pro-homosexual group that targets children as young as 12 years old! Please recall all license plates that have already been issued. Please let me know what you will do."

Senate: **317-232-9400** or 800-382-9467
House: **317-232-9600** or 800-382-9842

*Visit **AdvanceAmerica.com** to find out who your Senator and Representative are!

ADVANCE AMERICA

101 WEST OHIO ST., SUITE 660
INDIANAPOLIS, IN 46204
WWW.ADVANCEAMERICA.COM
(317) 684-3300

ERIC I. MILLER
FOUNDER & EXECUTIVE DIRECTOR

Fig. 175.1 Electronic flyer sent by AdvanceAmerica.com via Rapid Action Network regarding the Indiana Youth Group specialty plate. The only visual clue that this license plate is LGBT (lesbian, gay, bisexual and transgender)-themed is the rainbow-colored arms comprising the plate's logo (Source: Flyer released for public distribution by AdvanceAmerica.com.)

use latex devices for sexual behaviors and how to remove the stigma and perceived risk of doing so?" (Clark 2012). The harnessing of moral hazard is self-evident as the two conflated phrases "pro-homosexual group" and "targets children as young as 12 years of age" appear five different times on the flyer and are evenly distributed throughout the document to remind the reader in a quotidian fashion, of this most basic moral hazard.

The flyer's governing culture narrative is equally as illuminating. First, the phrase near the top of the flyer "This Should Not Happen In Indiana" is telling of Advance America's perception that the conservative Christian culture of Indiana functions as a bulwark to the geographical spread of secular humanism that is often imbricated with acceptance of a gay lifestyle; America's exceptionalist nature may be waning, but not Indiana's. Advance America's role in governing culture at the local scale is in "mobiliz[ing] reified notions of history and roots" (Edensor 2002: 25) whereby internal territorial space is closed down (Craib 1998) in a geographical process of "digging in" (Coleman 2004: 15) often associated with moral panic. Additionally, Advance Indiana engages the civil rights issue to make its case that gays are not deserving of protection, nor should they be regarded as a special interest group. Other groups who have specialty plates are worthy of support, but not a pro-homosexual group. In making this claim, gays are even further marginalized as a minority wishing inclusion in mainstream culture.

175.5.2 Governing Moral Hazard: State Actors

The first government anti-gay and governing culture salvo originated from Republican Representative Jeffrey Thompson who represents a largely rural district in the western exurbs of Indianapolis; Representative Thompson is an ex-high school teacher and member of a non-denominational mega-church. On January 30, Representative Thompson introduced an amendment to House Bill 1197 regulating moped registration and use to mandate that the BMV retroactively deny any organization's specialty license plates that "encourages, promotes or supports all of the following": (1) use of illegal drugs, (2) use of tobacco or tobacco products by a minor, (3) sexual activity involving a minor, (4) sexual activity outside of marriage, (5) use of pornography, (6) abortion, (7) disrespect for parents, the family, religion, healthy behavior, or good citizenship, (8) the undermining of the Constitution of the United States or the Constitution of the State of Indiana, (9) undermining of the law or the overthrow of the government of the United States, (10) a lifestyle that is contrary to the marriage of one man and one woman (www.in.gov/legislative/bills/2012/HAMF/MO119701.001.html). The first nine sections of the amendment are an attempt of course to hide the final section that addresses anti-gay sentiment. Representative Thompson's amendment was not accepted.

Simultaneous to Representative Thompson's amendment was a bill to redefine BMV practices that passed the Senate (SB0327) and was sent to the House Committee on Roads and Transportation chaired by Republican Representative

Edward Soliday. Much like Thompson, Soliday represents an exurban district in northwestern Indiana; Soliday is a military veteran and is a member of a conservative Baptist church. Soliday made an amendment to SB 0327 that in effect indirectly targeted the IYG specialty plate. The amendment attempted to eliminate all specialty plates approved by the BMV in 2011 that did not sell at least 1,000 plates. The existing threshold was 500 plates and the IYG plates had yet to reach 1,000 plates sold, as did other new specialty plates. The second part of Soliday's amendment was to transfer the decision making power of approving specialty plates from the BMV to the state legislature. This agenda is a common one among conservative politicians and is central to the governing cultures process because when compared to democratically elected representatives of the people, government bureaucracies populated with political appointees know little of popular sentiment. Indeed, when interviewed by a local radio station concerning this attempt to transfer decision making to those that 'represent the people,' Senator Long, President Pro Tempore, stated "[m]ost people think that the legislature ought to probably end up having control again" (WIBC 2008). The amendment passed on February 29 and came up for a second reading the next week. There were, however, an additional six amendments, one of which was a paring down of Representative Thompson's lengthy amendment in HB 1197. In part because of only a few days left in the already shortened 2012 legislative session, and the inability of Soliday to convince Thompson to further modify his amendment, Soliday let SB 0327 die without a vote.

With only one day remaining in the legislative year, Senator Long informed the press that an institutional solution to eliminate the IYG license plate had been discovered. Senator Long informed the press that a group of 20 Republican senators had discovered a violation of IYG's contract with the BMV. The senators claimed that IYG had used its low-numbered specialty plates as thank you gifts to some of IYG's donors in direct violation of BMV rules, although the rules read "sold" or "auctioned." That same day, the group of 20 senators sent a letter to the BMV requesting an investigation of this rules violation, but the BMV failed to act on this request because other organizations with specialty plates had done the same in the past. The following day, the BMV spokesperson who stated the reason for inaction to a local newspaper reporter was summarily fired. One week later, the BMV Counsel General informed the IYG that the sales of its specialty plate would be immediately suspended. To not give the appearance that IYG was specifically targeted, the BMV also revoked the specialty plates of the 4-H Foundation and Greenways organizations for the same reason.

Many of the 20 senators requesting that the IYG plate be revoked possess conservative Christian, if not Christian Right credentials. The following biographical information of the 20 senators is incomplete and relies on a variety of on-line sources using an ethnographic-like methodology. In terms of religious affiliation, 19 of the 20 senators are Protestant and eight of these are members of decidedly conservative or evangelical churches at either the denominational or congregational level; the face of governing culture is no longer one in which evangelical non-state actors network with government actors, but now evangelicals are state actors. For example, one worked for Focus on the Family, one introduced creation education in schools legislation, one is executive director of the Confessing Movement of the United Methodist Church, an evangelical movement to contest gays in church positions, one is a member of Gideons International, one attended a Christian University and is a member of both Advance Indiana and the American Family Association, one held an administrative position at a Christian school, and one legislatively supported the Ten Commandments on public property. In addition, there are other socio-demographic variables that are commonly associated with Christian Right activists (Burris 2001; Green 2000). Of the 20 senators, 11 are small business owners and some of these, plus other senators are or were associated with law enforcement; of the four in law enforcement, three possess law degrees and worked in county prosecutor's offices and one was a county sheriff. Five senators are military veterans. In addition, their average age is 60 with five being 65 or older with only one being younger than 40 years old. Lastly, of the 20 senators, only three are female. This last observation is important because regardless of the degree of religiosity, women are more pro-gay rights when compared to men (PRC 2012).

Equally as important, 17 of the 20 senators geographically represent largely rural districts located throughout the state. That 85 % of senators who signed the letter requesting the BMV to revoke the IYG specialty license plate are from rural districts is expected based on the "traditionalist" and "modernist" cultural framework that is commonly used in research addressing the intersections of the culture wars and politics (Airriess et al. 2012; Luebke 1990) and more specifically, the culture wars and anti-gay political behavior (Brown et al. 2005; Ormrod and Cole 1996). These competing moral visions of the world possess attributes that directly influence the degree of political activism toward gay civil rights. Traditionalist or social conservatives are characterized as rural or small town inhabitants, status quo social roles, and those who are generally suspicious of big government and egalitarian social movements; in the American context, they are often associated with fundamentalist Protestantism. In contrast, modernists are characterized by urban residents, secularization, higher educational attainment and the promotion of activist government and social change (Luebke 1990).

The remaining three senators who signed the BMV letter represent suburban districts of Indianapolis, two of which include exurban space. A suburban district is not surprising as Evangelical Protestantism has increasingly become mainstream in suburban America (Luhr 2009; Cavalcanti 2007). Importantly, the three senate districts are located in the eastern and southwestern edges of the urban region, and not the far more affluent suburbs of the northern half of the Indianapolis urban region. Lastly, the 20 senators felt politically comfortable to engage in this brazen antigay political opportunism in part because they represent relatively safe seats; in the last election cycle, four ran opposed, 11 won with margins greater than 60 %, and only five senators won their election with between 55 and 51 % of the vote.

The fallout from the action of the 20 senators to revoke the IYG specialty license plate illuminates the increased role of religious conservatives in governing the state's culture. The dependably staid Indianapolis Star, the state's largest newspaper, stated in an editorial "[c]an Indiana's state government get away with such uneven, heavy-handed actions? Maybe. Does it violate a basic sense of fairness?

Without question." (INDST 2012). Similarly toned editorials appeared in other newspapers around the state. Even the Republican governor, with a history of not aggressively pursuing social issues, responded tersely to a reporter's question about whether the state should change its specialty license plates policies with "I don't know and I don't care' (Schneider 2012). Some Republican senators claimed that they will review specialty license plate policies in the next legislative session. Rep. Soliday convened a study committee in the summer, but was unable to obtain a majority recommendation to transfer the decision making power of approving specialty plates from the BMV to the state legislature. IYG plans to legally challenge the BMV's decision and since the revocation of the license plate, donations to the IYG have dramatically increased to essentially challenge the sole right of Christian conservatives to govern culture in a pluralistic society.

175.6 Discussion

Unlike so many other Western secular democracies, religious institutions in the U.S. continue to possess great influence in effecting national scale cultural identities. In turn, politicized religious institutions comprising the networked Christian Right have successfully wielded their governing culture power in contesting gay civil rights. Their strident anti-gay movement, however, has been most successful at the local scale where political opportunities emerged as a result of recent Republican Party election dominance. Non-state and state actors of the Christian Right in Indiana, and indeed in many U.S. states have experienced almost unqualified political and legal success in their networked religious crusade to govern culture and eliminate perceived moral hazards associated with "homosexuality." Indeed, the revocation of the IYG specialty license plate illustrates the very local nature of the Christian Right's governing culture practices. Considering these broad achievements then, what is it about such a seemingly banal material object in the material form of a specialty license plate that prompted the orchestrated and panicked challenge by the Christian Right against the IYG plate? The simple answer is that the Christian anti-gay movement will take advantage of any political opportunity to deny movement towards normalizing gay culture; according civil rights protection to gays would in effect institutionalize acceptance of gay culture and introduce moral hazard incentives for immoral behavior.

For the Christian Right, the IYG specialty license plate symbolizes the normalization of gay civil rights, thus making gays moral equals to all the other stakeholder organizations possessing specialty license plates. It is the nature of auto culture in the U.S. that makes license plates the unexpected target of the Christian Right. More than in most any other country, autos are material representations of consumerism, individual identity, freedom, and mobility (Sheller and Urry 2000: 738) and are in a sense platforms for the reproduction of identity through a highly mobile form of public discourse (Airriess et al. 2012; Sculle and Jakle 2008; Edensor 2004). Whether LGBT or not, those that choose to purchase the IYG specialty plate are in effect unconsciously translating their feelings about a salient and personal social issue onto an external material object in public space (Kingsbury 2007). In doing so, they are publically stating that gays should possess stakeholder citizenship rights similar to other marginalized groups. For the Christian Right, the IYG license plate translates or constitutes the institutionalization of the right of gays to cultural citizenship by virtue of the banal, routine, or quotidian nature of the material object and the resulting inability to govern "immoral" culture in public spaces. Similarly, gay pride parades possess a similar moral hazard for Christian anti-gay groups because as Marston observes with reference to contesting gay participation in New York City's St. Patrick Day parade, governing the culture of public space should reflect "how the world is and how it ought to be" (2002: 387).

What constitutes moral hazard or risk, however, is not static in nature because its application changes based on large scale social, economic, and political structural transformations (Beck 1995) and are thus manufactured in the sense that they are subjectively constructed (Mythen 2004). Despite the process of recognizing gays as stake holding political citizens has been painstaking slow when compared to other affluent and secular countries, there does exist demographic evidence that points to the emergence of a less religiously informed and more gay friendly social environment in the U.S. What constitutes moral hazard and who possesses the power to govern culture will be called into question because public opinion polls tell us that America's younger generation, while not being overtly liberal on economic issues, are decidedly less religious and more accepting of same sex marriage and an activist government when compared to previous generations (PRC 2012). Some explain this generational transformation based on the fact that the younger generation are more culturally diverse or grew up in a more diverse social environment and that contrary to conventional wisdom, particularly with reference to social issues, individuals do not grow more politically conservative with age (Leonhardt 2012). This suggests a future that includes a less religiously informed moral hazard environment producing the moral opportunity to accommodate gays as a minority community deserving of civil rights protection.

References

Adam, B. D. (2003). The defense of marriage act and American exceptionalism. *Journal of the History of Sexuality, 12*(2), 259–276.

Airriess, C. A., Hawkins, M., & Vaughan, E. (2012). Situating banal nationalism, the culture wars, and civil religion: Governing localized geographies of national identity in Indiana. *Social and Cultural Geography, 13*(1), 49–68.

Baker, T. (2000). Insuring morality. *Economy and Society, 29*(4), 559–577.

Bang, H. P. (2003). Governance as political communication. In H. P. Bang (Ed.), *Governance as social and political communication* (pp. 7–23). Manchester: Manchester University Press.

Barkun, M. (1985). The awakening-cycle controversy. *Sociological Analysis, 46*(4), 425–443.

Barnett, C. (1999). Culture, government and spatiality: Reassessing the 'Foucault effect' in cultural-policy studies. *International Journal of Cultural Studies, 2*(3), 369–397.

Beck, U. (1995). *Ecological politics in an age of risk*. Cambridge: Polity Press.

Bennett, T. (1989). Culture: Theory and policy. *Media Information Australia, 53*, 9–11.
Berlet, C. (2008). The United States: Messianism, apocalypticism, and political religion. In R. Griffin, M. Feldman, & J. Tortice (Eds.), *The sacred in twentieth century politics: Essays in honour of Professor Stanley G. Payne* (pp. 221–257). Basingstoke: Palgrave Macmillan.
Berlet, C. (2012). Collectivists, communists, labor bosses, and treason: The tea parties as right-wing populist counter-subversion panic. *Critical Sociology, 38*(4), 565–587.
Boyer, P. S. (1992). *When time shall be no more: Prophecy belief in modern American culture.* Cambridge, MA: Belknap/Harvard University Press.
Brown, M., Knopp, L., & Morrill, R. (2005). The culture wars and urban electoral politics: Sexuality, race, and class in Tacoma, Washington. *Political Geography, 24*(3), 267–291.
Burack, C. (2008). *Sin, sex, and democracy: Antigay rhetoric and the Christian right.* Albany: State University of New York Press.
Burris, V. (2001). Small business, status politics, and the social base of new Christian right activism. *Critical Sociology, 27*(1), 29–55.
Cavalcanti, H. B. (2007). *Gloryland: Christian suburbia, Christian nation.* Westport: Praeger.
Clark, M. (2012). *Those who care to know, do care.* Retrieved June 21, 2012, from www.afain.net/those-who-care-to-know-do-care/
Clement, S., & Green, J. C. (2011). *The tea party, religion and social issues.* Retrieved June 10, 2012, from http://pewresearch.org/pubs/1903/tea-party-movement-religion-social-issues-conservative-christian
Coleman, M. (2004). The naming of 'terrorism' and the evil 'outlaws:' Geopolitical place-making after 11 September. In S. Brunn (Ed.), *11 September and its aftermath: The geopolitics of terror* (pp. 87–104). Portland: Frank Cass.
Craib, I. (1998). *Experiencing identity.* London: Sage.
Dean, M. (2003). Culture governance and individualization. In H. Bang (Ed.), *Governance as social and political communication* (pp. 7–23). Manchester: Manchester University Press.
Diamond, S. (1995). *Roads to dominion: Right-wing movements and political power in the United States.* New York: Guilford.
Diamond, S. (1998). *Not by politics alone: The enduring influence of the Christian right.* New York: Guilford.
Donovan, J. M. (1997). DOMA: An unconstitutional establishment of fundamentalist Christianity. *Michigan Journal of Gender and Law, 4*(2), 335–373.
Edensor, T. (2002). *National identity, popular culture and everyday life.* Oxford: Berg.
Edensor, T. (2004). Automobility and national identity: Representation, geography and driving practice. *Theory, Culture and Society, 21*(4–5), 101–120.
Ericson, R., Barry, D., & Doyle, A. (2000). The moral hazards of neo-liberalism: Lessons from the private insurance industry. *Economy and Society, 29*(4), 532–558.
Ewald, F. (1999). The return of the crafty genius: An outline of a philosophy of precaution. *Connecticut Insurance Law Journal, 6*(1), 47–79.
Fejes, F. (2008). *Gay rights and moral panic: The origins of America's debate on homosexuality.* New York: Palgrave Macmillan.
Fetner, T. (2008). *How the religious right shaped lesbianism and gay activism.* Minneapolis: University of Minnesota Press.
Fogel, R. W. (2000). *The fourth great awakening and the future of egalitarianism.* Chicago: University of Chicago Press.
Goldberg, M. (2007). *Kingdom coming: The rise of Christian nationalism.* New York: W.W. Norton.
Green, J. C. (2000). Antigay: Varieties of opposition to gay rights. In C. A. Rimmerman, K. D. Wald, & C. Wilcox (Eds.), *The politics of gay rights* (pp. 121–138). Chicago: University of Chicago Press.
Green, J. C., Guth, J. L., & Wilcox, C. (1998). Less than conquerors: The Christian right in state Republican parties. In A. Costain & A. McFarland (Eds.), *Social movements and American political institutions* (pp. 117–135). Lanham: Rowman & Littlefield.

Hacker, J. S., & Pierson, P. (2005). *Off center: The Republican revolution and the erosion of American democracy*. New Haven: Yale University Press.

Hardisty, J. (1999). *Mobilizing resentment: Conservative resurgence from the John Birch Society to the Promise Keepers*. Boston: Beacon.

Harrub, B., Thompson, B., & Miller, D. (2004). This is the way God made me-a scientific examination of homosexuality and the gay gene. *Reason and Revelation: A Monthly Journal of Christian Evidences, 24*(8), 73–80.

Herman, D. (2000). The gay agenda is the devil's agenda: The Christian Right's vision and the role of the state. In C. A. Rimmermann, K. D. Wald, & C. Wilcox (Eds.), *The politics of gay rights* (pp. 139–160). Chicago: University of Chicago Press.

Hicks, S. (2003). The Christian right and homophobic discourse: A response to 'evidence' that lesbian and gay parenting damages children, *Sociological Research Online, 8*(4), 1–9. Retrieved May 21, 2012, from www.socresonline.org.uk/8/4/hicks.html

Hicks, S. (2006). Maternal men-perverts or deviants: Making sense of gay men as foster carers and adopters. *Journal of GLBT Family Studies, 2*(1), 93–113.

INDST. (2012). *Unfair play in license plate game*. Retrieved March 23, 2012, from www.indystar.com/article/20120323/OPINION08/203230320/Unfair-play-license-plate-game

Kingsbury, P. (2007). The extimacy of space. *Social and Cultural Geography, 8*(2), 235–258.

Leonhardt, D. (2012). Old versus young. *The New York Times*, Sunday Review Section, Sunday, June 24, pp. 1 and 8.

Luebke, P. (1990). *Tar Heel politics: Myths and realities*. Chapel Hill: University of North Carolina Press.

Luhr, E. (2009). *Witnessing suburbia: Conservatives and Christian youth culture*. Berkeley: University of California Press.

Marston, S. A. (2002). Making a difference: Conflict over Irish identity in the New York City St. Patrick's. *Day parade, Political Geography, 21*(3), 373–392.

Marston, S. A. (2004). Space, culture, state: Uneven developments in political geography. *Political Geography, 23*(1), 1–16.

McLoughlin, W. G. (1983). Timepieces and butterflies: A note on the great awakening-construct and its critics. *Sociological Analysis, 44*, 103–110.

McPhee, L. (2008). *Evangelical lobbyist Eric Miller: The most powerful man in the state house*. Retrieved March 10, 2012, from www.nuvo.net/news/article/evangelical-lobbyist-eric-miller-most-powerful-man-in-indiana-state-house

Mythen, G. (2004). *Ulrich Beck: A critical introduction to the risk society* (p. 2004). London: Pluto Press.

Oldfield, D. M. (1996). *The right and the righteous: The Christian right confronts the Republican Party*. Lanham: Rowman & Littlefield.

Ormrod, R., & Cole, D. (1996). Tolerance and rejection: The vote on Colorado's Amendment Two. *The Professional Geographer, 48*(1), 14–27.

PRC. (2012). Partisan polarization surges in Bush, Obama years, trends in American values: 1987–2012, *Pew Research Center for the People and the Press*. Retrieved June 23, 2012, from www.people-press.org/2012/06/04/section-2-demographics-and-american-values/

Robinson, C. M., & Spivey, S. E. (2007). The politics of masculinity and the ex-gay movement. *Gender and Society, 21*(5), 650–675.

Rose, N., & Miller, P. (2008). *Governing the present: Administering economic, social and personal life*. Cambridge: Polity.

Schimitter, P. (2000, September). *Governance*. Paper presented at the conference democratic and participatory governance: From citizens to 'holders'. European University Institute.

Schneider, B. M. (2012). *Fallout over gay youth group's license plate continues*. Retrieved April 15, 2012, from www.indystar.com/apps/pbcs.dll/article?AID=2012203210313

Sculle, K. A., & Jakle, J. A. (2008). Signs in motion: A dynamic agent in landscape and place. *Journal of Cultural Geography, 25*(1), 57–85.

Sheller, M., & Urry, J. (2000). The city and the car. *International Journal of Urban and Regional Research, 24*(4), 737–757.

Smart, A. (2008, July 3–5). *Housing support for the 'undeserving': Moral hazard, stigma and laissez-faire in Hong Kong.* Presentation at Max Planck Institute for Social Anthropology invited workshop on "Who cares … and how? An anthropological inquiry into support", Halle, Germany.

Smith, M. (2008). *Political institutions and lesbian and gay rights in the United States and Canada.* New York: Routledge.

Smith, C., Emerson, M., Gallagher, S., Kennedy, P., & Sikkink, D. (1998). *American evangelicalism: Embattled and thriving.* Chicago: University of Chicago Press.

Stone, D. (2002). Beyond moral hazard: Insurance as moral opportunity. In T. Baker & J. Simon (Eds.), *Embracing risk: The changing culture of insurance and responsibility* (pp. 52–79). Chicago: The University of Chicago Press.

Swyngedouw, E. (2005). Governance innovation and the citizen: The Janus face of governance-beyond-the-state. *Urban Studies, 42*(11), 1991–2006.

WIBC. (2008). *UPDATE: Gay rights group gets to keep specialty tag, though battle may not be over.* Retrieved March 23, 2010, from www.wibc.com/news/Story.aspx?ID=1667580

Williams, R. (1996). Religion as political resource: Culture or ideology? *Journal for the Scientific Study of Religion, 35*(4), 368–378.

Wuthnow, R. (2007). Myths about American religion. *Heritage Lectures, 1049,* 1–9.

Chapter 176
Geographic Support for the Ordination of Same Sex Clergy by American Lutheran and Presbyterian Denominations

Bradley C. Rundquist and Stanley D. Brunn

176.1 Introduction

In the past decade four major denominations have held discussions on removing barriers for gays and lesbians serving as clergy or church leaders; Episcopalians, Lutherans (Evangelical Lutheran Church in America, or ELCA), Presbyterians (Presbyterian Church, USA or PCUSA), and the United Church of Christ (Burke 2011). The issues have been debated, sometimes rather heatedly, by clergy, members and those in support and opposed to such initiatives. The acrimony at times became so intense that some individual churches decided to withdraw, sometimes forming a new congregation, other times keeping the name, but moving to a different location. Property issues were often a major legal point for individual congregations wishing to sever relations with the larger national governing body. Below we examine the backgrounds and accounts of two denominations, ELCA and PCUSA. We investigate through a set of maps and tables the geographic patterns of support and opposition to noncelibate same-sex ordinations by ELCA at the congregational level and the presbyteries (a territorial governing body) and synods of PCUSA.

B.C. Rundquist (✉)
Department of Geography, University of North Dakota, Grand Forks, ND 58202, USA
e-mail: bradley.rundquist@und.edu

S.D. Brunn
Department of Geography, University of Kentucky, Lexington, KY 40506, USA
e-mail: brunn@uky.edu

© Springer Science+Business Media Dordrecht 2015   3361
S.D. Brunn (ed.), *The Changing World Religion Map*,
DOI 10.1007/978-94-017-9376-6_176

176.2 The Lutheran Story

On August 19, 2009, the 11th biennial Church Wide Assembly (CWA) of the ELCA adopted by a two-thirds majority (676 votes in favor to 338 opposed) a Social Statement on Human Sexuality (ELCA 2009a). The Social Statement addresses many issues including marriage, family, children, divorce, sexuality outside marriage, and friendship. It also covers social issues, including sexual abuse and global sex-trade exploitation. The document positions ELCA in opposition to all forms of discrimination against homosexuals and states that it welcomes all people into its congregations, regardless of sexual orientation. The CWA adopted resolutions that allow ELCA congregations, if they choose, to recognize and support lifelong, monogamous, same gender relationships and hold them publicly accountable; and to find a way for people in such relationships to serve as rostered leaders in the ELCA (ELCA 2009b). ELCA rosters include ordained (pastors) and lay church leaders (ELCA 2013). The decision repealed an earlier ELCA social statement that banned homosexual clergy unless they agreed to remain celibate.

In his closing remarks to the 11th CWA on Aug. 22, 2009, Presiding Bishop Mark S. Hanson said:

> When we began the process to develop the social statement on human sexuality, I said that one of my greatest fears was that we would spend more of our time and energy engaging people with whom we agree than those who have different perspectives. That is still my concern. Now – perhaps more than ever – we need to stay engaged with one another. I ask those of you wondering about your place in this church to let us be a part of that discernment. Take time with your decision. Step back. Understand the magnitude of the decision if you choose to leave, because we will be diminished by your absence, and the capacity for us to do the work God has gifted us through the Spirit and freed us in Christ to do will also be diminished. (ELCA 2009c)

A month later, on Sept. 23, 2009, Bishop Hanson penned a letter to ELCA pastors in which he wrote:

> I am disappointed that some are encouraging congregations and members to take actions that will diminish our capacity for ministry ... Although these actions are promoted as a way to signal opposition to churchwide assembly actions or even to punish the voting members who made them, the result will be wounds that we inflict on ourselves, our shared life, and our mission in Christ. And yet, as devastating as such actions could be for our shared life and for our global and ecumenical partners, my greatest sadness would be if we missed this opportunity: to give an evangelical and missional witness together to the world. (ELCA 2009d)

That letter was sent two days before the Lutheran Coalition for Reform's (CORE) convocation at Fishers, Indiana, to discuss a response to the CWA decisions. At the time, CORE was a coalition of ELCA members that worked within ELCA to uphold traditional teaching on marriage and sexuality. Soon after the 2009 CWA, CORE changed its name to the Lutheran Coalition for Renewal and in November 2009 CORE leaders announced they would begin a new church body called the North American Lutheran Church (NALC), which officially formed in Columbus, Ohio in August 2010 (NALC 2013). NALC now has about 380 congregations (NALC 2013).

At the time of the 2009 CWA, the ELCA had more than four million members in more than 10,100 congregations nationwide. By December 31, 2012, about 640 congregations, or 6 % of the 2009 total, had officially withdrawn from the ELCA. The geographic patterns of the congregations that officially withdrew were investigated by ELCA region, ELCA synod, U.S. state, and counties within states. The data were provided by the ELCA's Research and Evaluation group in Chicago.

Most of those withdrawing or departing are now affiliated with either the Lutheran Congregations in Mission for Christ (LCMC 38 %), an affiliation of autonomous Lutheran congregations originating in 2001 in response to the ELCA's 1999 adoption of the Called to Common Mission that established full communion between The Episcopal Church and the ELCA (LCMC 2013a), or NALC (33 %). Both LCMC and NALC have more conservative views than ELCA about human sexuality and clergy issues. The LCMC has more than 700 member congregations (LCMC 2013b). Many other ELCA congregations voted on the matter, but decided to remain a part of the ELCA.

The congregations that officially withdrew by December 31, 2012, had a combined membership of more than 240,000 baptized members or about 5 % of ELCA membership in August 2009. One cannot conclude that all were lost as ELCA members, because some may have joined a different ELCA congregation or banded together to start a new ELCA congregation. Bishop Hanson reported 53 new ELCA congregations started in 2011, half of which had multicultural or ethnic-specific bases (ELCA 2011a). However, congregations that did not leave the ELCA likely lost members. At the same time, ELCA congregations may have gained members who are supportive of the denomination's decision. The lasting impact of the social statement on the baptized membership in ELCA congregations is difficult to analyze even today.

176.2.1 Background

There is at least one ELCA congregation in every state, as well as Puerto Rico (24 congregations), the District of Columbia (13), the U.S. Virgin Islands (7), The Bahamas (1), Bermuda (1), and Ontario (1). The geographic distribution of ELCA churches is such that there is a dense concentration extending from eastern North Dakota through Pennsylvania and the northeastern coast of the U.S. Another concentration is found running from eastern North Dakota into south-eastern Nebraska. In addition, several clusters are associated with major metropolitan areas such as Seattle, San Francisco, and Los Angeles on the west coast and Dallas-Fort Worth and Houston in Texas and also Atlanta, Columbia, Charlotte, and Tampa-St. Petersburg-Orlando. On the state level, as of August 2009, Pennsylvania was home to the most ELCA congregations (1,263) with Minnesota second (1,130). North Dakota had, by far, the most ELCA congregations per capita (61.4 per 10,000 people), followed by South Dakota (30.3), Minnesota (21.3), Iowa (16.0), Montana (14.4), Nebraska (13.9), and Wisconsin (12.8) (Table 176.1).

Table 176.1 Top 10 states by ELCA congregation per 100,000 population prior to the August 2009 adoption of the Social Statement on Human Sexuality

State	Congregations	Congregations per 100,000 people
North Dakota	413	61.4
South Dakota	247	30.3
Minnesota	1,130	21.3
Iowa	488	16.0
Montana	142	14.4
Nebraska	253	13.9
Wisconsin	725	12.8
Pennsylvania	1,263	9.9
Ohio	604	5.2
Kansas	127	4.5

Source: Bradley C. Rundquist and Stanley D. Brunn, with data published by ELCA Research and Evaluation

The ELCA categorizes congregation size according to average Sunday service attendance (ELCA 2012). "Family" congregations are those with an average attendance of 75 or fewer. "Pastoral" congregations see an average attendance between 76 and 199, "Program" congregations are between 200 and 399, and "Resource" (or "corporate") congregations have an average Sunday attendance of 400 or more. Within the "resource" category, ELCA classifies "large" churches as those with 750 or more attendees on average and "mega" churches as those with 2,000 or more attendees (ELCA 2012). About 80 % of ELCA congregations are family- or pastoral-sized. According to data provided by ELCA for this research, as of August 2009 there were nine megachurches, five of which were in the Minneapolis-St. Paul metro area. The largest ELCA church, in terms of average Sunday service attendance, was in West Des Moines, IA (more than 7,000 in attendance on average). Other ELCA megachurches were in Fargo, ND, Glendale, AZ, and Columbus, OH.

Some congregations reacted quickly to the adoption of the Social Statement on Human Sexuality. The first congregation to leave the ELCA over the social statement did so during the 2009 CWA proceedings. According to data provided by ELCA, that pastoral-sized church, located in Holland, IN (a small town in southern Indiana north of the Ohio River), had voted 4 years earlier to leave ELCA if the social statement passed. The second congregation to leave because of the social statement did so on Sept. 27, 2009: a megachurch in Glendale, AZ, a Phoenix suburb, with an average Sunday attendance of more than 2,000. On June 10, 2010, ELCA lost a second megachurch, this time in Lakeville, MN, a suburban Minneapolis church with an average Sunday attendance of 4,000. Despite the loss of two megachurches, nearly 70 % of the congregations withdrawing from the ELCA during the study period are classified as family or pastoral. Figure 176.1 illustrates the temporal trend in official congregation withdrawals from ELCA during the study period. Note that it was not until January 2010 that large numbers of congregations began to withdraw from ELCA on a monthly basis. The peak withdrawal month during the

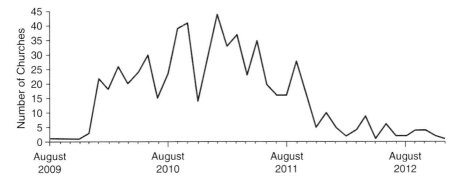

Fig. 176.1 Temporal trend showing the number of church withdrawals, August 2009–December 2012 (Source: Bradley C. Rundquist and Stanley D. Brunn)

study period was January 2011, 6 months after the CWA vote, when 44 congregations officially left the ELCA. The number of congregations withdrawing from the ELCA has been 10 or less per month since November 2011.

Congregations can terminate their relationship with ELCA through a process described in the Constitutions, Bylaws, and Continuing Resolutions of the ELCA (2011b). Termination requires two separate two-thirds majority votes of the congregation at two separate legally called and conducted congregational meetings. The first vote is on a written resolution to separate and can be held no sooner than 30 days after the bishop of the synod is notified of the meeting. The second vote is on termination of the relationship and can be held no sooner than 90 days after the approved resolution to separate is received by the bishop of the synod (ELCA 2011b).

176.2.2   Impact by ELCA Region

ELCA is organized into nine geographic regions that are numbered and without name (Fig. 176.2). Regional boundaries often, but not always, coincide with state boundaries. Not depicted on this map are Alaska (in Region 1), Hawaii (in Region 2), Ontario (in Region 7), and Puerto Rico, the Bahamas, Bermuda, and the U.S. Virgin Islands (in Region 9), all of which have small numbers of ELCA congregations.

Region 5, consisting of Illinois, Iowa, and Wisconsin, had the most ELCA congregations (1,805) in August 2009, followed closely by Region 3, which includes Minnesota, North Dakota, and South Dakota. These are the traditional "strongholds" of ELCA membership, along with parts of Region 7 (New England with 1,229 congregations), 8 (Pennsylvania with 1,263), and 6 (Michigan and Ohio with 1,071). Although Region 1, which is the northwestern U.S. and Alaska, had the fewest ELCA congregations in August 2009 (590), it was fourth in terms of the percentage (7.6 %) of congregations withdrawing from ELCA by December 31,

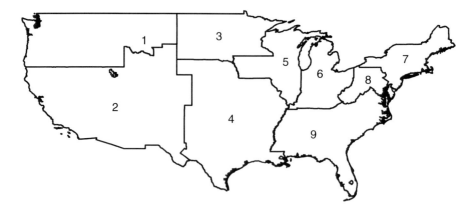

Fig. 176.2 Regions of the Evangelical Church of North America (Map by Bradley C. Rundquist and Stanley D. Brunn)

Table 176.2 ELCA congregations that withdrew between the August 2009 adoption of the Social Statement on Human Sexuality and December 31, 2012

Region No.	Congregations	Number withdrawing	Percent withdrawing
1	590	45	7.6
2	719	52	7.2
3	1,791	115	6.4
4	908	78	8.6
5	1,805	136	7.3
6	1,071	83	7.8
7	1,229	12	1.0
8	1,125	45	4.0
9	938	73	7.8

Source: Bradley C. Rundquist and Stanley D. Brunn, with data published by ELCA Research and Evaluation

2012. First was Region 4, the south central U.S., which had 8.6 % of its congregations withdrawing. Notable is that Region 7 (New England) saw only 12 of its 1,229 congregations (1 %) withdraw by December 31, 2012. Region 8 had only 4 % of its congregations withdraw. Region 8 includes Pennsylvania, the state with the most ELCA congregations (Table 176.2).

176.2.3   Impact by ELCA Synod

ELCA's nine regions are sub-divided into 65 synods of regional governing bodies. Synods are numbered and named (Fig. 176.3). Because this figure depicts the contiguous U.S., it does not illustrate Synod 1A: Alaska, or 9 F: Caribbean. Hawaii

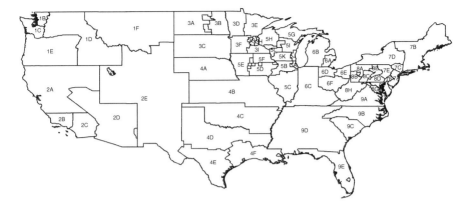

Fig. 176.3 Synods of the Evangelical Church of North America (Map by Bradley C. Rundquist and Stanley D. Brunn)

is included in Synod 2C: Pacifica, and the ELCA congregation in Ontario is in Synod 7G: Slovak Zion. The Slovak Zion Synod, founded by Slovak immigrants long before the establishment of the ELCA, is the only non-geographic ELCA synod (Slovak Zion Synod 2012). The boundaries of the 64 geographically based ELCA synods usually, but not always, follow state and county boundaries.

Between the adoption of the Social Statement on Human Sexuality and December 31, 2012, the Southwestern Texas Synod (4E) saw the highest percentage (20 %) and third highest number (33) of its congregations leave the ELCA. Western Iowa (5E, 16.8 % or 27 congregations), North Carolina (9B, 15.7 % or 36 congregations), and South Dakota (3C, 13.9 % or 34 congregations) were also high in terms of percentage and the number of congregations withdrawing from the ELCA. Four synods had no congregations withdraw following the adoption of the statement; three are included in Region 7: New England (7B, with 184 congregations), Metropolitan New York (7C, with 207 congregations), Slovak Zion (7G, with 26 congregations), and Metropolitan Washington D.C. (8G, with 76 congregations). Also low in terms of percent withdrawn are Northeastern Pennsylvania (7E, 0.35 % of 286 congregations), New Jersey (7A, 0.56 % of 180), Greater Milwaukee (5 J, 0.74 % of 135), Allegheny (8C, 0.79 % of 127), and Arkansas-Oklahoma (4C, 1.79 % of 56).

176.2.4   Impact by State

The state with the highest percentage of ELCA congregations leaving by December 31, 2012 was North Carolina (15.58 % or 36 congregations), followed by South Dakota (13.36 % or 33 congregations) and Idaho (12.82 % or 5 congregations)

Table 176.3 Top 10 states by percent of ELCA congregations withdrawing

State	Percent withdrawing	Number withdrawing
North Carolina	15.58	36
South Dakota	13.36	33
Idaho	12.82	5
Texas	12.76	49
Montana	12.68	18
Iowa	11.89	58
Kentucky	11.43	4
Arizona	10.23	9
Georgia	10.00	8
Missouri	9.33	7

Source: Bradley C. Rundquist and Stanley D. Brunn, with data published by ELCA Research and Evaluation

Table 176.4 Top 10 states by number of ELCA congregations withdrawing

State	Percent withdrawing	Number withdrawing
Minnesota	6.28	71
Iowa	11.89	58
Ohio	9.11	55
Texas	12.76	49
Pennsylvania	3.33	42
Wisconsin	5.52	40
North Carolina	15.58	36
California	8.28	36
South Dakota	13.36	33
Illinois	5.62	29

Source: Bradley C. Rundquist and Stanley D. Brunn, with data published by ELCA Research and Evaluation

(Tables 176.3 and 176.4). Minnesota lost 71 of its 1,130 congregations, 26 in the Northwestern Minnesota Synod alone. Iowa, Texas, North Carolina, and South Dakota stand out as states with relatively high numbers of ELCA congregations that lost more than 10 %. Several states saw no ELCA congregations withdraw, although most had relatively few congregations to begin with: Alabama (19), Arkansas (20), Connecticut (71), Delaware (13), Hawaii (11), Louisiana (18), Massachusetts (65), Maine (17), Mississippi (12), New Hampshire (14), New Mexico (18), Rhode Island (12), and Vermont (7). Note that the two "clusters" of states with no withdrawals are in New England and in the Mississippi Delta region. New Jersey had one withdrawal out of 182 congregations.

176.2.5   Impact by County

As of August 2009, there were three counties in the U.S. containing more than 100 ELCA congregations: Cook County (Chicago), IL, Los Angeles County, CA, and Hennepin (Minneapolis-St. Paul) County, MN. There were also five counties in Pennsylvania with at least 50 congregations (Table 176.5).

The geographic pattern of communities losing at least one ELCA congregation generally follows the spatial pattern of the overall distribution of ELCA congregations, especially through Minnesota, Iowa, Wisconsin, and into northern Illinois (Fig. 176.4). In Pennsylvania, another ELCA stronghold, withdrawals were concentrated in relatively few communities compared to the overall distribution of ELCA congregations in that state. The same can be said for eastern North Dakota, eastern South Dakota, and eastern Nebraska. Of note, there are other interesting concentrations of ELCA withdrawals in south central and southeastern Texas, southern California, and in Colorado along the front-range of the Rocky Mountains. Apparent too is a cluster of congregational withdrawals in the vicinity of Charlotte, NC, and a more dispersed cluster that spans south central Nebraska and north central Kansas.

There were 38 counties that saw 100 % of their ELCA congregations withdraw by December 31, 2012. Of these 28 of were home to only one congregation in August 2009. Other counties losing 100 % were Daniels (northeast Montana) and Pondera (northwest Montana) counties (4 congregations each); Shelby County (west central Iowa), Rusk County (Northwest Wisconsin), and Haakon County (west central South Dakota) (3 congregations each); and Jim Wells (south Texas) and Lee (northwest of Houston) counties, Boone County (south of Cincinnati),

Table 176.5 Top 10 counties by number of ELCA congregations, including ties (the congregational totals are for before August 2009 adoption of the Social Statement on Human Sexuality and December 31, 2012)

County	State	Population	Metro area	Congregations
Cook	Illinois	5,194,675	Chicago	136
Los Angeles	California	9,818,605	Los Angeles	109
Hennepin	Minnesota	1,152,425	Minneapolis-St. Paul	106
Allegheny	Pennsylvania	1,223,348	Pittsburgh	81
Berks	Pennsylvania	411,442	Reading	81
York	Pennsylvania	434,972	York-Hanover	69
King	Washington	1,931,249	Seattle	61
Maricopa	Arizona	3,817,117	Phoenix	54
Lehigh	Pennsylvania	349,497	Allentown	51
Dane	Wisconsin	488,073	Madison	50
Milwaukee	Wisconsin	947,735	Milwaukee	50
Montgomery	Pennsylvania	799,874	Philadelphia	50

Source: Bradley C. Rundquist and Stanley D. Brunn, with data published by ELCA Research and Evaluation

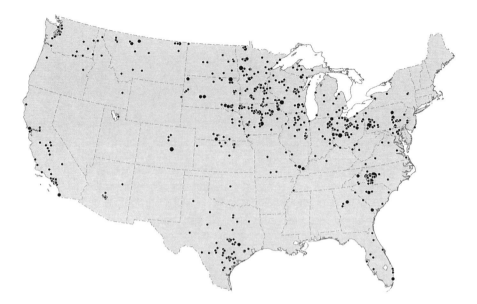

Fig. 176.4 Congregations that withdrew from the ELCA, August 2009 to December 31, 2012 (Map by Bradley C. Rundquist and Stanley D. Brunn)

KY, Atchison County (northwest Missouri), and Morrow County (southeast Ohio) (2 congregations each). Waupaca County, in east-central Wisconsin, lost eight ELCA congregations by December 2012, representing two-thirds of all ELCA congregations in the county in August 2009. Rowan County (west central North Carolina) lost 10 of its ELCA congregations while Maricopa County, AZ, lost eight (Table 176.6).

176.3 The PCUSA story

Presbyterians have been debating election and ordination of elders and pastors in regards to sexual orientation for most of the recent decade. The Presbyterian Church USA has 1.9 million members (compared to a peak of 4.2 for the Presbyterian denomination in the mid-1960s), less than half the number of ELCA. PCUSA has a little more than 10,000 churches, with the largest number in Pennsylvania (1,800). PCUSA and ELCA have approximately the same number of churches, but PCUSA has less than half the membership of ELCA.

As stated above the administrative organization of the denomination includes presbyteries and synods. There are 173 presbyteries that are essentially district governing bodies. The next higher level administrative units are synods; there are 16 which are regional governing bodies. Most synods adhere to state boundaries.

Table 176.6 Top 10 ELCA counties, including ties, by number of ELCA congregations withdrawing between August 2009 adoption of the Social Statement on Human Sexuality and December 31, 2012

County	State	Percent withdrawing	Number withdrawing
Rowan	North Carolina	31.25	10
Waupaca	Wisconsin	66.67	8
Maricopa	Arizona	14.81	8
Catawba	North Carolina	29.17	7
Orange	California	18.92	7
Los Angeles	California	6.42	7
Armstrong	Pennsylvania	26.09	6
Champaign	Illinois	50.00	5
Wright	Minnesota	26.32	5
Williamson	Texas	45.45	5
Gaston	North Carolina	33.33	5
Richland	Ohio	27.78	5
Butler	Pennsylvania	25.00	5
Franklin	Ohio	13.51	5
San Diego	California	13.51	5
Hennepin	Minnesota	4.72	5

Source: Bradley C. Rundquist and Stanley D. Brunn, with data published by ELCA Research and Evaluation

Examples of synods include: Synod of the Pacific, which includes Oregon, southern Idaho, Nevada and most of California except Southern California (which has 10 presbyteries), the Synod of the Trinity includes 6 presbyteries in West Virginia and Pennsylvania, and the Synod of the South Atlantic includes 16 presbyteries in South Carolina, Georgia and Florida. The number of presbyteries varies by the synod (Presbyteries, Synods 2011).

176.3.1   The Language

Passage of Amendment 10-A, which is discussed here, removed an amendment in the Book of Order that prohibited the ordination of sexually active unmarried Presbyterians from church offices. It replaces language in the 1996 Book of Order that stated that:

> Those who are called to the office of the church are to lead a life in obedience to Scripture and in conformity to the historical confessional standards of the church. Among these standards is the requirement to live either in fidelity within the covenant of marriage between a man and a woman (W-4.9001), or chastity in singleness. Persons refusing to repent of any self-acknowledged practice which the confessions call sin shall not be ordained and/or installed as deacons, elders or ministers of the Word and Sacrament. (Van Marter 2011)

The new language, G-6.0106b states:

Standards for ordained service reflect the church's desire to submit joyfully to the Lordship of Jesus Christ in all aspects of life. The governing body for ordination and/or installation shall examine each candidate's calling, gifts, preparation and suitability for the responsibilities of office. The examination shall include, but not be limited to, a determination of the candidate's ability and commitment to fulfill all the requirements as expressed in the constitutional questions for ordination and installation. Governing bodies shall be guided by the Scripture and the confessions in applying standards to individual candidates. (Van Marter 2011)

The General Assembly Stated Clerk, Gradye Parsons, informed the *Presbyterian News Service* in a press release May 11, 2011 acknowledging that "Clearly what has changed is that persons in a same-gender relationship can be considered for ordination." Also he added that "the gist of our ordination standards is that offers submit to the Lordship of Jesus Christ and ordaining bodes (presbyteries for ministers and sessions for elders and deacons) that have the responsibility to examine each candidate do so with no blanket judgments." www.pcusa.org/news/2011/5/11 pcusa-relaxes-cpmstitional-prhiibition-gay-and-lesbians

The sensitivity of the issue was echoed by Sharon Young, Communications Coordinator of PCUSA who stated:

Reactions to this change will span a wide spectrum. Some will rejoice, while others will weep. Those who rejoice will see the change as an action, long in coming, that makes the PC(USA) an inclusive church that recognizes and receives the gifts for ministry of all those who feel called to ordained office. Those who weep will consider this change one that comprises biblical authority and acquiesces to present culture. The feelings on both sides run deep. (Young 2011)

The decision whether Amendment 10-A would be approved was left up to the presbyteries where a majority had to pass for it to be accepted by the entire denomination. Thus the focus during late 2010 and early 2011 was on presbyteries, not individual congregations. However, congregations send delegations, which include clergy (senior and associate pastors) and elders, to presbytery meetings.

176.3.2 The Votes

Most votes (130) were taken in the first 5 months of 2011 (131); only 20 voted in 2010. Some presbyteries had not voted before the majority had already approved Amendment 10-A. Most presbyteries (153) reached their decision by formal ballot votes at presbytery meetings. Eight votes were recorded by hand, voice or standing (PCUSA, Presbytery Results 2011).

The largest number of votes by presbyteries were in February (48), March (33), and May (20). The numbers for the other months were 18 in January and 11 in April. Of the 130 presbyteries voting from January – May, 83 voted in favor of Amendment 10-A; 47 did not. As of May 11, 2011, the unofficial vote was 87–62 with 24 presbyteries still to vote. The change would occur July 10, 2012, 1 year after the 219th General Assembly adjourned.

The first presbyteries to vote on Amendment 10 did so in early 2010 where the results were divided: New Castle (Delaware and eastern shore of Maryland) on 18 January (70 % "Yes"), San Jose on 22 January (58 % "Yes"), Northern Waters (northern Minnesota with Duluth the key city) on 2 February (74 % "Yes"), Hudson River (southern New York adjacent to New York City) on 17 February (93 % "Yes"), Palo Duro (Texas panhandle) on 19 February (41 % "No"), San Jaoquin on 5 February 2010 (31 % "No") and Salem (central North Carolina: Raleigh, Chapel Hill, Durham area) on 9 April (63 % "Yes").

In the last 3 months of 2010 there were 14 presbyteries that voted on this amendment and 12 voted "No." These votes included Alaska presbytery on 15 October (31 % "No"), Albany on 16 October (85 % "Yes"), Boston on 1 November (64 % "Yes"), and Lake Michigan on 13 November (72 % "Yes") and Tropical Florida on 20 November (66 % "No").

A closer look at the votes is obtained by looking at two key months: February 2011 and May 2011. In February there were 49 presbyteries that voted on the amendment: 33 in support of Amendment 10A and 16 opposed. The *highest percentages* in support came from Hudson River Presbytery (93 %), Santa Fe (86 %), John Knox (southwest Wisconsin and northeast Iowa), Chicago (75 %), Arkansas (northern) (74 %) and Giddings-Lovejoy (southeast Missouri) (72 %). The *total absolute vote* of all the 49 presbyteries that voted in May in favor of Amendment 10-A was 4,056 Yes (64 %) and 2,597 No (36 %). The presbyteries with the *lowest percentages* voting in February were Flint River (southwest Georgia) (26 %), Pueblo (southeast Colorado) (31 %), Western Colorado (also 31 %), Olympia (southwest of Seattle) (33 %) and Inland Northwest (western Washington) (37 %). The *total absolute votes* for all the 16 presbyteries that did not support Amendment 10-A were much less than those supporting it: 733 No (56 %) and 572 Yes (44 %).

The May results also exhibited nearly the same number of presbyteries supporting Amendment 10-A (12) and those opposing it (9). The *highest percentages* in support were Twin Cities Presbytery (79 %), New York City (74 %) and Des Moines (69 %). The *lowest percentages* were Los Ranchos (north of San Diego) (28 %), New Harmony (northeast South Carolina) (also 28 %) and Prospect Hill (northwest Iowa) (33 %). In *total absolute numbers* for the 13 presbyteries that supported the amendment, there were 1,073 votes (62 %) in support and 656 against (38 %) it. The total absolute numbers for the nine presbyteries that opposed the amendment were 628 No (61 %) and 396 Yes (39 %).

That majority (half plus one) was reached in May 2011 when the 87th PCUSA presbytery voted in favor of the amendment. This favorable vote was cast by the Twin Cities Presbytery. Prior to that specific vote 70 had voted "No," in 2 presbyteries there was a tie and in 4 there were no data reported. It merits mention that there were 24 presbyteries that had not voted at the time the majority had voted in favor of Amendment 10-A. Fifteen voted subsequently; 10 voted Yes and 6 Voted No; there were no recorded votes for 6 other presbyteries. Those voting Yes were Boise, Charlotte, Des Moines, Hanmi, Homestead (central Nebraska, not Omaha), Missouri

Union (northeast Missouri), Sheppards and Lapsley (central Alabama), and Western New York. Those voting No were; Prospect Hill (northwest Iowa), Los Ranchos (between San Diego and Los Angeles), San Gabriel (a tie vote, which counted as No), Shenago (central Pennsylvania) and Western Kentucky.

There were 22 presbyteries that switched their votes from "No" to "Yes" and also four that switched from "Yes" to "No." Among those the switch from "No" to "Yes" where these presbyteries: North Alabama, Middle Tennessee, Central Nebraska (western and northeast Nebraska), Cincinnati, Missouri Union (central and northeast Missouri) and San Francisco. The four presbyteries that reversed initial "Yes" were de Cristo (Phoenix and southwest Arizona), Lake Huron (Saginaw and Bay City area), Minnesota Valleys (west central and southwest Minnesota) and West Jersey (southern New Jersey).

The margin of passage also varied significantly across the presbyteries. For some, such as Charlotte, Eastminster (northeast Ohio and south of Cleveland), East Oklahoma (eastern half of the state), Mission (southern Texas), Middle Tennessee, South Alabama, Tres Rios (El Paso area of west Texas), Yellowstone (eastern and southern Montana), and Sacramento (Sierra Mission Partnership), the amendment narrowly passed (50.1 – 53.0 %). Among the presbyteries with the lowest support (less than 30 %) were: Beaver-Butler (27 %; the Pittsburgh area), Kendall, Idaho (Snake River Mission Area) (23 %), Mississippi (southern half of the state) (19 %), Stockton (Sierra Mission Partnership) (12 %), and San Diego (25 %). Central Washington (18 %). The strongest support came from these presbyteries: Hudson River (93 %), Mid-Kentucky (92 %; Louisville south to the northern Tennessee state line), Santa Fe (86 %), Albany (85 %), Northern New England (81 %), Newton (northern New Jersey), (80 %), Twin Cities (97 %), John Knox (76 %; northeast Iowa), and Denver (78 %).

Because the number of members of churches in presbyteries varies significantly, it is important to consider this fact alongside the percentage of support. In some presbyteries there were very few votes cast, for example, in Cimarron (panhandle Oklahoma) there were only 28 votes cast (19 Yes, 9 No), in Kendall (Snake River Mission Area; eastern Idaho) in Idaho only 22 cast (5 Yes, 17 No), and in Eastern Oregon (Snake River Mission Area) (18 Yes and 9 No). Of the top 15 presbyteries in total votes cast, 13 supported Amendment 10-A.

The record of passage is not completely accurate as in the case of the Cayuga-Syracuse presbytery where there was a "voice estimate" of 60 Yes and 5 No; in Northern New York presbytery and also in Winnebago (central and east central Wisconsin) presbytery (recorded as passed), in Redwoods (northern coastal California) and San Fernando (both California presbyteries) "standing vote" (passed). In Shenago (west central Pennsylvania) the voice vote was "No."

An additional perspective on this issue is gained by looking at the presbyteries with the largest and smallest total votes and how they voted. There were 43 presbyteries that had more than 150 votes cast. Of these 23 voted Yes, 17 voted No and in 3 the votes were tied. Of the 15 largest votes, 11 were in support, 4 against (one was a tied vote that counted as No). The presbyteries with the largest total votes are shown in Table 176.7 as well as the percentage voting Yes. The presbyteries with the largest absolute voting totals were: Greater Atlanta (491), Mission (395; southern

Table 176.7 PCUSA votes by the largest presbyteries, 2010 and 2011

Presbytery	Yes vote	No vote	Percent yes	Total vote
Greater Atlanta	262	157	63	419
Mission	201	194	51	395
Grace (Dallas, Fort Worth)	243	149	62	392
San Francisco	198	143	58	341
Charlotte	162	154	52	316
Coastal Carolina	113	192	37	305
James (Central VA)	152	152	50	304
New Covenant (Houston)	144	156	48	300
Salem (Central NC)	186	107	63	293
Philadelphia	182	108	63	290
National Capital	204	80	72	284
New Hope (Northeast NC)	158	118	57	276
Twin cities	205	56	79	261
Chicago	185	61	75	246
Western North Carolina	145	99	59	244
Pittsburgh	80	163	33	243

Datasource:https://spreadsheets.google.com/pub?key=t-MAxyQn6VWgFOr83FqGGA&output=html

Texas), Grace (392; Dallas. Ft. Worth area), San Francisco (341), Charlotte (316); Coastal Carolina (305), James (304; central Virginia, south of Washington, DC), New Covenant (300), Salem (central North Carolina), and Philadelphia (293). Four of the latest presbyteries voted in 2010, 12 in 2011, mostly in February, which was the major 2011 month when all presbyteries voted on the amendment. For example, Salem presbytery (central North Carolina) voted "Yes" in April 2010. Grace Presbytery (Dallas-Fort Worth) voted "Yes" in March 2011 and Twin Cities and Charlotte voted "Yes" in May 2011.

Contrasting with largest presbytery votes are those 15 presbyteries where there were 60 total votes or fewer. These included rural areas in Alaska, Oklahoma, Oregon, Idaho, Wyoming, Minnesota, Virginia, Nebraska and Utah. The presbytery votes were split: eight voted for Amendment 10-A and seven voted against it. Among the presbyteries that voted in favor were: Cimarron (Oklahoma), Eastern Oregon, Yellowstone (Wyoming), Utah, Northern Waters (northern Minnesota, Duluth area), Palisades (New York city area), and Central Nebraska. These presbyteries were among those voting "No": Stockton, Mississippi, Alaska, Kendall (Snake River) and Northumberland; in each the support was less than 35 %. Of the seven that opposed the amendment, five voted "No" in 2010, of those that supported it, seven of the nine voted "Yes" in 2011. In short, the smallest presbyteries that voted No did so early; the smallest presbyteries that voted "Yes" did so in mostly in March and April.

The votes in synods also varied (Table 176.8). The strongest support came from the Northeast Synod (20 presbyteries in New England and adjacent states voting in favor of the amendment), Lakes and Prairies Synod (10 – Upper Middle West and

Table 176.8 Votes by presbyteries in synods

Synod	Number of votes		
	Yes	No	No record
Southern California (Los Angeles, San Diego) & Hawaii	2	5	1
Southwest (Arizona and New Mexico)	2	2	0
Alaska-Northwest (western Washington and Idaho panhandle)	0	7	0
Puerto Rico	0	1	2
Covenant (Michigan and Ohio)	9	2	0
Lakes & Prairies (Minnesota, Wisconsin, Iowa, North Dakota, South Dakota, east and central Nebraska)	10	5	1
Lincoln Trails (Illinois and Indiana)	7	0	1
Living Waters (Kentucky, Tennessee, Alabama and Mississippi)	7	5	0
Mid-America (Kansas and Missouri)	4	2	0
Mid-Atlantic (Maryland, Delaware, District of Colombia, Virginia and North Carolina)	8	6	0
Northeast (New England states, New York, Massachusetts, Connecticut, Rhode Island and New Jersey)	20	2	0
Pacific (Oregon, southern Idaho, Nevada, all California except Los Angeles and San Diego areas)	6	4	0
Rocky Mountains (Montana, Wyoming, Colorado, Utah and southwest Nebraska)	4	4	0
South Atlantic (South Carolina, Georgia and Florida)	5	9	2
Sun (Arkansas, Louisiana, Oklahoma and Texas)	7	3	0
Trinity (Pennsylvania and West Virginia)	4	11	1
Total	85	68	8

Data source: http://oga.pcusa.org/links/

northern Great Plains), Covenant (9 – Michigan and Ohio), and Mid Atlantic (8). Nearly half of the required number of votes to approve the amendment came from presbyteries in these four synods. The synods with the most opposition were, perhaps somewhat surprisingly, Trinity (11 presbyteries in Pennsylvania and West Virginia), South Atlantic (9 presbyteries in South Carolina, Georgia and Florida), Alaska-Northwest (7 presbyteries in Alaska and Washington) and Mid Atlantic (6 presbyteries in Maryland, Delaware, DC, Virginia and North Carolina).

176.3.3   Geographic Patterns

To understand the presbytery votes we examine 4 maps here. The first is the Yes-No vote by presbyteries (Fig. 176.5). There are two dominant patterns; those presbyteries that for and against Amendment 10-A. The majority or presbyteries voting against the amendment were in coastal northern California, Oregon and Washington, three states that have long been in the forefront of promoting gay and lesbian issues in the past 25 years. The complexities of this issue were evident in the support for Amendment 10-A in the majority of presbyteries in Alabama, Kentucky, Tennessee and Texas. Also many of the presbyteries in the rural and urban Middle West and

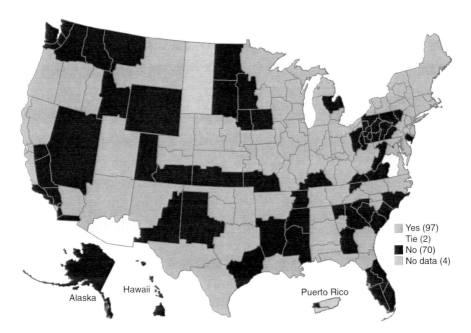

Yes (97)
Tie (2)
No (70)
No data (4)

Alaska   Hawaii   Puerto Rico

Fig. 176.5 Majority presbytery votes on approval of Amendment 10. Yes votes approve ordaining LBGT as deacons, elders and pastors (Map by Bradley C. Rundquist and Stanley D. Brunn, data from www.amendmenta10.org, July 2011)

New England were supportive. Opposition was evident across the U.S., including in some unsuspected places, such as presbyteries in southern California, much of peninsular Florida and Washington. Probably many outsiders, and insiders, would have predicted that presbyteries in these regions would have supported the amendment. The rural West had presbyteries both supporting and opposing the amendment.

The second map (Fig. 176.6) looked at the total absolute vote (yea and nay) for the amendment in each presbytery. This pattern is somewhat different than the previous one, but it also shows some widely varying patterns. There are three significant patterns on the map. First, there are many presbyteries where the total number of votes cast was less than 100. Most of these were in the western half of the country. Second, eight presbyteries had more than 300 total votes; these were, not surprisingly, mostly large urban areas: Houston, Dallas-Ft. Worth, San Francisco, Atlanta, San Francisco, Washington, D.C. and central North Carolina. Third, there mix of sizes was especially high in much of the Middle West and Mid-South.

The third map (Fig. 176.7) shows the percentage of each presbytery vote that was "Yes" or in favor of Amendment 10-A. There are nine categories, but five are for the following: hand vote, no vote, vote not submitted, a standing vote and a voice vote. The most significant are the first four categories. The presbytery voting results were also complex and varied, often within the same state or region. The highest support came from presbyteries in the Mid-South, including Alabama and parts of northern Florida, the Upper Middle West, especially in Wisconsin and Minnesota, New England also the New York City-New Jersey area and also most of Texas. There was

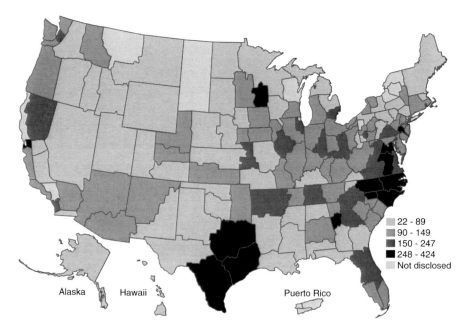

Fig. 176.6 Pastor, co-pastor, associate pastor, and elder votes on approval of Amendment 10. Yes votes approve ordaining LBGT as deacons, elders and pastors (Map by Bradley C. Rundquist and Stanley D. Brunn, data from www.amendmenta10.org, July 2011)

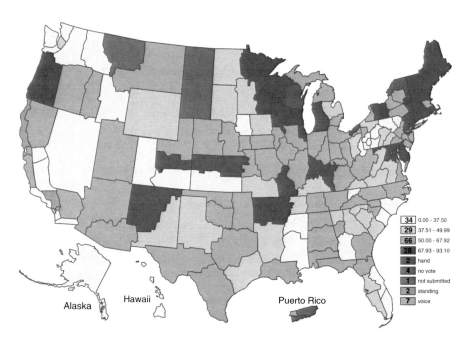

Fig. 176.7 Amendment 10 approval percentage, by Presbytery (Map by Bradley C. Rundquist and Stanley D. Brunn, data from www.amendmenta10.org, July 2011)

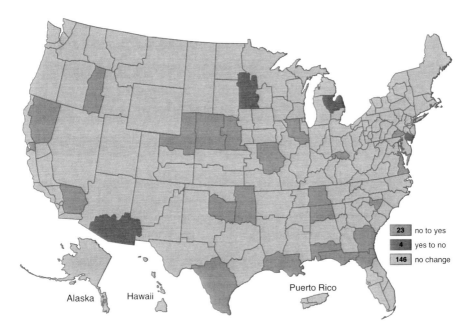

Fig. 176.8 Switched vote on Amendment 10 after 2009 (Map by Bradley C. Rundquist and Stanley D. Brunn, data from www.amendmenta10.org, July 2011)

less support in some presbyteries one would have expected strong support, for example, in those in the San Francisco, Seattle and Los Angeles areas. The opposition also had some very mixed regional results: central Pennsylvania, South Carolina-Georgia area; and much of the Rocky Mt. and northern Great Plains presbyteries. Presbyteries in some states had widely different degrees of support, for example, in Michigan, Ohio, New York and California.

The fourth (Fig. 176.8) map shows the changes from the vote on Amendment 8 in 2009 (which failed), only to be considered as Amendment 10-A in 2011. There were 146 presbyteries that did not change, 22 that switched from No to Yes and 4 that switched from Yes to No. The nearly two dozen presbyteries that changed from No to Yes were mostly in South (two in Alabama and Oklahoma and one each in Florida, Louisiana, Georgia, South Carolina and Tennessee). The four that switched from No to Yes were in four different states (New Jersey, Michigan, South Dakota-Minnesota area) and southern Arizona.

When we look at the vote with respect to the region or location of the presbytery and timing we find little regional consistency, except for strong support in the Northeast and a sizable number of presbyteries in the South opposing it. Basically, when looking at the votes at the national level, there was little strong uniformity. No region had consistently very high or very low support for Amendment 10-A. Such mixed results are not unexpected in a society where there is almost instant reporting about public opinion polling results, legal decisions and statewide votes on controversial issues. Clergy, elders and members of congregation can and do receive and exchange information daily about issues where there are strong sentiments for and against an institution's positions.

To illustrate this "fluidity" on this social issues, we can look at the presbytery votes in three groups: (a) the first twenty presbyteries that voted on the amendment, (b) those that voted on the amendment in March 2011 (the second largest month of presbyteries voting on this amendment in 2011), and (c) those voting in May 2011, the final month of voting. Eight presbyteries voted in the first 5 months of 2010. Five voted Yes and the other three voted No. The Yes votes came from San Jose Presbytery (California), Hudson River (just outside New York City), Northern Waters (northern Minnesota – Duluth the largest city), Pacific (Los Angeles) and Salem (central North Carolina) while the No votes came from San Jaoquin Presbytery (California's Central Valley), Lake Huron Presbytery (the thumb are around Flint, Bay City and Saginaw) and Prospect Hill Presbytery (in northwest Iowa). In short, in the early voting there was no regionality to the support for or against Amendment.

There were presbytery votes taken from June – September. From 15 October to 13 November 13 there were 5 that voted against the amendment and 3 that voted in favor. The No votes were mostly cast by presbyteries in the South: James (south central Virginia; the vote was tied, but this was regarded as a No vote), Mississippi (southern part of the state), Shenandoah (central Virginia), New Covenant (Houston area). Alaska presbytery also voted against the amendment. The supportive votes came from Boston, Lake Michigan (western part of the state) and Albany presbyteries.

In March 2011 there were 15 presbyteries voting on the amendment. Nine voted in favor; these were also in Eastminster (northeast Ohio, south of Cleveland), Geneva (western New York), Monmouth (northern New Jersey), Baltimore, Donegal (southeast Pennsylvania), Cascades (Oregon), Mission (southern Texas) and Grand Canyon (northern Arizona). These votes came from the northeastern states, but also a number of presbyteries outside the northeast. Three of negative votes were most from southern presbyteries, Abington (central Virginia), Coastal Carolina, Charleston Atlantic. Others came from presbyteries in southeast Ohio (Muskingam Valley), Central Washington, and south central Pennsylvania (Carlisle Presbytery).

The final votes were taken in May 2011. Seventeen states voted during the month; 10 voted Yes and 7 voted No. The votes were closely watched by those in the national denomination offices as the results showed both support and opposition, but always, as in previous months of 2011, slightly more in favor of Amendment 10A than opposed to it. The negative votes tended to come from southern presbyteries, for example, Foothills (northwest South Carolina), Peaks (south central Virginia), New Harmony (northeast South Carolina) and St. Andrew (northern Mississippi). But also presbyteries in northwest Iowa (Prospect Hill) and in between Los Angeles and San Diego (Los Ranchos Presbytery). While the opposition was mostly from southern presbyteries, there were presbyteries in the South that supported the amendment: Charlotte and Middle Tennessee. Additional support came from presbyteries in the West, for example, Plains and Peaks (Central Oregon), Pacific (Los Angeles), Nevada and several in the Middle West: Des Moines, Twin Cities, Homestead (eastern Nebraska, outside Omaha), and Union (central and northeast Missouri). In sum, there were presbyteries in the South and in the Midwest and West that both supported and opposed the amendment. We would expect similar results if we looked at the votes in other months, for example, December 2010 and January, February and April 2011.

In sum, the amendment passed not because of strong support in one, two or three regions but because of a "mix" of presbyteries voting for it in almost all synods. As we have seen, the Yes votes came from very large urban presbyteries and also that had very few members. Support came from, as expected Northeast states, but also from a significant number in the South, Southwest and rural Middle West. In short, there was nationwide support from rural and urban areas and presbyteries, large and small presbyteries, and presbyteries in all regions except the Northwest.

176.3.4   Churches Dismissed from the Denomination

Amendment 10A, as noted above, was a contentious issue for many clergy, members and individual churches. Strong emotions were expressed, undoubtedly, by members, elders, and non-elected members to church offices and position. Opposition was often based on scriptural interpretations, using certain biblical passages to support their positions, but also the unwillingness to accept the diversity that resulted from passage of the amendment. Within 10 months of the denomination approving the ordination of gay clergy, elders and deacons, 38 churches had left or started the process of leaving the denomination. This departure was the "largest decline since 2007 when over 57,000 members left" (May 2012).

The General Assembly (GA), the denominational wide both that meets biannually to discuss, debate and vote on issues, keeps records on discussions and votes of individual churches on this and other sensitive issues. The GA records official show that there were 22 churches that were dismissed from PCUSA in 2010 and 21 in 2011 (Fig. 176.9). While the map shows a scattered pattern of churches in all regions of the country, the most significant cluster was in the Mid-South Presbytery in western Tennessee. It is worth noting that Mid-South Presbytery voted on Amendment 10-A on February 12, 2011; the vote was 48 for the amendment and 49 against it. Two churches left the denomination in 2010 and eight in 2011. (We have no data for 2012 as of this writing.) The only other significant clusters of dismissals were four churches in California's Central Valley, three in west central Pennsylvania and three in western New York. Considering the number of presbyteries that voted No Amendment, it is surprising that so few had severed ties with the national denomination. However, this figure is not an accurate reflection of the current status as we still do not know the number of churches that left in 2012 or are contemplating doing so. A new group of Presbyterians has been formed, The Fellowship of Presbyterians (FOP) and ECO: A Covenant Order of Evangelical Presbyterians. ECO as of February 2013 has 28 congregations under its wings and another 48 are in the process of joining (La Berge 2013b). This information appeared in *The Layman*, a conservative PCUSA lay newsletter, which also stated another 75–100 congregations are in the discernment process. (The website for FOP and ECO is: www.layman.org.) In another

Fig. 176.9 Dismissed churches, 2010 and 2011 (Map by Jeff Levy, University of Kentucky Gyula Pauer Center for Cartography and GIS; commissioned by the editor)

article in the same issue of the newsletter, La Berge noted that PCUSA is losing about 60,000 members per year; at this rate "PCUSA would cease to exist in 32.5 years" (La Berge 2013a: 5).

176.4   Conclusions

Our purpose was to examine the geographic patterns of support and lack thereof by the ELCA (Evangelical Lutheran Church in America) and PCUSA (Presbyterian Church, USA 2011) for the ordination of same-sex clergy. While both denominations approved such language through churchwide voting (ELCA) and presbytery (PCUSA) voting, the issue is far from a settled issue as there are churches in both denominations which are considering or are in the process of severing ties with their national denominations.

The August 2009 adoption of the ELCA Social Statement on Human Sexuality was and remains a continuous issue. Between its adoption and August 31, 2011, nearly 8.5 % of member congregations held at least one vote on whether to leave the ELCA. By the end of December 2012, about 640 congregations, or about 6 % of the total number of ELCA congregations prior to the adoption of the social statement, had officially withdrawn. PCUSA reached its decision in May 2011 when the 87th presbytery (Twin Cities Presbytery) voted in favor of Amendment 10-A. In the analysis above we looked for areas of support and opposition and also the geographic pattern of those that withdrew from the national denomination.

In regards to ELCA analyses at the regional, synod, state and county level, the geographic pattern of withdrawals is consistent with the pattern of the distribution of ELCA churches. That is, many withdrawals came from traditional ELCA strongholds in the Upper Midwest Region. There are a few notable findings to highlight, however. New England states had very low rates of withdrawal (less than 1 %). On the other hand Iowa, Texas, North Carolina, and South Dakota showed relatively high numbers and percentages of withdrawals from ELCA. Texas, in particular, was highlighted in analyses at the regional, synod, state, and county scales. The county-level analysis shows that the area of Texas most impacted is in the south central and southeastern parts of the state, centered on the San Antonio metropolitan area. Multiscale analyses also indicated that parts of the northwestern U.S. witnessed relatively high rates of withdrawal, although there are relatively few ELCA congregations in that region and county-level analysis shows that the geographic pattern is dispersed. Analysis at the county level also shows that, overall, congregations in major urban centers tended to stay with the ELCA while those in suburban and rural counties were more likely to withdrawal.

The PCUSA results revealed a mixed record of support and opposition to Amendment 10-A. While there was strong support from presbyteries in the Northeast and Middle West especially, there were also presbyteries in the Great Plains, South and Southwest that supported the ordination of same-sex clergy and elders. The opposition was strongest in the South Atlantic and South, but the opposition was not uniform across presbyteries in these regions. Opposition was also strong in the Pacific Northwest states, California and also the rural Great Plains. The largest presbyteries supported the amendment. In short, it was a mix of rural and urban presbyteries, large and small presbyteries and those in all regions that contributed to passage of this amendment.

The research opens up a number of areas for future research by scholars in the social sciences and humanities. For example, future studies on GBLT issues could analyze the demographic and political make-up of the counties with high and low rates of withdrawal from following the 2009 adoption of the ELCA Statement on Human Sexuality. For example, it would be interesting to see how those counties and presbyteries voted in recent presidential elections. As this study has shown, there are some regional and urban similarities between the ELCA and PCUSA results. Pennsylvania, California and Florida would be interesting states to examine in greater detail because of regional differences in each. Also it would be valuable to compare the ELCA and PCUSA national, regional and metropolitan votes with the votes by the Episcopalian and United Church denominations and to investigate whether the withdrawing congregations are in a "core" region of a denomination or in a periphery where they would probably be competing with other churches for new and "recycled" members. Once we had the cartographic and demographic results it would be useful to look at any consistencies in the votes of these four progressive denominations and at potential votes on GBLT issues in other denominations.

Acknowledgements Brad Rundquist wants to thank Kenneth Inskeep, ELCA Executive for Research and Evaluation, and Jacqueline Skrypek, ELCA Senior Demographic Analyst, for providing the data set upon which our ELCA research is based and for answering the many questions we posed about the data set. Stan Brunn wishes to acknowledge the assistance provided by Kris Valerius of PCUSA; she provided the data on dismissed churches from the denomination. Also he wants to thank University of Kentucky undergraduate geography majors Charles J. Altendorf and Tiffany N. Mullins, both who took an interest in the subject on an independent research project in summer 2011 and also Jeff Levy of the University of Kentucky GIS Laboratory who aided them in their project and who also constructed the two maps of churches that left ELCA and PCUSA.

References

Burke, D. (2011). Even with the win, gay clergy likely to remain limited. *Religions New Service* http://www.religionews.com. Accessed 12 May 2011.

Evangelical Lutheran Church in America (ELCA). (2009a). *A social statement on human sexuality: Gift and trust.* Available at http://www.elca.org/What-We-Believe/Social-Issues/Social-Statements/JTF-Human-Sexuality.aspx. Accessed 2 July 2012.

Evangelical Lutheran Church in America (ELCA). (2009b). *Frequently asked questions about the 2009 churchwide assembly actions regarding human sexuality.* Available at http://www.elca.org/What-We-Believe/Social-Issues/Social-Statements/JTF-Human-Sexuality/cwafaqs.aspx. Accessed 23 Feb 2013.

Evangelical Lutheran Church in America (ELCA). (2009c). *Remarks to the eleventh churchwide assembly from presiding Bishop Mark S. Hanson.* Available at http://www.elca.org/Who-We-Are/Our-Three-Expressions/Churchwide-Organization/Office-of-the-Presiding-Bishop/Messages-and-Statements/090901.aspx. Accessed 23 Feb 2013.

Evangelical Lutheran Church in America (ELCA). (2009d). *Pastoral letter to rostered leaders from presiding Bishop Mark S. Hanson.* Available at http://www.elca.org/Who-We-Are/Our-Three-Expressions/Churchwide-Organization/Office-of-the-Presiding-Bishop/Messages-and-Statements/092309.aspx. Accessed 23 Feb 2013.

Evangelical Lutheran Church in America (ELCA). (2011a). *Churchwide assembly, report of the presiding Bishop, Part 1.* Available at http://www.elca.org/~/media/Files/Who%20We%20Are/Office%20of%20the%20Presiding%20Bishop/2011%20Message%20Archive/MSH%20 8%2013%2011%20Report%20of%20the%20Presiding%20Bishop%20Part%201.pdf. Accessed 23 Feb 2013.

Evangelical Lutheran Church in America (ELCA). (2011b). *Constitutions, bylaws, and continuing resolutions of the Evangelical Lutheran Church in America.* Available at http://www.elca.org/Who-We-Are/Our-Three-Expressions/Churchwide-Organization/Office-of-the-Secretary/ELCA-Governance/Constitutions-of-the-Evangelical-Lutheran-Church-in-America.aspx. Accessed 23 Feb 2013.

Evangelical Lutheran Church in America (ELCA). (2012). *Ministry: Frequently asked questions.* Available at http://www.elca.org/Growing-In-Faith/Ministry/Large-Membership-Congregations/Frequently-Asked-Questions.aspx. Accessed 2 July 2012.

Evangelical Lutheran Church in America (ELCA). (2013). *Rostered leadership.* Available at http://www.elca.org/Growing-In-Faith/Vocation/Rostered-Leadership.aspx. Accessed 23 Feb 2013.

La Berge, C. F. (2013a). Addressing the rumor that the PCUSA is going out of business. *The Layman, 46*(1), 3.

La Berge, C. F. (2013b). Three sneezes (and a gesundheit). *The Layman, 46*(1): 1.

Lutheran Congregations in Mission for Christ (LCMC). (2013a). *History.* Available at http://www.lcmc.net/history/232.html. Accessed 24 Feb 2013.

Lutheran Congregations in Mission for Christ (LCMC). (2013b). *Member congregations.* Available at http://www.lcmc.net/congregations.html. Accessed 24 Feb 2013.

May, D. (2012). *Churches continue exodus from PCUSA*. http://ecumenicalnews.com/article/churches-continue-exodus-from-pc-usa-1309 Accessed 23 Feb 2013.

North American Lutheran Church (NALC). (2013). *North American Lutheran Church history*. Available at http://thenalc.org/the-nalc-history.htm. Accessed 23 Feb 2013.

PC(USA)- Presbyteries, synods and other links of interest. http://oga.pcusa.org/links/. Accessed 13 May 2011.

Presbyterian Church (U.S.A.). (2011). *PC (USA) relaxes constitutional prohibitions of gay and lesbian ordination*. http://www.pcusa.org/news/2011/5/11/pcusa-relaxes-constitutional-prohibition-gay-and-l/. Accessed 2 Feb 2011.

Presbyterian Church (U.S.A.) Presbytery results on Amendment 10-A. https://spreadsheets.google.com/pub?Key=t-MAxyQn6VWgFOr836FqGGA&output=html. Accessed 26 May 2011.

Slovak Zion Synod. (2012). *Who we are*. Available at http://www.slovakzionsynod.org/webdirectory.html. Accessed 2 July 2012.

Van Marter, J. (2011). *PC(USA) relaxes constitutional prohibition of gay and lesbian ordination*. http://www.pcusa.org/news/2011/5/11/pcusa-relaxes-constitutional-prohibition-gay-and-l Accessed 22 Feb 2013.

Young, S. (2011). *A churchwide letter concerning Amendment 10-A*. http://www.pcusa.org/news/2011/5/10/churchwide-letter-concerning-amendment-10-/

Part XIV
Politics, Reconciliation and Advocacy

Chapter 177
Are High Levels of Existential Security Conducive to Secularization? A Response to Our Critics

Pippa Norris and Ronald Inglehart

177.1 Introduction

During recent decades, religion became increasingly prominent in politics around the world. The persistence of church-going in the United States, the growing importance of liberation theology in Latin America, and the resurgence of identity politics in the Balkans—all suggest that religion remains a potent force in contemporary politics. These observations were reinforced by the events of 9/11 in the United States, growing European tensions over religious identities, and sporadic outbursts of violent conflict dividing faith-based communities in Nigeria, Sudan, and India (Klausen 2009).

What links these disparate events? One popular assumption is that they reflect a new worldwide religious revival (Berger 1999: Thomas 2005). Hence Rodney Stark and Roger Finke (2000) claim that it is time to bury the secularization thesis: *"After nearly three centuries of utterly failed prophesies and misrepresentations of both present and past, it seems time to carry the secularization doctrine to the graveyard of failed theories, and there to whisper 'requiescat in pace.'"* But Sacred and Secular (Norris and Inglehart 2004) presented a theory of existential security, backed by empirical evidence from scores of societies, demonstrating a more complex picture; religious values remain strong in many developing societies, which also have rapidly growing populations, yet secularization is occurring in most advanced industrial

P. Norris (✉)
John F. Kennedy School of Government, Harvard University, Cambridge, MA 02138, USA

Government and International Relations, University of Sydney, Sydney, NSW, Australia
e-mail: pippa_norris@harvard.edu

R. Inglehart
Institute of Social Research, University of Michigan, Ann Arbor, MI 48104, USA

World Values Survey Association, Madrid, Spain
e-mail: RFI@umich.edu

© Springer Science+Business Media Dordrecht 2015
S.D. Brunn (ed.), *The Changing World Religion Map*,
DOI 10.1007/978-94-017-9376-6_177

societies. The erosion of church attendance, religious values, and beliefs has been most clearly established in Scandinavia and Western Europe (Aarts et al. 2008). But this development is not simply confined to this region, as similar developments are evident in Australia, New Zealand, Japan, and Canada; even in the U.S., there has been a trend toward secularization, partly masked by large scale immigration of people with traditional worldviews. Norris and Inglehart (2004) argued that *existential security* explained these divergent trends. This account emphasizes that the public's *demand* for transcendent religion varies systematically with levels of vulnerabilities to societal and personal risks and threats.

Nevertheless, with the evidence available at the time, *Sacred and Secular* was unable to address certain important issues. In particular, critics have argued that despite the wealth of indirect cross-national macro-level indicators, Norris and Inglehart did not establish a *direct* individual-level link between religiosity and either perceptual or experiential measures of existential insecurity. In addition, commentators have also often pointed to the apparent anomaly of the U.S., as one of the world's richest nation which, nevertheless, has relatively high levels of religious participation and values. Accordingly this chapter seeks to address these issues. *Part II* of this chapter summarizes the theory of existential security and reviews what we know about this from the previous research literature. *Part III* summarizes the data sources and methods used in this study. We draw upon new evidence concerning from the 2005–2007 World Values Survey conducted in 55 nations, as well as from the 2007 Gallup World Poll conducted in 132 nations. *Part IV* presents the results of the analysis examining the impact of experiential security on religiosity, utilizing the Gallup Lived Poverty index. Building upon this foundation, *Part V* pays particular attention to perceptual or subjective measures of security and risk, from the 5th wave WVS. The conclusion in *Part VI* suggests that the dynamics of secularization is much more complex than the simple decline of religion proposed by some early theories, or a universal revival of religion worldwide, as suggested by many contemporary commentators. Instead, as we will demonstrate, rising existential security brings declining emphasis on religion in many post-industrial societies worldwide, and thus a growing religiosity gap worldwide.

177.2   The Theory of Existential Security

One of the most popular contemporary approaches to explaining the strength of religiosity in any society draws upon rational choice theories which emphasize the 'supply-side' of religious markets (Stark and Bainbridge 1985, 1987; Finke and Stark 1992; Finke and Iannaccone 1993; Warner 1993; Stark and Finke 2000). These accounts posit that religious institutions can be regarded as equivalent to firms competing in the economic marketplace. Mass religious participation depends in this view upon the degree of pluralist competition among diverse religious organizations. This process actively recruits members of mass congregations, at least in types of faith based on communal forms of worship. Hence the strength of religiosity in the United States is explained by the rich diversity of churches, sects, temples, synagogues and mosques actively striving to attract members.

Yet this approach assumes that the mass demand for religion is constant. To develop an alternative perspective, *Sacred and Secular* developed a revised version of secularization theory that emphasizes variations in the "demand-side" of the equation. Norris and Inglehart (2004: 5–6) argued, in particular:

1. The publics of virtually all advanced industrial societies have been moving toward more secular orientations during the past 50 years. Nevertheless,
2. The world as a whole now has more people with traditional religious views than ever before—and they constitute a growing proportion of the world's population.

Though these two propositions may initially seem contradictory, they are not. As we will show, the fact that the first proposition is true helps account for the second—because secularization and human development have a powerful negative impact on human fertility rates.

The book theorized that transcendent religion is usually weakened by a sense of existential security—that is, the feeling that survival is secure enough that it can be taken for granted. We argue that feelings of vulnerability to physical, societal and personal risks are a key factor directly driving religiosity and we demonstrate that the process of secularization—understood as a systematic erosion of religious values and practices—has occurred most clearly among the most prosperous social sectors living in affluent and secure post-industrial nations. By "values" we mean the importance of religiosity in people's everyday lives. By "practices," we mean the common rituals which express religious values in different faiths, such as prayer and attendance at churches, synagogues, mosques and temples. The term "secularization" can refer to many developments. In this regard, we do not examine other forms of secularization (Casanova 1994), such as any decline in the authority of religious leaders and ecclesiastical institutions in social life and the public sphere.

As illustrated schematically in Fig. 177.1, our parsimonious model assumes a simple sequential pathway where human development and societal modernization gradually reduces exposure to both socio-tropic (societal) and ego-tropic (personal) risks. In turn, this process diminishes anxiety and stress, promoting feelings of psychological well-being and existential security. And in turn, this process usually reduces the central importance of religion in people's lives. The first steps in this model are widely established; inhabitants in poor nations often remain highly susceptible to unpredictable socio-tropic risks, exemplified by subsistence farmers

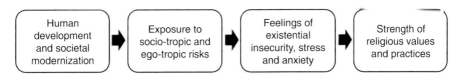

Fig. 177.1 Model of existential security and religiosity (Source: Pippa Norris and Ronald Inglehart)

facing sudden disasters from drought or flood, disease or landslides, crop failure or weather-related emergencies. Poor nations have limited access to the basic conditions of survival, including the provision of uncontaminated water and adequate food, access to effective public services offering basic healthcare, literacy, and schooling, and an adequate income. These societies also often have weak defense against external invasion, threats of internal coup d'état, and, in extreme cases, state failure. The final link in the model remains controversial, however we reason that virtually all of the world's major transcendent religions provide reassurance that, even though the individual alone cannot understand or predict what lies ahead, a higher power will ensure that things work out. This belief reduces stress, enabling people to shut out anxiety and focus on coping with their immediate problems. Without such a belief system, extreme stress tends to produce withdrawal reactions. Under conditions of insecurity, people have a powerful need to see authority as both strong and benevolent—even in the face of evidence to the contrary.

Through strengthening feelings of security, the processes of human development and societal modernization therefore have significant consequences for religiosity; as societies transition from agrarian to industrial economies, and then develop into postindustrial societies, the conditions of growing security that usually accompany this process commonly reduces the importance of religious values. The main reason, we believe, is that the need for religious reassurance becomes less pressing under conditions of greater security. These effects operate at both societal-level (socio-tropic) and at personal level (ego-tropic), although we suspect that the former is more important. Greater protection and control, longevity and health found in postindustrial nations mean that fewer people in these societies regard traditional spiritual values, beliefs and practices as vital to their lives, or to the lives of their community. This does not imply that all forms of transcendent religion necessarily disappear as societies develop; residual and symbolic elements often remain, such as formal adherence to religious identities and beliefs, even when their substantive meaning has faded away. But we expect that people living in secure advanced industrial societies tend to become less obedient to traditional religious leaders and institutions, and become less willing to engage in conventional religious practices. Contrary to the religious markets school, we, therefore, assume that the 'demand' for spirituality is far from constant; instead striking variations are evident due to experience of the basic living conditions common in rich and poor nations.

We can draw a further distinction in Western societies between transcendent religions and post-Christian forms of spirituality (Silver 2006). *Transcendent* religions acknowledge some supernatural forces or other-worldly powers, and emphasize belief in the existence of an afterlife or reincarnation. The way that faith provides reassurance against life-threatening risks is expected to be particularly important for these forms of religion. A wide range of post-Christian "New Age" alternative beliefs and practices, which have become popular in Western societies, promise spiritual or material improvement in this world, such as yoga, astrology, healing rituals, channeling, and self-awareness meditation (Pike 2006; Houtman and Aupers 2007). Theoretically these diverse forms of spirituality may also serve as a way to alleviate stress and anxiety arising from worldly suffering (Silver 2006). We suspect that similar orientations may lie behind both, although this chapter will not attempt to confirm whether motivations for transcendent religiosity and post-Christian forms of spirituality are indeed similar.

177.2.1   Previous Empirical Studies

What evidence from previous studies in the empirical literature helps to evaluate the existential security theory? *Sacred and Secular* presented a wide range of data drawn in particular from the first four waves of the World Values Survey 1981–2001. The analysis of the evidence demonstrated that the importance of religiosity persists most strongly among vulnerable populations, especially those living in poorer nations, facing personal survival-threatening risks. Yet in the original study much of the macro-level evidence demonstrating the links between existential security and religiosity was indirect, such as the significant correlations consistently linking objective macro-level indices of human development, societal modernization and economic inequality with the strength of religious values and practices in any society (see, for example, *Sacred and Secular, 2004,* Table 3.2).

Secondary evidence for the security thesis is also available from a wide range of studies published in social psychology, health care, and welfare studies which have demonstrated the final step in the model, confirming that religiosity helps individuals cope with stress and anxiety arising from uncontrollable life events. One thorough review of this literature by Pargament (1997) found that three-quarters of all empirical studies based on survey and experimental methods reported that religion usually serves to reduce life stresses, at least partially. Another meta-analysis comparing almost fifty published research studies concluded that people often turn to religion when coping with stressful events, such as severe ill-health and death, and this coping strategy proves effective in reducing anxiety and psychological well-being (Ano and Vasconcelles 2005). At the same time despite the wealth of empirical studies, most previous psychological research has examined religiosity within predominately Roman Catholic and Protestant communities, notably in the United States. Yet any effects may well prove situational; at the most extreme, for example, tensions among bitterly divided religious communities in countries emerging from deep-rooted conflict may be expected to exacerbate anxiety and stress (Pargament 2002). Although religious ties may bond adherents together within ethnic communities, such as the Bangladesh and Hindu communities in Manchester and Bradford, nevertheless, minorities often experience discrimination and prejudice in society at large. The effects of religion on psychological well-being may also vary by type of faith (Pargament 1997). Further systematic cross-national evidence is, therefore, needed to sort out whether competing claims for the effects of religiosity on psychological well-being hold in many different societal contexts, and across diverse social risks and types of religion.

Sociologists and economists have also reported similar research findings. Hence comparative survey research by Ruiter and van Tubergen (2009) used multilevel analysis to examine contextual and individual factors contributing towards religious attendance in 60 countries. The study concluded that religious attendance is strongly affected by personal and societal insecurities; in particular, financial insecurities at individual level (measured by household income and unemployment status) are associated with greater religious attendance. Another study by Gill and Lundsgaarde (2004) examined the link between macro-level welfare state spending and religious attendance in almost two dozen nations, confirming a strong relationship, even after controlling for per capita GDP. The research concluded that the most secure European welfare states with well-developed social safety-nets, exemplified by Scandinavian societies, tend to have the emptiest churches.

177.3   Evidence and Measures

Therefore a growing body of literature, drawing upon diverse disciplines, methods and approaches, has generated a large body of findings which are consistent with the existential security thesis. This lends further confidence in the core claims but, nevertheless, further work is needed to persuade doubters. In particular, this chapter will demonstrate more precisely the way that both attitudes towards existential security, and the experience of lived poverty, encourage faith in transcendent religiosity—and thus the search for security and reassurance in the after-life.

177.3.1   Measuring Religiosity

Fortunately both the 2005 wave of the World Values Survey (WVS), as well as the 2007 Gallup World Poll (GWP), contain suitable indicators of religiosity and security, allowing us to investigate these questions with individual-level evidence about cultural values in many different societies around the globe. The use of two independent surveys also strengthens replicability—and thus tests the robustness of the findings—as well as expanding the cross-national comparative framework.

The longest time-series coverage is available from the pooled World Values Survey/European Values Survey (WVS), a global investigation of socio-cultural and political change conducted in five waves from 1981 to 2007. This project has carried out representative national surveys of the basic values and beliefs of the publics in more than 90 independent countries, containing over 88 of the world's population and covering all six inhabited continents. It builds on the European Values Survey, first carried out in 22 countries in 1981. A second wave of surveys was completed in 43 countries 1990–1991. A third wave was carried out in 55 nations in 1995–1996, and a fourth wave, in 59 countries, took place in 1999–2001. The fifth wave covering 55 countries was conducted in 2005–2007.[1]

---

[1] Full methodological details about the World Values Surveys, including the questionnaires, sampling procedures, fieldwork procedures, principle investigators, and organization can be found at: www.worldvaluessurvey.com

The heart of our theory relates to religious *values*, understood as general motivational goals which transcend specific situations and which can be ordered in their relative importance. These are measured in the WVS survey by the importance of religion in people's lives, as monitored by the question: *"How important is God in your life?"* where responses use a scale ranging from not important (0) to very important (10).[2] The primary indicator of religious *practices* analyzed in this study is measured by the standard behavioral question that is widely used in the literature: *"Apart from weddings, funerals and christenings, how often do you attend religious services?"* Responses in the World Values Survey ranged on a 7-point scale from *"never"* (scored 1) to *"more than once a week"* (scored 7). Based on this item, *"regular"* religious participation is understood to denote at least weekly attendance (that is, combining either "once a week" or "more than once a week").

The 2007 Gallup World Poll also provides data on living standards, social deprivation, exposure to societal risks, and religiosity among the publics living in 132 societies worldwide.[3] The GWP survey is conducted using probability-based nationally-representative samples among at least 1,000 respondents of the adult population in each country, using a standard set of core questions. The total sample included 142,671 respondents. In developing societies, Gallup use face-to-face interviews, and they employ telephone interviews in countries where telephone coverage reaches 80 % of the population. Two items contained in the Gallup World Poll are most useful to gauge religiosity. Hence religious values are monitored by: *"Is religion an important part of your daily life?"* (measured as a dichotomous 'yes/no' response). Since this does not refer to any specific concept, meaning, or definition of "religion," or to any particular practices and beliefs, this item is the most suitable for cross-faith comparisons. In addition, religious practices are monitored by Gallup: *"Have you attended a place of worship or religious service within the last seven days?"* (also measured as a dichotomous "yes/no" response). This is the conventional measure of religious participation used in studies of religions involving communal forms of worship in churches, temples, synagogues and mosques, although it is less useful when comparing types of faith which do not follow these collective practices, such as Confucian and Shinto religions.

177.4   Experiential Security and Religiosity

To go further, we need to clarify and operationalize the core concept of "existential security" needs. *Sacred and Secular* compared the impact of objective developmental indices, such as those concerning health and wealth, but these are relatively blunt

---

[2] To test whether this is item suitable for use in non-monotheistic societies, we examined its national-level correlations with responses to the question, "How important is religion in your society?" which does not refer to God. The two tap essentially the same dimension, correlating at .92. Because the question about the importance of God has a longer time series, we use it here.

[3] More details about the methodology, fieldwork and sampling practices, and questionnaire can be found at https://www.gallup.com

and imprecise instruments of human security (Silver 2006). It could be argued that many social changes are associated with societal modernization—including growing individualism, and the expanded cognitive skills associated with the spread of education—which could provide alternative explanations to account for the decline of religious values and practices observed in post-industrial societies. Even at the individual level, levels of income are only proxy measures for personal feelings of security; low income households with close support networks from extended families and the local community may have informal means of coping with humanitarian catastrophes and personal risks which are unavailable to more affluent but isolated individuals. As the conclusion of *Sacred and Secular* acknowledged, psychological perceptions of risk and insecurity needed further analysis as the intermediary variables (see Fig. 177.1).

Yet the empirical challenges of operationalizing the concept of existential security remain challenging, especially for cross-national research. The core notion is complex, relating to multiple forms of vulnerability, whether arising from extreme poverty, hunger, disease, armed conflict, criminal violence, environmental degradation, state repression, natural disasters, or many other causes (UNDP 1994; King and Murray 2001; Newman 2010). The threat of extreme poverty, ill-health and malnourishment are most severe and widespread in the world's poorest societies. This is most commonly measured by the UNDP's Human Development Index (HDI), combining life expectancy, income, and educational attainment.[4] Yet this index may fail to capture exposure to varied societal risks arising from natural or humanitarian disasters, and even affluent post-industrial societies are not immune from certain terrifying forms of insecurity, exemplified by the perceived or actual threat of terrorist violence (Kinnvall 2004). We use two distinct indices; the Gallup lived poverty index can be understood as a summary objective measure of experiential insecurity, while the WVS measures attitudes towards security and risk, which reflects more subjective notions of perceived security.

177.4.1 Experiential Security: The Lived Poverty Index

One way to operationalize the experience of insecurity is by monitoring vulnerability to multiple risks and forms of social deprivation. Since cash income is only a poor proxy, especially in subsistence economies, the Afro-Barometer pioneered the use of a Lived Poverty scale which measures how far people go without a range of basic necessities during the course of a year (Mattes 2008). To construct a similar objective scale, the Gallup World Poll contains eight items which ask respondents to report how far they have enough money to buy food or shelter in the previous year, how far they are satisfied with their standard of living and state of health, whether their home has basic facilities such as running water, electricity and a

---

[4] UNDP. Human Development Reports. http://hdr.undp.org/en/statistics/indices/hdi/

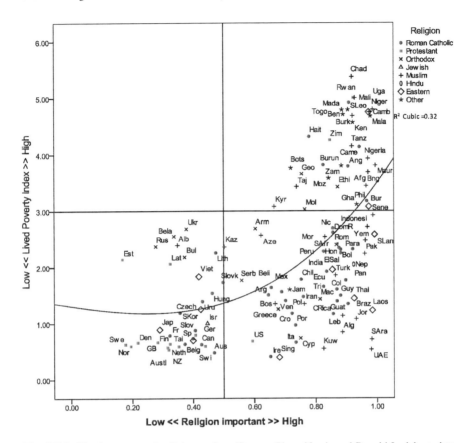

Fig. 177.2 Lived poverty and religious values (Source: Pippa Norris and Ronald Inglehart, data from Gallup World Poll 2007)

landline telephone, and whether they have experienced health problems.[5] These multidimensional items were summed and these proved to form a consistent scale of Lived Poverty (Cronbach's Alpha = 0.70), thus demonstrating a high level of internal consistency. The index it is also strongly correlated at macro-level with both per capita GDP (in PPP) (Pearson R = 0.884, P = 000 N = 120) and the UNDP Human Development Index (R = 0.673 P = 000 N = 123), suggesting high levels of external validity.

The scatter-plots presented in Figs. 177.2 and 177.3 illustrate the macro-level relationship between the Lived Poverty index and the distribution of religious values and practices across the nations where complete data are available. The results

---

[5] The Gallup Lived Poverty index is constructed from the following items: "Have there been times in the past 12 months when you did not have enough money... To buy food that you or your family needed? To provide adequate shelter or housing for you and your family?" "Are you satisfied or dissatisfied with your standard of living?" "Home has no running water, no electricity and no land-line telephone."

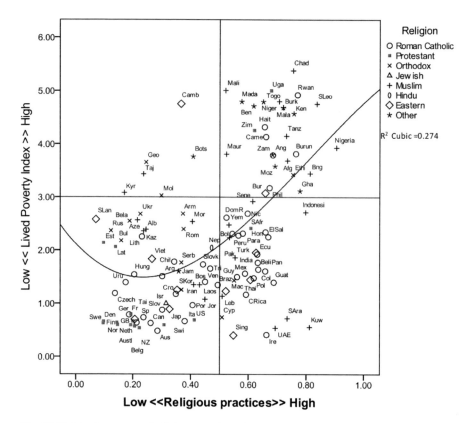

Fig. 177.3 Lived poverty and religious practices (Source: Pippa Norris and Ronald Inglehart, data from Gallup World Poll 2007)

confirm that the Lived Poverty Index was strongly correlated with religious values (R = 0.541 P = 000 N = 128); hence some of the poorest developing societies, such as Chad, Rwanda and Mali gave the greatest priority to religious values, while, by contrast some of the most affluent post-industrial societies in the world, led by Sweden, Norway, Denmark and Britain, proved the most secular. At the same time, the cubic fit shows that there remain many moderate-income countries in the bottom right-hand quadrant with low levels of lived poverty which are also highly religious in their values, notably the oil-rich Gulf states and some other countries with predominately Muslim cultures.

The comparison with religious practices (see Fig. 177.2) shows a similar almost equally strong relationship; thus the lived poverty index proved to be a significant predictor of participation in religious services (R = 0.497, P = .000, N = 127). Again the least developed nations such as Chad and Rwanda clustered together in the top right-hand quadrant, being the most religious and the poorest, while Scandinavian and West European Protestant societies were the least engaged in church-attendance. But there was also a cluster of countries which were outliers to these general

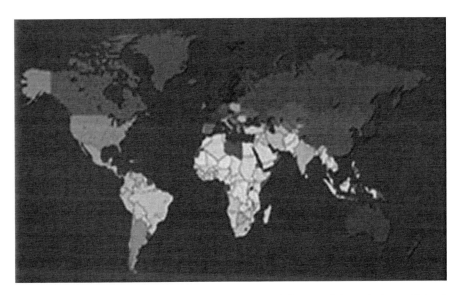

Fig. 177.4 The world map of religious values (*light color* indicates religious, *dark* nonreligious) (Map from http://en.wikipedia.org/wiki/Irreligion; Gallup World Poll 2007 data)

patterns; interestingly among the more affluent societies, although the U.S. is commonly regarded as a deviant case, it ranks high on religiosity in comparison with other rich countries, but much lower than most low-income societies. Religious market theory explains the relatively high levels of found in the U.S. in terms of religious pluralism—but, seen in a broader global perspective, the U.S. is considerably less religious than such countries as Ireland, Saudi Arabia and Kuwait—all of which have minimal religious pluralism.

The geographic world map illustrated in Fig. 177.4 presents the comparison in an alternative way visually, showing the strength of religious values across the African continent, as well as in South East Asia and Latin America. By contrast post-industrial societies are uniformly secular, including Scandinavia, Western and post-Communist Europe, and Australia, New Zealand and Canada.

Another way to examine the data is to consider the way in which both religious values and practices rise steadily across the Lived Poverty index, as shown in Fig. 177.5, where the trends show a remarkably uniform pattern. To make sense of these, they suggest that roughly nine out of ten people who lack the most basic necessities of life report that religion is an important part of their daily lives, but this proportion drops to just six out of ten people who have these basic needs met. Similar disparities can be observed for at least weekly religious participation, although behavioral indicators are always lower than the expression of religious values (see Fig. 177.5).

Moreover important debate continues to consider the reasons for the growing transatlantic rift dividing secular Western Europe from church-going Americans (Bellin 2008). One of the strongest challenge to secularization theory arises from

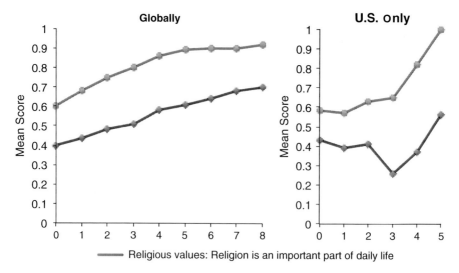

Fig. 177.5 Relationship between religious importance and poverty, globally and U.S. only (Source: Pippa Norris and Ronald Inglehart, data from Gallup World Poll 2007)

observers who commonly point out that claims of steadily diminishing congregations in Western Europe are sharply at odds with American trends, at least until the early 1990s (Berger 1999; Greeley 2003). The theory of existential security is probabilistic, seeking to explain cross-national patterns and to predict long-term trajectories of societal value change, although inevitably there are both leaders and laggards in any social processes. Particular anomalies in survey results can be attributed to many well-established issues associated with measurement error—such as the equivalence and translation of cross-national survey questions, national differences of field-work and sampling practices, and levels of freedom of expression across diverse societies. Nevertheless, we do not believe that religiosity in the United States is an anomaly; there is far too much cumulative evidence documenting the contrasts between America and Western Europe. Among affluent post-industrial societies, the theory of existential security can provide insights to account for certain important outliers, notably the persistence of relatively strong religious participation in Italy and Ireland, as well as the United States. What matters for societal vulnerability, insecurity, and risk, that we believe drives religiosity, are not simply levels of national economic resources, but also their distribution. The growth of the welfare state in industrialized nations insures large segments of the public against the worst risks of ill health and old age, penury and destitution, while private insurance schemes, the work of non-profit charitable foundations, and access to financial resources have transformed security in postindustrial nations and also reduced the vital role of religion in people's lives. Even relatively affluent nations have multiple pockets of long-term poverty and economic inequality. Populations typically most at risk in industrialized nations, capable of falling through the welfare safety-net, include the elderly and children, single-parent female-headed households, the long-term disabled, homeless and unemployed, and ethnic minorities. As Gill and Lundsgaarde (2004) demonstrate, cross-national patterns of welfare spending are significantly linked with religiosity. If feelings of vulnerability are driving religiosity, even in rich nations, then this should be evident by comparing levels of economic inequality and feelings of security within societies, for example to see whether religiosity is strongest among the poorer and least secure sectors of American society.

Figure 177.5 illustrates that the patterns already observed globally, where Lived Poverty helps to predict religious values and practices, also hold consistently in the United States quite well; far from an anomaly, the existential security thesis fits America as well. Hence, among the poorest segments of American society almost everyone reports that religion is important to their lives, but among the most affluent segment, only six out of ten do so. Patterns of church-going are less linear, but, nevertheless, the poorest group of Americans, according to the Lived Poverty Index, are also the most likely to fill the Sunday pews. Milanovic (2005) demonstrates that the U.S. has a remarkably skewed income distribution compared with most OECD countries. These sharp inequalities in American society help to explain its relatively high religiosity. Household income inequality, representing disposable income after taxes and transfers, is most simply compared by the Gini coefficient, ranging from 0 (the most equal) to 1.0 (the most unequal).[6] In the mid-2000s, for example, the OECD estimates that after taking account of taxes and transfers, the United States had a Gini coefficient of 0.38, (and 0.35 in both Italy and Ireland) compared with 0.23 for secular Sweden and Denmark, 0.27 for the Netherlands, and 0.28 for France.[7]

To subject these descriptive observations to more rigorous analysis, Table 177.1 presents the results of binary logistic regression models where the Lived Poverty Index is regressed on both religious values and practices (coded 0/1), controlling parsimoniously for some of the most common demographic characteristics which have commonly been found to predict religiosity, namely age and gender. Further controls for macro-level economic development (per capita GDP or HDI) and micro-level socioeconomic status were considered but rejected for inclusion due to issues of multicollinearity, since the aggregate factors and individual-level measures of education and income were strong correlated with the Lived Poverty index. The models were run both for all countries included in the Gallup World Poll, as well as just for the U.S., to see whether similar patterns held, as predicted, in the American case as well. It is well established that women tend to be more religious, and this is indeed what the results confirmed in the pooled model with all countries and concerning religious values (but not church-going) in just the U.S. case. We see this as a natural extension of the security thesis, since women

---

[6] The Gini coefficient is based on equivalised household disposable income, after taxes and transfers. The Gini coefficient is defined as the area between the Lorenz curve (which plots cumulative shares of the population, from the poorest to the richest, against the cumulative share of income that they receive) and the 45° line, taken as a ratio of the whole triangle. The values of the Gini coefficient range between 0, in the case of "perfect equality" (i.e. each share of the population gets the same share of income), and 1, in the case of "perfect inequality" (i.e. all income goes to the individual with the highest income).

[7] OECD Stats Extracts. http://stats.oecd.org/. Accessed November 2010.

Table 177.1 Models predicting religious values and practices

		All nations		U.S. only	
	Individual level	Religious values (A)	Religious practices (B)	Religious values (A)	Religious practices (B)
Demographic controls	Age (in years)	−.005†	−.002†	.013†	.008*
		(.000)	(.003)	(.003)	(.003)
	Sex (male = 1)	−.308†	−.037†	−.561†	−.115
		(.013)	(.012)	(.120)	(.118)
Socioeconomic resources	Lived Poverty 8-pt index	.272†	.146†	.129*	.114*
		(.004)	(.001)	(.014)	(.059)
	Constant (intercept)	.708	.362	.115	.663
	Nagelkerke R^2	.078	.028	.048	.011
	Number of respondents	121,658	120,394	1,198	1,203
	Number of nations	120	120	1	1
Percentage correctly predicted:		71.2	57.3	59.2	59.4

Data source: Gallup World Poll 2007
Note: Models present the results of the binary logistic regression models predicting religious values and practices including the beta coefficient (the standard error shown in parenthesis), and the significance. *p = .05 and †p = .001. (A) Religious values: *"Is religion an important part of your daily life?"* (B) Religious participation: *"Have you attended a place of worship or religious service within the last seven days?"*

tend to be disproportionately vulnerable to problems of poverty arising from child-care, old age and lower wages, as well as other security threats arising for victims from violence. The age profile in the pooled model for all countries proved more unexpected, with older generations more religious by both indicators in America, although this appears to reverse in the pooled model, a pattern which requires further exploration. After controlling for these demographic characteristics, the Lived Poverty Index remained strong and statistically significant as a predictor of religious values and practices, in the pooled model for all countries as well as in just the American sample.

It could be argued, however, that the patterns observed so far are actually confined to Catholic and Protestant Christianity, as the form of religion characteristic of most post-industrial societies, rather than a consistent trend across all forms of religion (Silver 2006). To examine this issue further we can also break patterns down by the individual's religious faith, and those without any religious beliefs at all. As illustrated in Fig. 177.6, the results demonstrate that the Lived Poverty Index predicts religious values across nearly all categories, including Muslims as well as Catholics, Buddhists and Confucian/Taoists, as well as Protestants and those of Orthodox faith. Therefore, across many comparisons, using an independent Gallup World Poll survey to corroborate the findings in *Sacred and Secular* based on the first four waves of the WVS, and across multiple nations and types of faith

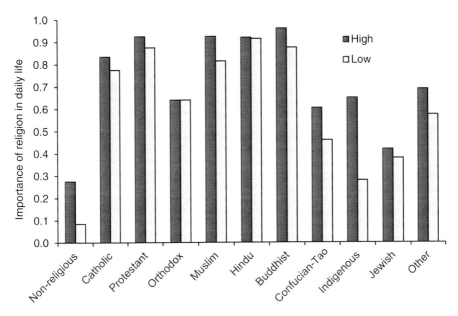

Fig. 177.6 Lived poverty and religious values by faith (Source: Pippa Norris and Ronald Inglehart, data from Gallup World Poll 2007)

worldwide, the results prove consistent and robust; the most vulnerable populations in the world—who lack the basic necessities of life such as food, running water and electricity—are also far more likely to feel that religion is important to their lives, as well as engaging more fully in religious practices.

177.5   Perceptual Security: Attitudes Towards Security and Risk

Still it could be argued that notions of lived poverty, while closer to capturing many important dimensions of human vulnerability than cash income or wealth, still do not tap fully into more subjective or psychological orientations towards threats. Do people believe that they live in a predictable and safe environment, or do they feel that the world is often a dangerous place? To examine these issues, we can analyze perceptual security, monitored by attitudes towards security and risk. This was measured in the 2005–2007 wave of the World Values Survey as part of the Schwartz (2001) value scales. These used the following questions:

> Now I will briefly describe some people. Using this card, would you please indicate for each description whether that person is very much like you, like you, somewhat like you, not like you, or not at all like you? V82: Living in secure surroundings is important to this person; to avoid anything that might be dangerous. V86. Adventure and taking risks are important to this person; to have an exciting life.

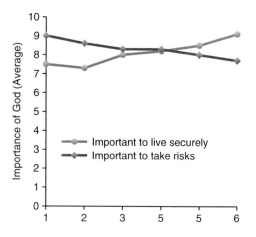

Fig. 177.7 Relationship between religious importance and security/risk-taking (Source: Pippa Norris and Ronald Inglehart, data from World Values Survey 2005–2007)

The results presented visually in Fig. 177.7 for the 55 nations included in this wave of the WVS show the remarkable linear relationship; religious values (monitored by the 10-point important of God scale) rise sharply with attitudes towards security values; those who give the greatest priority to living in secure surroundings and avoiding danger also regard religion as important to their lives. By contrast, those who feel that it is important to take risks and experience adventure are the least likely to see religion as important to themselves.

177.6 Conclusions and Implications

In recent decades public interest in religious differences around the world has grown tremendously, and the debate about secularization theory and its recent critiques has become increasingly relevant to contemporary concerns. The idea of secularization has a long and distinguished history in the social sciences with many seminal thinkers arguing that religiosity was declining throughout Western societies. Yet the precise reasons for this erosion of spirituality were never entirely clear. Weber attributed secularization to the spread of education and scientific knowledge; our own interpretation emphasizes the role of existential security. But by the mid-1960s the claim that religion was in a state of terminal decline rested upon flimsy evidence. Its proponents cited empirical evidence of declining churchgoing in Western Europe, and a handful of case studies that fit the thesis, rather than a systematic examination of empirical evidence from many countries (Hadden 1987).

It was not surprising, therefore, that during the last decade some American sociologists have mounted a sustained counterattack on the basic premises of secularization theory (Stark 1999; Stark and Finke 2000). This critique threw many former proponents on the defensive; Peter Berger recanted former claims, noting that many exceptions had accumulated that appeared to challenge the basic prophesies of Weber and Durkheim—pointing to the continuing vitality of the Christian Right in the United States, the evangelical revival in Latin America, the new freedom of religion in post-Communist Europe, the reported resurgence of Islam in the Middle East, or evidence that religious practices and beliefs continued to thrive throughout most of Africa and Asia (Berger 1999). Some of these reported phenomena may have been over-stated, but the simplistic assumption that religion was everywhere in decline, common in earlier decades, had become implausible to even the casual observer. Too many counter-examples existed around the world.

The religious market argument sought to reconstruct our thinking about the primary drivers in religious faith, turning attention away from long-term sociological trends in the mass public's demand for spiritual faith, and emphasizing instead institutional factors affecting the supply of religion, including the role of church leaders and organizations, and the role of the state in maintaining established religions or restrictions on freedom of worship for certain faiths (Warner 1993). The attempt to reconstruct the early twentieth century sociology of religion was long overdue, but the religious market theory was, we believe, fundamentally mistaken in trying to generalize from the distinctive American experience to the world as a whole. It is clear that the U.S. public remains far more religious than the publics of most other postindustrial societies, but we believe that this largely reflects other causes than those cited by religious market theory.

Early versions of modernization theory, from Marx to Weber, held that religious beliefs were dying out and would disappear with the spread of education and scientific knowledge. More recently, it has become apparent that religion continues to play a prominent role, leading to claims of a "Global Resurgence of Religion" (Thomas 2005). The truth lies between these two extremes. When examined in the global longitudinal perspective provided by the World Values Survey, it becomes evident that religion has indeed become more important in many countries—but it has continued to decline in many others. And one finds a clear pattern underlying these changes. As Fig. 177.8 demonstrates, in recent decades religion has become increasingly important in two types of countries: (1) developing countries and (2) ex-communist societies, where the collapse of communism has opened up an

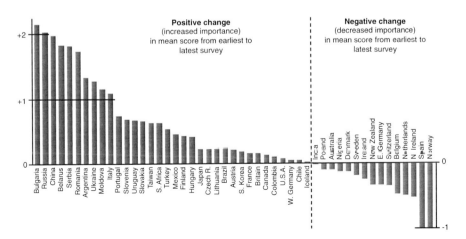

Fig. 177.8 Changes in level of religiosity from earliest to latest available survey, 1981–2007 (Source: Pippa Norris and Ronald Inglehart, data from World Values Surveys, 1981–2007)

ideological vacuum that is being filled, for many people, by increasing emphasis on religion. But this resurgence of religion is by no means universal. Among the publics of high income countries—who have grown up with high levels of existential security—the importance of religion is low and has continued to decline.

The results presented in this chapter serve to strengthen and further confirm the basic argument presented in *Sacred and Secular*, rather than modifying our initial ideas. The evidence has further demonstrated that, with rising levels of existential security, the publics of virtually all advanced industrial societies have been moving toward more secular orientations during at least the past 50 years. Earlier perceptions of this process gave rise to the mistaken assumption that religion was disappearing. "God is dead," proclaimed Nietzsche more than a century ago. A massive body of empirical evidence indicates that he was wrong. As a result of contrasting demographic trends in rich and poor countries, the world as a whole now has more people with traditional religious views than ever before—and they constitute a growing proportion of the world's population. The social and political divisions between those with religious and secular values, beliefs and identities are thus growing – leading to some of the tensions observed today in contemporary Europe.

What are the broader implications of this thesis? One important issue is the growing salience of the gap between those adhering to sacred and secular values. While hardly novel, this gap has grown in prominence in many post-industrial societies due to patterns of population migration combined with the aftermath of 9/11. Migration patterns have been particularly evident in Western Europe, where diverse populations have moved in pursuit of economic opportunities and political freedoms. The OECD (2005) estimates that 70.5 million international migrants living in Europe represent almost one-tenth of the region's total population. These patterns are particularly evident in urban areas, such as the population of Turkish guest workers in Berlin, Bremen and Frankfurt; the Moroccan, Turk and Sudanese communities in Rotterdam; Franco-Maghrebis in Marseilles; and Bangladeshis and Pakistanis in the East End of London, Bradford, or Leicester. While some migrants come from relatively secular countries, especially those from Central and Eastern Europe, many other migrants are drawn from highly religious cultures in Asia, the Middle East, and Africa. Due to these developments, European countries which used to be relatively homogeneous in their cultural heritage, historical traditions, ethnic composition, language, lifestyles, and religious faith—such as Denmark, France, and Sweden—have become far more socially diverse today. The close proximity of residents drawn from highly secular and religious societies heightens the importance of these differences in many contemporary post-industrial societies, with important consequences for politics.

Another contemporary issue concerns the regime changes associated with the Arab uprisings which have rocked the foundations of deeply-entrenched autocracies throughout the Middle East, generating after-shocks rippling out as far as Beijing and Moscow. In Tunisia, President Ben Ali fled to Saudi Arabia, replaced by an elected coalition government of the Islamist Ennahda and the left-wing Ettakatol parties. The Egyptian uprisings led to the Muslim Brotherhood's Freedom and Justice Party (FJP) victory in parliament, although government power remains with the military. In Yemen, President Ali Abdallah Saleh has agreed to stand down and the regime transition continues to unfold. Libya saw bloody civil war, the death of Gaddafi, and governance by the interim National Transitional Council. Seeking to prevent similar events in Bahrain, Sheikh Hamad bin Isa Al Khalifah unleashed Saudi security forces to repress Shia unrest. In Syria, President Bashar al-Assad uses even greater brutality, deploying mortar shells against civilians. For decades, authoritarian regimes in the Arab world had seemed immune from change, apparently impervious to the third wave of democratization sweeping the rest of the world, and untouched by the color revolutions in Eastern Europe. The uprisings caused seasoned observers to revise the standard perspectives in seeking to understand the causes—and forecast the consequences—of these developments. It remains too early to prognosticate upon their full implications, except to note that many predictions of democratic transitions were probably premature, and instead a process of regime change provides a more accurate way to conceptualize these events. The consequences for the balance of power among political parties will probably differ in each society; Tunisia, Libya, Egypt and Yemen differ sharply from each other, rather than providing a single model of regime change. Nevertheless it seems likely that the loosening of the grip of the old autocracies will probably facilitate the mobilization of religious forces in these states, generating new tensions between more traditional and more liberal groups.

At the same time, despite these observations, it would be mistaken to conclude that religiosity is strengthening in secure post-industrial societies, or indeed that these long-term developments will necessarily lead to growing conflict. If security is the driver of secularization, as we believe, then expanding human security through enabling the capacity for sustainable development around the world, and economic equality within societies, is also the key to reducing tensions over religious values. There is no inevitable "clash" but instead the outcome of the heightened contrasts between religious and secular values depend on our capacity for managing more diverse societies.

References

Aarts, O., Need, A., te Grotenhuis, M., & de Graafo, N. D. (2008). Does belonging accompany believing? Correlations and trends in Western Europe and North America between 1981 and 2000. *Review of Religious Research, 50*(1), 16–34.

Ano, G. G., & Vasconcelles, E. B. (2005). Religious coping and psychological adjustment to stress: A meta-analysis. *Journal of Clinical Psychology, 61*(4), 461–480.

Bellin, E. (2008). Faith in politics. New trends in the study of religion and politics. *World Politics, 60*, 315–347.

Berger, P. L. (Ed.). (1999). *The desecularization of the world*. Washington, DC: Ethics and Public Policy Center.

Casanova, J. (1994). *Public religions in the modern world*. Chicago: University of Chicago Press.

Finke, R., & Iannaccone, L. R. (1993). The illusion of shifting demand: Supply-side explanations for trends and change in the American religious market place. *Annals of the American Association of Political and Social Science, 527*, 27–39.

Finke, R., & Stark, R. (1992). *The churching of America*. New Brunswick: The University of Rutgers Press.

Gallup International. More details about the methodology, fieldwork and sampling practices, and questionnaire can be found at https://www.gallup.com

Gill, A., & Lundsgaarde, E. (2004). State welfare spending and religiosity – A cross-national analysis. *Rationality and Society, 16*, 399.

Greeley, A. M. (2003). *Religion in Europe at the end of the second millennium*. New Brunswick: Transaction Publishers.

Hadden, J. (1987). Toward desacralizing secularization theory. *Social Forces, 65*(3), 587–611.

Houtman, D., & Aupers, S. (2007). The spiritual turn and the decline of tradition: The spread of post-Christian spirituality in 14 western countries, 1981–2000. *Journal for the Scientific Study of Religion, 46*(3), 305–320.

King, G., & Murray, C. J. L. (2001). Rethinking human security. *Political Science Quarterly, 116*(4), 585–610.

Kinnvall, C. (2004). Globalization and religious nationalism: Self, identity, and the search for ontological security. *Political Psychology, 25*(5), 741–767.

Klausen, K. (2009). Why religion has become more salient in Europe: Four working hypotheses about secularization and religiosity in contemporary politics. *European Political Science, 8*(3), 289–300.

Mattes, R. (2008). The material and political bases of lived poverty in Africa: Insights from the Afrobarometer. In V. Møller, D. Huschla, & A. Michalos (Eds.), *Barometers of quality of life around the globe: How are we doing?* (pp. 161–185). Dordrecht: Springer.

Milanovic, B. (2005). *World apart: Measuring international and global inequality*. Princeton: Princeton University Press.

Newman, E. (2010). Critical human security studies. *Review of International Studies, 36*, 77–94.

Norris, P., & Inglehart, R. (2004). *Sacred and secular*. New York: Cambridge University Press.

OECD. (2005). Stats extracts. http://stats.oecd.org/

Pargament, K. I. (1997). *The psychology of religion and coping: Theory, research, practice*. New York: Guilford.

Pargament, K. I. (2002). The bitter and the sweet: An evaluation of the costs and benefits of religiousness. *Psychological Inquiry, 13*(3), 168–181.

Pike, S. M. (2006). *New age and neo-pagan religions in America* (Columbia contemporary American religion). New York: Columbia University Press.

Ruiter, S., & van Tubergen, F. (2009). Religious attendance in cross-national perspective: A multilevel analysis of 60 countries. *American Journal of Sociology, 115*(3), 863–895.

Schwartz, S. H. (2001). Extending the cross-cultural validity of the theory of basic human values with a different method of measurement. *Journal of Cross-Cultural Psychology, 32*(5), 519–542.

Silver, D. (2006). Religion without instrumentalization. *Archives Europeennes De Sociologie, 47*(3), 421–434.

Stark, R. (1999). Secularization, RIP. *Sociology of Religion, 60*(3), 270.

Stark, R., & Bainbridge, W. S. (1985). A supply-side reinterpretation of the 'secularization' of Europe. *Journal for the Scientific Study of Religion, 33*, 230–252.

Stark, R., & Bainbridge, W. S. (1987). *A theory of religion*. New York: Peter Lang.

Stark, R., & Finke, R. (2000). *Acts of faith* (p. 79). Berkeley: University of California Press.

Thomas, S. M. (2005). *The global resurgence of religion and the transformation of international relations: The struggle for the soul of the twenty-first century*. New York: Palgrave MacMillan.

UNDP. *Human development reports*. http://hdr.undp.org/en/statistics/indices/hdi/

United Nations Development Program. (1994). *New dimensions of human security*. New York: Oxford University Press.

Warner, R. S. (1993). Work in progress: Towards a new paradigm in the sociology of religion. *American Journal of Sociology, 98*(5), 1044–1093.

Chapter 178
The Religious Context in Political Place-Making

Herman van der Wusten

178.1 Introduction

The relation of religion and politics is a sensitive one. Changes in one of these spheres may well trigger significant effects in the other. A shift in religious allegiances tends to challenge the legitimacy of the political regimes in place. Think of the way the Reformation redid the European political order (Rokkan 2000: 178–184). From the other side a shift in political regime tends to challenge the institutional arrangements regulating the relations of church and state resulting in new positions. Remember how the communist revolutions in Russia and China resulted in abrupt shifts in the position of the Orthodox Church in Russia and of the various religious creeds in China, as did the demise of the Soviet Union and the important changes in China after Mao's death (Bourdeaux and Popescu 2006; Goossaert and Palmer 2012). Even where pronounced secularization has moved organized religion out of the limelight and churches are granted a secure though clearly subordinate position, remnants of old church-state arrangements may be maintained as continuations of a still powerfully felt tradition, e.g. the continued use of Reims cathedral as a French national shrine, where once upon a time French kings were inaugurated (Sauerländer 2012).

Within a polity attempts to control the inherent tensions between religion and politics have been made in different directions:

- the pursuit of maximum religious homogeneity so that religion can function as the basis of commonality with institutional provisions to control religious authorities (cuius regio eius religio as basis of state formation in early modern Europe, Russia under the tsar with intimate connection between political authority

H. van der Wusten (✉)
Department of Geography, Planning and International Development Studies, University of Amsterdam, P.O. Box 15629, 1001 NC Amsterdam, The Netherlands
e-mail: h.h.vanderwusten@uva.nl

© Springer Science+Business Media Dordrecht 2015
S.D. Brunn (ed.), *The Changing World Religion Map*,
DOI 10.1007/978-94-017-9376-6_178

and synodal church organization, contemporary Morocco with the king as highest religious authority in the country chairing the supreme council of ulemas),

- church-state separation with religion or religious communities to some degree autonomous in their own sphere under different degrees of civil freedoms (from the millet system in the Ottoman empire to variously regulated religious freedoms in secular constitutional regimes like France or the U.S.),
- suppressing traditional religion by the development of a state ideology that acquires religious features as in fascist, national-socialist and communist states plus later mixes in the decolonized world (Maier 2004) called political religion. In a version more complementary to traditional religion and more respectful of civil liberties, this was also identified in the US particularly in the 1950s and 1960s as civil religion (Bellah 1967).

The sensitive character of the religion-politics nexus has to do with the ultimate nature of their concerns. Religion provides beliefs about ultimate grounds for human existence, politics provides social mechanisms for ultimate decisions. Such beliefs and mechanisms necessarily touch upon each other by their ultimate nature and provoke highly charged reactions in case of tensions. These reactions tend to touch upon other spheres of life as well.

Religion comes in many forms, not only substantively as a wide variety of beliefs but also in terms of institutionalization: practices and organization. Religion as a rule expresses itself materially pretty massively and therefore leaves a lot of traces. Even if there is a reluctance to express the substance of religious beliefs in imagery, rituals are accompanied with material equipment, burial signs are crafted, meeting places are marked and constructed, offices and residences support the religious organization. Politics is markedly territorially organized, in the current historical period the land masses on the globe are still dominantly subdivided in the entities of the state system. States as a rule have distinct political centers within the national capitals and subcenters in various layers of government, another layer of nodes of trans – inter – and suprastate cooperation is in the making (Claval 2010). In this chapter I concentrate on the ways in which religion impacts upon the political centers of states as a rule situated within capital cities.

In order to do this I consider three basic attributes of political centers: their locations; their manifestations in town – and landscapes as a connected series of buildings and public spaces, an ensemble in architectural terms; and finally the ornaments, statues and other monuments that decorate this part of the town/landscape. Strictly functional considerations from the standpoint of the material needs of efficient and effective rule (how to enable the everyday functioning of a busy institution occupied with the management of large information flows and the crafting of a host of consequential decision-making including the care for its implementation) determine the shape of a political center to a considerable extent. But the intention to realize symbolic significance also conditions their form. To communicate the distinction of the political authority they accommodate is also an important function. These symbols refer to the power of the residing political agent, his particular functions (e.g. the parts of the trias politica he represents) and the basis and justification of his power giving rise to its presumed authority.

Decisions on the location, layout and decoration of political centers are to some extent driven by considerations of symbolism of patrons and executing artists, craftsmen and engineers. The inputs of the different participants in the realization of such projects is however difficult to reconstruct. One has also to recognize that these are often extremely long-term projects with lots of projects within projects and refurbishings and reconstructions over time that may well shift the originally intended meanings to be read in those centers. From the other side their symbolic content is read by frequent users of the facilities, the general public and critical professionals and their opinions may well change over time as well (e.g. on account of changes in aesthetic preferences) even if the material form of political centers does not (van der Wusten 2012b). Consequently the symbolic significance of political centers is a multifaceted phenomenon to say the least. In this paper I aim to provide a series of entries into the question how religion may drive the intended and unintended production of significant symbolisms related to the attributes of political centers that I have proposed and their various appreciations.

Religion is a social phenomenon that can among other things be expressed in terms of an individual, personal attribute (a series of beliefs and their cultural envelop). These individual attributes produce a connected public and captive audience for symbolisms as those in political centers if expressed in an appropriate idiom for that particular belief system. A longstanding example even in religiously adverse times is the shared familiarity of the expressions used in the King James Bible and the English language translations of the Christian holy scriptures that slightly preceded it which was then applied in the political rituals and public events of the United Kingdom and the United States from the early modern period to the present (Pogue Harrison 2012). The religious allegiances of the population amongst whom a polity takes shape possibly affect the different attributes of political centers (location, architecture and decoration) in this way. Religion, if organized, inevitably results in religious nodes and centers of various kinds with their material equipment (churches and temples of course, but also bishop seats, convents and monasteries, pilgrimage shrines, synod venues, religious schools). Religious centers may affect political centers in all sorts of ways and we will discuss such possible impacts again with reference to the location, the townscape and the decoration of the ensuing political centers.

178.2   The Religious Idiom and the Political Center

Religious idioms have a presence on the map. They are distributed across space by people that are positioned in longtime residential patterns and hold on to a particular faith. They are spread out by others who have moved elsewhere (for example, Greek colonies across the Mediterranean and the Black Sea; the Puritans and then countless additional immigration waves in the U.S. all holding on to different creeds). And also by the missionary zeal of religious specialists and even larger social groups outside the core areas where a particular faith originated: Christ's disciples and their successors pushing for Christian conversions across the Roman empire and then further, the Arabs fanning out from the Arabian peninsula to bring the Quran across the Middle East, North Africa and the Northern shores of the Mediterranean from where the Islam travelled still farther out (Barraclough and Overy 1999: 92–93, 98–101).

Religious idioms may have a particularly strong and sharply profiled presence in the channels and nodes of large scale social networks and gradually become less distinctive along gradients toward peripheral areas. But there are also cases of relatively pure survival for extended periods of time in isolation (e.g. Amish). Where different religious idioms meet they may sharply demarcate their differences and then either accommodate to each other's presence more or less graciously (the tradition of Protestant and Catholic pillars during a large part of twentieth century Netherlands) or engage into sustained conflict ('clashes of civilization') or they may merge into syncretistic versions of the two idioms where in the end the original idioms may get lost. Sharp demarcation or syncretism may well result from the substance of different faiths (e.g. the nature of the revelations on which they are based) but the position of the different groups either in well-connected positions within larger units (in nodes or on network links) or in peripheral parts of the various religious realms may also be important.

The Arabs and their Islam may in the initial stages of its expansion have militarily overpowered the religious belief systems that they encountered with amazing speed but the eventual acceptance of the Muslim faith by the conquered populations could take centuries and would take very different forms. The religious infrastructure of ulema's, mosques and religious schools had to be developed and the Muslim faith was practiced in the midst of traditional local customs with all kinds of alternative religious intent (Brown 2011, 27). Similar stories can be told for other religions.

Political centers have traditionally emerged in the context of such religious landscapes and they have then in their turn contributed to their further transformation. The Roman realm was stamped by the myths and deities of the antique world and republican and then imperial Rome showed it. In one of the corners of this political construct, a Christian creed came to fruition that was then slowly and with many ups and downs transmitted through the available channels of the empire across its vast domains until the political entity itself became Christianized. In this last process Constantine was one pivotal figure. He temporarily reunited the then fragmenting empire by force in 312–313 and accepted and supported the Christian religion that gradually was to become the dominant religion of the realm. In Rome he stressed the Christian presence by building an important, episcopal church close to his palace and founded a few Christian churches commemorating local martyrs.

A decade later Constantine decided to shift the political capital of the empire to the East. Rome had already lost importance as a prime political center as the military organization on which the empire now mainly rested had shifted its principal command posts closer to the prime areas of action along the imperial borders (for example in the West to Milan and Trier). In the East were in general the better developed parts of the realm and Christianity had its center of gravity there despite

Rome's special position as the seat of Saint Peter's successors. Constantine's origins were also in the East, but it should be added that he had also long stayed in
more Western parts of the empire, not only in Rome but repeatedly also in France
and Britain. In the end Constantine opted for Byzantium, an old Greek colony
alongside the connection between the Mediterranean and the Black Sea with excellent long distant connections, a defensible harbor and nearby important food producing areas.

In 6 years 324–330 the city was converted into the imperial capital called the new
(second) Rome, later called the city of Constantine: Constantinople. The reconstruction of the urban landscape at this stage is very difficult. Archeological evidence of the period is massively destroyed, written sources are few and far between
and often highly hagiographic. A more or less shared current version goes like this
(Mango 1985; Harris 2007) (Fig. 178.1).

The city situated on the eastern tip of a narrow isthmus was delimited by a new
wall far extending the existing urban surface still largely empty to the West. Its
public core consisted of two squares and a few 100 m of the main thoroughfare in
between. To the West just across the erstwhile wall of old Byzantium, was the

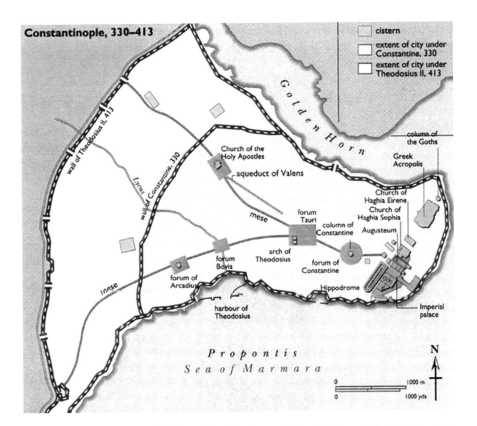

Fig. 178.1 Constantinople (Map from a historic atlas, available at HistoryandCivilization.com.)

Forum of Constantine, a round space with the Senate House to the North and a huge statue of Constantine in the middle. This Forum was opened up from two sides by the East-West main thoroughfare the Mese. Further West the Mese was in later times interrupted by further fora dedicated to later emperors and ended in two stretches at the main city entrances. Going from the Forum of Constantine through the Mese to the East one reached the Augusteion square. Entering the square one passed an arc with another statue of Constantine and a milestone indicating the distances to the major places in the empire. The Augusteion was surrounded by the imperial palace, another senate accommodation and a basilica – a familiar item of Roman cities. It was a roofed hall with arcades and law chambers where business transactions could be settled and accompanying legal issues solved. There were also a horse race track and a bath house.

Somewhat further to the Northeast was the already long existing acropolis, an elevation where the already existing temples of the Olympian gods remained intact. Between the acropolis and the Augusteion, Constantine built the Christian episcopal church of Saint Eirene. Apart from this church building Constantine built two others for two martyrs, one inside and one outside the city walls. A modern commentator deems the Christian presence at this stage as hardly outspoken (Mango 1985: 36). His policy in this respect resembled the line he had earlier followed in Rome. There was an apparent presence of old Olympian gods in a prominent elevated, long established position, the references to a godlike emperor in statues at the most central spots in the city and the spread-like presence of Christian churches across built up urban space and beyond. This changed profoundly with emperor Justinian's final shaping of the Hagia Sophia church in the sixth century, the third and magnificent effort to found an overpowering Christian symbol along the Augusteion that could dominate the urban landscape. This followed a serious urban riot in which existing buildings along the Augusteion square were severely damaged. The Hagia Sophia became the principal church of the Orthodox creed, temporarily in the thirteenth century an important church of the Latin Christianity, subsequently a mosque and then a museum.

That is to say, the location of the new capital was perhaps slightly influenced by the religious map of the empire. Constantine followed a very cautious policy in ornating his new capital with religious symbols largely following the contemporary ambiguities in the makeup of the religious landscape. The dominance of Christian beliefs was still highly uncertain and so was his own missionary zeal. In the initial plan of Constantinople the residence of the religious function was not particularly emphasized. This was to be a political center par excellence with the new Augustean emperor in the very center and public facilities with a political role emphasized. While the church was not a first priority, it eventually got a prominent position when the Christian presence was more fully established.

In the expanding Islamic world from the seventh century onward new religious allegiances were accompanied by the construction of mosques in an enormous variety of forms everywhere making use of local building traditions and aesthetic formal repertoires (Brown 2011) (Fig. 178.2). Mosques in the Arabic Middle East tend to differ strongly from those in the Persian tradition, in Central Asia and India and

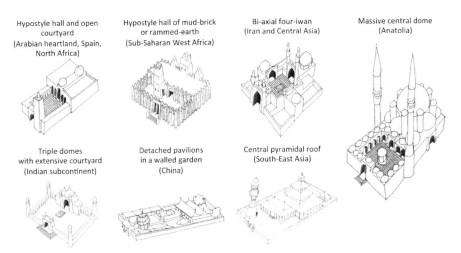

Fig. 178.2 Mosque types (Modified from Frishman and Khan 1994: 13)

then again from those in Southeast Asia. Very generally there is the preference for repetitive geometric decorations, sometimes showing birds and foliage but without other living figures. Religious imagery, if at all existent is flat, not three-dimensional just as in the Orthodox tradition with its icons. In Islamic decoration Arabic calligraphy expressing texts from the Quran is of cardinal importance. The political centers of the Islamic rulers, their palaces and forts were also deeply impregnated by these cultural variants within an overarching Muslim world. So were the funerary monuments that the ruling elites left behind as part of the expression of their enduring power in the context of their faith.

The Red Fort in Old Delhi (Kaul 1985, particularly 41–47), home of South Asia's Mughal rulers from the mid-seventeenth to the mid-nineteenth century, their Islamic beliefs refracted in the surrounding Hindu culture, had a military, a governmental and a private residence part on its spacious grounds. The different pavilions of the private residence contained elements of Hindu building styles and were decorated with descriptions of paradise in the Quran. An artificial flow connected its different parts, the surroundings were magnificently shaped and the words of a famous poet of the sixteenth century were repeated again and again: If paradise there is, this is it. Some 200 km (120 miles) away the founding father of the Red Fort had built the Taj Mahal, a funerary monument for his deceased wife, but also the luxurious expression of a powerful reign full of Muslim symbolisms. In some of these symbols (for example, on top of the dome) Islamic imagery merges subtly with references to the Hindu gods. This Mughal ruler had then moved his political center and built the Red Fort. Mughal rule later declined and finally ended in 1857 when the British forces conquered the Red Fort and put up their headquarters there. On their way they had defaced the Taj Mahal building. Still during colonial times it was extensively restored and the gardens were redrawn with British-style lawns. The British decided to move their political center from Calcutta to Delhi in 1911, one important motive being the continuation of government from a site that had earlier been home to Moslim/neighbourliness rulers who had replaced different periods of Hindu reign in which the place had played a prominent role (Kaul 1985: p. xxix).

For all the very different and numerous instances where a political center came successfully to life through history, the question arises how its profile was affected by its origin in a distinctive religious realm. Constantinople's foundation is one case. The numerous forts and palaces in the extensive Muslim world just mentioned are other ones. But it is also relevant how political center formation adapts to or finds itself confronted with changing religious realms over time. Constantinople, Byzantium, Istanbul is also an extremely interesting case of this last type. Repeatedly shifts in religious allegiances had to be accommodated in the old political center (1204, 1261, 1453). When the Ottomans finally conquered the city the sultan found the old Byzantine palace abandoned and destroyed. After he had spent a few years in a temporary residence later called "the old palace" nearby he moved to his "new palace" 6 years later. It had been built on the elevated site of the old acropolis and this meant the final disappearance of the pre-Byzantine Olympian gods from the city. The palace became later famous as Topkapi, now a museum. The main Christian church Hagia Sophia was made a mosque. After the demise of the Ottoman empire the entire political center was replaced to Ankara from 1923 as part of the efforts by Atatürk to modernize the religion-state relation in an intended secularizing context. The vibrant cultural and economic center in the former capital is still intact (see Nobel Prize for Literature 2006 Orhan Pamuk's loving commemorations and recollections in several fiction and half-fiction books, the latest one Pamuk 2005).

In the next section I confront political and religious center formation and their mutual impacts in a number of very different settings. Constantinople/Istanbul is one extreme case in that part of its history from 1453 onward when its long-established religious center function in the context of the Orthodox faith found itself confronted with the ruler of the Ottoman caliphate. The sultan as proclaimed supreme leader of the Islamic world established a new political center with the obligation to protect the ultimate religious center of Islam in Mecca. The existing religious center function of the Orthodox faith, replaced toward a more modest location in the city, but hanging on to its formal position still exists in today's Istanbul in sharply reduced circumstances.

178.3 Religious and Political Centers: Variations in Cohabitation

Politics cannot exist without centers, at least in the sense of network nodes to which information flows are sent and from where steering efforts emanate. These nodes tend to materialize in spatial concentrations to be read as centers on the map. Now that the means for telecommunication have multiplied and have become generally accessible, these materializations in geographical centers may somewhat subside, but they have an extremely long history. Face to face contacts seem to remain irreplaceable in politically relevant instances and the symbolic importance of a geographically concentrated site where politics is situated, is not easily given up. In sum, political centers as places are here to stay, still (Deutsch 1963; Castells 1996; van der Wusten 2012a). The symbolic significance of political centers and the capital cities in which they are situated has been communicated to a wider audience for a long time (for the Berlin example, see Biskup and Schalenberg 2008).

Religious centers in the sense of geographical sites may have various functions. They can be places where the sacred materializes and thus function as shrine for pilgrimage or at least attract attention from afar. They may also be places where the religious institutions that administer a certain faith have their residences and therefore attract the attention of the believers. These functions can also be mixed. It is intriguing in this respect that the Roman basilica was residence to commercial transactions and a legal institution and then was transformed as the basic shape of the Christian church functioning as the place of congregation of the believers, but also as the place where clerical authority was displayed, all with an air of nearness to the sacred. Religions obviously differ in the ways they are institutionalized and, therefore, there is a wide variety of types of religious centers.

How can political and religious centers be considered as connected in terms of location? Two variants can be distinguished: succeeding each other in the same location and be part of one ensemble. In addition, to get a sense of the mechanisms involved, it is useful to explore cases where there is no apparent connection at all in terms of location. There is still another possibility of connection that I do not consider in any detail. Efforts can be made to bridge existing distances between these different types of centers by extra physical infrastructure or enforcement of accessibility. For example, the pope for a long time tried to ensure accessibility of political centers elsewhere through his local representatives (bishops and special envoys), the different khalifats facilitated the annual hajj from their capitals (Baghdad, Cairo, Istanbul) by extra infrastructure (secured roads and regularly spaced resting places for the annual caravans) in the direction of Mecca (Porter 2012).

178.3.1   Sequencing

In the first case political centers succeed religious centers in one location by direct replacement or after some time interval. Examples of shifts in the other direction are rare: Rome as the political capital of an empire succeeded by the center of western Christianity might be seen as such notwithstanding the papal political pretensions. An interesting and open question is if the succession is due to motives concerning the properties of the first facility or to the more general qualities of the location. Relevant properties of the first facility might be at the symbolic level: the intention to benefit from the religious center's fame, to suppress its memory or to show the predominance of the second facility. They might also be at the level of presumed functional suitability: the expectation to satisfy the needs of the second user in terms of appropriate spaces and technical facilities as well as it has suited the first. Relevant locational qualities might be: exposed position, good general connections, accessibility of needed assets.

Examples of this "invasion and succession" type of connection are to be found in Catholic Europe where formerly religious buildings though not churches were often reconverted into state buildings. This was sometimes inspired by anticlerical motives and to underline the liberal nature of the current regimes but also by the fact that these buildings and their central locations were considered particularly well suited for their new functions. The parliaments of Portugal and Italy sit in a former monastery and a papal law court (Fig. 178.3), Spain's senate sits in a former convent school (the lower house has nineteenth century purpose built accommodation) and some of Spain's regional governments also sit in former religious buildings. It seems that generally in Protestant, northern Europe such buildings had already passed into secular, communal hands during the Reformation. They had then been converted into other functions like hospitals, alms houses and orphanages. Practical considerations and a less convincing symbolic potential may have prevented their use as political trophies for state glory purposes later on (Schilling 2004).

Fig. 178.3 Montecitorio Rome: Papal law court reconverted into a parliament. The building was selected as parliament after Italian independence, but its interior was soon found to be inconvenient. In the early twentieth century the Chamber of Deputies was completely rebuilt, only the façade was left (Photo by Manfred Heyde, http://en.wikipedia.org/wiki/File:Palazzo_Montecitorio_Rom_2009.jpg)

178.3.2   *More or Less Harmonious Cohabitation*

Political and religious centers may also share positions in a larger architectonic ensemble, thus being located in each other's immediate vicinity. Sharing a location or at least being both part of an architectural ensemble necessitates some form of accommodation in the longer term. This can be based on power differences and terms dictated by the stronger partner that have to be considered legitimate or by the mutual acceptance of different competence domains with the ability to sort out the occasional transgressions. In addition, following a standard argument developed in economic geography, concentration in a more or less single place may be stimulated by general site qualities and connectivities that attract public functions. As political and religious institutions are so sensitive to each other, mutual harmony – if not attraction – is practically indispensable for long term shared geographical positions.

The blending of spiritual and political authority is particularly obvious in cases where religious and political top functions are merged in a single person like Gottmann's early bishops. The Japanese and Chinese emperors in still recent times were in that same position. The religious and political centers of their realms were fused in their palaces. Another famous, only marginally less closely tied example is the Latin American plaza, often constructed on top of the ruins of the earlier town centers with their religious and public buildings lining the central square with a cathedral, a government structure and possibly additional public buildings for leisure and display (Low 2000).

Jean Gottmann, a founding father of modern political geography, suggested a kind of evolutionary path in which situational qualities and an initial religious central function give rise to a harmonious process of place-making to establish political authority. Prehistoric man apparently was a frequent traveler using existing paths. Fear for the unknown struck him particularly at crossroads where religious signs were then erected to protect him and restore his calm: monuments, crosses, graves of holy men, white or black stones. These can be read as the early stages of a higher spiritual authority. Later in history a demand for political authority grew to create order, particularly along the main traffic arteries (some of the erstwhile walking paths). Lords (often simultaneously bishops) established their castles at the crossroads, started to take toll and the legitimation of their power was assisted by the presence of the spiritual authority and the iconography that it engendered. In this way the human movement factor and the established iconography of spiritual authority at the crossroads helped locate and strengthen the centers of political authority (Gottmann 1952: 517–519).

The blending of spiritual and political authority is particularly obvious in cases where religious and political top functions are merged in a single person like Gottmann's early bishops. The Japanese and Chinese emperors in still recent times were in that same position. The religious and political centers of their realms were fused in their palaces. Another famous, only marginally less closely tied example is the Latin American plaza, often constructed on top of the ruins of the earlier town centers with their religious and public buildings lining the central square with a cathedral, a government structure and possibly additional public buildings for leisure and display (Low 2000).

A classic European example of the harmonious togetherness of the political and a religious center is the set of buildings lining Parliament Square in London. The most important are the Palace of Westminster, home of the mother of all parliaments and Westminster Abbey, a very prominent national church. The palace is the successor building of one of the earliest residences of the English king, the House of Commons gathers in a copy of his private chapel. The palace was initially established close to an existing community of Benedictine monks who inhabited an abbey close to a long established river crossing. In other words, the religious center was first and even older was the network node (remember Gottmann). The abbey then came under the special protection of the English king and was later reconstructed as a French style Gothic cathedral, the current 'abbey' (a case of power differences and apparent accommodation, that has been repeatedly questioned later). It became the coronation church of what eventually became the United Kingdom, it has a whole collection of tombs of royals and national luminaries. It should be stressed that the Church of England eventually established at the age of the Reformation has important formal, organizational links with the British crown and Parliament but at the same time is thoroughly decentralized and has its primate bishop and his cathedral in Coventry (the long term form of accommodation found). The Palace of Westminster eventually lost its royal occupants and became the seat of parliament that after a devastating fire in the nineteenth century got a completely redone special purpose facility with a host of historical references in an appropriately nationalized neo-Gothic style.

The Berliner Dom, a huge Protestant church in central Berlin has a comparable background. Its location was adjacent to the Stadtschloss of the Hohenzollern Prussian princes who eventually became the emperors of the post-1871 (second) German realm. As the rulers of Brandenburg they had in the fifteenth century moved their residence to this site, close to an existing Dominican monastery, but also on a spot eminently suitable to control traffic and take toll. The princes then constructed a chapel that became a church. After a number of changes in religious faiths and architectural transformations of both palace and church, an impressive ensemble resulted in the early twentieth century. It consisted of a large mainly neo-classical palace more or less definitely built in the early nineteenth century by Schinkel, a massive evangelical domed church in neo-baroque, newly built around 1900. These core buildings complemented with a sizable French style public park and at some distance away the new parliament, the Reichstag, and in a different direction close by the emerging state bureaucracy (Fig. 178.4).

It was not to last. Karl Liebknecht proclaimed the doomed socialist republic from one of the palace balconies after it was stormed and occupied by revolutionaries. The building was no longer used as a political site during Weimar and the Nazi regime. The park however was, for open air rallies. The whole ensemble was severely damaged in World War II, the palace was destroyed by the new communist regime in 1950 and replaced by a new parliament cum cultural facility as part of the East German political center. The old Reichstag was with difficulty preserved by the other side of the Cold War contestation. The restoration of the church was finally allowed. After the demise of the East German regime the future of its new parliament building was for years a major bone of contention in German and Berlin politics. Finally it was levelled and a copy of the Stadtschloss is in the process of resurrection at the time of writing. It will get a museum function. The church attracts masses of tourists. In the meantime the German political center has moved westward to the gloriously restored Reichstag, now the seat of the federal parliament adjacent to the new large chancellery. There is no church in sight there.

Many noble houses tended to have their personal chapel adjacent to or even inside their palaces. Ever since the oldest manuals on diplomatic practice a point was made of the admissibility of a religious facility of any creed on the protected

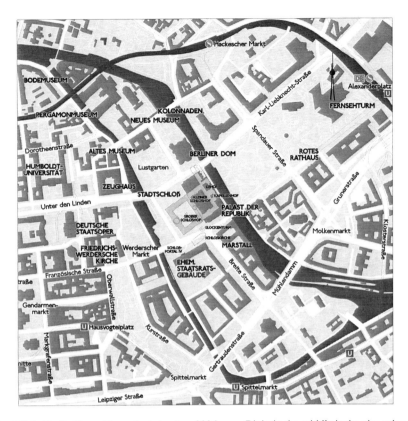

Fig. 178.4 Berlin Ensemble projected onto a 2006 map: Right in the middle is the city palace of the Prussian king then German emperor in central Berlin with the clocktower and private chapel in the south already removed in the eighteenth century and one wing of the palace in the north removed in 1930 to allow traffic to pass through the park. The ensemble consisted of the palace (Stadtschloss), the park (Lustgarten) and the church (Berliner Dom). The rest of the map shows the buildings existing in 2006. At the location of the Stadtschloss was then the Palast der Republik, the parliament annex leisure building of the GDR (removed since then), nearby also the building of the State Council, the major administrative organ of the GDR (part of its façade was a partial front and balcony of the former palace where the communist Liebknecht tried to start a revolution in late 1918) and the building of the GDR ministry of Foreign Affairs that had already disappeared by 2006 (Map from http://commons.wikimedia.org/wiki/File:Karte_berlin_stadtschloss.png)

territory of the diplomat provided it did not have very visible towers and loud noise that could annoy the local population. Thus, the importance of the expression of the religious identity by the esteemed representative was underlined at the same time granting the possibly delicate nature of that identity on foreign ground. A harmonious ensemble of political and religious center may result from the preference of political rulers to seek the vicinity of existing religious establishments (as in London). Probably more often, major churches are positioned close to the rulers' residence or the government buildings, apparently to enhance the prestige of the ensemble.

A particularly interesting ensemble of a longtime complicated, but mutually harmonious relationship of political and religious centers is found in Dresden.

The capital of Saxony, it knew a tumultuous religious history. Luther started his version of the Reformation here. Saxony eventually had an overwhelmingly Protestant population and initiatially its rulers also converted to the new faith. But eventually their descendants moved back to Catholicism at the beginning of the eighteenth century (in order to improve their chances at the Polish throne). Central Dresden had a longstanding Residenz-schloss where the reigning family resided. In the early eighteenth century an impressive building program started in the baroque style. First a big palace à la Versailles, the Zwinger, was constructed largely built by Italians. Then two major churches in the same sumptuous style were projected and eventually nearly simultaneously realized. First at some distance from the political quarters the Protestant Frauenkirche (only recently rebuilt after its destruction in the Dresden bombardment of 1945) appeared and then in the immediate vicinity of the palaces the Catholic Hofkirche. Apparently a careful balancing act was done to satisfy the different religious preferences.

Yet another political effort to align a central religious institution as part of an architectural ensemble to an existing political center was and has again become the reconstructed building of the enormous (largest Orthodox church ever built) Cathedral of Christ the Saviour close to the Moscow Kremlin (Sidorov 2000). The project originated from the tsarist wish to celebrate Russian victory after Napoleon's disastrous military campaign directed at Moscow in 1812. At the time of Napoleon's Russian campaign and the following celebratory project Moscow of course was not the actual national political center, but the Kremlin had been retained as the place of coronation of the tsars and had been provided with a number of palaces, other public buildings, convents and churches and a garrison from Catherine the Great onward. The main office holder of the autocephalous Russian Orthodox Church was the patriarch of Moscow and all Rus. He has resided in different places across Moscow in the course of time but has some space available in the Kremlin as well (not in Soviet times obviously).

The first effort (a less ambitious venture than the finally realized building planned far from the Kremlin) came to nothing. The next tsar Nicolas I thoroughly changed the project into a larger version of Constantinople's Hagia Sophia Byzantine Church/mosque thus underlining the Russian ambition to make Moscow into the third Rome. He also changed the location to a prominent spot close to the Kremlin. In this way his freshly crafted imperial ideology of Orthodoxy, autocracy, nationality materialized. Nicolas I initiated the project in 1839, but the cathedral was only consecrated in 1883 under a successor. It was not only richly decorated with icons and frescoes but also contained large marble tablets with inscriptions of the units and military commanders who had at the time successfully fought Napoleon's army.

In 1931 the Cathedral was completely destroyed on Stalin's orders only to be replaced by his intended celebratory and staggeringly large building of the Bolchevik regime, the Palace of the Soviets. This project ended as an open air swimming pool, the cathedral was finally resurrected in its seemingly former shape at the same spot by Turkish construction crews using lots of concrete in the 1990s. The Cathedral now plays an important public role and also acts as a venue in the rituals of the current Russian regime. In response it has also been 'invaded' recently by political opponents who staged and filmed demonstrative rock performances before they were forcefully arrested (http://en.wikipedia.org/wiki/Pussy_Riot).

Another weird ensemble is the Royal Palace in Amsterdam with the neighbouring New Church in the Netherlands (Goverde 2002). The Palace was built as the local town hall and only became a palace in the early nineteenth century at the initiative of the French temporary occupier. The church had at the time for centuries been one of the major protestant churches of the city. As the kingdom of the Netherlands was instituted in 1813, Amsterdam was named its capital. The political center was and is in the Hague, the royal palace has hardly ever functioned as such. The only practical manifestation of Amsterdam's capital city status is the constitutionally prescribed role of the city as the place where a royal successor takes the oath and is inaugurated. The new incumbent following tradition shows himself on the palace balcony after the abdication of his predecessor followed by a ceremony in the church with the parliament attending, while this is explicitly not deemed to be a religious ceremony. This is a rare ritual. It has only been performed seven times in two centuries.

In many historic cases of political center formation an existing religious institution representing vivid religious beliefs, acquires a distinctive place of honour within a political center (for example, Westminster Abbey). In the twentieth century the political sphere has generated ideologies aiming at the realization of an ideal in this world. Such ideologies can be preached and practiced as religions while traditional religious faiths are sidelined, ignored or suppressed. Such political religions have emerged as the guiding spirits of communist, fascist and national socialist regimes that had their heyday in the interwar years. In these political religions the very center of attention is politics itself and the religious features are made part of the political rituals that sustain the various regimes.

To illustrate the above, massive public celebrations in the open air were organized: Mussolini's speeches on the Venezia square in Rome represent one early instance where sound amplification was used to great effect. Other examples include parades to commemorate the anniversary of the revolution and May 1 popular manifestations on the Red Square with the leaders on top of the mausoleum and party gatherings in Nuremberg after carefully choreographed marches of uniformed sections to the field (see Riefenstahl's artful propagandistic documentary 'Triumph des Willens') with light shows, flags and also bombastic speechmaking (Figs. 178.5, 178.6, and 178.7). The functional equivalents of churches where the political religion can be celebrated have hardly materialized as these regimes were short lived and perhaps did not prioritize such facilities. The huge intended accommodations to celebrate these political religions in Moscow and Berlin were never realized (Van der Wusten 2000). The pseudo-cathedrals and the open air manifestations were all, Nuremberg excepted, situated close to the political centers of these various regimes. In the case of the nazis the park, part of the Stadtschloss/Dome ensemble in Berlin that was earlier mentioned, was also intensively used for mass demonstrations and even especially refurbished with new stone pavement for the purpose.

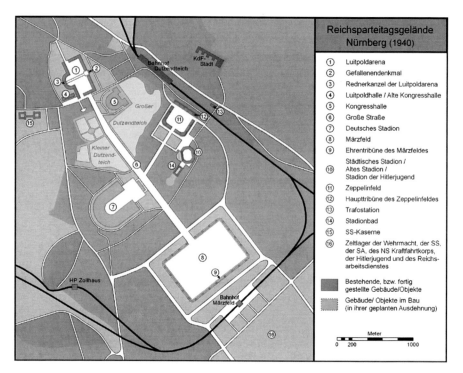

Reichsparteitagsgelände Nürnberg (1940)

① Luitpoldarena
② Gefallenendenkmal
③ Rednerkanzel der Luitpoldarena
④ Luitpoldhalle / Alte Kongresshalle
⑤ Kongresshalle
⑥ Große Straße
⑦ Deutsches Stadion
⑧ Märzfeld
⑨ Ehrentribüne des Märzfeldes
⑩ Städtisches Stadion / Altes Stadion / Stadion der Hitlerjugend
⑪ Zeppelinfeld
⑫ Haupttribüne des Zeppelinfeldes
⑬ Trafostation
⑭ Stadionbad
⑮ SS-Kaserne
⑯ Zeltlager der Wehrmacht, der SS, der SA, des NS Kraftfahrtkorps, der Hitlerjugend und des Reichsarbeitsdienstes

Bestehende, bzw. fertig gestellte Gebäude/Objekte

Gebäude/ Objekte im Bau (in ihrer geplanten Ausdehnung)

Meter
0   200   1000

Fig. 178.5 Political religion in Nuremberg. This 11 sq km area was used earlier for trade fairs at the outskirts of Nuremberg. Hitler and Speer intended (and only partly realized) the area to become 'the Temple city of the Movement' where public manifestations of Hitler's party (NSDAP) and its different organizations were held (Map from http://en.wikipedia.org/wiki/File:Karte_Reichspartei tagsgel%C3%A4nde_N%C3%BCrnberg_1940.png)

Fig. 178.6 The first NSDAP annual party manifestation in Nuremberg in 1927. Left of Hitler are Himmler and Hess. The very start of mass organized ritual was in this place (Photo from http://de. wikipedia.org/w/index.php?title=Datei:Bundesarchiv_Bild_146-1969-054-53A,_N%C3%BCrnberg,_ Reichsparteitag.jpg&filetimestamp=20111201203822 from Wikimedia Commons: Bundesarchiv, Bild 146-1969-054-53A/CC-BY-SA)

Fig. 178.7 Hero worship during the annual party manifestation in Luitpoldarena in 1934. Central figures: Hitler (chancellor), Himmler (leader SS), Lutze (leader SA) (Photo from http://en.wikipedia.org/wiki/File:Bundesarchiv_Bild_102-04062A,_N%C3%BCrnberg,_Reichsparteitag,_SA-_und_SS Appell.jpg from Wikimedia Commons: Bundesarchiv, Bild 102-04062A/CC-BY-SA)

American civil religion, especially as perceived in the 1950s and 1960s is similar to political religion in the sense that political values and themes are celebrated in ways that are derived from religious practice. Rituals with that background are practiced in an atmosphere that also suggest a nearness to the sacred. This is e.g. quite visible in and around Lincolns Monument in Washington D.C. with its texts that resound the biblical rhythms, its enormous statue of the man himself and the surrounding temple like building (Vale 1992: 4–7). The monument was inaugurated in 1922 long before the perception of an American civil religion had gained ground.

It is very intriguing that in fact mostly after the period in which American civil religion was most frequently debated, Washington's Mall has been increasingly occupied with monuments that touch the national nerve by commemorating recent wars and their outcomes in various ways. They can be read as set pieces for celebrating the American civil religion (Savage 2009: 4–5 in particular). American civil religion differs significantly from political religion in that civil religion is for most people not contrary to the religious practices in which they are involved, nor is the state in any sense trying to monopolize the sphere of religion for its own purposes. It is, however, clear that beliefs from these various origins in this instance co-exist generally harmoniously, with some instances of quite vocal dispute and contestation added.

178.3.3 Single or Living Apart Together

Finally, there is the possibility that political and religious centers are locationally disconnected or unconnected. The pure type would be where no sharing of locations or neighbourliness either in sequence or simultaneously can be discerned. Some cases come to mind. Early European cities tended to have several public squares often with market functions for different kinds of products and public buildings are often scattered among them. Town halls and main churches may find themselves at different squares. In early Italian cities a ruler could establish himself in one defensible corner of the city including a private chapel (the entire ensemble defensible also against the occasional ire of fellow townsmen) whereas the main church was located at a central position within the city.

More modern cases seem to be the result of exceptional circumstances. Upon Italy's unification starting in 1859 from Turin as the seat of the Piedmont state, the capital city function first shifted provisionally to Florence and then moved to Rome. The Pope as the local ruler who did not want to give up his status as the holder of a territorial sovereignty, had to withdraw to the Vatican. The newly established Italian state took over many of the buildings that had functioned as parts of the seat of the Papal state (for example, the Italian parliament took over the papal law court, the king occupied a major papal residence), a new government center took shape along Via XX September, named after the day when in 1870 the royal troops had entered the city. The history of the third Rome in situ started that day succeeding the histories of the Rome of the empire and the Rome of the Pope (Kostof 1973). As earlier mentioned, ambitions to succeed elsewhere as a new Rome were attempted in Constantinople and then in Moscow. From 1870 on, the religious center in real Rome was, if anywhere, in the Vatican, across the river and clearly at a distance from the parliament, the palace and the new government departments. But the circumstances were obviously exceptional: a complicated tangle of different jurisdictions in which the two centers had to be located expressing the political issues at stake.

There are some similarities with the situation in Jerusalem (Fig. 178.8). After the cease fire in 1948 Israel constructed a national political center in the western part of

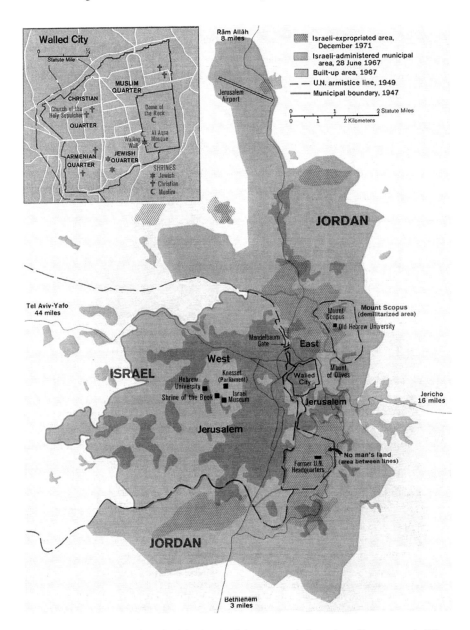

Fig. 178.8 Spatial separation of political and religious center in Jerusalem. Government buildings including the Israeli Parliament Knesset are to the West; religious establishments are within the Walled City to the East. This situation resulted from the 6 Days War in 1967 and was put on the map in 1973. What is currently known as the West Bank was at the time still nominally part of Jordan though under Israeli administration (Map by U.S. Central Intelligence Agency, 1973)

the city against the wishes of the international community whose members do not want to station their diplomats in the city. The center is also at a distance from the extraordinary multi-religious center in the old city in the eastern part of Jerusalem. The very diverse set of religious stakeholders present there, keep each other under close watch and are constantly at least at the edge of contestation. The Jewish part, the remainder of the temple of classical times, is a globally important shrine for religious Jews. Since the 1967 war this Wailing Wall and East Jerusalem at large is accessible from the rest of Israel though the status of the connection in international law is still disputed. In this case as well the separation of political and religious center is closely related to the complicated political and military situation in which the territories of the relevant state units were shaped.

Still another set of circumstances in which political and religious centers are in clearly separate places apparently result from conscious planning. Looking at the typical capital city projects of the modernist age (in an architectural sense) like Ankara, Chandigarh, Brasilia (Gordon 2006) the political centers are designed to be at the apex of the local experience. No religious elements are prominently positioned. Apparently at that historical juncture government was deemed strictly secular unconnected with already existing religious practice. One could interpret the projects to build these new capital cities in themselves as expressions of civil/political religion. More recently this seems to have changed if the following examples warrant a more general impression.

Kazakhstan's new capital city Astana (Yacher 2011) has a very special, large scale political center annex civic center with several dominating elements: the presidential palace, a huge leisure center, a mega building for the management of the country's vast reservoir of natural resources and a gigantic glass pyramid, over 60 m (197 ft) high called the Palace of Peace and Reconciliation designed by British star architect Norman Foster. It has accommodations for various religions but also organizes interfaith symposiums encouraging religious tolerance. Special sessions occur in a room modeled after the UN Security Council meeting room in New York. It has also a museum, library and research center. This structure gives the impression of a return to the religious impulse with a number of contemporary elements added. What this will mean in practice is still unclear. The building was opened in 2006 (Fig. 178.9).

Burma/Myanmar also has a new capital city in the early years of the new millennium (Preecharushh 2011). The military regime moved the government from Rangoon/Yangon to the brand new city of Naypidaw twinned to the already existing city of Pyinmana that was also importantly extended. The political center and its military function are concentrated in the North and Central sector of Naypidaw with huge statues (10 m or 33 ft high) of three ancient kings to stress the continuity of the governmental function and refer to the heritage that the current office holders carry (Naypidaw means Seat of the King or Royal City). At a hilltop well visible from all sides but not closely adjacent to the political center has been erected the replica of the most famous pagoda of Burma. The gilded dome of the original (100 m or 328 ft high) stands in the outskirts of the former capital Rangoon, a temple with important Buddhist relics. Buddhist monks congregate here. It was the major site first of national protest against the British colonizer in the 1940s and then of repeated religious and popular protest against the military regime since the 1980s. The rep-

Fig. 178.9 The ecumenical pyramid is a palace of peace and reconciliation in Astana (Photo by Nicolle K. Hamilton, http://en.wikipedia.org/wiki/File:PyramidPeaceReconciliation.jpg from Wikimedia Commons)

lica in Naypidaw is apparently meant to also get a sacred function (Preecharushh 2011: 1040) as a corollary of its older example but to be considered as sacred place of Burmese orderly development under the wise guidance of the government, not as a sacred place where contestation of the order could be justified.

A final example of the same kind, where new capital cities and their political centers are now again willingly provided with a complementary religious center, could also eventually be Ivory Coast's Yamoussoukro. The first president's natal village was soon after independence in the 1960s designated as the future capital to be. The initial emphasis was on the construction of large homes for the Muslim, the Protestant and the Catholic faith. In particular, the Catholic home, modeled after the Saint Peter basilica in Rome with similarly vast dimensions drew much attention. One of its many stained glass windows shows the face of the president. An institute with its own large building named after the president was aimed to facilitate inter faith relations. One overall ambition was to make the future capital also a place of ecumenism. A presidential palace and a parliament (the last largely paid for by the Chinese) were more recently achieved. In the urban planning was a regulation disapproving any building higher than the dome of the basilica (Mataillet 2006). Meanwhile the government has not moved from the existing political center in Abidjan. The president died and there was repeated violent conflict that only seemed to die in 2011. The existing buildings are quickly dilapidating and the future of this project is in doubt as it has been on earlier occasions. What will eventually come of it is unclear at the time of writing.

178.4   Concluding Comments

The religious idiom in which political authority emerges tends to affect the ways in which that authority is displayed in a political center. Some arrangement in which religious and political centers co-habitate, also in locational terms, is normal. Religious and political centers are occasionally unconnected in terms of location. Such cases seem to be concentrated either in early stages of European urban development after prolonged overall dominance of religious authorities if any and a gradual re-assertion of civic political authority, or in situations of strong secularism dominating decisions concerning the place-making for political centers or during severe political conflict involving religion that prohibits the consideration of harmonious cohabitation. In a few instances the one follows the other at the same location. That also indicates strained relations.

As societies secularize the location of their main religious facilities becomes a less pressing concern, but they nevertheless tend to be maintained as part of a valued context. Their position shifts from being a prominent part of an inalienable patrimony to being a significant part of a self-selected and accepted heritage (Cosgrove 2000). A secularized context occasionally provokes the demand for new meaning and expected salvation in this world to be satisfied by an ideology. The attempts at realization can give it the appearance of a (political or civil) religion. This is also readily apparent in the efforts to create quasi-religious centers that show significant similarities with the ways in which religious centers traditionally have been located and shaped taking into account the changes in building styles and constructional opportunities. The fear of religious strife now also seems to induce religiously inspired constructions as part of political centers. They seem mainly intended to defuse possible hostile encounters that might unhinge the political regime.

All across the world and in many epochs efforts have been made to religiously homogenize populations living under a single public authority, often to no avail. Religious heterogeneity had in very different ways, often grudgingly, to be accepted. A politically predominant religion can be backed up by a display of prominence close to the political leadership. A politico-religious ensemble in the heart of a capital city may for the adherents of that particular faith further underline the legitimacy of the political regime. But in many cases adaptations to this pure type have to be made, for example, a second central church in Dresden, Hindu inspired decorations on the Taj Mahal funerary monument of pre-British India's Muslim rulers. The seeming neutrality of modernist political center designs with respect to religion has now apparently given way to other efforts to accommodate religious heterogeneity in Astana and Yamoussoukro. In Naypidaw the question centers around the accommodation of anti-government feeling within the predominant Buddhist religion.

Fig. 178.10 Heritage by theft: Venice, circa 1870. The horses were stolen by the Venetians in Constantinople in 1204 during the Fourth Crusade and put on display at the entrance of Saint Mark's church as part of an invented tradition. The current statues are replicas (Photo by Carlo Naya, 1816–1882,   http://commons.wikimedia.org/wiki/File:Naya,_Carlo_(1816–1882)_-_n._047_-_ Venezia_-_Cavalli_di_S._Marco.jpg from Wikimedia Commons)

Emulation is a frequent mechanism in the production of political and religious centers when the aim is to at least equal the radiance of an erstwhile predecessor. It shows in the church building in Westminster taking the example of the glorious French cathedrals and in the nineteenth century neo-styles as used for the latest version of the church in central Berlin. Not to speak of the Yamoussoukro version of the Saint Peter. It can be seen in the appropriation of classic Greek and Roman forms from the Christian basilicas to Lincoln's Monument in Washington D.C. and in the reconstruction of antique Rome in the second Rome of the Pope that then was used as a model all across Europe to embellish the palaces of the new sovereigns of the early modern period. Rome itself was remodeled in at least three versions. The next one always emulated the earlier versions. Rome was externally emulated by Constantine's new capital city, and then by Moscow at the time that it initially functioned as the Russian capital. These claims were accompanied by outbursts of building activity to realize a splendid politico-religious center.

"A policy of visual splendour (…) seems to be an indispensable element of the exercise of sovereignty" as the editors of a recent volume on palaces and power conclude. "Governing is not only, as Machiavelli wrote, a question of make believe, it is also a question of display (faire croire et faire voir)." (Auzépy and Cornette 2003, 370). This is right. A graphic example of the combined result is the display in their city of sacred and secular items stolen from Constantinople by the Venetians in

1204 as part of the Fourth Crusade. Statues, marbles and columns were transported back to represent and make people believe the physical and moral renewal of the city as a bulwark of Christendom in the face of Islam but also to underline the city's autonomy with respect to Rome and Byzantium (Cosgrove 2000: 8). Religious and political elements were consciously mixed in the effort (Fig. 178.10).

References

Auzépy, M.-F., & Cornette, J. (2003). *Palais et pouvoir. Paris: De Constantinople à Versailles.* Saint-Denis: Presses Universitaires de Vincennes.

Barraclough, G., & Overy, R. (Eds.). (1999). *The Times history of the world (atlas).* London: HarperCollins. New Edition.

Bellah, R. N. (1967). Civil religion in America. *Journal of the American Academy of Arts and Sciences, 96*(1), 1–21.

Biskup, T., & Schalenberg, M. (Eds.). (2008). *Selling Berlin. Imagebildung und Stadtmarketing von der preussischen Residenz bis zur Bundeshauptstadt.* Stuttgart: Franz Steiner Verlag.

Bourdeaux, M., & Popescu, A. (2006). The Orthodox church and communism. In M. Angold (Ed.), *Eastern Christianity. The Cambridge history of Christianity* (Vol. 5, pp. 558–579). Cambridge: Cambridge University Press.

Brown, P. (2011, December 8). On the magic Carpet of the Met. *New York Review of Books, LVIII*(19), 26–28.

Castells, M. (1996). *The rise of network society.* Oxford: Blackwell.

Claval, P. (2010). *Les espaces de la politique.* Paris: Colin.

Cosgrove, D. (2000). *Return to Delphi. Heritage: Cultures and politics.* London: The Eleni Nakou Foundation.

Deutsch, K. W. (1963). *The nerves of government: Models of political communication and control.* New York: Free Press.

Frishman, M., & Khan, H.-U. (Eds.). (1994). *The Mosque: History, architectural development and regional diversity.* London: Thames & Hudson London.

Goossaert, V., & Palmer, D. A. (2012). *The religious question in modern China.* Chicago: University of Chicago Press.

Gordon, D. L. A. (Ed.). (2006). *Planning twentieth century capitals.* London: Routledge.

Gottmann, J. (1952). The political partitioning of our world: An attempt at analysis. *World Politics, 4*(4), 512–519.

Goverde, H. J. M. (2002). *Politics and public space. The Amsterdam Dam Square urban ensemble in cultural geographic perspective* (Research Team Governance and Places GAP Department Geography and Planning Working Paper Series 2002/15, unpublished, 35 p). Nijmegen: Radboud University.

Harris, J. (2007). *Constantinople: Capital of Byzantium.* London/New York: Hambledon Continuum.

http://en.wikipedia.org/wiki/Main_Page (lemma's on buildings, cities and persons mentioned in this paper, also occasionally in French, German and Dutch versions of Wikipedia)

http://en.wikipedia.org/wiki/Pussy_Riot

Kaul, H. K. (1985). *Historic Delhi: An anthology.* Delhi: Oxford University Press.

Kostof, S. (1973). *The third Rome 1870–1950. Traffic and glory.* Berkeley: University of California Press.

Low, S. M. (2000). *On the plaza: The politics of public space and culture.* Austin: University of Texas Press.

Maier, H. (Ed.). (2004). *Totalitarianism and political religions.* London: Routledge London.

Mango, C. (1985). *Le développement urbain de Constantinople (IV-VIIe siècles)*. Travaux et mémoires du centre de recherche d'historie et civilisation de Byzance. Collège de France. Monographies 2. Paris: de Boccard.

Mataillet, D. (2006, June 6). Quand l'Áfrique s'écrit au futur. *Jeune Afrique*. http://www.jeunea-frique.com/Article/LIN04066quandrutufu0/?art_cle=LIN04066quandrutufu0

Pamuk, O. (2005). *Istanbul – Memories and the city*. New York: Knopf.

Pogue Harrison, R. (2012, February 9). The book from which our literature springs. *New York Review of Books*, LIX(2),40–45.

Porter, V. (Ed.). (2012). *Hajj: Journey to the heart of Islam*. Cambridge: Harvard University Press. Catalog exhibition British Museum.

Preecharushh, D. (2011). Myanmar's new capital city of Naypyidaw. In S. D. Brunn (Ed.), *Engineering earth; the impacts of megaengineering projects* (pp. 1021–1044). Dordrecht: Springer.

Rokkan, S. (2000). *Staat, Nation und Demokratie in Europa. Die Theorie Stein Rokkans aus seinen gesammelten Werken rekonstruiert und eindgeleitet von Peter Flora*. Frankfurt am Main: Suhrkamp.

Sauerländer, W. (2012, March 22). Reims: The queen of cathedrals. *New York Review of Books*, LIX(5), 25–26.

Savage, K. (2009). *Monument wars. Washington D.C., The national mall, and the transformation of the memorial landscape*. Berkeley: University of California Press.

Schilling, H. (2004). Calvinist and Catholic cities – Urban architecture and ritual in confessional Europe. *European Review, 12*(3), 293–312.

Sidorov, D. (2000). National monumentalization and the politics of scale: The resurrection of the Cathedral of Christ the Savior in Moscow. *Annals of the American Association of Geographers, 90*(3), 548–572.

Vale, L. J. (1992). *Architecture, power and national identity*. New Haven: Yale University Press.

van der Wusten, H. (2000). Dictators and their capital cities: Moscow and Berlin in the 1930s. *GeoJournal, 52*(4), 339–344.

van der Wusten, H. (2012a). Political global cities. In B. Derudder, M. Hoyler, P. J. Taylor, & F. Witlox (Eds.), *International handbook of globalization and world cities* (pp. 40–50). Cheltenham: Edward Elgar.

van der Wusten, H. (2012b). Symbols in political centres. Where they are and what they mean. *Belgeo* 1–2. http://belgeo.revues.org/6110

Yacher, L. (2011). Astana, Kazakhstan: Megadream, megacity, megadestiny. In S. D. Brunn (Ed.), *Engineering earth: The impacts of megaengineering projects* (pp. 1001–1020). Dordrecht: Springer.

Chapter 179
The Geography of Religious Freedom

Daniel McGowin and Gerald R. Webster

> *"We can say that the religious environment is in its best period since the founding of new China. It is an indisputable fact that the number of religious Chinese is on the rise, which is evidence of the freedom of religion." Xinping Zhuo, director of Institute of World Religions, Chinese Academy of Social Sciences (State Administration for Religious Affairs of PRC 2012)*

179.1   Introduction

Religious belief systems have existed throughout human history. Hoult (1958: 3) goes so far as to assert that "Wherever there is man (*sic*) there is religion," and Nottingham (1954: 1) claims that "religion . . . is a cultural universal." Rooted in concepts of morality, ethics, spirituality, mysticism and explanations of creation, religious belief systems form the core of societal world views and have dictated the daily lives of billions of adherents. Davis (1949: 509) argues that "so universal, permanent, and pervasive is religion in human society that unless we understand it thoroughly we shall fail to understand society." Though arguably universal, religious belief systems are expressed in myriad forms in different parts of the world through rituals, festivals, prayers, birth rites, funeral procedures, and other manners of private and sometimes public expressions (Cohn 1962; Sopher 1967).

Not all expressions of religiosity occur at the private personal or even small group levels. While certainly many festivals take place in public spaces, society as a whole may also choose to publically express its religious beliefs through legal and political means. And throughout history various religions have been officially adopted by governments, from the Buddhist missionaries of the Asoka kingdom to the adoption of Christianity in Armenia to the spread of Islam to the various sultanates of Southeast Asia (Barro and McCleary 2005). Additionally, some governments

D. McGowin (✉)
Department of Geology and Geography, Auburn University, Auburn, AL 36849, USA
e-mail: dam0027@auburn.edu

G.R. Webster
Department of Geography, University of Wyoming, Laramie, WY 82071, USA
e-mail: gwebste1@uwyo.edu

© Springer Science+Business Media Dordrecht 2015
S.D. Brunn (ed.), *The Changing World Religion Map*,
DOI 10.1007/978-94-017-9376-6_179

such as the People's Republic of China and the former Soviet Union have chosen to take an official nonreligious (or atheistic) stance, while others have emphasized an official position of not favoring one religion over another. This latter concept speaks to notions of religious freedom, and even some countries adopting an official religion may formally guarantee freedom of religion for other faiths (U.S. Department of State 2011).

According to Article 18 of the United Nations Universal Declaration of Human Rights, religious freedom includes the "freedom to change his (*sic*) religion or belief, and freedom, either alone or in community with others and in public and private, to manifest his (*sic*) religion or belief in teaching, practice, worship and observance" (United Nations 2012). Such a human rights perspective was foreshadowed by U.S. President Franklin Roosevelt, who in his 1941 State of the Union address referenced a freedom of religion where "every person [could] worship God in his (*sic*) own way – everywhere in the world" (Roosevelt 1941). Thus, today religious freedom is generally viewed as a fundamental human right.

Religious freedom is not always fully guaranteed or protected by the world's governments. In fact, while concepts of religious freedom predate the United Nations Declaration and President Roosevelt's "Four Freedoms," it is still a relatively new "right." Various religious wars throughout the world, in particular in Europe during the sixteenth and seventeenth centuries, led to more calls for religious tolerance (Richardson 2006). Some of those who faced religious intolerance and conflict fled Europe for North America. This experience can in part explain the drafting of the "Establishment Clause" in the U.S. Constitution (Kramnick and Moore 1997; Richardson 2006).

Since the late eighteenth century several countries have drafted similar constitutional guarantees or laws that provide for religious freedom, even if an official religion is adopted by the state. Despite the formal codification establishing religious freedom, both governments and societies vary greatly in the actual defense of such freedoms (Richardson 2006). Governments may choose to create laws that make blasphemy, apostasy and vilification a crime. For example, Iran established laws that outlaw blasphemy and apostasy in such a way that it can be used in a discriminatory fashion (Daftari 2012). On the other hand, some countries decriminalized blasphemy laws, holding that such laws conflict with religious freedom. However, while governments may overturn blasphemy laws, society may enforce punishments for blasphemous expression. An example of this is the codification of blasphemy laws by the Italian Football Federation (FIGC) (Gianfreda 2011). According to Article 66 of the Regulations Internal of the FIGC (2012a, translated from original), players must "not use offensive, abusive, threatening or blasphemous language" or otherwise be subject to disciplinary action as outlined in the Italian Sports Code of Justice's Article 19 (FIGC 2012b). In other cases, states resort to claims of protecting public safety and well being when limiting or restricting religious activities (Cumper and Lewis 2010; Emilianides et al. 2011; Bowen 2011; Zhang and Zhu 2011).

Recognition of certain religions, or lack of recognition, as well as declarations of a state religion may also bring about restrictions on religious freedom. For example, the Republic of Georgia constitutionally guarantees the freedom of religion and avoids establishing an official religion. However, the Georgian constitution recognizes the historical importance of the Georgian Orthodox Church and societal attitudes towards "other" religions tend to be suspect (Tsintsadze 2008). Russia similarly notes the importance of the Russian Orthodox Church that pays homage to the past connections between church and state, but also imparts favoritism (Krindatch 2006). China officially recognizes only five religions – two of which are branches of Christianity – while labeling all other religions as "evil cults," predictably leading to harsher treatment of non-recognized religions. This harsher treatment is exemplified by the plight of the Falun Gong, a religious group not recognized by the Chinese government and whose practitioners face interrogation, torture and even death for adhering to their faith (United Nations 2010). But it is worth noting that despite their recognition by the state, adherents to Buddhism, Catholicism, Islam, Protestantism, and Taoism can also face discrimination and harsh treatment in China. Such examples include the arrest and punishment of Uyghur Muslims for "illegal gathering," the confiscation of passports for unauthorized pilgrimages to Mecca, the incarceration of Catholic bishops who profess allegiance to the Pope, and the disruption of Buddhist ceremonies at temples due to the local government's "fear" of a riot (Tong 2010; United Nations 2010; Zhang and Zhu 2011).

Similarly, while Kosovo guarantees religious freedom and declares no state religion, it specifically lists five religions in its religious freedom law insinuating official status or recognition (Woods 2008). Countries such as Nigeria have allowed its northern states to adopt Sharia law, though leaders in these areas insist that it applies only to Muslims living there and that the state has not officially adopted Islam. It is important to mention that while there are conflicts between Muslims and Christians living in Nigeria, many of these conflicts run deeper than religious divisions and often include social and ethnic tensions (Ludwig 2008). And, although countries such as Denmark allow for freedom of expression that includes the criticism of religion, actions such as the printing of the "Muhammad Cartoons" in the Danish newspaper *Jyllands-Posten* can lead to social protests both at home and abroad (Cobb 2006; Hassner 2011).

The purpose of this chapter is to analyze the geography of religious freedom around the world. Using data gathered by the Hudson Institute's Center for Religious Freedom, we map the Center's indicator of religious freedom for over 100 countries and territories using the categories adopted by the Center: "free," "partly free," and "unfree" for the religions with the greatest number of adherents in the world. We further analyze the spatial patterns of religious freedom by major religious category and world region. Finally, we provide examples of the levels of religious freedom by region and faith, as well as possible explanations for the major patterns that emerge.

179.2 Methodology

As noted above, we use data compiled by the Hudson Institute's Center for Religious Freedom. The Center collected data on religious freedom for 101 different countries and territories spanning all populated continents and major religious groups. The Center's Index on Religions Freedom (IRF) focuses on "the denial to anyone of rights of a particular kind, those connected with practicing one's religion, and the denial of rights for a particular reason, because of the religious beliefs of those who are persecuted and/or those who persecute" (Marshall 2008: xiii). Using the categories provided by the Center, we first map these data to analyze the geography of religious freedom in different parts of the world. We further analyze these data in terms of the dominant religion in each country using these same three Hudson Institute categories of religious freedom. Lower values on the Center's numeric scale indicate greater religious freedom with high numbers reflecting lesser levels of religious freedom. According to the Center, a country with a score of 1–3 is considered to be "Free," a "Partly Free" country has a score of either 4 or 5, while "Unfree" countries have scores of 6 or 7.

The Center's numeric values for each country are arrived at through a survey including 119 questions completed by experts on each country. The experts were contracted by the editor based upon knowledge of the specific country, with a smaller number of country profiles being written by the staff of the Center. For example, the editor completed the profile of Egypt. In other cases those writing the original drafts of some country profiles were selected based upon referrals by other experts. The states of the former Soviet Union were undertaken by the staff of Forum 18 due to the confidence of the editor in their work. In all cases the country profiles were works for hire and unsigned with the editor reserving the right to edit the draft profiles (Marshall 2012). The questions posed to the country experts include among others whether citizens are free "to have or not have a religion or belief of their choice?," and whether religious groups "have the right to solicit and receive voluntary financial and other contributions from individuals and institutions, either domestically and internationally?" (Marshall 2008: 451–476).

It is also important to point out that the scores are not for specific governments which may or "may not be particularly blameworthy," but rather for countries and territories. As stated by the Center (Marshall 2008: 450): "We are interested in the practical situation, not per se in the culpability of any government.... The agents of religious repression might be terrorists or, perhaps most commonly, 'society,' as when religious groups are attacked by mobs or face pervasive discrimination" (see also Stump 2000: 213–229).

We also examine three other measures employed by the Center: government regulation of religion (GRI), government favoritism of religion (GFI) and social regulation of religion (SRI). These indicators were first developed by Grim and Finke (2006) using the U.S. State Department's International Religious Freedom Reports. The Center for Religious Freedom calculated the same index by providing Grim and Finke's questions to the country experts noted above (see Marshall 2008: 495–498). By taking into account these three measures, we are able to provide a more thorough and rich investigation into the geography of religious freedom. Additional explanation for these measures is included below.

There are other organizations that have created measures of "religious freedom." For example, the United States Department of State annually publishes a report on religious freedom as part of its foreign policy analysis (U.S. Department of State 2012a), while non-profit organizations such as the International Coalition for Religious Freedom (2012) and the Pew Research Center also publish their own reports (Pew Forum on Religion & Public Life 2011). However, we opted to use the Hudson Institute's data because of its quantitative measures, its multiple measures, and its wide-ranging examples across religions and regions (Marshall 2008).

179.3   Results

The countries and territories examined by the Hudson Institute are first mapped based on dominant religion (Fig. 179.1). Each of the major religions of the world is represented, with Christianity subdivided into branches (Catholicism, Orthodox, and Protestantism). We also mapped the data for religious freedom, government regulation of religion, government favoritism of religion, and social regulation of religion. On each of the latter four maps the countries are labeled based on their score on the four observed measures.

The countries were organized based on religion and region (Table 179.1). Africa is represented by sixteen different countries (five each for Islam, Protestant, and

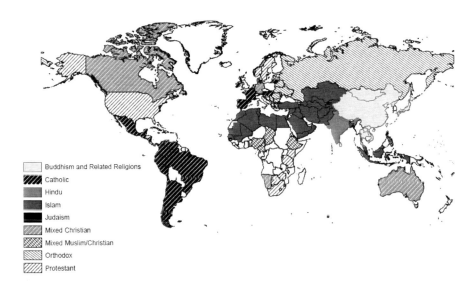

Fig. 179.1 Dominant religion, by country (Map by Kelly Koenig and Craig Remington, University of Alabama Cartographic Research Laboratory, commissioned by the author)

Table 179.1 Religions, by region

Religion	Africa	Asia	Former USSR and Eastern Europe	Latin America	North Africa and West Asia	Western Europe and North Atlantic
Buddhism and related religions	–	10	–	–	–	–
Catholic	–	1	3	11	–	7
Hindu	–	2	–	–	–	–
Islam	5	4	7	–	19	–
Judaism	–	–	–	–	1	–
Mixed Christian	1	–	1	–	–	3
Mixed Muslim/Christian	–	–	–	–	–	–
Orthodox	–	–	10	–	–	1
Protestant	5	–	1	–	–	4
Total	11	17	22	11	20	15

Compiled by Daniel McGowin and Gerald R. Webster with data from Marshall (2008)

Table 179.2 Regions, by categorization

Region	Free (%)	Partly free (%)	Unfree (%)
Africa	38	44	19
Asia	24	35	41
Former USSR and Eastern Europe	32	55	14
Latin America	73	18	9
North Africa and West Asia	5	62	33
Western Europe and North Atlantic	100	0	0

Source: Compiled by Daniel McGowin and Gerald R. Webster from data drawn from Marshall (2008)
Note: The percentages were rounded to the nearest whole number and may not equal 100 %

Mixed Muslim/Christian, and one for Mixed Christian). The seventeen Asian countries represented include ten dominated by Buddhism, while North Africa and West Asia includes 21 countries with nineteen being mostly Muslim. Both the territory of the Former USSR and Eastern Europe region (22 countries) and Western Europe and North Atlantic region (15 countries) are mostly religiously mixed, while all eleven Latin American countries are Catholic.

In terms of regions, the Western Europe and North Atlantic region is the only one with all countries and territories falling into the "Free" category (Table 179.2). The only other region where more than 50 % of its countries are labeled "Free" is Latin America (8 of 11 countries, or 73 %). This region is followed by Africa (37.5 %), the territory of the Former USSR and Eastern Europe (32 %), Asia (23.5 %), and North Africa and West Asia (only 5 %). The latter region has the highest percentage of its countries and territories falling into the "Partly Free" (62 %), with the territory of the Former USSR and Eastern Europe region placing 54.5 % into this category. Asia has the highest percentage of countries and territories in the "Unfree" category (42 %), with North Africa and West Asia second (33 %).

Table 179.3 Religions, by categorization

Religion	Free (%)	Partly free (%)	Unfree (%)
Buddhism and related religions	30	20	50
Catholic	82	14	5
Hindu	0	100	0
Islam	5	57	37
Judaism	100	0	0
Mixed Christian	100	0	0
Mixed Muslim/Christian	0	83	17
Orthodox	36	55	9
Protestant	80	20	0

Source: Compiled by Daniel McGowin and Gerald R. Webster from data drawn from Marshall (2008)
Note: The percentages were rounded to the nearest whole number and may not equal 100 %

The data also cuts across a wide variety of religions and religious branches (Table 179.3). The largest sample comes from Islam, with 35 countries represented (34 % of all countries examined). Although 43 Christian countries are examined, the data are divided into branches: Catholic (22 countries), Orthodox (11), and Protestant (10). Ten countries classified as Buddhist and Related Religions are included, followed by Mixed Muslim and Christian (6), Mixed Christian (5), Hindu (2), and Judaism (1).

As for categorization, the only religions with 100 % of its samples in the "Free" range are Judaism (Israel) and Mixed Christian (Australia, Canada, Germany, Latvia, and Namibia). The only other religions above 50 % are other Judeo-Christian religions, Catholicism (82 %) and Protestantism (80 %). Both Hindu countries (India and Nepal) fall into the "Partly Free" category, while 88 % of Mixed Muslim and Christian states are also placed in this category (Cameroon, Chad, Ethiopia, Lebanon, and Nigeria). Additionally, 20 Muslim countries (57 %) had scores placing them in the "Partly Free" category. Islam also has a significant percentage of its countries in the "Unfree" category (37 %). However, 50 % (5 countries) of countries dominated by Buddhism and Related Religions are categorized as "Unfree." Hinduism, Judaism, Mixed Christian, and Protestant have no countries in the "Unfree" category.

Forty-one countries are listed as "Free." Only four of those countries scored a one (most free) and include Estonia (Protestant), Hungary (Catholic), Ireland (Catholic) and the United States (Protestant). At the opposite end of the spectrum, 21 countries were labeled "Unfree," with 11 countries receiving the worst possible score (seven on the Institute's scale). These 11 countries include seven Muslim countries including Iran, Iraq, Maldives, Saudi Arabia, Sudan, Turkmenistan, and Uzbekistan. Three of the other four are Buddhist countries and territories (Myanmar, North Korea and Tibet), while the last one is a Mixed Muslim and Christian country (Eritrea) (Fig. 179.2).

In addition to the Religious Freedom Index, we also examined three other measures – the government regulation Index, the government favoritism index (GFI), and the social regulation index (SRI) calculated by the Institute. By examining the

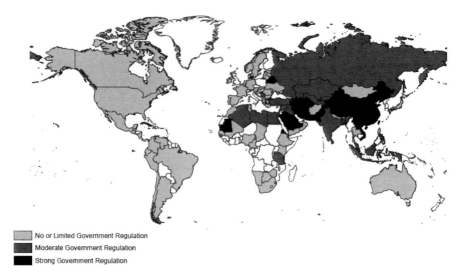

No or Limited Government Regulation
Moderate Government Regulation
Strong Government Regulation

Fig. 179.2 Religious freedom index (Map by Kelly Koenig and Craig Remington, University of Alabama Cartographic Research Laboratory, commissioned by the author)

U.S. State Department's 2003 International Religious Freedom Report, Grim and Finke (2006) developed a questionnaire that targets key words in the report. For the GRI, the measure included each "state's specific actions for regulating religious mission work, proselytizing, preaching, conversion, and worship, as well as more general legal and policy actions" (Grim and Finke 2006: 13). The GFI "refers to the actions of the state that provide one religion or a small group of religions special privileges, support, or favorable sanctions" and focuses on "a broad range of indicators on the governments' favoritism of religion, measuring formal religious establishment as well as the various forms of government funding for religion" (Grim and Finke 2006: 15–16). Finally, the SRI focuses on societal regulations rather than state policies, examining "restrictions religious groups face from the larger culture and other institutions." Each of these measures provides a deeper understanding of the levels of religious freedom associated with each dominant religion and each country (Tables 179.4 and 179.5).

Protestant countries tend to have the least amount of government regulation of religions, averaging 0.78 in the GRI. This average is slightly ahead of Mixed Christian countries (0.78)[1] and Judaism (0.83). The religious category with the highest level of government regulation includes those countries which are dominated by Buddhism or Related Religions (5.71), followed by Islamic countries (5.61). In terms of regions, Western Europe and the North Atlantic subdivision has the lowest level of government regulation (0.63) while Asia has the highest (5.33), with the latter's number coinciding with restrictions on Buddhism (Fig. 179.3).

---

[1] The GRI for Protestantism is officially 0.777801, while the number for Mixed Christian is 0.777808.

Table 179.4 Regulation and favoritism indices, by religion

Religion	GRI	GFI	SRI
Buddhism and related religions	5.7	5.9	4.0
Catholic	1.1	4.1	1.6
Hindu	3.8	7.8	6.9
Islam	5.6	6.5	6.7
Judaism	0.8	4.5	7.0
Mixed Christian	0.7	3.5	1.1
Mixed Muslim/Christian	3.2	4.6	4.6
Orthodox	3.8	5.7	7.0
Protestant	0.7	3.2	1.0

Source: Compiled by Daniel McGowin and Gerald R. Webster from data drawn from Marshall (2008)

Table 179.5 Regulation and favoritism indices, by region

Region	GRI	GFI	SRI
North Africa and West Asia	4.8	7.3	6.9
Latin America	1.7	2.9	1.8
Former USSR and Eastern Europe	4.1	4.9	5.4
Western Europe and North Atlantic	0.6	5.4	1.7
Asia	5.3	6.1	4.9
Africa	2.8	3.6	3.7

Source: Compiled by Daniel McGowin and Gerald R. Webster from data drawn from Marshall (2008)

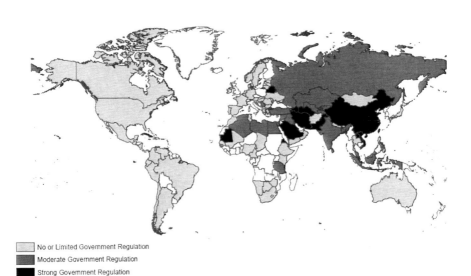

No or Limited Government Regulation

Moderate Government Regulation

Strong Government Regulation

Fig. 179.3 Government Regulation Index (GRI) of religion (Map by Kelly Koenig and Craig Remington, University of Alabama Cartographic Research Laboratory, commissioned by the author)

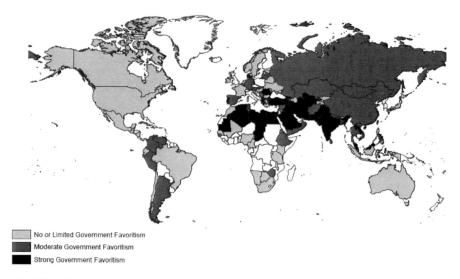

No or Limited Government Favoritism
Moderate Government Favoritism
Strong Government Favoritism

Fig. 179.4 Government Favoritism Index (GFI) toward religion (Map by Kelly Koenig and Craig Remington, University of Alabama Cartographic Research Laboratory, commissioned by the author)

Protestant countries also tend to have the least amount of government favoritism towards a particular religion with an average GFI of 3.2, followed by Mixed Christian (3.53) and Catholic countries (4.11). Hindu countries have the highest level of government favoritism (7.83) with Islamic countries displaying the second-highest levels of government favoritism (6.59). Latin American countries generally lack government favoritism (2.92), while North Africa and West Asia countries have the highest level of favoritism (7.39) (Fig. 179.4).

With regards to social regulation, Protestant countries once again display the lowest levels of regulation of religions (1.07), followed by Mixed Christian (1.33) and Catholic (1.63) countries. At the other end of the spectrum is another Christian branch – Orthodox – with an average SRI of seven. Orthodox countries are closely followed by Jewish (7.0),[2] Hindu (6.92) and Islamic (6.79) countries. The Western Europe and North Atlantic regional subdivision once again tops the regions with the lowest levels of social regulation (1.74), while North Africa and West Asia exhibits the highest such regulation (6.92) (Fig. 179.5).

Thirty-eight countries scored under one on the GRI, with 28 countries scoring zero (that is, no government regulation). Countries with no government regulation include several which are Protestant (Botswana, Sweden, United States), Catholic (Brazil, Ireland, The Philippines), Orthodox (Georgia, Ukraine), Muslim (Mali, Senegal), and Buddhist and Related Religions (Japan, Thailand). Countries scoring over seven on the GRI tended to be Muslim and include Bangladesh, Mauritania, and Saudi Arabia. However, other religions were represented here as well such as

---

[2] Officially, Orthodox Christianity has an average SRI of 7.00043, while Judaism's SRI is 7.0004.

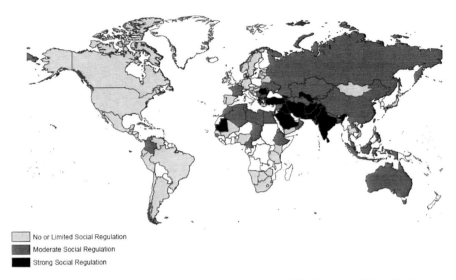

Fig. 179.5 Social Regulation Index (SRI) of religion (Map by Kelly Koenig and Craig Remington, University of Alabama Cartographic Research Laboratory, commissioned by the author)

Buddhism (China, Viet Nam), Catholicism (Cuba), Orthodox Christian (Belarus), and Mixed Muslim and Christian (Eritrea).

   Government favoritism tended to follow a similar pattern, though only 12 countries scored less than one on the GFI. These countries and territories represent a variety of religions once again, such as Buddhism (Tibet), Catholicism (Cuba, Ecuador, Guatemala, Mexico, The Philippines), Islam (Tajikistan), Mixed Christian (Namibia), Mixed Muslim and Christian (Cameroon), Orthodox (Ukraine), and Protestant (Botswana). However, unlike government regulation, government favoritism tends to be stronger with 42 countries and territories scoring over seven on the measure. This grouping cast a wide net and includes countries which are Protestant (Denmark), Catholic (Peru, Italy, Portugal), Orthodox Christian (Serbia, Greece, Romania), Islamic (Syria, Egypt, Maldives), Buddhist (Viet Nam, Laos, Myanmar), and Hindu (India).

   Finally, countries with low levels of social regulation of religion tended to be dominantly Christian such as The Philippines (Catholic),[3] Norway (Protestant), and Germany (Mixed Christian). Buddhism (Thailand, Mongolia) and Islam (Senegal) were also represented. On the other end of the scale, 15 of the 27 countries scoring over seven on the SRI are dominantly Muslim, including Indonesia, Afghanistan, and Saudi Arabia.[4] Orthodox countries such as Serbia, Armenia and Greece also scored high on the SRI, as did India (Hindu), China (Buddhism), and Israel (Judaism).

---

[3] The Philippines is the only country to score a zero on all three indices.

[4] Saudi Arabia was the only country to score over 10 on all three measures.

Israel presents an interesting, though by no means unique (for example, Colombia, Georgia, Afghanistan) case as it has a very low level of government regulation of religion (0.83), but displays a high level of social regulation (7.0). Israel's GFI (4.5) places it in the middle of that ranking. While Israel displays a generally strong commitment to religious freedom, including the continuation of Ottoman-era allowance of local religious authorities to determine personal status, favoritism towards Judaism remains high. Visitors to Israel are occasionally detained and even deported if there is suspicion that the individuals will attempt missionary work in the country. Furthermore, while Christians and Muslims face discrimination and bias, non-Orthodox Jews also face similar obstacles as Orthodox interpretations and laws tend to take precedent. In addition, the Palestinian areas under either Israeli or Palestinian Authority control exhibit restrictions on religious freedom (Marshall 2008; U.S. Department of State 2012b).

Other examples of countries with differing levels of regulation include Ukraine (Orthodox), which has both low GRI and GFI scores but high a SRI value, and Azerbaijan (Islam) with high GRI and SRI scores, but low GFI score. There are very few countries and territories with low SRI scores but high GRI indicators. Tibet is one such example with a 0.67 SRI and a 6.94 GRI. Among other examples is the Islamic country of Bahrain (SRI: 2.5; GRI: 7.8), Buddhist Viet Nam (2.7 and 8.6), and Belarus, an Orthodox Christian country (2.7 and 9.4). Countries with low SRI's and high GFI's include Denmark (Protestant), Venezuela (Catholic), Germany (Mixed Christian), Thailand (Buddhism), and the aforementioned Bahrain (Islam).

179.4 Discussion

While certainly the geographies of "Free" and "Partly Free" countries are worth further examination, we primarily focus on the geography of those classified as being "Unfree." This examination will also include those countries with high levels of religious regulation either from governments or society. In all, there are 21 countries listed as "Unfree." According to the Hudson Institute's measures, nearly two-thirds of these countries are dominantly Muslim. Five countries are categorized as being dominated by Buddhism and Related Religions, while there is one each labeled Catholic, Mixed Muslim and Christian, and Orthodox Christian.

Bangladesh is one of the Islamic countries considered to be "Unfree." Part I Section 2A of Bangladesh's constitution states that "the state religion of the Republic is Islam, but other religions may be practiced in peace and harmony in the Republic" (Constitution of the People's Republic of Bangladesh 1972), while Part III Section 41 guarantees the freedom of religion. The government attempts to maintain its country's secular status and has taken steps such as banning the use of religion in election campaigns, rejecting calls to adopt sharia law, and even ruling fatwas illegal (Marshall 2008: 87). However, laws such as the now-repealed Vested Property Act (Drishtipat 2012) target "enemies" of the state, which tends to be a reference to non-Muslims in Bangladesh. The Act allowed for the government to take the property from non-Muslims. "Minority rights groups in Bangladesh claim that 60 % of the property belonging to members of minority communities has been illegally confiscated by individuals in positions of authority" (Marshall 2008: 88). Marshall (2008: 88) also notes that harassment and violence including beheadings and rapes have occurred as many non-Muslims have been targeted by Muslim groups, leading some to seek asylum abroad (see also Humphreys 2012).

Saudi Arabia provides another example of a Muslim country labeled "Unfree." The Southwest Asian country is ruled by the Basic Law of Governance (Royal Embassy of Saudi Arabia 2012), which states that "Saudi Arabia is a sovereign Arab Islamic State" with its religion being Islam and the "constitution is Almighty God's Book." The Basic Law protects "Islamic and Arab heritage" and encourages families in Saudi society to "be raised in the Islamic Creed, which demands allegiance and obedience to God, to His Prophet and to the rulers, respect for and obedience to the laws, and love for and pride in the homeland and its glorious history." The law is in accordance to sharia and the Saudi government "shall form armed forces and equip them to defend the Islamic Creed, the Two Holy Mosques, the society and the homeland" (Marshall 2008: 346–349; Royal Embassy of Saudi Arabia 2012).

While in Saudi Arabia "the State shall protect human rights in accordance with the sharia," there is no guarantee of religious freedom. And it is not just a severe restriction of religions other than Islam as "the Wahhabi brand of Sunni Islam is adopted as state ideology and vigorously propagated by the state" (Marshall 2008: 347). Shia Islam also faces discrimination in Saudi Arabia, such as refusals by the courts to hear testimony from Shiite citizens and the restriction of Shiite holidays and festivals. As for other religions, Marshall (2008: 348) notes that Christians and Jews are held as being inferior while Hinduism and Buddhism are considered "polytheistic and therefore suffer even greater discrimination." Because of the application of the sharia, punishments such as stoning, amputation and beheading are practiced and the penalty for apostasy is death (Marshall 2008: 348).

This latter punishment for apostasy is highlighted in the case Hamza Kashgari. On 4 February 2012, the day of observance for the birth of Muhammad, Mr. Kashgari used the social networking site Twitter to send out three tweets directed towards the Islamic prophet:

> On your birthday, I will say that I have loved the rebel in you, that you've always been a source of inspiration to me, and that I do not like the halos of divinity around you. I shall not pray for you.
> On your birthday, I find you wherever I turn. I will say that I have loved aspects of you, hated others, and could not understand many more.
> On your birthday, I shall not bow to you. I shall not kiss your hand. Rather, I shall shake it as equals do, and smile at you as you smile at me. I shall speak to you as a friend, no more.
> (Hopkins 2012)

Mr. Kashgari also tweeted that "No Saudi women will go to hell, because it's impossible to go there twice," referencing the social restrictions on women in Saudi Arabia (Keyes 2012). After calls by Saudi King Abdullah for the arrest of the young Saudi man, Kashgari attempted to flee to New Zealand. However, he was detained in Malaysia and deported back to Saudi Arabia. Since then, Mr. Kashgari repented for his "crime" while another user of Twitter, Mohammed Salama, has also been accused of blasphemy for his tweets (Nikolas 2012).

A final example of an "Unfree" Muslim country is Egypt, which according to a Pew Research report on religious freedom experienced an increase in the late 2000s in government restrictions on religion, while also ranking very high in social restrictions (Pew Forum on Religion & Public Life 2011). Like Bangladesh and Saudi Arabia, Egypt declares that the religion of the country is Islam. Yet, unlike Saudi Arabia but like Bangladesh, the Provisional Constitution of Egypt "guarantees … the freedom to religious rites" and that the "law applies equally to all citizens … [and] may not be discriminated against due to race, origin, language, religion, or creed" (Egyptian Government Portal 2011). Despite this, Egypt prohibits ridicule and insults directed towards Islam.

Social regulation of religion in Egypt is of great concern, particularly towards the country's Coptic Christians. Churches have been attacked by Muslim mobs, Coptic women have been kidnapped and forced to convert to Islam, and some Copts refuse to leave home for fear of attacks (Marshall 2008: 156–160). In addition, tension between Muslims and Copts has increased since the fall of the regime of former Egyptian president Hosni Mubarak. Despite relative solidarity during the uprising against Mubarak, clashes between Copts and Muslims involved both sides attacking one another. It is important to note that not all clashes have been started by Muslims. For example, in 2011 clashes between Copts and Muslims began after a Coptic woman who had converted to Islam and married a Muslim man was allegedly abducted by a group of Copts. Because divorce is strictly prohibited in the Coptic Church, the woman converted to Islam in order to escape her marriage to a Coptic priest. Word of the abduction spread and a mob of Muslims descended on the church where the woman was held. Fighting soon broke out between both sides, eventually leading to the Muslim crowd setting fire to a Christian-owned apartment building and a church (Michael and El Deeb 2011). In another case of social regulation, a Coptic family was forced by a tribal council to leave their home town despite having their family business burned down; none of the alleged attackers were prosecuted (Batrawy 2012). And, following the death of Pope Shenouda III, the leader of the Egyptian Coptic Church, many Copts in the North African country fear that discrimination and attacks will only increase (Fahim 2012).

Islam is not the only religion with a categorization of "Unfree." Belarus provides an example of a Christian country with a high level of governmental favoritism (GFI: 7.2). While the constitution guarantees the "right to determine independently [one's] attitude towards religion … to express and spread beliefs connected with [one's] attitude towards religion, and to participate in the performance of acts of worship and religious rituals and rites" (President of the Republic of Belarus 2012), there is a high level of favoritism towards the Belarusian Orthodox Church, a branch of the Russian Orthodox Church. Belarusian President Aleksandr Lukashenko pushed forth the law "On Freedom of Conscience and Religious Organisation," which gives preferential treatment towards religious organizations that have an "influence upon forming spiritual, cultural and state traditions of the Belarusian people," which is recognized as the Orthodox Church (Republic of Belarus 2002). The law also requires religious organizations to register with the state, though some registrations are rejected. Unregistered religions face stiff fines for practicing their faith. Hindus in Belarus face arrests for practicing their religion (Wines 2002), while a pastor for a Church of God congregation in Zhodino was fined for "unregistered religious activity," even though the church repeatedly and unsuccessfully attempted to register with the state (Allen 2011).

The Hudson Institute lists five Buddhist countries and territories as being "Unfree," including Myanmar, China, North Korea, Tibet, and Viet Nam. In all five control over religions and religious organizations is seen as a means to further the state's agenda. For example, Myanmar and Viet Nam both guarantee religious freedom, but Myanmar uses Buddhism as a conduit for the dissemination of state propaganda while Viet Nam requires registration and strictly controls registered religions (Marshall 2008).[5]

Among these countries, China offers one of the most intriguing cases. China guarantees the freedom of religion in as much as "no state organ, public organization or individual may compel citizens to believe in, or not to believe in, any religion; nor may they discriminate against citizens who believe in, or do not believe in, any religion" (People's Daily Online 1982). However, while the Chinese constitution protects "normal religious activities," these activities must not "disrupt public order, impair the health of citizens or interfere with the educational system of the state" (People's Daily Online 1982). In addition, the Chinese government recognizes five main religions – Buddhism, Catholicism, Islam, Protestantism and Taoism (Embassy of China 1997) – yet promotes the concepts of Marxist atheism (Marshall 2008: 125–132). Unregistered groups regardless of religious affiliation face intimidation and discrimination (Marshall 2008), while religious sites are often destroyed (Schiller 2010). Buddhist and Taoist groups also face discrimination, with the most prominent group targeted being Falun Gong (Beaumont 2009).

While unregistered religions and religious branches face discrimination and intimidation, recognized religions are not immune to such hardships. Muslims living in the autonomous regions of Ningxia and Xinjiang face repression and regulation. Xinjiang's Uyghurs make up a plurality of the region's population, but Chinese efforts to disrupt the demographic composition of Xinjiang include the movement of ethnic Han Chinese to the western region. Chinese regulation in Xinjiang includes restriction of mosque construction, censoring the Quran, the closing of Arabic-language and Islamic schools, and even the forced imposition of female imams on the Uyghur communities (Davis 2008; Marshall 2008). In addition, the Chinese government used the terrorist attacks on the United States on 11 September 2001 as justification for further crackdowns on Uyghur Muslims, including the classification of some groups as terrorists (Blanchard 2011).

---

[5] In the case of Myanmar, democratic and liberalization movements between 2010 and 2012 could lead to a change in attitude towards religious freedom. Indeed, a report on religious freedom released in 2012 by the U.S. Department of State notes some improvement in Myanmar.

China's other western autonomous region also faces instances of religious persecution and discrimination – Xizang, also more commonly known as Tibet. Chinese crackdowns on Tibetan Buddhists include restrictions on schools, limiting the number of monks and nuns at religious institutions, and labeling the Dalai Lama, the religious leader of Tibetan Buddhism, as the "root of Tibet's instability" (Marshall 2008: 135). In addition, the Chinese government re-educates Tibetans on its territory's history by teaching a narrative that Tibet has always been part of China (Marshall 2008). Religious rites and history are also restricted. While the succession of the Dalai Lama is traditionally achieved through reincarnation, the current Dalai Lama (Tenzin Gyatso) has suggested that due to his exile he may choose his successor. However, China claims that the Dalai Lama must be reincarnated and that the reincarnation must occur on Chinese soil, something Tenzin Gyatso has stated cannot happen since he will not return as long as China occupies Tibet (Wee and Blanchard 2011). Furthermore, much like Chinese attitudes towards progressive Uyghurs, Buddhist monks and nuns in Tibet who self-immolate as a means of protest against the Chinese government are often labeled as "terrorists" (Martina and Rhoads 2012).

179.5   Conclusion

In this chapter we provided an overview analysis of the geography of religious freedom around the world. We note that the Muslim world had one of the smallest proportions of countries with high levels of religious freedom. Of the 35 countries dominated by Islam only Mali and Senegal were classified as "free" on the religious freedom index. In comparison, the Catholic world included 23 countries with 19 being classified as "free." The other three indices that we examined confirm this pattern.

Considering all 49 Christian countries whether Catholic, Protestant, Orthodox or mixed, 36 were classified as "free." Countries dominated by Buddhism and Related Religions were divided with half being "free" and half being either "partly free" or "unfree." Finally, there are six countries classified as "mixed Muslim/Christian," all of which are labeled as "partly free" (Cameroon, Chad, Ethiopia, Lebanon, and Nigeria) or "unfree" (Eritrea).

In terms of regions, there is a contiguous collection of religiously "unfree" states stretching from North Africa to East Asia, including Southwest Asia and South Asia. This area includes countries such as Mali, Libya, Saudi Arabia, Pakistan, India, China, and North Korea. There is some variance with regards to the three other indices. Southwest Asian states and Communist countries in East Asia display the highest levels of government regulation, while government favoritism tends to be highest in North Africa, South Asia, and Southwest Asia. Social regulation is most prevalent in South and Southwest Asian states. States in developed regions such as Europe and North America, as well as the countries of Australia and Japan, tend to be labeled "free" with exceptions receiving a "partly free" designation. With regards to this latter region, there is no significant difference with regards to the three additional indices.

It is fair to note that a country's status in such religious freedom reports is subject to change, especially if an internal conflict or regime change occurs. For example, the U.S. Department of State's International Religious Freedom Report for 2011 (2012c) highlights the increase in religious freedom in Egypt, Libya, and Myanmar, crediting political transitions in those countries. However, the report is cautious to note that discrimination towards religious minorities in those three countries still exists, and internal unrest in countries such as Bahrain, Iraq, Nigeria, and Russia continue to restrict religious freedoms. The report also notes that religious freedom deteriorated in countries designated "Countries of Particular Concern" such as China, Eritrea, Iran, and Uzbekistan. Also, labeling countries as being restrictive towards religious freedom has led to some governments questioning the motives of such reports. For example, the Chinese government released a statement in response to the 2011 U.S. Department of State report claiming it is based on "groundless accusations" and is a politically-motivated attempt to "meddle in Chinese affairs." The statement also noted that strict application of the law towards religious groups occurs when "religion is used engage in criminal activity," insinuating activities by "separatists" and "evil cults" (The Central People's Government of the People's Republic of China 2012, translated from original).

Explaining the above patterns is a complex task. Certainly, countries that are dominated by Islam demonstrate a general lack of religious freedom, yet exceptions do exist. The close connection in many Muslim countries between "mosque and state" may explain such a restriction. Though the prospect of separation of the two may seem unlikely, Bernard (2008) suggests that it is possible to separate religion from politics in the Muslim world without sacrificing the importance of Islam, drawing parallels to a similar debate which took place during the infancy of the United States. Beyond Islam, while several Buddhist countries also display a lack of religious freedom, these states tend to be ruled by Communist governments. Indeed non-Communist Buddhist countries such as Japan, Mongolia and Thailand are labeled as "free" in the Center's categorization. Finally, while there is a general divide between developed countries being "free" and developing countries being "unfree," there are several of the latter countries that are labeled as "free" or at least "partly free."

Through this chapter, we have mapped the various levels of religious freedom demonstrated throughout the world. However, this chapter only scratches the surface of religious freedom and offers a brief global view of religious rights and restrictions. A more thorough examination of individual countries that promote or restrict religious freedom, as well as a deeper investigation into the regional patterns of religious freedom are certainly topics worthy of pursuit. Other potential topics highlighted in this research include a comparison of religious freedom in secular and religious Muslim countries, regional variations of social regulation of religion in the United States (Brunn et al. 2011), implications of the "Arab Spring" on religious freedom in the Middle East, and the role religious diversity plays in promoting religious freedom.

Although many countries, including some of those labeled as "unfree," constitutionally guarantee religious freedom, reality suggests otherwise. To paraphrase Chan (2004: 347, emphasis added), government regulations and social control tend to affect "the freedom of *religious belief*, not the freedom of religion." Countries will tolerate religion "so long as it poses no threat to the ruling regime and no challenge to established social institutions, such as education and marriage" (Chan 2004: 347). Thus, while countries may tolerate religions, being able to actually practice one's religious belief – be it Protestantism in Belarus, Buddhism in China, or Islam in France – may face many obstacles if such beliefs clash with established "norms. Yet, by restricting religious freedom, governments and religious groups may very well be "fail[ing] to understand [human] society" (Davis 1949: 509) (Table 179.6).

Table 179.6 Countries and territories, by religious freedom measure

Country	Religious freedom	GRI	GFI	SRI	Religion
Estonia	1	0	3.1668	0	Protestant
Hungary	1	0	2.1668	3.5	Catholic
Ireland	1	0	1.8335	0	Catholic
United States	1	0	1.1667	0	Protestant
Australia	2	2.2224	3.1668	4.5004	Mixed Christian
Austria	2	0.8333	7.167	1.5	Catholic
Botswana	2	0	0	2.6668	Protestant
Brazil	2	0	1.1667	3.1668	Catholic
Canada	2	0.8333	2.3334	0.6668	Mixed Christian
Chile	2	0	1.5	1.1668	Catholic
Denmark	2	0.8333	9.5006	0	Protestant
Ecuador	2	0.8333	0.3334	1	Catholic
Guatemala	2	0	0	1.1668	Catholic
Italy	2	0.8333	7.667	0.5	Catholic
Japan	2	0	2.3335	3.1668	Buddhism and related religions
Latvia	2	0	5.8335	0	Mixed Christian
Lithuania	2	3.889	6.5002	0.668	Catholic
Mali	2	0	1.5	2.6668	Islam
Namibia	2	0	0.6668	0	Mixed Christian
Norway	2	0	3.6669	0	Protestant
Portugal	2	0	8.6672	1.5	Catholic
Senegal	2	0	2.8335	0.6668	Islam
South Africa	2	0	1.8335	0.6668	Protestant
Spain	2	0	7.167	2.1668	Catholic
Sweden	2	0	3.5002	0	Protestant
Ukraine	2	0	0	6.8336	Orthodox
Argentina	3	3.3333	6.1671	0.6668	Catholic
Belgium	3	0.8333	7.167	1.5	Catholic

Country	Religious freedom	GRI	GFI	SRI	Religion
Bulgaria	3	2.2223	1.6667	4.1668	Orthodox
France	3	0	2.5	4	Catholic
Germany	3	0.8333	5.667	0.5	Mixed Christian
Greece	3	2.2223	10.001	9.334	Orthodox
Israel	3	0.8333	4.5003	7.0004	Judaism
Kenya	3	0	2.3335	2.3336	Protestant
Mongolia	3	3.889	5.5002	0.6668	Buddhism and related religions
Nicaragua	3	0.8333	1.8336	1.1668	Catholic
Peru	3	0	7.5003	0	Catholic
Philippines	3	0	0	0	Catholic
Romania	3	2.2223	10.001	9.334	Orthodox
Thailand	3	0	7.3337	0	Buddhism and related religions
Venezuela	3	3.0557	6.1671	0	Catholic
Armenia	4	6.1113	5.5003	8.1672	Orthodox
Cameroon	4	0	0	4.3336	Mixed Muslim/Christian
Georgia	4	0	9.3338	8.6672	Orthodox
Jordan	4	2.5	7.3337	6.5004	Islam
Kosovo	4	N/A	N/A	N/A	Islam
Kyrgyzstan	4	4.7223	1	7.6672	Islam
Lebanon	4	0	3.167	3.8336	Mixed Muslim/Christian
Macedonia	4	3.889	8.0005	8.1672	Orthodox
Malaysia	4	5.5557	9.0006	7.834	Islam
Mexico	4	2.2223	0	2.3336	Catholic
Moldova	4	3.6113	3.167	5.5004	Orthodox
Morocco	4	2.5	3.5	5.3336	Islam
Oman	4	2.5	7.3336	5.3336	Islam
Russia	4	5.5557	6.5003	6.6672	Orthodox
Serbia	4	6.9447	7.5004	7.5004	Orthodox
Slovakia	4	0	7.5004	0	Catholic
Tanzania	4	5.5557	1	3.8336	Protestant
Algeria	5	4.7223	8.0004	7.0004	Islam
Azerbaijan	5	8.6113	1.5	6.8336	Islam
Bahrain	5	7.778	9.5006	2.5004	Islam
Chad	5	3.0557	8.3338	4.3336	Mixed Muslim/Christian
Colombia	5	0.5557	7.5003	7.0004	Catholic
Comoros	5	6.9447	7.8338	9.334	Islam
Egypt	5	5.5557	8.167	7.5004	Islam
Ethiopia	5	3.0557	6.167	7.5004	Mixed Muslim/Christian
India	5	4.7223	8.3338	8.334	Hindu
Indonesia	5	4.4447	2.5	7.5004	Islam
Kazakhstan	5	6.9447	6.167	6.8336	Islam
Kuwait	5	3.3333	7.3337	5.3336	Islam
Laos	5	8.6113	8.8338	5.997	Buddhism and related religions

(continued)

Table 179.6 (continued)

Country	Religious freedom	GRI	GFI	SRI	Religion
Libya	5	4.1667	9.1672	5.3336	Islam
Nepal	5	3.0557	7.3337	5.5004	Hindu
Nigeria	5	3.889	3.1669	2.0004	Mixed Muslim/Christian
Sri Lanka	5	6.1113	5.5004	9.5008	Buddhism and related religions
Syria	5	5.5557	7.5005	6.8336	Islam
Tajikistan	5	6.9447	0	5.8336	Islam
Tunisia	5	2.5	6.5002	5.3336	Islam
Turkey	5	6.9447	9.5006	8.834	Islam
Yemen	5	2.5	6.8337	3.3336	Islam
Zimbabwe	5	1.389	5.8336	1.1668	Protestant
Afghanistan	6	3.6113	4.0003	8.834	Islam
Bangladesh	6	7.778	7.667	6.1672	Islam
Belarus	6	9.4447	1.5	2.6668	Orthodox
China	6	8.6113	6.8337	7.6672	Buddhism and related religions
Cuba	6	8.6113	0	2.8336	Catholic
Mauritania	6	9.4447	10.001	8.834	Islam
Pakistan	6	8.6113	9.6672	9.5008	Islam
Palestinian Areas	6	7.778	8.3338	10.001	Islam
Vietnam	6	8.6113	7.5004	2.6634	Buddhism and related religions
Burma	7	8.6113	9.5006	6.334	Buddhism and related religions
China-Tibet	7	6.9447	0	0.668	Buddhism and related religions
Eritrea	7	9.4447	6.8338	5.8336	Mixed Muslim/Christian
Iran	7	8.3337	9.3338	10.001	Islam
Iraq	7	7.778	8.1672	10.001	Islam
Maldives	7	8.3337	10.001	6.8336	Islam
North Korea	7	N/A	N/A	N/A	Buddhism and related religions
Saudi Arabia	7	10	10.001	10.001	Islam
Sudan	7	2.2223	0.5	3.5004	Islam
Turkmenistan	7	9.167	8.8338	6.8336	Islam
Uzbekistan	7	7.5003	8.0005	9.334	Islam

Source: Compiled by Daniel McGowin and Gerald R. Webster from data drawn from Marshall (2008)

References

Allen, B. (2011). *Baptists fined for unregistered worship in Belarus.* Retrieved April 12, 2012, from www.abpnews.com/content/view/6859/53/

Barro, R. J., & McCleary, R. M. (2005). Which countries have state religions? *Quarterly Journal of Economics, 120*(4), 1331–1370.

Batrawy, A. (2012). *Christian family evicted from Egypt town.* Retrieved April 10, 2012, from www.ajc.com/news/nation-world/christian-family-evicted-from-1392821.html

Beaumont, P. (2009). *China's Falun Gong crackdown*. Retrieved April 15, 2012, from www.guardian.co.uk/world/2009/jul/18/china-falun-gong-crackdown

Bernard, A. (2008). The advantage to Islam of Mosque-State separation. *Policy Review*, (147), 65–75.

Blanchard, B. (2011). *Insight: China's war on terror widens Xinjiang's ethnic divide*. Retrieved April 15, 2012, from www.reuters.com/article/2011/09/09/us-china-xinjiang-idUSTRE7880CE20110909

Bowen, J. (2011). How the French state justifies controlling Muslim bodies: From harm-based to values-based reasoning. *Social Research, 78*(2), 325–348.

Brunn, S. D., Webster, G. R., & Archer, J. C. (2011). The bible belt in a changing South: Shrinking, relocating and multiple buckles. *Southeastern Geographer, 51*(4), 513–549.

Chan, K. K. (2004). China's socioeconomic changes and the implications for the religion-state dynamic in China. *Brigham Young University Law Review, 2004*(2), 325–356.

Cobb, K. (2006). The Muhammad cartoons. *Conversations in Religion and Theology, 4*(1), 1–8.

Cohn, W. (1962). Is religion universal? Problems of definition. *Journal for the Scientific Study of Religion, 2*(1), 25–33.

Constitution of the People's Republic of Bangladesh. (1972). *Constitution of the People's Republic of Bangladesh*. Retrieved April 10, 2012, from www1.umn.edu/humanrts/research/bangladesh-constitution.pdf

Cumper, P., & Lewis, T. (2010). Last rites and human rights: Funeral pyres and religious freedom in the United Kingdom. *Ecclesiastical Law Society, 12*(2), 131–151.

Daftari, L. (2012). *12 Christians in Iran await verdicts after Easter Sunday apostasy trial*. Retrieved May 31, 2012, from www.foxnews.com/world/2012/04/12-christians-stand-trial-on-easter-sunday-in-iran/

Davis, K. (1949). *Human society*. New York: Macmillan.

Davis, E. (2008). Uyghur Muslim ethnic separatism in Xinjiang, China. *Asian Affairs, 35*(1), 15–29.

Drishtipat. (2012). *Vested Property Act, 1974*. Retrieved April 10, 2012, from www.drishtipat.org/HRLaw/vestedprop.htm

Egyptian Government Portal. (2011). *Constitutional declaration 2011*. Retrieved April 10, 2012, from www.egypt.gov.eg/english/laws/constitution/default.aspx

Embassy of China. (1997). *White paper – Freedom of religious belief in China*. Retrieved April 15, 2012, from www.china-embassy.org/eng/zt/zjxy/t36492.htm

Emilianides, A., Adamides, C., & Eftychiou, E. (2011). Allocation of religious space in cyprus. *The Cyprus Review, 23*(1), 97–121.

Fahim, K. (2012). *Coptic Pope's death adds to fears in Egypt's time of transition*. Retrieved April 10, 2012, from www.nytimes.com/2012/03/21/world/middleeast/coptic-pope-shenouda-iiis-death-adds-to-fears-in-egypt.html

FIGC. (2012a). *Norme Organizzative Interne della FIGC, Parte II: Le Funzioni*. Retrieved July 23, 2012, from www.figc.it/Assets/contentresources_2/ContenutoGenerico/20.$plit/C_2_ContenutoGenerico_3817_Sezioni_lstSezioni_numSezione_1_lstCapitoli_numCapitolo_2_upfFileUpload_it.pdf

FIGC. (2012b). *Codice di Giustizia Sportiva, Titolo II: Sanzioni*. Retrieved July 23, 2012, from www.figc.it/Assets/contentresources_2/ContenutoGenerico/44.$plit/C_2_ContenutoGenerico_3815_Sezioni_lstSezioni_numSezione_0_lstCapitoli_numCapitolo_1_upfFileUpload_it.pdf

Gianfreda, A. (2011). Religious offences in Italy: Recent laws concerning blasphemy and sport. *Ecclesiastical Law Society, 13*(2), 182–197.

Grim, B., & Finke, R. (2006). International religion indexes: Government regulation, government favoritism, and social regulation of religion. *Interdisciplinary Journal of Research on Religion, 2*, 1–40.

Hassner, R. (2011). Blasphemy and violence. *International Studies Quarterly, 55*(1), 23–45.

Hopkins, C. (2012). *Malaysia may repatriate Saudi who faces death penalty for tweets*. Retrieved April 5, 2012, from www.csmonitor.com/World/Global-News/2012/0210/Malaysia-may-repatriate-Saudi-who-faces-death-penalty-for-tweets

Hoult, T. F. (1958). *The sociology of religion*. New York: Dryden Press.

Humphreys, A. (2012, January 27). Let Buddhist man remain in Canada, court rules; Temple worker still in danger in Bangladesh. *National Post*, p. A7.

International Coalition for Religious Freedom. (2012). *Religious freedom country reports*. Retrieved April 3, 2012, from www.religiousfreedom.com/index.php?option=com_content&view=article&id=47&Itemid=29

Keyes, D. (2012). *Saudi writer Hamza Kashgari faces charge of blasphemy after tweets about Muhammad*. Retrieved April 5, 2012, from www.washingtonpost.com/opinions/saudi-writer-detained-after-tweets-about-muhammad/2012/02/09/gIQApsgW2Q_story.html

Kramnick, I., & Moore, R. L. (1997). *The godless constitution*. New York: W.W. Norton.

Krindatch, A. (2006). Changing relationships between religion, the state, and society in Russia. *GeoJournal, 67*(4), 267–282.

Ludwig, F. (2008). Christian-Muslim relations in Northern Nigeria since the introduction of Shari'ah in 1999. *Journal of the American Academy of Religion, 76*(3), 602–637.

Marshall, P. A. (2008). *Religious freedom in the world*. Lanham: Rowman & Littlefield Publishers.

Marshall, P. A. (2012, July 26). Personal Communication with the authors.

Martina, M., & Rhoads, B. (2012). *China gets restive taste of post-Dalai Lama era*. Retrieved April 16, 2012, from http://in.reuters.com/article/2012/02/06/china-tibet-idINDEE81501S20120206

Michael, M., & El Deeb, S. (2011). *Church burning deepens tumult of Egypt transition*. Retrieved April 10, 2012, from www.usatoday.com/news/topstories/2011-05-08-1652241639_x.htm

Nikolas, K. (2012). *Saudi Arabia accuses another twitter user of apostasy*. Retrieved April 5, 2012, from www.digitaljournal.com/article/321896

Nottingham, E. K. (1954). *Religion and society*. New York: Random House.

People's Daily Online. (1982). *Constitution of the People's Republic of China*. Retrieved April 15, 2012, from http://english.people.com.cn/constitution/constitution.html

Pew Forum on Religion & Public Life. (2011). *Rising restrictions on religion*. Washington, DC: Pew Research Center.

President of the Republic of Belarus. (2012). *Constitution of the Republic of Belarus*. Retrieved April 12, 2012, from http://president.gov.by/en/press19330.html#doc

Republic of Belarus. (2002). *The new Belarusian law on religion*. Retrieved April 12, 2012, from www.unhcr.org/refworld/pdfid/4c2217f12.pdf

Richardson, J. (2006). The sociology of religious freedom: A structural and socio-legal analysis. *Sociology of Religion, 67*(3), 271–294.

Roosevelt, F. (1941). *Message to congress*. Retrieved March 20, 2012, from www.fdrlibrary.marist.edu/pdfs/ffreadingcopy.pdf

Royal Embassy of Saudi Arabia. (2012). *The basic law of governance*. Retrieved April 5, 2012, from www.saudiembassy.net/about/country-information/laws/The_Basic_Law_Of_Governance.aspx

Schiller, B. (2010). *Christians come under attack in China*. Retrieved April 15, 2012, from www.thestar.com/news/world/china/article/84527--hristians-come-under-attack-in-china

Sopher, D. E. (1967). *Geography of religions*. Englewood Cliffs: Prentice Hall.

State Administration for Religious Affairs of PRC. (2012). 中国新闻网——卓新平:宗教参与中国社会建设空间正在扩大. Retrieved March 16, 2012, from www.sara.gov.cn/xwzx/mtjj/12689.htm

Stump, R. W. (2000). *Boundaries of faith: Geographical perspectives on religious fundamentalism*. Boulder: Rowman and Littlefield.

The Central People's Government of the People's Republic of China. (2012). 外交部就美国"国际宗教自由报告"涉华内容答问. Retrieved on August 2, 2012, from www.gov.cn/gzdt/2012-08/02/content_2197045.htm

Tong, J. (2010). The new religious policy in China: Catching up with systemic reforms. *Asian Survey, 50*(5), 859–887.

Tsintsadze, K. (2008). Legal aspects of Church-state relations in post-revolutionary Georgia. *Brigham Young University Law Review, 2007*(3), 751–774.

United Nations. (2010). *Report of the Special Rapporteur on freedom of religion or belief: Summary of cases transmitted to governments and replies received*. Retrieved July 24, 2012, from http://daccess-dds-ny.un.org/doc/UNDOC/GEN/G10/107/83/PDF/G1010783.pdf?OpenElement

United Nations. (2012). *Universal declaration of human rights*. Retrieved March 20, 2012, from www.un.org/en/documents/udhr/

U.S. Department of State. (2011, August 17). *What is religious freedom?* Fact Sheet: Bureau of Democracy, Human Rights and Labor, Office of International Religious Freedom. Retrieved June 5, 2012, from www.state.gov/j/drl/rls/fs/2011/170635.htm

U.S. Department of State. (2012a). *Religious freedom*. Retrieved March 31, 2012, from www.state.gov/j/drl/irf/

U.S. Department of State. (2012b). *Israel*. Retrieved July 28, 2012, from www.state.gov/documents/organization/171736.pdf

U.S. Department of State. (2012c). *International religious freedom report for 2011, executive summary*. Retrieved August 2, 2012, from www.state.gov/j/drl/rls/irf/religiousfreedom/index.htm#wrapper

Wee, S. L., & Blanchard, B. (2011). *China says Dalai Lama has to reincarnate*. Retrieved April 16, 2012, from www.reuters.com/article/2011/03/07/us-china-npc-tibet-idUSTRE72624L20110307

Wines, M. (2002). *New Belarus law codifies rising religious repression*. Retrieved April 12, 2012, from www.nytimes.com/2002/11/23/international/europe/23MINS.html?pagewanted=all

Woods, K. (2008). Religious freedom in Kosovo: Prenatal care to a new nation. *Brigham Young University Law Review, 3*, 1009–1055.

Zhang, Q., & Zhu, Y. (2011). Religious freedom and its legal restrictions in China. *Brigham Young University Law Review, 3*, 783–818.

Chapter 180
Geographies of Cosmic War: Comparing Secular and Religious Terrorism in Space and Time

Steven M. Radil and Colin Flint

180.1 Introduction

The agenda of social scientists and practitioners with interest in international security issues has become dominated by the topic of terrorism. Catalyzed by the attacks of September 11, 2001 and given a dubious academic crutch by the revived notion of a "Clash of Civilizations," it appeared as if new geopolitical actors and agendas had suddenly materialized to reshape the very essence of international security (Huntington 1993; Esposito 2002; Stern 2004; Hoffman 2006; White 2011). No longer was the focus upon inter-state competition and the possible resort to "old wars" in a Clausewitzian sense (Black 2010). Instead, it was claimed we had moved to an era of "new wars" in which an ugly assortment of criminal, religious, ideological and state groups fought for the ability to keep fighting and to attain more ill-gotten gains (Kaldor 2001). Globalization, the military dominance of the United States, cultural differences, and terrorism as the chosen method of conflict were seen to underlie security practitioners' interest in "asymmetric warfare" (Kaldor 2001; Kahler and Walter 2006) and academic interest in religious terrorism (Juergensmeyer 2000; Stern 2004). Though the contemporary impact and threat of religiously motivated terrorism is hard to deny, our approach is to look at the geography of religious terrorism in order to critically engage its "newness" (Rapoport 2001; Flint 2003; Flint and Radil 2009). In particular, we engage whether religious terrorism is just a new representation of old political motivations or whether new geographies of terrorism are emerging.

S.M. Radil (✉)
Department of Geography, University of Idaho, Moscow, ID 83844, USA
e-mail: sradil@uidaho.edu

C. Flint
Department of Political Science, Utah State University, Logan, UT 84322, USA
e-mail: colin.flint@usu.edu

© Springer Science+Business Media Dordrecht 2015
S.D. Brunn (ed.), *The Changing World Religion Map*,
DOI 10.1007/978-94-017-9376-6_180

Our investigation of these questions is organized in the following way. First, we introduce terrorism as a form of geopolitical agency that occurs, but is not determined by, geographic settings. Religious terrorist groups are geopolitical actors and their actions are part of the complex process of the dynamic social construction of space, a construction that is situated within existing social and geographic settings but may change existing patterns. Second, we move from the general idea of the social construction of space to the key geopolitical actors in this analysis, religious terrorist groups. We introduce contemporary religious terrorism using the historical framework of Rapoport's (2001) waves of terrorism, and discuss their geographical expression. Third, we discuss why religious terrorism is expected to produce a new geography of violence. In the fourth section we introduce our case study of terrorism in India to investigate whether terrorist activity is indeed becoming increasingly motivated by religious belief and, if so, whether it actually displays new geographic patterns. Tentative conclusions are offered to close the chapter.

180.2   Terrorists as Geopolitical Agents

Though theoretically and methodologically eclectic, the discipline of human geography may be defined by one overarching axiom: that the built environment is socially constructed, or it is the product of individual and collective action (Massey 2005). Moreover, the process is recursive in that the geography of the built environment is the setting which, to some degree, limits the possibility of future actions while subsequent actions maintain but also alter the geographic setting and political and social possibilities. The recursive process is one that identifies a range of possible or likely actions by individuals and groups, but is not deterministic (Agnew 1987).

Geopolitics is a particular form of the social construction of space that focuses on political actions and the political geographies that actors form and operate within (Flint 2011). Historically, geopolitics was the preserve of leaders of states with normative agendas for their own state's foreign policy agendas (Dodds and Atkinson 2000). However, the revival of geopolitics has reclaimed the topic for social scientists wishing to analyze rather than proscribe the world. In addition, the cast of actors involved in geopolitics has expanded beyond the singular concentration upon states to include social movements, multinational companies, religious organizations, and the media (for a review see Cox et al. 2008).

Terrorism, and religious terrorism in particular, is readily accommodated in the new geopolitical research agenda. Terrorism is a form of political activity that operates within existing political geographic settings with an agenda to disrupt them and impose new spatial political arrangements (Flint 2003). Hoffman (2006: 37) provocatively refers to terrorists as "altruists" to illustrate how they develop a belief system that frames their violent actions as necessary to right the wrongs imposed upon a marginalized social group. Such rhetoric claims that all conventional peaceful political avenues have been closed and the resort to violence is the only means left to pursue a political agenda that will offer a long sought after justice for an oppressed group. In other words, the geopolitical agency of terrorist groups is one in which the violence is calculated to have a lasting political impact.

From a geographic perspective, Hoffman's terminology requires a further element that helps us to understand the reason for the sense of injustice and oppression and also highlights the implications of a successful terrorist campaign. The perceived goals of terrorist groups can be more fully identified as political geographic altruism (Flint 2011: 177). The conceptual framework of the social construction of space requires us to think of politics and geography as mutually constructed. Hence, any sense of political injustice is associated with a geographical arrangement that creates and perpetuates the injustice. In turn, any proposed political alternative is likely to require the construction of a new geographic arrangement. The inseparable connection between politics and geography, with regard to terrorism, is probably most clear in the arena of national separatism. For example, the demands of the Irish Republican Army or the Basque separatists Euskadi Ta Askatasuna (ETA) are framed within an understanding that the existing territorial expressions of the United Kingdom and Spain, respectively, are unjust, in that they do not allow the full expression of Irish and Basque national identity. In turn, the goals of these groups are a re-drawing of political boundaries in a way that is seen to enable national liberation. In sum, the political solution requires a new geography and the existing political injustices are an expression of a "flawed" or oppressive geographical arrangement.

Geography also plays a role in the way that terrorists strategize. The choice of target is, partially, a geographic decision based upon the symbolic value of geographic settings. Timothy McVeigh's 1995 attack on the Alfred P. Murrah Federal Building in Oklahoma City was not based solely on the politics of identifying the Bureau of Alcohol, Tobacco and Firearms as unconstitutional oppressors of the people, but also as a statement that this was a battle in and for the geographic realm of the "American heartland." The decision of the IRA to enact a campaign on the British "mainland" sent a message that the question of Irish republicanism was a British problem and not just Ulster's. In each case, the geographic setting of terrorist activity has a symbolic value connected to a general understanding of the cultural and historic significance of particular spaces or places (Flint 2005: 200). With regard to religious terrorism, Al Qaeda's representation of the Arabian Peninsula as the spiritual home of Islam has provided the basis for its justification of the U.S. as both an invading and infidel force.

There is one more component of geopolitics that is useful for our discussion. Geopolitics is both practice and representation (Flint 2011: 35) in that it is not only the actions of geopolitical actors that are important but the way they are represented. Representation is the discourse used by an actor in an attempt to portray their actions as necessary and just. Of course, in the competitive world of geopolitics opponents offer their counter-representations. In the classic statement "one person's terrorist is another's freedom fighter," terrorist groups portray themselves as liberators while the state authorities they fight portray them as "murderers." The identification of geopolitics as both practice and representation is not simply a semantic exercise or a call for a discursive analysis (Ó Tuathail 1996). The issue is that we should not identify the goals and identification of terrorist groups just by what they say, but also, arguably primarily, by what they do. More plainly, geopolitical actions speak louder than geopolitical words in identifying the purpose of a terrorist group.

We take this stance in order to critically investigate contemporary terrorism and the argument that we are living through something fundamentally new: an era of religious terrorism. Moreover, we believe that the mutual construction of geography and politics provides a means to tackle this question. Religious terrorism is believed to have fundamentally different objectives than previous expressions of terrorism, motivated by nationalist and ideological agendas (Rapoport 2001). The geographic perspective leads to hypotheses that changing agendas are likely to create new geographies of terrorist activity, including intrastate patterns of terrorist attacks. Our case study of India is an empirical examination of trends in terrorist activities to see if geographic patterns of activity expected from a rise in religious terrorist activity are actually occurring. But before we can focus on a particular case study we must introduce the broad historical trends of terrorist activity to situate and understand contemporary religious terrorism.

180.3   Waves of Terrorism

Rapoport (2001) has identified four waves of terrorism in the modern era. These broad categorizations are useful in situating contemporary terrorism and assessing the argument that we are currently experiencing a fundamentally new period dominated by religiously motivated terror (Juergensmeyer 2000). Each of these periods has a related geographic expression (Flint 2011: 180–184). However, one geographic issue that is still to be addressed is the change in the geographical location of terrorist attacks that is implied in the conceptual understanding of how and why terrorism is inspired by religious belief. To understand the intra-state geography of terrorism that is the focus of our case study we must first situate religious terrorism in broader historical trends.

The first wave of modern terrorism occurred, generally speaking, between 1880 and 1914 (Rapoport 2001). The main motivating force was the ideology of anarchism and its goals of revolutionary change against monarchies and aristocratic regimes. The gradual electoral and social reform that European governments were, reluctantly, making were seen by the anarchists as a mere sop, but one that could dampen the revolutionary ardor of the masses. Hence, anarchists believed that dramatic events – such as the assassination of prominent political elite figures – would act as a trigger to tip society across a threshold and into revolution. The geographic focus of this wave of terrorism was intrastate, or the overthrow of particular national governments to be replaced by a new form of national political organization (Flint 2011: 180). The key event that marked the end of this period, the assassination of Archduke Franz Ferdinand in Sarajevo in June 1914, illustrates that the political motivation of terrorism was already changing.

The second wave of terrorism, roughly 1920–1960, occurred within a context of decolonization: Specifically, the dismantling of the formal European empires in the aftermath of the First and Second World Wars and the rise of U.S. power that emphasized an open world economy (Rapoport 2001; Smith 2003). Following on from the assassination of Franz Ferdinand, the inspirational power of anarchism was replaced by the ideology of nationalism: The belief was that individuals were members of national groups that had a right to independent sovereign and territorial states. The geographic strategy of this wave of terrorism was aimed at creating new nation-states, rather than changing the form of government in existing states as in the first wave (Flint 2011: 180). Examples of terrorism in this wave include Irgun's activities and the desire to create the state of Israel and the Mau Mau campaign for an independent Kenya.

The third wave of terrorism, generally speaking 1960–1990, maintained the nationalist motivations of the previous wave, but added a twist of political ideology and a new geographical expression (Rapoport 2001). Political ideology of the extreme left and right became part of the terrorist lexicon. In some cases this was the sole focus – such as the Red Brigade and Baader Meinhof Gang – while in others it blended with nationalist goals – such as the Irish National Liberation Army (Laqueur 1987: 214). More significantly for the argument we make in this chapter, the organization and strategy of terrorist organizations became transnational (Flint 2011: 181). Groups cooperated with each other for training purposes. Furthermore, terrorist actions were undertaken to ensure a global impact. The PLO's killing of Israeli athletes at the Munich Olympic Games in 1972 was staged in a manner that the whole world could no longer claim ignorance of the nationalist aspirations of the Palestinian people. The practice of aircraft hijacking also ensured that terrorist activities were transnational: a plane of a particular national airline being hijacked en route from one country to another and being diverted to yet another country transnationalized acts of terror. Hence, though the geography of terrorist aims remained largely focused on territorial states and the politics of nationalism, the geographic expression of terrorist activity was loosening its bonds to specific national territories.

The fourth wave of terrorism, roughly the 1990s to the present day, displays a continuation of this geographic trend, but with a dramatic change in the motivation for terrorist activity. Rapoport (2001) claims that we are living in an era of religious terrorism; described as a fundamental shift and an unprecedented situation. Though focus in the U.S. and Europe may simply identify religious terrorism with Islamic extremism, it is actually identified as a global phenomenon. Extremist Christian, Jewish, Sikh, and Buddhist groups have also been identified as advocating and committing acts of terrorism (Juergensmeyer 2000).

The trend towards pan-religious extremism is suggested by mapping terrorism events associated with religion. As seen in Fig. 180.1, we have mapped the number of events by country using data from the Global Terrorism Database (GTD) available from the National Consortium for the Study of Terrorism and Responses to Terrorism (START 2011). The GTD contains information on terrorist events around the world from 1970 through 2010, including a variety of information

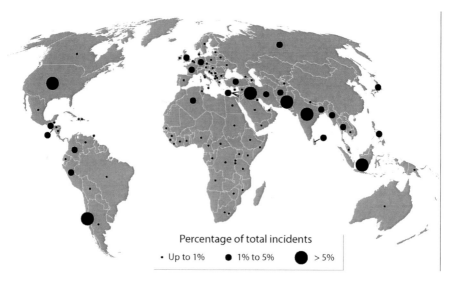

Fig. 180.1 Religiously-motivated terrorism by country, 1970–2010 (Map by S. Radil)

about incident locations, groups involved in attacks, and the types of targets. Mapping incidents where religion was a primary targeting motive suggests that religious terrorism is a global phenomenon with incidents present in numerous countries all around the world. While the specific motivations behind the acts of violence are not universal (for example the high number of incidents in the U.S. is attributable to abortion-related violence which is nearly unique to the U.S.), this global geography bolsters Juergensmeyer's (2000) claim that religious terrorism is more than just Islamic extremism.

180.4   The Geography of Religious Terrorism

The geographic expression of religious terrorism, as conceived by Western academics, is possibly a fundamental change and one with worrying implications. Juergensmeyer (2000) claims that religious terrorists are fighting what they perceive as a "cosmic war" – a war over good against evil adjudicated by a supreme deity. The implication is that religious terrorism will provoke a greater level of violence than terrorism motivated by nationalist or ideological goals. The reasoning behind this logic rests on the assumption that secular inspired terrorists realize that at some stage they are likely to engage in a political negotiation with state authority. Hence, secular groups balance a belief that they must commit atrocities in order to be placed on the political agenda with a strategy to keep the level of violence within certain bounds to enable a government to justify the morality of talking with them. Religious terrorists, fighting for their understanding of what a supreme deity demands of them, have no interest in potential negotiation with secular authorities and instead are compelled to commit massive acts of violence that will usher in a period of final judgment for humankind. In addition, religiously motivated terrorism may also be seen as a performance intended to reach a broad audience, beyond state elites, to communicate the ideals such groups are fighting for (Juergensmeyer 2000: 124). Accordingly, constructing the broadest possible audience demands increasingly spectacular acts of violence.

However, going back to our definition of geopolitics that identifies the interaction of practice and representation (Flint 2011: 35), what a geopolitical actor claims may just be a form of justification or public image that belies underlying motivations and strategies. Most scholarly studies of terrorism, like Jurgenmeyer's (2000), focus almost exclusively on the representations of the groups that use this particular form of violence but pay little heed to potential inconsistencies between the practices of these groups and their representations. Rather than focus on representations, we instead consider the actual practices of terrorism across Rapoport's (2001) third wave of secular terrorism and fourth wave of religious terrorism. We consider practices because we would expect the geography of secular terrorism to be quite different from that of religious terrorism. For instance, terrorism motivated by national separatism is more likely to have a geographic expression in the areas that are at the heart of the territorial claim, with also some activity in urban centers associated with political control of the claimed areas, such as a national capital or provincial capital. On the other hand, if Juergensmeyer's (2000) claims are correct, we expect that religious terrorism would target population centers in an attempt to inflict maximum casualties and to communicate to the broadest possible audience.

By theorizing the material geographies of terrorism in this way and drawing on geopolitics as both practice and representation, we also provide an opportunity to consider if the practices of religious terrorism actually differ from those of secular terrorism. If the practices do differ, then our findings would support the arguments that religious terrorism is indeed a unique phenomenon with a correspondingly unique set of issues at its core as argued by Juergensmeyer (2000) and Rapoport (2001). However, if we see little difference in the practices of religious and secular terrorism, we may be left to conclude that all that is really new is the choice of representation to portray large scale violence as necessary and just.

As our approach is to consider the practices of terrorism, we have chosen to focus our analysis on a particular state where both types of motivations for terrorism are present. While the global geography of religiously motivated terrorism suggests more than a few candidates, we focus on the geography of terrorism within the state of India. Since independence in 1950, India has suffered from protracted communal and terroristic violence between Hindus, Muslims, and Sikhs. While much of the violence has direct roots with the partition of India and the creation of Pakistan and East Pakistan (which in 1971 gained full independence and is now called Bangladesh), religion has been central to many of these conflicts as a basis for collective identity. Additionally, India has also seen a long-running Maoist insurgency. The Naxalite movement has engaged in violence since the mid-1970s, aimed at ushering in Marxist-Leninist style political and economic systems within India. Put succinctly, India has a long history of both religiously- and secularly-motived terrorism and constitutes an ideal case to consider our argument.

The specifics of our approach are as follows. First, by drawing on the above arguments about differing geographies of secular terrorism and religious terrorism, we would expect religious terrorism to be a largely urban phenomenon within India. The perceived need to commit spectacular acts of violence (meaning large number of victims and witnesses) is expected to lead to a strategic focus on cities as the likeliest stage for the "performance." In contrast, secular terrorism is less likely to be an exclusively urban phenomenon due to the perceived need to challenge state control in all contested areas, including rural areas, smaller villages and minor cities or towns in the urban hierarchy in India (for example, Lohman and Flint 2010). Second, by drawing on Rapport's arguments about shifting motivations for terrorism between historic waves, we would expect a corresponding shift in the geography of terrorist attacks over time within India, moving beyond contested areas and into major population centers.

Taking on these questions requires a robust set of data on the geography of terrorism over time and for that we have used the GTD (START 2011). The GTD is a longitudinal dataset, covering terrorism from 1970 to 2010. As such, it spans the two most recent waves of terrorism in Rapoport's formulation, including what Rapoport sees as the nationalist/ideological motivations for terrorism through the 1970s and 1980s and the religious motivations for terrorism from 1990 on. The GTD is also a geographically disaggregated dataset as it records the geographic location of incidents within states as precisely as possible. As a practical matter, this often means that incidents are assigned to the smallest administrative areal unit in a particular county when more detailed information is not available. However, even when incidents are geographically aggregated in this way the GTD also codes for incident location based on an urban/rural typology. These twin features of the GTD allow an examination of changing space-time patterns of terrorism within India across two arguably distinct eras.

To be included in the GTD, incidents have to meet a combination of several criteria, including intentionality, the use or threat of violence (including property violence and sabotage), political, religious, or social motivations, an intent to coerce local or state officials, or the deliberate targeting of civilians and/or non-combatants (START 2011). However, the GTD does not record violence by state forces against opposition forces or civilians and, therefore, may provide only an incomplete presentation of the geography of terroristic violence within an individual state. With these criteria and limitations in mind, the GTD yielded 5,355 incidents within India from 1970 through 2010 with incidents coded as occurring either within rural areas or within urban areas at the level of a District (an administrative unit below that of States and Union Territories in India). This represents a robust longitudinal set of data with enough geographic specificity for the examination of our key questions.

Table 180.1 Trends in the geography and lethality of terrorism in India during Rapoport's fourth wave

	All incidents (1970–2010)		Third wave (1970–1990)		Fourth wave (1991–2010)	
	Rural	Urban	Rural	Urban	Rural	Urban
Incidents	24.8 %	75.2 %	6.8 %	93.2 %	29.4 %	70.6 %
Fatalities/Event	2.57	2.42	2.60	2.49	2.56	2.40

Source: Steven M. Radil and Colin Flint

First, we drew on the rural/urban coding of incidents to consider the overall geography of terrorism in India within the urban hierarchy. For instance, the GTD codes 75 % of all incidents from 1970 to 2010 as occurring within urban settings, which is defined by India's census as inhabited places with 5,000 residents or settled areas with a population density of 400 people/km^2 (Ministry of Home Affairs 2011). At first glance, such an overwhelmingly urban geography of terrorism could be seen as support for the Juergensmeyer's cosmic war thesis (2000) as we have hypothesized that the geography of religious extremism would likely be focused on cities. However, grouping the data using Rapoport's typologies suggests that this geography is dynamic over time but in a way that complicates the cosmic war thesis. From 1970 to 1989, the GTD codes only 7 % of all incidents as rural while from 1990 on, 29 % are coded as rural (Table 180.1). While the geography of terrorism is clearly dynamic across Rapoport's third and fourth waves, the trend is precisely opposite of what we might expect from a cosmic motivation. Terrorism in India has become more rural over time rather than more urban.

Second, this dynamic geography of incidents also confounded our expectations of the cosmic war thesis when we considered the average lethality of incidents in the GTD. Based on fatality data present in the GTD, we calculated the mean number of fatalities per incident in both rural and urban settings. As shown in the table, the mean number of fatalities per incident was slightly higher among rural incidents than urban incidents in both time periods (1970–1989 and 1990–2010). Further, the mean lethality of incidents has declined overtime, whether rural or urban. Put plainly, terrorism in India was slightly more deadly outside cities in either era, but has become less deadly in the era of religiously motivated violence.

Finally, we considered the overall geographic distribution of incidents within India based on the expectation that religiously-motivated violence would have a broader geographic expression than would secular terrorism. By mapping the percentage of the total number of incidents by state/union territory in each period, the overall geography of terrorism was largely unchanged (Fig. 180.2). Indeed in both periods, terrorism was focused within disputed areas associated with partition and the creation of India and Pakistan/Bangladesh; specifically Jammu and Kashmir and Punjab states in the northwest and Assam, Bihar, and West Bengal states in the east. Notably, terrorism was less common in the second period in Delhi state (the national

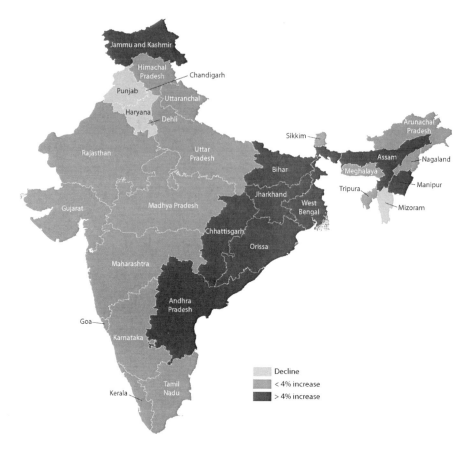

Fig. 180.2 Change in incidents from third to fourth wave (Map by S. Radil)

capital) and more common in Chhattisgarh and Jharkhand states. Both outcomes are related to the expansion of the Naxalite movement, as terroristic violence by various Naxalite groups spread to these largely rural states in the second period. Rather than religiously-motivated groups being responsible for a broader geography, the expansion of terrorism within India had to do with a deepening secular resistance.

In sum, we found little in the changing geography of terrorism within India that is consistent with the expectations drawn from the "cosmic war" thesis. We theorized that a shift toward religious motivations for terrorism would produce a shifting geography of terrorism that was more urban in form and more violent. We also theorized that in a "wave" of religious terrorism attacks would be less focused on historically-disputed territory or places that represent centralized state authority. Despite our expectations, we found that terrorism became more rural rather than more urban and no more violent that it had been previously, while remaining concentrated in disputed areas.

180.5 Religious Terrorism as Geopolitical Representation

As our analysis shows, the changing geography of terrorism within India raises questions about the validity of the cosmic war thesis and the associated claims that terrorism is now motivated by issues beyond earthly politics. We recognize that a study of India does not allow for universal claims. Rather, our case study is suggestive that the claims to a "cosmic war" need to be tempered and that more analysis of the actual changing pattern of terrorism is necessary to interrogate the motivations underlying contemporary terrorism. Claims to a new "cosmic war" suggest that terrorist groups now seek to inflict the maximum number of casualties to bring about a period of final judgment for humankind and to communicate with the broadest audience possible. We have hypothesized that such new motivations will also necessarily produce new geographies, with terrorism shifting to population centers in search of more victims and more witnesses. By considering the case of India, a state with a long history of terroristic violence associated with both secular and religious issues, we have indeed uncovered new geographies of terrorism. However, counter to our expectations, terrorism within India has become less urban and less deadly over time while simultaneously concentrating over time within areas associated with long-standing territorial disputes.

Because our findings are based on terrorism data, one possible explanation for this unexpected result may be the nature of terrorism data itself. Because terrorism is a subjective and contested phenomena, there is a potential for large sample, cross-sectional datasets (such as the one we have used) to be biased geographically. The GTD, like most similar terrorism databases, are largely created from media reports of incidents and attacks. Reporting agencies and news organizations may bias coverage toward large urban areas, especially if an attack produces no fatalities. The sum effect of such geographic bias could be to depress the reporting of rural incidents, particularly in the early years of the dataset which covers Rapoport's secular third wave.

Given that we partially base our findings on the increasing of percentage of rural incidents between the third and fourth waves, we assessed how many additional unreported third wave rural incidents would be necessary to present the same rural/urban proportion as observed in the fourth wave. To reach the same proportion of rural incidents as found in the fourth wave, there would have to be over four unreported rural incidents for every reported rural incident during the third wave. This is the level of underreporting required to produce an unchanged geography of terrorism between Rapoport's secular (third) and religious (fourth) waves. Accordingly, far more underreporting would be required to produce a geography where terrorism matched the expectations of the cosmic war thesis by shifting into urban areas and population centers. Bias toward urban incidents may indeed be present, but such a markedly high level of underreporting seems unlikely.

What then can we attribute this counter-intuitive geography of terrorism? For this, we return to our argument about geopolitical practice and representation. By focusing on the material geography of terroristic attacks in India we have revealed the geopolitical practices of intrastate terrorism during eras with presumably differing motivations. These practices give no indication of the pursuit of "cosmic war" (Juergensmeyer 2000) in India. Rather, we may see the rhetoric of religious terrorism as part of the geopolitical representation of the use of terroristic violence. By connecting the use of violence against civilians or non-combatants to larger struggles of good versus evil, the leadership of groups can attempt to portray their actions as necessary and just to both internal and external audiences. Further, this kind of representation is connected to the process of demonizing one's opposition, a process that Juergensmeyer (2000: 171–178) himself argues is a necessary step in engaging in organized violence of all stripes. There is a need for further analysis to see if such conclusions find support in other parts of the world.

Despite the adoption of the language of cosmic war to represent the necessity and justness of violent struggle, the practices of terrorism in our example suggest far more earthly concerns. As we described earlier, a key point regarding Juergensmeyer's notion of cosmic war is the assumption that evil cannot be negotiated with and must be destroyed. This assumption, in combination with the desire to communicate to a mass audience, is thought to lead to a lack of restraint in the use of violence. In other words, religious terrorism is not about creating opportunities to force state elites to negotiate settlements that create new political geographies on the ground. That is an earthly outcome and presumably of no interest to those who see their cause as an "all-or-nothing struggle against whom [they are] determined to destroy" (Juergensmeyer 2000: 148). And yet, the geographies of terrorism we have uncovered in India suggest a persistent concern with more terrestrial motivations. For instance, the high levels of violence in the main areas of partition (Punjab and Bengal) are artifacts of the creation of geographical arrangements that have been seen as persistent sources of social injustice. And while violence in these areas is indeed connected to religion, the Punjabi and Bengali partitions were primarily attempts to create new political geographies to resolve long-standing communal clashes between Hindus and Muslims (and to a lesser extent Sikhs) during decolonization. In other words, issues associated with religiously-motivated nationalism did not lead to a geography of cosmic war. Instead, it led to a geography that appears little changed from the previous waves of terrorism in which the construction of new geographic arrangements was dominant.

By concerning ourselves with more than what terrorist groups say, we believe we have exposed what may actually be fundamentally new in Rapoport's (2001) wave of religious terrorism and Juergensmeyer's (2000) era of cosmic war. What could be new about contemporary terrorism is the set of geopolitical representations that have adopted the rhetoric of religious symbolism to justify and moralize the use of violence in ongoing conflicts. And just as terroristic violence is one tactic among many available to geopolitical actors, the geopolitical representations of conflicts often shift over time and can be quite disconnected to actual practices. This disconnect can occur as geopolitical elites search for effective ways to motivate their rank and file to continue the struggle, counter the representations of their opposition, and influence external parties. As such, it is possible that contemporary terrorism has actually changed little over time as it is still concerned with political geographic altruism and the construction of terrestrial political geographies aimed at redressing perceived injustices and grievances (Flint 2011), even in a state like India where representations that draw on religious arguments regarding good and evil might resonate due to the presence of religiously-defined communal conflicts.

The rush to find something new about terrorism after the attacks of September 11, 2001 has led to an imbalanced understanding in the terrorism literature that has overly emphasized the religious themes in the geopolitical representations of terrorism. A more balanced and sensible approach is one that considers geopolitical practices as well. Given the persistence of groups that employ terrorism as a tactic and of the earthly issues that motivate them, terrorism scholars would do well to consider both representation and practice. Doing so is an ideal way to consider what's fundamentally new the next time terrorism seems to reshape the very essence of international security.

References

Agnew, J. A. (1987). *Place and politics*. London: Allen and Unwin.

Black, J. (2010). The western way of war. In G. Kassimeris & J. Buckley (Eds.), *The Ashgate companion to modern warfare* (pp. 11–16). Farnham: Ashgate.

Cox, K., Low, M., & Robinson, J. (2008). *The Sage handbook of political geography*. Los Angeles/London: Sage.

Dodds, K., & Atkinson, D. (2000). *Geopolitical traditions: A century of geopolitical thought*. London: Routledge.

Esposito, J. L. (2002). *Unholy war: Terror in the name of Islam*. Oxford: Oxford University Press.

Flint, C. (2003). Geographies of inclusion/exclusion. In S. L. Cutter, D. B. Richardson, & T. J. Wilbanks (Eds.), *The geographical dimensions of terrorism* (pp. 53–58). New York/London: Routledge.

Flint, C. (2005). Dynamic metageographies of terrorism: The spatial challenges of religious terrorism and the "War on Terrorism.". In C. Flint (Ed.), *The geography of war and peace* (pp. 198–216). Oxford/New York: Oxford University Press.

Flint, C. (2011). *Introduction to geopolitics* (2nd ed.). Abingdon/New York: Routledge.

Flint, C., & Radil, S. (2009). Terrorism and counter-terrorism: Situating al-Qaeda and the Global War on Terror within geopolitical trends and structures. *Eurasian Geography and Economics, 50*(2), 150–171.

Hoffman, B. (2006). *Inside terrorism*. New York: Columbia University Press.

Huntington, S. (1993). The clash of civilizations. *Foreign Affairs, 72*(1), 22–49.

Juergensmeyer, M. (2000). *Terror in the mind of God: The global rise of religious violence*. Berkeley: University of California Press.

Kahler, M., & Walter, B. F. (Eds.). (2006). *Territoriality and conflict in an era of globalization*. Cambridge/New York: Cambridge University Press.

Kaldor, M. (2001). *New and old wars. Organized violence in a global era*. Stanford: Stanford University Press.

Laqueur, W. (1987). *The age of terrorism*. Boston/Toronto/London: Little, Brown and Co.

Lohman, A., & Flint, C. (2010). The geography of insurgency. *Geography Compass, 4*(8), 1154–1166.

Massey, D. (2005). *For space*. London/Thousand Oaks: Sage.

Ministry of Home Affairs. (2011). *Census terms*. http://censusindia.gov.in/Data_Products/Library/Indian_perceptive_link/Census_Terms_link/censusterms.html

Ó Tuathail, G. (1996). *Critical geopolitics*. Minneapolis: University of Minnesota Press.
Rapoport, D. C. (2001). The fourth wave: September 11 in the history of terrorism. *Current History, 100*(650), 419–424.
Smith, N. (2003). *American empire: Roosevelt's geographer and the prelude to globalization*. Berkeley/Los Angeles: University of California Press.
START (National Consortium for the Study of Terrorism and Responses to Terrorism). (2011). Global terrorism database [Data file]. Retrieved from www.start.umd.edu/gtd
Stern, J. (2004). *Terror in the name of God: Why religious militants kill*. New York: Harper Perennial.
White, J. R. (2011). *Terrorism and homeland security*. Beverly: Wadsworth Publishing.

Chapter 181
A Content Analysis of Session-Opening Prayers in the U.S. Congress

Fred M. Shelley

181.1 Introduction

On July 28, 1787, Benjamin Franklin rose to address the Constitutional Convention in Philadelphia. At 81 years of age, Franklin was the oldest delegate to the Convention. Franklin expressed concern over the slow pace of deliberations. He observed that his fellow delegates were "groping as it were in the dark to find political truth, and scarce able to distinguish it when presented to us."

Franklin encouraged his fellow delegates to call upon the aid of the Almighty to guide their deliberations and recommended that each session of the Convention begin with a prayer. He said:

> I have lived, Sir, a long time and the longer I live, the more convincing proofs I see of this truth – that God governs in the affairs of men. And if a sparrow cannot fall to the ground without his notice, is it probable that an empire can rise without his aid? We have been assured, Sir, in the sacred writings that "except the Lord build they labor in vain that build it." I firmly believe this; and I also believe that without his concurring aid we shall succeed in this political building no better than the Builders of Babel.[1]

Franklin concluded his address by introducing a formal motion to begin each day's session with a prayer. He moved that "henceforth prayers imploring the assistance of Heaven, and its blessings on our deliberations, be held in this Assembly every morning before we proceed to business, and that one or more of the Clergy of this City be requested to officiate in that Service." Franklin's motion carried, and daily prayers were held at the beginning of each remaining session of the Convention.

---

[1] The original quotation in Franklin's handwriting is preserved by the Library of Congress and can be viewed at www.loc.gov/exhibits/religion/vc006642.jpg.

F.M. Shelley (✉)
Department of Geography and Environmental Sustainability, University of Oklahoma, Norman, OK 73019, USA
e-mail: fshelley@gcn.ou.edu

© Springer Science+Business Media Dordrecht 2015
S.D. Brunn (ed.), *The Changing World Religion Map*,
DOI 10.1007/978-94-017-9376-6_181

The outgrowth of these discussions among the delegates to the Constitutional Convention was, of course, the Constitution itself. Article I, Section 1 of the Constitution states that "All legislative Powers herein granted shall be vested in a Congress of the United States, which shall consist of a Senate and House of Representatives." Since Congress met first in 1789, daily sessions of each House of Congress have begun with prayers. On many occasions, individual Members of Congress have called on their colleagues to pray for, or to issue prayer proclamations, for particular individuals, groups of persons, or localities. Others have asked Congress to pass resolutions or legislation in support of public prayer for various individuals or groups of people.

What places are mentioned in these prayers and prayer requests? And what is the content of these prayers? The purpose of this paper is to shed light on these questions by examining prayers and prayer requests made on the floor of the Senate and the House of Representatives between 1989 and 2010.

181.2 "Clergy of this City" in Congress

In his motion to begin each session of the Constitutional Convention in Philadelphia with a prayer, Benjamin Franklin asked that "Clergy of this City be requested to officiate" at these prayers. Since the beginning of the First Congress in 1789, each House has maintained the office of Chaplain. Early Senate and House Chaplains were, as Franklin recommended, local members of the clergy. The first Senate Chaplain was the Rt. Rev. Samuel Provoost, the Episcopal Bishop of New York who was appointed when the First Congress met in New York in 1789. In the following year, Congress moved to Philadelphia and the Chaplain of the Senate was that city's Episcopal Bishop, the Rt. Rev. William White. After Congress moved to the new city of Washington, D. C. in 1800, local clergy continued to be selected as chaplains. In the early years of the U.S. government, Senate Chaplains served short terms but the average term of service increased after the Civil War. Forty three clergy members held this office between 1789 and 1865, but only 19 have served in this office since 1865.

The Senate Chaplain's position was a part-time office until the 1940s. At that time, the office of the Senate Chaplain became a full-time salaried position, with the Chaplain and the Chaplain's staff paid by Congressional funds. Since 1949, only five clergymen have served as the Senate Chaplain. Senate policy does not require the Senate Chaplain to be a Christian. However, all but one of the 52 Senate Chaplains have been Protestants.[2] The office has been dominated by Episcopalians with 16 Chaplains, Presbyterians with 13, and Methodists with 12. Of the others, five were Baptists, two were Unitarians, one was Lutheran, and one was Congregationalist.

---

[2]The official list includes 62 Senate Chaplains. However, eight served two non-consecutive terms and one, Rev. Henry Slicer, served three non-consecutive terms. Thus the office of Senate Chaplain has been held by 52 individuals.

The only Roman Catholic to have served as Senate Chaplain was Father Charles Constantine Pise, who served in 1832 and 1833. The current Chaplain, Rev. Barry C. Black, is a Seventh-Day Adventist and the first African-American to hold the position.

The majority of Senate chaplains were born in the Middle Atlantic states and only one was born west of the Mississippi River.[3] Maryland has produced the most Senate chaplains with ten, followed by Pennsylvania with eight and New Jersey and New York. Other states in which Senate chaplains were born included Vermont with four, Massachusetts with three, Kentucky with two, and Connecticut, Delaware, North Dakota, Ohio, Rhode Island, Virginia, and Wisconsin with one each. Eight were born outside the United States, with four from Ireland, three from England, and one from Scotland. The preponderance of Chaplains who were born along the Eastern Seaboard reflects the fact that the large majority of Chaplains served before the Civil War. Thus, they were born before large-scale Western expansion beginning in the early nineteenth century.

The House of Representatives also established the office of House Chaplain in 1789. The first Chaplain was a Presbyterian minister from New York City, Rev. William Linn. As with the Senate, early Chaplains of the House served short terms and the length of the House Chaplains' service increased after the Civil War. Fifty two members of the clergy have served as House Chaplains.[4] Of these, two were Roman Catholics and the other 50 were Protestants. The 50 Protestants included 16 Methodists, 15 Presbyterians, 7 Baptists, 4 Episcopalians, 2 Lutherans, 2 Unitarians, 1 Congregationalist, 1 Disciples of Christ, 1 Universalist, and 1 "Christian." As is the case with Senate Chaplains, all of the House Chaplains have been men. The current Chaplain of the House of Representatives, Father Patrick J. Conroy, is a Roman Catholic as was his immediate predecessor, Father Daniel P. Coughlin (Brudnick 2011). Six persons have served as chaplains of both Houses of Congress, although none served in both capacities simultaneously.

The Chaplains of the Senate and the House of Representatives have many duties. They provide pastoral care for Members of Congress and their families. They also supervise and maintain the Capitol Prayer Room, which is located near the Rotunda of the Capitol Building and is available for use by Members who desire silent prayer and contemplation. The Chaplains also meet American and foreign religious dignitaries who visit Congress. On occasion, they are called upon to officiate at funerals or memorial services for deceased Members.

The Chaplains are also responsible for the conduct of prayers said at the beginning of each meeting of their respective Houses of Congress. Usually, the Chaplains lead the prayers themselves. On occasion, however, the prayers are delivered by

---

[3] Information about the birthplaces of Senate Chaplains was obtained from Wikipedia through individual biographies linked to http://en.wikipedia.org/wiki/Chaplain_of_the_United_States_Senate.

[4] The official list includes 59 House Chaplains. However, four served two non-consecutive terms and two, Rev. William Henry Milburn and Rev. Thomas H. Stockton, served three non-consecutive terms. Thus the office of House Chaplain has been held by 52 individuals. Milburn was one of six persons who served as the Chaplain of both the Senate and the House.

guest chaplains. Over the years, these guest chaplains have included some women and have represented many different religious faiths and denominations, both Christian and non-Christian.

Generally, chaplains have been selected without regard to partisan politics. The last three Senate chaplains were nominated by bipartisan search committees whose recommendations were ratified by the entire Senate (Brudnick 2011). Politics have not entered the search for a Congressional chaplain since 1947, when the newly elected Republican-majority Senate replaced Rev. Frederick Brown Harris, who had served since 1942, with Rev. Peter Marshall. However, Marshall died in 1949 and the Senate reinstated Harris, who continued to serve until he retired in 1969.

The constitutionality of paying clergy members to open legislative sessions with prayers has been challenged through the Federal court system. As early as the 1850s, some Members of Congress objected to the selection of chaplains on the grounds that this practiced violated the principle of separation of church and state. The practice of selecting chaplains was abandoned in the late 1850s but was resumed during the Civil War (Brudnick 2011). In 1983, a Nebraska state legislator challenged this practice on the grounds that it violates the Establishment Clause of the First Amendment to the U.S. Constitution, which states that "Congress shall make no law respecting an establishment of religion." A Federal district court ruled that the practice of paying chaplains does violate the Establishment Clause. Noting that the same chaplain had served the Nebraska legislature for 16 years and that his prayers were published as part of the official records of the legislative sessions, an appellate court issued an injunction prohibiting Nebraska from continuing the practice.

In 1983, the U.S. Supreme Court reversed the appellate court's ruling in *Meyer v. Chambers* (463 U.S. 783). In doing so, the Court called attention to the long history of opening legislative sessions with prayers with the words "The opening of sessions of legislative and other deliberative public bodies with prayer is deeply embedded in the history and tradition of this country. From colonial times through the founding of the Republic and ever since, the practice of legislative prayer has coexisted with the principles of disestablishment and religious freedom" (p. 786). In fact, the Court's opinion pointed out that its own sessions begin with the words "God save the United States and this honorable Court." The Court concluded that "In light of the unambiguous and unbroken history of more than 200 years, there can be no doubt that the practice of opening legislative sessions with prayer has become part of the fabric of our society. To invoke Divine guidance on a public body entrusted with making the laws is not, in these circumstances, an "establishment" of religion or a step toward establishment; it is simply a tolerable acknowledgment of beliefs widely held among the people of this country" (p. 792).

181.3   Text of Prayers and Prayer Resolutions in Congress

Geographers and political scientists have obtained valuable information about the expressed views of prominent public officials and the impacts of these views on public policy through systematic examination of speeches, textual material, and other documents. For example, Ragsdale (1984) and O'Loughlin and Grant (1990) examined the politics and political geography of annual State of the Union addresses and other major speeches delivered by U.S. Presidents. Sutherland and Webster (1994) examined transcripts of the presidential debates held prior to the 1992 U.S. Presidential election to identify statements indicative of political geography. At the international level, Brunn (1999) undertook a content analysis of addresses made by ambassadors from small countries to the United Nations in order to interpret the views of these small states concerning their place in the contemporary world. This paper follows in this tradition by examining prayers and prayer-related debates in the United States Congress between 1991 and 2011.

Article I, Section 5 of the U.S. Constitution requires each House of Congress to keep a journal of its proceedings, and these records have been kept since the first session of the First Congress met in 1789. The *Congressional Record* has been published in its current form since 1873, and the daily prayers have been published as part of the *Congressional Record* since 1914. A searchable database consisting of each issue of the *Congressional Record* published since the beginning of the first session of the 101st Congress in 1991 is available for public use. This database can be accessed at http://thomas.loc.gov/home/LegislativeData.php?&n=Record&c=111. This database was examined systematically using the keywords "pray and prayer" in order to find the information used for the research presented in this paper.

181.4   Prayer and Place in the U.S. Congress

Using the words "pray" and "prayer" as keywords, the database was searched for the 101st through the 111th sessions of Congress meeting between 1989 and 2010. In part because the prayers are brief, they very seldom mention specific places. For example, the record for the 109th Congress, meeting in 2005 and 2006, contains exactly 2,000 entries containing the words "prayer" or "prayers." Of these entries, 914 or nearly half were to the daily prayers as published in the *Congressional Record*, including the introduction of the Chaplain or a guest chaplain by a Member of Congress.

Of the remaining 1,086 entries, many involved recognizing the passing of various individuals who had died during this period. In some cases, the Senator or Representative delivers the comment on the floor of his or her House of Congress during its formal session. More often, however, these comments are printed in the *Congressional Record* as part of its session entitled "Extensions of Remarks." Members can and often do "revise and extend" the comments that they have made more briefly on the floor. This is done after the day's session ends, but before the *Congressional Record* covering this session is published. As a result, Extensions of Remarks are generally longer and more detailed than the original comments made during the sessions themselves.

Some of those persons for whom prayers were delivered or advocated by Members of Congress were well-known public figures including Pope John Paul II, Chief Justice William Rehnquist, civil rights activist Rosa Parks, and former Secretary of

Defense Caspar Weinberger. Others recognized in this way were former Members of Congress including Senator Lloyd Bentsen of Texas and Representatives Robert Matsui of California, Shirley Chisholm of New York, and Carroll Campbell of South Carolina. Other Members of Congress recognized members of the Armed Forces from their districts who had been killed in action in Afghanistan or Iraq. For example, on May 18, 2005 Representative Mike McIntyre of North Carolina stated "Mr. Speaker, I rise today to pay tribute to Sgt. John "Mac" Smith of Wilmington, North Carolina, for serving his country valiantly with the 11th Armored Cavalry Regiment in Operation Iraqi Freedom. On May 11, 2005, Sgt. Smith lost his life when a roadside bomb hit his convoy. He was courageously serving his second tour of duty in Iraq, and our heartfelt thanks and prayers go out to his family and friends in this time of grief" (*Congressional Record,* House of Representatives, 109th Congress, 1st Session, May 19, 2005, p. E1009).[5] Still other Members recognized persons of local prominence in their districts who had passed away.

In some cases, Members of the Senate or the House called for prayers for the victims of natural disasters in the United States or abroad. During the 109th Congress, prayers were asked for the victims of the Indian Ocean tsunami of December 2004, the victims of Hurricane Katrina in 2005, the victims of a devastating earthquake in India and Pakistan in November 2005, and tornadoes which caused considerable damage in Tennessee also in November 2005. Most of these prayers and prayer request were initiated by individual Members of Congress, although the Chaplains themselves have called attention occasionally to foreign disasters. For example, 8 days after the earthquake that devastated Haiti on January 12, 2010, House Chaplain Coughlin delivered the following prayer:

> Lord God, Creator of heaven and Earth, Eternal Shepherd of the living and the dead, as a Nation we unite with other nations of the world and pray for our suffering brothers and sisters in the poverty-stricken and Earth-shaken nation of Haiti. Have mercy on us all.
>
> We beg You to help all the people of Haiti in all their needs. Come to the aid of the afflicted. Take pity on the helpless and the most vulnerable. Raise up the fallen as well as the ruins where human life and human remains may be still hidden. Restrain the wayward and sustain the brokenhearted. Bring compassion to those who mourn and eternal life to those who are buried in anonymity. (Congressional Record, House of Representatives, 111th Congress, 2nd Session, January 20, 2010, p. H195)

Similarly, in the first meeting of the Senate after the Indian Ocean tsunami, the following prayer was delivered by Senate Chaplain Black:

> Sovereign God, Creator and sustainer of us all, our hearts ache for the tsunami victims and for all touched by this tragedy. In our sadness, our eyes turn to You. We confess that we do not fully understand why bad things happen to good people. Nonetheless, by faith, we believe that You can carve tunnels of hope through mountains of despair.
>
> Lord, make us Your eyes, ears, feet, and hands to bring solace to those who suffer. Guide us as we seek to help the hurting and empower the global humanitarian effort. Comfort those who mourn, and strengthen the widows and orphans.

---

[5] Pages in the Congressional Record are numbered with the prefix S, H, or E representing the Senate, the House of Representatives, and extensions of remarks respectively. These pages are numbered within each session of Congress. Thus page H195 associated with the quotation refers to the 195th page of the House proceedings for the second session of the 110th Congress in 2010.

Today, bless our 109th Congress. Give our new Senators wisdom and courage as You order their steps. Help them trust You so completely that Your will may be done on Earth. (Congressional Record, Senate, 109th Congress, First Session, January 4, 2005, p. S1)

Some other requests for prayers were delivered by Members of Congress in response to accidents such as an airplane crash in Armenia in May 2006. Other Members asked for prayers for the victims of terrorist acts or other deliberate violent acts. For example, a few days after a terrorist attack in London in 2005, Senator Saxby Chambliss of Georgia stated as follows:

Mr. President, my wife Julianne and I express our deepest sympathies to those who lost loved ones and those injured in the terrorist attacks in London last Thursday. Our thoughts and prayers are with them.

The terrorists who claim allegiance to al-Qaida undertook these atrocious acts in response to the United Kingdom's unflinching, courageous support for the global war on terrorism. Prime Minister Tony Blair and the British people have stood along side [sic] the United States and the other members of the coalition in the war on terrorism.

This is a reminder that we must always be vigilant against those who wish to attack our freedom and our way of life. We must not waiver [sic] in our resolve to pursue and bring to justice those who commit these heinous crimes. [Congressional Record, Senate, 109th Congress, First Session, p. S8207]

Thus, the formal prayers offered by the Chaplains seldom mention specific places, but remarks made by individual Members often do mention them. In addition to the places associated with natural disasters and terrorist activities, foreign places mentioned specifically in calls for prayer in the 109th Congress included Darfur, Jerusalem, China, and the Philippines.

Individual members of Congress have often called upon their colleagues to promote prayers on behalf of the people of parts of the world outside the United States, and/or on behalf of American troops and diplomatic personal stationed in places of crisis. In some cases, in their comments these Members advocate the establishment of a National Day of Prayer on behalf of the people of a particular country. For example, in 1990 Representative Ileana Ros-Lehtinen of Florida called attention to a Worldwide Day of Prayer for Freedom and Democracy in Cuba on June 14, 1990. Representative Ros-Lehtinen's remarks were as follows:

Mr. Speaker, I would like to focus the attention of this legislative assembly to the efforts of a group of Cuban-Americans from my congressional district of Miami, FL, who are organizing a Worldwide Day of Prayer for Freedom and Democracy in Cuba to be observed on July 15, 1990.

This group of Cuban-Americans are a collection of lay people of different faiths who are united by two main ideas. First, the desire for absolute freedom and democracy for the island of Cuba; and, second, the fervent belief in the power of prayer to produce such change.

The objective of the group is to contact 50,000 congregations of different denominations around the world requesting their prayers to implore the Almighty for devine [sic] intervention for the bloodless liberation of Cuba. They hope to achieve increased international pressure against Cuba's Communist regime, as well as to deliver a message of good will to the people of Cuba from freedom-loving people of around the world which have the capability to deliver such a message. (Congressional Record, June 14, 1990, 101st Congress, Second Session, pp. E1981–E1982)

Prayer requests and prayer advocacy of this sort involves an identifiable group of people, with the prayer request involving divine intervention in order to achieve a specific objective. In this example, Representative Ros-Lehtinen called for prayers for the people of Cuba. The objective was freedom and democracy for the Cuban people and the overthrow of the Communist regime of Fidel Castro. In the last section of the remarks, she compared the situation in Cuba with that of countries in Eastern Europe, which at that time were in the process of overthrowing their Communist governments. Significantly, Representative Ros-Lehtinen represented (and continues to represent) a House district in Miami that contains a large population of Cuban-Americans, many of whom had left Cuba and moved to southern Florida in order to escape the Castro regime.

181.5 Prayer Resolutions in Congress: Introduction, Debate, and Resolution

The foregoing examples illustrate that prayer-related speeches and discussions in Congress can be divided into two categories. The daily prayers by the Senate and House Chaplains are direct prayers to the Almighty. Most ask for God's blessings on the Congress and His guidance to its Members as they debate and resolve the important issues of the day, and/or for God's blessings upon and guidance to the American public. In other cases, Members of Congress call upon Americans to pray for specific people or places. Some of these prayer requests are proposals to enact formal resolutions calling for national days of prayer.

A National Day of Prayer has been observed annually since 1952. The National Day of Prayer is held on the first Thursday in May each year, and is proclaimed by the President. On this day, Americans are exhorted to "turn to God in prayer and meditation." However, the National Day of Prayer does not identify specific prayers or subjects of prayers.

Prayer resolutions have involved people and nations throughout the world. Some have been controversial, while others were enacted with little or no opposition. Some resolutions were passed by Congress with unanimous or near-unanimous support. For example, on July 11, 2005, Republican Representative Christopher Smith of New Jersey introduced House Concurrent Resolution 333. The text of the resolution was that "That the House of Representatives (1) supports the goals and ideals of a National Weekend of Prayer and Reflection for Darfur, Sudan; (2) encourages the people of the United States to observe that weekend by praying for an end to the genocide and crimes against humanity and for lasting peace in Darfur, Sudan; and (3) urges all churches, synagogues, mosques, and religious institutions in the United States to consider the issue of Darfur."

In advocating House Concurrent Resolution 333, Representative Smith pointed out that the House and passed a resolution in the previous year declaring the situation in Darfur to be genocide. However, Smith called attention to continuing atrocities in Darfur. In his words:

> For its part, the U.N. Security Council has passed no less than six resolutions addressing the situation in Darfur. "Still, the crisis, the genocide, continues. Out of a pre-conflict population of 6.5 million, anywhere between 300,000 and 400,000 Darfurians have perished, and an estimated 10,000 continue to die each and every month. Over 2 million have been forced from their homes. Entire villages have been looted and destroyed, and countless men, women and children have been murdered, abducted, abused or raped" (Congressional Record, House of Representatives, 109th Congress, 1st Session, July 11, 2005, p. H5611). He continued "Mr. Speaker, I believe it is our Nation's richest and most important tradition that we turn to God in prayer, in good times and bad, to thank Him for His blessings and to ask and to petition Him for His help. I believe we delude ourselves if we think that we possess the wisdom or the courage or the skill to solve the myriad of vexing problems that we face. We needed God's help." (Congressional Record, House of Representatives, 109th Congress, 1st Session, July 15, 2005, p. H5611)

Several Members from both parties spoke in favor of House Concurrent Resolution 333, which was passed by the House by a vote of 364-2, with one Democrat and one Republican opposed.[6]

Many of the remarks made by Members of Congress in support of prayer for non-Americans have been entered into the *Congressional Record* with very little opposition, as in the case of House Concurrent Resolution 333 calling for prayers for the people of Darfur. However, in some cases resolutions calling for a national day of prayer have been debated vigorously. Much of this debate has involved the larger question of the role of religion in public life, especially in public schools.

Shortly after the September 11, 2001 terrorist attacks, a prayer resolution was offered by Representative Walter B. Jones, Jr., a Republican from North Carolina. In his proposed House Concurrent Resolution 239, Representative Jones called for "Expressing the sense of Congress that schools in the United States should set aside a sufficient period of time to allow children to pray for, or quietly reflect on behalf of, the Nation during this time of struggle against the forces of international terrorism." Resolutions of this sort express the sense of Congress, but they do not carry the force of law and they require a two-thirds majority to be passed.

Much of the debate involving House Concurrent Resolution 239 involved questions of constitutionality, associated especially with the Supreme Court's 1962 decision in *Engel* v. *Vitale* (370 U.S. 421), in which the Court ruled that official prayers in public schools are unconstitutional. In expressing his opposition to the resolution, Democrat George Miller of California said "I am afraid the resolution is really about, once again, trying to introduce some form of content or prayer into the schools under the guise of the tragedies of Sept. 11 and the events that have occurred since then. We should really not do that" (quoted by Marus 2001). In response,

---

[6] The only Republican who voted against House Concurrent Resolution 333 was Representative Ron Paul of Texas, who ran unsuccessfully for the Republican Party's nomination for President in 2008 and 2012. The only Democrat to vote in the negative was Representative Gary Ackerman of New York.

Representative Johnny Isakson, a Republican from Georgia, said "This resolution encourages and does not require the schools of America to set aside a sufficient period of time for children in America to pray for or reflect on our nation."

On November 15, the House passed House Concurrent Resolution 239 by a vote of 297 to 125.[7] Republicans in the House supported the motion by a margin of 212 to 3. The Democrats were divided more evenly, with 84 in favor, 121 opposed, and one voting "present." A month after the debate, Jones criticized Miller and two of his Democratic colleagues for their opposition to the Resolution. He also identified and criticized various organizations that had opposed House Concurrent Resolution 239. He singled out the National Parent-Teacher Association (PTA) and quoted its President as having said "[B]ecause the legislative intent is clearly to endorse religious expression, it does not conform with current constitutional standards" (*Congressional Record*, 107th Congress, 1st Session, December 13, 2001, p. H9751). In response, Jones said "Mr. Speaker, that is not what it did. What it said was that the children of America should have a moment of prayer or a moment of reflection. But, again, my point is, I am very disappointed in the National PTA, which is supposed to strengthen families, encourage education and encourage families to be together. Why they would take this type of position, I do not know. But, again, I was very surprised and disappointed that they would" (*Congressional Record*, 107th Congress, 1st Session, December 13, 2001, p. H9751).

In this discussion, Jones also expressed his support for restoring prayers to public school. He pointed out that the resolution did not require prayer, and also allowed for silent and possibly secular reflection and meditation. However, he also pointed out that the United States was founded on Judeo-Christian principles (Marus 2001). In concluding his defense of the Resolution, Jones said:

> Mr. Speaker, prior to 1962, we had prayer in this Nation. I think the children of this country, and since September 11, I think there have been more adults in the churches, the synagogues, the mosques, than there have been in a long, long time. Again, for these groups that are supposed to help educate our children like the National PTA, I was very disappointed that they would oppose a resolution that was only the sense of the Congress. When governors, when the President, when other leaders of State and local and national government are asking people to pray for America and to pray for our men and women in uniform, I just felt like I needed to come to the floor and say "thank you" to those who voted for this resolution on November 15. Again, it passed with 297, only 125 in opposition. They are the kind of messages, Mr. Speaker, in my opinion, we need to be sending to the American people, because every survey I have seen over the last 2 years, better than 70 % of the American people, say they would like to see prayer returned to the school systems of America. (Congressional Record, 107th Congress, 1st Session, December 13, 2001, p. H9754)

Thus Representative Jones framed the question of prayer in schools for the victims of the September 11, 2001 terrorist attacks on the United States in terms of the larger question of school prayer. On the other hand, Representative Miller argued that the resolution would lead toward reinstituting official prayer in public schools.

---

[7] House of Representatives resolutions do not carry the force of law. House rules require that resolutions, as opposed to pieces of legislation, be passed by a two-thirds majority of those Representatives present and voting.

The framing of the resolution in this manner may have contributed to the substantial partisan division of the House on this question, with Republicans nearly unanimous in favor of the resolution and the Democrats divided. On many other occasions, Members of Congress arguing in support of the National Day of Prayer, and more specific prayers, have also brought up support for overturning *Engel* v. *Vitale*. For example, in 2010 Representative Mike Pence of Indiana made a brief speech in support of that year's National Day of Prayer. Representative Pence said "Sadly, voluntary prayer has been under attack of late. It has been driven from our public schools and from our graduation ceremonies by activist courts. That ruling ignored our history, our traditions, and it should be overturned" (*Congressional Record,* House of Representatives, 111th Congress, 2nd Session, May 6, 2010, p. H3206).

Why were Americans asked to pray for these places? Why did Members of Congress introduce or support resolutions calling officially for these prayers? In all three examples quoted above, (Cuba, Darfur, and the September 11 terrorist attacks), prayers by Americans were requested in response to atrocities associated with these places respectively, including atrocities and violations of human rights in Cuba and Darfur. But levels of Congressional support were influenced in some cases by larger questions involving the role of religion in American public life.

A third type of prayer-related resolution involves the use of prayer in public buildings and memorials. On January 24, 2012, Republican Representative Bill Johnson of Ohio introduced the World War II Memorial Prayer Resolution, which became House Resolution 2070. House Resolution 2070 directed the Secretary of the Interior to "install in the area of the World War II Memorial in the District of Columbia a suitable plaque or an inscription with the words that President Franklin D. Roosevelt prayed with the nation on June 6, 1944, the morning of D-Day." President Roosevelt's D-Day prayer, which has been referred to as the "Let Our Hearts Be Stout" prayer, read in part:

> [I]n this poignant hour, I ask you to join with me in prayer: Almighty God: Our sons, pride of our Nation, this day have set upon a mighty endeavor, a struggle to preserve our Republic, our religion, and our civilization, and to set free a suffering humanity.
>
> Lead them straight and true; give strength to their arms, stoutness to their hearts, steadfastness in their faith.
>
> They will need Thy blessings. Their road will be long and hard. For the enemy is strong. He may hurl back our forces. Success may not come with rushing speed, but we shall return again and again; and we know that by Thy grace, and by the righteousness of our cause, our sons will triumph.

Representative Johnson noted that the official position of the Administration and of the Department of the Interior was in opposition to adding this prayer to the World War II Memorial, although Secretary of the Interior Ken Salazar had testified that he was personally in support of the idea. Despite opposition from the Administration, the resolution was passed by the House by a vote of 386-26, with 21 Members abstaining. All 233 Republicans voting voted for the resolution, with 8 abstaining. Democrats supported the resolution by a margin of 153-26, with 13 abstaining. As in the case of the 2001 prayer resolution, opposition among some Democrats was based on concerns about the role of religion in public life.

181.6 Conclusion

Prayer has played a significant role in the United States Congress since the United States Constitution went into effect in 1789. Through official prayers and through exhorting Americans to pray, Members of Congress have often discussed and in some cases have enacted formal resolutions calling upon God to guide Americans at times of crisis. Many of these times of crisis have involved consideration of the role of the United States in the world, and its relationships with other countries. In general, these prayers and prayer requests have generated little or no formal opposition. However, the role of religion in American life remains a matter of considerable controversy. Thus some Members of Congress have opposed prayer resolutions on the grounds that they imply, or would leave the door open, for an increased role of religion in the American polity that they regard as unwarranted and perhaps unconstitutional. This opposition may reflect the important larger question of how Americans have reacted, favorably and unfavorably, to an increased secularization of American public life relative to the past.

Prayers in American public life are a rich source of further research. As O'Loughlin and Grant (1990) and Sutherland and Webster (1994) reported in their analysis, maps of places mentioned in requests for prayers could provide a useful source of information for future analysis, especially if these maps are prepared sequentially and can be compared over time and linked to the changing importance of varying places in global affairs.

In addition, further analysis could be devoted to the tabulation and mapping of support and opposition to various prayer resolutions by state or district. Is there a consistent pattern of geographical as well as partisan opposition to proposals that can be interpreted as unwarranted expression of religion in public life? Issues such as prayer in public schools, the celebration of Christian holidays, the placement of the Ten Commandments and other religious inscriptions on public buildings, and other such topics remain controversial. The analysis presented in this paper represents a valuable starting point for such future research.

References

Brudnick, I. A. (2011, May 26). House and Senate chaplains: An overview. Congressional Research Service. http://chaplain.house.gov/chaplaincy/ChaplainHistoryCRS.pdf

Brunn, S. D. (1999). A content analysis of 1995 U.N. speeches. *Geopolitics, 4*, 17–33.

Marus, R. (2001, November 26). House resolution sparks new debate on school prayer. *Baptist Standard.* www.baptiststandard.com/2001/11_26/pages/resolution.html

O'Loughlin, J., & Grant, R. (1990). The political geography of presidential speeches, 1946–1987. *Annals, Association of American Geographers, 80*, 504–530.

Ragsdale, L. (1984). The politics of Presidential speechmaking, 1949–1980. *American Political Science Review, 78*, 971–984.

Sutherland, C. L., & Webster, G. R. (1994). The geography of the 1992 presidential debates. *The Geographical Bulletin, 36*(2). Available at http://www.gammathetaupsilon.org/the-geographical-bulletin/1990s/volume36-2/article3.pdf

Chapter 182
Political Pilgrimages: American Presidents and Religious Communities, 1933–2012

Kevin Coe, David Domke, and Anthony Schmidt

182.1 Introduction

In 1933, Franklin Roosevelt began what would turn out to be a lengthy stay in the White House. Elected four times, Roosevelt served more than 12 full years before dying in 1945. During this period, on only four occasions did Roosevelt leave the White House to give a speech to a distinctly religious audience or at a site imbued with religious meaning. More than 60 years later, in 2009, George W. Bush finished his second term in the White House. Over the course of his 8 years in office, Bush had made such religious-cum-political trips *79 times*. The striking increase in these moments of presidential bread-breaking—what we call political pilgrimages—underscores the perceived value of such events to those who engage in them, as well as the rapidly changing relationship between religion and politics in America.

With an eye toward better understanding this relationship, this chapter examines in detail the political pilgrimages that have taken place in the modern U.S. presidency. Building on a concept we introduced in our book, *The God Strategy: How Religion Became a Political Weapon in America* (Domke and Coe 2010), we develop a conceptual framework for understanding political pilgrimages and their importance in American politics. We then track the contours of these pilgrimages across 13 presidencies over 8 decades. Three specific questions guide our analysis: (1) To what extent

K. Coe (✉)
Department of Communication, University of Utah, Salt Lake City, UT 84112, USA
e-mail: kevin.coe@utah.edu

D. Domke
Department of Communication, University of Washington, Seattle, WA 98195, USA
e-mail: domke@uw.edu

A. Schmidt
Department of Communication Studies, Edmonds Community College, Edmonds, WA 98036, USA
e-mail: tony.schmidt@email.edcc.edu

© Springer Science+Business Media Dordrecht 2015
S.D. Brunn (ed.), *The Changing World Religion Map*,
DOI 10.1007/978-94-017-9376-6_182

do presidents engage in pilgrimages? (2) Where do presidents go when undertaking pilgrimages? (3) How do presidents talk about God and faith when undertaking pilgrimages? Answering these questions is an important step in forwarding scholarship on the role that politicians play in American civil religion (for example, Bellah 1967; Marty 1974; Pierard and Linder 1988; Roof 2009), on presidents' religious rhetoric (for example, Chapp 2012; Domke 2004; Hart 2005; Shogan 2006), and on the geography of presidential travel and discourse (see Brunn et al. 2011; Coe and Neumann 2011; Doherty 2007; Ellis 2008; O'Loughlin and Grant 1990).

182.2 Political Pilgrimages: A Conceptual Framework

The idea of a pilgrimage is deeply rooted in the American experience. Families take vacations to national parks, seashores, and military battlefields. Sports fans visit mythic stadiums, while others might visit famous birthplaces or homes. In all instances, the symbolism of the site transcends the location itself. As Campo (1998) explains in his study of "American pilgrimage landscapes," by traveling to these iconic places Americans insert themselves into broader "national narratives" that function as connective cultural tissue (see also Chidester and Linenthal 1995). The importance of such pilgrimages is not lost on politicians. For officeholders or candidates interested in signaling support for a key constituency, a powerful means of doing so is to visit people or places that have elevated significance to the group. By taking the time and trouble to visit a specific group at a specific site, politicians make it clear with whom they wish to be identified—and the targeted groups appreciate this show of solidarity.[1]

Such pilgrimages have particular importance when they are religious in nature; that is, when a politician travels to deliver a speech at a religious institution or before a distinctly religious audience. In these journeys, political leaders insert themselves into religious narratives in subtle yet understood ways. Indeed, the very idea of a "pilgrimage" conjures up an image of the faithful trekking across the globe to a sacred location (see Eade and Sallnow 1991; Turner and Turner 1978). This section of the chapter develops a framework for understanding political pilgrimages and their meaning for America's religious politics, focusing on the context of the U.S. presidency. Our thinking is based on three precepts that we consider in turn: (1) Presidential travel is a powerful symbolic activity; (2) Presidents are crucial to the construction and maintenance of American civil religion; and (3) Presidents are strategic actors and are, therefore, deeply concerned about public perceptions.

---

[1] For example, when George W. Bush was campaigning for the presidency in 2000, he visited Bob Jones University, one of the nation's most conservative religious institutions which, at the time, prohibited interracial dating. As Bush took some criticism for the visit, Ralph Reed, former director of the Christian Coalition, highlighted the visit's religious importance: "I don't think it was an endorsement of racial exclusion, it was an endorsement of the inclusion of people of devout faith" (see Ainsworth 2000).

Travel by high-level politicians is nearly always electorally significant (Doherty 2007; Shaw and Gimpel 2012); when the politician in question is the president, travel also takes on a very important symbolic dimension (Hart 1987; Ragsdale 1996). Ellis (2008), in his extensive historical analysis of presidential travel, concludes that the necessary pomp and circumstance accompanying modern presidential travel symbolically reinforces the notion of a "regal presidency." In other words, when a president shows up somewhere, people notice. But much of the symbolic power of presidential travel is derived from exactly *who* notices. When a president gives a major speech to the entire nation, he is "broadcasting" his views to the mass public. More is needed, though, for a constituency to feel a deep connection with a politician or a political party. All groups want to feel special, to sense that they are more than a cog in an electoral machine. Presidential travel delivers this feeling because it is a kind of "narrowcasting." When political leaders narrowcast, they put forth constituency-targeted actions and words that are public but fly below the radar of most Americans (see Jacobs 2005; Jacobs and Shapiro 2000; Overby and Barth 2006; Wattenberg 2004). Narrowcasting, in contrast to broadcasting, is an effective political strategy because it creates a sense of *relationship* between sender and receiver—an especially meaningful feeling for many in America's large Christian communities, who often speak of a "personal relationship" with God, and who make "fellowship" activities with other people of faith a central part of their lives.[2] Thus a political pilgrimage to a religious site is a very concrete way for presidents and other politicians to demonstrate an understanding and appreciation of religious voters' key concerns and values, thereby practicing what scholars have called "the social embodiment of religious beliefs" (Kellstedt et al. 1996).

This manner of presidential religious engagement plays a substantial role in sustaining and shaping what Bellah (1967) identified as America's "civil religion." Bellah, borrowing a phrase from Jean-Jacques Rousseau and building from a theoretical foundation developed by Alexis de Tocqueville, Emile Durkheim, and others, conceived of civil religion as "a set of beliefs, symbols, and rituals" through which a society "interprets its historical experience in light of transcendent reality" (Bellah 1967: 4, 1975: 3; see also Marty 1974; Pierard and Linder 1988). American presidents are expected to be the primary overseers of this civil religion. They fulfill this role every time they make a political pilgrimage because their presence at a religious site signals a bond between God and country. What presidents say while they are engaged in such pilgrimages also contributes to this bond. Presidents regularly invoke broad religious themes

---

[2] Consider a few examples. The Southern Baptist Convention's web site at www.sbc.net has a section that addresses the issue of "How to become a Christian." The answer begins: "You're not here by accident. Jesus loves you, and He wants you to have a personal relationship with Him." The National Association of Evangelicals, meanwhile, "promotes fellowship, cooperation, networking, and dialogue as means of evangelical witness." This according to their web site at www.nae.net. In the words of Black (2004), a professor at Wheaton College, "[O]ne of the things that makes an evangelical faith look different, perhaps, than others, [is] there's an emphasis on....a personal relationship with Jesus Christ."

K. Coe et al.

to build and reinforce a sense of national community, thereby functioning as "the high priest of the national faith" (Hart 2005: 34). These presidential expressions of religiosity have important consequences for public perceptions. In an extensive empirical study drawing on both experimental and survey data, Chapp (2012: 16) found that invocations of civil religion by presidential candidates reliably influenced public attitudes, concluding that "religious rhetoric is a central force responsible [for] shaping the contours of American political culture. Religious rhetoric is also electorally consequential and culturally significant, with important implications for how we interpret American political representation" (see also Calfano and Djupe 2009). Clearly, where presidents go, and what they say while there, matters for America's religious politics.

Given the stakes involved, it is no surprise that political pilgrimages are highly strategic. Presidents and their advisors devote considerable attention to determining the appropriate site and message for each presidential pilgrimage. This is *not* to say that presidential displays of faith are inauthentic. They may be perfectly authentic—and we have no reason to doubt the sincerity of any president's faith—but must nonetheless be expressed in the context of a broader political calculus. Consider one example: political pilgrimages to the state of Indiana have been popular among presidents. There have been 11 over the past 8 decades, placing Indiana among the 5 most-visited states. This may seem surprising, given that Indiana is neither close to Washington, DC nor especially populous. It becomes less surprising, however, when one notes that Indiana houses the University of Notre Dame—the specific site of every one of the 11 presidential pilgrimages to Indiana. For religious symbolism in America, especially for Catholics, it is hard to top this location. A speech delivered there is simply different than the same speech delivered anywhere else in the United States and, of course, presidents understand this. Going to Notre Dame is, therefore, a specific and valuable political strategy. Thus both *where* presidents go and *what they say* are carefully planned and deeply consequential. With this in mind, we turn our attention to mapping the contours of the political pilgrimages that presidents have undertaken over the past eight decades.

182.3 Method

Our analysis focuses on the modern presidency, which most scholars define as beginning with the administration of Franklin Roosevelt (for example, Greenstein 2004; Leuchtenburg 1988). During Roosevelt's lengthy term the United States and especially the presidency changed significantly. Beginning with Roosevelt will, therefore, allow multiple presidents to be included in the analysis while still providing a fair amount of consistency in the cultural position of these presidents. We examined all public remarks about religion made by presidents over the past eight decades, from Roosevelt's inauguration in 1933 to the end of Barack Obama's third full year in office, January 19, 2012. Via extensive manual and computer searches of the National Archives' *Public Papers of the Presidents* (the definitive record of presidents' public communications; available online at www.americanpresidency.org), we identified every instance in which a president left the White House to deliver a speech at a religious location or to a distinctly religious audience. Four types of pilgrimage became apparent in this analysis. First, presidents during foreign trips occasionally met with religious leaders or spoke at religious sites. Second, on U.S. soil presidents delivered public addresses at religiously symbolic schools, often at commencement or convocation ceremonies. Third, presidents spoke in churches in a range of contexts, including stump speeches, sermons, and funeral services. Fourth, presidents met in formalized settings with religious leaders and groups to discuss policy ideas, social developments, and broader goals. In all, this collection procedure yielded 373 speeches.[3]

We recorded several characteristics about these speeches, most notably the location (abroad or, for domestic speeches, the state in which the speech was delivered). In 11 cases, the location where the speech was given could not be determined. For analyses that required location, these cases were excluded. We then used the computer content-analysis program TextQuest (www.textquest.de) to track two themes in these speeches: *invocations of God* and *invocations of faith*. The first of these themes tracked specific mentions of a higher power, such as God, Jesus, Lord, Divine Providence, and the like. The second tracked mentions of broader terms that signal a commitment to religious principles, such as faith, angel, heaven, pray, and worship. Our measurement of these concepts follows Domke and Coe (2010), except that our measure of God is more conservative. This is because we rely here on computer-assisted content analysis rather than manual coding. This makes it difficult to capture personal references to God, such as "He" and "His." Our measure of God invocations is therefore based on just the following terms and phrases (with all relevant variants also included): Almighty, Christ, Creator, Dios, Divine Providence, the Divine, Eternal Father, Father and Preserver, God, Good Shepherd, Jesus, Lord, Messiah, Prince of Peace, Redeemer, Savior, Son of God, Supreme Being. A list of the several dozen terms included in our analysis of faith invocations is available in Appendix B at www.thegodstrategy.com. Table 182.1 provides a selection of presidents' invocations of God and faith.

---

[3] Three additional notes should be made about the sample. First, when presidents made comments to the press after attending their home churches, we did not consider this to be a pilgrimage. This was an uncommon occurrence among presidential remarks. Second, we focused on Christian-themed pilgrimages because these constituted the vast majority of pilgrimages and because of the political importance of Christian conservatives in recent decades (see Domke and Coe 2010). Finally, we included as pilgrimages the few cases where a president made live telephone or satellite remarks at a religious event.

Table 182.1 Selected presidential invocations of God and faith (Source: Kevin Coe, David Domke and Anthony Schmidt)

President	Invocation
Roosevelt	I am very sure that the spirit in which we are approaching those difficult tasks…are going to be exemplified in the lives of all the people calling themselves Christians who believe in God and uphold the works of the Church. (Hyde Park Methodist Episcopal Church, Hyde Park, NY, 9/29/1933)
Truman	Men can build a good society, if they follow the will of the Lord. Our great Nation was founded on this faith. (Gonzaga University, Spokane, WA, 5/11/1950)
Eisenhower	We must remember the spiritual base that underlies man's existence, and the spiritual base that underlies all free government; else we shall surely fail. (World Christian Endeavor Convention, Washington, DC, 7/25/1954)
Kennedy	I think this country will continue its commitments to support the world of freedom, for as we discharge that commitment we are heeding the command which Brigham Young heard from the Lord more than a century ago. (Mormon Tabernacle, Salt Lake City, UT, 9/26/1963)
Johnson	I find for myself … a sustaining strength from the moments of prayer, whether we assemble together or whether we pray silently alone. (Presidential Prayer Breakfast, Washington, DC, 2/4/1965)
Nixon	I can also tell you America would not be what it is today … if this were not a nation which has made progress under God. (Billy Graham's East Tennessee Crusade, Knoxville, TN, 5/28/1970)
Ford	Although our religious denominations are different, I have long admired the missionary spirit of Baptists and the fact that you strive to keep the Bible at the center of your lives. (Southern Baptist Convention, Norfolk, VA, 6/15/1976)
Carter	There's no incompatibility between the constraints and the shackles on our lives by standards prescribed by God on the one hand, and the ultimate freedom that can come when the spirit of the Lord is present. (National Prayer Breakfast, Washington, DC, 1/18/1979)
Reagan	Think of it: the most awesome military machine in history, but it is no match for that one, single man, hero, strong yet tender, Prince of Peace. His name alone, Jesus, can lift our hearts, soothe our sorrows, heal our wounds, and drive away our fears. (National Religious Broadcasters, Washington, DC, 1/31/1983)
G.H.W. Bush	We are one Nation under God. We must remember that. We must advocate that. We must continue to state that we are one Nation under God. (Mount Zion Missionary Baptist Church, Los Angeles, CA, 5/7/1992)
Clinton	We have a wonderful array of people of faith here. We have Christians who are Catholic and Protestant; we have American Jews here; we have American Muslims here. (United Michigan Clergy, Detroit, MI, 10/21/1996)
G.W. Bush	I'm grateful for your prayers…Thanks for having me today. May God bless your organization, and may God continue to bless our great country. (Knights of Columbus, Dallas, TX, 8/3/2004)
Obama	As I look out at this audience, I'm reminded of the power of faith in America; faith in God, and a faith in the promise of this great country. (National Hispanic Prayer Breakfast, Washington, DC, 6/19/2009)

182.4   Results

Our first question dealt with the extent to which modern presidents have undertaken political pilgrimages.[4] Since 1933, presidents have made 373 pilgrimages, averaging nearly five per year. Every president since Franklin Roosevelt has undertaken at least a few political pilgrimages, but presidents have varied considerably in the extent to which they have done so. Figure 182.1 reveals this variation by showing how many pilgrimages each president made during his average year in office. It is clear that there has been a steady and marked increase in the number of pilgrimages taking place. Roosevelt averaged not even half a pilgrimage per year, whereas George W. Bush averaged nearly 10. The rise throughout time has been quite steady, excepting a dramatic spike during Gerald Ford's brief presidency.[5] Importantly, although some of this rise is to be expected given a general increase in the amount of presidential travel over time, it is nonetheless clear that some presidents have particularly focused their energy on presidential pilgrimages.

Consider, for example, that Ronald Reagan traveled to give general public remarks about 25 % less than did his immediate predecessor, Jimmy Carter (see Doherty 2007). Yet Reagan's number of political pilgrimages was about 20 % *higher* than Carter's (see Fig. 182.1). In other words, relative to Carter, Reagan focused a

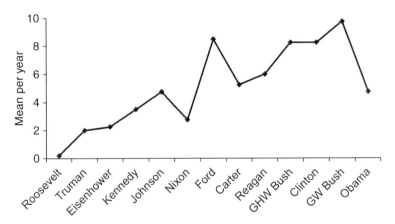

Fig. 182.1 Presidential pilgrimages per year, 1933–2011 (Source: Kevin Coe, David Domke and Anthony Schmidt)

---

[4] Throughout the results section we report no statistical tests. We are working with the census of political pilgrimages that have taken place in the modern presidency, so inferential tests are unnecessary.

[5] It is not entirely clear why Ford engaged in so many pilgrimages, but it may have to do with the fact that his reelection campaign pitted him against Jimmy Carter, whose religious faith was on full display during the campaign. This possibility is supported by the fact that more than half of Ford's pilgrimages occurred in the 10 months leading up to the 1976 election.

large proportion of his presidential travel on visiting religious audiences and sites. Nowhere is this kind of variation in average pilgrimages per year more striking than in the case of Barack Obama. Obama is the only president since Gerald Ford to make fewer than 5 pilgrimages in the average year, and one of only two (along with Nixon and Carter) to decrease his pilgrimages from the standard set by his immediate predecessor. What is remarkable is not just that Obama broke from the standard set by George W. Bush, but how dramatically he did so. Obama engaged in pilgrimages *not even half as often* as Bush.[6] It is hard to imagine a more striking departure from presidential norms. Clearly, presidents vary in their eagerness to embrace the religious faithful via political pilgrimages.

Our second question focused on where presidents went on their religious pilgrimages. Far and away the most common type of pilgrimage was to speak to a religious group or organization (53.6 % were of this kind), with visits to churches (20.1 %) and schools (19.8 %) somewhat less common. Pilgrimages abroad were rare (6.4 %), with just 24 total since 1933—nearly all of which took place in the past three decades (and more than a third of which were visits to the Vatican to meet with the Pope). That the majority of pilgrimages have been to religious meetings as opposed to churches or schools reflects two things. First, presidents since John Kennedy in 1961 have spoken annually at an ecumenical prayer breakfast in Washington, DC, making this religious meeting the most consistent and familiar political pilgrimage that exists in America today. Second, the increased political mobilization of religious organizations in recent years has given presidents—especially Republican ones—a wide range of potential hosts for pilgrimages. For example, the National Religious Broadcasters Association and the National Association of Evangelicals are umbrella organizations that represent millions of Americans who lean conservative. Since Ronald Reagan took office in 1981, presidents have traveled to address these organizations 14 times (and in every case, it was a Republican president making the trip). Democrats have found sites for pilgrimages (Bill Clinton was a regular in churches, for example), but have had fewer options than their Republican counterparts during the past few decades.

Domestic pilgrimages—those to meetings, schools, and churches—were quite limited geographically. Table 182.2 reveals this by reporting the frequency of presidents' domestic pilgrimages since 1933. Consistent with what is typical in other forms of presidential travel (see Doherty 2007), presidents prefer that their pilgrimages take place close to home. Indeed, roughly half of all domestic pilgrimages over the past eight decades took place in Washington, DC, Virginia, and Maryland, with DC in particular dominating the geography of presidential pilgrimages. DC is so dominant largely because many of the most consistent pilgrimage events—including the national prayer breakfast and other prayer breakfasts that are more recent additions (e.g., the national Hispanic prayer breakfast, the national Catholic prayer

---

[6] Notably, Obama does not have a full election year in this dataset, which could depress his overall number of pilgrimages (given that some presidents increase their pilgrimages during election years). However, even if Obama in his election year doubled his average, he would still not reach Bush's yearly average.

Table 182.2 Locations of presidents' domestic pilgrimages

Location	Count	%
District of Columbia	152	45.0
New York	26	7.7
Texas	20	5.9
Illinois	11	3.3
Indiana	11	3.3
Virginia	9	2.7
Pennsylvania	9	2.7
Maryland	8	2.4
Ohio	8	2.4
California	7	2.1
Utah	7	2.1
Georgia	7	2.1
Massachusetts	7	2.1
New Jersey	7	2.1
Louisiana	6	1.8
Missouri	6	1.8
Florida	5	1.5
Tennessee	5	1.5
Michigan	5	1.5
Colorado	4	1.2
Arkansas	3	.9
Washington	2	.6
North Carolina	2	.6
Kansas	2	.6
Wisconsin	2	.6
Hawaii	1	.3
Oregon	1	.3
Mississippi	1	.3
Oklahoma	1	.3
South Carolina	1	.3
Minnesota	1	.3
Connecticut	1	.3
Total	338	100

Source: Kevin Coe, David Domke and Anthony Schmidt

breakfast)—take place there. Additionally, several major religious organizations and schools that presidents like to address, including the National Association of Evangelicals and Georgetown University, are located in DC. Once presidents get beyond DC, only a few states are regularly targeted: Maryland and Virginia for the sake of proximity, New York, Texas, Illinois, Pennsylvania, and Ohio because of their large cities. And, as discussed, Indiana is a draw because of Notre Dame. Outside of a few regular locations, however, presidents are quite selective: In eight decades, more than half of the states in the union have been visited just once or not

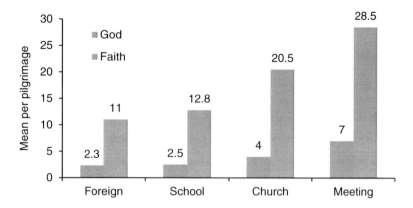

Fig. 182.2 Invocations of God and faith by context (Source: Kevin Coe, David Domke and Anthony Schmidt)

at all. This reality underscores why political pilgrimages matter so much to the local populations who host them.

Our final question focused on the degree to which presidents invoke God and religious faith when making their political pilgrimages. We considered three possible types of variation in such rhetoric: variation across speaking context, regional variation in pilgrimages to different U.S. locales, and variation among individual presidents. Figure 182.2 provides insight into how context matters, showing presidential invocations of God and faith in pilgrimages abroad, to schools, to churches, and at meetings with religious audiences or organizations. It is evident that presidents elevate their religious rhetoric when speaking in churches, and then elevate it even more so when speaking at religious meetings. Total religious language is more than 60 % higher in churches than it is when presidents speak abroad or in schools, and it increases another 45 % over that already-elevated level when they speak at religious meetings.[7] Variation by region is much more subdued.

Figure 182.3 shows presidents' religious rhetoric in each of the census bureau's traditional regional divisions (with Washington, DC listed separately).[8] DC and the South lead the way in religious rhetoric, but the differences are mostly small. Further, the marginal difference between the South and some other regions does not

---

[7] Importantly, these trends could be sensitive to the average word count in speeches across different contexts. We chose not to control for word count in these analyses, however, because the decision a president makes about how much to say during a pilgrimage is also a strategic and important one. It turns out in these data that average word count is higher in school settings (3,081), where religious rhetoric is less used, than it is in churches (2,082) and at meetings (2,010). Foreign addresses were typically shorter (629).

[8] These divisions are as follows: Northeast (Maine, New Hampshire, Vermont, Massachusetts, Rhode Island, Connecticut, New York, Pennsylvania, New Jersey); Midwest (Wisconsin, Michigan, Illinois, Indiana, Ohio, Missouri, North Dakota, South Dakota, Nebraska, Kansas, Minnesota, Iowa); South (Delaware, Maryland, District of Columbia, Virginia, West Virginia, North Carolina, South Carolina, Georgia, Florida, Kentucky, Tennessee, Mississippi, Alabama, Oklahoma, Texas, Arkansas, Louisiana); West (Idaho, Montana, Wyoming, Nevada, Utah, Colorado, Arizona, New Mexico, Alaska, Washington, Oregon, California, Hawaii)

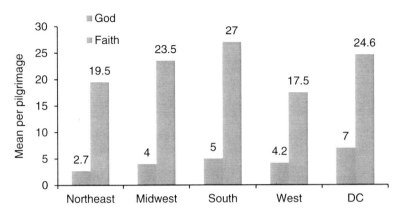

Fig. 182.3 Invocations of God and faith by region (Source: Kevin Coe, David Domke and Anthony Schmidt)

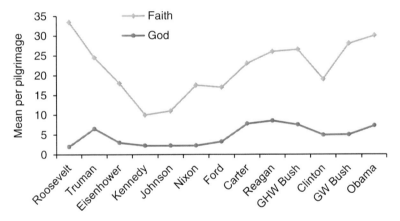

Fig. 182.4 Invocations of God and faith by president (Source: Kevin Coe, David Domke and Anthony Schmidt)

appear to be caused by presidents targeting states with a higher percentage of church-going Christians, because limiting the analysis to just "Bible-belt" states does not strengthen the trend (nor are most of these states consistently higher than others when considered individually). It appears, then, that presidents do elevate their religious rhetoric to fit the likely comfort level of their audience, but that they find context (as opposed to region) a better guide in making this determination.

Examining presidents individually shows further variation in the degree to which they invoke God and faith during their pilgrimages. Figure 182.4 shows each president's average number of invocations of God and faith per pilgrimage. Three trends are noteworthy. First, there is a somewhat low period—from Dwight Eisenhower to Gerald Ford—when presidents were using relatively less religious rhetoric than they were before or since. The combined God and faith average during that period (17.05) is only slightly more than half the presidential average before (32.88) and since (32.22). The recent high period, which begins with Carter, corresponds to the increased mobilization of Christian conservatives in American politics. Second, party does not appear to be a strong determinant of this rhetoric. During the recent high period, the three Democrats have averaged 30.66 mentions per pilgrimage; the three Republicans have averaged a very similar 33.78. Finally, Barack Obama is once again an interesting case. We saw above that he had broken with recent presidential history by sharply decreasing his presidential pilgrimages. But in terms of what he says when he is there, Barack Obama is the most religiously out-spoken president in modern history.

Consider a few examples of Obama's invocations of God and faith. Speaking at his first National Prayer Breakfast, Obama said: "We come to break bread and to give thanks, but most of all to seek guidance and to rededicate ourselves to the mission of love and service that lies at the heart of all humanity. St. Augustine once said: 'Pray as though everything depend on God, then work as though everything depended on you'" (Washington, DC, 2/5/09). A year later, during a church service honoring Dr. Martin Luther King, Jr., Obama greeted the audience with "Good morning. Praise be to God," and later during the same speech said: "So let us hold fast to that faith, as Joshua held fast to the faith of his fathers, and together, we shall overcome the challenges of a new age. Together, we shall seize the promise of this moment. Together, we shall make a way through the winter, and we're going to welcome the spring. Through God all things are possible" (Washington, DC, 1/17/10). And in 2011, speaking at Trinity College in Dublin, Ireland, Obama asked that "God bless the eternal friendship between our two great nations" (5/23/2011). Clearly, both undertaking a pilgrimage and invoking religious language while there are signals of support for the religious faithful. Obama has been much more likely to send the latter signal than the former.

182.5 Discussion

This chapter focused on an aspect of America's religious politics that had received only scant attention in the past, but one that is clearly an important and growing part of American political culture. Several points warrant discussion.

First, the increasing presence of political pilgrimages has made them—for the last few decades, at least—a remarkably common feature of presidential politics. Since Gerald Ford took office in 1974, it has been more likely in the average month to see a president making a pilgrimage than to not. There is little doubt that presidents see value in undertaking these pilgrimages—and that the audiences are usually grateful for the attention. For example, in 2003 George W. Bush visited the National Religious Broadcasters Association to speak at its annual convention. The association was so pleased with this attention that they issued a resolution praising the president. As a tax-exempt organization, the NRB had to stop short of explicitly endorsing Bush's reelection—but only barely. The resolution concluded this way: "We recognize in all of the above that God has appointed President George W. Bush to leadership at this critical period in our nation's history, and give Him thanks" (see Blake 2003). But there are also political risks associated with too brazenly mixing religion and politics. As the 2012 election cycle got into full swing, a plurality of Americans (38 %) reported feeling that there was "too much expression of religious faith and prayer from political leaders" in America—a number that had grown from only 12 % a decade prior (Pew Research Center 2012). In many ways, the political value of a pilgrimage is that it can reap electoral rewards without usually risking such backlash. As a narrowcast message, pilgrimages go unnoticed by many in the population, but still speak powerfully to those being targeted.

Second, these pilgrimages are narrowly focused, geographically speaking, and often target recurring events and major religious organizations or institutions. This is consistent with the reality that presidential travel is dependent primarily on two things: practical considerations and political calculations (see Doherty 2007; Ellis 2008). Attending events in Washington, DC is relatively easy for the president, and those scheduling major religious gatherings understand this. In the much rarer cases when the president leaves Washington to make a speech, he wants to maximize the strategic value of that visit. Consequently, major population centers and well-known religious institutions receive considerable attention—as do organizations with clear political leanings and substantial political clout. For example, recent Republican presidents spoke on multiple occasions to the National Association of Evangelicals, the National Religious Broadcasters Association, the Southern Baptist Convention, the Knights of Columbus, and other similar organizations. The political calculus that underlies these patterns is also at work in determining what presidents will say when they undertake a pilgrimage. It is far from happenstance, for instance, that presidents in their pilgrimages increased their rhetorical emphasis on God and faith just as Christian conservatives in America were beginning to mobilize politically (see Domke and Coe 2010; Kellstedt et al. 2007). With this mobilization, religion entered American politics with greater force than in past decades; presidents of both parties clearly responded by increasing their pilgrimages and upping the religious rhetoric present in those pilgrimages.

Finally, these data cast America's most recent president in a very interesting light. Barack Obama did something that presidents very rarely do: he broke with tradition. Coming into office at a time when political pilgrimages (and America's religious politics more generally) had reached a modern-era apex, he cut pilgrimages in half. His strategy was simple: he continued the familiar events, such as national prayer breakfasts, but rarely made time for additional ones. It was not simply the case that Obama differed from his Republican predecessor in lacking a host of conservative religious organizations to visit. Obama certainly could have done as Bill Clinton did: make up for this deficit by speaking frequently in churches. But he did not. It may be the case that Obama's shift was meant to send a signal of support to America's growing secular population. At the same time, Obama made moves that suggest he grasped the electoral importance of engaging with the faithful. Specifically, he led all presidents in total religious language used during pilgrimages. Put simply, Obama's technique was to do more with less: More religiosity in fewer pilgrimages. Time will tell if this strategy was an effective one. What is clear so far is that Obama's faith has not been well understood by a sizable segment of the public. Indeed, three and a half years into his presidency, roughly 1 in 6 Americans incorrectly believed that Obama was a Muslim (Markoe 2012). It may be the case that these persistent questions about Obama's faith led him to emphasize religious language so strongly when making his pilgrimages. It may also be the case that undertaking additional pilgrimages could have altered these misconceptions. At a minimum, they probably could not have hurt.

182.6   Future Research

The present study suggests a few avenues for additional research. For one, it will be useful for future studies to employ manual content analysis to the texts of presidential pilgrimages to achieve a finer level of detail than was possible using computer-assisted analysis. It is plausible that presidents' rhetoric in such contexts has substantial variation beyond just what was revealed here with measures of God and faith. It would be useful, for instance, to examine the degree to which presidents talk about political policy during their pilgrimages. Given the obviously religious settings, this would be an effective way to measure the willingness of presidents to explicitly mix religion and politics. Additionally, future research could examine via case studies how individual presidents used pilgrimages strategically. Consider that even though in the aggregate there was little regional variation in presidents' religious rhetoric during pilgrimages, there were some cases where individual presidents exhibited potentially meaningful variation. For example, two presidents with especially strong ties to religious conservatives, Ronald Reagan and George W. Bush, both noticeably elevated their religious rhetoric when making pilgrimages to the South. Detailed analysis of these presidents' rhetorical choices in different contexts could be revealing. In a similar vein, it would be interesting to track whether or not the past two presidents' pilgrimages correlate at all to the allocation of funds for the faith-based initiatives program started under George W. Bush and continued under Obama. Finally, future research might incorporate measures of presidents' total domestic travel as a means of understanding the proportion of travel that presidents devote to pilgrimages. It would be interesting to know, for example, if religious pilgrimages are more or less common than pilgrimages to speak to other social groups. All of these avenues would provide additional insight into the many ways that geography, rhetoric, and strategy interact in the U.S. presidency.

References

Ainsworth, B. (2000, February 4). Bradley rips Bush for stop at Bob Jones U.; South Carolina school prohibits interracial dating. *San-Diego Union Tribune*. Retrieved October 11, 2006, from Nexis database.

Bellah, R. (1967). Civil religion in America. *Daedalus, 96*, 1–21.

Bellah, R. (1975). *The broken covenant: American civil religion in time of trial*. New York: The Seabury Press.

Black, A. (2004, April 29). *Interview for The Jesus factor* [Television broadcast]. Boston: Public Broadcasting Services.

Blake, M. (2003, May/June). Stations of the Cross: How evangelical Christians are creating an alternative universe of faith-based news. *Columbia Journalism Review*. Retrieved July 4, 2007, from http://cjrarchives.org/issues/2005/3/blake-evangelist.asp

Brunn, S. D., Webster, G. R., Morrill, R. L., Shelley, F. M., Lavin, S., & Archer, J. C. (Eds.). (2011). *Atlas of the 2008 elections*. Lanham: Rowman & Littlefield.

Calfano, B. R., & Djupe, P. A. (2009). God talk: Religious cues and political support. *Political Research Quarterly, 62*, 329–339.

Campo, J. E. (1998). American pilgrimage landscapes. *The Annals of the American Academy of Political and Social Science, 558*, 40–56.

Chapp, C. B. (2012). *Religious rhetoric and American politics: The endurance of civil religion in electoral campaigns*. Ithaca: Cornell University Press.

Chidester, D., & Linenthal, E. T. (Eds.). (1995). *American sacred places*. Bloomington: Indiana University Press.

Coe, K., & Neumann, R. (2011). International identity in theory and practice: The case of the modern American presidency. *Communication Monographs, 78*, 139–161.

Doherty, B. J. (2007). The politics of the permanent campaign: Presidential travel and the Electoral College, 1977–2004. *Presidential Studies Quarterly, 37*, 749–773.

Domke, D. (2004). *God willing? Political fundamentalism in the White House, the "war on terror", and the echoing press*. London: Pluto.

Domke, D., & Coe, K. (2010). *The God strategy: How religion became a political weapon in America* (Updated ed.). New York: Oxford University Press.

Eade, J., & Sallnow, M. J. (Eds.). (1991). *Contesting the sacred: The anthropology of Christian pilgrimage*. London: Routledge.

Ellis, R. J. (2008). *Presidential travel: The journey from George Washington to George W. Bush*. Lawrence: University Press of Kansas.

Greenstein, F. I. (2004). *The presidential difference: Leadership style from FDR to George W. Bush* (2nd ed.). Princeton: Princeton University Press.

Hart, R. P. (1987). *The sound of leadership: Presidential communication in the modern age*. Chicago: University of Chicago Press.

Hart, R. P. (2005). The original argument. In R. P. Hart & J. L. Pauley II (Eds.), *The political pulpit revisited* (pp. 13–95). West Lafayette: Purdue University Press.

Jacobs, L. R. (2005). Communicating from the White House: Presidential narrowcasting and the national interest. In J. D. Aberbach & M. A. Peterson (Eds.), *Institutions of American democracy: The executive branch* (pp. 174–217). New York: Oxford University Press.

Jacobs, L. R., & Shapiro, R. Y. (2000). *Politicians don't pander: Political manipulation and the loss of democratic responsiveness*. Chicago: University of Chicago Press.

Kellstedt, L. A., Green, J. C., Guth, J. L., & Smidt, C. E. (1996). Grasping the essentials: The social embodiment of religion and political behavior. In J. C. Green, J. L. Guth, C. E. Smidt, & L. A. Kellstedt (Eds.), *Religion and the culture wars: Dispatches from the front* (pp. 175–192). Lanham: Rowman & Littlefield.

Kellstedt, L., Green, J., Smidt, C., & Guth, J. (2007). Faith transformed: Religion and American politics from FDR to George W. Bush. In M. A. Noll & L. E. Harlow (Eds.), *Religion and American politics: From the colonial period to the present* (pp. 269–295). New York: Oxford.

Leuchtenburg, W. E. (1988). Franklin D. Roosevelt: The first modern president. In F. I. Greenstein (Ed.), *Leadership in the modern presidency* (pp. 7–40). Cambridge, MA: Harvard University Press.

Markoe, L. (2012, May 10). One in six voters still thinks Obama's a Muslim: Why? *Washington Post*. Retrieved May 17, 2012, from http://www.washingtonpost.com

Marty, M. E. (1974). Two kinds of civil religion. In R. E. Richey & D. G. Jones (Eds.), *American civil religion* (pp. 139–157). New York: Harper & Row.

O'Loughlin, J., & Grant, R. (1990). The political geography of presidential speeches, 1946–87. *Annals of the Association of American Geographers, 80,* 504–530.

Overby, L. M., & Barth, J. (2006). Radio advertising in American political campaigns: The persistence, importance, and effects of narrowcasting. *American Politics Research, 34,* 451–478.

Pew Research Center. (2012, March 22). More see "too much" religious talk by politicians. Retrieved October 15, 2012, from http://pewresearch.org

Pierard, R., & Linder, R. (1988). *Civil religion and the presidency*. Grand Rapids: Academie [sic] Books.

Ragsdale, L. (1996). *Vital statistics on the presidency: Washington to Clinton*. Washington, D.C.: Congressional Quarterly.

Roof, W. (2009). American presidential rhetoric from Ronald Reagan to George W. Bush: Another look at civil religion. *Social Compass, 56,* 286–301.

Shaw, D. R., & Gimpel, J. G. (2012). What if we randomize the governor's schedule? Evidence on campaign appearance effects from a Texas field experiment. *Political Communication, 29,* 137–159.

Shogan, C. J. (2006). *The moral rhetoric of American presidents*. College Station: Texas A&M University Press.

Turner, V. A., & Turner, E. (1978). *Image and pilgrimage in Christian culture: Anthropological perspectives*. New York: Columbia University Press.

Wattenberg, M. P. (2004). The changing presidential media environment. *Presidential Studies Quarterly, 34,* 557–572.

Chapter 183
Walking on the Razor's Edge: Religious Groups and the 2011 Arab Spring

Ghazi-Walid Falah and Laura J. Khoury

183.1 Introduction

The so-called "Arab Spring" of December 18th 2010 (first in Tunisia) and its ongoing aftermath have caught many observers by real surprise within and outside the Arab world. One of the many demands for the youth who initiated the revolt—and we are using the term revolt instead of uprising in order to emphasize the notion of a break from one political era into a new one—was the establishment and consolidation of true democracy, however defined. What is even more surprising is the non-participation by few Islamic groups, spreading ideas and distributing flyers saying the values of democracy violate the law of God and rejecting all voices advocating for a civil state (see Halawa 2011) and, in addition, condemning protestors who were causing *fitna* (strife) demanding that Muslims should oppose all attempts at toppling the legitimate rulers of their countries.

The purpose of this paper is to examine the reaction of those key religious scholars, or *'Ulamā* [1] in the Arab world and certain religious groups (for example, the Salafis in Egypt) in the region to the "spring" in their respective countries and beyond (that is, the Salafis exist in many Arab countries). We are excluding from our discussion two groups which are more political organizations in their actions and organizational structure and less religious ones, namely the Lebanon-based Hezbollah and the Hamas

---

[1] 'Ulamā or religious scholars are Muslim scholars who arbitrate the sharia law and or/ and hadith and tafsir for many years or those who studied at an Islamic schools or madrasah and teach Islamic discipline ranging from a mufti, as a higher rank, to imam, as the lowest rank.

G.-W. Falah (✉)
Department of Public Administration and Urban Studies, University of Akron,
Akron, OH, USA
e-mail: falah@uakron.edu

L.J. Khoury
Department of Sociology, Birzeit University, P.O.Box #14, West Bank, Palestine
e-mail: ljkhoury@birzeit.edu

© Springer Science+Business Media Dordrecht 2015
S.D. Brunn (ed.), *The Changing World Religion Map*,
DOI 10.1007/978-94-017-9376-6_183

in Gaza Strip (see Da'na 2009; Khoury and Da'na 2009, 2012). Although Hezbollah adheres to the Shiai' Islamic ideology, it was very much in support of regime change in Tunisia, Egypt, Libya, Yemen, Bahrain, but not in Syria. Syria has been the only Arab country that did not have a peace treaty with Israel, housed all Palestinian resistance movements, and most importantly, any regime change targets the resistance core in the Arab World. Hezbollah identified this as problematic and sought the Israel-U.S. intervention as a threat to resistance. In addition, the facts on the ground are more and more uncovering this U.S.-Israel-Gulf-led plan to change the regime. As for Hamas, they were under siege by Israel in Gaza and were very happy to see the end of Mubarak's regime, but stayed neutral and were very careful watching as events unfold, especially in Syria where the Hamas headquarters are located and where many of its top leaders live.

First of all, we should be clear about the great variety of interpretations of the discourse of political Islam among different groups and individuals in the Moslem Arab world and those who call themselves Islamist or publicly identify themselves as adhering to the Islamic *Shariaa*, that is, the faith of Islam grounded in the Quran and the *Hadith* (the sayings and deeds of Prophet Mohammad during his life time). These differences in interpretation and practice clearly surfaced in their reactions to the dramatic events brought with the "Arab Spring." For example, the Moslem Brotherhood in Egypt announced formally (aired on national and international media outlets) at the beginning of the revolt in January 2011 that they would not be participating in the revolt. A point we will discuss later, because it was so ironic to find them winning the elections of the Egyptian new presidency (Muhammad Morsy) on 1/24/2012.

According to An-Nai'm (1999:103) "[p]olitical Islam can be broadly defined as the mobilization of Islamic identity in pursuit of particular objectives of public policy, both within an Islamic society and in its relations with other societies." Such mobilization can of course be simultaneously both positive and negative. As will be shown below, Islamic groups had to change their ideological position from one edge of the continuum to another in order to accommodate themselves to the changing circumstances as the revolt unfolded. The wave of political and social change in the region powered by spontaneously revolting masses who were characteristically disorganized at the beginning, lacking both charismatic leadership and an ideological underpinning, have shaken the ideology of many religious groups and individuals' beliefs to the very core, as we will show through their statements and *fatwas*.

First we discuss how mainstream Islamists perceive democracy. This is followed by a section looking at two cases of how Arab regimes have treated religious movements and groups. A third section examines responses of key religious scholars or *'Ulamā* to the Arab Spring. Section 183.4 looks at how religious groups view the Arab Spring.[2] There we focus on two groups, namely the Muslim brotherhood and

---

[2] Most of the Arab Spring groups were named after the day the revolt broke out, for example, December 14 in Tunisia, January 25th in Egypt, and February 14th in Bahrain. So, you find social media using the terms like The Youth of January 25th and this, in our analysis, denotes the independent spirit of this revolution, its spontaneity, and disorganization. However, this does not mean that certain groups, like the Nasserites in Egypt, were not organizing and raising awareness well before the revolution began.

Salafi group. This is followed by a brief conclusion and then some thoughts on the future impacts of these developments.

183.2   Islam and Democracy

Democracy is today seen by many societies, governments and individuals as the ultimate goal for each person on the planet. This taken for granted conventional wisdom ignores two basic realities: (1) that each society has the right to choose its own type of ruling government and (2) what is appropriate for one society is not necessarily good for another. In the case of the Arab world, where Islamic culture and religion are dominant, Western democracy and secularism are concepts brought to the region at the time of European colonization. They basically sprang from the Christian experience and the long European struggle for separation of Church from the state.

There is nothing parallel in the Muslim world which is called separation of "mosque and state." Even in the most secular regions such as that of Iraq under Sadam Hussein or Libya under Gadhafi, their respective regimes occasionally relied on the Qur'an and molded religion in one way or another in governing their people. It is true that many citizens in the Arab world object to Islamic fundamentalist ideology and call for separation (or minimizing the role or religion) in government but as An-Ana'm claims "If presented with European secularism as the only alternative to the so-called Islamic state and application of Shari'a, Islamic societies will clearly prefer the latter, however serious its conceptual faults and practical difficulties" (1999:119).

Al Qadimi (2012) summarizes the Islamists' objections to a democratic system in five key points:

1. Democracy is a Western concept and has nothing to do with our heritage of political Islam.
2. This form of government is essentially based on the principle of the rule of the people whereas in Islam Sharia rules, not the people.
3. Democracy places the power to decide on those who rule in the hands of the citizenry (including, the scholar, the ignorant, the good and bad, the cheats and impostors—all have an equal vote whereas in the Islamic system, the decision should be in hand of Ahel al-Hal and al Aqid (the problem-solver and maker of contracts, that is, those versed in the Sharia),
4. In democracy, people elect an executive authority whereas in Islamic polity it is the function of the Showra (that is, an advisory council) to enlighten the ruler and the Showra's decision is not binding. There is no mechanism for Showra in Islam so some countries appointed by the ruler while others inherit it.
5. Democracy makes a formal opposition legal.

These objections to democracy have spawned much debate scholars trying to prove or disprove that Islam is compatible with democracy. Da'na (2011) critiques

Western views of democracy in Islam (especially the work of Bernard Lewis) and characterizes their orientalists views as a clear "mismeasurement of Arabs." Voll explored how "some Islamic scholars and intellectuals argue that Islam has within its historic and theological repertoire of symbols and concepts the resources to create democratic discourses and to provide the foundations for systems of Islamic democracy" (2005:92). There is a continuum between this version of Islamic democracy and synthesizing the two. For a synthesis between democracy and Islam Turkey is a good example of political system of a secular-republic tradition with an Islamic vision or the return to the Quran when needed. Iran on the other hand would be considered at the other end of the synthesis where public participation in politics is advocated and a version of Islamic democracy is advanced. The latter does not mean it is a position by the Shi'a sect because you find Shia'a leaders like Ayatollah Muhammad Baqir al Sadr from Iraq advocating a synthesis. Space limitation does not allow us to expand on this fundamental debate further.

183.3 Religious Groups in the Pre-revolutionary Phase

Since the beginning of the 1980s religious fundamentalism[3] has strengthened worldwide, with increased political influence on state politics and public space. Some argue that it is in many ways the successor of failed nationalist programs because the founders of many Islamic movements were formerly in nationalist movements like Hasan-al-Banna of the Muslim Brotherhood in Egypt and Rachid Ghannoushi of *al-Nahda* or Renaissance Party in Tunisia (see Choueiri 2010; Esposito 1994). This has led many regimes to try to contain, suppress or even co-opt key religious leaders and fundamentalist groups in order to maintain their regime (for example, Egypt and Saudi Arabia). Despite this, Islamic fundamentalism[4] remains a factor of Western capitalism and cannot be explained except by situating it in the context of anti-Western sentiment due to its penetration in the social and political lives of these countries and the perceived threat of Islam by the West, especially since the fall of the Eastern Block. This analysis is incomplete if it was not addressed as a universal reaction to the changes in the global power structure and the wider religious resurgence. This is the context in which the Arab spring appeared.

In most Arab countries rocked by the Arab spring, civil society has been (or was) closely controlled. Hellyer (2011:12) quotes an activist in Cairo saying that during

---

[3] The periodization of the development of Islamic type of fundamentalism according to Choueiri specifies them as Islamic Revivalism (1744–1885), Islamic Reformism (1839–1954), and Islamic Radicalism (1945 and onward). However we will use Da'na's (2009:217) categorization.

[4] The definition of Islamic fundamentalism varied but the core idea is in what to Sayyid Qutb, an ideologue of the Muslim Brotherhood in Egypt, defined as a restoration of Islamic life governed by Islamic conception and Shari'ah and system. The variations are more related to how best to achieve this goal. Some sought vilayat el-faqih model of a renewed caliphate by Khomeni, others implemented an ideal Islamic state like the Taliban Afghan model or Sudanese Hassan al-Turabi model of sovereignty.

the Mubarak's time, the authorities would not allow individuals to organize a recycling initiative in his neighborhood, let alone organize political protest. It is well known that university students were not allowed to demonstrate outside their respective campus walls. According to Hashem (2011), the campaign in Syria "to instill in the nation secular-socialist and anti-Islamic programs took a sectarian meaning in the 1970s after the Alawites assumed power. Symbolic violence increased as insults to ordinary Muslims flared up periodically, including a sacrilegious drawing in a regime-sponsored magazine and the president's brother's militia forcibly removing the head scarves women in the streets of Damascus." Hashem (2011:72–73) wrote: "As a stark example, the greeting of "Asalam Alaikum" in a formal setting not only became unacceptable, but in certain situations could lead to questioning by the security forces." Hashem further describes the situation in Syria by saying that "by the late 1970s … the regime cracked down on Islamic activists in general and on the Muslim Brotherhood in particular, since its leadership abroad had endorsed the rebellion. Muslim activists were jailed and tortured, some disappeared, and many fled the country. The city of Hama was bombed, leaving tens of thousands dead" (Hashem 2011:73).

The case of Egypt was not much different. According to Khalil al-Anani (2012:10), "Mubarak dealt with religion as if it had a monopoly over the Egyptian state and its official institutions, and made sure that its use in the public sphere did not exceed the limits of the symbolic and ritualistic. Religion was employed to support the regime's legitimacy and reinforce it internally and externally…. Mubarak used religious movements against each other in order to weaken them, and prevented any rapprochement among them that could threaten the existence of his regime."

In the early 1980s, Mubarak allowed the Muslim Brotherhood to play a more significant role in order to be able to confront violent groups like Egyptian Islamic Jihad and Al-Gamaa al-Islamiyya who threatened his rule. Having succeeded in this, he tightened the screws on the Muslim Brotherhood throughout the 1990s, when many of the group's prominent leaders were tried in court for the first time since the era of Abdel Nasser. Over the last decade, the regime used the Salafi movement to marginalize the Muslim Brotherhood, particularly after its surprise victory in the parliamentary elections of 2005 (see more details in al-Anani 2012).

In sum, we agree with al-Anani (2012) that Mubarak sought to manipulate Islamic movements in a way that would guarantee that they did not become too powerful or expand too far into the community. He did not uproot them in a way that would have led to a violent reaction that might have threatened the stability of his regime. The Muslim Brotherhood in Egypt committed itself to the rules of the game as set by the regime, since whenever they tried to increase their political influence; they were met by repression from the security forces, political exclusion, and social harassment. Previous Egyptian presidents Abdel Nasser and Sadat have also treated members of Brotherhood group in a similar manner and have imprisoned them for an extended period of time.

183.4 Reaction of Key Religious Scholars to the Arab Spring

As a general rule, religious scholars were very hesitant to support the Arab Spring when it first erupted. When Mohamed Bouazizi set himself blaze on December 17, 2010 in Tunisia in protest against a local police officer's decision to confiscate his fruit cart, religious scholars in the Arab world refused to consider him a martyr, arguing that suicide is never a legitimate means of political protest. Al-Azhar University in Cairo issued at the time a stern reminder that suicide is categorically prohibited in Islam, and was prepared to exonerate suicides only for reasons of mental illness or defect (see further discussion in Fadel 2011).

Early on in the Egyptian revolt, Shaykh Ali Gomma, the official mufti of Egypt, issued a *fatwa* (religious edict) in which he stated that "it was permissible for individuals not to attend Friday prayers in the present circumstances on the grounds that to do so would entail unreasonable exposure to risk of losing life or property". Another fatwa of his asked demonstrators to return to their homes after the first of Mubarak's speeches in which he promised to reform the government and set it on a course toward genuine democracy. Overall, Gommaa's fatwas were consistent with historical Sunni views that regard revolution with skepticism, if not outright terror at the prospect of public disorder (see Fadel 2011: 18).

*In addition, part of the 48-page fatwa that t*he spiritual leader of Algeria's influential Salafist movement Sheikh Abdelmalek Ramdani (who moved to Saudi Arabia due to threats) has issued is: *"As long as the commander of the nation is a Muslim, you must obey and listen to him. Those who are against him are just seeking to replace him, and this is not licit" (Chikhi, Reuters* 2011*).* Thus urging Muslims to ignore calls for change because democracy is against Islam and that they can only "pray and be patient" when faced with an unwanted ruler. For the demonstrators though patience after 30 years is impossible though.

The Grand Imam of Al-Azhar, Sheikh Ahmed al-Tayyib of Egypt tried to "hold the stick from the middle" so as not to upset the regime or appear to be in the position of defending the revolution. Instead of Al-Azhar issuing a statement condemning the killings of the demonstrators and protesters in Tahrir Square, it asked both sides to calm down for the sake of the country's unity, something that pushed the revolutionary youth to demand al-Tayyib's resignation (al-Anani 2012:7). The position of the Egyptian Coptic Church was worse as Pope Shenouda III, Patriarch of Saint Mark's Episcopate, announced his support for Mubarak to remain in power, rejected the demonstrations of January 25, and called for Copts not to participate in them. The Pope's remarks prompted anger and complaints among many Coptic priests (al-Anani 2012:7).

The chief mufti of Saudi Arabia, Sheikh Abdul Aziz Abdullah, condemned the Egyptian demonstrators with extremely harsh language, and the famous Syrian scholar, Muhammad Said Ramadan al-Buti, has repeatedly condemned the Syrian demonstrators (Fadel 2011: 18). In the wake of pro-democracy Arab spring movement in Bahrain in March of 2011, new laws were created in Saudi Arabia criminalizing the freedom of expression. The military intervened in Bahrain to quell protests –protecting the headquarters of the US Navy 's fifth fleet-and as the Saudi ruling al-Khalifa family are cohorts with the regime of Ali Abdullah Saleh in Yemen, another trusted U.S. ally; they supported the Gulf Cooperation Council (GCC) initiative for the Yemeni crisis by transferring power to the vice president Abd Rabu Mansur Hadi. This came to be called cosmetic change (Alwazir 2011).

Of special interest is the stance of Shaykh Yusuf al-Qaradawi, an Egyptian religious scholar in Qatar, who frequently appears on al-Jazeera and has his own Program 'al-shaiaa and life.' He was invited to give the first Friday sermon at Tahrir Square after Mubarak's resignation. Reflecting on Bouazizi's death, he affirmed that suicide was generally a major sin (*kabira*), but blamed the Tunisian state for Bouazizi's sin and prayed that God would absolve him of any blame for that sin. Al-Qaradawi was outspoken in support of the Egyptian revolution and he called for Libya's military to assassinate Col. Moammar Gadhafi on the grounds that he was waging war on his own people. Killing him in these circumstances would be a legitimate act of self-defense (Fadel 2011: 18).

Al-Qaradawi's reputation for moral courage in the face of Arab dictators, however, suffered a significant blow as a result of his refusal to condemn the actions of the Bahraini and Saudi governments in violently suppressing the peaceful protests in Manama's Pearl Square. His attempts to distinguish the Bahraini protests on the ground that they were sectarian in character rather than national hardly seemed plausible at the time (Fadel 2011:18).

183.5 The Reaction of Religious Groups to the Arab Spring

With the exception of the Libyan religious establishment, the 'ulama,' who generally threw their weight behind the rebels, showing that in some circumstances, religious authorities in the Muslim world, contrary to popular belief, do advocate physical resistance to an unjust ruler (Hellyer 2011, 14), the two groups who were vocal in their statements against the revolution are: the Salafi group who based on Da'na's categorization belong to the pre-nation state and the Muslim Brotherhood who belong to the nation state as in his table below (2009: 217). We explained earlier why Hamas and Hizbollah were excluded in our study and is justified by Da'na as they belong to the post-nation state (Table 183.1).

Obviously, religious establishments and individuals who do not advocate the manner in which the Arab spring was evolving were grounding their objections on some phrases from the Qur'an which basically state that a citizen of a state should

Table 183.1 Trends in Islam

Wahabi	Muslim brotherhood	Muslim brotherhood first renewal (Hamas)
Iraqi/Arab Shii'sm	Iraqi Shi'ism First Renewal	Iraqi/Arab Shi'ism – Second Renewal (Hizbollah)

Data source: ABS 2011 Census Community Profiles

not revolt against his ruler. The idea appears in the Quran (Verse 54 of Suret al Nisa'a), but there are many interpretations, some argue that it is absolute loyalty and that if a person commits any sin, scholars may advise him but only God can punish him. Others restrict submission to authority as long as the ruler is governing according to the rules and commits no sin. The interpretation of the Wahabis, Salafis and conservative Islamists, usually cites and builds on Ibn Taimiya's interpretation in Minhaj Al Sunna (See Salem 1986 [1295]). Ibn Taimiya (1263–1328 CE), of course, is the source of most conservative interpretations and the father of Salafism. But Islamic rulers have manipulated this against any opposition and authority. Scholars use this all the time, especially in Saudi Arabia. Al Qadimi (2012) states that the Islamist stance of the Arab Spring can best be seen at two levels; one is that related to Muslim Brotherhood groups and movements " beyond the Brotherhood" that include the Turkish and Moroccan political party of Justice and Development, Tunisia's Ennahda Party, Egypt's Wasat Party, and other independent groups and individuals; the second level includes the Salafi groups.

183.5.1 Brotherhood Groups

This group is well organized and traditionally adheres to a low-profile education of its members, advocates gradual reform and calls for consolidating resistance to repressive regimes. They express their resistance by patience when under oppression; the means of absorbing oppression tactics emanates from values like being patient. A religious education to adhere to patience in whatever situation of repression leads them to denying any idea of using violence in order to change regimes. Such collective ideology made the group resilient in confronting repressive regimes, but at the same time hindered them for making a fast decision when protesters of the Arab Spring started pouring into the streets and squares of their capitals. In other words, they were confused and hesitant to make hasty decisions or reveal their stance. Alqadimi (2012) has monitored the behavior of this group throughout the revolution. He claims that despite the facts that the Brotherhood had been constantly the victim of the regime, their stance in the protests of January 25 (better known as the "day of anger") were marked by hesitation and confusion. They initially were reluctant to participate in demonstrating with others, claiming at the time that they were not invited to participate or they were still evaluating the situation. In another statement they said that their participation will be limited to some of its key Brotherhood leaders. Another statement said that they do not mind if Brotherhood members go out and demonstrate. This graffiti simply says "Egypt is the tomb of Brotherhood" (Fig. 183.1). Tellingly, this means that they did not request their young members to participate in demonstrations against the regime. All this was oral and not written; it was disseminated by word of mouth.

In the arena of documented position via written statements, the Moslem Brotherhood's positions were consistent with those of other political parties who were accused of coordinating with the Mubarak regime and were refusing to sign a

Fig. 183.1 This graffiti simply says, "Egypt is the tomb of Brotherhood" (Photo by Christine Smith, used with permission)

call on January 21st requesting the masses to go into the streets and demonstrate on January 25th. On January 19th the Brotherhood issued a written statement calling on the Egyptian regime to carry out some reforms, including abolishing the Emergency Law, disbanding the People's Assembly and carrying out certain constitutional amendments. On January 23rd the Brotherhood issued another written statement reminding the regime of the need for reforms. But it never mentioned or gave any hint of the upcoming day of anger on January 25.

After the huge demonstration of masses on January 25 the Brotherhood apparently realized that they needed to move on to different strategies, lest they be left behind and loose possible support from the demonstrating masses. On January 26 they issued another statement in writing confirming the legality of demonstrations on the day of anger, and that their groups are an integral part of the people and they should participate in demonstration. On January 29, the Brotherhood restated their previous call for reforms and stressing there was a need for constituting a transitional government whose members should not come from Mubarak's National Democratic Party. By that time the demonstrators were already calling for the fall of the Mubarak regime. At this point, the Brotherhood failed to call for the fall of the regime. It was only on January 31st that they called for regime change and they refrained from using the word "fall" in their statement and instead said the government should "leave." Moreover, while demonstrators used the term "revolution" (*thawra* in Arabic) to describe their revolt against the regime, the Muslim Brotherhood used the softer descriptor and called it Intifadah (uprising). Clearly, the Palestinian Intifada is not on their minds either as young revolutionaries in fact made the connection more concrete as in this graffiti that has a picture of Muhammad El-Durra who was protected by his son but killed by Israeli forces (which became iconic of the Intifada) and the graffiti says the struggle is the same and the border is

Fig. 183.2 Graffiti showing Muhammad El-Durra, who was protected by his son but killed by Israeli forces (which became iconic of the Intifada). The graffiti says the struggle is the same and the border is just simply made of earth (Photo by Christine Smith, used with permission)

just simply made of earth (Fig. 183.2). It was only on February 2nd that the Brotherhood used the word 'revolution' in their statements.

It is clear that the Brotherhood was somewhat hesitant in putting all of their political cards with the demonstrators, and was essentially pragmatic in evaluating the political change in Egypt day after day, as the above chronology indicates. After the removal of Mubarak from reign, Muslim brothers who are well organized have entered Egyptian political life with full power and capacity, although they experienced some splintering into smaller subgroups that have formed independent political parties. Their ability to mobilize political power among the masses after the fall of Mubarak allowed them to win the presidential election. Interestingly, the graffiti here shows how the thoughts about whoever came after Mubarak are responding to a call from him "the one who missions others did not die" (Fig. 183.3).

In Syria, "The Muslim Brotherhood has insinuated itself into Syria's popular uprising against that country's Ba'athist dictator. The Muslim Brotherhood in Syria has a score to settle with the Assad dynasty over the massacre of thousands of Ikhwan faithful in the 1980s; their strategic plan is to ride the Syrian revolt to the very end" (Fares 2011). Lund (2012:5) in his study shows how Assad is the perfect Islamist enemy and contends that two factors led to their rise: first the descent into sectarian conflict between Sunni Moslems against the secular Alaweits, and

Fig. 183.3 Graffiti showing how the thoughts about whoever came after Mubarak are responding to a call from him "the one who missions others did not die" (Photo by Christine Smith, used with permission)

secondly, "the foreign support pouring in from regional governments and non-state organizations which is disproportionately empowering Islamist groups."

183.5.2 Salafi Awakening?

In the home of the Arab Spring (Tunisia), arch-conservative Salafis have emerged on the streets and in the public eye as a major challenge to the traditional more secularist tradition in the country and the elected government headed by Ennahda, a moderate Islamic group. They are pressing to ban alcohol, have raided bars and stores in Sidi Bouzid, where the Arab Spring in effect began.

Many Salafi Muslims were behind bars or undergrounds in the pre-revolutionary Tunisian state, under heavy oppression. They played a muted role in the revolt, but now are pushing for a more *sharia* as the basis of law in the new Tunisia that is a-borning. They have also attacked a TV station and motion picture theater that showed films they regarded as celluloid blasphemy. In Egypt, there was one Salafi group called the Salafi Movement Towards Reform who stood by the revolution from the start like Sheikh Muhamed Abdel Maksud and his followers are the exclusion. Therefore, there are Salfis like sheikh Muhamad Jibreel who praise the first speech by president Mubarak after the revolution, he once greets the youth and another time calls them to go withdraw, Mustafa al-Adawi did the same but on TV, Yaser Burhami released a fatwa that forbids demonstrating, and Sheikhs Mahmud el-Masri and Muhamad Hassan Yakoob who called for not participating in the *fitna* and stay at home.

The Salafi anti-democratic flyers in Egypt from the onset of the revolution read things like the civil state "equates between citizens in their rights and duties, even if God has created differences in them" and democracy "allows the people to govern themselves even if they are violating the rule of God" (Halawa, Egypt Independent 2011). The Salafis in Egypt, they were long repressed under Mubarak and have emerged as a small but visible ultra-fundamentalist group pressing for a strict Islam close to its early origins. The Muslim Brotherhood has decided to allow them to press their agenda. Some think the MB is indeed allowing the Salafis to assume the role of pressing openly for a more fundamentalist Islamic state in Egypt, while the MB pursues what appears to be a more "moderate" course. In the election, the Salafi bloc won about a quarter of the seats, astonishing many. They have a very conservative stance on women's and minority rights. For the moment, the largest Salafi party in Egypt has struck a more pragmatic line in respect to Egyptian foreign policy, but Salafis are known for their particular hostility to Israel and the U.S. presence in the region. What will accompany this "Salafi awakening," as a recent article from the Brookings Institution termed this growing impact in Egypt, remains to be seen (see www.brookings.edu/research/articles/2012/06/28-egypt-byman-gold).

In Syria, the revolution has had a different switch to it, especially of what Lund (2012) called the new foreign fighters phenomenon. He argues that the role of two of the most extremist salafists in the Syrian rebel movement are foreign: Ahrar el-Sham Brigades and Jabhet el-Nosra. The former accepts non-Syrian volunteers and "according to their website, they are made of nearly fifty jihadi groups, such as the Qawafil el-Shuhada (Idleb), Ansar el-Haqq (Idleb), and Salaheddine Brigade (Hama)" (2012:29). The latter, in specific, has strong links to the international salafi-jihadi community and it is "the one most likely to gain official approval by al-Qaida" (2012:23).

183.6   Conclusion

In conclusion, the religious/political picture regarding the Arab Spring in many countries remains highly diverse and in some flux. It is clear that in Tunisia and Egypt, the extreme fundamentalist religious forces like the Salafis, and even the Muslim Brotherhood, did not play a major role in the revolt, for various reasons. But especially in Egypt, and to a visible degree in Tunisia, the Salafis have emerged from repression to a militant role in public space, and in the democratic process of election in the Egyptian case. We exhibited in this paper how religious leaders that were reluctant and or denounced any support the revolt now speak with many voices. However, now that powerful circles in the U.S. and European governments are probably seeking in many ways to shape the post-revolt situation across the region, their concern is doubtless to strengthen the hand of moderate Islamic groups and to isolate the fundamentalists. But as mentioned in the paper's beginning, public opinion on the street may be little enamored of certain aspects of Western democracy, and the strong secularist currents know they have a struggle ahead in shaping the new societies and polities that are emerging.

Nonetheless, it seems clear as well that religious groups and individuals may move in "moderate" directions if they think that traditional strict views of Islam will restrict them from adapting to new dynamic circumstances. The Muslim Brotherhood shows that even a major force can rethink its positions and be ready to compromise when they see that they will be left behind. They can make or invent new Fatwas in order to adjust to the new situation if they see fit. For example, in August (2012) a condemnation of a religious fatwa legislating the killing of demonstrators in mass protest against the new president Muhammed Morsi in Egypt caused much controversy (see: http://gulfnews. com/news/region/egypt/death-fatwa-against-protesters-causes-a-stir-1.1063071). Soon after the president's spokesman Yasser Ali repudiated the Fatwa and Al-Azhar dissociated itself from the *fatwa* (saying the clerk was not a member of their committee). It is so ironic to see how the same position that was taken by some of their *ulama* at the start of the revolution is now rejected as the Moslem Brotherhood affiliated president rules Egypt. Denouncing demonstrators' actions in both cases may reflect consistency of how the interpretation of Islam to those religious leaders concerning public involvement is, but there has been no consistency since what was viewed as wrong early on in the revolution, but it is repudiated now. May we conclude that the Arab spring unleashed some aspects of Islamic fundamentalism, or awakened it, or simply allowed to become what it is?

References

Al Qadimi, N. B. A. R. (2012). *Al-Slamyoon wa Rabia' al Thawrat: al Mumarasah al Muntijah lil Ifkar* (Islamists and spring of revolutions: Creative practice of Thouts). Doha: Arab Center for Research & Policy Studies (in Arabic).

al-Anani, K. (2012). *The role of religion in the public domain in Egypt after the January 25 revolution, case analysis*. Doha: Arab Center for Research & Policy Studies.

Alwazir, A. (2011, November 30). Yemen's GCC initiative: cosmetic or comprehensive change?(In English al-Akhbar). Retrieved September 1, 2012, from http://english.al-akhbar. com/node/2088

An-Nai'm, A. A. (1999). Political Islam in national politics and international relations. In P. L. Berger (Ed.), *The desecularization of the world. Resurgent religion and world politics* (pp. 103–121). Washington, DC: Ethics and Public Policy Center.

Chikhi, L. (2011, March 15). Top Algerian Salafist's Fatwa says Unrest Un-Islamic. *Reuters Africa*: Retrieved September 1, 2012, from http://af.reuters.com/article/topNews/ idAFJOE72E0B620110315

Choueiri, Y. M. (2010). *Islamic fundamentalism* (Vol. 3). London: Continuum International Publishing Books.

Da'Na, S. (2009). Islamic resistance in Palestine: Hamas, the Gaza war and the future of political Islam. *Holy Land Studies, 8*(2), 211–228.

Da'Na, S. (2011). The mismeasure of Arabs: Culture and revolution. *The Arab World Geographer, 14*(2), 145–152.

Esposito, J. L. (1994). Political Islam: Beyond the green menace. *Current History: A Journal of Contemporary World Affairs, 93*(579), 19–25.

Fadel, M. (2012, September 15). Religion and the Arab Spring: Between opposition, equivocation and Liberation. *The Islamic Monthly,* Summer/Fall, 17–19 Retrieved September 1, 2012, from http://theislamicmonthly.com/articles/religion-and-the-arab-spring-between-opposition-equivocation-and-liberation/

Fares, W. (2011). Muslim brotherhood riding the crest of Arab Spring. *Ion Newsmax*. Retrieved September 20, 2012, from http://www.newsmax.com/WalidPhares/muslimbrotherhood-arabspring-gadhafi/2011/06/03/id/398700

Halawa, O. (2011, March 28). Salafi anti-democracy flyers handed out around Cairo. http://www.egyptindependent.com/news/salafi-anti-democracy-flyers-handed-out-around-cairo

Hashem, M. (2011). Syria exceptionalism and revolution. *The Islamic Monthly*. Summer /Fall, pp. 71–74.

Hellyer, H. A. (2011). The genie of change is out of the bottle. *The Islamic Monthly*, Summer/Fall, pp. 11–15.

Khoury, L., & Da'Na, S. (2009). Hizbollah's war of positions: The new Arab-Islamic praxis. *Arab World Geographer, 12*(3–4), 136–149.

Khoury, L., & Da'Na, S. (2012). Decolonizing the geographies of resistance: Imperial cartography of the arab world. *Arab World Geographer, 15*(3), 189–225.

Lund, A. (2012). Syrian Jihadism. In Ulbrief. Swedish Institute of International Affairs. No. 13: (pp. 1–45).

Salem, M. R. (1986 [1295]). *Ibn Taymiyah: Minhaj as-Sunnah an-Nabawiyyah* (The Pathway of as-Sunnah an-Nabawiyyah). Beirut: Dar Al-Kutob al-Ilmyyah (in Arabic).

Voll, J. O. (2005). Islam and democracy: Is modernization a barrier. In S. Hunter & H. Malik (Eds.), *Modernization, democracy and Islam* (pp. 82–97). Westport: Praeger.

Chapter 184
The Role of Religion in the Formation of a New State on the World Map: South Sudan

Rainer Rothfuss and Yakubu Joseph

184.1 Introduction: Roots of the North-South Conflict

As history shows from ancient times of slave raiding onwards, the conflict fault line between the northern and southern parts of Sudan have always had – even when the conflict manifested in struggles around land or mineral resources, economic or socio-political deprivation – a racial, cultural and religious component (cf. Herbst and Mills 2012). When the Numeiri government, under the political pressure of the Muslim Brotherhood, declared Islamic Sharia law in 1983 as the basis for all legal affairs in the whole multiethnic, multicultural and multireligious country[1] religious identity was symbolically reconfirmed to be the core issue of the north-south conflict which Jok (2007: 171) addresses as a "battle over identity." As Kaiser (2009: 191) reminds us, once conflicts reach a stage where they concentrate on differences between religious or racial identities and are based on mutual mistrust due to consistent and increasing violent clashes, such conflicts can hardly be resolved through negotiations like in cases of competing socioeconomic or power-sharing interests. To put it in the words of Stump (2008: 267): When the opposing parties' positions are "grounded in incontrovertible religious truths […] such conflicts often resist easy resolution and may persist over long periods of time."

This chapter examines the role of religion in the formation of the Republic of South Sudan, which gained independence on 9 July 2011. The youngest internationally recognized State on the world map gained independence after an unprecedentedly long civil war in the history of the African continent between the Muslim-Arab dominated central government in the north and the mainly Christian-Black African

---

[1] According to Deng (2010: 394) Sudan is among the most diverse country in Africa with 181 indigenous languages, 500 dialects and, according to Mathok (2009: 147), with more than 500 tribes.

R. Rothfuss (✉) • Y. Joseph
Geographisches Institut, University of Tübingen, D-72070 Tübingen, Germany
e-mail: rainer.rothfuss@uni-tuebingen.de; yakubu.joseph@uni-tuebingen.de

© Springer Science+Business Media Dordrecht 2015
S.D. Brunn (ed.), *The Changing World Religion Map*,
DOI 10.1007/978-94-017-9376-6_184

3515

liberation movement in the south. The civil war had two phases with the first beginning in 1955, a year prior to Sudan's Independence from British rule. It was prompted by the south's desire to achieve regional autonomy from the hegemonic central government in the north (Cole and De Blij 2007). The mediation efforts of the World Council of Churches (WCC) and the All Africa Conference of Churches (AACC), which have not been deservedly acknowledged, led to the Addis Ababa Agreement (AAA) in 1972 and the granting of regional autonomy to the south (Kobia 2008 and Nyang and Johnston 2003). The second phase started in 1983 as a direct consequence of: (a) the failure to fully implement the peace agreement; (b) the discovery of oil in the south in 1978 and the central government's order, in a bid to control the resources, that the oil should only be refined in the north and (c) the imposition of Sharia law on the whole country in 1983 under President Nimeiri's highhanded government, despite the south's unique multiethnic and multireligious character (Rader 2012: 47–48; Cole and De Blij 2007: 561). The liberation struggle, therefore, can be interpreted as a struggle for political and socioeconomic self-determination and religious freedom. As Nyang and Johnston (2003: 215) observe, postcolonial Sudan has been "under continual contestation" as it never enjoyed the "universal approval of those living within its 'jurisdiction'." The core problem of state-building for them was the "general absence of widely shared principles by which the state is to be governed, such as the principle of political equality for all citizens."

The chapter analyzes the religious symbolisms as they are reflected in the national anthem, independence celebrations, political discourse and foreign relations. Furthermore, it systematizes the South Sudanese experience by drafting a concept for a form of geopolitics of the subaltern that may serve as an example for promoting the liberation of other marginalized peoples suffering from severe and enduring oppression, especially as they relate to identity issues.

184.2 Territorialization Through Identity and Nation Building

De Blij (2009: 52) argues that "for the faithful, religion is the key to identity. And such identity is part of the impress of place. Religion and place are strongly coupled." Any kind of conflict in Africa – where religious identity plays a central role in public life – can only be understood when acknowledging that "religion is in some ways the most powerful among the powers of place" (De Blij 2009: 80). The authors of this chapter subscribe to the hypothesis that the process of territorialization driven by the factor of cultural and religious identity in itself is not necessarily problematic. Violent conflicts between territorially rooted identity groups will only evolve if their normative system is not based strongly enough on universal values like justice, tolerance, fairness and openness for dialogue to settle disputes, often also traditionally inscribed in autochthonous cultures. The lack of such basic values can easily lead to processes of stereotyping and "othering" towards minority or neighboring groups with a distinct cultural and religious identity (Sauer 2009) eventually leading to violent identity based conflicts.

Identity based territorialization is a common process taking place among nations and ethnic groups around the world. Occurring on a larger scale, it has laid the foundation for the formation of the present system of nation states which even in times of pronounced globalization processes remains an important framework for any kind of inter – and transnational interaction. Whereas secular Western societies tend to become more and more footloose through spatially and ever more dispersed and far reaching patterns of social and economic interaction, land as a source of livelihood and faith as a sense of social belonging, plays a central role for collective territorially based self-identification in Sub-Saharan Africa where societies are usually more strongly rooted in traditions and religion.

In the case of Sudan, fundamentalist movements had already achieved in the 1980s the formation of a totalitarian Islamic state[2] (Stump 2008: 288). "External expressions of religious territoriality" (Stump 2008: 266) are probably most pronounced in predominantly Muslim countries as "Islam does not recognize the distinction of the secular and the religious in human affairs: it prescribes rules for nearly all contingencies in life, and sets the aim of introducing the comprehensive Islamic state on earth" (Chapman 1990: 114). The Sudanese conflict can be rendered more intelligible when the strong influence of religion is taken into account (Williams 2011: 137). As already stated, when Islamic law was forcefully imposed as the source of legislation and jurisprudence on the whole territory of the multiethnic and multireligious country in 1983, establishing the Sudanese "Islamic apartheid system" (Both 2003), violence and struggle for secession became the automatic consequences for the southerners. Stump (2008: 267) describes how this constellation of oppressive power and minority resistance conflate into armed encounter: "For religious minorities, conflict often emerges from patterns of resistance to the hegemony of dominant groups, while the latter have often provoked conflict through the suppression or persecution of minorities." From the perspective of critical geopolitics, the discursive and legal construction of Sudan as a purely Muslim territory, despite its multireligious demography, proves that the hegemonic religious actors and the respective political beneficiaries of that country had subscribed to the Islamist georeligious agenda of subjugating southern Sudan, by means of Jihad, to the "dar al-Islam," the "sacral geography" (Heidenreich 2010) of Islamic rule, an extreme form of religious territoriality. Bona Malwal, the author of *People & Power in Sudan: The Struggle for National Stability*, noted that as far back as 1966, the Islamists had declared that "the failure

---

[2] Hassan al-Turabi, the spiritual and ideological architect behind political Islam in Sudan and the countrywide introduction of Sharia law, defined this Islamic type of statehood and governance as follows (cited in Dau 2011: 27): "An Islamic state is part of Islam's comprehensive, integrated way of life where there is no division between the private and the public, between the state and the society."

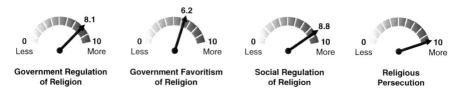

Fig. 184.1 Religious freedom country profile of Sudan before the independence of South Sudan (Source: Rainer Rothfuss and Yakubu Joseph, data from the Association of Religion Data Archives, 2012)

of Islam in southern Sudan would be the failure of Sudanese Muslims to the international Islamic cause. Islam has a holy mission in Africa and southern Sudan is the beginning of that mission" (cited in Cole and De Blij 2007: 561). The religious freedom country profile of Sudan before the independence of South Sudan reflected the reality of this ambition (Fig. 184.1); although the situation has got even worse in the north since the separation.

But, in order to capture the full spectrum of relevant causes and factors behind the Sudanese conflict it has to be considered, of course, that "conflicts are always provoked by several, mostly inter-related factors, there is always a specific set of other criteria, such as poverty, ethnicity, resource access, corruption and other forms of government dysfunction which play a role in the formation and development of a conflict that shows as a common baseline the religious identification of the involved groups" (Rothfuss and Joseph 2010: 49).

184.3 Territory, Identity and Terrorization: The Historical Struggle of Cultural Domination vs. Self-Determination Between Northern and Southern Sudan

Herbst and Mills (2012: 1) refer to the independence of South Sudan as "one of the most dramatic events of the twenty-first century [...] after years of bloodshed along racial and religious fault lines." Rader (2012: 53–54), in contrast, reduces this "fault line" to an "artefact of unequal economic relations" and highlights "economic and political marginalization" as the "principal driver of conflict in Sudan." She suggests that the territorial division between Sudan and the newly independent Republic of South Sudan should be seen as a consequence of the construction of "robust identity categories with a hard 'inside/outside' boundary" through economic and political marginalization rooted in "colonial interpretations of Sudan." From a historical perspective, it has to be doubted, however, whether colonialism played such an important role in the development of the north-south "fault line" as the century-old problem of Arab slave-raiding in the whole Black-African Sahel belt and Sudan,[3] existed long before the British arrived and even sought to protect the southern communities against the stiff resistance of slave traders (Dau 2011: 6–7). Thus, the implication of the Arab slave raiding in the conflict is too weighty to be underplayed and, to date, remains a vital historical antecedent of the north-south fault line. Ethnically, religiously and geographically, the group and terrains of victims have always been sharply defined: Arab slave raiders, due to their Muslim faith, were only allowed to enslave non-Muslim tribes (Euler 2008; Flaig 2009: 120–123) who, in the case of Sudan, happened to be the Black Africans in the south. As Breitlid et al. (2010: 93) put it "the slave raids in the South created an atmosphere of domination and subjugation, establishing the big divide and gap between the North and the South." By the middle of the twentieth century, southerners did not fear the British colonialists as much as their looming new 'masters' from the Arab-Muslim north (Jok 2007: 79) who, even before independence, usually referred to them disparagingly as "abiid" (or "our slaves") and seemingly had no respect for their autochthonous culture and identity (Breitlid et al. 2010: 93; Dau 2011: 19; Mathok 2009: 151; Mückusch 2008: 40).

In a historical perspective, Herz and Heide (2010: 425) underline that the distinct and territorially defined self-identification of the black Africans in the upper Nile basin dates back to 2000 BC when the area was called "Cush," "land of black people." After antiquity, over hundreds of years, the region was influenced by the Orthodox-Christian kingdom of Aksum with its center of power in Ethiopia. From the seventh century onwards, right after the foundation of Islam, Muslim Arabs expanded into the northern parts of Sudan. As Breitlid et al. (2010: 30) underline "for 1,000 years (from the sixth century to the fifteenth century AD) Christianity was the official religion of the three Sudanese kingdoms, resisting successfully the southward expansion of Islam." Only in 1560, the Black African Nuba region in the southern part of northern Sudan was conquered by Ottoman-Egyptian forces, whereas the southern regions still successfully defended their Ethiopian-Christian identity and respective socio-economic interdependencies (Smidt 2008: 22).

From 1821 to 1885, for the first time in history, the south had been conquered and administered – even though as a separate region – as part of modern Sudan through the Ottoman-Egyptian regime. From 1872 onwards, attempts were made by British governors to stop slave raiding[4] in southern Sudan (Dau 2011: 10; Fradin 2003: 90). Muhammed Ahmad ibn Abdallah, the self-styled *Mahdi*, declared Jihad against the external rulers, reconquered Sudan in 1885 and established a rigid Islamist regime under which slave raiding among Black African communities flourished again (Dau 2011: 10; Deng 2010: 66) until the British regained control in

---

[3] Flaig (2009: 148) shows how Islamic Jihad over the centuries served the purpose of raiding slaves from that region and that 17 million slaves were captured to meet the demands of mostly Islamic countries, by far exceeding the number of slaves within transatlantic trade (10.06 million). Flaig reminds that for one traded slave, on average one more person had been killed during the raids, doubling the overall figures of victims.

[4] This policy can be seen as an integral part of the then general abolition policy of the British and French colonial powers against the fierce opposition of the Islamic countries and traders, active in the respective regions (Flaig 2009: 210–212).

R. Rothfuss and Y. Joseph

1898. Muhammad Ahmad ibn Abdallah proclaimed himself to be the *Mahdi*, that is, the messianic redeemer or reformer of Islam (Cockett 2010: 13). From 1899 to 1955 the mainly "Arab-Muslim" northern and "Black African-Animist" southern parts of the colonial construct of "Sudan" were governed under the same colonial Anglo-Egyptian condominium, but under separate administrative units. To protect the vulnerable and "underdeveloped" south from further relentless exploitation and slave-raiding, visas were required to cross the north-south border. This "Closed District Ordinance" also restricted free access of Christian missionaries and consequently further deepened the isolation and "backwardness" (Dau 2011: 49; Mathok 2009: 11–12) of the south. Among northern proponents of Sudanese independence this colonial policy nurtured fears that the south shall gain independence separately from the north. In a geopolitical maneuver during the Juba conference in 1947, the British gave into the demand of northern Sudan and Egypt to incorporate the southern part of Sudan into a new and independent country. While Egypt sought to "safeguard the headwaters of the Nile by unifying Sudan under Arab leadership" (Nyang and Johnston 2003: 211), the British, in return, needed the support of Egypt for continued control over the Suez Canal. Jok (2007: 79) summarizes that the "British had created a mess by ruling the south and the north as separate entities without setting up two different countries."

184.4 Independence of Sudan in 1956: From External to Internal Colonialism and Civil War

When the new Sudanese administration was formed just before independence in 1956, out of 800 higher positions only six were conceded to southerners (Collins 2008: 65). Already in 1955 south Sudanese soldiers started an uprising against the colonial construct of a united Sudan, dominated by the "Muslim-Arab" north which feared that any concessions to the south concerning federalism or semi-autonomy were a threat to national unity (Dau 2011: 21). The Aboud military regime from 1958 to 1964 intensified its repressive military and Islamization agenda against the south, aggravating the "racial and religious polarization" which in the long run turned out to become an effective "politics of disunity" (Jok 2007: 1). Friday was declared the only public holiday also in the south and Sunday a usual working day. Christian missionary schools were converted into Islamic schools, Arabic replaced English as language of instruction and became the only language of administration (Nyang and Johnston 2003: 211). In 1964 all foreign Christian missionaries were forced to leave the country. Dau (2011: 22) summarizes that "the flood of Islamic schools and teachers that poured into the south after the expulsion of Christian missionaries convinced many southern Christians that the real reason behind the expulsion of missionaries was the imposition of Islam and Arabic culture on the south."[5]

---

[5] Kröpelin (2010), in contrast, upholds the counter-narrative, stating that "Christianity has been spread by western missionaries for a century, laying the foundation for today's problems."

Nyang and Johnston (2003: 212–213) interpret "the north's policy of 'Arabization'" in a way that "the Sudanese elite have tried to remake their non-Arabic-speaking populations in the image of Sudanese Arabism."

The internationalization of the first Sudanese civil war in 1967 and 1968 unveils the evolving geopolitical architecture behind the conflict scene. Besides some pragmatic "the enemy of my enemy is my friend" constellations, it reveals some cultural bonds between allied states that are, to date, of major importance in the ongoing struggle between Sudan and South Sudan. The regime in Khartoum oriented itself more and more toward Egypt and supported it in the Six-Day War against Israel. In return, Egypt and Libya provided weapons to the Sudanese government to fight the rebels in the south.[6] A military assistance agreement between Sudan with the Soviet Union provided improved access to advanced weapon technology and revealed the extent to which relationships between the regime in Khartoum and the West had been disrupted. Under the government of Dschafar al-Numeiri, Sudan broke off its cooperation with the Soviet Union and resumed relationships with the West. Under the patronage of the Ethiopian Emperor Haile Selassie I in 1972, he achieved, in negotiations with the rebel movements of southern Sudan, the Addis Ababa Agreement (Mathok 2009: 80–141), which was brokered by the World Council of Churches (WCC) and the All Africa Conference of Churches (AACC), as mentioned earlier. After 17 years of civil war there were more than half a million casualties mainly among civilians in the south. The end brought a period of relative autonomy and peace to the southern region of Sudan.

Unfortunately, the relative calm was short-lived when under the increasing political influence and pressure of the Muslim Brotherhood, headed by Hassan al-Turabi,[7] who had become Minister of Justice in 1979, President Numeiri revoked the semi-autonomous status of southern Sudan in 1983 and introduced strict Islamic Sharia law[8] as the sole source of jurisprudence and custom for the whole country (Burr and Collins 2010). This policy caused a "social tsunami" in the whole country (Deng 2010: 176).[9] Only a few years earlier oil resources had been detected in the

---

[6] These three countries were engaged from 1969 to 1974 in a geopolitical integration project with the vision of a joint Islamic state to form the core of an envisaged greater future pan-Arabic state.

[7] Hassan al-Turabi, one of the first Sudanese holding a doctor's degree from Sorbonne University in Paris, had considerable influence for many years in Sudanese politics as spiritual and party leader and became Speaker of the Parliament in 1996. As founder of the Popular Arab and Islamic Conference (PAIC), an extremist counter-organization to the Organization of the Islamic Conference (OIC), in the 1990 he contributed to the profile of Sudan as one of the worldwide hotbeds of Muslim extremism, joining over the years under the umbrella of PAIC jihadist and terrorist movements that strived for the further radicalization of global Islamist politics and transnational terrorism (Collins 2008: 219; Deng 2010: 187).

[8] On 1 August 2010, the National Assembly even called for the punishment of stoning to be introduced into national public order legislation, proving that the recent trend points toward an ever stricter implementation of Sharia law (Women Living Under Muslim Laws 2011: 22).

[9] Sharia, as the "sole source of legislation" has even been confirmed by 96 % of the (northern) Sudanese citizens in a referendum in 1998 (Collins 2008: 224).

southern part of Sudan. Thus any kind of federalism or semi-autonomy was seen as detrimental to the Islamist northern regime's quest to secure access to those important resources.

The de facto imposition of Islamic law even on all non-Muslims in the south led to the immediate break-out of the second civil war in 1983. In 1989 Islamist military forces around Omar al-Bashir, supported by the Muslim Brotherhood under the leadership of Hassan al-Turabi, usurped power. Al-Bashir established the Popular Defense Forces (PDF) which forcefully recruited young northern men in schools, universities and even on open streets to carry out a Jihad[10] against the "Black African Infidels" in the regions of southern Sudan and the Nuba Mountains (Collins 2008: 247; Weber 2008: 84). In the second Sudanese civil war, under the leadership of Dr. John Garang and with the support of some East African states and Israel within its Periphery Doctrine, the rebel forces largely joined under the umbrella of the Sudanese People's Liberation Army (SPLA) and evolved into a rather well organized underground army (Mathok 2009: 142–213).

Through Khartoum's military forces, the PDF and government sponsored nomadic militias, the so called "murahilin" (sometimes referred to as "Janjaweed" as in the Darfur civil war), which were mainly from the Arabic "misseriya" or "baqqara" (cattle herder) groups, the north carried out genocidal attacks in the south which caused over two million fatalities, displaced four to six million people (Collins 2008: 258) and, as "inherent part of the war tactic" (Jok 2007: 181), brought approximately 300,000 southerners into slavery who mostly were detained in Arab households or on farms in northern Sudan (Gerber 2006: 68). During the 22 years of civil war, vast traditional settlement areas were deserted and the autochthonous social systems of many tribes have been severely disrupted.

184.5 The Comprehensive Peace Agreement in 2005: Breakthrough in the Struggle for Liberty and Self-Determination

Largely united by the spirit of aspired self-determination, the well-trained and highly motivated troops of the SPLA finally managed to defeat the northern government. Taking into consideration Khartoum's access to sophisticated weapons and, through the support of China, to oil revenues since 1997, this victory only became possible in 2003 when a second front against the central government was opened by rebel movements from the Darfur region. Following a remarkable initiative of neighboring East African states, organized within the regional development organization of the Intergovernmental Authority on Development (IGAD), the SPLA and

---

[10]According to Williams (2011: 136) in 1992 "a fatwa was issued by ulama (scholars of Islamic doctrine and law) which defined the war against the SPLM as a jihad."

Fig. 184.2 Ubiquitous advertisements for the independence referendum (Photo by R. Rothfuss)

the government in Khartoum agreed on a Comprehensive Peace Agreement (CPA) in 2005.[11] The core of the CPA was a referendum which foresaw the possibility of independence of the south.[12] On 9 January 2011, despite repeated attempts of the northern government to delay and circumvent the referendum, under the pressure of mainly the U.S. and under the scrutiny of international electoral observers, the southern Sudanese population finally got the chance to vote for independence (Fig. 184.2).

Dau (2011: 58) mentions that on 21 September 2010 the President of Southern Sudan Salva Kiir had issued a call for national prayer for a breakthrough in referendum issues as he was on a visit to the United States[13] to meet world leaders and ask for their support for a timely and transparent referendum as stated in the CPA (Fig. 184.3). According to Dau (2011: 59) "senior government officials attended the prayer" and "it was declared [by church leaders] that we would go for 101 days of prayer and fasting for the nation and [...] peaceful conduct of the referendum." As Dau mentioned in an expert interview on 6 July 2011 in Juba, for most South

[11] As Collin (2008: 268) highlights "the role of the international community, including the "troika," consisting of the US, the UK and Norway, has often been exaggerated by an ebullient Western press."

[12] Egypt, supported by Libya, opposed this option of secession, fearing the loss of control of Nile water supply, and started a counter-initiative for negotiating peace between north and south following the premise of unity (Both 2003: 71; Johnson 2011: 24).

[13] According to Ben (2011: 13) on 1 November 2010 the US extended sanctions against Sudan in order to pressure the government to stick to the referendum deadline and, at the same time, offered to drop Sudan from the list of state-sponsors of terrorism if the referendum was held on time and the result was respected.

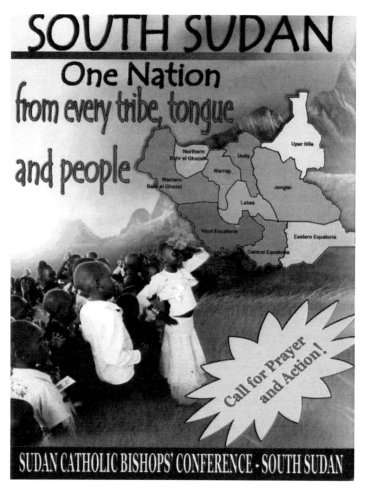

Fig. 184.3 Independence prayer week of the South Sudan Catholic Bishop's conference (Image courtesy of Caritas Internationalis, used with permission)

Sudanese the day of the referendum was the most important and longingly awaited day in their lives. In his book, *Free at Last* (Dau 2011), he bared his emotional experiences on the referendum day:

> January 9th came and it was a very special day that many of us thought would never come. We were unbelieving believers in that regard. [...] On the day itself, I was reduced to tears of excitement and joy as I witnessed a sea of jovial humanity, singing, dancing and ready to cast their votes and decide their own destiny. They came in their thousands to the polling stations as early as 1 AM and some, I later learned, never slept at all. I arrived at the polling station at 6 AM, thinking I was early. The day of voting being Sunday, I had intended to vote before 8 AM and then go to preach in a church but this was going to be hard with thousands of people queuing ahead of me. However, the voters were generous to let me jump the queue because they recognized me as their pastor. Consequently, I was able to vote shortly after the President and other dignitaries had voted. It was very emotional for me to

vote for the first time in 52 years of my life and in such a historic event. I stood still holding my ballot paper before the ballot box, for a minute or two prior to casting it. I whispered a prayer of gratitude to God and cast my vote in the name of the Father, of the Son and of the Holy Spirit. The poll attendant said "Amen." She had been watching my lips!

184.6   Independence of South Sudan in 2011: (Dis-)integration Between State Secularism, Christian Identity and Tribalism

On 9 January 2012, an overwhelming 98.83 % of the South Sudanese voted for independence. After decades of suffering and disappointing attempts to find a common way of living together within one nation state, South Sudanese used their first opportunity in history for fair elections to choose the "extreme fault-line management option" (Rader 2012: 45): To fall apart from the Muslim-Arab north, despite all looming difficulties and challenges as "one of the least developed places on earth" (Cockett 2010: 253) after "the civil war deprived an entire generation of the right to an education" (Downie and Kennedy 2011: 18).

The population of South Sudan had never experienced a well functioning statehood, granting to its citizens equal rights and opportunities for development. Therefore, in historical perspective, no strong national identity could evolve in relationship to a nation-state. Ethnic groups, tribes and clans have traditionally been the focus of self-identification (Lacher 2011; Jok 2011). Whereas the fight against the oppressive Muslim-Arab regime in the north unified most of the different tribes for several decades, now the recently established Republic of South Sudan faces the challenge of creating a common identity for all southerners, regardless of many persevering tribal disputes[14] (Fig. 184.4). As Downie and Kennedy (2011: 17) state, however, "ethnic rivalry is rarely the sole cause of violence" but an "extremely powerful mobilizing tool, used by politicians and conflict entrepreneurs to manipulate existing grievances over such issues as contested internal borders or the allocation of scarce resources."

There are no official numbers concerning the proportions of different religious groups in South Sudan. Based on earlier estimations, scholarly works often still assume that a majority of South Sudanese adhere to traditional African religions (Kaufmann 2004: 45). Williams (2011: 137–138) shows that allegiance to the "new found brand of Christianity" comprised only 20 % of southerners in 1980 and reached between 60 and 70 % in the year 2000. The World Council of Churches (WCC) reports the following shares for the total population of South Sudan of an estimated 10.8 million[15]: 60.5 % Christians, 32.9 % adherents to African traditional

---

[14] In early 2012 these erupted into violent conflicts between the Lou Nuer and Murle tribes over raided cattle in the northeastern part of South Sudan, costing the lives of 3,000 people.

[15] The government of South Sudan estimates that the real size of the population amounts to 11 to 13 million in contrast to the official figures of the 2008 census (8.26 million) under the northern Sudanese administration (www.southsudaninfo.com/Population_of_South_Sudan; 21 May 2012).

Fig. 184.4 On Independence Day, the government uses banners to remind people of South Sudan to discard tribal cleavages and unite in (Christian) faith (Photo by R. Rothfuss)

religions and 6.2 % Muslims.[16] Bishop Michael Taban Toro, the chair of the Sudan Council of Churches (SCC), however, estimates that Christians already make up 80 % of the population of South Sudan, by far outnumbering Animists (10 %) and Muslims (10 %).[17] The common trend in Africa that animist religions are more and more being overshadowed and replaced by the dominating monotheistic religions can also be witnessed in South Sudan. In a representative poll, carried out shortly after Independence in September 2011 among 2,225 South Sudanese from randomly chosen households from the urban and rural areas of all districts of the country, showed the astonishing result of 93 % identifying themselves as Christian, 2 % as Muslim and only 1 % as other (IRI, 2011: 67).[18] This information coincides with sky-rocketing membership figures of the established churches organized within the SCC, but also of the Pentecostal churches of South Sudanese origin, such as the Cush International Church which has emerged in recent years (Fig. 184.5). Interestingly, this autochthonous church considers itself as the spiritual trailblazer for the "rising nation of Cush (South Sudan) as the nucleus for the nation of Greater Cush" which "will be made up of all the neighboring states found in the ancient land of Cush from Lake Victoria in Uganda up to Alexandria in Egypt, the ancient Biblical land of Exodus and Babylon" (Cush International Church, 2011 cited in Zillinger 2013: 18).

Natsios (2012: 220–221) underlines that the "value that churches inculcate, and the extent to which they use their influence to pursue the public good, will determine

---

[16] WCC quotes as source the World Christian Database 2011. According to this statistics, Christian denominations rank as follows: Catholics: 3,579,937; Anglicans: 1,554,000; Protestants: 1,163,600; Pentecostals: 40,482; Orthodox: 3,000; Independent: 48,850 (www.oikoumene.org/en/member-churches/regions/africa/south-sudan.html; 20 May 2012).

[17] Interview with Bishop Taban on 6 July 2011 in Juba, Southern Sudan.

[18] It has to be assumed, however, that this survey was not truly representative, as the population practicing traditional African religions usually spreads over the most remote areas of the rural parts of South Sudan which are physically hardly accessible for interviewers.

Fig. 184.5 Baptism ceremony of the Pentecostal Cush International Church at the shores of the White Nile in Juba (Photo by R. Rothfuss)

the South's future. […] In South Sudan, Christian churches are *the* central private institution, exercising a powerful influence in the development of the emerging social order." Christianity as the uniting faith of South Sudanese citizens is being seen by the government as an important source of common identity and shared values. In a meeting with 70 delegates of the Sudan Council of Churches (SCC) on 6 July 2011 in Juba, the Minister of Information, Dr. Barnaba Marial Benjamin stressed the role of Christian churches to unite all tribes of South Sudan and serve as common source of identity instead of tribalism which too often has lead to violent confrontation, paternalism and corruption in the past. However, the IRI (2011: 63) poll shows that identification among the South Sudanese citizens with their new nation is already making tremendous progress: only 3 % of the respondents considered themselves exclusively as member of a tribe and 6 % self-identified more strongly as tribe than South Sudanese citizen. An astonishingly high 49 % identified themselves only through South Sudanese citizenship whereas tribal affiliations did not matter at all, and 23 % indicating that they "feel more South Sudanese than tribe." There may be a difference between a personal statement being made during an interview in times of euphoria shortly after independence and the often unconscious behavior and action of people in the struggles of daily life. Yet, a similar poll from Nigeria showed a very different pattern of self-identification through nationality. In that country struggling with a

Fig. 184.6 Independence celebrations in Juba, July 9, 2011: "We are not worse Arabs but better Africans" (Photo by R. Rothfuss)

north-south fault line conflict as well, only 5 % of Muslims and 9 % of Christians see nationality as their most important source of identity whereas 91 % of Muslims and 76 % of Christians see religion in that position (Pew Forum 2006) (Fig. 184.6).

In its struggle for self-determination, be it within a federal state of Sudan or an independent Republic of South Sudan, the SPLA always made clear that it rejected the adoption of any religion as state religion as was the case under successive Islamist governments in Khartoum, the most fundamentalist being the present one. In the 2011 transitional constitution of the Republic of South Sudan, regardless of the reference to the "Almighty God" in its preamble,[19] the separation of state and religion and religious freedom rights have been duly anchored in part I, article 8:

Religion:

(1) Religion and State shall be separate.
(2) All religions shall be treated equally and religion or religious beliefs shall not be used for divisive purposes.

---

[19] "Preamble: We, the People of South Sudan, grateful to the Almighty God for giving the people of South Sudan the wisdom and courage to determine their destiny and future through a free, transparent, and peaceful referendum […]." The full text of the transitional constitution can be downloaded from www.sudantribune.com/IMG/pdf/The_Draft_Transitional_Constitution_of_the_ROSS2-2.pdf (21 May 2012).

Fig. 184.7 Spiritual guidance from the Sudan Council of Churches for the geopolitical reorientation of the 54th state in Africa (Photo by R. Rothfuss)

According to Bishop Michael Taban Toro, the chair of the Sudan Council of Churches (SCC), the governmental Directorate of Religious Affairs in dealing with the SCC and the Islamic Council seemed to even miss out on the principle of equity in its attempt to uphold the practice of nondiscrimination. For example, if the government gave out equal amount of subvention or financial support to both faith groups, for example, for their contributions to the program during independence celebrations – despite their enormous difference in size: "This is where our government has to be very careful."[20]

Despite fully implemented state secularism it is no secret that 95 % of all key persons in the present South Sudanese government are strongly rooted in Christian faith.[21] In a meeting between Dr. Barnaba Marial, Minister of Information and Broadcasting and church leaders assembled in an SCC meeting in Juba on 6 July 2011, he highlighted the role of the church to unite the fractious tribal society of South Sudan. He stated, however, that despite the central contribution that the government expects from Christian churches in the long-term unification and national identity formation process, credibility of church and state organizations needed to be upheld by maintaining a due institutional distance (Fig. 184.7).

The national anthem surely is one of the most powerful elements contributing to a common sense of national identity. In August 2010, all citizens were invited by the then semi-autonomous government to compose the future South Sudanese national

[20] Interview with Bishop Taban on 6 July 2011 in Juba, Southern Sudan.

[21] Newspaper article in the Sudan Tribune edition of 24 February 2012: www.sudantribune.com/spip.php?iframe&page=imprimable&id_article=41707 (24 May 2012)

anthem. After three rounds of competition, students from the University of Juba won with their lyrics titled "South Sudan Oyee" (Hurray)[22]:

Oh God
We praise and glorify you
For your grace on South Sudan,
Land of great abundance
Uphold us united in peace and harmony.

Oh motherland
We rise raising flag with the guiding star
And sing songs of freedom with joy,
For justice, liberty and prosperity
Shall forever more reign.

Oh great patriots
Let us stand up in silence and respect,
Saluting our martyrs whose blood
Cemented our national foundation,
We vow to protect our nation
Oh God bless South Sudan.

In contrast to this joyful, yet historically sensitive anthem, reflecting the hope and aspirations of the youth of the young nation South Sudan, the formerly shared national anthem of Sudan "We are the army of God and of our land" and now anthem of the remaining rump state of the northern Republic of the Sudan breaths a burdensome spirit of fighting and suffering to uphold its imaginary supremacy[23]:

We are the army of God and of our land,
We shall never fail when called to sacrifice.
Whether braving death, hardship or pain,
We give our lives as the price of glory.

May this Our land, Sudan, live long,
Showing all nations the way.

Sons of the Sudan, summoned now to serve,
Shoulder the task of preserving our country.

Whereas the northern Sudanese state motto "Victory is ours" does not imply any ethical values and completely disregards the suffering of the unmentioned "others" who supposedly shall have been defeated, the official motto of the Republic of South Sudan highlights the humanistic ideals of a modern secular state that the deprived people of the Black African regions of Sudan had so long been longing and fighting for: "Justice, liberty, prosperity" (Fig. 184.8).

The South Sudanese flag summarizes some of the most important ingredients of national identity of the young nation, including a reference to the over two million people who died during the liberation struggle. Both in the anthem and the flag, the

---

[22] Insiders say that President Salva Kiir supported the proposed anthem so strongly that he even threatened critics of it, that he would not be available as the first president after independence if the religious elements were skipped. Listen to the anthem at: www.goss-online.org/magnoliaPublic/en/about/symbols.html (21 May 2012).

[23] Listen to the anthem at: www.nationalanthems.info/sd.htm (21 May 2012).

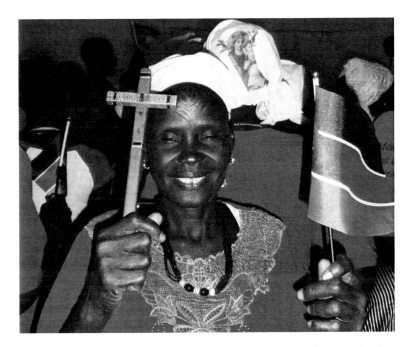

Fig. 184.8 Jubilant woman with two of the most important symbols of the new national identity: the Christian cross and the South Sudanese flag (Photo by R. Rothfuss)

remembrance of this national trauma is given a prominent place, lifting up to martyr status the departed liberation fighters, family members and friends who every single South Sudanese had to mourn. According to the constitutional state secularism, the government deliberately avoided any religious symbols or interpretations concerning the flag. Initially it has been used by the SPLM/A during the decade long liberation struggle. According to the government its colors symbolize the following[24]:

Red: Blood that was shed by the liberation struggle martyrs
White: Peace attained after many years of the liberation struggle
Blue: Waters of the Nile River, a source of life for the country
Green: The country's natural resources
Black: Black African skin
Yellow: Star guiding the country and its citizens.

In a gesture of forgiveness and reconciliation with the north, southern President Salva Kiir spontaneously declared in presence of the northern President Omar al-Bashir and a cheering crowd of half a million people attending the independence celebrations on 9 July 2011 in the new capital of Juba[25] that the Sudanese flag that

---

[24] Please refer to the section on "State Symbols" under the chapter "About South Sudan" on the government website: www.goss.org (25 May 2012).

[25] On 2 September 2012, the Council of Ministers decided to build a new planned capital at Ramciel, over 100 miles north of Juba, together with an international airport for large cargo planes, allowing to capitalize on South Sudan's strategic location between eastern, central and northern

had just been lowered would not be given back, as stated in the diplomatic protocol, but it would be kept in the archives of the Republic of South Sudan in remembrance of the long way that both countries had gone together. Already during the announcement of the referendum results on 9 February 2011, Salva Kiir had pointed out in his speech "Hope for a Better Future" that South Sudan was willing to seek amicable relations with its northern neighbor[26]: "To President Bashir […], we will work together to help Sudan regain its pride. We will assist in persuading the world to remove sanctions from Sudan." During Sunday service in the Saint Theresa Roman Catholic cathedral in Juba on 16 January 2011 the President even called southerners to forgive the north for the death of two million killed during the second civil war: "For our deceased brothers and sisters, particularly those who have fallen during the time of the struggle, may God bless them with eternal peace and, like Jesus Christ on the cross, forgive those who have forcibly caused their death."[27]

The Independence of South Sudan with its major and unusually tall Dinka tribe, its symbolic flag raising ceremony and fanfare play on Independence Day was seen by many South Sudanese, as the fulfillment of the prophecies of Isaiah, chapter 18 from the eighth century BC, concerning the fate of the Black African land of Cush alongside the upper Nile river[28]:

Sudan Will Be Punished[29]

[1]How horrible it will be for the land of whirring wings which lies beyond the rivers of Sudan. [2]It sends messengers by sea in boats made of reeds skimming over the surface of the water. Go, swift messengers, to a tall and smooth-skinned people, a people who are feared far and near, a strong and aggressive nation, whose land is divided by rivers. [3]Look when someone raises a flag on the mountains. Listen when someone blows a ram's horn, all you inhabitants of the world who live on the earth. [4]This is what the Lord says to me: I will keep quiet and watch from my dwelling place. My presence will be like scorching heat in the sunshine, like heavy dew in the heat of the harvest. [5]Before the harvest, when blossoms are gone and grapes are ripening from blossoms, he will cut off the shoots with pruning shears and chop off the spreading branches. [6]They will be left for the birds of prey on the mountains and the wild animals. The birds of prey will feed on them in the summer, and all the wild animals on earth will feed on them in the winter. [7]At that time gifts will be brought to the Lord of Armies from a tall and smooth-skinned people, a people who are feared far and near, a strong and aggressive nation, whose land is divided by rivers. They will be brought to Mount Zion, the place where the name of the Lord of Armies is.

It seems that with Isaiah, chapter 18 South Sudan has found another unifying "foundation myth" that obviously every nation has and needs to have to join its yet fractious society around a heroic narrative or a bright destiny. As a natural consequence of believing that this verse refers to the new nation of South Sudan, church

---

Africa to form a new international trade hub (www.sudantribune.com/Ramciel-s-survey-as-South-Sudan,42133; 25 May 2012).

[26]The full text of the speech can be downloaded from the government website under "Press Releases": www.goss.org (24 May 2012).

[27]www.sudantribune.com/Salva-Kiir-urges-Southerners-to,37650 (25 May 2012).

[28]The first time I was confronted with the belief that Isaiah 18 was prophesying the future of South Sudan was on a flight from Addis Ababa to Juba on 2 July 2011 when a pastor was sitting next to me reading the Bible.

[29]God's Word Translation (GW).

leaders started getting active in making the prophecy a self-fulfilling one. On 23 February 2012 several church representatives presented to Vice-President Dr. Riek Machar their plan to undertake a pilgrimage to Israel and asked him for financial support and advice for an appropriate gift to be brought to the Lord at Mount Zion.[30]

The government of South Sudan itself has found its own way to express its strong bonds with Israel and the Jewish culture symbolically. Repeatedly, it has confirmed that, as the newest state in the world, it will establish its embassy in the historical Jewish capital, the "Holy City" of Jerusalem, and not in the modern political capital of Tel-Aviv. While the Israeli government seems to appreciate this politically delicate expression of gratitude for its decade long support of the South Sudanese cause within its Periphery Doctrine, the Arab nations which had strongly opposed independence of the South (Both 2003: 70) presumably will see it as an affront against their cultural, religious and finally political aspirations projected onto this probably most intensively contested and – for three world religions – most mean-ingful place on earth. In a speech in early May 2012 at a rally for the presidential election campaign of the Muslim Brotherhood and now President of Egypt, Dr. Muhammad Mursi, an Imam made clear in the presence of tens of thousands of spectators who were chanting "Millions of martyrs march toward Jerusalem!" that the vision of a Pan-Islamic Caliphate, uniting Egypt with other Arab countries of the region will be realized with Mursi in power, and the capital of the "United States of Arabs" shall be the reconquered city of Jerusalem.[31] This religious symbolism may be qualified as just being a politician's attempt to please voter's emotions. Yet, the radicalization of the masses should not be underestimated in its potentially devastat-ing political effect for the Middle East if any unexpected incidents should deepen the fractions along this major sociopolitical fault line in the region.

184.7   Independence: From Renewed Calls for Jihad to External Reorientation

Because of current oil disputes, arising from disagreement over Khartoum's over-pricing of its pipeline and port for shipment of oil from South Sudan and alleged failure to account for crude worth hundreds of millions of dollars, South Sudan has completely stopped its oil production since the end of January 2012, despite its 98 % dependency of state revenues on oil (Natsios 2012: 210). As a consequence, tensions on the north-south fault line have been increasing rapidly again. On 20 April 2012, Sudanese President Omar Al-Bashir[32] delivered a hate speech in Port

---

[30] www.sudantribune.com/Machar-says-independence-of-South,42151 (21 May 2012).

[31] Watch film section 1:51 to 3:30: www.youtube.com/watch?v=ZjmT1uBgrG0 (24 May 2012).

[32] An arrest warrant for President Omar al-Bashir was issued on 4 March 2009 by the International Criminal Court due to crimes against humanity (murder, extermination, forcible transfer, torture and rape) and war crimes (pillaging and attacks against civilians) committed against Black African Muslim tribes in Darfur since 2003.

Sudan, directed mainly against Black African minorities and SPLM-North rebels in the Nuba Mountains, where an estimated 30 % of the population is Christian, but also threatening the South Sudanese government[33]:

> Our Lord, open up a new gateway for us – a gateway to Paradise, Allah willing. Our casualties go to Paradise and theirs go to the Hellfire. [...] If anybody dares to lay his hand on Sudan, we will chop it off. [...] We don't want any vermin left in the Blue Nile Province. We don't want any insect left in the Nuba Mountains. We want [Defense Minister] Abd Al-Rahim here to deliver us the land in its entirety before the end of autumn and leave the land of Sudan free of any poisonous vermin. [...] We want to liberate our brothers in the South from those vermin.

The Governor of the province of North Kordofan, Mu'tasim Zaki Al-Din further fueled hate against the neighboring country by resorting to conspiracy theory while addressing the political alliance between Israel and South Sudan and calling for Jihad against the infidels:

> South Sudan [...] wants to backtrack and continue in the path of enmity, and of subservience to global Zionism, which remains hostile to our country. But we praise Allah for allowing us these moments so that we could get dust on our boots on the path of Allah, in response to the call to Jihad, from which we took a respite for a while because we wanted peace. Peace is the basis of Islam. [...] We were previously in the forests of the south – steadfast "mujahideen." The path of Jihad continues, even if it took a few days respite for the sake of peace. Jihad, however, must return. Today, we in the Province of North Kordofan, just like in any other province of Sudan, have declared a state of alert and mobilization. Mr. President, these are the sons and daughters of Kordofan who have responded to the call of Jihad.[34]

When a "Jihad" is declared it allows the government to extract funds from Sudan's Muslim evangelical agencies, the so called "da'wa" (Williams 2011: 137). Moreover, it enables the Sudanese regime to raise support from other Muslim countries for the "Holy War."[35] To use De Blij's (2009: 80) terminology, we would qualify these statements as another sign of the alleged "twenty-first century Endarkenment" that Islam, seemingly, is not undergoing in Sudan alone. Extremist movements, like the Muslim Brotherhood that has nurtured Islamist totalitarianism in Sudanese politics throughout the past 60 years and the al-Qaeda related Ansar al-Suna, the mainstream Sudanese movement of Salafi Islam which incited recent

---

[33] The speech has been aired on Sudan TV: www.aijac.org.au/news/article/sudanese-president-s-all-out-call-for-genocide (25 May 2012).

[34] This statement is clear evidence of another present round of dangerous mobilization of religion and race in a struggle over resource control. Yet, some international observers tend to neglect the fundamentalist character of the Bashir regime (for example, Kröpelin, 2010) and the international community seems to have capitulated concerning the aim to preventing the northern regime from committing further war crimes and genocide.

[35] It is interesting, however, that in May 2012 the government of the Islamic monarchy of Qatar provided hundreds of millions of dollars to the Bank of South Sudan as a credit line for importing essential commodities which require hard currency (www.sudantribune.com/US-based-billionaires-pledge-to,42677; 25 May 2012).

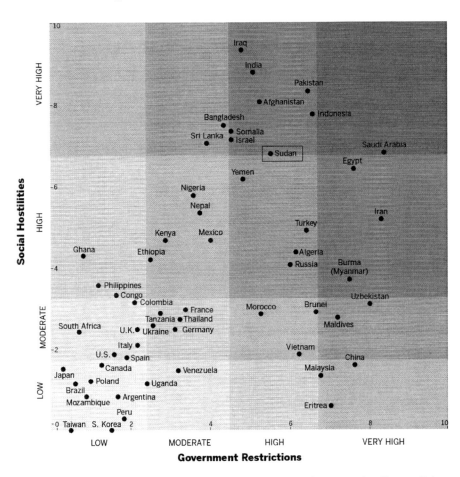

Fig. 184.9 Restriction of religious freedom in the 50 most populous countries (Source: Rainer Rothfuss and Yakubu Joseph, data from Pew Forum, 2009: 28)

attacks against Christians and the torching of a church in Khartoum[36] are on the rise in almost every country of the world. North Africa and the Middle East were religiously much more diverse a hundred years ago than they are today (Johnson and Kenneth 2009) (Fig. 184.9).

---

[36] According to a South Sudanese newsblog "On 21 April, a 300-strong mob attacked a Presbyterian church compound in Khartoum's Al-Jiraif District" just 3 days after "President Omar al-Bashir vowed to 'liberate' South Sudan from its government, labeling it an "insest" regime." (http://south-sudannet.activeboard.com/t48833714/muslims-storm-burn-and-loot-church-in-khartoum; 25 May 2012). According to the International Institute for Religious Freedom (IIRF) the atrocities were committed by members of the Salafi Ansar al-Suna.

In a comparative analysis, the Pew Forum's (2009: 14, 24) report *Global Restrictions on Religion* showed that Muslim governments and civil societies are on average the strongest opponents of the religious freedom rights laid down in article 18 of the United Nations Universal Declaration of Human Rights. Non-Muslim minorities faced, throughout the twentieth century, and still face discrimination and even systematic persecution in many predominantly Muslim countries of the world. The long-term results of historical genocide and forced eviction, as well as persevering systematic discrimination can be observed in a country like Turkey, widely hailed for having promoted a moderate version of Islam within a secular state: Christian population in Turkey dropped from approximately 20 % in 1900 to 0.15 % at present (Backes 2005: 128). A similar development has been sought by the Khartoum government in southern Sudan through a decade long civil war including genocide, ethnic cleansing of traditionally non-Muslim settlement areas and abduction into slavery combined with forced conversion to Islam. It is estimated that even today over 35,000 Black African citizens remain enslaved under northern Sudanese Arabs.[37] On 8 April 2012, the al-Bashir regime, apparently in another attempt to finally achieve the pure Arab-Muslim territory at least in the remaining rump state of the north, has set an ultimatum of 30 days for the remaining 500,000 to 700,000 Black Africans, out of previously over two million southerners, to leave the country or otherwise loose citizenship. Sudanese of southern origin, willing to move south, even if many of the younger ones have never lived there, reportedly have been kept from getting aboard airplanes to Juba due to allegedly missing special documents for travelling abroad.[38] The government of South Sudan, contrasting diplomatic habits of equal mutual treatment in the relationship with other nations, has offered all northern Sudanese living in the south dual citizenship rights. Furthermore, it has pleaded for a soft north-south border to allow the traditional cattle herders from the north, to maintain their traditional livelihoods and transhumance patterns.[39]

Optimistic observers had expected that independence leading to the formation of a sovereign state, recognized by the United Nations and the African Union, and the imposition of an international boundary would halt violent military action against the people and territories of South Sudan.[40] There is evidence that the government of northern Sudan, besides having charged from the south exorbitant pipeline usage

---

[37] The faith-based NGO Christian Solidarity International still repeatedly reports the liberation and transfer to South Sudan of hundreds of slaves every year (http://csi-usa.org/slave_liberation.html). Further details on the fates and living conditions of contemporary slaves in Sudan were documented by Gerber (2006) and Walzer (2009).

[38] It is worth noting that 42.35 % of the southern Sudanese living for decades as refugees or migrants in northern Sudan voted against independence on 9 January 2011.

[39] Statement by the South Sudanese ambassador to Canada, Joseph Moum Majak Ngor Malok at the Stand Leadership Conference, 15–16 October 2011 in Toronto (www.goss-online.org; 25 May 2012).

[40] As Wagschal et al. (2010: 19) have shown, only 14 % of all violent conflicts worldwide between 1945 and 2007 evolved across international borders.

fees,[41] has seized considerable amounts of crude oil by underreporting transduction volumes,[42] illegal shipping and clandestine secondary pipelines. Unsettled territorial disputes in oil rich regions and the lack of demarcation alongside 20 % of the 1,800 km (1,240 miles) – of common border incite the north to attack both nearby South Sudanese communities and the Black African ethnic minorities in the northern Sudanese Nuba mountains, the latter facing literally genocide with daily aerial bombings of villages. A local rebel commander from the Nubian SPLA-North reflects the seemingly unbridgeable gap between the Islamist regime in Khartoum and the marginalized peasant communities in the Nuba region, facing a major famine, by declaring[43]: "We will fight the Omar Bashir regime until we die. Then our children will fight. And if they die, then the Arabs will make babies with Nuba women until the population becomes Arab."

184.8   Conclusion: Geopolitics of the Subaltern

Schroer (2006: 214–215) explains the "renaissance of the regional and the local," in extreme cases, even leading to secession and hence the reconstitution of container spaces, as a growing demand of people for self-segregation and preservation, driven by globalization. The people of South Sudan opposed over decades the incorporation into the "geographical imagination" (Gregory 1994) of a culturally homogenizing container space of an artificially construed Muslim and Arab Sudan. For South Sudan, secession does not mean to constitute a new container space in order to escape the challenges of globalization. For the historically marginalized South Sudanese it means to escape from a post-colonial container-like prison and to achieve through statehood an appropriate framework and the right of self-determination and participation as dignified human beings in African and global society.

When Routledge (1998: 236) coined the term "anti-geopolitics" to describe the resistance within civil society that "challenges the notion that the interests of the state's political class are identical to the community's interests," he sticks to a mere organizational delimitation of the forces behind "anti-geopolitics." It is too simplistic to see the political, economic, and cultural hegemony of the state on the *dominant* side alone and the subaltern position of civil society, including religious institutions, the media and educational institutions, on the *dominated* side. The lat-

---

[41] Usual pipeline transduction fees vary between 60 and 80 cents of a US Dollar per barrel for the distance between the north-south border and Port Sudan. Khartoum, however, charged the South $32.20 per barrel.

[42] The NGO Global Witness has revealed discrepancies of 9-26 % between the oil production figures published by the Khartoum government and those reported by Chinese state-owned CNPC (www.globalwitness.org/campaigns/corruption/oil-gas-and-mining/sudan-and-south-sudan; 25 May 2012).

[43] New York Times documentary by Nicolas D. Kristof, 22 February 2012: www.nytimes.com/2012/02/23/opinion/kristof-dodging-bombers-in-sudan.html?_r=2&ref=opinion (25 May 2012).

ter often are mere instruments in the hands of powerful institutions, often at the service of the state but sometimes also of powerful segments of civil society. The case of South Sudan exemplifies very well that a state and military-like organized movement like the SPLA can be in the subaltern position over decades. On the contrary, an extremist religious movement stemming from civil society like the Muslim Brotherhood can serve to establish the cultural and religious hegemony of a totalitarian government that oppresses non-conforming parts of civil society. Therefore, we wish to speak of the "geopolitics of the subaltern" when oppressed groups, nations or even states strive in a systematic and strategic way for the recognition of their rights and for full emancipation. Both civil society groups and states, respectively state-like organizations, as we have seen, can be perpetrators as well as victims of oppression. Therefore, there is a need to develop appropriate strategies and concepts for a more broadly defined "geopolitics of the subaltern."

References

Association of Religion Data Archives. (2012). Sudan: national profile. Retrieved June 10, 2012, from www.thearda.com/internationalData/countries/Country_211_1.asp

Backes, R. (2005). *Sie werden euch hassen: Christenverfolgung heute*. Augsburg: Sankt Ulrich Verlag.

Ben, A. (2011). The road to independence of the Republic of South Sudan. *South Sudan Business Review, 1*(1), 9–14.

Both, P. L. (2003). *South Sudan: Forgotten tragedy*. Bloomington: 1st Books Library.

Breitlid, A., Said, A. A., & Breitlid, A. K. (2010). *A concise history of South Sudan*. Kampala: Fountain Publishers.

Burr, J. M., & Collins, R. O. (2010). *Sudan in turmoil. Hasan al-Turabi and the Islamist state*. Princeton: Markus Wiener Publishers.

Chapman, G. (1990). Religious vs. regional determinism: India, Pakistan and Bangladesh as inheritors of empire. In M. Chisholm & D. M. Smith (Eds.), *Shared space: Divided space. Essays on conflict and territorial organization* (pp. 106–134). London: Unwin Hyman.

Cockett, R. (2010). *Sudan: Darfur and the failure of an African state*. New Haven/London: Yale University Press.

Cole, R., & De Blij, H. J. (2007). *Survey of Subsaharan Africa: A regional geography*. New York: Oxford University Press.

Collins, R. O. (2008). *A history of modern Sudan*. Cambridge: Cambridge University Press.

Dau, I. M. (2011). *Free at last. South Sudan independence and the role of the church*. Kijabe: Kijabe Printing Press.

De Blij, H. (2009). *The power of place. Geography, destiny, and globalization's rough landscape*. New York: Oxford University Press.

Deng, F. M. (2010). *New Sudan in the making? Essays on a nation in painful search of itself*. Trenton/Asmara: The Red Sea Press.

Downie, R., & Kennedy, B. (2011). *Sudan: Assessing risks to stability*. Washington, DC: CSIS.

Euler, H. (2008). Scharia, Verfassung, Menschenrechte. Zur Problematik eines gemeinsamen Rechtsverständnisses im Sudan. In B. Chiari & D. H. Kollmer (Eds.), *Wegweiser zur Geschichte: Sudan* (pp. 161–169). Paderborn: Schöningh.

Flaig, E. (2009). *Weltgeschichte der Sklaverei*. Munich: Beck.

Fradin, M. S. (2003). *Jihad. The Mahdi rebellion in the Sudan*. Lincoln: Authors Choice Press.

Gerber, D. (2006). *Fünfzehn Dollar für ein Leben*. Basel/Gießen: Brunnen Verlag.

Gregory, D. (1994). *Geographical imaginations*. Cambridge, MA: Blackwell.

Heidenreich, E. (2010). *Sakrale Geographie. Essay über den modernen Dschihad und seine Räume*. Bielefeld: Transcript.

Herbst, J., & Mills, G. (2012). Introduction: Managing fault lines in the twenty-first century. In J. Herbst, T. McNamee, & G. Mills (Eds.), *On the fault line. Managing tensions and divisions within societies* (pp. 1–16). London: Profile.

Herz, D., & Heide, M. (2010). Sudan. In W. Gieler (Ed.), *Afrika-Lexikon: Geographie, Geschichte, Kultur, Politik und Wirtschaft* (pp. 423–433). Frankfurt am Main: Lang.

International Republican Institute (IRI). (2011). Survey of South Sudan public opinion, September 6–27, 2011. Retrieved May 19, 2012, from www.iri.org/sites/default/files/2011%20December%205%20Survey%20of%20South%20Sudan%20Public%20Opinion%2C%20September%206-27%2C%202011.pdf

Johnson, H. (2011). *Waging peace in Sudan. The inside story of the negotiations that ended Africa's longest civil war*. Brighton/Portland/Toronto: Sussex Academic Press.

Johnson, T. M., & Kenneth, R. R. (2009). *Atlas of global Christianity*. Edinburgh: Edinburgh University Press.

Jok, J. M. (2007). *Sudan: Race, religion and violence*. Oxford: Oneworld Publications.

Jok, J. M. (2011). Which way South Sudan? Cultural Diversity and the fundamentals of nation-building. Retrieved May 19, 2012, from http://africanarguments.org/2011/03/28/which-way-south-sudan-cultural-diversity-and-the-fundamentals-of-nation-building/

Kaiser, W. (2009). Entwicklungszusammenarbeit, Religionen und Konflikte. In J. Wilhelm & H. Ihne (Eds.), *Religion und globale Entwicklung. Der Einfluss der Religionen auf die soziale, politische und wirtschaftliche Entwicklung* (pp. 187–193). Berlin: Berlin University Press.

Kaufmann, E. P. (2004). *Rethinking ethnicity: Majority groups and dominant minorities*. London: Routledge.

Kobia, S. (2008, April 2). Keynote Speech at the end of the ecumenical solidarity visit to Sudan and the Sudan Church Leaders' conference, Juba, World Council of Church. Retrieved May 25, 2012, from www.oikoumene.org/en/resources/documents/general-secretary/speeches/at-the-sudan-church-leaders-conference.html

Kröpelin, S. (2010). *No land in sight*. Retrieved October 8, 2014, from http://www.theeuropean-magazine.com/stefan-kroepelin--2/6114-the-secession-of-south-sudan

Lacher, W. (2011). Staatsaufbau im Südsudan. Rahmenbedingungen, Erfolgsaussichten und Grenzen internationalen Statebuildings. Retrieved May 19, 2012, from www.swp-berlin.org/fileadmin/contents/products/studien/2011_S19_lac_ks.pdf

Mathok, D. (2009). *Politics of ethnic discrimination in Sudan. A justification for the secession of South Sudan*. Kampala: Netmedia Publishers.

Mückusch, A. (2008). Der erste Bürgerkrieg und die schwierige Unabhängigkeit 1956 bis 1983. In B. Chiari & D. H. Kollmer (Eds.), *Wegweiser zur Geschichte: Sudan* (pp. 39–49). Paderborn: Schöningh.

Natsios, A. S. (2012). *Sudan, South Sudan, and Darfur. What everyone needs to know*. Oxford: Oxford University Press.

Nyang, S., & Johnston, D. (2003). Conflict resolution as a normative value in Islamic law. In D. Johnston (Ed.), *Faith-based diplomacy: Trumping realpolitik* (pp. 210–227). Oxford: Oxford University Press.

Pew Forum. (2006). Nigeria's presidential election: The Christian-Muslim divide. Retrieved May 19, 2012, from www.pewforum.org/Politics-and-Elections/Nigerias-Presidential-Election-The-Christian-Muslim-Divide.aspx

Pew Forum. (2009). Global restrictions on religion. Retrieved May 19, 2012, from, http://pewforum.org/uploadedFiles/Topics/Issues/Government/restrictions-fullreport.pdf

Rader, A. (2012). Overcoming the past: War and peace in Sudan and South Sudan. In J. Herbst, T. McNamee, & G. Mills (Eds.), *On the fault line. Managing tensions and divisions within societies* (pp. 45–68). London: Profile.

Rothfuss, R., & Joseph, Y. (2010). The spatial dimension of Muslim-Christian conflict in the middle belt of Nigeria. *International Journal for Religious Freedom, 2*(3), 47–71.

Routledge, P. (1998). Anti-geopolitics. In G. Ó Tuathail, S. Dalby, & P. Routledge (Eds.), *The geopolitics reader* (pp. 236–248). London/New York: Routledge.

Sauer, C. (2009). The religious other as a threat: Religious persecution expressing xenophobia – A global survey of Christian-Muslim convivience. *International Journal for Religious Freedom, 2*(2), 45–71.

Schroer, M. (2006). *Räume, Orte, Grenzen. Auf dem Weg zu einer Soziologie des Raums.* Frankfurt am Main: Suhrkamp.

Smidt, W. (2008). Schwarze Königreiche von der Antike bis zur kolonialen Unterwerfung. In B. Chiari & D. H. Kollmer (Eds.), *Wegweiser zur Geschichte: Sudan* (pp. 16–25). Paderborn: Schöningh.

Stump, R. (2008). *The geography of religion: Faith, place, and space.* Lanham: Rowman & Littlefield.

Wagschal, U., Croissant, A., Metz, T., Trinn, C., & Schwank, N. (2010). Kulturkonflikte in inner- und zwischenstaatlicher Perspektive. *Zeitschrift für internationale Beziehungen, 17*(1), 7–39.

Walzer, C. (2009). *Out of exile. Narratives from the abducted and displaced people of Sudan.* San Francisco: McSweeney's Books.

Weber, A. (2008). Machstrukturen und politische Lager. In B. Chiari & D. H. Kollmer (Eds.), *Wegweiser zur Geschichte: Sudan* (pp. 75–85). Paderborn: Schöningh.

Williams, P. D. (2011). *War & conflict in Africa.* Cambridge: Polity Press.

Women Living Under Muslim Laws. (2011). Mapping stoning in Muslim contexts. Retrieved July 7, 2012, from www.wluml.org/sites/wluml.org/files/Mapping%20Stoning%20in%20 Muslim%20Contexts_Final.pdf

Zillinger, M. (2013). Christliche Modernisierungen. Erfolg und Misserfolg religiöser Reinigungsarbeiten und ihrer Medientechniken im Sudan. *Zeitschrift für Kulturwissenschaften, 7*(1), 1–21 (accepted paper, retrieved July 7, 2012, from www.uni-siegen.de/phil/medienwissenschaft/personal/lehrende/zillinger_martin/zillinger_reinigungsarbeiten.pdf).

Chapter 185
Quaker Lobbying on Behalf of the New START Treaty in 2010: A Window into the World of the Friends Committee on National Legislation

Stephen W. Angell

185.1 Introduction

The Friends Committee on National Legislation (FCNL), founded in 1943, is the "oldest registered ecumenical lobby in Washington, D.C." and "A Quaker lobby in the public interest," according to its website. The nonpartisan Friends Committee on National Legislation was established by, and is still governed by, a broad range of meetings and churches of the Religious Society of Friends (Quakers). This essay is especially concerned with one of its many lobbying successes in the year 2010,[1] its work on behalf of Senate ratification of the New Strategic Arms Reduction Treaty (START) between Russia and the United States. FCNL has more than 30 staff persons, and is governed by a General Committee of about 200 Quakers, hailing from 35 states and the District of Columbia. About 130 of these are appointed by 33 Yearly Meetings and other Quaker organizations; the others are appointed at large. Among the representatives appointed to the General Committee, the state with the largest number of General Committee members reside is North Carolina with 17, followed by Indiana with 14, Iowa and Ohio with 12, and

---

[1] FCNL's website listed eleven successes for 2010: New START ratification; payment by the US government to Native Americans for land use; a Senate resolution on genocide prevention; improvement of Native American health care; a boost for U.S. efforts to prevent nuclear materials from falling into the hands of violent extremists, increased support for prevention of deadly conflict through the U.S. State Department; stronger public safety on Native American reservations; grassroots support for new priorities for the federal budget; and for a new Afghanistan policy; and health insurance reform. http://fcnl.org/about/history/past_successes/ (June 8, 2012)

S.W. Angell (✉)
Earlham College, School of Religion, Richmond, IN 47374, USA
e-mail: angelst@earlham.edu

© Springer Science+Business Media Dordrecht 2015
S.D. Brunn (ed.), *The Changing World Religion Map*,
DOI 10.1007/978-94-017-9376-6_185

Table 185.1 Home states of members of the 2012 FCNL General Committee (Data sources: www.fcnl.org and communications with Alicia McBride)

Number	State
17	North Carolina
14	Indiana
12	Iowa, Ohio
11	Pennsylvania
10	California
9	Maryland, New York
8	Washington
6	Massachusetts, Oregon, Virginia
5	Colorado, Florida, Illinois, Michigan, Texas
4	District of Columbia, Nebraska
3	Maine, Minnesota, New Jersey, Oklahoma
2	Alabama, Alaska, Connecticut, Kansas, Tennessee, Utah, Wisconsin
1	Delaware, Montana, New Hampshire, New Mexico, South Carolina

Pennsylvania with 11. Each of these states is historically an area where there have been many Quakers and Quaker meetings[2] (Table 185.1).

185.1.1   FCNL's Legislative Policies

FCNL's work is governed by its "Statement of Legislative Policy." Its current policy statement was approved by its General Committee in 2003. While remarkably broad in scope, the policy statement does not include all issues, and it specifically does not take policy stands on issues on which Quakers themselves are not in agreement. One such area, listed as a "challenge," is abortion policy. The statement of legislative policy observes that "members of the Society of Friends are not in unity on abortion issues. Therefore, FCNL takes no position and does not act either for or against abortion legislation. On occasion, FCNL may appeal to lawmakers not to use the abortion debate to paralyze action on other legislation" (FCNL 2003: 3).

"Our legislative policy grows out of our basic belief that there is that of God in every human being and that God's love endows all creation with worth and dignity," according to the FCNL Policy Statement. The opening section of the policy statement is that "We seek a world free of war and the threat of war." This has been a major

[2]Communication from Alicia McBride, July 12, 2012; and "Friends Committee on National Legislation: Appointed Members of the 2012 General Committee, 11-06-2011," and "2012 At-Large members of the General Committee, 11-04-2011," http://fcnl.org/assets/about/who/generalcommittee11_12.pdf

focus of FCNL's work since it was founded in the midst of World War II. Its 1943 Policy Statement had expressed two "obligations" that Quakers had:

> 1. To work toward our goal of a non-military world order based so firmly on justice, spiritual unity, and voluntary cooperation that there is no place for war or armaments for war; [and] 2. To view sympathetically steps toward the goal, such as increasing international cooperation, progressive disarmament, and the bringing of national force under the control of world organization, while mindful that these fall short of our ideal. (Wilsm 1975:18)

Both the idealism and the practicality that were expressed in this initial Policy Statement have been of great importance for the development of FCNL, although later development would cause, of necessity, if Friends desired to be effective, a somewhat greater degree of emphasis upon the practical dimension of the work and a reflective contextualization of the idealism. Thus the 2003 statement affirms that "Friends are called to help build a framework for world peace." It emphasizes the positive roles to be played by the United Nations, treaties and covenants, and international frameworks for law, justice, and human rights. Promotion of "educational, cultural, scientific, commercial, and other exchanges among nations and peoples" are encouraged for the purpose of building mutual understanding. It urges a move toward comprehensive disarmament, including unilateral steps by the United States "toward its own disarmament, believing that other nations will respond affirmatively to this example." As a challenge, the Policy Statement asks: "In situations of genocide or intense conflict, how should Friends respond to the use of UN or other multilateral military forces to impose order or settlement?" (FCNL 2003:1–4).

In Part II of this Policy Statement, "We seek a society with equity and justice for all," FCNL commits itself to reducing the influence of money in the electoral process, holding "government institutions and officials to rigorous ethical standards of fairness, honesty, openness, and avoidance of even the appearance of conflicts of interest," criminal justice reform that would emphasize restoration over retribution, abolition of the death penalty, promotion of human rights, and support for the separation of church and state, among other issues. There are subsections ("A Society Free of Racism," "Relationships with Native Americans," and "Immigration and Refugees") which declare Quaker concerns with the treatment of minorities within the United States. Part III, "We seek a community where every person's potential may be fulfilled," provides a detailed vision for national and international economic justice, including such dimensions as employment, health care, housing, food production, and transportation. The final section, "We seek an earth restored," provides a strong ecological vision that addresses problems of climate change, population growth, and overconsumption, among others (Wilson 1975: 18).

185.1.2   The Expanding Scope of FCNL's Work

From its strong initial orientation toward addressing issues of war and peace, FCNL gradually branched out into these other areas over its almost seven decades of existence. In its early years, FCNL naturally moved into lobbying to support efforts to combat world hunger. In the late 1940s, much of those efforts concerned the peoples of war-torn Europe, but in the 1950s and 1960s, FCNL lobbying was devoted to directing American food aid to the peoples of such occasionally famine-stricken Asian countries as India, Japan, and China. Edward F. Snyder, Executive Secretary Emeritus of FCNL,[3] identified these major themes of FCNL lobbying toward assisting people in the developing world: "Channel most U.S. aid through the United Nations and other international programs; Focus first of all on basic human needs – especially food and hunger; Stress the importance of sound motivation for aid workers; and Stress the need for adequate capital as well as technical assistance in the development process." FCNL urged a greater American generosity in foreign aid, urging the U.S. to contribute 2 % of its gross national product. Snyder noted, however, that "this did not happen, in fact the trend has been toward even less U.S. true economic assistance" (Snyder and Mullen 1995: 151–153).

As a matter for FCNL concern, fairness toward victims at home of America's wars began as early as 1947, with lobbying on behalf of Japanese Americans displaced during World War II, and 1960, with lobbying on behalf of Native Americans, in this case, Senecas, whose reservation was threatened by a planned dam for the purposes of flood control. Also significant was FCNL participation in the broad coalition of religious, civil rights, and labor organizations that lobbied successfully in support of the landmark civil rights legislation for African Americans in the mid-1960s (Wilson 1975: 127–145). Edward Snyder stated that "not until the cities began burning in the 1960s – Watts in 1965, Detroit and Newark in 1967, and nationwide after Martin Luther King's assassination in 1968 – did FCNL seriously undertake a domestic legislative agenda beyond issues traditionally associated with Friends' testimonies for peace, civil liberties, civil rights and American Indians. The impact of urban and racial turmoil and its close relation to the Vietnam war virtually mandated a significant FCNL response." Eventually, FCNL turned to working on such domestic U.S. issues "within a much broader context, as FCNL has increasingly made connections between domestic and international priorities." From 1975 to 1976, the author of this present piece worked for FCNL as a legislative assistant, and previously an intern, focusing on domestic affairs with a particular emphasis on the areas of full employment and criminal justice reform (Snyder and Mullen 1995: 166).

In 1974, a $5,000 pledge from Nebraska Yearly Meeting of the Religious Society of Friends inaugurated a "Friend in Washington" program on behalf of Native Americans. FCNL's first two lobbyists devoted to Native American issues, Bryan Michener and Diana Payne, succeeded in persuading Congress to enact a Indian Health Care Improvement Act before the end of the Congressional term in 1976 (Snyder 1995: 119–121). FCNL's intensified involvement in lobbying on behalf of Native Americans has strongly continued in the almost 4 decades since, and Native American issues were an important part of FCNL's work in the year 2010.

---

[3] Snyder began working for FCNL in 1955. He was Executive Secretary when he retired in 1990. www.swarthmore.edu/library/peace/DG226-250/dg241esnyder.html

FCNL took the longest to begin work on the fourth and, so far, final section of its policy statement, "We seek an earth restored." In the mid-1970s, it was often observed by FCNL staff that FCNL could afford to focus its work elsewhere, leaving legislative advocacy on environment issues to the many fine, capable organizations already working in that area. As the imperatives to change behavior worldwide under the reality of climate change sunk in, and as Quakers at the grass-roots increasingly became involved in environmental activism, FCNL began putting more emphasis on work in this area. One way that it did so was by leading through example. FCNL's historic building at 245 Second St. NE in Washington, D.C., is across from the Hart Senate Office Building, had developed serious structural deficiencies by the 1990s, and it became clear that a major renovation would be necessary. After long and careful discussions, the FCNL General Committee decided to create a green building, with one consideration being that such a building would be "an important witness to FCNL's goal 'We seek an earth restored.'" Also of concern was to provide a compelling witness to Quaker "testimonies," or ideals, of simplicity and good stewardship. An important goal was for FCNL not to use fossil fuels to heat or cool its building; among other green measures, a geothermal loop was installed to assist with managing the building's temperature. The fundraising and construction on the FCNL building renovated with the most up-to-date green technology occurred between 1990 and 2005 (Fig. 185.1). In addition to receiving a variety of awards for modeling environmental best practices, the closeness of a Senate Office Building "has served to raise consciousness of green building among

Fig. 185.1 FCNL "Green Building" (Photo by Jim Morris, used with permission)

US Senators…Congressional staff and visitors have regularly taken pictures of FCNL's vegetated roof from the Hart Building roof and comment on this new addition to the local landscape." A problem arose in 2010 when the geothermal loop proved to be too warm, but FCNL promptly made appropriate adjustments to maintain its green building (Abbott, 2008; Volk 2010a; see also Veila 2008).[4]

FCNL does not engage in many of the tactics that other Washington lobbies utilize. For example, FCNL has never computed ratings for members of the Congress, either in the House or the Senate. In the past, FCNL published an annual newsletter which provided voting records for all members of Congress, with the vote of each Representative or Senator listed on selected issues, along with the position that FCNL would have preferred, but they no longer regularly publish Congressional voting records in their newsletters.[5] Instead, the advent of the Internet has enabled FCNL to tailor access to voting information about members of Congress in a manner they apparently see as preferable to a newsletter. Now up-to-date information on how one's own Representative or Senator voted can be found by entering one's zip code on the FCNL website http://capwiz.com/fconl/officials/congress/. Unless one recalls a zip code in another district, however, there is no convenient way to check on the voting records of members of Congress, other than one's own, through the FCNL website. (www.fcnl.org) And at no time has FCNL converted its voting records into percentages, as many other Washington lobbies do.

FCNL takes legislative policy stands on many more issues than it can reasonably work on in a concerted fashion at any one time. Thus, in addition to establishing policies, FCNL's General Committee, in consultation with Quaker meetings throughout the United States, also establishes legislative priorities, and it does so every 2 years. It attempts to discern where Friends can lead and where Friends' testimonies are especially relevant. For the 111th Congress, which included the year 2010, FCNL's legislative priorities included:

- promoting arms control, disarmament, and nuclear nonproliferation;
- building U.S., government capacity for peaceful prevention and resolution of deadly conflict;
- working for demilitarization of U.S. foreign policy and aid programs;
- working to remove U.S. bases and combat forces from Iraq; opposing torture and secret prisons;
- supporting immigration reform;
- promoting tribal sovereignty of indigenous peoples within the United States; supporting programs to alleviate poverty and address pressing human needs; and
- supporting bold action to counter global climate change.

---

[4]A fascinating video of the opening of FCNL's "green building" can be found at: "FCNL Green Building Ice Cream Social," YouTube, Feb. 8, 2008: www.youtube.com/watch?v=daxuz{TREl4

[5]According to David Culp, (private communication, August 13, 2012) FCNL is "considering 'publishing' a voting record for the 112th Congress (2011–2012) on our website in early October. We haven't made a decision yet."

- "As way opens," FCNL would continue working in areas where Quakers have a long-standing witness, for example, in favor of a just and lasting peace in the Middle East. The priorities are based in part on the issues that FCNL staff believe will be most prominent in the upcoming Congress, thus giving FCNL its maximal opportunity to help shape current legislation.

Given the limited scope of this article, it would be helpful to focus on one of FCNL's lobbying campaigns.

185.2   Anatomy of a Successful Lobbying Effort: The New START Treaty

185.2.1   Historical Background for FCNL's Lobbying for Nuclear Disarmament

In December, 2010, the U.S. Senate ratified the New START Treaty signed by President Barack Obama of the United States and President Dmitri Medvedev of Russia in April of that year. Treaty ratification requires a two-thirds supermajority, and the Senate, which had been tied up in bitterly partisan conflict for much of the year, mustered a decisive bipartisan 71-26 tally in favor of ratification. Yet the treaty ratification was a much closer prospect than the final vote would have indicated, and FCNL was deeply involved in much of the strategy that led to this ultimately successful outcome.

As we have seen, a call for "progressive disarmament" was part of FCNL's original 1943 policy statement. FCNL indeed has been active in working for "general and complete disarmament" since the 1940s. Shortly after nuclear bombs were dropped by the U.S. on Hiroshima and Nagasaki in 1945, FCNL's first executive secretary, E. Raymond Wilson, joined 33 other religious leaders in urging President Harry Truman "to take immediate steps to discontinue...production [of nuclear weapons], and to press for commitments by all nations outlawing the atomic bomb and also war" (Wilson 1975: 73–83). FCNL was actively involved in every subsequent decade in the disarmament effort, joining with other peace groups, for example, to achieve Senate ratification of the Partial Test Ban Treaty in 1963 (Wilson 1975: 227–229; Snyder 1995: 73–83).

At the beginning of 2010, FCNL had prepared to make arms control and disarmament a major focus of its work, but Executive Secretary Joe Volk believed that the Comprehensive Test Ban Treaty would be the main arms control measure that FCNL would be working on during the year. The high bar of a two-thirds supermajority meant that Senators from both the Republican and Democratic Parties would have to be committed to vote "yes" and Volk had secured at least one former member of Congress who was willing to help out this cause (Volk 2010b). Adopted by the United Nations General Assembly in 1996, the U.S. Senate had once before, in 1999, considered the CTBT for ratification and had rejected it by a 48-51 vote. In Prague, President Obama identified ratification of this treaty as a key objective of his administration: "To achieve a global ban on nuclear testing, my administration will immediately and aggressively pursue U.S. ratification of the Comprehensive Test Ban Treaty. After more than five decades of talks, it is time for the testing of nuclear weapons to finally be banned."[6]

185.2.2   FCNL Shifts Arm Control Work to New START Treaty

At the same time, however, the treaty between the United States and Russia limiting strategic weapons was soon to expire, and in the same speech in Prague, Obama mentioned his determination to sign a new START treaty with Russia. Table 185.2 presents a chronology of the strategic arms reduction treaties between the United States and the former Soviet Union and Russia since 1991. In April 2010, Obama and Medvedev signed the New START treaty in Prague, and FCNL shifted its arms control work to ensuring the ratification of this new treaty.

Table 185.2   Important dates for the START and New START treaties

Date	Event
July 31, 1991	President George H. W. Bush and USSR President Mikhail Gorbachev signed the first Strategic Arms Reduction Treaty (START I)
January 3, 1993	President George H. W. Bush and Russian President Boris Yeltsin signed the START II Treaty
December 5, 1994	The START I treaty went into force.
January 26, 1996	The U.S. Senate ratified the START II Treaty. However, Russian ratification was stalled until 2000.
May 24, 2002	President George W. Bush and Russian President Vladimir Putin signed the Treaty on Strategic Offensive Reductions. (SORT)
June 14, 2002	Russia formally withdrew from the START II Treaty, 1 day after the United States withdrew from the Anti-Ballistic Missile Treaty.
June 1, 2003	SORT went into force.
December 5, 2009	The START I Treaty expired.
April 8, 2010	President Barack Obama and Russian President Dmitri Medvedev signed the New START Treaty in Prague.
December 22, 2010	The U.S. Senate ratified the New START Treaty.
January 26, 2011	Following ratification by the Russian Duma, the New START Treaty went into force. The SORT treaty expired the following month, having been superseded by the New START Treaty.

Compiled by S. Angell from various sources

---

[6] Remarks by President Barack Obama, April 5, 2009, Hradcany Square, Prague, Czech Republic. www.whitehouse.gov

According to Joe Volk, FCNL staff soon "realized that the increasingly partisan atmosphere in Washington could make ratification of this important treaty much more difficult, perhaps impossible. Not ratifying the New START Treaty with a strong majority is unthinkable. If opponents of arms control defeat this treaty, then the future ratification of the CTBT would be 'toast'" (Volk 2010b). FCNL General Committee member Welling Hall states that "there had been real enthusiasm for working on CTBT, but that turned out to be beyond reach" (Private Communication, August 13, 2012).

Some preceding arms agreements between the United States and the old Soviet Union had actually allowed for increases in nuclear weapons on both sides, although such agreements may well have modestly restrained the growth of such stockpiles. The New START agreement does better than this. It required both the U.S. and Russia to reduce their deployed strategic nuclear weapons by about 30 %; from the current level of 2,200 such weapons for each county, the new START treaty mandated a reduction to 1,550 for each signatory ("Provisions," 2010:2804). These reductions would leave the United States with fewer nuclear warheads than any time since the early 1960s (Schaub and Forsythe 2010). When asked why FCNL was willing to put so much effort into this treaty ratification, Joe Volk responded, "The easy answer is that of the 24,000 nuclear weapons in the world today, some 22,000 are in the United States and Russia. With 90 % of the world's nuclear weapons, these two countries have a responsibility to take another step to reduce their stockpiles of nuclear weapons" (Volk 2010b).

On August 11, Joe Volk (2011) issued a statement on the New START Treaty:

> FCNL strongly endorses the New START treaty and urges the U.S. Senate to ratify it expeditiously…Our religious experience convinces us that it is not God's will for us to threaten each other with weapons of mass destruction. We unite with the 1955 statement of the Meeting for Sufferings of London Yearly Meeting (of Friends), "To rely on the possession of nuclear weapons as a deterrent is faithless; to use them is a sin.' … Threat reduction and confidence building between God's children in the U.S. and Russia may help to prevent a new nuclear arms race. What could please God more?"

Actually, although FCNL was in this instance able to give its strong endorsement to this treaty, FCNL almost always is involved in a process of weighing the good in a given piece of legislation, and its shortcomings as regards to Friends' principles. Thus, it is relevant to recollect here that FCNL's ultimate goal has always been "general and complete disarmament," and this treaty, however laudable, of course falls short of that aim. This dynamic is reflected in David Culp's measured praise of the Obama administration's nuclear disarmament efforts, as embodied in its Nuclear Policy Review, constituting "a modest incremental step in the right direction." That bolder actions to reduce nuclear weapons have not yet been forthcoming merely shows that "nuclear weapons supporters still control the Pentagon" (Weyl 2010:882). Or, as Volk observed specifically about the New START treaty, "Our ultimate goal is a world free of nuclear weapons. Senate ratification of this treaty is the next logical step toward that goal" (Volk 2010b).

185.2.3 Organizing Efforts at the State Level

FCNL's next step was to determine in which states it needed most to work. FCNL ended up focusing its work in support of the treaty in Tennessee, Ohio, Maine, Alaska, Georgia, and Indiana. Many of these states were represented by "swing senators," who could be persuaded to support this treaty. (One listing by FCNL's David Culp itemized 21 Republican Senators who fit into the favorable, undecided, or possibly persuadable categories; ratification would not succeed unless treaty advocates convinced at least eight or nine of these Republican Senators to vote for ratification, assuming that all Democrats and Independents voted for ratification, as indeed would be the case.)[7] One Senator who was especially crucial to ratification prospects than Tennessee and its Republican Senator Bob Corker. FCNL worked with Rotary Clubs in Tennessee to arrange presentations on behalf of those treaties, on two occasions flying a leading national expert to Tennessee. A leading newspaper in the state, the *Nashville Tennessean,* published what FCNL staff regarded as "an all-important editorial" supporting ratification. They also paid the airfare of "a former administrator of the nation's nuclear weapons complex to fly to Knoxville, Tennessee, to brief local business-people, clergy, and other opinion leaders on the importance of the treaty," which led to favorable coverage in the Knoxville newspaper. The results of all of this low-key, mostly behind-the-scenes work away from Washington, D.C. was just what supporters of the treaty desired.

FCNL facilitated a letter on nuclear disarmament by religious leaders in Indiana to Richard Lugar, the ranking Republican member of the Senate Foreign Relations Committee. Legislative Representative David Culp engaged in a weeklong speaking tour in Alaska, home to another swing Republican Senator, Lisa Murkowski. (Culp was FCNL's leading lobbyist on this issue; a representative of another peace organization, one who was familiar with his work, called Culp "indefatigable." Welling Hall describes Culp as "one of the unsung heroes of the arms control and disarmament community" ("Peace Action" 2010; Hall 2012). Joe Volk made a 3 day trip to Maine, where he met with newspaper editorial boards in an effort to persuade those boards to publish editorials favor to ratification of the treaty; Maine had two moderate Republican Senators, Susan Collins and Olympia Snowe.[8] In general, FCNL convened a network of state-level organizers on the START Treaty,

---

[7] The math changed slightly when Republican Mark Kirk replaced Democrat Roland Burris as Senator from Illinois in December. The enumeration of 21 Republican Senators as persuadable is drawn from David Culp, "Memo to NGO Colleagues," December 15, 2010, incorporated in "Getting Close on New Start ratification in Senate, Action Still Urgently Needed!" Peace Action Peace Blog, December 16, 2010. http://peaceblog.wordpress.com/2010/12/16/getting-close-on-new-start-ratification-in-senate-action-still-urgently-needed/

[8] "Coinciding with [Volk's] visit, three Maine newspapers endorsed New START: The Kennebec Journal, The Morning Sentinel, and The Times Record": Elise Connor, **"Friends Committee on National Legislation** on the Road to Push New START," August 5, 2010, www.ploughshares.org/blog/2010-08-05/friends-committee-national. See also: "Legislative Action Message: August 5, 2010: START in Danger as Senate Delays," in FCNL Archive 2010.

hoping to energize opinion leaders in key states to advocate on behalf of the treaty. Much of this work took place in half-hour conference calls every week. FCNL co-chaired the START and CTBT lobbying group, maintaining its email list. FCNL also chaired the Interfaith Nuclear Disarmament Group and maintained its email list as well.

The Senate Foreign Relations Committee backed the treaty on September 16 by a vote of 14-4, and Senator Corker and two other Republican Senators, Richard Lugar and Johnny Isakson, were among those who supported the treaty, reporting it to the whole Senate (Volk 2010b).

185.2.4   Responding to Concerns of Treaty Opponents

After this promising step, FCNL did not let up. It was apparent in the fall that there would be no vote prior to the midterm elections, but FCNL hoped that the treaty would be ratified in the "lame duck" session following the elections. The 2010 elections themselves, in which the Republicans gained six Senate seats, increased the urgency for ratification, as it was less likely that the more conservative Senate in the 112th Congress would ratify the treaty. For many months, the Obama Administration had worked tirelessly with a leading Republican Senator, Jon Kyl, who was raising criticisms of the New START treaty, and the Administration made substantial concessions to Kyl and Senate Republicans in the areas of missile defense and modernization of the nuclear weapons stockpile. At least one FCNL supporter regretted that political realities did not permit the administration to negotiate a New START treaty that would have cut back on missile defense programs.[9]

Yet Kyl and many other Senate Republicans responded to these concessions, not with belated support of the treaty, but with a call to delay its consideration until the new Congress convened. A longtime Republican Senator, Richard Lugar of Indiana, with decades of arms control experience, strongly favored the treaty, but throughout much of 2012 it was unclear how many other Republican Senators would follow Lugar's lead rather than that of Kyl. Some Republicans questioned the urgency of the treaty, seeing the nuclear aspirations of North Korea and Iran as the real issues to be addressed. Republican experts were divided, with five former Republican Secretaries of State (including Colin Powell and Henry Kissinger) urging its ratification, and experts from the Heritage Foundation urging delay (Cadei 2010a, b, c).

FCNL's first action message on the subject, that of August 5, 2010, gave three reasons for supporting the treaty: First, "START is needed to further reduce the number of deployed nuclear weapons in the arsenals of our country and Russia." Second, a point of particular concern to Senator Lugar of Indiana and other Republican moderates, inspections of Russian nuclear stockpiles had ceased in

---

[9] Dan Schlitt, "A Nuclear Weapon-Free World," http://nebraskansforpeace.org/nuclear-weapon-free-world. Schlitt is a member of FCNL's General Committee.

2009 with the expiration of the old treaty, and this new treaty needed to be ratified so that inspections could be resumed. Finally, if START was not ratified,
"prospects dim for future ratification of the Comprehensive Test Ban Treaty"
("LAM" 2010a). Three more action alerts on the same subject followed during
the fall, mentioning the substantial support for ratification from Powell, Kissinger,
and other members of the national security establishment from both parties, and
the "due diligence" of the Senate, which had held 21 hearings or briefings since
the treaty had been signed in April ("LAM" 2010b, c). FCNL's fourth (and, as it
turned out, final) action alert addressed the role of Senator Kyl in the process:
"[Kyl] wants to prevent a vote until next year when it will be much less likely the
treaty will pass." If FCNL constituents could persuade their Senators to speak
out right away, "your senators can make sure that Senator Kyl does not get a
'one-man veto'" ("LAM" 2010d).

185.2.5 A Final Push

The activities around treaty ratification in the fall of 2010 were extensive. David
Culp was in weekly contact with the White House strategy team. FCNL was also in
extensive contact with other colleagues in "NGO" organizations (that is, other organizations from a variety of constituencies lobbying on behalf of the treaty), and
FCNL freely shared its many materials with them. (A Friends' Meeting in Illinois
surely spoke for many Quakers and non-Quakers when it called FCNL's materials
on this issue "excellent" (Thomas 2010)), FCNL also arranged for a toll-free telephone number to the Capitol switchboard, and this, too, was shared with other organizations. Of course, Culp and other FCNL staffers attended the numerous hearings
and engaged members of Congress and their staff. Coalition action, in which like-
minded lobbying groups support each other on the causes they agree on, is a constant feature of work on Capitol Hill, and it is a method in which FCNL is very
experienced and uses with alacrity.

On November 14 and 15, FCNL sponsored a New START Lobby Weekend, and
this event was co-sponsored by Citizens for Global Solutions, Physicians for Social
Responsibility, and Women's Action for New Directions. This turned out to be the
largest lobbying weekend which FCNL had ever held, with 100 activists from 25
carefully selected states in attendance. Thirty-five Senators were lobbied, including 16 Republicans. The speaker was the chief negotiator of the New START
Treaty, Rose Gottemoeller. She is the Assistant Secretary of State and, within the
Department of State, heads the Bureau of Arms Control, Verification and
Compliance. FCNL's Deanna Boyd called the lobbying weekend "incredibly successful," and she was "particularly impressed that many of [the Senators'] constituents continued to lobby their senators on the importance of New START long after

the Lobby Weekend" (Volk 2010c). The larger significance is that all of the arms control and religious organizations lobbying together with FCNL had come to see the wisdom of its strategy of targeting the states of swing Republican Senators and had adopted the same strategy.

In December, as Senator Jon Kyl tried to have treaty consideration delayed until the 112th Congress, FCNL joined with more than 30 other organizations, including the American Friends Service Committee, the Hoover Institute, and the League of Women Voters, to push for a ratification vote in the lame duck session prior to adjournment of the 111th Congress. Writing to Democratic Senators Harry Reid, Richard Durbin, and Charles Schumer, they argued that "failure to act on the New START treaty *this year* would undermine the country's national security interests, as both our military leadership and numerous former Republican officials have noted" (Biencke 2010). When the Democratic Senate leadership did bring the treaty to the floor of the Senate, citizen lobbyists continued to contact their Senators in support of the treaty. According to David Culp, over 1,700 constituents used FCNL's toll-free number to contact their senators, requesting them to back the New START treaty (Volk 2010c).

A successful cloture vote (67-28) on Dec. 21 presaged the even larger margin (71-26)[10] for ratification on the following day, Dec. 22. The cloture vote attracted "yes" votes from 11 Republican Senators; 13 Republicans voted for ratification. (All of the Democratic and Independent Senators present had supported cloture and ratification.) Republican Senators in Maine, Alaska, and Tennessee, where FCNL had expended so much time and effort, all ended up voting for the treaty. As the accompanying map shows, there was widespread support for the treaty among Senators in all sections of the country, with opposition concentrated mostly in the Sunbelt and Rocky Mountain regions. In states President Obama had carried in the 2008 election, 55 % (6 of 11) of the Republican Senators voted for the treaty, whereas in states carried by his opponent Senator John McCain, only 23 % (7 of 31) of the Republican Senators voted in favor. Of course, those 7 Republican Senators voting "aye" from states that Obama had not carried provided the crucial margin for the constitutionally mandated two-thirds supermajority required for treaty passage. Adding in Democrats and Independents, 91 %

---

[10]The cloture vote was Senate Roll Call Vote 292, Dec. 21, 2010; the ratification vote, Senate Roll Call Vote 298, Dec. 22, 2010, "Adoption of the resolution of ratification for the New Strategic Arms Reduction Treaty (START) with Russia," http://library.cqpress.com/cqweekly/document.php?id=floorvote111-227090000&type=hitlist&num=0& Three Republican Senators, Kit Bond (MO), Sam Brownback (KS), and Jim Bunning (KY), all of whom were to retire from the Senate at year's end, were absent and did not vote on Roll Call Vote 298. These three Senators likely would have opposed the treaty if they had voted; see, for example,, Timothy Carpenter, "Brownback anxious on START," Topeka Capital-Journal, Dec. 21, 2010; Conn Carroll, "Bond on New START: A Dangerous Path," The Foundry, Oct. 13, 2010. http://blog.heritage.org/2010/10/13/bond-on-new-start-a-dangerous-path/

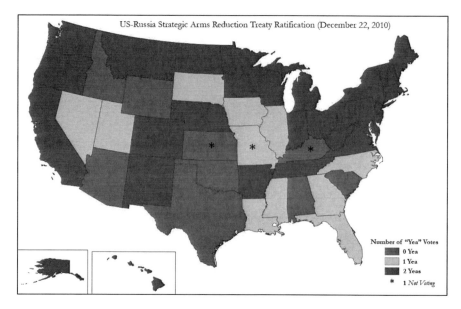

Fig. 185.2 Map of Senate Roll Call Vote 298, December 22, 2010, on adoption of the resolution of ratification for the New Strategic Arms Reduction Treaty (START) with Russia. (Map by J. Clark Archer, commissioned by the editor)

of the Senators in the 28 states that Obama carried voted in favor of the treaty, and 45 % of the Senators in the 22 states that his opponent carried voted in favor of the treaty.

Or, to look at this vote another way, 82 % of the Senators (36 of 44) in states that are home to three or more FCNL General Committee members supported treaty ratification; in the remaining states, only 63 % of the Senators (35 of 56) supported ratification (Fig. 185.2).

The *New York Times* justly declared this result to be "an important foreign policy victory" for President Obama,[11] and surely Obama would want to share credit with the grassroots lobbyists who helped to make the result possible. In a letter to Quakers throughout the United States, FCNL proclaimed that "the final Senate ratification vote of 71 to 26 was a huge success. It mobilized a Senate super majority in favor of international engagement as the path to greater security for our country. It sets the stage for work to prevent cuts in nuclear nonproliferation funding and ratification of the Comprehensive Test Ban Treaty" (Green 2011).

---

[11] The Obama Administration's extensive efforts on behalf of the New START Treaty were detailed in Peter Baker, "Obama's Gamble on Arms Pact Pays Off," *New York Times*, Dec. 22, 2010; and idem, "Arms Treaty with Russia Headed for Ratification," *New York Times*, Dec. 21, 2010 (quotation in text from Dec. 21 article).

185.3   Conclusion

With Republican gains of six Senate seats in the 2010 elections, it appeared that FCNL's work would become more difficult, and that the room for bipartisan action would be lessened. Joe Volk, FCNL's Executive Secretary at the end of 2010 (already having announced his retirement for March 2011), made just these kinds of observations a few days prior to the elections: "The outcome of a more conservative Congress will, of course, impact FCNL's work...In a 112th Congress that is more conservative, we would have to shift back to work that tries to impede or stop what we see as destructive initiatives" (Volk 2010d).

While there is considerable truth to Volk's prognostications, it is also true that, in FCNL's search for common ground, it has helped to find bipartisan majorities for sensible policies in the area of nuclear disarmament, no matter the political complexion of the Congress. FCNL staff report mixed results in the 112th Congress, some heartening and some deeply worrying. Successes in Congress, as of June 2012, include a strongly bipartisan 328-89 vote in the House to increase funding for the Global Threat Reduction Initiative, which assists in securing "vulnerable bomb-grade nuclear material from countries around the world."[12] Supporting this initiative were 192 Republicans and 136 Democrats; the opposition was evenly split between the two parties, with 44 Republicans and 45 Democrats voting "nay." The accompanying map (Fig. 185.3) shows strong support for the initiative in all regions of the country, with the small minority of opponents concentrated somewhat in the Sunbelt, and a scattering of opposition elsewhere, including the major metropolitan areas of New York, Chicago, and Houston.

However, Alicia McBride writes that "Congress is at a turning point . . . [and] progress could be slowing." She points to worrying legislation from the House of Representation, legislation that would block the implementation of the New START Treaty. "Even though this legislation isn't likely to become law, it's disturbing how much support it got in Congress" (McBride 2012). The *FCNL Newsletter* for May-June 2012 points also to attacks that legislators favoring nuclear disarmament are facing on the campaign trail. Specifically, Senator Lugar, the most energetic Republican advocate for the New START Treaty, "lost a primary in part because of his support for international treaties to reduce the numbers of nuclear weapons" ("A World Free," 2012).

Some Quakers are vocal about the urgent need for further action in nuclear disarmament. With considerable reason, physicist and FCNL General Committee member Dan Schlitt sees the 1,550 nuclear warheads still permissible for both Russia and the United States under the New START Treaty as "beyond any con-

---

[12] House Roll Call Vote 325, June 6, 2012, 4:51 PM, Fortenberry, R-Neb., amendment that would shift $17 million within the Defense Nuclear Nonproliferation account, with the aim of increasing funding for the Global Treat Reduction Initiative. http://library.cqpress.com/cqweekly/document.php?id=floorvote112-235608000&type=hitlist&num=0& David Culp, "Victories in the House: New Nuclear Bomb Plant Blocked," June 7, 2012, http://fcnl.org/blog/2c/Nuclear_bomb_plant_blocked/

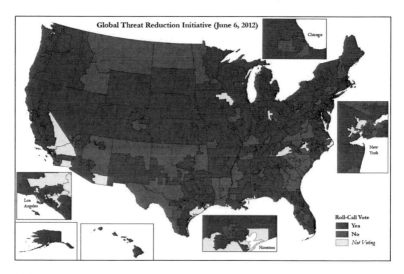

Fig. 185.3 Map of House Roll Call Vote 325, June 6, 2012, on Fortenberry, R-Neb., amendment that would shift $17 million with the Defense Nuclear Nonproliferation account, with the aim of increasing funding for the Global Threat Reduction Initiative. (Map by J. Clark Archer, commissioned by the editor)

ceivable rational need." He also concurs with Volk that it should be urgent for the United States to ratify the Comprehensive Test Ban Treaty, which "bans all nuclear explosions on Earth whether for military or peaceful purposes" (Schlitt 2012). Schlitt's views are shared by some academicians at military universities, such as the Air War College's Gary Schaub and James Forsyth of the School of Advanced Air and Space Studies. They believe that the United States "could address its conceivable national defense and military concerns with only 311 nuclear weapons" (Schaub and Forsythe 2010).

Unfortunately, these bold steps are exactly what would make conservative Republicans like Jon Kyl most nervous, and over the past two decades, Kyl and some other Republicans have gone to great lengths to forestall ratification of the CTBT (Cadei 2010c). On the other hand, there are many who strongly support the CTBT in the present and past national security leadership. Henry Kissinger and George Schulz, both former Secretaries of State, and former Defense Secretary William Perry can be numbered among the CTBT supporters. And, indeed, there has not been much progress on CTBT ratification since the ratification of the New START Treaty in 2010, although it is possible that Congress may vote on CTBT ratification in 2013 ("Test Ban" 2012).

But one thing seems certain. The kind of patient organizing in crucial states that was critical to the success of the New START Treaty ratification will be the most likely route to further treaty ratifications in the nuclear disarmament area. The Friends Committee on National Legislation, the dean of religious lobbies in Washington, D.C., has put its long experience to very good use, and in the effort for

New START treaty ratifications and similar lobbying campaigns, has developed a useful blueprint for lobbying in the public interest in Washington, D.C. Other public interest lobbyists would do well to take heed.

Acknowledgments   I would like to thank J. Clark Archer, Department of Geography, University of Nebraska, Lincoln, for his generous assistance with the maps that accompany this essay; Alicia McBride of the Friends Committee on National Legislation for her generous assistance with my research; and David Culp and Welling Hall for commenting on a draft of this essay.

References

A World Free. (2012, May/June). A world free of nuclear weapons: Congress at a turning point. FCNL Washington Newsletter No. 753, 1.

Abbott, M. P., & Abbott, C. (2008). Redefining Quaker simplicity: The Friends Committee on National Legislation, 2005. *Quaker Studies, 12*(2), 230–232, 241, 246.

Arms Control Association. (2012). www.armscontrol.org/factsheets/start1, www.armscontrol.org/factsheets/start2chron, www.armscontrol.org/factsheets/NewSTART

Atom Arsenals. (1993, January 4). It would reduce Atom Arsenals about 75%: Summit in Moscow. *New York Times.*

Beincke, F. (2010, December 9). Letter from Frances Beinecke et al. to Harry Reid, Richard Durbin, and Charles Schumer. www.lwv.org/content/league-urges-senators-bring-new-start-treaty-senate-floor

Cadei, E. (2010a, May 17). For GOP, a Cold War Vet is not hawk enough. *CQ Weekly,* pp. 1191–1193. http://library.cqpress.com/cqweekly/weeklyreport111-000003664040

Cadei, E. (2010b, October 25). Facing a world without START. *CQ Weekly*, pp. 2420–2422. http://library.cqpress.com/cqweekly/weeklyreport111-000003755061

Cadei, E. (2010c, December 6). Measuring Kyl's Clout. *CQ Weekly*, pp. 2800–2806. http://library.cqpress.com/cqweekly/weeklyreport111-000003773456

FCNL. (2003, November/December 1–4). FCNL Policy Statement (Approved 11/03), *FCNL Washington Newsletter.*

Friends Committee on National Legislation website. http//fcnl.org

Green, G. D. (2011, March). Clerk, FCNL General Committee, To friends everywhere: The importance of friends lobbying in the public interest in 2010–2011, In *FCNL Archive 2010.*

Hall, W. (2012, August 13). Culp "unsung hero": Private communication, Welling Hall.

LAM. (2010a, August 5). Legislative action message. START in danger as senate delays. In *FCNL Archive 2010.*

LAM. (2010b, September 16). Legislative action message. START's next Stop. The senate floor. In *FCNL Archive 2010.*

LAM. (2010c, November 4). Legislative action message. Senate should listen to the american people. In *FCNL Archive 2010.*

LAM. (2010d, November 18). Legislative action message. Don't let Sen. Kyl stand in the way on START. In *FCNL Archive 2010.*

Landei, M., & Erlanger, S. (2011, February 11). U.S. and Russia activate a new arms treaty. *New York Times.*

McBride, A. (2012, June 15). Nuclear weapons: Congress at a turning point. Email communication.

Obama. (2010, April 8). Obama, Medvedev sign treaty to reduce nuclear weapons. *The Washington Post.* Available at www.washingtonpost.com/wp-dyn/content/article/2010/04/08/AR2010040801677.html

Peace Action. (2010, December 16). Culp "indefatigable": "Getting close on new start ratification in senate, action still urgently needed!" Peace Action Peace Blog.

Provisions. (2010, December 6). Provisions of New START. *CQ Weekly*, p. 2804. http://library.cqpress.com/cqweekly/weeklyreport111-000003773462

Ratification. (2011, January 26). With Russian ratification of New START, what's next for US-Russia relations? *The Christian Science Monitor*. Available at http://www.csmonitor.com/World/Europe/2011/0126/With-Russian-ratification-of-New-START-what-s-next-for-US-Russia-relations. Accessed 9 June 2013.

Schaub, G., Jr., & Forsyth, J., Jr. (2010, May 23). An arsenal we can all live with. *New York Times*. www.nytimes.com/2010/05/24/opinion/24schaub.html

Schlitt, D. (2012, May/June) A nuclear weapon-free world. http://nebraskansforpeace.org/nuclear-weapon-free-world. The comprehensive test ban treaty: Still a good idea. *FCNL Washington Newsletter No. 753,* pp. 2–3.

Snyder, E. F., & Mullen, T. (Eds.). (1995). *Witness in Washington: Fifty years of friendly persuasion*. Richmond: Friends United Press.

Test Ban. (2012, May/June). The comprehensive Test Ban Treaty: Still a good idea. *FCNL Washington Newsletter*, pp. 2–3.

Thomas S. (2010, November 11–14). Report on annual meeting of friends committee on national legislation. www.oakparkfriends.org/friends-committee-on-national-legislation.html

Treaty. TREATY: Bush, Gorbachev Salute Treaty as new foundation of peace. *CQ Weekly*, August 3, 1991, pp. 2192. Available at http://library.cqpress.com/cqweekly/WR102403999 and http://library.cqpress.com/cqweekly/document.php?id=floorvote111-227090000&type=hitlist&num=0. Accessed 9 June 2013.

Veila, M. (2008, August 28). Building a Greener America. *Bloomberg Businessweek*.

Volk, J. (2010a, October). *FCNL's green building at 5*. Washington, DC: Friends Committee on National Legislation.

Volk, J. (2010b, October 15). Memo to FCNL key supporters. In *FCNL Archive 2010*.

Volk, J. (2010c) Executive secretary's report – Fourth quarter 2010. In *FCNL Archive 2010*, 39–40.

Volk, J. (2010d). Executive secretary's report – Third quarter 2010. In *FCNL Archive 2010*, 2–3.

Volk, J. (2011, August 11). FCNL executive secretary. FCNL executive secretary's statement on the New START Treaty. http://fcnl.org/issues/nuclear/fcnl_executive_secretarys_statement_on_the_new_start_treaty/

Weyl, B. (2010, April 12). Some evangelicals agree with Obama on nuclear arms control. *CQ Weekly*, p. 882. http://library.cqpress.com/cqweekly/weeklyreport111-000003638549

Wilson, E. R. (1975). *Uphill for peace: Quaker impact on congress*. Richmond: Friends United Press.

Chapter 186
Interpreting the Transforming Geographic Mosaic of Religion in America: Its Impact on Congressional Representation and Increasing Political Polarization

Josiah R. Baker

186.1   Introduction

This chapter examines the changing geographic and political dynamics of religious composition within the United States Congress for the period 1959–2009. Of the three U.S. government branches, Congress best reflects the will of the broader populace because they are directly elected (unlike the President or the Supreme Court). Since its members are directly elected by the people, democracy within the U.S. governmental system is best observed in the voting behavior and representation witnessed in the House of Representatives and in the Senate. This chapter focuses on the contributing factors that have helped alter the religious make-up of the House and Senate over the past 50 years. The findings indicate that the increasing diversity of religious thought has contributed to the increasing polarization of the two parties, as observed in the voting patterns, and the increasing dysfunctionality of both elected bodies.

To what extent have there been changes in religious composition of the U.S. House and Senate since 1959? In particular, does the change of religious affiliation correspond to changes in the religiosity of the greater U.S. population? Are American Democrats and Republicans represented by people who represent their religious affiliation? How often and under what circumstances can religious affiliation influence voting behavior more than partisanship? What geographic trends are apparent by religious representation from state to state?

In *Religion on Capitol Hill*, Peter Benson and Dorothy Williams (Benson and Williams 1982) used a specially designed random sample from 535 House and Senate members to evaluate the religious beliefs, values, attitudes, and behavior of 80 members of Congress. From this effort, Benson and Williams concluded the

J.R. Baker (✉)
Methodist University, Fayetteville, NC, USA

Economics and Geography, Methodist University, Fayetteville, NC 28311, USA
e-mail: josiah.r.baker@gmail.com

© Springer Science+Business Media Dordrecht 2015
S.D. Brunn (ed.), *The Changing World Religion Map*,
DOI 10.1007/978-94-017-9376-6_186

following: (1) nearly all Congress members believed in God; (2) Congress members were no less religious than the people they represented (Benson and Williams 1982: 81), (3) political conservatives and liberals perceived themselves as nearly equally religious, and (4) Evangelical Christians, because of black Americans, were not a united conservative political force.

According to Benson and Williams, "Three-quarters of the members of the U.S. Congress were genuinely religious, to the point where religion was a definite influence on their thoughts and actions, an integral part of their lives" (Benson and Williams 1982: 72). In addition, they wrote that a majority of the 80 members that they interviewed acknowledged that religious beliefs and values influenced their votes. The significance of religion in U.S. Congressional voting behavior was also supported by E.J. Dionne and Alan Wolfe (2006).[1] They provided detailed analysis of religion's role in American politics which surmised, "When evangelicals began to flex their political muscle in the 1970s, almost no political scientists – and very few journalists– paid them much attention. That we could ignore something so important was…one of the biggest mistakes our fields could have made" (Nivola and Brady 2006: 206).

Further research shows that "religious affiliation significantly affects specific and general roll call voting behavior, and that religious groups in Congress differ from one another, both within and across parties" (Fastnow et al. 1999: 687). House abortion votes between 1959 and 1994 were examined in their article, "Holy Roll-Calls: Religious Tradition and Voting Behavior in the U.S. House." They concluded that "results not only show that religion affects congressional voting, they also suggest the way religion plays a role in congressional decision making." In addition, Fastnow, Grant, and Reynolds wrote, "Religion stands up to the competition of a variety of other theoretically powerful predictors of voting behavior, including party, constituency preferences, and important demographic indicators" (Fastnow et al. 1999: 697).

Another reason that can explain the unexpected significant influence of religion in Congressional voting is that lobbyists can ignore morality issues because they often lack financial and political considerations that would otherwise generate interest to the special interest groups that they represent. Representatives and Senators feeling less pressure are more able vote according to their conscience, whereas in areas such as defense spending and tax bills lobbyists clearly possess reasons to want to exert influence on voting (Marriott 2005). In a paper presented at the Midwest Political Science Association in April 2005, David Mariott studied legislative roll call voting in the U.S. House on morality issues. He compared gay rights and abortion roll call votes in the Senate and in the House over 3-years. Marriott observed that accountability can change according to the difference in length of terms. Therefore, Senators are more independent than Representatives on morality votes.

---

[1] Nivola and Brady (2006). E.J. Dionne, Chapter 4, and Alan Wolfe, "Myths and Realities of Religion in Politics."

186.2   Religion's Contribution to Polarization

Since the rise of the Religious Right during President Ronald Reagan's 1980 election campaign, a culture war has wracked American politics.[2] According to Hunter (1991, 1994) and other supporters of the culture war hypothesis, the conflict has emerged over such social and moral issues as abortion, homosexuality, affirmative action, and school prayer. Evangelical Christians and liberal Jews as they have become more dominant within the Republican and Democratic Parties have contributed to an uncompromising political environment. True political moderates in the House and Senate have become less common. *Polarization*, which is best defined as the overall proportional voting variance from one party to the other on a particular vote, is the ultimate result of the inability of Congressional factions to find a common ground necessary for compromise.

A contrast in the vision of a "good society" according to Hunter (1991, 1994) is the blame for the emergence of such political divisions on issues like abortion. One view is grounded in an orthodox, transcendent understanding of the world, based on "God-ordained" fundamental beliefs, values, and norms. This view postulates that people must obey the exact laws given to them by God as they are written in the Bible. A fundamental or literal interpretation of the Bible is the basis of this world view (Hunter 1991, 1994). In this study Evangelical Christians[3] and conservative Catholics generally follow this approach.

The other perspective, often called "progressive" views life as a process that is forever unfolding or evolving and that "truth" is best interpreted by the use of contemporary scientific theories and reason. The progressive perspective does not necessarily deny the existence of a God but it relies on a broader interpretation of traditions. Philosophically, this approach is rooted from ideas of the Enlightenment Era in the eighteenth century. Reformed Jews, Mainline Protestant, and liberal Catholics mainly subscribe to this approach in blending science, reason and religion into a form of faith.

These two approaches are more than just a difference of opinion. Specific issues such as abortion have become "anchoring points" for two distinct visions of the good society. Aside from abortion, the two groups differ on taxes, the size and scope of government, welfare concerns for the poor and the disadvantaged, and a variety of social programs. While fundamentalists may view many of these policies as futile, socialistic, or morally wrong, progressives often believe that government is not only effective in addressing these issues, but that there is a moral prerogative to promote society's general welfare.

---

[2] DiMaggio et al. (1996), Evans et al. (2001), Evans (2002a, b), Gitlin (1995), Hall and Lindholm (1999), Hunter (1991, 1994), Luker (1984), Mouw and Sobel (2001), Smelser and Alexander (1999), Wuthnow (1988), Williams (1997).
[3] Black Protestants are often an exception due to party loyalty.

186.3   Historical Overview

Benson and Williams (1982) observed that there are very few historical studies on religion and the American Congress. Despite such little attention more scholars are realizing that religion does influence Congressional voting (Hertzke 1988; Mariott 2005). Senators and House Representatives hold a wide range of personal religious beliefs, and may at times question their beliefs. The First Congress (1787–1788) consisted of Episcopalian, Congregational, and Presbyterian churches, with smaller numbers belonging to Christian Reformed, Catholic, and Quaker denominations. Though there were 66 members, only 46 declared a religious affiliation. In contrast, no House member of the 111th U.S. Congress (2009) was identified as having no religious affiliation and only one Senator.

In colonial times, political leaders in America were heavily influenced by philosophical trend of the times: the Enlightenment. A sometimes competing, but sometimes complementary philosophy found in Puritanism also dominated colonial political philosophy (Benson and Williams 1982: 86). American ideology was, indeed, shaped by a fusion of Enlightenment and Puritanical thinking. Benson and Williams referred to Puritanism as a "nation-building religion" because of its vast influence.[4] This view contributed to a rejection of a central religious authority such as the Catholic Church or the Church of England. Puritans and Congregationalists rejected the Pope and the King of England. Religion was supposed to be practiced at the local, not the national level. This belief contributed to the separation of Church and State and the desire for members of Congress to shy away from religious issues when crafting legislature.

Early American Protestants played an active role in contributing to the conditions that led to the First Great Awakening, a spiritual movement that focused on making the Bible easy to understand, and more active worship practices. In the early to middle decades of the nineteenth century, the Great Awakening was greatly aided by the ability of the distribution of printing press technology through the construction of the railroad and other infrastructure projects such as the Erie Canal which made the mass production of Bibles affordable for low income farmers. Indeed, technological innovations that allowed the individual to gain direct access to the writings of Christianity enabled the emergence of new Christian-based movements such as the Seventh Day Adventists, the Church of Christ, and Mormons. Today, these churches are classified as "Evangelical."

In the eighteenth century and earlier, based on Biblical references to a "Promised Land," the Puritans and Pilgrims created their colonies with the belief that God had a special plan for developing America (1982: 89–93). This mentality of a divine role for the United States became widely accepted by other denominations. The phrase, "God Bless America," which Presidents and other politicians frequently invoke when trying to unify the country is a reflection of the widespread acceptance or

---

[4] Many Puritans were Presbyterians, while others were Congregationalists. Much of the difference had to do with how the groups were governed.

perception of God's special role for America. Those who believed the latter felt united in their belief that the ideals of freedom, hard work, prosperity, and promise would lead them toward a national destiny in keeping with divine intent. According to Benson and Williams, "It was a world view that bound the nation together, cutting through the theological and cultural divisions of the day" (1982: 93). Though not all citizens in the early years of the United States agreed, the idea of a providential God with a divine plan for the new nation served as a centripetal force for a vast number of Americans (Hertzke 1988: 23–26).

186.4   Religious Diffusion

As previously stated, the United States Congress was dominated by Mainline Protestants from pre-Revolutionary Period until the mid-twentieth century. Much of the present-day American societal norms and practices are still a reflection of Mainline Protestant cultural values and beliefs. In reviewing the diffusion of religious representation in the United States Congress from 1959 to 2009, four main religious groups (Mainline Protestants, Evangelicals, Catholics, and Jews) are classified according to their beliefs and perspectives.[5]

The next four maps will show that since 1959, the religiosity of the Senate and the House of Representatives has considerably changed. From the Almanac of American Politics, Congressional membership data shows that after 1969, the once dominant Mainline Protestants gradually declined while Republican Catholics and Democratic Jews effectively replaced them. Evangelicals initially declined in overall numbers, but eventually rebounded but with a partisan shift (shifting from Democrats to Republicans). The 86th Congress (1959) consisted of a greater number of Mainline Protestants of both parties and more Evangelical Democrats (particularly in the South) (Fig. 186.1). Mainline Protestants were the largest group with – only Rhode Island and Utah did not have an elected Mainline Protestant in their Congressional delegation. In 36 states, Mainline Protestants represented 50 % or more of the Congressional delegation. Evangelical Christians were overall the second largest group with elected officials in 38 states and with 50 % or more majorities in Alaska, Nevada, Utah, Arkansas, South Carolina, and Maine.

In 1959, Catholics, far fewer in number than later, were in 24 states but were more concentrated in the Mid-Atlantic and Northeast. For example, California (the most populated state) had almost no Catholics in its Congressional delegation.

---

[5] Mainline Protestant incorporates Liberal and Moderate Protestant groups as they are categorized by Wade Clark Roof and William McKinney. Liberal Protestants include Episcopalians, Presbyterians, and Congregationalists (United Church of Christ); Moderate Protestants include Methodists, Lutherans, Northern Baptists, Disciples of Christ, and Christian Reformed. Evangelical Christians include Southern Baptists, Church of Christ, Evangelicals/Fundamentalists, Nazarenes, Pentecostals/Holiness, Assemblies of God, Church of God, and Adventists. Mormons are also counted as part of the Evangelical group, although they do not consider themselves Protestants (Roof and McKinney 1987).

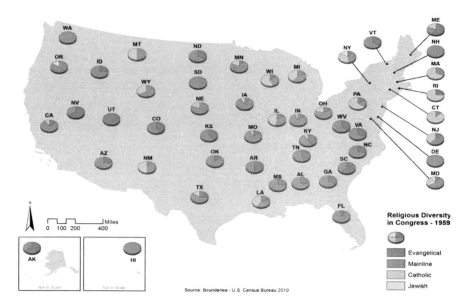

Fig. 186.1 Religious diversity, U.S. Congress, 1959 (Map by C. Kraemer, Gainesville State University, Gainesville, Georgia, 2012)

Other than Louisiana there were virtually no Catholics in Congress in the American South. Florida, for example, was almost completely represented by Mainline Protestants with no elected Jews or Catholics. Only Connecticut, Massachusetts, Pennsylvania, New Mexico, Montana, and Rhode Island had a Catholic-majority of elected members of Congress. Jews were the smallest group having elected members of Congress in only four states (New York, Maryland, Oregon, and Pennsylvania). Of these, Jews consisted of a minority of each congressional delegation. The 94th Congress (1975) represents the beginning of the changes in religious diversity (Fig. 186.2). While Mainline Protestants actually increased in their dominance from 1959 having 50 % or more in 38 states, Jews and Catholics began to be represented in more states. Only New Mexico did not have an elected Mainline Protestant in 1975's Congress. By then, the number of Catholics had increased and exceeded Evangelical Christians in their overall representation with elected members in 33 states. Of these, eight states (an increase from six) had 50 % or more representation. Catholics increased their regional dominance in the Northeast and Mid-Atlantic. They also had increased representation in the South and Southwest (Florida, Virginia, Missouri, Oklahoma, Arizona, Nevada, New Mexico, Oregon, and Washington State).

By 1975, Evangelicals declined in their overall representation to 33 states. Instead of six states, only five (Idaho, Utah, Kentucky, Wet Virginia, and North Carolina) had 50 % or more of Evangelicals representing them in Washington, DC. Evangelicals weakened in the Northeast (Maine, Vermont, New Jersey, and Pennsylvania) and in the Mountain West (Utah, Nevada, Colorado, and Alaska). While still the smallest

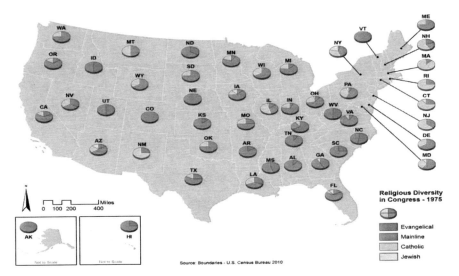

Fig. 186.2 Religious diversity, U.S. Congress, 1975 (Map by C. Kraemer, Gainesville State University, Gainesville, Georgia, 2012)

group, the Jewish presence in Congressional delegations doubled from four to eight states. Jews were no longer confined to the Northeast or Mid-Atlantic, but had diffused to Florida, Georgia, Iowa, and Arizona. The South began to show increased religious diversity by 1975. New York remained the center of Jewish congressional representation. The 103rd (1991) Congress reveals the continuing trend of increasing religious diversity. Twenty-nine states (down from 38 states) had 50 % or more Mainline Protestant representation (Fig. 186.3). Three states (an increase from one) did not have an elected Mainline Protestant. Most important, Mainline Protestants only represented a quarter of California and New York's delegations (down from 75 % in 1959 for CA, and down from nearly 50 % for NY). In the South, religious diversity continued to increase with more states having more than two groups for the first time (such as Alabama Mississippi). States with Catholic members of Congress continued to increase with a total of 40 (up from 33). The Catholic increase is observed across the country all regions. Eight states had Catholics representing 50 % or more of their delegation (the same amount, but less concentrated in the Northeast – i.e., Minnesota, South Dakota, Nevada, and New Mexico). Catholics became much more geographically dispersed in all regions.

By 1991, Evangelicals had elected members in 37 states (an increase from 33), but dominance (50 % or more) was to only Utah, Kentucky, and New Hampshire (down from six). In some cases, Evangelical representation increased by modest amounts due to their increasing participation within the Republican Party in the West (Colorado, California, Utah, Oregon, Washington State) and with increases of black Democrats within the Democratic Party in the Northeast and Mid-west (New Jersey, Maryland, Wisconsin, and Illinois) after 1980. The largest increase was with

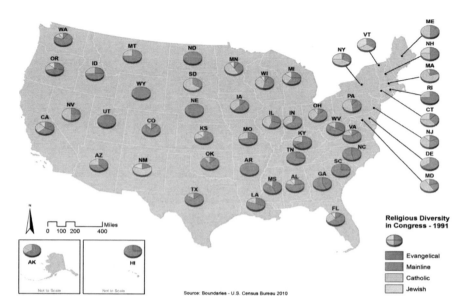

Fig. 186.3 Religious diversity, U.S. Congress, 1991 (Map by C. Kraemer, Gainesville State University, Gainesville, Georgia, 2012)

Jewish representation going to 21 states (up from eight). Though no states had 50 % or more Jewish representation, Jews were now nationally distributed with members in the South, the West, Mid-West, Mid-Atlantic, and the Northeast. New York remained the center of Jewish congressional influence, but the nearby and neighboring states of Vermont, Massachusetts, New Hampshire, and New Jersey increased in their proportion of Jewish representatives. The 111th Congress (2009) revealed a further decline in the numbers of elected Mainline Protestants (Fig. 186.4). In New York, Mainline Protestants virtually disappeared. Mainline Protestants in California are now at the same proportion as Jews, with both being the smallest group. While only two states (Vermont and Nevada) do not have Mainline Protestants in their delegation, only 16 states (down from 29 states) have 50 % or more elected Mainline Protestants. Only in the South do Mainline Protestants still have a regional dominance. The West and the Mid-West became much more diverse. States with Catholic members of Congress dropped very slightly to 39 (from 40), but 11 states had 50 % or more Catholic representations (an increase from 8). Catholic dominance increased within the Northeast, but also expanded southward and westward. New York, Ohio, Pennsylvania, New Jersey, Louisiana, and Alaska all had majority Catholic legislative delegates.

By 2009, Evangelicals had elected Congress members in 40 states (up from 37) and had a proportion of 50 % or more in 4 states (up from three). The numbers of Evangelicals has increased, but they seem to be more evenly geographically distributed than Catholics. Evangelical representation has proportionally increased in the middle of the U.S. (Oklahoma, Kansas, Colorado, Nebraska, South Dakota), and in migration destination states such as California, Texas, and Nevada with the exception

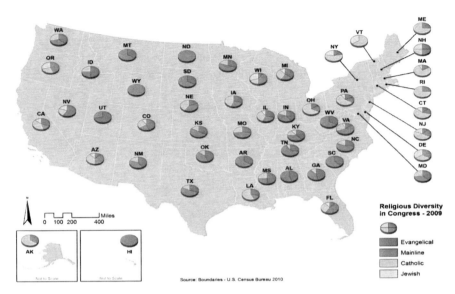

Fig. 186.4 Religious diversity, U.S. Congress, 2009 (Map by C. Kraemer, Gainesville State University, Gainesville, Georgia, 2012)

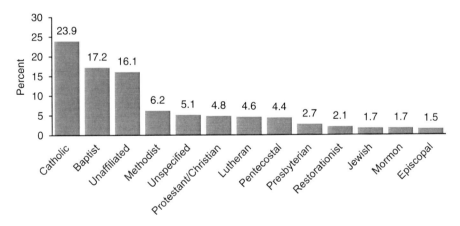

Fig. 186.5 Religious affiliations of U.S. population, 2009 (Source: Josiah R. Baker)

of Florida. Jews had elected members in 20 states (down from 21) but showed a growing presence in the West (Colorado, Arizona, Nevada) while losing some ground in the middle of the country (Kansas) and in the South (Alabama). Overall, there are fewer Jews that are proportionally concentrated in and around New York State.

Figure 186.5 shows the current percentage of the U.S. population that is identified with each major religious denomination and illustrates the apparent American religious diversity among the general population. The Catholic Church claims the highest proportion of Americans with just under 25 %. Baptists are second with

17 %, followed by Methodists at 6 %. In recent years, the fastest growing group has been "unaffiliated," which may include people who no longer personally identify with a denomination but are still believers in addition to those who consider themselves as "non-religious." Despite that trend, members of Congress are more likely to have an affiliation than in years before. Next, we have a general description of each major religious group with a brief review of their beliefs and historical background.

186.5 About Mainline Protestants

Mainline Protestants consist of the theologically liberal (Episcopalian, Presbyterian, and Congregationalist/United Church of Christ) and moderate denominations (Methodist, Evangelical Lutherans, Northern Baptist, Disciples of Christ, and Christian Reformed). The Episcopal Church advocated a relationship between religious and civil life adhering to practices of the Church of England. Under this model, "the church and state cooperate…to promote good governance and sound religion, but the state reign(s) supreme, even on issues of doctrine and clerical leadership" (Anderson 2009: 93). Despite 45 % of the original House members belonging to the Episcopal faith, and a greater proportion having ancestral roots in it, political independence from England directly affected the passage of the First Amendment to the U.S. Constitution which separated state and church (Anderson 2009: 101).

The First Amendment's purpose was to guarantee avoiding the institutional replication of the Church of England in the United States, which was the primary reason for why the Puritans, Pilgrims, and many Presbyterians left England (Demerath 1995). As Maps 1 through 4 (Figs. 186.1, 186.2, 186.3, and 186.4) demonstrated Mainline Protestants have declined in recent decades, which followed a sharp drop in overall membership and an even sharper decrease in regular attendance. Nevertheless, Mainline Protestant principles have interjected themselves into most American religions (Smith and Snell 2009). Two examples are the belief of personal autonomy, which means that everyone is accountable for their own behavior, and the value of following one's conscience. In his famous visit to the United States, De Tocqueville observed how personal autonomy and how it directly connected to responsibilities to family members and to the greater community was evident throughout the country in its early days. While these Mainline Protestant values are still influential, Joseph Bottum and others have noted the decline in Mainline Protestant membership:

> Over the past thirty years, Mainline Protestantism has crumbled at the base, as its ordinary congregants slip away to evangelicalism, on one side, or disbelief, on the other. But it has weakened at the head, too, as its most serious theologians increasingly seek community — that longed-for intellectual culture of people who speak the same vocabulary, understand the same concepts, and study the same texts — in other, stricter denominations. (Bottum 2008, Section IV)

Bottum argued that American culture is negatively affected by the decline in Mainline Protestantism: "Protestantism…gave America something vital: a social unity and cultural definition that did not derive entirely from political arrangements and economic relations. And America gave Protestantism something in return: a chance to flourish without state interference, a freedom to fulfill the human desire for what lies beyond the material world" (Bottom, 2008: Sect. VII).

186.6   About Evangelical Christians

In contrast to Mainline Protestants, state and church have a more integrated role for Evangelical Christians. This justified government involvement in greater society because the state is perceived as a means to an end with respect to religion. Since almost all Evangelical denominations are relatively new and lack a lengthy history involving of maintaining relations between state and church, its members are less mindful of historical problems related to having government involvement with religion. Evangelical churches emerged many years after the colonization process and beyond the American Revolution. In the minds of many Evangelical Christians, the church and state are "separate and unequal, with an elevation of church over the state" (Joireman 2009: 77). The government's prerogative is to maintain peace and stability for society in the background. The church is supposed to play an active role in addressing most social ills. Evangelical Christian include Southern Baptists, Churches of Christ, Evangelicals/Fundamentalists, Nazarenes, Pentecostals/Holiness, Assemblies of God, Churches of God, Adventists, and Mormons.[6]

Evangelical Christians view the state's role as more expansive, believing the state to be capable of both providing order and supporting moral growth. President George W. Bush's administration had an office on Faith-based Initiatives which exemplified the perspective that although separation of church and state should exist, the state can possess positive spiritual functions. Timothy Samuel Shah described this perspective:

> On the one hand, evangelicals generally favored a sharply limited political role for themselves, as well as a sharply limited political role for the state. On the other hand, evangelicals believed that the state was an appropriate and necessary instrument for the redress of moral and social evils they perceived to be significant. (Shah 2009: 124)

186.7   About Roman Catholicism

As shown in Maps 1–4 (Figs. 186.1, 186.2, 186.3, and 186.4), Catholic influence has greatly expanded in the United States in recent decades. Catholic influence in American politics has its origins from Spanish (in the Southwest), the French (parts of New England), and the English Catholic colony located in Maryland. Later, in the nineteenth century German, Irish, and Italian Catholics arrived to the United States during the "great white migration." Catholic priests and bishops built the Catholic

---

[6] Mormons are included Conservative Protestants due to their theology and voting patterns.

Church and ministered to the growing numbers of new citizens, which included their own school system. Catholic leaders rejected the Mainline Protestant teachings of a "common core" found in public education. The perceived attempts by public schools to socially engineer Catholic children into good Anglo-Saxon protestants led to resentment and resistance.

Church leaders were determined that education should not include "Protestantizing" the growing waves of Catholic immigrants. In the Third Plenary Council held in Baltimore in November 1884, the bishops (71 in number) mandated the establishment of a Catholic school in every parish, with pastors and parishioners responsible for funding the schools and requiring all parents to send their children to these parochial schools. Gradually, ethnic parishes built their own variations of Catholic schools with the intent of perpetuating the Catholic faith while also retaining core aspects of their community's culture and language, at least for a couple generations. Catholic schools had the further effect of re-enforcing the ethnic factor because it helped to solidify Catholic immigrant and their children as a political constituency. The Catholic school system continued to increase in its influence as a key centripetal force for American Catholicism and reached its peak in the 1950s, with almost half the Catholic student population enrolled in some 13,000 Catholic schools (Shah 2009: 68).

Politically, Catholic lobbying efforts from the United States Conference of Catholic Bishops (USCCB) have played a significant role in promoting public policies that reflect the church's positions. Professing to "promote the greater good which the church offers humankind" (U.S. Conference of Catholic Bishops), the USCCB draws upon the Catholic Church's overall aim to benefit the "common good" and the "general welfare" of society. The Catholic Church does not see itself as operating separately from the state, but especially since Vatican Council II in 1964, rather in connection with it. Viewing the state as both a necessary and formative part of culture and society, the Catholic Church finds the state to be a useful means for realizing its primary aim, promoting the "common good" (Shelledy 2009:17).

Robert Shelledy (2009) describes the Catholic perspective as being based on social teachings that view the purpose of humanity as understanding people and their relationships to each other (2009: 18). Thus, the Catholic perspective on government and the state is a natural extension of this understanding. The state has a responsibility not just to safeguard its citizens, but also to positively influence their conduct. More recently, the Catholic Church has sought to realize this principle through the standards outlined in the Second Vatican Council (1962–1965) and the resulting publication of *Gaudium et Spes*, both of which were designed to illuminate the Catholic Church's position in contemporary politics (Shelledy 2009: 22). As a result of these examinations, "the church moved toward a tolerance of pluralism and the promotion of religious freedom," and its ideological position became supported by the Catholic Church as long as they foster the common good (Shelledy 2009: 22).

The sharp division within Catholicism is based on how Catholicism addresses social issues such as poverty and welfare spending. The conditions of the global economic depression in 1930 led the Church to further examine the role of the individual with the state. In 1931, Pope Pius XI further developed this principle in his encyclical *Quadragesimo Anno*, which was followed by *Economic Justice for All* by the United States Conference of Catholic Bishops. Pope Pius XI's "Quadragesimo Anno," page 79 states, "*It is a fundamental principle of social philosophy, fixed and unchangeable, that one should not withdraw from individuals and commit to the community what they can accomplish by their own enterprise and industry.*"

In recent decades, American conservative and libertarian Catholics such as Reid Buckley have pursued the principle of subsidiarity, which is a Catholic social teaching that postulates that government should only manage those initiatives that exceed the ability of individuals or private sector groups' independent efforts. Pope Leo XIII first formally developed the principle of subsidiarity in the 1891 encyclical *Rerum Novarum*, as a means to formulate a compromise between *laissez-faire* capitalism and various forms of communism, which sought to subject the individual to the state. Subsidiarity as defined by American conservative and libertarian Catholics is fundamentally opposed to increased centralized or federal governmental efforts. Government functions are believed to be best handled at the local level whenever possible. Republican Catholics such as Speaker of the House John Boehner and Rep. Paul Ryan adhere to this approach, which helps to justify their reluctance to increase federal government social spending.[7]

186.8   About Jews

Jewish influence on American culture, following a similar pattern as Catholic political influence, has substantially grown during the period of this study (1959–2009). Political scientists Kenneth Wald and Michael Martinez explain that "the diversity within Judaism is a function of religious development as abetted by Diaspora," and they pointed to "revivalist and reformist protests" within Judaism over the last few centuries were key factors within this process (Wald and Martinez 2001: 379).

The Jewish position in the United States is indeed unique.[8] Paul Burstein pointed out that Jews as a group "are much more successful educationally and economically than other ethnic, racial, and religious groups in the United States" (Burstein 2007: 209). The explanations for this disproportionate success is attributed to education, a

---

[7] Rep. Paul Ryan of Wisconsin told the Christian Broadcasting Network that it was his Catholic faith that helped shape the budget plan. In his view, the Catholic principle of subsidiarity suggests the government should have little role in helping the poor. Source: http://zionica.com/2012/04/16/was-jesus-for-small-government/, "Was Jesus for small government?", *Zionica Magazine*.

[8] Since 1959, the American Jewish population has declined in proportional and in absolute terms (from 5 to 2 % of the population and from more than five million to less than four million) yet within the House Jewish representation has tripled to 30 seats and there are 12 Jewish Senators. Within the next 10–20 years, further declines as a proportion and in absolute terms are expected due to low birthrates and the increase of Jews marrying or having children with non-Jews.

strong work ethic, traditional Jewish values and lifestyle, including mutual aid societies where Jews offer assistance to one another (Burstein 2007: 214).

Judaism faces a different situation from Catholicism in terms of its ability to prescribe and maintain a unified set of principles to its believers (Wald and Martinez 2001: 378–379). Judaism, while possessing a strong religious tradition, lacks the hierarchical structure of Catholicism, and its doctrine is often subject to a variety of interpretations. In terms of political involvement, Jews have largely identified with the Democratic Party since the era of FDR (Greenberg and Wald 2001: 163).[9] In Congress, this partisan attachment to the Democratic Party is even more pronounced. Wald and Martinez explain their circumstances: "As a minority that has thrived in a multicultural state that disclaims a formal religious identity, American Jewry has developed a distinctive liberal political culture" (2001: 381). The Jewish identity – how they view themselves in the world – contributes to a form of liberalism that supports government possessing a responsibility to promote welfare programs while also maintaining the liberal view that heavily supports the separation of church and state.

186.9   Changes in the House: Increasing Religious Diversity

When reviewing the patterns of religious affiliation that characterize Democrats and Republicans in the U.S. House of Representatives, several questions arise: Do the two parties attract members more or less randomly from the array of religious groups in the U.S. population? Which religious groups, regardless of their size among the general population, have tended to dominate either one or both parties in the House? Are there geographic trends due to migration, immigration, or the aging population?[10]

The next two figures illustrate changes in membership of religions within the House and the Senate from 1959 to 2009. Figure 186.6 shows the changes in major religious groups that have occurred in the U.S. House of Representatives since 1959 (the 86th Congress). As indicated, Mainline Protestants in the House have steadily decreased since the 86th Congress when they held 250 of 435 seats. This number declined to 148 by 2009, from constituting 57 % of the House in 1959 to 33 % in 2009 at a time when Mainline Protestants were just under 13 % of the total U.S. population. There were shifts in partisanship. For example, in 1959, there were more House Episcopalian Democrats than Republicans. While Episcopalian Republicans held steady at around 20, the number of Episcopalian Democrats declined from

---

[9] See page 163 and pages 170–171 for more information. In terms of Jewish self-image in America, Greenberg and Wald note the Jewish Public Opinion Study, in which "35 % of Jews call themselves liberal compared to 18 % of non-Jews; 8 % of Jews call themselves conservative compared to 26 % of non-Jews." (Greenberg and Wald 2001: 170).

[10] Some of the religious groups identified here, such as Christian Scientists, are classified as "Mainline Protestant." Eastern Orthodox and Greek Orthodox groups are included in this category for the purposes of this study.

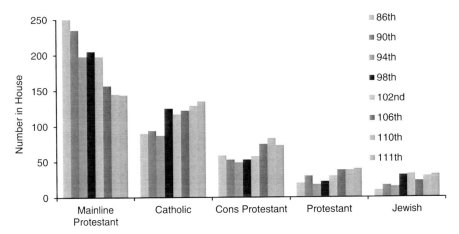

Fig. 186.6 Religious affiliations of members of the U.S. House of Representatives, 1959–2009 (Source: Josiah R. Baker)

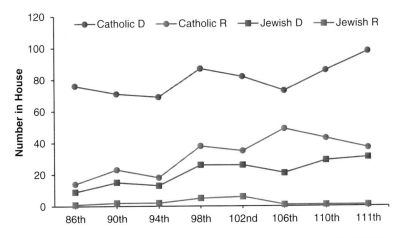

Fig. 186.7 Catholics and Jews in the U.S. House of Representatives, by party, 1959–2009 (Source: Josiah R. Baker)

29 to 13. A similar pattern of decline was observed among Presbyterian Democrats (declining from 28 to 15) with Presbyterian Republicans also decreasing. Methodist Democrats in the House also lost more membership than did their Republican counterparts. With the exception of Lutherans among Mainline Protestant groups, Democrats experienced a much steeper decline in numbers than did Republicans.

In effect, Mainline Protestants were replaced by every other group, and most significantly by Roman Catholics, who had almost matched the proportion of Representatives with 31 % up from 21 % in 1959. As shown in Fig. 186.7, major growth in both Catholic and Jewish groups in the U.S. House occurred among Democrats. Though Catholic Republicans also increased in number, Catholic

Democrats in 2009 still outnumbered Catholic Republicans by a margin of more than 2 to 1.

Though Catholics had the highest overall gain, Jews experienced the highest proportional increase, with their share more than tripling (from a mere 2 % in 1959 to just over 7 %. As noted previously with Maps 1 through 4 (Figs. 186.1, 186.2, 186.3, and 186.4), Jewish Congressional members have spread to all regions of the country and are no longer largely limited to New York and the immediate surrounding states.

In contrast to Mainline Protestants, all major Evangelical Protestant groups increased in overall House membership between 1959 and 2009, owing largely to greater Republican numbers. While Evangelicals (listed in Fig. 186.1 as "Cons Protestant" and "Protestant") have always been more prominent in the South, among Republicans they have increased in more Western and Plains states showing a modest growth of about 24 % of the total House membership in 2009 (up from 14 % in 1959). Major Evangelical Protestant groups in the U.S. House of Representatives include Baptists (primarily Southern Baptists) and numerous smaller evangelical groups.[11] Since the 86th Congress,[12] Baptists (formerly almost all white) have increased in number and in racial diversity by about 20 %, largely due to the growing presence of recently elected blacks. By 2009, Baptists alone comprised about 13 % of the total House membership and about 16 % of the total U.S. population.[13] Aside from the Baptists, other Evangelical groups grew in House membership overall; between 1959 and 2009, their physical numbers have roughly doubled.

186.10   Changes in the Senate: Increasing Religious Diversity

Figure 186.8 displays the major religious denominations and their membership trends in the U.S. Senate since 1959.[14] Over the 50-year period, the Senate followed a similar pattern of change of composition as did with the House. For example, Mainline Protestants declined significantly after 1959 (after peaking at 65 % in 1967), Mainline Protestants decreased to a low of 38 % of the Senate in 2009. Despite that, Mainline Protestants are still over-represented in the Senate, having about 13 % of the U.S. population in 2008.

---

[11] For charting purposes, "evangelical" here encompasses those Evangelical Protestants who self-identify as Evangelical/Fundamentalist as well as Brethren in Christ, Pentecostal/Holiness, Pietist, Adventist, Assembly of God, Church of Christ, Nazarene, Church of God, and Mormons.

[12] Only the major Conservative groups are assessed here, Evangelicals include Pentecostal and Assemblies of God.

[13] Most of the growth among Southern Baptists is from the increase in the number of African Americans in Congress since the Voting Rights Act of 1965. Though their theology regarding homosexuality and abortion is conservative, African Americans generally are among the most liberal in voting patterns dealing with abortion.

[14] "Denomination" is used in broad terms to distinguish between religious groupings.

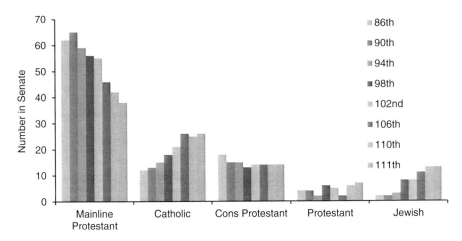

Fig. 186.8 Major religious groups in the U.S. Senate, 1959–2009 (Source: Josiah R. Baker)

The decline of Mainline Protestants, in conjunction with the rise of Catholics and Jews, constitutes the most dramatic shift in the Senate within the half-century between 1959 and 2009. Catholics, in particular, grew from being a group with modest membership in 1959 to becoming the largest religious group by 2009. In the process, the number of Catholic Republicans in the Congress increased unevenly, but by 2009 they were a significant part of the Republican Party. A similar pattern did not appear among American Jews because they were overwhelmingly dispro-portionately Democratic by 2009.

Thus, Mainline Protestant Senators decreased in their total numbers while being replaced by Catholics and Jews. However, their decline happened later in the Senate (after 1991) than in the House. Party divisions explain some of these changes. Among Episcopalian Senators, both Democrats and Republicans declined in numbers after 1959, although Republicans embarked on a more dramatic course, reaching a high of 16 % in 1981 (98th Congress) and then declining to 2 % by 2009. This trend was not evident within all Mainline Protestant groups. For example, Presbyterian Democratic Senators declined by 50 % from 1959 to 2009, while Presbyterian Republicans more than doubled their numbers during the same period of time.

Due to having only 100 members instead of 435, proportional changes in the Senate are expected to be more dramatic (and they were) than for the House. For example, Jewish gains in the Senate increased to 13 % in 2009 (they were only 2 % in 1959). Also, Catholics more than doubled their proportion to 26 % in 2009 (they held 12 % in 1959). Despite these similar trends, there was a difference regarding Evangelical Protestant Senators. Their numbers, decreased slightly since 1959 (unlike in the House, where they experienced modest increases). Baptists, as the largest Evangelical group, lost members, shrinking from 14 % in 1959 to 8 % in 2009. It is interesting that Evangelicals claim to have about 25 % of such adherents nationally, and between 70 % and 80 % vote Republican yet Evangelicals are significantly under-represented in Congress, which is the opposite circumstance for Jews.

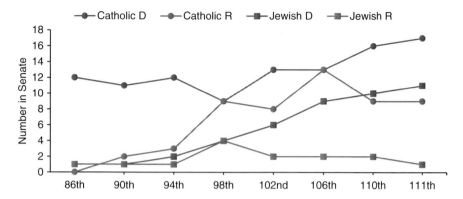

Fig. 186.9 Catholic and Jewish members of the U.S. Senate, by party, 1959–2009 (Source: Josiah R. Baker)

Figure 186.9 shows that most of the growth of Catholic and Jewish Senators was within the Democratic Party. While Catholic Republicans experienced some gains in the Senate, Jews have practically disappeared within the Republican Party.[15] Jewish Republicans were the only group that declined in membership; they went from a high of 4 % of the Senate in the 98th Congress to merely 1 % (1 seat) of the Senate in 2009.

As shown above, Senators who were Catholics and Jews increased their numbers, with Catholics rising from 12 % in the 86th Congress (1959) to 26 % by 2009. The number of Jewish Senators also sharply increased, going from 2 to 13 % over the 50-year period. In proportion to their representation in regard to population, these increases were inversely related to Evangelicals in the Senate, whose proportion increased, even as their share of the population rapidly increased. Many Mainline Protestants had converted or switched to Evangelical dominations: upwards of one-third of the adult Catholic population had also become Evangelical Christians.[16] The growth of Jewish proportional representation is especially evident because the Jewish percentage of the U.S. population dropped from 5 to 1.2 % during this period of time. The Jewish influence in Congress is overwhelmingly partisan (within the Democratic Party), but is also increasingly geographically diverse. Though the Jewish population has spread to other metropolitan areas beyond the Northeast, they have climbed the ranks within the Democratic Party. Thus, in the arena of domestic policy, Jews have been instrumental in initiating and supporting Democratic Party legislation.

Figure 186.10 illustrates the average voting differences of religious groups within the party, for Mainline Protestant Democrats. The differences were greater for House Mainline Protestant Democrats than for their Republican counterparts, with the glaring exceptions of 1987 and 1989 (100th session and the 110th session).

---

[15] Currently, there are no Jewish Republican Senators and there is only one Jewish Republican in the House.

[16] D'Antonio et al. (2012).

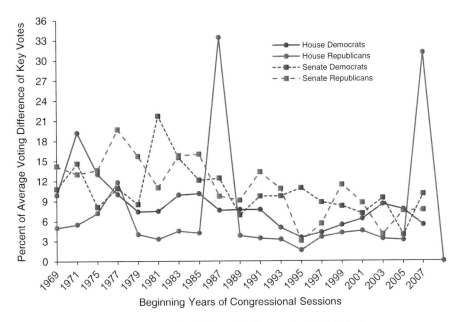

Fig. 186.10 Mainline Protestant voting differentials (Source: Josiah R. Baker)

The differential in voting behavior among House Mainline Protestants compared to the Senate may also reveal how their local (district) interests are separate from party interests as a whole. For example, while a Democratic House member might vote to keeping a military base in their district, their party as a whole could support shutting down the base as part of cuts in defense spending.

The chart indicates that Mainline Protestant House Democrats remained the most consistent in their independent voting patterns. For sessions 100 (1987) and 110 (2007), these two exceptions had votes that included the 100th (1987) session, by Mainline Protestant House Republicans voted much more differently than non-Mainline Protestant House Republicans. Examples within the 100th session (1987–1988) included votes on restricting money for the homeless (81 % difference), supporting a deficit reduction (78 % difference), and supporting nuclear testing (77 % difference). For the 110th session (2009), the exceptional votes of Mainline Protestant House Republicans in contrast to non-Mainline Protestant House Republicans included the following: no military operations in Iraq (94 %), repeal of DC's gun law (81 %), foreign aid abortion ban (66 %), and ban gay bias in the workplace (63 %). These are exceptional percentages because most of the key votes in this analysis were less than 10 % in difference. In any case, Democrats showed greater variability based on affiliation to Mainline Protestantism than did Republicans. In this regard, religiosity is stronger within Democrats when examining Mainline Protestantism's influence.

Trends within the Senate between Mainline Protestant Democrats and Mainline Protestant Republicans had inconsistent patterns, with a change in leads 11 times

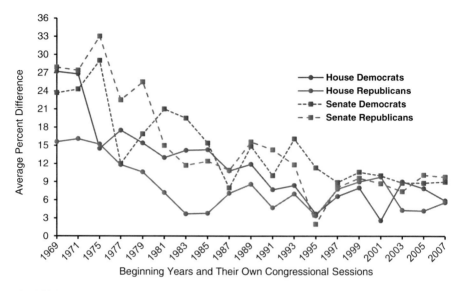

Fig. 186.11 Catholic voting differentials (Source: Josiah R. Baker)

within 20 sessions. Note that differences generally declined over time, indicating that religious affiliation had a decreasing role in explaining votes that varied from the party line. Since there are far fewer Senators (100 vs. 435), this would imply that variability could have been easier to achieve. Yet, Fig. 186.10 shows that it did not. Party loyalty was stronger for Democrats and Republicans Senators who represented their states rather than a district. Thus, Mainline Protestant Senators appear to have experienced far less influence from their religious affiliation.

Regarding average voting differences of religious groups within parties, for Catholic Democrats, the differences within the House were also compared consistently greater for Catholic Republicans within the House, the Senate. In the Catholic case, the actual numbers of Catholic members of Congress greatly influenced the interpretation of results. For example, Fig. 186.11 shows that variability was far greater from the 91st (1969) through the 96th (1979) sessions, when there were very few Catholic Senators. Though the differential decreased, Catholic Senate Democrats voted consistently more different than their party, compared to their Republican co-religionists. In all cases, the differences generally declined, which indicates that religion as a factor declined in determining a difference of voting, whether one was Democrat or Republican, or in House or Senate. Nonetheless, the factor did not disappear, and could make a critical difference if a key vote is close, particularly in the House and especially for Catholic House Democrats. Aside from the size of the House membership, the role of the filibuster in the Senate changes the importance of voting when considering smaller differences, because House votes require a simple majority plus one. Therefore, Senators can vote more independently than House members, often without fear of upsetting the election outcome

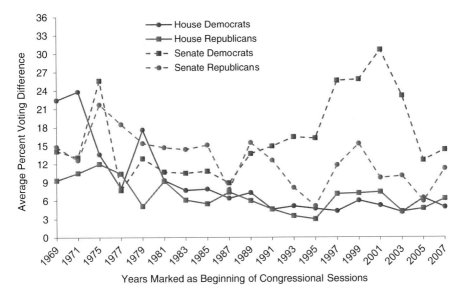

Fig. 186.12 Evangelical voting differentials (Source: Josiah R. Baker)

(due to the higher voting requirements to end cloture). However, as shown, Senators do not vote as independently as do members in the House. Catholic Democrats, like Mainline Protestant Democrats, are again more likely to vote independently of their party than their Republican counterparts.

Regarding average voting differences of religious group within party, voting behavioral differences between Evangelical House Democrats and Evangelical House Republicans after 1981 was remarkably flat (Fig. 186.12). Note that from the 97th (1981) session to the 110th (2007) session the overall differences between Evangelical Democrats and Evangelical Republicans from their party were virtually the same. Another consistent trend was that House Evangelical Republicans from the beginning of this study (91st session) remained at the lowest of all four groups.

The greatest differences when evaluating Evangelicals is found with Senate Evangelical Democrats. However, this with the exception of the 94th (1975) session, is a relatively new development: not until the 102nd (1991) and onward did Evangelical Senate Democrats emerge as a much more independent voting group from their party than the Evangelical Senate Republicans. Only with Evangelical Senate Democrats did the differences generally increase, indicating that with this exception, religion as a factor declined in determining a difference of voting, whether one was a Democrat or a Republican in the House or a Republican in the Senate. Again, with Evangelical Senate Democrats the factor actually increased which was unexpected. This would indicate that Evangelical Senate Democrats do occasionally vote against their party which strengthens the Republican ability to launch a successful filibuster. Surprisingly, though Evangelicals are often seen as the most religious group, they actually exerted the least independent voting behavior compared to their Mainline and Catholic colleagues.

186.11 The Influence of Southern Democrats

These last two figures illustrate the effect of Southern Democrats as a separate voting group. The decrease in the number of Southern Democrats after the civil rights era has contributed to an increase in voting polarization (Figs. 186.13 and 186.14). The rise of the Republican Party (using Nixon's Southern strategy) may have benefitted the party on the whole, enabling it to take control of the House in 1994, but it was at the expense of bipartisanship. Continually, Republicans, regardless of geographic and religious differences, vote with far less independence from the party than Democrats. These figures help to illustrate a key proportion of the differences: the impact of the Southern Democrat.

Figure 186.13 shows the distribution of voting polarization between Southern Democrats and other House Democrats. These points represent the proportion of agreed key votes (Barone's Almanac of American Politics, 1969 to 2009) between Southern Democrats and other Democrats in the House of Representatives. The y-axis represents the differential percentage of voting independently of the party. The x-axis represents Congressional sessions, beginning with the 91st (1969) and ending with the 110th (2006). Low points reflect Southern Democrats agreeing with other Democrats on key votes. High points represent votes when Southern Democrats did not agree with other Democrats. Overall, Southern Democrats exhibited a polarity hovering around 50 %, meaning that half of them did not vote with other Democrats.

Figure 186.14 shows the distribution of voting polarization between Republican House members and House Southern Democrats. As with the prior figure, the lower

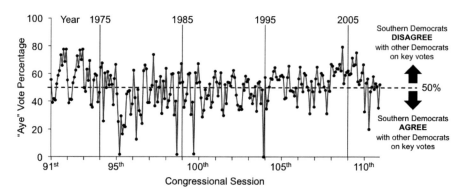

Fig. 186.13 Voting polarization between Southern Democrats and other House Democrats (Source: Josiah R. Baker; data from Barone's Almanac of American Politics from 1969 to 2009)

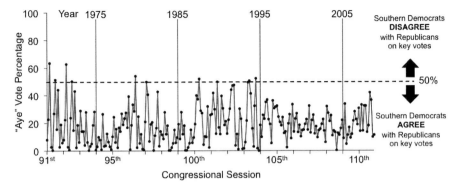

Fig. 186.14 Voting polarization between Southern Democrats and House Republicans (Source: Josiah R. Baker; data from Barone's Almanac of American Politics from 1969 to 2009)

points indicate agreement, meaning that Southern Democrats tended to vote with Republicans for those bills. The higher points indicate increasing disagreement with the Republican Party. The proportion of the votes and the general differential pattern did generally change much between 1969 and 2006 for either figure. However, there were substantially fewer members of House Democrats in the South by 2006. Note that the voting polarity between Southern Democrats and Republicans was less than it was when comparing votes with other Democrats. Instead of the differential hovering around 50 %, the differences between House GOP and Southern Democrats was about 20 %, indicating that in some ways this group acted almost as a third political party.

186.12   Summary

Today, the American perspective on religion and politics is sharply divided. One side believes that religion serves as a guideline for civic behavior and that morality is subject to scientific and historical adaptations. The other side believes that faith is attached to an individual's prosperity and growth, including the concept that the Bible is without error (Benson and Williams 1982: 94). Recent decades have brought changes to how different religious groups participate in politics, and where they participate. Baptists who were Democrats in the 1970s and 1980s, if they are in Congress now, are more likely to be white Republicans or black Democrats. Evangelicals, such as Baptists, were once conservative Democrats; they are now almost exclusively Republicans. On the other hand, in the 1950s, being Catholic meant being a Democrat, because Catholic religious beliefs and political ideologies overlapped. But by 2009, there were almost as many Catholic Republican representatives as Democratic House members. Thus, it is fair to say that within Catholicism, two political ideologies have emerged.

This chapter examined how the changing geopolitical landscape is affected by the dynamics of religious composition within the United States Congress for the period 1959–2009. Democrats, particularly in the House, are more likely than Republicans to vote independently of their party. Mainline Protestants generally proved more likely to vote the party line than Catholics or Jews. The decline of Mainline Protestantism correlates with increased polarization. These findings indicated that the increasing diversity of religious thought has contributed to the increasing polarization of the two parties. Last, Southern Democrats did play a larger role in moderating the voting differences between Republicans and Democrats. The decline of Southern Democrats in number and in prominence has also contributed to polarization. Therefore, regional differences did have a moderating past influence that however, seemingly has declined. Geographic influences regarding independent voting behavior have declined, especially in the Senate. Thus, the culture war based on two separate visions of America as observed by Hunter and others is partially attributed to increasing religious diversity. Religious differences do contribute to an increasingly dysfunctional House and the Senate.

Further research should explore why Texas did not experience a significant change in religious composition compared to large populated states such as California, Florida, and New York. Why did Texas maintain the status quo when so many other states transformed during this 50 year period? Another issue that should be addressed is to measure the impact of the black Evangelical vote on the Democratic Party and if there are circumstances when they would break from the party and join Republican Evangelicals. This study would have to concentrate on House voting due to the limited number of blacks and Evangelicals in the Senate. Nonetheless, to what degree and how frequent do black Evangelicals exhibit a separate voting pattern as attributed to their religiosity? Last, current trends point to ever increasing religious diversity within the United States, such as the House, now having one Islamic member. In the future, it is expected there will be further increases in religious diversity, including the continual rise of the population (now at around 15 %) that identifies as having no religious affiliation. How will the rise of secularism and the proportion of non-Judeo-Christian religions influence Congressional voting behavior?

References

Anderson, L. S. (2009). The Anglican tradition: Building the state, critiquing the state. In S. Joireman (Ed.), *Church, state, and citizen* (pp. 93–101). New York: Oxford University Press.

Benson, P. L., & Williams, D. L. (1982). *Religion on capital hill: Myths and realities*. New York: Oxford University Press.

Bottum, J. (2008, August). The death of Protestant America: A political theory of the Protestant mainline. *First Things*. Available at http://www.firstthings.com/article/2008/08/001-the-death-of-protestant-america-a-political-theory-of-the-protestant-mainline-19

Burstein, P. (2007). Jewish educational and economic success in the United States: A search for explanations. *Sociological Perspectives, 50*(2), 209–214.

D'Antonio, W., Tuch, S., & Baker, J. R. (2012). *Congress and religion*. Washington, D.C.: Rowman & Littlefield.

Demerath, N. J. (1995). Cultural victory and organizational defeat in the paradoxical decline of liberal Protestantism. *Journal for the Scientific Study of Religion, 34*, 4–6.

DiMaggio, P., Evans, J., & Bryson, B. (1996). Have Americans' social attitudes become more polarized? *American Journal of Sociology, 102*, 735–36.

Dionne, E. J., & Wolfe, A. (2006). Polarized by God: American politics and the religious right. In P. S. Nivola & D. W. Brady (Eds.), *Red and blue nation: Characteristics and causes of American's polarized politics* (pp. 23–28). Washington, DC: The Brookings Institution.

Evans, J. H. (2002a). Polarization in abortion attitudes in U.S. religious traditions, 1972–1998. *Sociological Forum, 17*(3), 17–19.

Evans, J. H. (2002b). *Have Americans' attitudes become more polarized? – An update* (Working Paper 24). Working Paper Series of the Center for Arts and Cultural Policy Studies at Princeton University: Princeton University. Princeton, New Jersey.

Evans, J., Bryson, B., & DiMaggio, P. (2001). Opinion polarization: Important contributions, necessary limitations. *American Journal of Sociology, 106*, 4–7.

Fastnow, C., Grant, J. T., & Rudolph, T. J. (1999). Holy roll-calls: Religious tradition and voting behavior in the U.S. House. *Social Science Quarterly, 80*(4), 687–701.

Gitlin, T. (1995). The twilight of common dreams: Why America is wracked by culture wars. New York: Henry Holt/Metropolitan Books.

Greenberg, A., & Wald, K. D. (2001). Still liberal after all these years? In L. Sandy Maisel & I. N. Forman (Eds.), *Jews in American politics* (pp. 161–193). Lanham: Rowman and Littlefield.

Hall, J. A., & Lindholm, C. (1999). Is America breaking apart? Princeton: Princeton University Press.

Hertzke, A. D. (1988). *Representing God in Washington: The role of religious lobbies in the American polity*. Knoxville: University of Tennessee Press.

Hunter, J. D. (1991). *Culture wars: The struggle to define America*. New York: Basic Books.

Hunter, J. D. (1994). *Before the shooting begins: Searching for democracy in America's culture wars*. New York: The Free Press.

Joireman, S. F. (2009). Anabaptists and the state: An uneasy coexistence. In S. Soiremen (Ed.), *Church, state, and citizen* (pp. 77–79). New York: Oxford University Press.

Kraemer, C. (2012). *U.S. religious composition maps*. Gainesville, GA: Gainesville State University.

Luker, K. (1984). *Abortion and the politics of motherhood*. Berkeley: University of California Press.

Mariott, D. (2005, April). *Righteous roll calls: Religion, choice, and morality politics in the U.S. House and Senate, 1999—2002*. Paper presented at the Midwest Political Science Association.

Mouw, T., & Sobel, M. E. (2001). Culture wars and opinion polarization: The case of abortion. American Journal of Sociology, 106, 913–943.

Nivola, P., & Brady, D. (Eds.). (2006). Polarized by God? American politics and the religious divide. In *Red and blue nation: Characteristics and causes of America's polarized politics*: Hoover Institution on War, Revolution, and Peace, Stanford University, 2006.

Roof, W. C., & McKinney, W. (1987). *American mainline religion: Its changing shape and future*. New Brunswick: Rutgers University Press.

Shah, T. S. (2009). For the sake of conscience: Some evangelical views of the state. In Joireman, S. (Ed.), *Church, state and citizen* (pp. 67–69 and 122–125). New York: Oxford University Press.

Shelledy, R. (2009). The catholic tradition and the state: Natural, necessary, and nettlesome. In S. Joireman (Ed.), *Church, state and citizen* (pp. 16–23). New York: Oxford University Press.

Smelser, N. J., & Alexander, J. C. (Eds.). (1999). *Diversity and its discontents: Cultural conflict and common ground in contemporary American society*. Princeton: Princeton University Press.

Smith, C., & Snell, P. (2009). *Souls in transition: The religious and spiritual lives of emerging adults*. New York: Oxford University Press.

Wald, K. D., & Martinez, M. D. (2001). Jewish religiosity and political attitudes in the United States and Israel. *Political Behavior, 23*(4), 378–381.

Williams, R. (Ed.). (1997). *Cultural wars in American politics*. New York: Aldine de Gruyter.

Wuthnow, R. (1988). *The restricting of American religion. Society and faith since World War II*. Princeton: Princeton University Press.

Chapter 187
Interfaith Advocacy Groups in American Politics

Katherine Knutson

187.1 Introduction

The energy level was high among the 850 people assembled in a St. Paul convention center located a few blocks from the Minnesota State Capitol in February 2011. Men and women, old and young, from four distinct religious traditions—Protestant, Catholic, Jew, and Muslim—gathered around tables with others from their home communities for the 2011 Joint Religious Legislative Coalition Day on the Hill lobbying event (Fig. 187.1). Interfaith political advocacy groups, like the Joint Religious Legislative Coalition, unite individuals or religious institutions from multiple faith traditions to influence public policy and politics. The development of interfaith advocacy groups is a relatively new phenomenon in American politics, though advocacy groups representing religious voices have a long history of political participation. As this chapter contends, interfaith groups, made up of Christians, Jews, Muslims, and members of other faith traditions, are an important part of the body politic.

In this chapter, I explore the rise of interfaith advocacy groups in American politics. Section two provides a conceptual framework for understanding the role of advocacy groups. Section three briefly summarizes of the history of religious advocacy groups. Section four describes the rise of interfaith activism and uses case studies of two interfaith groups, The Interfaith Alliance (TIA) and the Joint Religious Legislative Coalition (JRLC). These two groups share a commitment to advocacy from an interfaith perspective, however, there are also significant differences between the two groups, which allows for a broad view of the interfaith work happening in the United States. Aside from the different level of focus (national versus state policy),

K. Knutson (✉)
Department of Political Science, Gustavus Adolphus College, St. Peter, MN 56082, USA
e-mail: knutson@gustavus.edu

© Springer Science+Business Media Dordrecht 2015
S.D. Brunn (ed.), *The Changing World Religion Map*,
DOI 10.1007/978-94-017-9376-6_187

Fig. 187.1 2011 Joint Religious Legislative Coalition Day on the Hill (Photo copyright © 2011 from http://csjjusticematters.blogspot.com/2011/02/joint-religious-legislative-coalition.html by JRLC. Reproduced by permission of Taylor and Francis Group, LLC, a division of Informa pic)

the two interfaith groups differ in their view of the role religious beliefs ought to play in the political process. TIA advocates a separation of religion and politics whereas the JRLC argues that religious values ought to influence political decision-making. Finally, section five analyzes challenges facing interfaith advocacy groups and opportunities of interfaith advocacy.

187.2   What Are Political Advocacy Groups?

Political advocacy groups, also known as interest groups or pressure groups, represent likeminded individuals in the political process and help to mobilize people to participate in the political process. Unlike social movements, advocacy groups have a formal organizational structure, and often employ professionals to organize group activity and to lobby government. Unlike political parties, advocacy groups do not run candidates for political office. Advocacy groups have a history in American politics dating back to the founding of the nation. Constitutional framer, James Madison warned of the potential harm of political advocacy groups in his essay, Federalist 10. He argued that the governmental structure codified in the Constitution would help protect against the "mischief" of factions.

Throughout most of the eighteenth and nineteenth centuries, advocacy groups in the U.S. represented economic interests such as farmers, business, and labor. The number of advocacy groups active in the political system and the types of groups changed dramatically beginning in the 1960s and 1970s. Scholars term this the "advocacy explosion" (Berry and Wilcox 2009). Today there are literally tens of thousands of advocacy groups active in national politics, and more at the state and local levels.[1] Many of the new groups formed around particular political issues, ideologies or aspects of individual identity.

Scholarship of advocacy group mobilization can be roughly divided into three broad theoretical perspectives: *pluralism*, *selective incentives*, and *resource mobilization*. Pluralists, such as Truman (1971), emphasize the ways in which groups form in response to other groups. When an imbalance develops in the interest group universe, new groups form to represent the viewpoints of unrepresented interests. As Olson (1965) pointed out, however, this perspective overlooks the difficulties inherent in overcoming collective action problems. Olson emphasized the importance of the provision of selective incentives, benefits available only to group members, to induce individuals to join an organization. Related to Olson's theory, Clark and Wilson (1961) argued that groups might offer multiple types of incentives to potential group members. In addition to the material incentives emphasized by Olson, groups could offer solidary or purposive incentives. *Solidary incentives* include social benefits such as socializing with peers and a sense of group membership, while *purposive incentives* involve the feeling of making a difference on issues that matter to the individual. Group leaders play an important role in finding the appropriate combination of benefits to offer group members (Salisbury 1969). Resource mobilization theories focus on the important role played by wealthy patrons, including charitable foundations and the government, to provide the resources necessary to form and maintain groups (Walker 1991).

187.3   A Brief History of Religious Advocacy Groups in American Politics

Religious advocacy groups have a long history of political activism in the United States. They established politically-minded committees to protect their own political interests and to lobby on behalf of policies deemed important. One of the first faith-based groups, the International Religious Liberty Association, was established

---

[1] Over 150,000 nonprofit groups are listed in the Encyclopedia of Associations: National Organizations of the US, and over 100,000 additional groups are listed in the Encyclopedia of Associations: Regional, State, and Local Organizations.

in 1893 by the Seventh-day Adventists to protect religious freedom. Credit for the first faith-based Washington lobbying office goes to the Methodist Church, which opened in 1916. Other religious groups followed suit, opening Washington-based offices and engaging in political advocacy for causes ranging from the prohibition of alcohol to peace and justice. An encyclopedia of religious advocacy groups in the United States published in 1994 identified 120 religious advocacy groups active in national politics (Weber and Jones 1994). A 2011 study found more than 200 such groups (Pew Forum on Religion and Public Life 2011). In addition to working at the national level, religious advocacy groups are also active at the state and local levels.

As a basic definition, religious advocacy groups are organized groups of people with policy interests that result from a shared religion or set of religious beliefs. These groups may take the form of a lobbying office of a national religious denomination (like the Baptist Joint Committee or the U.S. Conference of Catholic Bishops), a coalition of religious organizations or denominations (like the National Council of Churches or the National Association of Evangelicals) or a think tank or research organization (such as the Acton Institute or the Center for Public Justice). Religious groups may also be issue-based and comprised of individual citizen members (like the Anti-Defamation League, the Council on American-Islamic Relations, or Bread for the World) or they may simply be a religious organization or movement that has more of a mailing list than a list of official group members (like the American Center for Law and Justice).

Religious advocacy groups have played important roles in many political debates, though political scientists were slow to note their influence. The increased political power of groups on the right in the 1980s helped spur academic interest in the role of religious advocacy groups. Hertzke (1988) finds that religious interest groups lobby in ways that are similar to non-religious groups, but concludes that religious groups are less successful at it because they lack experience and the ability to compromise. Hofrenning (1995) also finds that religious groups are constrained in their ability to compromise. He describes them as operating under a "principled and moralistic vision" that he terms "prophetic politics" (1995: 6). Other scholars, however, point to successes enjoyed by religious interest groups especially in electoral politics and in framing debates over issues such as abortion and gay rights (Adams 1970; Rozell and Wilcox 1995, 1997; Wood and Davis 1991).

187.4   The Rise of Interfaith Activism

In contrast to groups representing those with shared religious beliefs, *interfaith advocacy groups* represent religious individuals or institutions from different religious traditions who share a commitment to specific political goals or ideologies without a shared set of religious beliefs. This type of group has received relatively little attention in the subfield of religion and politics, but it is useful to differentiate interfaith groups from other religious advocacy groups.[2] Interfaith groups take one of two primary forms: individual membership groups and coalitions. Some interfaith groups are comprised of individual citizens from various faith traditions who join together to advance a particular issue or set of issues. The Interfaith Alliance is a prime example of this type of group. Other groups involve a coalition of religious groups from various faith traditions who join together on an ad hoc or permanent basis to advance a particular issue or set of issues. The Joint Religious Legislative Coalition, discussed in detail below, is an example of a permanent coalition. Both types of groups are noteworthy because they attempt to bridge theological differences in the interest of achieving shared political goals.[3]

Interfaith political advocacy is a relatively new phenomenon because Protestantism dominated the American religious landscape until the mid-1800s and, thus, there was less religious pluralism from which to draw an interfaith movement. Beginning in the 1840s, Catholic and Jewish immigrants began arriving in the United States in significant numbers. The increased religious diversity initially resulted in growing tensions in the political realm, particularly between Protestants and Catholics. For example, education became a heated topic because public schools often reflected Protestant biases by using a Protestant version of the Bible and textbooks with obvious anti-Catholic biases. These tensions stifled cooperation between Protestants and Catholics.

Strains of anti-Semitism among both Protestants and Catholics also inhibited the formation of interfaith groups (Feldman 1990, 2001). While some theologically liberal Protestants initiated interfaith dialogue with Jewish leaders in the late 1800s, increases in Jewish immigration following World Wars I and II primarily resulted in backlash against the new immigrants.

As a result of these tensions, the interfaith advocacy movement in the U.S. did not began in earnest until the 1960s, when alliances developed between progressive Protestants, Catholics, and Jews in political debates over Civil Rights and the Vietnam War. Arguably, the most important events facilitating the formation of interfaith alliances in the U.S. were changes instituted within the Catholic Church. In 1961, Pope John XXIII called for a meeting of the world's bishops for the purpose of modernizing the Catholic Church. One document that emerged from the meetings, titled *The Church in the Modern World*, "encouraged Catholics to involve themselves

---

[2] Chapters on advocacy groups in two texts on religion and politics (Corbett and Corbett 1999; Fowler et al. 2010) include overviews of the lobbying strategies of Jewish, mainline Protestant, Catholic, evangelical Protestant, and Muslim groups but do not specifically discuss interfaith groups. Zwier (1989) discusses some interfaith alliances in his study of coalition strategies of religious groups. Interfaith advocacy groups have received more attention from scholars in the field of religious studies (see, for example, Pedersen 2004; McCarthy 2007).

[3] The interfaith advocacy alliances I discuss in this chapter are distinct from ecumenical alliances. The ecumenical movement is a theological movement emphasizing the development of relationships between churches within the Christian tradition. Many religious advocacy groups, such as the National Council of Churches, are the result of ecumenical cooperation. In contrast, interfaith alliances involve relationships between faith traditions, such as between Christian and non-Christian individuals or groups.

in the daily social and political struggles around them. No longer should they think of Catholicism as the one true religion and all other branches of Christianity as heresies. Instead, they should now think of Protestants as 'separated brethren' within the Christian family" (Allitt 2003: 81). The change in tone resulting from the Second Vatican Council helped bring to an end the long era of Protestant-Catholic tensions in America and helped to begin dialogue between Jewish and Catholic leaders.

During the 1950s and 1960s, many religious leaders from various faith traditions became involved in the Civil Rights Movement. Religious interest groups such as the National Council of Churches, the National Conference of Catholics for Interracial Justice, and the American Jewish Congress endorsed civil rights demonstrations, which led many lay people associated with them to participate in protests. As relationships developed as a result of this advocacy work, leaders realized there may be benefits to more structured interfaith cooperation.

The networks forged in the fight for civil rights carried over into the battle against U.S. involvement in the Vietnam War. Several interfaith groups formed, including the Interreligious Committee on Vietnam and Clergy Concerned About Vietnam (later renamed Clergy and Laity Concerned). The Interreligious Committee on Vietnam sponsored a series of newspaper advertisements against the war in 1965 and later organized a delegation of Christian and Jewish clergy to lobby Congress directly (Allitt 2003). Clergy Concerned About Vietnam, formed in 1965 and led by Lutheran minister Richard Neuhaus, Rabbi Abraham Heschel, and Catholic priest Daniel Berrigan, brought together 100 clergy members opposed to the Vietnam War.

The formation of interfaith advocacy groups carried on throughout the latter half of the twentieth century and the policy interests of these new groups expanded beyond the issues of civil rights and war. The Washington Interreligious Staff Council, an informal coalition of Protestant, Catholic, and Jewish groups, formed in 1968 to provide coordination between progressive groups. The Interfaith Center on Corporate Responsibility and the Interfaith Impact for Justice and Peace both formed in the mid-1970s and the Interreligious Health Care Access Campaign formed in the early 1990s. Many of these interfaith groups were initiated by the National Council of Churches.

While interfaith advocacy groups tend to form among individuals and groups on the liberal end of the political spectrum, there are also examples of conservative interfaith groups. The Moral Majority, founded in 1979, was an attempt by evangelical leader Jerry Falwell to unite conservatives from various religious traditions for political action. Falwell hoped to include conservative Roman Catholics, Jews, and Mormons in the coalition, but the attempt failed to gain traction among the dominant Baptist and fundamentalist members of the group.

More successful efforts at interfaith cooperation on the Religious Right have centered around particular issues such as support for Israel, opposition to abortion, opposition to the proposed Equal Rights Amendment, and opposition to same-sex marriage. Butler (2006) documents an effort by a coalition of conservative evangelicals, Catholics, and Mormons to influence international policies at the United Nations. The Protect Marriage campaign organized Evangelical Christians, Mormons, Catholics, and Jews in support of Proposition 8, a 2008 ballot measure in California to ban same sex marriage (Garrison 2008). Conservative interfaith advocacy groups tend to form on an ad hoc basis and dissolve once the issue is resolved.[4]

As these examples suggest, most interfaith advocacy groups active today involve three dominant religious groups in America: Protestants, Catholics, and Jews. Among Protestants, it is useful to distinguish between evangelical Protestants, who make up approximately 26 % of the U.S. population, and mainline Protestants, who comprise approximately 18 % of the population (Pew Forum on Religion and Public Life 2007). Catholics make up approximately 24 % of the population, while Jews represent just under 2 % of the population. Some interfaith advocacy groups also include other religious minorities, who make up a small segment of the U.S. population. In the wake of the terrorist attacks of September 11, 2001, many interfaith groups made a particular effort to include Muslims in their coalition. Finally, some religious groups, such as the Jehovah's Witnesses and some fundamentalist Protestants, are unlikely to be included in interfaith advocacy efforts because their theology eschews political involvement.

187.4.1   The Interfaith Alliance

One of the largest and most well-known interfaith advocacy groups is the Interfaith Alliance (TIA). The Interfaith Alliance is a national interfaith advocacy group made up of individual members from various religious traditions that advocates for a strict separation of church and state. The group draws strength from the fact that the call for keeping faith out of politics comes from self-professed religious people from a range of religious traditions. This section provides a brief history of TIA, examines the group's size and structure, and provides an overview of the main political issues of concern to the group.

Three political operatives, Robert Norris, Jeff Kleuter, and Jill Hanauer, founded TIA in 1994 out of concern for what they perceived as the growing infusion of religious and political extremism in American politics throughout the 1980s. After an initial meeting, the organizers enlisted the support of the National Council of Churches, which shared an interest in countering the political power of the Christian Coalition. The group's first president, Presbyterian minister Herb Valentine, led the launch of the new group on July 14, 1994 with a high profile news conference and an appearance on NBC's *Today Show*. The Interfaith Alliance outlines its mission as "celebrating religious freedom by championing individual rights, promoting policies that protect both religion and democracy, and uniting diverse voices to challenge extremism."[5]

---

[4] In contrast, research by Kevin Hula (1999) suggests that conservative advocacy coalitions, on the whole, are more likely than liberal advocacy coalitions to form long-term coalitions.

[5] www.interfaithalliance.org/about. Accessed: February 29, 2012.

Since the group was formed to contest the growing power of the Religious Right, it should come as no surprise that its initial efforts focused attention on the, then powerful, Christian Coalition. During its first year, TIA appealed to religious leaders from across the country encouraging them not to distribute voter guides produced by the Christian Coalition. In the next election cycle, in 1996, TIA produced and distributed its own voter guides. As the group continued to evolve, emphasis shifted, first to encouraging civil political discourse, and then to taking specific policy positions on certain controversial issues.

The Interfaith Alliance operates both at a national level and through affiliates at the state and local level, with an emphasis on individual (rather than institutional) members. Within 10 years of its founding, TIA claimed 150,000 members, a cyber-network of 10,000 activists and 75 local activist groups (Freeman 2003). By 2012, the group claimed 185,000 members from over 75 faith traditions. Individuals can join the national group by signing up on the group's website. Members are encouraged to donate money to the organization and to communicate with elected officials regarding issues of concern to the group. In return, they receive a quarterly newsletter and access to policy information and other online resources.

According to Greg Lebel, the group's Field and Political Director during the 1996 campaign, TIA "didn't really exist as a national organization that could move the agenda at that point. We were more of a clearing house [that could] tell the folks in Iowa what the folks in Washington state were doing" (Freeman 2003: 10). Over 15 years later, the national group still does not maintain a publicized list of issue priorities, but current president Rev. Dr. C. Welton Gaddy summarizes the mission of TIA as being "dedicated to promoting mutual respect, cooperation and civility … [and striving] … to promote religion as a positive and healing force in the life of the nation."[6] Though the group does not have a set of issue priorities, it does respond to government decisions that involve religion through issuing press releases and email alerts to members. Press releases from 2012 focus on topics ranging from President Obama's decision to require health insurance plans to include contraceptive coverage to concern over the Federal Bureau of Investigation's training material dealing with Muslims.

Unlike the national organization, many state and local affiliates do have lists of specific issue priorities. For example, the Iowa chapter includes an "Issues" page listing five of the group's key issues: "Democracy Not Theocracy," "Respect Iowa Courts," "Strengthening Civil Rights," "Principles for a Strong Public Education," and "Civility."[7] The Interfaith Alliance of Colorado lists several ballot initiatives they oppose and highlights comprehensive immigration reform as a key issue.

---

[6] www.interfaithalliance.org/about/meet-our-president Accessed: January 15, 2012.

[7] www.interfaithallianceiowa.org/index.php?option=com_content&view=article&id=77&Ite mid=56 Accessed: February 29, 2012.

187.4.2   The Joint Religious Legislative Coalition

In contrast to the Interfaith Alliance, which relies on individual members and operates at a national level, the Joint Religious Legislative Coalition is a permanent interfaith coalition comprised of institutional members, active in state-level politics. This section provides a brief history of the JRLC, examines the group's size and structure, and provides an overview of the main political issues of concern to the group.

The Joint Religious Legislative Coalition is an interfaith advocacy group that seeks to influence public policy in the state of Minnesota. The group is comprised of institutional members representing four religious traditions: The Minnesota Council of Churches,[8] the Minnesota Catholic Conference, the Jewish Community Relations Council of Minnesota and the Dakotas, and the Islamic Center of Minnesota. In addition to these four sponsoring bodies, individual citizens from churches, synagogues, and mosques affiliated with the four sponsors form a network of activists across the state.

JRLC was founded after a chance meeting between two Minnesota lobbyists representing the Catholic Conference and the Council of Churches. The run-in and their subsequent correspondence led the two religious groups to begin laying the groundwork for a formal organization. In the fall of 1969, a formation committee was established to formalize the relationship between the Catholic Conference and the Council of Churches and to articulate a joint legislative agenda for the 1971 session. Throughout the summer, the committee worked to organize task forces on the issue areas identified by the group. Task forces were comprised of members of the formation committee and community members who had expertise in the various subject areas.

Within months of formation, the Minnesota Rabbinical Association asked to participate. The new group joined the coalition in September 1969. In early 1974, the Jewish Community Relations Council replaced the MRA as the institutional body representing the Jewish community on the JRLC board.

The most recent organizational change to the JRLC involved the addition of Muslims to the coalition. Adding Muslim representatives to the coalition was something that board members were excited about, but it was also challenging because the Muslim community did not have a formal statewide organization analogous to the other three sponsors until the formation of the American Muslim

---

[8] Denominations represented by the Minnesota Council of Churches include the African Methodist Episcopal Church, American Baptist Churches USA, Christian Church (Disciples of Christ), Church of God in Christ, Church of the Brethren, The Episcopal Church, Evangelical Lutheran Church in America, Greek Orthodox Archdiocese of America, Mennonite Church, National Association of Congregational Christian Churches, National Baptist Convention, Presbyterian Church (USA), United Church of Christ-Minnesota Conference, and the United Methodist Church-Minnesota Annual Conference.

Council-Minnesota Chapter in 1995. The process of joining the JRLC began with the granting of "observer status" to the American Muslim Council-Minnesota Chapter in October 1995. Obtaining "observer status" meant that the Muslim representatives of the AMC-MC to the JRLC board could participate in all board discussions and have access to all written records of the group. The observers, however, did not have a binding vote in board meetings during this initial period. In 2004, the Islamic Center of Minnesota replaced the AMC as the institutional representative of Muslims to the JRLC board and the group transitioned into a full sponsoring member in April 2004.

A 16 member board comprised of four representatives from each of the sponsoring bodies governs the group. Each sponsor has its own method for selecting board members to serve, though most report that they seek people who are committed to interfaith advocacy and who represent the broad diversity of their constituents. The board rotates leadership positions annually between religious traditions, with the Chair and Vice-Chair representing separate traditions. Decision-making requires unanimous approval from all four sponsoring bodies, which means that the group only lobbies on behalf of issues agreed to by the state leaders of all four religious traditions.

Although only these four institutional groups are considered members, the JRLC also encourages individuals to be involved with the group. Individuals can sign up to receive information from the JRLC and they are encouraged to participate in the group's primary lobbying event, the annual legislative Day on the Hill.

The JRLC entered the 1971 legislative session with 15 position papers and a legislative guide of 85 specific recommendations on a range of issues including taxation, agricultural workers, consumer affairs, criminal justice, alcohol and drug abuse, ecology, education, housing, human rights, senior citizens, and welfare. The group maintains a position until it is replaced by a new position paper, and so many of these position papers from 1971 are still in effect. More recent position papers include "Human Trafficking in Minnesota: A Violation of Human Dignity" and "Health Care and Justice in Minnesota." The position papers form rough guidelines for political advocacy, but the JRLC Board and staff members narrow the focus by selecting a handful of issues on which to concentrate each year. In 2011, the JRLC focused on protecting the safety net for low income Minnesotans, changing the selection process for state judges, and creating a more progressive taxation structure.

While the Interfaith Alliance focuses on one central issue, religious liberty, the JRLC takes a stand on a wide range of issues. While some of these issues might not seem to relate to religious values, JRLC leaders argue that they involve questions of justice and fairness, which are values at the heart of the four religious traditions involved in the coalition. Also unlike TIA, leaders of the JRLC argue that religious voices and values ought to play a role in the political process. Their claim is that people of faith can make important contributions to political deliberations and play a critical role in advocating on behalf of the poor and powerless. Rather than trying to establish a clear division between religious faith and political decision making, they encourage decision makers to consider the teachings of their faith traditions in making decisions.

187.5   Challenges and Opportunities of Interfaith Advocacy

187.5.1   *Challenges*

As the examples of the Interfaith Alliance and the Joint Religious Legislative Coalition demonstrate, interfaith political advocacy is possible, but these types of groups also face a number of substantive challenges. In this section, I outline two of the most significant challenges facing interfaith groups: overcoming theological differences and deciding whom to include.

Perhaps most significantly, interfaith advocacy is challenging because it requires individuals or groups to overcome important theological differences and historical baggage. Most religious advocacy groups are rooted in a common historical and theological base. In contrast, interfaith groups—by definition—do not have this shared understanding and experience from which to draw. Individuals or institutions from different faith traditions must overcome historical prejudices and reach acceptance regarding differences in fundamental beliefs about human nature in order to participate effectively in an interfaith group. This tends to be easier for some religious groups than others, which helps explain why interfaith groups frequently include a regular cast of characters. In particular, Protestants, Catholics, and Jews, are regular participants in interfaith initiatives because they draw from a shared faith tradition.

As interfaith groups expand to include individuals or institutions from different religious traditions, they are forced to deal with the challenge of overcoming theological differences that impact practical political decisions. For example, when Muslims joined the JRLC, the group was forced to recalibrate some of its established positions. The JRLC expressed clear opposition to the death penalty in the 1992 position paper "Crime and Justice," however, Muslim tradition allows for capital punishment in some situations. In response, the JRLC altered the language of the position paper in order to focus more on the moral injustices in the judicial system. Thus, the position became more of a critique of the justice system than a critique of capital punishment itself.

Interfaith groups can also respond to differences in practice resulting from differences in theology by avoiding those issues altogether (Zwier 1989). For example, the JRLC deals with issues that might be controversial as a result of theological differences through, what one board member described as "self-censorship." Board members are aware of where the four faith traditions stand on these issues and so they choose to focus on issues where there is common ground rather than on issues where they know disagreement exists. As Executive Director Brian Rusche remarked, there are not many controversial items that come before the JRLC board because people have "self-discipline" and avoid bringing those issues to the table.[9]

This challenge of overcoming theological differences is more pronounced among interfaith groups with institutional members, like the JRLC, than it is among those

---

[9] Brian Rusche, in discussion with the author, June 16, 2010.

with individual members, such as The Interfaith Alliance. Individuals from various faith traditions who associate with TIA are free to accept or reject whichever aspects of their faith might prohibit interfaith action, if they so choose. In contrast, institutional members experience more constraint in determining their participation in interfaith organizations.

In addition to the challenge posed by theological and historical differences, interfaith advocacy groups also face a challenge faced by other types of advocacy groups: deciding whom to include in the group. For the most part, political advocacy groups live by the mantra "more is better." There is substantial political strength when a group can point to a large and active membership base. However, among interfaith groups, especially those that operate as coalitions, there is reason to suspect that a group may want to restrict membership (Zwier 1989). In discussing one long-term coalition, Graziano found that "their strength lies in prestige rather than numbers" (2001: 52). Hojnacki argues that individual groups are more accountable when a coalition is small and that a small coalition decreases incidences of free riding (1998). Furthermore, certain groups may be viewed as "pivotal players," making them more desirable coalition members while other groups may be protective of their group autonomy making them less likely to want to join a coalition (Hojnacki 1997). Simply having more members does not mean that the group will necessarily be perceived as more powerful. Similarly, having group members from a variety of religious traditions does not automatically grant an interfaith group any authority over political questions. Interfaith groups must weigh the benefits and drawbacks of unrestricted membership.

The Interfaith Alliance has not limited its membership and includes individuals from any religious tradition as well as those who claim no religious affiliation. The group can do this easily because it represents individuals rather than institutional members. While this provides TIA with the ability to claim a large membership base—over 185,000 members at last count—there are challenges to this model. For example, members may not be equally committed to the work of TIA or members may not be distributed throughout the country in ways that give political advantage to the organization. Although the group does not specify where its members live, the website only lists active state groups in 17 states, suggesting that its presence in the other 33 states may be minimal. Additionally, the group gives no indication of how their membership is distributed among the 75 faith traditions it claims or of how strongly those religious beliefs are held by its members. Members may claim a religious tradition they do not actively practice. Thus, having a large membership base representing a range of religious traditions does not allow for much control on the part of the group and may lead to collective action problems.

In contrast, groups comprised of institutional members have more control over membership. The JRLC has faced the question of who should be included several times in its 40-year history. Soon after its initial founding, representatives from the Minnesota Rabbinical Association requested inclusion in the coalition. More recently, representatives from the Muslim population requested representation. In both of these cases, the JRLC moved to become more inclusive of these non-Christian faiths, but did so though an introductory period where the new groups were included as "observers" for a period of time before receiving full membership.

To be successful in the political arena, interfaith advocacy groups must bridge historical and theological differences that threaten to divide religious groups and they must overcome challenges in organization and structure facing most advocacy groups.

187.5.2 *Opportunities*

Despite these significant challenges faced by interfaith groups, I contend that interfaith groups will continue to play an important role in the political process. Interfaith groups can be particularly influential because they function as coalitions, bringing together unlikely allies to address political problems. Perhaps more significantly, the social capital developed through interfaith advocacy groups allows for quick and effective responses to challenges faced by a pluralistic society.

The adage, "politics makes strange bedfellows," reflects a fundamental principle of politics: advocacy group leaders know that the groups they oppose on one issue might be allies on another. The most influential coalitions in politics are often those that are formed by groups that usually find themselves on opposite sides of issues. By their very nature, religious groups operate largely in opposition to other religious groups (Finke and Stark 1992). Thus, any efforts on the part of religious groups from different religious traditions are especially significant because they carry the weight of being an unlikely alliance. Simply stated, interfaith advocacy groups draw political strength from the "surprise factor" of having individuals or institutions representing various faith traditions working together toward a common political goal.

Working together in interfaith groups also allows the practical benefit of sharing scarce resources. Research on religious interest groups suggests that coalitions are especially important for religious interest groups because religious groups tend to have fewer financial resources than other types of interest groups (Zwier 1989; den Dulk and Hertzke 2006).

Finally, interfaith advocacy groups represent a unique opportunity for the building of *social capital* within local, state, and national communities. There are many ways of conceptualizing social capital, but the primary meaning as it relates in this case involves the norms of trust and reciprocity that develop because of regular interactions with a diverse group of people. Putnam (2000) distinguishes this form involving relationships between heterogeneous groups of people as "bridging" social capital and suggests that this type of social capital is especially critical in maintaining a healthy democracy. As Bretherton argues, "...politics properly understood is the process through which people maintain community and recognize and conciliate conflict with others in pursuit of shared goods" (2011: 364).

Interfaith advocacy groups help develop bridging social capital by bringing individuals and institutions from different faith traditions together for regular interaction and action on common issues. The individual relationships developed through these groups facilitate creative problem solving and provide an avenue for addressing challenging issues. For example, the interfaith relationships developed among JRLC board members are particularly useful when new problems or challenges emerge within local communities. As one board member said, when contentious issues, such as cases of discrimination, arise, the different religious groups can move quickly to address the issue jointly, "because we all know each other."[10]

Similarly, these established interfaith relationships can help interfaith advocacy groups to begin to address issues where there is less agreement between the religious traditions. For the JRLC, one such issue involves civil rights for gays and lesbians. Despite the four sponsoring institutions expressing very different views on gay rights, the JRLC has discussed civil rights for the GLBT community since 1972, and the group issued a formal position in 1993 in a document titled "Human Rights with Regard to Sexual Orientation." In this document, the JRLC affirmed its support of human rights for all people and the need for government to protect the dignity of all people. As these examples suggest, one of the most significant benefits of interfaith advocacy groups may be the fact that their very existence can help to address many challenges stemming from pluralism the United States will most certainly face in years to come.

187.6   Conclusion

The Interfaith Alliance and the Joint Religious Legislative Coalition represent only a small fraction of the work of interfaith advocacy groups in the United States, yet they help highlight many of the important dimensions of interfaith groups, including the challenges and opportunities for interfaith advocacy. Interfaith advocacy groups are related to, but substantively different from religious advocacy groups and merit increased attention on the part of scholars of religion and politics. These groups face significant theological, historical, and organizational challenges, but when these challenges are overcome, interfaith advocacy groups have the potential to both influence public policy in significant ways and to promote the health of American democracy through the creation of social capital.

References

Adams, J. (1970). *The growing church lobby in Washington*. Grand Rapids: Eerdmans.
Allitt, P. (2003). *Religion in America since 1945: A history*. New York: Columbia University Press.
Berry, J., & Wilcox, C. (2009). *The interest group society* (5th ed.). New York: Pearson Longman.
Bretherton, L. (2011). A postsecular politics? Inter-faith relations as a civic practice. *Journal of the American Academy of Religion, 79*(2), 346–377.
Butler, J. (2006). *Born again: The Christian right globalized*. London: Pluto Press.

---

[10] JRLC board member, in discussion with the author, July 7, 2010.

Clark, P., & Wilson, J. (1961). Incentive systems: A theory of organizations. *Administrative Science Quarterly, 6*(3), 129–166.

Corbett, M., & Corbett, J. M. (1999). *Politics and religion in the United States.* New York: Garland.

Den Dulk, K., & Hertzke, A. (2006). Conclusion: Themes in religious advocacy. In E. Cleary & A. Hertzke (Eds.), *Representing God at the statehouse: Religion and politics in the American states* (pp. 225–241). Lanham: Rowman & Littlefield.

Feldman, E. (1990). *Dual destinies: The Jewish encounter with Protestant America.* Urbana: University of Illinois Press.

Feldman, E. (2001). *Catholics and Jews in twentieth-century America.* Urbana: University of Illinois Press.

Finke, R., & Stark, R. (1992). *The churching of America 1776–1990: Winners and losers in our religious economy.* New Brunswick, NJ: Rutgers University Press.

Fowler, R. B., Hertzke, A. D., Olson, L. R., & Den Dulk, K. (2010). *Religion and politics in America: Faith, culture, and strategic choices* (4th ed.). Boulder, CO: Westview Press.

Freeman, M. (2003). *Answering the call: The Interfaith Alliance's first decade.* Retrieved April 8, 2010, from www.interfaithalliance.org/images/PDF_DOCS/10year_2003.pdf

Garrison, J. (2008). Churches plan a big push against same-sex marriage. *Los Angeles Times,* August 24. http://articles.latimes.com/2008/aug/24/local/me-faith24

Graziano, L. (2001). *Lobbying, pluralism and democracy.* London: Palgrave.

Hertzke, A. (1988). *Representing God in Washington: The role of religious lobbies in the American polity.* Knoxville: University of Tennessee Press.

Hofrenning, D. (1995). *In Washington but not of it: The prophetic politics of religious lobbyists.* Philadelphia: Temple University Press.

Hojnacki, M. (1997). Interest groups' decisions to join alliances or work alone. *American Journal of Political Science, 41*(1), 61–87.

Hojnacki, M. (1998). Organized interests' advocacy behavior in alliances. *Political Research Quarterly, 51*(2), 437–459.

Hula, K. (1999). *Lobbying together: Interest group coalitions in legislative politics.* Washington, DC: Georgetown University Press.

McCarthy, K. (2007). *Interfaith encounters in America.* New Brunswick, NJ: Rutgers University Press.

Olson, M. (1965). *The logic of collective action: Public goods and the theory of groups.* Cambridge, MA: Harvard University Press.

Pedersen, K. (2004). The interfaith movement: An incomplete assessment. *Journal of Ecumenical Studies, 41*(1), 74–94.

Pew Forum on Religion and Public Life. (2007). *U.S. religious landscape survey.* Retrieved March 7, 2012, from http://religions.pewforum.org/reports

Pew Forum on Religion and Public Life. (2011). *Lobbying for the faithful: Religious advocacy groups in Washington, DC.* Retrieved March 7, 2012, from www.pewforum.org/Government/Lobbying-for-the-faithfu--xec.aspx

Putnam, R. (2000). *Bowling alone: The collapse and revival of American community.* New York: Simon & Schuster.

Rozell, M., & Wilcox, C. (1995). *God at the grass roots: The Christian right in the 1994 elections.* Lanham: Rowman & Littlefield.

Rozell, M., & Wilcox, C. (1997). *God at the grass roots: The Christian right in the 1996 elections.* Lanham: Rowman & Littlefield.

Salisbury, R. (1969). An exchange theory of interest groups. *American Journal of Political Science, 47*(1), 1–32.

Truman, D. (1971). *The governmental process: Political interests and public opinion* (2nd ed.). New York: Knopf.

Walker, J. (1991). *Mobilizing interest groups in America: Patrons, professions, and social movements.* Ann Arbor: University of Michigan Press.

Weber, P., & Jones, W. L. (1994). *U.S. religious interest groups: Institutional profiles.* Westport: Greenwood Press.

Wood, J., & Davis, D. (1991). *The role of religion in the making of public policy.* Waco: Baylor University Press.

Zwier, R. (1989). Coalition strategies of religious interest groups. In T. Jelen (Ed.), *Religion and political behavior in the United States* (pp. 171–186). New York: Praeger.

Chapter 188
The Election of a Lesbian Mayor in a Religiously Conservative City: The Case of Houston, Texas

Nancy Palmer Stockwell and Ira M. Sheskin

188.1 Introduction

The United States is the most religiously diverse nation in the world, an anomaly among industrialized nations. Despite this diversity, geographers have identified pockets of religious homogeneity in which a dominant religion reinforces social capital, influences political ideology, and may serve as a predictor of voting behavior among its adherents. The American South remains a pocket of predominantly conservative, Baptist and Evangelical Christian hegemony; however, metropolitan areas of the South display more diversity and pluralism as a result of foreign and domestic migration into the region, changes in religious affiliation, an increasing population choosing no religious affiliation, and the growing influence of popular culture on religious participation. These forces act within a paradigm of neosecularization in which personal conscience exerts more influence than religious authority.

Houston, Texas, within the context of the South, serves as the case study for this chapter which examines the temporal and spatial changes to Houston's religious and political landscape and the manner in which these change s influence, or do not influence, voter behavior. It has become almost cliché to equate conservative fundamentalist Christians with the Republican Party and liberal Black Protestant and mainline Protestants with the Democratic Party. These stereotypes may hold

N.P. Stockwell (✉)
Senior Contract Administrator, Enerfin Resources, Houston, TX, USA
e-mail: nancypstockwell@gmail.com

I.M. Sheskin
Department of Geography and Regional Studies, University of Miami,
Coral Gables, FL 33124, USA
e-mail: isheskin@miami.edu

© Springer Science+Business Media Dordrecht 2015
S.D. Brunn (ed.), *The Changing World Religion Map*,
DOI 10.1007/978-94-017-9376-6_188

true at the national level. This stereotype may apply in Houston in a national or statewide election, but a more complex voter profile appears in city elections in which the candidate slate includes a lesbian accountant, a Black attorney, a Hispanic firefighter, and a white anti-gay activist. In 2009, Houston became the first major American city to elect an openly homosexual mayor. The election of Democrat Annise Parker garnered national coverage not only because of her sexual orientation, but also because Houston is home to a large, active Christian community. Official statements issued by some of these Christian groups express disapproval of homosexuality. Parker is not the first GLBT elected to public office in America. Since Kathy Kozachenko's 1974 election to the Ann Arbor, MI City Council, many GLBT candidates have won municipal, state and congressional positions (Gay and Lesbian Victory Fund 2012). Houston's status as the nation's fourth largest city elevates the significance of Parker's election.

Following introductory remarks, this chapter opens with an outline of the transition in Houston's religious and political landscapes and mention of the agents driving this transition. Theories of secularization and neosecularization explain religion in the United States and the South as a pocket of religious homogeneity. Historic and contemporary examples of deterritorialization and reterritorialization of religious space in the South are explored. This is followed by a discussion of Houston's demographics and its diversity following Richard Florida's concept of the "creative class." An examination of national trends and local exceptions in the composition of religious populations and religion, race, and sexual orientation in politics precedes the conclusions.

188.2   Religion and Politics in Transition

This chapter demonstrates the following features of Houston's religious and political landscape:

- Houston's religious landscape has changed over the past three decades.
- Houston is following the national trend that shows an increasing number of people professing no religious affiliation.
- Houstonians are exhibiting an increased tolerance for diversity and acceptance of gays, lesbians, bisexuals and transgenders (GLBT).
- Religion in Houston does not serve as a reliable predictor of voting behavior in local mayoral elections.

The overriding evidence in this study is that Houston's religious landscape has changed amid a climate of increasing diversity and tolerance, thus creating the environment for Annise Parker's political success. In general, this study examines the changes that have occurred in Houston's religious landscape and in attitudes toward gays and lesbians by analyzing data at the zip code level from the Houston Area Survey (Klineberg 2010). Selected survey variables pertaining to religion, political preference, attitudes toward gays, and demographic characteristics are used as independent variables to analyze voting results in the 2009 and 2011 mayoral elections. This research will demonstrate the process of neosecularization, also termed detraditionalization, by which people of faith rely more upon their personal wisdom than upon the guidance of religious authorities when making decisions about secular matters. As will be shown, the high level of religious affiliation, primarily Protestant, among Houstonians does not serve as a reliable predictor of voting behavior, and in the case of the 2009 and 2011 elections, the minority secularist population and the "creative class" as defined by Richard Florida (2002) determined the outcomes.

188.3   Religion in the United States and the South

The complex religious landscape of the United States has intrigued and perplexed geographers and other social scientists for decades, Halvorson and Newman (1994), Stump (2008) and Zelinsky (1961) among others. From its inception when colonists of many different faiths sought a space within which they could worship without fear of persecution, Protestant Christianity exerted hegemonic power in the United States (Kosmin and Lachman 1993). Through a process Kosmin and Lachman call Protestantization,

> the collective religion became America's civil religion, incorporating secular as well as religious values and emphasizing a national purpose... Eventually, the collective Americanized Protestant religion made its peace with other religions, such as Catholicism and Judaism, as long as they recognized the United States as being 'different' from other lands and the spiritual descendent of biblical Israel. (p. 22).

188.3.1   Secularization and Neosecularization

The process of secularization that has significantly reduced the impact of religion in most of the industrialized Western nations has facilitated religious diversity in the United States, where choosing a religious affiliation becomes a response to market competition (Kosmin and Lachman 1993; Lindsay 2008, Warf 2006). This paradigm of secularization describes at the societal, organizational and local scales the transformation and persistence of religion rather than its decline and eventual disappearance (Kosmin and Lachman 1993; Kosmin et al. 2009; Lindsay 2008; Smith 2007; Warf 2006; Warf and Winsberg 2008; Yip 2002); however, emphasis has shifted from religious authority to individual spirituality described by the new paradigm of neosecularization facilitating analysis at the individual scale (Yip 2002).

188.3.2 Religious Homogeneity in the South

In his seminal study, Zelinsky (1961) mapped religious distribution in the United States in a "first approximation of the nation's religious regions" (p. 139) in the hopes of providing insight into America's cultural regions. Overcoming variations among religious bodies in methods of counting membership, lack of data for some religious bodies, and variations in denominational definitions, Zelinsky identified three primary classifications within which the major Christian denominations fall. *The British Colonial Groups* include Protestant Episcopal, Presbyterian bodies, Methodist bodies, Congregational Christian, Baptist bodies and Friends (Quakers). The *Immigrant European Groups* include Roman Catholic, Jewish congregations, Lutheran bodies, Evangelical and Reformed, Reformed bodies, Brethren churches, Mennonite bodies and Moravian bodies. The *Native American Groups*[1] include Disciples of Christ, Unitarian and Universalist churches, Latter-Day Saints (Mormons), Adventist bodies, Church of the Nazarene, Assemblies of God, Churches of God and Evangelical United Brethren. Within these classifications, Zelinsky mapped the membership of each group and the "latitudinal zonation in the western movement" (Zelinsky 1961: 159) of most groups across the American continent. Three major pockets of religious homogeneity persisted through the migration: a concentration of Roman Catholics in the Northeast, a concentration of Mormons in Utah and Idaho, and a high concentration of Baptists in the South (Zelinsky 1961). Lindsay (2008) posits that the secularized, pluralist environment of the U.S. nourishes a higher level of religious commitment in the South where social capital is reinforced through strong adherence to the Evangelical Christian, primarily Baptist, worldview. This homogeneity is in part a reflection of the South's historically less advanced economic condition and weaker migration into the area (Lindsay 2008; Zelinsky 1961). Following the Civil War, the evangelical church was the only civic institution left intact to which Southerners could turn (Webster 1997).

188.3.3 Deterritorialization and Reterritorialization of Religious Space in the South

Two decades before the outbreak of combat, the Baptist Church split over the issue of slavery, and the Southern Baptist Convention was formed in 1845 and remains the largest Protestant denomination in the country (Warf and Winsberg 2008; Webster 1997; Winter 2000). Before the Civil War, Blacks in the South "generally worshipped in white churches, balcony sections being reserved for their presence" (Webster 1997: 159). During Reconstruction, Blacks were discouraged or barred from attending their former Southern Baptist Churches, thus leading to the development

---

[1] In Zelinsky's terminology "Native American" refers to Christian denominations originating in the United States, not the religious beliefs of indigenous peoples.

of exclusively Black churches with Black clergy who became the religious and political leaders of their communities (Webster 1997). Baptists were not the only denomination to fracture over the issue of slavery. Presbyterians divided in 1857 and Methodist Episcopals in 1844 (Goen 1983). The divide remained deep well into the twentieth century. Not until 1983 did the Presbyterian Church in the United States, based in the South, join the northern-based United Presbyterian Church in the U.S.A. to form the Presbyterian Church (U.S.A.). Many Southern conservatives opposed the reunion and broke away to form the Presbyterian Church in America and the Evangelical Presbyterian Church (Winter 2000). Beginning in the late twentieth century and continuing into the twenty-first century, the debate over the role of gays, lesbians, bisexual and transgender individuals within the religious community has ignited heated debate and denominational division (D'Emilio 2012; ECO: Evangelical Covenant Order of Presbyterians 2012; Jenkins 2011; The Fellowship of Presbyterians 2012), leading to renewed deterritorialization and reterritorialization of religious space. Unlike divisions over the issue of slavery with its definite North-South geography, congregations across the country are realigning themselves according to acceptability or disapproval of the active participation in leadership roles by homosexuals. As these religious bodies recreate their worship space, Houston finds itself at the vanguard of change on both sides of the debate.

188.4   Demographics and Diversity: Counterforces Against Protestant Hegemony

Current census data reveal economic improvement and demographic changes in the South (Macken et al. 2011) prompting changes in the South's religious landscape. Between 2000 and 2010, Houston's population increased significantly with changes in its racial and ethnic composition (Humes et al. 2011; Macken et al. 2011). The distribution of the three largest racial and ethnic groups in Houston shows the widespread dispersal of non-Hispanic Whites (Fig. 188.1), the clustering of Blacks (Fig. 188.2) and the clustering of Hispanics (Fig. 188.3). Increased demographic diversity, as shown by Richard Florida's "diversity index," in metropolitan areas across the South has attracted highly-educated individuals seeking "low barriers to entry for human capital" (Florida 2002: 743). Florida terms this group the "creative class." According to Florida, the presence of a large gay community, such as the one established in Houston (Gates and Ost 2004), indicates that an area is open and accepting of a wide range of nationalities, races, ethnicities, sexual orientation and gender identities, thus increasing the area's diversity index (Florida 2002).

This chapter will demonstrate how this segment of Houston's population may have influenced the outcome of the 2009 election and will contribute to the body of knowledge established by geographers and social scientists who have analyzed voting patterns on the national and regional scales in elections involving gays and gay rights referenda (Chapman et al. 2007; O'Reilly and Webster 1998; Quinton and Webster 2011; Webster et al. 2010). Despite the demographic changes,

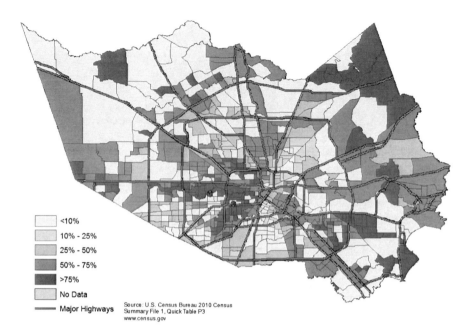

Fig. 188.1 Non-Hispanic whites in Harris County, by Census Tract, 2010 (Map by Nancy Palmer Stockwell and Ira M. Sheskin, data from the U.S. Census Bureau, 2010 Census)

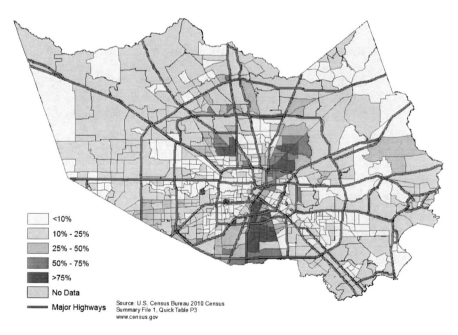

Fig. 188.2 Blacks in Harris County, by Census Tract, 2010 (Map by Nancy Palmer Stockwell and Ira M. Sheskin, data from the U.S. Census Bureau, 2010 Census)

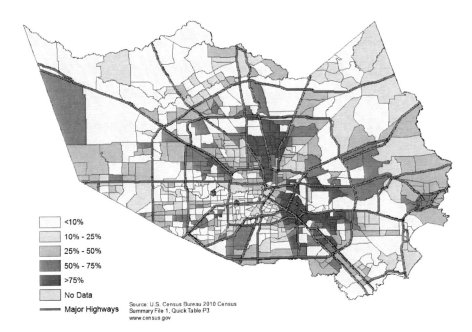

Fig. 188.3 Hispanics in Harris County, by Census Tract, 2010 (Map by Nancy Palmer Stockwell and Ira M. Sheskin, data from the U.S. Census Bureau, 2010 Census)

more recent studies of religion in America (Kosmin and Lachman 1993; Kosmin et al. 2009; Putnam and Campbell 2010; Warf 2006; Warf and Winsberg 2008) indicate that these homogeneous regions persist when data are analyzed on a national level. However, these studies also shed light on the forces countering the Protestant Christian hegemony: domestic migration and foreign immigration, religious "switching" in a competitive market, increases in the number of people claiming no religious affiliation, and the influence of popular culture. To this list, Smith (2007) adds the exposure to other religions through people or services and certain religious doctrines that are more tolerant. Bellah (2005) and Kong (2001) describe civil religion, moral geographies and social movements that may have a religious basis, but may be influenced by other secular agents of morality, that is, the state.

188.5   Religion and Politics: National Trends, Local Exceptions

Religion, though, by its very nature is more than data to be analyzed and mapped on a national or regional scale. Religion is personal. Religion is "intimately related to human identity, serving to cement political, cultural, and social mores in ways that at times appear to defy logical argument or rational action" (Buttimer 2006: 200).

Religion and belief systems form the basis for humans to form bonds with each other, their lived space, and the natural world in ways that build moral convictions shaping the believer's everyday life (Buttimer 2006). Observed at the local scale using the Houston Area Survey (Klineberg 2010), this study examines the lived expression of religious beliefs and whether it impacts the political climate of the city.

188.5.1 Baptists in the South and Houston

Even though the Baptist Church, "the state church of the ex-Confederacy" (Warf and Winsberg 2008: 420) remains strong in the South, recent data compiled by The Association for Religion Data Archives (ARDA) (InfoGroup, Social Explorer 2011) show a broader range of religious affiliation in Southern metropolitan areas. Zelinsky's 1952 data ranked Houston second to Atlanta in its concentration of Baptists (Zelinsky 1961). The 2009 ARDA data for Harris County, Texas, which encompasses Houston, indicate a drop in the percentage of religious adherents declaring Baptist affiliation (InfoGroup, Social Explorer 2011).

188.5.2 Composition and Trends of Houston's Religious Population

Categorizing and defining religious bodies presents challenges. One method found useful by researchers in grouping the myriad Protestant denominations follows the biological taxonomic system in which religious traditions are grouped by genus and individual denominations are species (Putnam and Campbell 2010). The three primary genera are Evangelical, Mainline, and Black Protestant (Putnam and Campbell 2010). Evangelicals (a) emphasize a personal relationship with Christ gained through a conversion "born-again" experience leading to salvation through the redeeming crucifixion and resurrection of Christ; (b) believe in the divine inspiration of the Bible and its ultimate religious authority; and (c) encourage followers to share their faith with non-believers (Noll 2001). Evangelical Protestants uphold a traditional, conservative interpretation of scripture, and the term as used today "encompasses all theologically conservative Protestants (except Black Protestantism…)" (Putnam and Campbell 2010: 13). Mainline Protestants generally are more liberal in their interpretation of scripture and emphasize social justice as a Christian's priority rather than "personal piety" (Putnam and Campbell 2010: 14). Mainline Protestants, who historically represented the dominant Christian tradition in America, value scholarly Biblical criticism and deemphasize supernaturalism (Putnam and Campbell 2010). Black Protestantism incorporates racial identity into its "theology, iconography, and worship" and blends "an evangelical focus on personal piety with a strong dose of Social Gospel" (Putnam and Campbell 2010: 15).

One of the fastest growing groups in the country is the "Nones," also referred to as "unaffiliated" (Kosmin et al. 2009; Pew Forum on Religion and Public Life 2008; Putnam and Campbell 2010). Kosmin describes "Nones" as people who do not identify with any religious group or organization. Although by definition atheists are included in this group, they comprise a very small proportion (7 %) of respondents who self-identified as "Nones" in the 2008 American Religious Identification Survey (ARIS 2008). Deists (people who believe in a higher power, but not a personal God) and theists (people who believe in a personal God) comprise 24 % and 27 % respectively in ARIS 2008. Agnostics comprised 35 % of the "Nones" in ARIS 2008 (Kosmin et al. 2009). The Pew Foundation uses the term "Unaffiliated" (Pew Forum on Religion and Public Life 2008) for this group. Data showing numbers of churches and adherents categorized by denomination at the county level for 2009 are archived and available from ARDA (Association of Statisticians of American Religious Bodies 2009). ARDA adjusted the total number of all adherents in the data to include an estimate of adherents in the Black denominations according to a formula presented by Finke and Scheitle (2005). However, the adjustment does not apply to individual denominations or theology groups within the general category of "Black denominations." The 2009 data were collected by the InfoGroup, organized by ARDA following the procedures described above, and tabulated and processed by Social Explorer, a web-based research tool designed to provide access to demographic information and historical census data. The InfoGroup does not include data on the unaffiliated (InfoGroup, Social Explorer 2011).

The U.S. Religious Landscape Survey (Pew Forum on Religion and Public Life 2008) was used for comparison of the 2009 Harris County data to the state and national rates of adherence. Christianity is the dominant religion, with Houston and Harris County showing a higher percentage of Evangelical Christians (38 %) than the state (34 %) and the country (26 %). Mainline Protestants, Historically Black Protestants, and Catholics bring the total Christian population in Harris County to 76 % of the general population, compared to 81 % in the state and 75 % in the country (Association of Statisticians of American Religious Bodies 2009; InfoGroup, Social Explorer 2011; Pew Forum on Religion and Public Life 2008).

Figure 188.4 illustrates a snapshot of Houston's religious landscape; however, analysis of the Houston Area Survey data reflects temporal trends in Houston's religious landscape (Fig. 188.5). Most notable is the increase in "Nones" from less than 5 % in 1982 to 11 % in 2010, following the national trend.

188.5.3   *Religion and Socio-political Attitudes*

Although Zelinsky (1961) hesitated to make a connection between religion and socio-political attitudes and behaviors of adherents because of inadequate data, current researchers have sought to fill this void (Klineberg 2010; Kosmin and Lachman 1993; Kosmin et al. 2009; Putnam and Campbell 2010; Webster 1997). Putnam's and Campbell's (2010) analysis of data collected in their *Faith Matters Survey 2006*

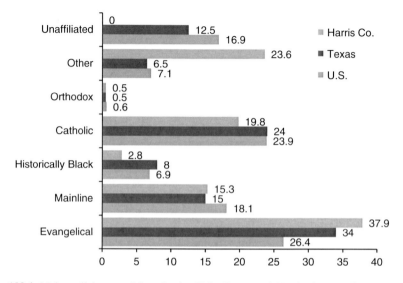

Fig. 188.4 Major religious traditions in the U.S., Texas, and Harris County (Source: Nancy Palmer Stockwell and Ira M. Sheskin, using the Association of Religion Data Archives database (www.thearda.com) and the U.S. Religious Landscape Survey)

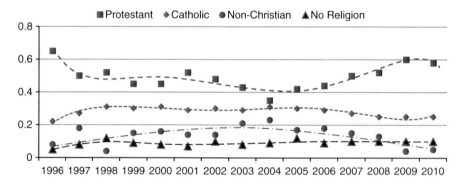

Fig. 188.5 Trends in Houston's religious participation, 1982–2010 (Source: Nancy Palmer Stockwell and Ira M. Sheskin, using data from the Houston Area Survey)

show that Evangelicals and individuals with a high level of religious participation tend to be conservative and Republican. Mainline Protestant and Black Protestant tend to be Independent or more liberal (Putnam and Campbell 2010).[2]

Studies of religion at the national and regional scale indicating a high level of religious participation leading to conservative Republican dominance appear inapplicable in Houston, where a shift toward a more liberal political atmosphere

---

[2] For added insight into national and regional voting patterns refer to Leib's recent analysis of Appalachia and the South (Leib 2012).

has occurred over the past 15 years. Houston is the "archetype laissez-faire city in the U.S.A." (Vojnovic 2003: 19). It is also entrepreneurial and individualistic with a flourishing religious market. The phenomenon of megachurches appearing on the religious landscape in the 1990s (Putnam and Campbell 2010) manifests itself in Houston with the presence of Lakewood Church, the largest Christian church in the country, and Second Baptist Church, the largest Baptist congregation and fifth largest Christian church (Stetzer 2009). Lakewood's pastor, Joel Osteen, influences a large congregation of approximately 40,000 who attend weekly services in the building that formerly housed the National Basketball Association Houston Rockets. Osteen's influence extends beyond Houston through his nationally syndicated broadcasts that reach over 200 million households in the US and around the world (Lakewood Church 2011). Second Baptist Church pioneered the multi-site church in Houston with five locations around the Greater Houston Area and a combined membership of approximately 54,000 (Marks 2011).

A vibrant Black Protestant presence is exemplified by Windsor Village United Methodist Church (WVUMC), the largest United Methodist congregation in the country (Windsor Village United Methodist Church 2004). Membership is derived from 16,000 households drawn from a 10–12 mile (16.4–19.7 km) radius around the church; weekly attendance averages 6,500 (Haynes 2011). Houston is home to the largest Catholic diocese in Texas and the twelfth largest in the United States (Archdiocese of Galveston-Houston 2010). The head of the Archdiocese, Daniel Cardinal DiNardo, is a member of the College of Cardinals and the first to be appointed from an archdiocese in the South (Archdiocese of Galveston-Houston 2010). There is also a small but stable Jewish population of approximately 45,000 (Lytle 2002; Sheskin and Dashefsky 2011). Virtually all population growth over the past 20 years can be attributed to immigration from Latin America, Asia, Africa and the Caribbean (Macken et al. 2011). The results of this study show that it is within this religious and ethnic milieu that Houston's religious and moral geography has been constructed in an environment of neosecularization on the micro scale. A phenomenon appearing in Houston is the clustering of gay-affirming churches (Fig. 188.6) (Sundby 2011) in neighborhoods found to have a concentration of gay and lesbian residents (Gates and Ost 2004). Rodriguez and Ouellette (2000) distinguish between gay-friendly churches that welcome participation of gays and lesbians, and gay-positive churches that "enable(s) gays and lesbians to alleviate the conflict between their religious beliefs and their homosexuality while increasingly enjoying identity integration" (p. 335). Studies have illuminated the desire of many gays and lesbians to openly participate in Christian worship (Gross and Yip 2010; Yip 2002) regardless of dogmatic teachings in some denominations on the sinfulness of homosexuality (Dudley 2006; Fryrear 2005; Piers Morgan Tonight 2011; United States Conference of Catholic Bishops 2006; Van Geest 2008) and the opposition of special interest groups (Dowland 2009; Van Geest 2008).

At the urging of Kong (2001), this study explores how "different religions may inform the constructions of different moral geographies ... and how these constructed moral geographies contradict or are negotiated or reinforced by other secular agents of morality (for example, the state)" and how these constructions are

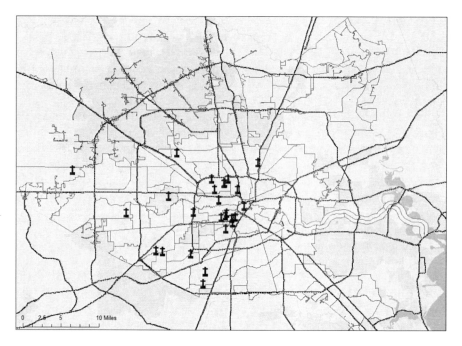

Fig. 188.6 Houston's GLBT-affirming churches (Map by Nancy Palmer Stockwell and Ira M. Sheskin, data from G.A.L.I.P. Foundation, www.gaychurch.org)

"played out in space" (p. 228). In Houston, these constructions appear to be played out in the changing voting patterns in presidential elections and, more significantly, in the election of the openly lesbian Parker as mayor in 2009. Voters in Harris County cast the plurality of votes for the Republican candidate in each presidential election since 1992 until favoring Democrat Barack Obama in 2008 (Texas Secretary of State and Elections Division 2010). Democrats have controlled the mayor's office in Houston since 1997 (Harris County Clerk's Office and Elections Division 2009). Houston, the fourth largest city in the nation, is the first major city to elect an openly homosexual mayor (Olson 2009). The homosexuality of the new mayor was "certainly not a positive in the campaign, but we can make it not a negative" (Parker 2011). There was an outcry from a small segment of the Christian community just before the run-off election, but it had no apparent effect on the outcome (Sappenfield 2009). According to Sherrill (1996), although there have been minor victories in local elections, gays, lesbians and bisexuals (GLB)[3] lack political clout as a voting bloc on a national level because of gays' wide and random dispersal

---

[3] At the time of Sherrill's writing, the term "transgender" was not commonly used. The term has since become part of the lexicon used when describing people of non-heterosexual orientation, and therefore, is used throughout most of this chapter.

throughout the population,[4] anti-gay attitudes leading to a lack of safety, and the lack of cohesion and collective identity. For these reasons, GLB voters must depend on support of heterosexuals to win elections. Parker served 12 years on Houston's City Council and as City Comptroller before becoming mayor, thus establishing a record of service. In an interview with the senior author, Mayor Parker explained her campaign strategy (Parker 2011). Comments in quotations in the following paragraph were made during the interview. Parker acknowledges that her sexual orientation is a negative in any campaign in which she is involved, but she and her staff "work hard to neutralize it" by emphasizing her experience in city office and her civic involvement in the GLBT community before her first term on City Council. She earned the endorsement of the conservative Houston Police Officers Union based on her "decades-long track record of working with them on policing issues" as the police liaison from the GLBT community. She also received the endorsement of the *Houston Chronicle*, which at the time of her first election in 1997, was a more conservative publication than it is today. In her words, those endorsements confused people and framed the election in 1997 in a way that begged the question, "do you want the good government lesbian who is endorsed by the cops, or do you want the lobbyist for the solid waste industry?" Parker fired back with the "lobbyist" label again in 2009 after Locke used a negative campaign piece about her sexual orientation. Apparently to the voters of Houston "somehow lobbyist is not a good thing." By combining her non-lobbyist position with her financial background working for an oil and gas company, her endorsements from two strong conservative organizations, and her experience in city government and civic organizations, she finds herself in the "odd dynamic that I end up being the conservative choice." She is the familiar candidate able to "pull votes from everywhere, even communities that might not have been expected to vote for me." Parker builds her campaigns with Houston's demographics in mind, and she states that there is "clearly ethnic and racial voting. Although the Hispanics voting for Hispanics is much less of an impact." According to Parker, Hispanics vote at "probably a quarter of what their demographics would indicate." She said there is a "very strong African-American shift always to the Black candidate." Parker feels strongly that religion does not influence votes for or against her, and the data seem to support this assertion. "At each level there was at least an appeal to voters who might be motivated to vote against me because of my sexual orientation. But I don't know that I would call that necessarily a religious factor, because I think that it transcends race; it transcends ethnicity." The GLBT community is a small portion of the electorate, which requires Parker to rely on coalition politics. However, her core of supporters locally and nationally is "committed to the campaign above how most people are engaged in politics."

Following a national trend toward increased tolerance of homosexuality detected by Putnam and Campbell (2010), Houstonians are becoming more accepting of homosexuality (Klineberg 2010) amid a neosecular religious environment in which

---

[4] Refer to The Gay and Lesbian Atlas, compiled by Gary J. Gates and Jason Ost in 2004 for updated information on the dispersal of gays in the United States.

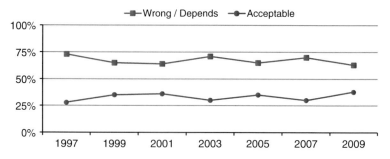

Fig. 188.7 Houston's evolving acceptance of gays (Source: Nancy Palmer Stockwell and Ira M. Sheskin, using data from the Houston Area Survey)

"the *self* of the believers, rather than authority structures, primarily steers the course" (Yip 2002). Although disapproval of homosexuality persists among most Houstonians, the gap is narrowing (Fig. 188.7). Research suggests that as the younger generation comes of age, the level of acceptance of homosexuality will increase, thus diminishing the negative perception of non-heterosexual identity (Putnam and Campbell 2010).

188.5.4 Religion, Race and Sexual Orientation in Local Politics

Houston's 2009 election for mayor highlighted the interplay between religion, race, and sexual orientation in local politics. Protestants still dominate Houston's religious landscape and shape the spiritual community; however, they had minimum influence on the outcome of the election even after factoring in the Black Protestants who voted for Gene Locke, the Black candidate. People claiming no religious affiliation, the "Nones," categorized as "Secularists" in the Houston Area Survey, comprise the smallest percentage of the religious landscape but had the strongest impact on the election results due to their support of Annise Parker. The geographic distribution of votes for Parker and votes for Locke closely followed the distribution of Blacks in the city and the distribution of "Secularists." The Houston Area Survey follows a matrix (Table 188.1) to categorize respondents into "Fundamentalist," "Progressive," and "Secularist" by computing respondents' answers to the questions "Which one of these three statements comes closest to describing your feelings about the Bible?" and "How important would you say religion is in your life? Would you say: very important, somewhat important, or not very important?"

The distribution of Fundamentalists (Fig. 188.8), Progressives (Fig. 188.9) and Secularists (Fig. 188.10) across the City of Houston overlaid with the precincts won by Annise Parker and Gene Locke illustrate the interplay between religion

Table 188.1   Houston Area Survey computed variable matrix

	Religion is very important	Religion is somewhat important	Religion is not very important
The Bible is the actual word of God and contains no errors	Fundamentalist		
The Bible is the inspired word of God written by humans and contains errors		Progressive	
The Bible is merely a book of myths and legends			Secularist

Source: Nancy Palmer Stockwell and Ira M. Sheskin

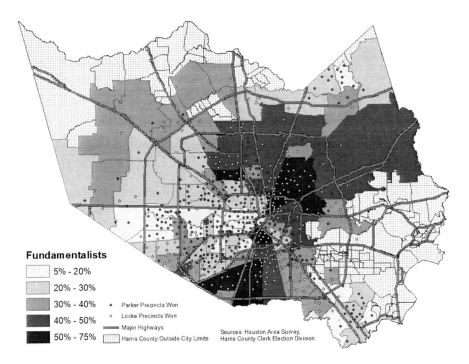

Fig. 188.8 Distribution of Fundamentalists, showing precinct winners (Map by Nancy Palmer Stockwell and Ira M. Sheskin, data from the Houston Area Survey, Harris County Clerk Election Division)

and voting.[5] Fundamentalists comprise 50 % or more of the respondents in more zip code areas of the city. Progressives and Secularists comprise 50 % or more of the respondents in very few zip code areas; most zip codes contain less than 50 % of respondents who are Progressive or Secularist. The precincts Parker won are located

---

[5] See Morris (2012) for a regression analysis that supports the conclusions in this paragraph.

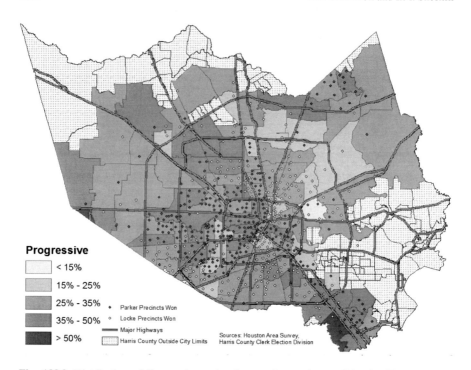

Fig. 188.9 Distribution of Progressives, showing precincts winners (Map by Nancy Palmer Stockwell and Ira M. Sheskin, data from the Houston Area Survey, Harris County Clerk Election Division)

in areas with high concentrations of Secularists; precincts Locke won are located in areas with high concentrations of Fundamentalists. There are 29 zip codes where more than 30 % of the respondents classify as Secularist and only one of these zip codes has more than 50 %. Parker received more votes than Locke in 26 of these zip codes. These zip codes are concentrated near Houston's central business district and extend west over an area that includes Montrose, recognized as one of the nation's gayest neighborhoods (Gates and Ost 2004). Ironically, while this area has the highest concentration of gay-friendly churches (see Fig. 188.5), Lakewood Church and Second Baptist Church also are located here. Conversely, 62 zip codes have more than 30 % Fundamentalists, of which 13 zip codes are more than 50 % Fundamentalist (see Fig. 188.7). Most of these zip codes are located in areas with high Black (see Fig. 188.2) and Hispanic (see Fig. 188.3) populations. Locke received more votes than Parker in 44 of these zip codes. Zip codes with more than 30 % Progressives are more widely dispersed across the city (see Fig. 188.8), and there is some overlap, primarily with Secularist zip codes. Parker received more votes than Locke in 23 of the 37 Progressive zip codes.

Results of a stepwise multiple regression with votes for Parker as the dependent variable indicate that the model in which the independent variables "Religion is not

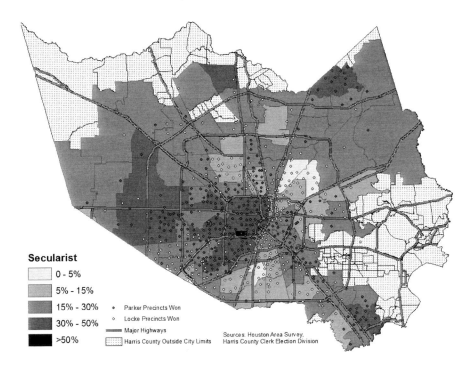

Fig. 188.10 Distribution of Secularists, showing precinct winners (Map by Nancy Palmer Stockwell and Ira M. Sheskin, data from the Houston Area Survey, Harris County Clerk Election Division)

very important" and "Blacks" best explain geographic variations in the dependent variable. The first variable is very strongly positively correlated ($R=0.626$) with Parker 2009 votes; the second very strongly negatively correlated ($R=-0.590$) with Parker 2009 votes. The adjusted R^2 is 47.7 % with an F-value of 50.7.[6]

It is recognized that the ecological fallacy is in play in this study. For example, a correlation is found between "Black" and votes for Locke at the zip code level. This does not necessarily prove that Blacks were more likely to vote for Locke. That is, the correlations at the zip code level may not transfer to the voting behavior of individuals, although in this case, it would seem to be highly likely.

Both candidates are Democrats, as were the previous three-term mayors, Bill White first elected in 2005 and Lee Brown first elected in 1997, evidence that religious preference does not reliably predict party affiliation in Houston mayoral elections. At the national and state level, researchers have noted a general tendency for Fundamentalists to align with conservative, Republican candidates (Kosmin and Lachman 1993; Kosmin et al. 2009; Putnam and Campbell 2010; Webster 1997).

---

[6] See Morris (2012) for a more complete explanation of the correlation testing and regression analysis.

188.6 Conclusions

Houston is a vibrant, entrepreneurial city in which a highly competitive religious marketplace thrives. The megachurches attract thousands of members locally by offering encouraging messages delivered by charismatic leaders and entertaining programs that appeal to younger audiences. These churches expand their influence nationally and internationally through the use of media and technology. As popular as the megachurches are, they do not appeal to everyone. With over 2,000 congregations, religious consumers in Houston can shop for the faith home that best suits them. In Houston's laissez-faire environment, choosing to be religiously unaffiliated is a viable choice for many Houstonians. Increasingly, homosexuality simultaneously has become both the lightning rod for denominational devolution and the rallying cry for clerical tolerance. These election results illustrate that despite a strong evangelical, fundamentalist Christian presence in Houston, a connection exists between the increased participation in other Christian, non-Christian and non-religious groups and the change in Houston and Harris County voting patterns. The influences of other religious and secular belief systems prompt a weakening of the hegemonic grip of Christianity on Houston culture. Parker's support from a coalition of Secular voters and others who, according to Peterson (2012), are able to "compartmentalize their faith and other elements of their lives, so they were able to see the mayor's office as something other than an extension of what they happen to believe when they go to church on Sunday" contributes to her local political success. David Peterson, senior pastor of Houston's Memorial Drive Presbyterian Church, has been a long-time opponent of the ordination of GLBTs in the Presbyterian Church U.S.A. and expressed his views in an interview with the author.

The ability to compartmentalize may be considered a manifestation of the concept of neosecularization in which personal wisdom guides an individual's spirituality from within rather than religious authority from without. It is within this construct that individuals reinterpret traditional orthodox views while maintaining their belief in the sacred and divine. In the "culture wars," generational differences highlight a growing acceptance of homosexuality among young people born in the post-baby boomer years. There remains opposition to homosexuality among young people of faith; however, the level of opposition is lower relative to baby boomers and pre-baby boomers. Putnam and Campbell (2010) predict that the issue of gay rights will follow the pattern of women's rights and eventually become less viable as a political issue.

For some who hold to the traditional Christian orthodox view of human sexuality, Christianity appears to have become marginalized in the process of neosecularization. In Houston, some have continued to openly express opposition to gay rights issues and same-sex marriage. Others have resigned themselves to the cultural tide and have taken the position of holding on to the traditional orthodox Christian lifestyle as an alternative to the cultural mainstream until the time they predict the world will rediscover Christianity. Until that time, they plan to remain friendly and work within Houston's diversity.

Parker (2011) acknowledges her advantage as the incumbent and expects to be reelected in the 2013 election, barring anything that would anger Houston's voters enough to come to the polls. As she says, "If people were truly unhappy I think there would have been more of a turnout." If Peterson (2012) is correct in his assessment of Houstonians, then Parker's chances of a third term are good. As he says, there is an attitude in Houston that keeps people from igniting around issues while managing to live creatively with enormous diversity. "You can't get people mad at each other. We don't want to be mad at each other…This city likes being a diverse plurality, and with that being the case and with a vibrant economy, then I just don't know that there's a heart in most cases for people to begin arguing, fighting with each other and giving an impression to the wider world that Houston's just an angry Southern city."

References

Archdiocese of Galveston-Houston. (2010). *Fast facts about the archdiocese*. Retrieved April 4, 2011, from www.archgh.org/Newsroom/Fact-and-Backgrounders/Fast-Facts-about-the-Archdiocese/

Association of Statisticians of American Religious Bodies. (2009). *County membership report – Harris County, Texas*. Retrieved April 13, 2011, from www.thearda.com/mapsReports/reports/counties/48201_2000.asp

Bellah, R. N. (2005). Civil religion in America. *Daedalus, 134*, 40–55.

Buttimer, A. (2006). Afterword: Reflections on geography, religion, and belief systems. *Annals of the Association of American Geographers, 96*(1), 197–202.

Chapman, T., Leib, J. I., & Webster, G. (2007). Race, the creative class, and political geographies of same sex marriage in Georgia. *Southeastern Geographer, 47*(1), 27–54.

D'Emilio, F. (2012, January 2). Vatican unveils home for Anglicans here; City to be hub for Episcopalian priests and their flocks that join Catholic Church. *Houston Chronicle*, p. A1.

Dowland, S. (2009). "Family values" and the formation of a Christian right agenda. *Church History, 78*(3), 606–631.

Dudley, S. (2006). *Forum on homosexuality and the Bible*. Retrieved February 7, 2012, from www.fellowship-pres.org/wp-content/uploads/Forum-on-Homosexuality-and-the-Bible.pdf

ECO: Evangelical Covenant Order of Presbyterians. (2012). Retrieved February 5, 2012, from www.fellowship-pres.org/evangelical-covenant-order/

Finke, R., & Scheitle, C. P. (2005). Accounting for the uncounted: Computing correctives for the 2000 RCMS data. *Review of Religious Research, 47*(1), 5–22.

Florida, R. (2002). The economic geography of talent. *Annals of the Association of American Geographers, 92*(4), 743–755.

Fryrear, M. (2005). *Homosexuality: Your questions answered*. Retrieved June 23, 2011, from http://erlc.com/article/homosexuality-your-questions-answered/

Gates, G. J., & Ost, J. (2004). *The gay & lesbian atlas*. Washington, DC: The Urban Institute Press.

Gay & Lesbian Victory Fund. (2012). *Election scorecard*. Retrieved June 18, 2012, from www.victoryfund.org/our_results/election_scorecard

Goen, C. C. (1983). Broken churches, broken nation: Regional religion and North-south alienation in Antebellum America. *Church History, 52*(1), 21–35.

Gross, M., & Yip, A. K. (2010). Living spirituality and sexuality: A comparison of lesbian, gay and bisexual Christians in France and Britain. *Social Compass, 57*(1), 40–59.

Halvorson, P. L., & Newman, W. M. (1994). *Atlas of religious change in America, 1952–1990*. Atlanta: Glenmary Research Center.

Harris County Clerk's Office, Elections Division. (2009). *Election results archives-canvass report.* Retrieved May 1, 2011, from www.harrisvotes.com/_eresults.aspx

Haynes, M. (2011, July 19). *Director of Ministries, administrative staff* (N. Morris, Interviewer), Houston.

Humes, K. R., Jones, N. A., & Ramirez, R. R. (2011). *Overview of race and Hispanic origin: 2010.* Retrieved May 5, 2011, from www.census.gov/prod/cen2010/briefs/c2010br-02.pdf

InfoGroup, Social Explorer. (2011). *Religion 2009: Religion by tradition (InfoGroup).* Retrieved April 13, 2011, from www.thearda.com/DemographicMap/displayIGMap.asp?ZipCode=77002

Jenkins, J. (2011, July 14). Presbyterians officially to allow gay clergy. *Houston Chronicle,* p. Belief 3.

Klineberg, S. L. (2010). *Data archive, Houston Area Survey 1982–2010.* Retrieved April 4, 2011, from www.thearda.com/Archive/browse.asp

Kong, L. (2001). Mapping 'new' geographies of religion: Politics and poetics in modernity. *Progress in Human Geography, 25*(2), 211–233.

Kosmin, B. A., & Lachman, S. P. (1993). *One nation under God: Religion in contemporary American society.* New York: Crown.

Kosmin, B., Keysar, A., Cragun, R., & Navarro-Rivera, J. (2009). *American nones: The profile of the no religion population.* Hartford: Trinity College, Program on Public Values.

Lakewood Church. (2011). *Home.* Retrieved April 2011, from www.lakewood.cc/Pages/Home.aspx

Leib, J. (2012). Appalachia and the American South. In S. D. Brunn, G. R. Webster, R. Morrill, F. Shelley, S. Lavin, & J. C. Archer (Eds.), *Atlas of the 2008 election* (pp. 126–128). Lanham: Rowman & Littlefield.

Lindsay, D. M. (2008). Mind the gap: Religion and the crucible of marginality in the United States and Great Britain. *The Sociological Quarterly, 49,* 653–688.

Lytle, D. E. (2002). *2001 Study of the Houston listed Jewish community.* Retrieved August 23, 2011, from www.jewishdatabank.org/Archive/C-TX-Houston-2001-Main_Report.pdf.

Macken, P., Wilson, S., Fischetti, T., & Goworowska, J. (2011). *Population distribution and change: 2000 to 2010.* Retrieved April 25, 2011, from www.census.gov/prod/cen2010/briefs/c2010br-01.pdf

Marks, J. (2011, July 19). *Minister on call* (N. Morris, Interviewer), Houston.

Morris, N. S. (2012). *The election of a lesbian mayor in a religiously conservative city: The case of Houston, Texas.* Open Access Theses. Paper 315. Retrieved April 27, 2012, from http://scholarlyrepository.miami.edu/oa_theses/315

Noll, M. A. (2001). *American evangelical Christianity: An introduction.* Malden: Blackwell.

O'Reilly, K., & Webster, G. R. (1998). A sociodemographic and partisan analysis of voting in three anti-gay rights referenda in Oregon. *The Professional Geographer, 50*(4), 498–515.

Olson, B. (2009, December 12). *Houston makes its choice; Parker makes history.* Retrieved September 25, 2011, from www.chron.com/news/houston-texas/article/Houston-makes-its-choice-Parker-makes-history-1589385.php

Parker, A. (2011, December 17). *Mayor of Houston* (N. Morris, Interviewer), Houston.

Peterson, D. (interview by Nancy Morris). (2012, March 12). *Senior Pastor, Memorial Drive Presbyterian Church,* Houston.

Pew Forum on Religion & Public Life. (2008). *U.S. religious landscape survey religious affiliation: Diverse and dynamic.* Washington, DC: Pew Research Center.

Piers Morgan Tonight. (2011). Retrieved August 18, 2011, from www.cnn.com/video/#/video/bestoftv/2011/01/24/piers.osteen.homosexuality.cnn?iref=allsearch

Putnam, R. D., & Campbell, D. E. (2010). *American grace.* New York: Simon & Schuster.

Quinton, N., & Webster, G. R. (2011). Electoral alignments and place-based cleavages in statewide votes in Alabama. In B. Warf & J. Leib (Eds.), *Revitalizing electoral geography* (pp. 195–218). Burlington: Ashgate.

Rodriguez, E. M., & Ouellette, S. C. (2000). Gay and lesbian Christians: Homosexual and religious identity integration in the members and participants of a gay-positive church. *Journal for the Scientific Study of Religion, 39*(3), 333–347.

Sappenfield, M. (2009, December 13). *Houston mayor race shows progress – and limits – of gay rights.* Retrieved April 18, 2011, from http://search.proquest.com/docview/405583215?accoun tid=14585

Sherrill, K. (1996). The political power of lesbians, gays and bisexuals. *PS: Political Science and Politics, 29*(3), 469–473.

Sheskin, I. M., & Dashefsky, A. (2011). Jewish population in the United States, 2011. In *Current Jewish population reports, number 4–2011*. Storrs: Mandell L. Berman North American Jewish Data Bank, The Association for the Social Scientific Study of Jewry, and The Jewish Federations of North America.

Smith, B. G. (2007). Attitudes towards religious pluralism: Measurements and consequences. *Social Compass, 54*(2), 333–353.

Stetzer, E. (2009). *100 largest churches in America for 2009.* Retrieved April 5, 2011, from http://churchrelevance.com/100-largest-churches-in-america-for-2009/

Stump, R. W. (2008). *The geography of religion: Faith, place, and space.* Lanham: Rowman & Littlefield.

Sundby, E. (2011). *Welcoming gay friendly churches in Texas – TX.* Retrieved August 25, 2011, from www.gaychurch.org/Find_a_Church/united_states/us_texas.htm

Texas Secretary of State, Elections Division. (2010). *General election county by county canvass report.* Retrieved May 18, 2012, from www.elections.sos.state.tx.us/elchist.exe

The Fellowship of Presbyterians. (2012). *About.* Retrieved February 7, 2012, from www.fellowship-pres.org/about/

United States Conference of Catholic Bishops. (2006). *Issues and action: Human life and dignity.* Retrieved September 24, 2011, from www.usccb.org/issues-and-action/human-life-and-dignity/homosexuality/upload/minstry-persons-homosexual-inclination-2006.pdf

Van Geest, F. (2008). Christian denominational and special interest political action on public policy issues related to sexual orientation. *Sociology of Religion, 69*(3), 335–354.

Vojnovic, I. (2003). Laissez-Faire governance and the archetype laissez-faire city in the USA: Exploring Houston. *Geografiska Annaler Series B, Human Geography, 85*(1), 19–38.

Warf, B. (2006). Religious diversity across the North American urban system. *Urban Geographer, 27*(6), 549–566.

Warf, B., & Winsberg, M. (2008). The geography of religious diversity in the United States. *The Professional Geographer, 60*(3), 413–424.

Webster, G. R. (1997). Religion and politics in the American South. *The Pennsylvania Geographer, 35*(2), 151–172.

Webster, G. R., Chapman, T., & Leib, J. (2010). Sustaining the 'societal and scriptural fence': Cultural, social, and political topographies of same-sex marriage in Alabama. *The Professional Geographer, 62*(2), 211–229.

Windsor Village United Methodist Church. (2004). *Meet the pastor.* Retrieved July 19, 2011, from www.kingdombuilders.com/templates/cuskingdombuilders/details.asp?id=23260& PID=126545

Winter, R. M. (2000). Division & reunion in the Presbyterian church, U.S.: A Mississippi retrospective. *Journal of Presbyterian History, 78*(1), 67–86.

Yip, A. K. (2002). The persistence of faith among nonheterosexual Christians: Evidence for the neosecularization thesis of religious transformation. *Journal for the Scientific Study of Religion, 41*(2), 199–212.

Zelinsky, W. (1961). An approach to the religious geography of the United States: Patterns of church membership in 1952. *Annals of the Association of American Geographers, 51*(2), 139–193.

Chapter 189
Moral Imperatives: Faith-Based Approaches to Human Trafficking

Martha Bettis Gee and Ryan D. Smith

189.1   Introduction

Children in the fishing industry on Lake Volta in Ghana or forcibly conscripted to serve as soldiers – men enslaved in agriculture or in mining – women in forced domestic servitude or in the sex trade. Though slavery is illegal in every country in the world,[1] legal statute has not eliminated the practice, as many assume. Rather, slavery has moved to the margins. In the shadows, in places where its legal abolition is not enforced effectively, modern slavery is flourishing in myriad forms.

In the campaign to eliminate chattel slavery in the late nineteenth century, people of faith were at the forefront of the battle. In the early twentieth century, the faith community took up the campaign to eliminate so-called white slavery. In both instances, the impetus for addressing these grievous social ills was a deep commitment to ending an abomination—though it is important to note that the faith community has at times accepted and perpetuated the use of slaves. The movements to end the practice thus took the form of a moral crusade characterized by rhetoric that would appeal to the sensibilities of people of faith. While hyperbole and emotional appeals often oversimplified complex issues and employed stereotypes that caricatured enslaved persons as helpless victims, these approaches were considered to be justified in order to achieve the greater good—eliminating slavery.

---

[1] The last country to abolish legal slavery was Mauritania, where it was officially abolished by presidential decree in 1981.

M. Bettis Gee (Retired) (✉)
Compassion, Peace and Justice, Peace and Justice Ministries, Presbyterian Mission Agency,
Presbyterian Church (USA), Louisville, KY 40202, USA
e-mail: mbgee@bellsouth.net

R.D. Smith
Compassion, Peace and Justice Ministries, Presbyterian Ministry at the U.N., Presbyterian
Mission Agency, Presbyterian Church (USA), New York, NY 10017, USA
e-mail: ryan.smith@pcusa.org

© Springer Science+Business Media Dordrecht 2015   3623
S.D. Brunn (ed.), *The Changing World Religion Map*,
DOI 10.1007/978-94-017-9376-6_189

Today, institutionalized religious groups and private non-profits founded by individuals motivated by their faith are at the forefront of the non-governmental movement to eliminate modern slavery. While faith-based advocacy and action is by no means limited to Christian individuals or organizations, Christians have played a key role in shaping how the United States has responded. In this chapter, we first frame the conversation by defining human trafficking using both the understandings of U.S. federal statutes and established international law as well as employing a practical definition used by networks doing anti-slavery work on the ground. We will examine the justification for identifying human trafficking as modern slavery and delineate the scope of slavery today, positioning it in the globalized context that continues to fuel the phenomenon. Then we will examine both the history out of which U.S. Christian faith-based responses grew and the more recent efforts, whereby evangelical Christian organizations and the partnerships that coalesced around the issue were able to influence the narrative and shape current federal legislation to address trafficking.

This chapter will then look at the current state of anti-trafficking policy and the work presently underway. We will examine the evangelical moral crusade and its crusaders—those whose focus is primarily to address sex trafficking as well as those evangelicals whose approach is grounded in an understanding of the free market. Then we turn to the perspective of some other faith-based organizations seeking to ground the work in an explicitly human–rights based approach, including a few illustrative case studies. Finally the chapter will look at trends and questions emerging in the ongoing faith-based movement to combat slavery.

The scope of modern slavery is continually changing, and effective responses must evolve in order to address this complex human rights violation. Any attempt to examine the issue must necessarily be viewed as a snapshot in time. This chapter attempts to present a nuanced picture, one that can serve as a benchmark for developing a comprehensive and effective approach that matches the strengths of the faith-based community with the skills of secular organizations.

189.2 What Is Human Trafficking?

Human trafficking is a term fraught with ambiguity and defined in a variety of ways. For purposes of clarity, we can consider human trafficking alongside the other most common forms of trafficking—drug and arms trafficking. These forms garner huge profits from the sale of a commodity, and in fact, human trafficking is also about selling a commodity. It is all about the enslavement and commodification of persons, who are then the property of the traffickers to use as they choose. To term this practice as modern slavery is, therefore, both more precise and more descriptive of the stark realities that face its victims.

The hallmark of modern slavery is the use of force, fraud or coercion to first enslave and then to keep the person enslaved. This ranges from actual physical violence and imprisonment, to threats of physical harm for the victim or family members, to more subtle forms of intimidation or control. Conventional wisdom to the contrary, the key variable in trafficking is neither gender, nor race, nor country of origin, but rather vulnerability.

189.2.1   International Law

International law defines human trafficking broadly and highlights its non-consensual nature in the UN Protocol to Prevent, Suppress and Punish Trafficking in Persons, Especially Women and Children:

> The recruitment, transportation, transfer, harboring, or receipt of persons, by means of the threat or use of force or other forms of coercion, of abduction, of fraud or deception, of the abuse of power or of a position of vulnerability or of the giving or receiving of payments or benefits to achieve the consent of a person having control over another person, for the purpose of exploitation. Exploitation includes, at a minimum, the exploitation of the prostitution of others or other forms of sexual exploitation, forced labor or services, slavery, servitude, or the removal of organs. (United Nations 2003: 2)[2]

189.2.2   Federal Statute

The Trafficking Victims Protection Act (TVPA), the major piece of federal legislation in the United States aimed at addressing trafficking, defines "severe forms of trafficking" as:

> a. sex trafficking in which a commercial sex act is induced by force, fraud, or coercion, or in which the person induced to perform such an act has not attained 18 years of age; or
> b. the recruitment, harboring, transportation provision, or obtaining of a person for labor or services, through the use of force, fraud, or coercion for the purpose of subjection to involuntary servitude, peonage, debt bondage, or slavery.[3]

While this definition implicitly reflects the widespread belief that sex trafficking is the most common form of trafficking, current research does not support this understanding. Labor trafficking, including both forced and bonded labor, is by far the most prevalent form of modern slavery worldwide. The 2012 *Trafficking in Persons Report* includes the following types of modern slavery: sex trafficking, child sex trafficking, forced labor, including involuntary domestic servitude, debt bondage, and child labor. Child soldiers are considered a category of trafficked in persons.[4] The Optional Protocol to the Convention on the Rights of the Child on the sale of children, child prostitution and child pornography also classifies any person under the age of 18 who works in prostitution as a victim of human trafficking.

---

[2]Yvonne C. Zimmerman reports that a 2010 multi-lingual review of the literature on human trafficking reveals that most definitions of human trafficking reference this definition.

[3]Trafficking Victims Protection Act of 2000.

[4]The Optional Protocol to the Convention on the Rights of the Child against the use of children in armed conflict sets the standard in international law for dealing with this special form of modern slavery.

189.2.3   A Working Definition

Although the technical definitions of both TVPA and the Palermo Protocol are necessary to ground the interpretation of the statutes, a working definition can also serve as a guide. The Freedom Network USA, a network of thirty organizations and individuals doing anti-trafficking work on the ground, defines human trafficking pragmatically in this way:

> … the recruitment, harboring, transporting, providing or obtaining, by any means, any person for forced labor, slavery or servitude in any industry or site such as construction, prostitution, manufacturing, begging, domestic service or marriage.[5]

189.2.4   Quantifying the Scope

Because modern slavery breeds in the shadows and trafficked persons often are not aware that they are in the process of being trafficked until they are enslaved, estimates of the numbers of enslaved persons are very difficult to either generate or verify. Dr. Kevin Bales, President of Free the Slaves and an acknowledged expert in the field of anti-trafficking work, cites the figure of 27 million enslaved persons (Bales 2007:1, 8). In its 2012 report, The International Labor Organization estimates that modern slavery around the world claims 20.9 million victims at any time and that 55 % of forced labor victims are women and girls, as are 98 % of sex trafficking victims. Yearly income lost by people in forced labor worldwide due to unpaid wages and recruitment costs surpasses 20 billion dollars. (ILO 2012 Global Estimate of Forced Labor Executive Summary 2012) The United Nations Children's Fund (UNICEF) reports that an estimated 1.2 million children are trafficked within and across borders (UNICEF 2012). Other estimates range widely and are more or less reliable, depending upon the source of the figures, the research tools used, and sometimes even the motivations of those providing the estimate. Regardless of the figures, human trafficking is a growing global phenomenon affecting millions of men, women and children and one that generates huge profits.

189.2.5   Contemporary Slaves: Disposable People

Kevin Bales notes that two characteristics distinguish slavery of today from that of the past: today's slaves are cheap, and they are disposable.[6] Along with the world population explosion has come an increase in potential slaves—a source that no

---

[5] Freedom Network (USA) uses this definition in trainings for FBI, law enforcement, and social service agencies doing anti-trafficking work.

[6] For a thorough discussion, see Chapter 1, Ending Slavery: How We Free Today's Slaves, by Kevin Bates (1999).

longer costs the equivalent of $40,000 per individual, as the enslaved field hand did prior to 1860, but rather the equivalent of less than $100 (Bales 2007:12). Another factor contributing to the rise of slavery is the global economy that has led to an extraordinary increase in the wealth of some countries, leaving others to languish in poverty of an unprecedented scale. In many parts of the world, endemic corruption, especially of law enforcement officials, has also both fueled slavery's growth and impeded a comprehensive and effective response.

189.3 Historical Context

189.3.1 Early History

For much of the 5,000 years that slavery has existed, it was an accepted institution, part of the social fabric of cultures, peoples and even faiths. There were slaveholders as well as slaves in the earliest Christian communities. While there were voices within the early Christian movement who challenged the values on which slavery rested (notably Gregory of Nyssa), official church policy in the Christian Empire supported the rights of slaveholders (Glancy 2011).

In the early Middle Ages and forward, the Church condoned slavery (unless it was Christians enslaved by the "infidels"—Muslims). With the advent of the transatlantic slave trade by the Portuguese, an era of exploration and conquest led to an increased need for slave labor. Some 900,000 African slaves were imported by the 1600s (New Internationalist 2012).

189.3.2 Abolitionist Movement

In the past 150 years, public opinion gradually shifted as individuals began to question the institution. While the slave trade in Britain had been abolished in 1807 and slavery was illegal there, it flourished in the British colonies. William Wilberforce, who in 1785 had undergone a conversion experience, joined other evangelicals and British politicians in a campaign that led to the Slavery Abolition Act of 1833, which abolished slavery in most of the British Empire.

In the United States as in Britain, many American abolitionists were evangelical Christians. Quakers in both countries were at the forefront of the abolitionist movement, adding weight to the moral imperative to end slavery.(Choi-Fitzpatrick 2006) In the United States, the prohibition of trading in slaves had taken effect in the year following Britain (1808), but slaves were still the backbone of the southern agricultural economy. Scholars observe that Christian slaveholders justified the practice using the theological teachings of their faith, arguing that slavery was part of God's plan to Christianize Africa (Boles 1988; Peach 2000. As cited in Zimmerman 2011:571). Others point out the influence of the Protestant work ethic, the idea of dedicated industry and thrift, what Zimmerman calls profit making as equated with divine sanction (Weber 1992. as cited in Zimmerman 2011:571). White slaveholders were able to view their economic success by virtue of slave labor as a justification for the institution of chattel slavery itself, a similar theological position to the contemporary understanding of the prosperity gospel.

189.3.3 "White Slavery" and the Social Purity Movement

Contemporary anti-trafficking work also has roots in the series of social purity movements of the late nineteenth century. With the view that women's moral nature was somehow purer than men's, evangelical Christian anti-vice crusaders targeted prostitution and what they viewed as sexual immorality (Zimmerman 2011:570). Their work coalesced around the fear of "white slavery"—defenseless white women forced into prostitution by evil men. Most historians agree that white slavery more accurately portrayed white, middle class fears than it did an actual reality, and there was an underlying racist narrative valuing the morality of white women above that of the black women who had been sexually exploited in chattel slavery. The Mann Act (White Slave Act) of 1910 marked a new perception of forced prostitution as a social problem in need of solving. It became a convenient excuse for targeting those perceived as engaging in practices differing from the Victorian moral sensibility, notably black men, Jews and immigrants (Doezema 2000; Irwin 1996; Roberts, 1993, as cited in Zimmerman 2011:571).

189.3.4 Persecution of Christians

In the mid-twentieth century, Christians became concerned about reports of the religious persecution of Christians in many less developed countries, leading to the passage of the International Religious Freedom Act in 1998. Many evangelical Christians saw religious freedom and human trafficking as parallel issues; indeed, conservative activist Michael Horowitz, among others, considered that the faith-based anti-trafficking movement was as much about expanding women's rights issues to include the protection of victims of sex trafficking as it was about redefining the human rights agenda to include the rights of religious believers (Cromartie 1999:54 as cited in Zimmerman 2011:572). Prior to the 1990s, the secular NGOs then at the forefront of anti-trafficking work focused primarily on labor trafficking and the constructs of exploitative labor out of which labor trafficking arose. At this writing human trafficking is a high profile issue and its rise to that status is to a large extent due to the reframing of the issue to focus on abusive and exploitative sex (Bernstein 2010; Ditmore 2002; Jakobsen and Bernstein 2009; as cited in Zimmerman 2011:575).

189.4   Shaping Current Federal Statues to Address Trafficking

Absent comprehensive federal legislation, any prosecutions that took place prior to 2000 to address labor and sex trafficking were implemented under more limited statutes or statutes passed during the emancipation era. While Ambassador-at-Large Luis CdeBaca of the Office to Monitor and Combat Trafficking gives credit to then-First Lady Clinton, along with Attorney General Janet Reno and Secretary of State Madeline Albright, for being instrumental in bringing this issue to the attention of policymakers in Washington (Transcript, special briefing 2011; TIP Report 2011) other forces, and other players, were also at work.

On March 11, 1998, then-President Bill Clinton issued a directive calling for legislative action to combat human trafficking. The Clinton administration sought congressional support for a new non- immigrant visa classification for victims of trafficking, as well as preventative measures and humanitarian assistance for undocumented victims who were ineligible for any other form of aid due to their status. The administration urged Congress to adopt harsher penalties for traffickers and for individuals who profit from trafficking schemes (Yeomans 2000:77, as cited in McReynolds 2008:36). But this action on the federal level was preceded by a decade or more of advocacy and action that involved a strange coalition: evangelical Christians and feminists.

189.4.1   Feminist and Evangelical Coalitions

Gretchen Soderlund, Assistant Director of the Center for the Study of Communication and Society and Lecturer in Sociology at the University of Chicago, notes that combating trafficking, specifically the trafficking of women, has become a common denominator that has united people across the political and theological spectrum (Soderlund 2005). Even so, an alliance of secular feminist activists and evangelical Christians seems unlikely at best. Just as evangelical Christians turned to human rights issues, seeing human trafficking as a natural fit with concerns about the religious freedom of Christians in the developing world, so feminists came to focus on introducing women's sexual and reproductive rights into the arena of human rights dialogue and action, with sex trafficking as its focus.

This issue became a priority for the U.N. in the 1990s as feminists and women's rights organizations sought to bring the broader issue of violence against women into the discussion. In international conferences such as the 1993 Vienna World Conference on Human Rights and the 1995 Beijing Conference on Women, sex trafficking and prostitution leapt to the forefront as "the most egregious form of violence against women imaginable and thus trafficking emerged as the centerpiece of

M. Bettis Gee and R.D. Smith

the campaign" (Soderlund 2005:70).[7] Factions grew out of the debates that raged around the issue of pornography, with some adhering to a strict abolitionist model that considered all sex work as sexual slavery and others adopting the position that these intensive campaigns undermined work to secure sex worker rights (Doezema 1998, as cited in Sunderland, 2005). Such organizations as CATW (the Coalition Against Trafficking in Women) favored laws on trafficking broad enough to encompass a prostitution, and others, notably GAATW (Global Alliance Against Traffick in Women), argued that sex trafficking was a part of a continuum of forced migrant labor. Yet these players in the earlier debate were largely absent from the coalition subsequently formed in the late 1990s.

In this new coalition, Christian evangelicals were joined by feminist groups such as Equality Now and the Protection Project. Laura Lederer, founder of the Protection Project and editor of *Take Back the Night* (and who was to become a State Department appointee), functioned as a liaison between feminists and evangelicals. Lederer remarked that religious organizations had introduced a "fresh perspective and biblical mandate to the women's movement (Crago 2003, as cited in Soderlund 2005:68.). This perspective and the moral clarity it brought were highlighted in June 1999 when the National Association of Evangelicals issued a statement signed by more than 130 religious leaders calling on Congress to pass a human trafficking law to end the "sinister trade that profits ruthless businessmen, criminals, and corrupt public officials at the expense of millions of women and children.... " The God-given dignity and integrity of each individual compels us to take action to combat this evil." (Schliecher 1999, as cited in Zimmerman 2010).

The fact that the 2000 TVPA codifies trafficking in specific ways is key to its effectiveness. Yet Yvonne Zimmerman argues that the law's particular conception of trafficking is one of the statute's chief weaknesses. Specifically, she argues that although the statute covers all forms of forced labor, a special concern for sex trafficking is not difficult to detect. For example, while acknowledging that "trafficking in persons is not limited to the sex industry," sex trafficking is nevertheless the first type of trafficking that the text specifies, and the only type of trafficking whose characteristics are explicated at length (Zimmerman 2010).

The TVPA of 2000 did not include a category for unforced prostitution. But the Clinton administration did distinguish voluntary prostitution from forced, resisting linking prostitution to trafficking. Further, administration officials did not make the claim that legal prostitution increased trafficking (Weitzer 2007:461).

189.4.2 Bush Administration Policies

New institutional partners had emerged out of the work of the coalition to pass anti-trafficking legislation, among them former U.S. Representative Linda Smith of Shared Hope International, Gary Haugen of International Justice Mission, and Kevin

---

[7] Then-First Lady Clinton was a leader in the U.S. delegation to the Beijing Conference.

Bales of Free the Slaves. This emerging anti-prostitution crusade was rebuffed in the effort to influence the Clinton administration's position. But all that was to change when the Bush administration took office in 2001. Despite the Clinton administration's strong commitment to the legislation, its passage only days before the presidential election meant that President Clinton could only oversee its most basic elements. It was left to the Bush administration to implement the law, and President Bush and his policy advisors welcomed the crusade's views with open arms.

Ronald Weitzer posits that this receptivity is reflected in public pronouncements of government officials, official positions of government agencies such as the Department of State, Department of Justice, USAID and HHS, and in the State Department's annual Trafficking in Persons Report (TIP) itself. The State Department also produced a seminal document, *The Link Between Prostitution and Sex Trafficking,* and the 2003 and 2005 reauthorizations of the TVPA show the stamp of the *End Demand for Sex Trafficking* bill, where leading feminist and faith-based activists played a key role (Weitzer 2007:461). Other indications of the influence of crusaders were seen in the 2002 presidential directive on trafficking that defined prostitution as "inherently harmful and dehumanizing" and in Bush's references to the sex trade in a speech at the United Nations.

189.4.3   The Prostitution Loyalty Oath

A significant amendment to the 2003 reauthorization of TVPA also reflected the positions of the crusaders: the Prostitution Loyalty Oath, a set of rules that stipulated that no federal funds could be used to implement any program where the recipient had not stated explicitly that it did not promote, support or advocate the legalization or practice of prostitution. Not only did this reflect a de facto conflation of prostitution and sex trafficking, it actually impeded the work of many reputable organizations, both in anti-trafficking work and in addressing HIV and AIDS.

189.4.4   Faith-Based Initiatives and Unintended Consequences

In the expansion of Clinton's charitable choice initiative by the Bush administration, faith-based social service programs now gained unprecedented access to federal funding through the newly established Office of Faith-based and Community Initiatives. In signing the landmark executive order establishing the Office, President Bush explained that "the days of discrimination against religious groups just because they are religious is coming to an end." (George W. Bush quoted in Anne Farris, et al., as cited in Zimmerman 2010). Bush referred to a policy whereby faith-based recipients of federal funds were required to create entities with separate incorporation and were restricted in using the funds to secular or non-religious activities (Zimmerman 2010).

The new office was quite open to funding anti-trafficking work. In the 8 years of the Bush administration, faith-based groups gained increasingly more funding for this work.[8] Shared Hope International and the International Justice Mission each received million dollar grants. Others receiving funding were the Salvation Army, with its Initiative Against Sexual Trafficking, the United States Conference of Catholic Bishops, and World Vision.

Aside from concerns about which groups were funded, another criticism leveled at the funding mechanism had to do with the questionable capacity of some groups to engage in effective anti-trafficking work. Yvonne Zimmerman illustrates this critique with the example of the Crossing the Bridge Project, an anti-trafficking initiative of Concerned Women for America that received some $300,000 over 2 years. Project coordinator Brenda Zurita reported that when she was hired in early 2005, she knew nothing about trafficking. Her training consisted of reading one of the TIP Reports, doing some Internet research and spending 2 days at a State Department meeting where there were representatives of NGO groups doing trafficking work. Following this "training," she designed a State Department – funded seminar to educate Mexican leaders (Zurita 2005, as cited in Zimmerman 2010). Zurita remarked that God is "the most effective weapon against the slave trade" and that prayer is "the ultimate ant-sex trafficking program" (Zurita 2006, as cited in Zimmerman 2010).

An unintended consequence of the availability of federal funds to faith-based groups was to place an undue emphasis on their religious credentials at the expense of the actual technical expertise and capacity a group might have. Claude d'Estree, Special Rapporteur for the Global Initiative to Fight Trafficking at the United Nations Office of Drugs and Crime has pointed out that groups like the CWA received sizeable grants for anti-trafficking work, which they were ill-prepared to execute (from a conversation with the author, Zimmerman 2010).

Zimmermann herself observes, with no small touch of irony:

> ...neither God nor prayer appear among internationally recognized best practices for the elimination of trafficking—practices that include strategies like enforcing labor laws, implementing witness protection programs for those who testify against traffickers, and enhancing the responsiveness of local governments to constituents' concerns. (Zimmerman 2010)

189.4.5   A New Administration Addresses Trafficking

With the Obama administration there has been a shift away from the ideological position of the Bush administration that reflected the institutionalization of the ideology of the moral crusade. One can detect that shift even in small but significant changes in rhetoric in the annual *Trafficking In Persons Reports*. To cite one example, in the introductory letters in the 2005–2008 TIP Reports, Secretary of State Condaleeza Rice

---

[8]These groups were almost exclusively Christian evangelical groups whose theological positions were aligned with the Bush administration's aims.

refers to the abolitionist movement, terminology used primarily by those in the moral crusade. Letters in subsequent reports from Secretary of State Hilary Rodham Clinton, while referring to modern slavery primarily as a grave human rights violation, have moved away from language explicitly reminiscent of faith.

Another indicator of the shift has been the defunding of the anti-trafficking work of the US Conference of Catholic Bishops because of their refusal to offer or refer persons to services that differ from their pro-life stand, including abortion services and contraception for victims of trafficking.

On September 25, 2012, President Obama issued an executive order that signaled a new commitment to addressing human trafficking. In a speech at the annual Clinton Global Initiative meeting in New York, Obama announced initiatives to ensure the U.S. would "lead by example" on trafficking-free government contracting, and new efforts to help victims of human trafficking recover, with better access to treatment, legal services and job searches, as well as a simpler visa process for victims brought to America against their will. The President praised the work of Christian groups, noting that they are "answering the Bible's call to 'seek justice' and 'rescue the oppressed'" (christianpost.com 2012).

Although the language of rescue reflects the earlier position of the Bush administration, Obama's remarks represent a significant shift away from the abolitionist rhetoric employed by that previous administration to frame the issue of modern slavery as primarily concerning sexual purity. Instead, the language of faith here underscores a different, but equally compelling call—the mandate of justice.

189.5   The Current State of Faith-Based Activism

189.5.1   Freedom as Economic Participation

In the current context, Christian activism and advocacy against modern slavery is far from monolithic. For many evangelical Christians, anti-slavery activism is grounded in an understanding of morality that presumes that nothing is more harmful to human dignity than sexual activity that occurs outside of marriage. But not all Christian anti-trafficking activists are motivated by these concerns. Some evangelicals self-described as socially moderate and progressive include David Batstone, President and Co-Founder of Not for Sale; Tony Compolo, evangelical leader of the Evangelical Association for the Promotion of Education; Richard Cizik, President of the New Evangelical Partnership for the Common Good; Jim Wallis of Sojourners; and Gary Haugen of the International Justice Mission (Zimmerman 2011:573–574).[9]

---

[9]Yvonne Zimmerman reports that in 2004 IJM partnered with an MSNBC producer to stage raids to free young girls being sold for sex in brothels in Cambodia, documented by a hidden digital camera and subsequently aired as a 2004 Dateline episode entitled "Children for Sale." The episode has served as an implicit endorsement of the tactics of the abolitionists, tactics that often have many unintended consequences, among them endangering other victims and driving the problem further underground.

For these more globally oriented social justice evangelicals, freedom from enslavement is framed in terms of formal economic participation. Activists encourage purchases that will contribute to faith-based organizations or items crafted by women who have been freed from sexual slavery. Elizabeth Bernstein identifies this as a 'business as mission' model, whereby slaves become free by entering the free market as consumers, and Western consumers embody freedom through their ability to purchase the products of these former slaves (Bernstein 2010, as cited in Zimmerman 2011). These and other more progressive evangelicals reject the idea that modern slavery is a by-product of capitalistic models, asserting that freedom consists of the ability to participate in the free market.

189.5.2   What Motivates the Activist? A Congregation on a Moral Crusade

What happens when the macro level work of an organization like IJM is translated to the micro level, viz., to a local church? Claire M. Renzetti, Ph.D., Judi Conway Patton Endowed Chair and Professor of Sociology at the University of Kentucky shares some preliminary results of ongoing ethnographic research to explore transcendent motivation. At an interdenominational evangelical church in Cincinnati, she attends group meetings and anti-trafficking events, as well as analyzing brochures and other publications. Most recently, Renzetti has been engaged in one-on-one interviews.[10] There is a strong concern with salvation and a personal relationship with Jesus Christ to attain salvation, as well as a belief in the inerrancy and authoritativeness of the Bible. The members of the church affirm the commandment to love God and neighbor, which to them means active engagement in solving social problems, including human trafficking.

Human trafficking is currently one core focus of the church's social justice ministry. Some of the most active members in the work are high school students who became involved after hearing Gary Haugen, founder of IJM, speak about the problem and the organization's work to address it.

The church itself has committed to financially supporting IJM's field offices in Mumbai and Kolkata, India as well as aftercare homes and reintegration programs for rescued trafficking victims. Fundraising materials for this effort state that they are committed to this "uphill battle for the long haul" and that they take seriously God's call, in Isaiah 1:17 to "seek justice and encourage the oppressed." They believe, they say, that God has "positioned [them] uniquely to make a real difference, one girl at a time."

This congregation has also organized educational events to raise public awareness, distributed written materials, organized and led prayer gatherings, and held fair trade fairs (to draw attention to the problem of labor trafficking and slave labor,

---

[10] The name of the church and of individuals interviewed is withheld here because the research is ongoing.

especially child victims). Some members, including two 17 year olds, have traveled abroad to work in the aftercare homes.

One of Renzetti's research participants summed up when he said, "Jesus left heaven, came into this crappy world and suffered. In doing the work we do, we are being like Jesus. We are the body of Christ. Jesus was willing to get his heart broken, too. Let the girls into your heart. Let Jesus into your heart, too" (Renzetti 2011).

189.5.3   Freedom as a Human Rights Issue

There is yet another perspective emerging among Christian activists that more explicitly affirms that to truly address modern slavery, it is necessary to view it as both a cause and a consequence of profound violations of human rights. On the one hand, people become enslaved because of a complex and interrelated set of factors characteristic of exploitative labor practices that can become a tipping point propelling persons into slavery. The basic rights of a person to be able to have a livelihood, to basic health care and education and housing, are thus violated.

But a person's human rights can potentially be violated by well-meaning Christian activists who seek to free enslaved persons without taking into account those persons' own desires, dreams and wishes. One of the realities of so called "rescue" operations of adult victims of sex trafficking is that reports of escapes from shelters are quite common. While some victims see the rescue as an opportunity to leave the sex industry, others see it as imprisonment—just another form of the violation of their human rights. When Phil Marshall of the United Nations Project on Human Trafficking in Southeast Asia's Mekong Region was asked by journalist Maggie Jones what he thought of the rehabilitation strategies in place at that time (2003), he remarked that he had "never seen an issue where there is less interest in hearing from those who are most affected by it" (Jones 2003:1, as cited in Soderlund 2005:66).

189.5.4   Three Case Studies

What follows here are three case studies of faith-based organizations engaged in advocacy and activism from an explicitly human rights based perspective, one that seeks to keep the perspective of the victim at the center of the work. The first is an NGO in India established by a person of faith, the second is a U.S. denominational initiative, and the last is an extensive and well-known international Catholic organization of organizations.

Case Study 1: The National Domestic Workers' Movement (NDWM)
The National Domestic Workers' Movement (NDWM) is an NGO working with domestic workers, child domestic workers and migrant domestic workers in 17

states in India and with Indian migrants working abroad. Domestic workers, particularly child workers and live-ins, comprise one of the most vulnerable and exploited groups of the unorganized work sector in India. Their vulnerability is exacerbated by the Indian caste system.

NDWM works on issues such as globalization and its effect on domestic workers; human rights and domestic work; the worst forms of child labor; trafficking for forced labor and domestic work; and domestic work as a contemporary form of slavery. NDWM had its genesis in the work of founder Sr. Jeanne Devos, in Dindigul District of Tamil Nadu, where she assisted in the formation of small groups of domestic workers. With no means to stand up for their rights or to bring to light their own stories of ill treatment, domestic workers and their needs were virtually invisible. Paid little or no salary, with poor working conditions, workers experienced mental and physical abuse and isolation stemming from little or no interaction with co-workers, whom they viewed more as rivals who could snatch away their jobs than as friends.

Bishops of the tribal areas, concerned about the number of women and girls who were leaving villages to go to towns, approached Sr. Jeanne. With prior experience with founding student movements, she agreed to come to Mumbai to start a few groups of domestic workers. Out of this work, the objectives of the movement were shaped into the policies of the NDWM. The movement extended operations to Patna (Bihar), Chennai (Tamil Nadu), Varanasi, (Uttar Pradesh).

NDWM states that its vision is to

> create a just society for all domestic workers, where they are treated with dignity and justice, their rights are upheld, their contribution recognized and their voices heard. It also envisions a society where child domestic work is completely abolished and children in work enjoy mainstream education and a fuller childhood.

To the end of enabling the voice of the workers, NDWM encourages their full participation and places leadership in the hands of domestic workers themselves. It empowers workers through leadership and capacity building programs and informing them of their rights. Various awareness, advocacy and lobbying campaigns are conducted to sensitize the public, governing bodies and policy makers to the plight of domestic workers. The Movement also intervenes in crisis situations and helps trafficked women and child domestic workers to be re-instated with their families before they are given necessary trauma counseling and support and medical aid.

To help prevent and curb the trafficking of children and women from rural areas, the National Domestic Workers' Movement undertook active campaigns in 2004 at both the source areas, the states of Jharkhand, Bihar, Chhattisgarh and Orissa, and the receiving areas of Mumbai, Delhi, Bangalore and Goa. Over time, NDWM realized that trafficking was spreading to other areas as well. After the tsunami, trafficking became quite rampant in the southern states, with traffickers targeting very small children who were lost, orphaned or abandoned.

To deal with the problem, NDWM launched a systematic campaign in 2005–2006. In Chhattisgarh, a delegation of NGOs, politicians, domestic workers and civilians went from village to village on foot and held meetings with the villagers using street plays, speeches and interactions to communicate their message. Meetings were also held at schools so that children would be aware of the myths associated with domestic work and that trafficking of children is for real. This campaign, an ongoing activity of the Movement, helped establish that forced migration was serious concern in need of appropriate action.

NDWM has witnessed domestic workers working for, rather than against, each other. Workers have gained awareness of their rights and the confidence to challenge employers and the general public. Practical changes such as fair pay, shorter working hours and using first names have created a relationship of justice between the employer and the employee. Fewer incidences of violence have been recorded.

But the process continues. In the words of Founder and National Coordinator, Sr. Jeanne Devos, "We do not give in, we do not give up, we do not grow tired."[11]

Case Study 2: The Presbyterian Church (U.S.A.) Human Trafficking Roundtable

As human trafficking has emerged as one of most critical justice issues of the day, most Christian denominations have sought to engage in work to address modern slavery. Women in organizations such as United Methodist Women, Presbyterian Women and the Women of the ELCA have long had a strong commitment to issues such as violence against women and pornography. But in recent years, some denominational efforts have moved beyond the arena of exclusively women's concerns.

The Presbyterian Church (USA) is a mainline Protestant denomination with some 10,400 congregations and just under two million members. Formed in 1983 with the reunion of the northern stream, the United Presbyterian Church in the USA (UPUSA), and the Presbyterian Church (US), the denomination brings a history of engagement in the public square that dates to the work of John Calvin at the time of the Reformation and a long standing commitment to justice issues.

Staff work on particular justice issues is grounded in and grows out of actions of the denomination's General Assembly, and human trafficking is no exception. Not surprisingly, actions of the Assembly on trafficking focused primarily on sex trafficking, an overture in 2006 calling the church to respond to child sex trafficking, and action by the subsequent Assembly in 2008 expanding that work to include the sex trafficking of women. But other actions of earlier assemblies had focused on the Fair Food Campaign and the work of the Coalition of Immokalee Workers on the exploitation of agricultural workers in Florida. That work, in which the PC (USA) was a partner, clearly revealed that trafficking for forced labor existed on a continuum of exploitative labor practices. Actions on immigration issues expand the policy base related to the issue.

As staff began their work in response to assembly actions on trafficking, they quickly found other staff partners whose portfolios in some way intersected with modern slavery: women's issues and advocacy, disaster assistance work, immigration

---

[11] Information excerpted from http://ndwm.org, website of the National Domestic Workers Movement. Retrieved: September 25, 2012.

services, hunger, peacemaking work, and even work with the men's organization. It also became clear that mission workers in many countries were uniquely positioned to both identify persons who were potentially vulnerable to being trafficked and the contexts ripe for slavery, as well partners who might address it. Staff in both the denomination's Office of Public Witness in Washington, DC and in the Presbyterian Ministry at the United Nations brought the perspective of national and international advocacy. And there were non-staff constituency partners from the larger church: a member of the domestic violence network and a leadership team member of the women's organization, Presbyterian Women.

From the beginning, the work of this roundtable was characterized by the partnership of staff persons with diverse, and sometimes divergent, views on modern slavery. Because the roundtable included staff whose work addressed labor as well as sex trafficking and the enslavement of men as well as women and children, there developed a more nuanced understanding of the scope of the problem. With input from those dealing with undocumented persons, a deeper understanding ensued of the particular vulnerabilities of these persons, but also of the distinction between smuggling, where a person is moved voluntarily (at least initially) from one country to another, and trafficking. Staff involved in disaster assistance further increased the group's understanding of the vulnerabilities of persons displaced by disasters. Staff at the UN and those with connection to mission personnel reminded the group of both the international scope of slavery and the myriad unique contexts in which it occurs. The staff person for Mission Responsibility Through Investment provided a connection to the use of shareholder power to impact practices of the tourism industry that contribute to trafficking.

The work of the Roundtable is to educate, raise awareness and advocate around the issue of modern slavery, thereby to equip individuals and congregations to respond in effective and responsible ways. Partnership with those doing the work on the ground is the watchword here, and secular partners include the Coalition of Immokalee Workers, ECPAT-USA, and the Freedom Network (USA). Particular staff persons also have strong connections to the work of the U.S. Department of Justice and Amber Alert.

Presbyterian congregations and other groups (that is, Presbyterian Women's groups, for example) present varying degrees of interest and commitment to activism, ranging from participation in purchasing ECPAT-USA's Tassatag luggage tags, produced by survivors of trafficking, the proceeds of which support the implementation of ECPAT's Code of Responsible Tourism, to participation in the Red Hand project, an initiative originally put forth by Human Rights Watch that sought to get organizations to create red handprints with messages urging an end to the use of child soldiers. The messages are then sent to the United Nations missions of countries that have not ratified the Optional Protocol against the use of children in armed conflict, with the goal of building international consensus against the use of child soldiers. The Roundtable also encourages more extended commitments that involve advocacy and partnerships to provide services.

One initiative that requires extensive logistical work is the sponsoring of professional trainings for law enforcement, FBI, and social services agencies and clergy in a community. These comprehensive trainings are offered by the Freedom Network (USA), a network of some 30 organizations doing anti-trafficking work on the ground. To date, several trainings have taken place, with more in the offing.

Awareness raising and training equips congregations to seek out other community partners, including other congregations, in order to provide what the on-the-ground providers of services and law enforcement agencies in their communities perceive to be the greatest needs and gaps in services. Often, the most critical need identified is safe transitional and long-term housing for trafficked persons.

What Motivates the Activist? A Congregation Addresses Safe Housing that Respects Victim Rights

In an urban area in the southeast, the Roundtable partnered with staff in its regional governing body (called a presbytery) to offer a Freedom Network (USA) training. One attendee was a member of a social justice committee of a local church searching for a ministry in which to engage. The training sparked a strong sense of commitment to respond to the issue of human trafficking, and connections made with law enforcement personnel underscored the urgent need for safe transitional housing for victims. Nationally, there is a severe shortage of beds available for persons who have been victims of trafficking, particularly transitional housing for which the need may be immediate and critical.

The social justice committee chair of the congregation speaks of the training as a time when she felt a clear sense of God's call to respond to this need. On the church's property are two houses that, while in need of rehabilitation, offered the ideal location for safe housing. But the tasks involved were daunting: not only getting the houses renovated and equipped to be ready for guests, but also securing the necessary funding, recruiting a board of persons with the necessary skills and capacity to run such a program successfully, and, most important, putting in place carefully thought out procedures that would represent best practices grounded in a victim-centered approach. Presbytery social justice staff and the national staff person who worked with Fair Food concerns worked with the congregation's committee, and the Freedom Network (USA) provided technical assistance. The first house was only just ready when the FBI notified the group that a guest needed housing. Renovation and equipping of the second house was subsequently completed. Persons involved in this work have a larger vision for the future—a network similar to Habitat for Humanity that locates and furnishes longer term housing for those attempting to rebuild their lives.

Members of the congregation engaged in this ministry understand that they are providing housing for persons in need of a sanctuary. They are aware that issues of privacy and confidentiality make it necessary for only board members and those who are part of the ministry to have more in-depth knowledge. For this congregation, it comes down to answering God's call for justice for the vulnerable.[12]

---

[12] Case study is based on conversations with members of the social justice ministry of the congregation and with presbytery staff, and on a site visit to see the housing and meet a guest.

Case Study 3: Caritas Internationalis

Caritas Internationalis is a global confederation of 165 Catholic organizations working in humanitarian emergencies and international development. Caritas splits its work into six functional areas: (1) Peace and Reconciliation, (2) Emergencies, (3) Economic Justice, (4) Climate Change, (5) HIV and AIDS, and (6) Women and Migration.

Caritas houses its human trafficking work within the areas addressing women and migration function. The Caritas Internationalis Commitment on Combating Trafficking in Human Beings states, "Combating trafficking in human beings is of relevance to the principle of human dignity, to the fight against poverty and to the promotion and defense of human rights. All these elements are at the very heart of the mission and work of Caritas" (Caritas document on Trafficking).

Caritas Internationalis and its Member Organizations strongly condemn trafficking in human beings as a criminal act, which violates basic human rights and the inviolable dignity and integrity of the human person. Caritas Internationalis and its Member Organizations acknowledge the need and commit themselves jointly to take decisive and effective actions:

- in raising public awareness and enabling people to take necessary actions with regard to trafficking in human beings, and thus preventing trafficking;
- in advocating for alternatives for vulnerable groups
- in assisting and protecting trafficked persons;
- in advocating for the rights and for the protection of trafficked persons, for effective anti-trafficking legislation and measures, for effective enforcement of such legislation and measures;
- in advocating for migration policies and economic policies that reduce vulnerability of people to trafficking;
- and in networking with authorities and with Churches and relevant actors in international civil society to jointly elaborate effective partnerships to challenge the human trafficking phenomenon. (Caritas Commitment on Trafficking 2005)

Sexual exploitation and the exploitation of women are of particular concern to Caritas. In their work, Caritas recognizes that "trafficking in human beings knows no boundaries in terms of age and gender." In their anti-trafficking work, the organization identifies the following strategies: (1) Prevention and awareness raising, (2) Assistance to trafficked persons, (3) Advocacy- focusing on alternatives for vulnerable groups and migration policy that reduces the risk of trafficking, and (4) Networking.

Member organizations of Caritas Internationalis may do advocacy and service work in their own local, national and regional contexts. This work varies depending on the specific priorities of individual Caritas organizations. Caritas Internationalis focuses its human rights approach on the international human rights mechanisms and the universal declaration of human rights. Caritas works within these frameworks to advocate for change.

Caritas Europa has been a convener and key player in COATNET- Christian Organizations Against Trafficking Network. COATNET membership has largely been representatives from Caritas member organizations around the world. COATNET has also partnered with the ecumenical community in Europe, particularly with CCME (Churches Commission on Migrants in Europe) with members from Protestant, Anglican and Orthodox communities. At present, COATNET and CCME work as a model for Christian cooperation on the difficult task of coordinating advocacy and service work across different theological divides.[13]

189.6   Conclusion

189.6.1   Unexamined Assumptions and Collaborative Work

People of faith and faith-based organizations have been engaged in action and advocacy to eliminate slavery for a century and a half. In the past three decades, the issue of modern slavery has moved onto center stage for many more Christians of all theological persuasions, as well as those of other faiths. But while modern slavery is a social issue that can unite people of faith across the spectrum of theological belief, the assumptions that underlie how faith-based groups address the problem and the approaches that groups take to address the issue can and do vary widely. Because these assumptions are sometimes not articulated clearly or worse, are even unspoken, comprehensive and coordinated action can be difficult at best.

An example of how this happens can be seen in the 2008 "Ecumenical Conference on Human Trafficking: Multiple Perspectives to Address This Complex Issue," a conference sponsored by the National Council of Churches Justice for Women Working Group and the United Methodist Women's Division United Methodist Seminar Program on National and International Affairs. The conference, as one might expect given its sponsorship, focused more on the trafficking of women and girls, and more intentionally on sex trafficking, though there was at least one presentation that addressed labor trafficking. Those in attendance were from denominations and faith-based groups engaged in work on trafficking from a variety of perspectives. Yet specific theological perspectives and approaches were not articulated among participants (and in fairness, the time frame and the need for presenting basic knowledge probably precluded that this would happen in that setting).

To build the consensus necessary for collaborative ecumenical work, surfacing differences in the moral grounding for anti-trafficking work would allow groups to identify points of agreement on which such work could be built. At this writing work is underway for more ecumenical work to address modern slavery. A group of staff from several member communions of the NCCC have initiated plans for an upcoming conference on modern slavery. Yet without addressing these same unexamined assumptions, collaborative work will face an uphill battle.

---

[13] For more information, see "Created in the image of God, treated like slaves…." The Caritas Internationalis commitment on combating trafficking in human beings (2005).

189.6.2 The Consequences of Actions Built on Myth

Another dynamic that serves as a possible roadblock to effective collaborative action is the tendency to move quickly to a clearly defined, concrete and attractive common action without examining critically whether that action is grounded in evidenced-based research. One case in point is the mobilization of initiatives against sex trafficking in advance of large sporting events such as the World Cup, the Olympics or the Super Bowl. These initiatives have been bolstered by rhetoric claiming that large numbers of women and girls would be trafficked in to the cities where these events are to be held. In fact, evidence-based research does not bear out that assertion. A 2011 study by GAATW found that there is a very large discrepancy between claims made prior to large sporting events and the actual number of trafficking cases found, with no evidence found that there is an uptick in trafficking as a result of these events (Global Alliance Against the Traffic in Women 2011). Anti-trafficking campaigns based on such unsubstantiated claims can negatively impact the very groups they are intending to protect. Groups like GAATW and others have urged instead that anti-trafficking efforts address other forms of trafficking or exploitation connected to these events, like exploitation of migrant workers in construction, workers' rights in the production of sporting equipment, and the like.

189.6.3 Faith-Based Anti-trafficking Work: The Way Forward

The face of modern slavery continues to evolve. As more is known about the scope and breadth of the problem, and as initiatives to address slavery become better grounded in what research shows to be best practice, groups engaged in the work will be able to make more nuanced and effective responses. Faith based groups and individuals have a vital role to play in this work. One emerging arena for response is in addressing supply chain issues—raising awareness among consumers about which products may be contributing to modern slavery. Websites such as www.slaveryfootprint.com allow persons to see how their lifestyle choices and purchasing habits contribute to slavery.

   While this chapter has focused on anti-slavery work within Christian faith-based groups, interest in and work to address slavery is by no means confined to Christians. In the religious texts and traditions of most faiths, there are strong prohibitions against many of the key aspects of slavery. Information about organizations in other faiths engaged in anti-trafficking work is less easily obtained, however. *Religion Link*, a web-based source of religion story ideas and sources for journalists, notes that Jewish groups in particular are raising awareness among Jewish congregations, some of which celebrate Freedom Shabbats (www.religionlink.com/tip_110615.php. Retrieved: October 23, 2012). The October 2008 issue of *Sh'ma: A Journal of Jewish Ideas* focused on human trafficking. It notes that American Jewish World Service (AJWS), a grant-making organization with almost 400 partners in 36 developing countries, funds both organizations assisting and promoting the human rights of trafficked persons and those that address underlying vulnerabilities that fuel trafficking (www.shma.com/2008/10/. Retrieved: September 25, 2012). Rabbis for Human Rights- North America has also been active with the Freedom Network as well as the Coalition of Immokalee Workers in advocating for a human rights based approach to address the issues of human trafficking.

In the Muslim world a number of organizations address root cause issues, notably Grameen Bank in Bangladesh, the microfinance organization. (http://met0.wordpress.com/organizations. Retrieved: November 12, 2012.)

In her blog for the *Huffington Post,* Katherine Marshall, who heads up the World Faiths Development Dialogue, notes that in a study WFDD undertook in Cambodia where some 60 Christian groups were doing anti-slavery work, they identified just one Buddhist organization (www.huffingtonpost.com/katherine-marshall/sex-trafficking-religion_b_1924000.html. Retrieved: November 12, 2012). In India, where most adherents of the Hindu faith live and where slavery has traditionally been connected to the caste system, Hindu activists were instrumental in the government's decision to end bonded labor slavery in 1976. (www.freetheslaves.net/Page.asp?pid-484. Retrieved: September 25, 2012). Clearly, more research is needed on how these faith groups and others are working to address modern slavery.

In the Christian community, deep theological differences will continue to cause divisions, and these differences manifest themselves in anti-slavery work. It will behoove people of faith to identify their differences and to seek to find congruence where they can. Increased partnerships with organizations doing the hard work on the ground will help guide the most effective responses.

189.7 Faith-Based Organization Websites

Web-based information and web addresses frequently change. If the addresses listed here are no longer current, use a search engine to enter organization names and obtain updated information.

In order to examine the nature of the moral imperatives that grounds a particular organization and to evaluate the effectiveness of its work, it will be necessary to spend time exploring its website and related resources in some depth.

189.7.1 U.S. Denominations/Agencies

America Baptist Church

- International Ministries—Information on missionaries working to confront human trafficking. Includes contact information for ordering a newsletter on human trafficking. www.internationalministries.org/ministry_focus_areas/human_trafficking

- American Baptist Women and Girls: "Break the Chain" website—Resources to address violence against women and girls, including some anti-trafficking resources. www.abwministries.org/resources/resource.cfm?ResourceID=1227

Christian Church (Disciples of Christ)—www/disciples.org

- Disciples Women—Information about Human Trafficking: Breaking the chains of modern day slavery, 2009–2014 Quadrennial Assembly Social Action Project. www.discipleshomemissions.org/pages/DW-SocialActionBreakChains

Christian Reformed Church in North America—www.crcna.org/

- Office of Social Justice—Information and steps to take action. www.crcna.org/pages/osj_trafficking.cfm

Cooperative Baptist Fellowship— www.thefellowship.info

- Justice and Peace Ministries—Downloadable resources, including human trafficking fact sheet and video. www.thefellowship.info/Missions/Communities/justice

Evangelical Covenant Church—www.covchurch.org

- Human trafficking—Information about the Break the Chains initiative, including downloadable resource, *Protect.* www.covchurch.org/trafficking/

Evangelical Lutheran Church in America—www.elca.org

- Women of the ELCA—Downloadable resource, *Understanding and Responding to Human Trafficking.* www.womenoftheelca.org/
- Justice for Women www.elca.org/Our-Faith-In-Action/Justice/Justice-for-Women/Resources.aspx **Downloadable resources, including** "Human Trafficking: A Faith Response" **brochure,** Human Trafficking Awareness Day worship resources, bulletin inserts, and a **Human Trafficking Curriculum Series** to be led for and by young people.
- Cherish Our Children—A national Lutheran ministry of prayer, education, relationship-building and action to prevent child sexual exploitation. www.cherish-children.org

Free Methodist Church World Ministries—http://fmcusa.org/fmwm/

- International Childcare Ministries—Advocates for the spiritual, educational, physical and social development of children by partnering with the global Free Methodist Church through sponsorships, scholarships and creative initiatives. Downloadable brochure for Freedom Sunday targeted at combating child trafficking in India. www.childcareministries.org

Presbyterian Church (USA)—www.presbyterianmission.org/

- Human Trafficking Roundtable—Cross divisional roundtable working collaboratively to raise awareness, educate, advocate and equip churches and individuals to respond to modern slavery. Downloadable resources, including a *Toolkit for Action: Modern Slavery*. www.presbyterianmission.org/ministries/human-trafficking/

Reformed Church in America—https://www.rca.org

- Social Justice: *Moving from Witness to Action*—Theological rationale and web links, including ordering information for *Hands That Heal*, a community-based international curriculum to engage churches against human trafficking and to equip them to provide transformational care to survivors. https://www.rca.org/Page.aspx?pid=4812

The Salvation Army—www.salvationarmyusa.org/

- Combating Human Trafficking—Extensive web resources and ways to help, including the Promise initiative to combat the trafficking of children. http://www.salvationarmyusa.org/usn/www_usn_2.nsf/vw-dynamic-index/8081A407 9639D55A802573E000530965?Opendocument

U.S. Conference of Catholic Bishops—www.usccb.org/about/anti-trafficking-program/

- Catholics Confront Global Poverty—An initiative of the U.S. Conference of Catholic Bishops and Catholic Relief Services. Represents the official voice of the Roman Catholic Church in the United States on policy issues. Downloadable resource, "Catholics Confront Human Trafficking: Answering the Church's Call to End Human trafficking" www.confrontglobalpoverty.org/

The Vineyard USA—www.vineyardusa.org/

- Justice Response—Resources for learning, getting equipped and responding to sex trafficking, forced labor and child trafficking. www.vineyardusa.org/site/task-forces/justice-response

The Wesleyan Church—www.wesleyan.org/

- Human Trafficking Awareness Day—Video downloads. www.wesleyan.org/freesource/r?id=221&tag=29&lang=1

The United Methodist Church: General Board of Church and Society—Actions include a national conference call to combat domestic violence and human trafficking; action alert to urge calls to support reauthorization of TVPA. http://umcgbcs.org/faith-in-action

- United Methodist Women—Information and downloadable resources, including *Human Trafficking: Preventing, Protecting, Prosecuting*. http://new.gbgm-umc.org/umw/act/trafficking/?search=human trafficking
- UMCOR (United Methodist Committee on Relief)—News stories of UMCOR projects and responses related to human trafficking. www.umcor.org

The United Church of Christ—www.ucc.org/

- Women's Ministries—Web links to resources, including liturgical resources for Human Trafficking Awareness Day. www.ucc.org/justice/womens-issues/human-trafficking.html
- Global Ministries—Resources for Human Trafficking Awareness Day. http://globalministries.org/news/news/human-trafficking-awareness.html

189.7.2 Global Faith-Based Organizations (A Selected List)

- CHASTE (Churches Alert to Sex Trafficking Across Europe)—Works for the global eradication of sex trafficking, the provision of practical care and support for trafficked persons, and the suppression of demand which drives the supply of women and the girl child. www.chaste.org.uk/
- COATNET -Christians Organised Against Human Trafficking—Links together many Christian groups which are fighting human trafficking. www.caritas.org/Resources/Coatnet/Coatnet.html
- National Domestic Workers Movement—Movement and a NGO working with domestic workers, child domestic workers and migrant workers in 23 states of India. http://ndwm.org/
- Hagar International Christian organization dedicated to recovery of extreme human rights abuses, particularly human trafficking, gender-based violence, and sexual exploitation, working with *women and children in Afghanistan, Cambodia and Vietnam.* http://hagarinternational.org/international
- World Council of Churches www.oikoumene.org A worldwide fellowship of 349 churches seeking unity, a common witness and Christian service.
- Human trafficking www.oikoumene.org/en/news/news-management/eng/a/article/1724/human-trafficking-violen.html Contact: A Publication of the World Council of Churches. No 84, May 2007 Human Trafficking. www.oikoumene.org/fileadmin/files/wcc-main/documents/p4/contact/con-184.pdf

189.7.3 Other Faith-Based Organizations

There are literally hundreds of organizations and groups that focus on anti-trafficking work, many of them founded by persons of faith. The following are some of the organizations with the most visibility.

- Free the Slaves—Dedicated to ending slavery worldwide, working on the ground with liberators around the world to free slaves and change the systems that allow slavery to exist. www.freetheslaves.net
- International Justice Mission—Brings rescue to victims of slavery, sexual exploitation and other forms of violent oppression. Lawyers, investigators and aftercare

professionals work with local officials to secure immediate victim rescue and aftercare, to prosecute perpetrators and to ensure that public justice systems effectively protect the poor. www.ijm.org

- Not For Sale—Through international work on the ground and in mainstream supply chains, proactively targets the root causes of slavery while engaging and equipping the movement for freedom. www.notforsalecampaign.org
- Shared Hope—Rescues and restores women and children in crisis. www.shared-hope.org/

References

Bales, K. (1999). *Disposable people: New slavery in the global economy.* Berkeley: University of California Press.

Bales, K. (2007). *Ending slavery: How we free today's slaves.* Berkeley: University of California Press.

Bernstein, E. (2010). Militarized humanitarianism meets carceral feminism: The politics of sex, rights and freedom in contemporary anti-trafficking campaigns. *Signs: The Journal of Women in Culture and Society, 36*(1), 45–71.

Boles, J. (1988). Introduction. In J. Boles (Ed.), *Masters and slaves in the house of the Lord: Race and religion in the American South 1740-1870* (pp. 1–18). Lexington: University Press of Kentucky Press.

Choi-Fitzpatrick, A. (2006). *Free indeed: the modern church's opportunity to end slavery again.* Retrieved October 1, 2012, at https://www.freetheslaves.net/SSLPage.aspx?pid=362

Crago, A. -L. (2003, May 21). Unholy collaboration. *Rabble.*

"Created in the image of God, treated like slaves…." The Caritas Internationalis commitment on combating trafficking in human beings (2005). 89th Session CI Executive Committee Document 6.07, CI Secretariat – Vatican City.

Cromartie, M. (1999). The Jew who is saving Christians. *Christianity Today, 43*(3), 50–55.

Ditmore, M. H. (2002). *Trafficking and se work: A problematic conflation.* Unpublished Ph.D. dissertation, Department of Sociology, City University of New York.

Doezema, J. (1998). Forced to choose: Beyond the voluntary v. forced prostitution binary. In *Global sex workers: Rights, resistance, redefinition* (pp. 34–50). New York: Routledge.

Doezema, J. (2000). Loose women or lost women? The re-emergence of the myth of 'white slavery' in contemporary discourses of 'trafficking in women'. *Gender Issues, 18*(1), 23–30.

Glancy, J. A. (2011). *Slavery as moral problem in the early church and today.* Minneapolis: Fortress Press.

Ham, J. (2011). *What's the cost of a rumour?: A guide to sorting out the myths and the facts about sporting events and trafficking.* Bangkok: Global Alliance Against Trafficking in Women (GAATW).

ILO 2012 Global Estimate of Forced Labor Executive Summary (2012). www.ilo.org/wcmsp5/groups/public/---ed_norm/---declaration/documents/publication/wcms_181953.pdf. Retrieved September 25, 2012.

Irwin, M. A. (1996). 'White slavery' as a metaphor anatomy of a moral panic. *Ex Post Facto: The History Journal, V*, 1–35.

Jakobsen, J. R., & Bernstein, E. (2009). *Religion, politics and gender equality: Country report USA.* Geneva: United Nations Research Institute for Social Development.

Jones, M. (2003, November/December). Thailand's brothel busters. *Mother Jones.*

McReynolds, M. (2008). The Trafficking Victims Protection Act: Has the legislation fallen short of its goals? *Policy Perspectives, 15*, 33–56.

New Internationalist. (2012). *A brief history of slavery*. www.newint.org/features/2001/08/05/history. Retrieved November 8, 2012.

Obama announces new effort against human trafficking, praises work of faith groups. www.christianpost.com/news/obama-announces-new-effort-against-human-trafficking-praises-work-of-faith-groups-82214/#TzTR3rQc6jTYUdSw.99. Retrieved November 2, 2012.

Peach, L. J. (2000). Human rights, religion and (sexual) slavery. *Annual of the Society of Christian Ethics, 20*, 65–87.

Renzetti, C. M. (2011, April 25). *Thy will be done: transcendent motivation and the fight against human trafficking*. Presidential Lecture Series, Morehead State University.

Schliecher, A. (1999). Unity against sexual trafficking. *Sojourners, 34*(9), 17.

Soderlund, G. (2005). Running from the rescuers: New U.S. crusades against sex trafficking and the rhetoric of abolition. *NWSA Journal, 17*(3) Fall, 64–87.

Trafficking Victims Protection Act (TVPA), Division A Public Law 106-386, 106th Congress (October 28, 2000), [section] 112(a); [section] 103(8).

Transcript, special briefing (2011). TIP report. www.state.gov/j/tip/rls/rm/2011/167149,htm. Retrieved September 25, 2012

UNICEF. (2012). www.unicef.org/protection/57929_58005.html Retrieved September 25, 2012.

United Nations. (2003). *UN Protocol to Prevent, Suppress and Punish Trafficking in Persons, Especially Women and Children*, supplementing the *United Nations Convention against Transnational Organized Crime*, Article 3 (2003) New York, New York.

Weber, M. (1992). *The Protestant ethic and the spirit of capitalism* (T. Parsons, Trans.). (1905 reprint). New York: Routledge.

Weitzer, R. (2007). The social construction of sex trafficking: Ideology and institutionalization of a moral crusade. *Politics and Society*. Sage Publications. http://pas.sagepub.com/cgi/content/refs/35/3/447. Retrieved September 25, 2012.

Yeomans, W. (2000). U.S. Department of Justice, U.S. Congress, Senate Committee on Foreign Relations, Sub Committee on Near Eastern and South Asian Affairs, International Trafficking of Women and Children, 106th Congress, 2nd session. April 4, 2006.

Zimmerman, Y. C. (2010). From Bush to Obama: Rethinking sex and religion in the United States' initiative to combat human trafficking. *Journal of Feminist Studies in Religion, 26*, 79–99, March 22, 2010.

Zimmerman, Y. C. (2011). *Christianity and human trafficking*. Retrieved September 25, 2012, at http://onlinelibrary.wiley.com/doi/10.1111/j.1749-8171.2011.00309.x/abstract/

Zurita, B. (2005). *Passion vs. apathy*. www.beverlylahayeinstitute.org/articledisplay.asp?id=7959&department=BLI& categoryid=reports&subcategoryid=blitraf. and BLI unites Mexico and the U.S. (2005).

Zurita, B. (2006). *Prayer, the ultimate sex-trafficking program*. www.cwfa.org/articles/10983/BLI/family/index.htm.

Chapter 190
Violence, Tolerance and Religious Peacebuilding in Northern Ireland

John D. Brewer and Francis Teeney

190.1 Introduction

Northern Ireland, sometimes also known as six-county Ulster, is a small country, geographically attached to the Irish Republic but territorially partitioned from it and thus remaining part of the United Kingdom (on the special condition of the "border areas" see Creamer et al. 2011), with a population in the 2001 census of just over 1.6 million, just below one-third of the Island of Ireland's total population (Fig. 190.1). Irish independence from Britain in 1921 left Northern Ireland with a contested status (for a short historical introduction to Northern Ireland see Mulholland 2003), with the population roughly split in the 2001 census between Catholics (40 %) and Protestant (46 %), with "not stated" (9 %) and "neither" (5 %) comprising the rest. The Catholic population is disproportionately concentrated in the Western parts of Northern Ireland (Fig. 190.2).

Northern Ireland's ambiguous position as British or Irish, part of a united Ireland or in Union with Britain, spurred a conflict, known colloquially as "the Troubles," a phrase to our minds that has always undervalued the level of trauma the conflict caused. In its last phase it lasted for 30 years, between the civil rights marches of 1968 and the signing of the Good Friday Agreement in 1998, although conflict existed for long periods before 1968 as part of the Irish war of independence and has broken out sporadically since the 1998 Agreement (for a history of Catholic-Protestant relations in Ireland see Brewer and Higgins 1998). Over 3,500 people were killed in the last phase, which per capita is more than the number of U.S. soldiers killed in the Vietnam War. Many tens of thousands were injured, roughly one in four of the population. Details of the violence are well known (see Edwards and McGrattan 2010 for what is styled a "beginner's guide to the conflict"). Our attention

J.D. Brewer (✉) • F. Teeney
Institute for the Study of Conflict Transformation and Social Justice,
Queen's University Belfast, Belfast BT7 1NN, UK
e-mail: j.brewer@qub.ac.uk; f.teeney@qub.ac.uk

© Springer Science+Business Media Dordrecht 2015
S.D. Brunn (ed.), *The Changing World Religion Map*,
DOI 10.1007/978-94-017-9376-6_190

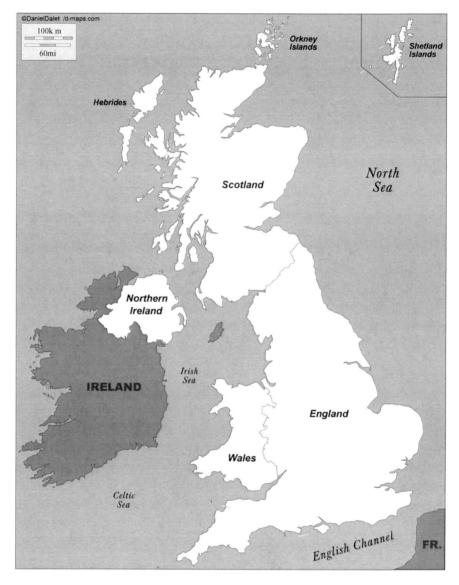

Fig. 190.1 Map of Britain and Ireland, showing Northern Ireland (Map by Daniel Dalet, /d-maps. com, http://d-maps.com/pays.php?num_pay=218&lang=en)

here is on the peace. Our chapter will address three issues: first, it will offer some reflections on the relationship between religion, peace, tolerance and co-existence; secondly, it will use these reflections to critique the new field of study called religious peacebuilding; and finally, it will address the dynamics of religious conflict and peacebuilding in Northern Ireland as a case study of religious peacebuilding.

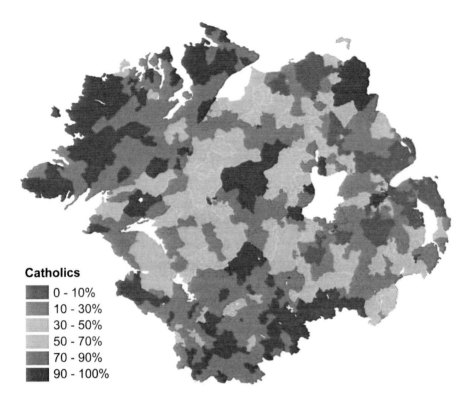

Fig. 190.2 Religious demography in Northern Ireland. Percentage of Catholics based on census figures from 2001 and 2006 (Map from Wikipedia, http://en.wikipedia.org/wiki/File:Scaoileadh_Creidimhin_in_UlaidhReligious_Division_of_Ulster.jpg)

190.2   Religion, Tolerance and Co-existence

Religion is a site well suited to the process of cultural "othering," the process of perceiving humans in group terms and believing that some belong to less deserving categories than one's own. However, religious "othering" presupposes processes of religious "belonging" as two sides of the same Janus face, for religion differentiates between those who belong and those who do not, marking both the "insider" and the "outsider" or "other." If religion serves as a scared canopy gluing society together (Berger 1967), it does so only for those who belong and in practice it reinforces the exclusion of the marginalized religious "other." It is easy to see why religion is so suited to this process. Religion involves making and believing truth claims. The inherent tendency for each world faith to see itself as the font of all truth can among some believers turn religious righteousness into self-righteousness.

The point about self-righteousness is the conviction of the untruth of others. This makes *in*tolerance rather than tolerance a virtue. For example, in the early modern period in Britain and its colonies, roughly the fifteenth to the eighteenth centuries, when religious conflicts were vivid, toleration was perceived pejoratively as a threat both to God and politics, since it reflected doubt in one's own religious truths and uncertainty in government. Religious *in*tolerance was valued; some historians have referred to this as a kind of "charitable hatred" (Walsham 2006). This is quite the reverse to the way in which today we like to see religion as carrying the ethic of liberalism, pluralism and tolerance In the past, religious toleration posed the same dilemmas and problems as religious *in*tolerance does today.

But the respect for religious *in*tolerance is not a just a historical quirk restricted to the distant past. Religion is still often a site of conflict, which belies its perception as a carrier of tolerance and pluralism. Conflict between religious groups is common, even in recent history. Indeed, sociologists of religion who otherwise denude religion of significance in their obsession with secularization, argue that religion retains its saliency in modernity only when it stands in as a surrogate for ethno-national and political conflict (for example Bruce 2011). In Bruce's view, the sociology of religion is really the sociology of secularisation and the substantive focus is on its decline, denuding it as marginal and irrelevant, save in one exception, the ancient association of religion, politics and conflict (on which see Norris and Inglehart 2004). There are several dimensions to the historic link between religion and conflict.

Religion is clearly wrapped up in the 'problem of large numbers,' where cultural majorities have the ability to impose exclusion on minorities in the form of persecution, competition or indifference. This is what we understand commonly by intolerance. Intolerance naturally leads to "external othering" for religious majorities assert their differences in order to exclude minorities who do not 'belong.' External "othering" of religious minorities can show itself in religious persecution, but most often neglect and indifference. Violent persecution is more likely to occur when the majority feels threatened and beleaguered, perhaps because of the size of the religious minorities or their political and cultural assertiveness, with indifference the likely result when the majority is not an unsettled community and there is no sense of threat.

It is in the context of this neglect and indifference that religion also gets wrapped up in what is popularly called today the "problem of small numbers." This concept describes the demand from small religious and cultural minorities for social, political and economic recognition. We might call this "self othering," for cultural differences are asserted by groups themselves to facilitate their recognition as a minority. This is not necessarily a demand that they become absorbed into the majority religion and made to feel they "belong;" it is a demand for their religious difference to be accepted as a legitimate minority status. In this regard, cultural and religious minorities are asserting difference in order to better separate themselves. Sometimes this demand for religious and cultural separatism is pursued by violent means, deepening the association between religion and violence; on other occasions by political mobilization. Note that in the first instance, "othering" is imposed on minorities from the outside, in the latter appropriated internally, hence, our terminology of "external" and "self othering." In both cases, however, religious differences can be the measure by which cultural distinctiveness is reproduced.

However, religious difference can be asserted for the purposes of social inclusion as much as exclusion. While "external othering" can manifest itself in terms of religious persecution at worst and neglect at best, "self othering" does not have to involve separatism but can encourage religious toleration. Recognition of their cultural and religious difference is merely the precursor to demands for a minority's social inclusion. It is fuller participation they demand, not separatism. This breaks the link between religion, conflict, and intolerance for religion can be a site for reconciliation, an arena for articulating the demand for toleration and mobilizing on its behalf. This uncouples "othering" from "unbelonging," for some religious minorities can assert a demand for acceptance in order to "belong."

Religious "othering," in short, does not have to provoke intolerance. In this instance, religious toleration is *not* premised on the demand to eliminate all religious differences, but to denude these differences of contestation. It is the wish to pacify, that is, make peaceful, obvious and clearly recognizable, but enduring religious differences. This effort requires that we understand what peace means. Peace is never about eliminating differences, the merging of the world religions in rainbow ecumenicalism, it is about the way *dis*agreements are handled better in the future. It is also about the reproduction of continued religious difference but in non-violent ways, encouraging all the religious groups to feel they belong despite their remaining religious differences.

Over the last two decades an interest has developed, particularly in the U.S., in religion as a site of reconciliation (Coward and Smith 2004; Hadley 2001; Johnston 2003; Little 2007; Schlack 2009; Shore 2009; Smock 2001, 2002, 2006, 2008). The U.S. is particularly suited as a cultural space for this kind of work. There is a plurality of religions in the U.S. as part of its racial and ethnic mix, but the country has never witnessed a religious or holy war and thus has no historical memory of religious hatred and violence of the kind that affects most of Europe. Its separation of church and state ensures no one religion has become the established faith and accorded privileged political status as a result. It is also a society where religious practice remains high, against the trend toward secularisation in the West, which encourages people to take religion seriously. It is also a society, for example, in which religion is recognized as a rich resource in politics, part of political diplomacy (for example, Johnston 2003) and incorporated into peace and civic education.

Religious peacebuilding is a rapidly growing field in the U.S. where it is distinguished by three defining characteristics: (a) an emphasis on inter-faith dialogue as the primary form of religious peacebuilding; (b) the commensurate privileging of ecumenism as the chief peace strategy; and (c) eschewing comparative research in favor of the case study method. Methodologically, the single case study approach dominates. Single case studies get set alongside each other in endless edited collections within this new literature, but there is no conceptual apparatus with which to compare the cases systematically.

There is another weakness in this literature. It concentrates on positive cases, situations where religious bodies, para-church organizations and faith-based NGOs bring warring factions together and where religion is above the fray and considered

Fig. 190.3 A typical anti-Catholic wall mural in a Protestant area of Belfast (Photo by John D. Brewer and Francis Teeney)

neutral so that religious actors have legitimacy as peacemakers (for example, Johnston 2003; Smock 2006). Celebrate as we might the odd case or two where this outcome has occurred, such as the short-lived peace deal in Sudan in 1972 or the stable settlement in Mozambique in 1992, we need a conceptual apparatus that focuses on the more numerous cases where religion is part of the problem.

This background introduces the problem of Northern Ireland. Religion was perceived to be wrapped up in the conflict to such an extent that it was difficult for religious peacebuilders to be seen as neutral. Religion was thought to be part of the problem in Ireland and incapable of turning itself into part of the solution (Fig. 190.3). Ireland has a legacy of conflict between Catholics and Protestants that goes back four centuries and historical memories that involve instances of extreme atrocity and depravity on both sides (see Brewer and Higgins 1998). Processes of religious othering and belonging ensured marked cultural and political differences between Catholics and Protestants. The national narrative of both the Irish Republic and Northern Ireland invokes religious symbolism to portray the struggle for nationhood as a religious one. Therefore, religious peacemakers challenge these stereotypes by going against history, culture and politics. Many commentators deny the churches a role in Northern Ireland's peace process or belittle it, focusing on the few well-known events of church involvement and the small number of high profile religious peacebuilders.

190.3   Religion and Politics in Northern Ireland's Conflict

In this section we address the nature of the conflict, assessing the contribution religion made to it. In the final sections we discuss the activities of religious peacebuilders in effecting peace and tolerance. By so doing, we seek to correct various misapprehensions about the ineffectiveness of the role of the churches by pointing to their major achievements in both the social and political dimensions of the peace process. We also develop a conceptual framework to understand religious peacebuilding, allowing the Northern Irish case study to speak to other conflicts where religion is thought to be problematic.

Jonathan Swift, the well-known eighteenth century satirist and Irishman, once said that Ireland had enough religion to make its citizens hate one another, but not enough to make them love. WB Yeats, another well-known literary Irishman, wrote of the Irish that there is more substance in their enmities than in their love. It appears strange that a society noted in the distant past for the conversion of Europe, a land of saints and scholars, and known today for maintaining very high levels of religiosity against the modern secular trend, should be associated with enmity, hatred and conflict. This picture is no paradox.

Religion, while not the cause of conflict, is the social boundary marker that demarcates the groups between whom there is conflict. While some observers seek to denude the conflict of any religious hue, preferring to present it as ethno-nationalist conflict for which religion is just an inconvenient surrogate (for example, McGarry and O'Leary 1995), others highlight the residual religious dimension at least at the symbolic level (for example, Barnes 2005; Mitchell 2006a, b). The truth is it is both political *and* religious, for religion maps onto and represents both real material and political differences. The conflict is over the legitimacy of the state and access to its political, economic and cultural resources, but religious affiliation defines the boundaries of the groups who are in competition. Religion provides some of the cultural resources for drawing moral boundaries between the ethnic groups in political competition, religious symbols become associated with political contestation and the churches also took sides in the war. The religious affiliations of protagonists once had strong theological meaning for most people involved in the conflict (see Brewer 2003a).

For most people today however, their religious affiliation has no substance in the conflict, such that "Protestant" and "Catholic" are merely labels representing contrasting positions on the legitimacy of the state (Fig. 190.4). Protestantism is understood by protagonists mostly in terms of its political and constitutional stance rather than theology; the same would be so for Catholicism, which is why people's sense of belonging to one or the other can be matched with low Christian observance and religious practice (see Brewer 2003b). People are *cultural* rather than *religious* Catholics and Protestants. Those sensitive to the political nature of the conflict prefer to use alternative nomenclature to describe these positions, distinguishing, on a continuum from the moderate to the more radical position, between Nationalists and Republicans within the Catholic community, and between Unionists and Loyalists within Protestantism.

Fig. 190.4 Wall mural in Protestant East Belfast showing the Protestant reformers, Calvin, Luther and Wesley (Photo by Keith Ruffles, http://www.geolocation.ws/v/P/16215658/protestant-reformation-mural-east/en)

This nomenclature is a fine distinction and it is easy to see how outsiders to Ireland can mistakenly perceive the conflict as a religious one, caused by religion, fought to defend religious principles, and contested by people for whom religious affiliation is the master status. Some protagonists in Ireland assist in the perpetuation of this error. For a small minority of Protestants see the conflict in religious terms and reinforce this belief by killing Catholics, burning churches or portraying Irish Republicanism as Catholic, in pursuit of a strategy directed by the antichrist in the Vatican (on the antichrist belief see Barkley 1967 and on its use in Northern Ireland see Higgins 2000, and for a discussion of anti-Catholicism in Northern Ireland see Brewer and Higgins 1998; Rafferty 1994). These beliefs are so marginal to mainstream Protestantism as to appear even to the majority of Northern Irish Protestants as extreme, although at moments of most threat, these ideas temporarily often gained much wider popularity among ordinary Protestants. And Republicans, while thoroughly political and criticized by the Catholic Church, were not averse to utilizing Catholic symbolism (Fig. 190.5).

But if the situation is not a religious war, it is nonetheless puzzling that religion still represents the boundaries of the groups between whom there is conflict. Religion survives as an important social cleavage primarily in the Two-Thirds

Fig. 190.5 A wall mural in Catholic West Belfast uses religious symbolism – the rosary and the Virgin Mary – to characterize the political protest of the Hunger Strikes in 1980–1981 (Photo by John D. Brewer and Francis Teeney)

World, but it is unusual in the modern industrial world for conflict to be socially marked by religion, or at least, for religion to remain important after the country has modernized and industrialized, since historically religion was once a powerful source of social cleavage and conflict throughout Europe and North America. The conundrum, then, is why religion in Northern Ireland retains its saliency as the critical social cleavage around which social division coheres. Marx, in the *18th Brumaire of Louis Bonaparte* gave us a glimpse of the explanation when he wrote that the tradition of the dead generation weighs like a nightmare on the brains of the living. Put another way, Northern Ireland has not transcended the social divisions of its colonial past, ensuring the perpetuation of the process of religious othering and belonging.

What religious peacemakers in Northern Ireland confront, therefore, is a situation in which ancient religious differences have ensured the survival of separate religious communities through such methods of same-religion marriages, residential segregation, distinct cultural organizations and segregated schools. The social structure of the two communities ensures the effortless perpetuation of distinct and separate groups marked by religious differences. Catholics and Protestants live in separate areas, they hold to separate symbols, they contest rather than share territory. Belfast is a divided city whose geography and physical space give vivid portrayal to the conflict. Those working for peace and for reconciliation in the churches thus had three obstacles to overcome: (i) the legacy of the past that created social division; (ii) the impact of a social structure that reproduces separateness; and (iii) the perception that religion was part of the problem and could not possibly become part of the solution.

190.4 Conceptualizing Religious Peacebuilding

The real contribution of sociology in explicating the subfield of religious peace-building is threefold: (1) to focus on those problematic instances where religion is wrapped up in the conflict so that we can more sharply see the potential for religion to assist in reconciliation, tolerance and co-existence; (2) to devise a theoretical framework that moves us beyond the case study method; and (3) to deploy this conceptual apparatus in cross-national comparative research. What follows is one such model.

While this model has been garnered and honed during a 4-year study of the role of the churches in Northern Ireland's peace process, it is proffered as a conceptualisation of religious peacebuilding that facilitates comparative research (what follows is a summary of Brewer et al. 2010, 2011). Figure 190.6 represents the model in diagrammatic form. We see these distinctions as sedimentary layers. At the base of the conceptualisation are three critical distinctions. The first is between active and passive peacemaking. The former lives out commitments to peace as a social practice, so that peacemaking is enacted rather than just talked about; the latter is full of an idealistic commitment, but lacking in application. Some faith-based NGOs, churches and para-church bodies talk peace, but are passive when it comes to its

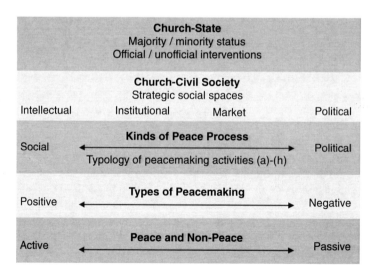

Fig. 190.6 The sedimentary layers of religious peacebuilding (Source: John D. Brewer and Francis Teeney)

practice. To flesh-out the forms that active peacemaking can take, we utilize Galtung's famous contrast between negative and positive peace (Galtung 1969). Negative peace is desirous of an end to the killings, in which peacemaking involves working to end violence. Positive peace involves working toward establishing (or reintroducing) wider principles of justice, equality, fairness and social redistribution as well. The former we can call conflict transformation, the latter social transformation. Some religious peacebuilders can be active when it comes to resolving particular incidents of violence, for example, in demanding military groups desist from killing and dialoguing with them to this end; a smaller number advocate and mobilize to achieve positive peace via the social gospel.

The final distinction is between the social and political peace processes. This antinomy needs more elaborate explanation (see Brewer 2010 for fuller details). All too often peace processes are understood to describe the negotiation process that results in a settlement and the monitoring of conformity to the accord afterwards. Negotiated compromise peace deals, in which parties opt for (or are forced by third parties to accept), are second-best preferences in order to resolve conflict. These are the foundation of peace processes. We refer to this as the *political peace process*. However, the negotiated settlement is never the end of peacemaking, for accords mostly leave unresolved the processes for realizing social healing. By this we mean reconciliation between erstwhile protagonists, social relationship-building and repair across a communal divide, and the replacement of brokenness by the development (or restoration) of people's feelings of wholeness. These concerns are either ignored by negotiators in the political peace process or assumed to follow naturally from the signing of the agreement itself. The *social peace process*, however, fills this void and deals directly with societal healing. It constitutes an important dimension to peacemaking, going on well after the new political institutions resulting from the accord are bedded in.

With this process in mind, we now have a conceptual field that shows what active (as distinct from passive) peacebuilding looks like. As informative as this is for fleshing out the content of *active peacemaking*, the broad sweeps by which the cells are painted need finer detail in order to complete the picture. A typology of examples of active peacemaking fills-in the cells. We suggest active religious peacemaking involves the following kinds of activity:

Social activity (+indicates positive, – indicates negative):

+ Ecumenical activity (breaking down barriers, stereotypes and developing contact in a religious context)
– Mediation (conflict resolution and prevention)
+ Cross-community activities (entry into secular spaces to try to break down barriers)
+ Peace initiatives (espousing peace and monitoring the conflict)
+ Anti-sectarianism/anti-racism etc. (challenging the conflict and redefining it)
+ Dealing with the problems of post-violence (assisting with post-conflict adjustment)

Table 190.1 Active peacemaking in practice

	Positive	Negative
Social	Involves civil society and grassroots groups working in areas of expertise to focus on social transformation and societal healing, whether in pre or post-agreement phases	Involves civil society and grassroots groups working in areas of expertise to focus on conflict transformation by intervening as mediators in specific instances of violence and/or campaigning to end the violence generally
Political	Involves political parties, negotiators and politicians incorporating social transformation and societal healing into the terms of the accord and/or using the new political structures to address social transformation and societal healing	Involves political parties, negotiators and politicians negotiating ceasefires and campaigning for all factions to desist from killing

Source: John D. Brewer and Francis Teeney

Political activity (+indicates positive, – indicates negative):

- – The churches as back channels of communication (provision of 'safe' private political spaces)
- + Churches' participation in negotiations over political settlements and their iterations and contributions to selling the deals (the churches' public political role).

It is useful to plot this typology in order to synchronize the elements of our conceptualization (Table 190.1). This conceptual apparatus enables us to see the variety of activities religious peacebuilding compromises and the diversity of activities religious peacebuilders can engage in.

As sociologists we do not see institutions as independent actors autonomous from the rest of civil society or the state; the key to this model is placing churches within the civil society-state matrix. Civil society is an idea much in vogue in peace studies (see Brewer 2010: 44–67; van Leeuwen 2009). We emphasize four socially strategic social spaces as important to a peace process that go well beyond the now familiar distinctions in the civil society literature between "good" and "bad," or progressive and regressive, civil society (Chambers and Kopstein 2001). These strategic spaces help us understand further the differences between kinds of religious peacebuilding. These strategic spaces are: *intellectual spaces* (as places for discussion of peace, development of visions for peace, ideas for conflict resolution, new ideas for reconciliation work, envisioning the new society, etc.); *institutional spaces* (religious organizations putting peace into practice in their own activities and behaviors); *market spaces* (their employment of social, symbolic, cultural and material resources to actively support peace and peace work); and *political spaces* (their engagement with the political peace process, engagement with political groups and their armed wings, with governments, etc.).

Table 190.2 Civil society's strategic social spaces

Intellectual: Ecumenism, anti-sectarianism	Institutional: Cross community activities, peace initiatives
Market: Transitional justice work with prisoners and families, social gospel	Political: Mediation, back channel political communication, formal political representation

Source: John D. Brewer and Francis Teeney

If we combine this typology with those above, we have four strategic social spaces for the reproduction of active religious peacebuilding, represented in Table 190.2. We suggest that in *intellectual spaces* we find the churches doing many forms of ecumenical activity and anti-sectarianism/anti-racism. In *institutional spaces* we see the churches engaging in cross-community activities and involving themselves in national and local peace initiatives. In *market spaces* the churches were involved in those post-violence adjustment problems that involved expending material and cultural resources, notably transitional justice work with prisoners and their families, and other forms of faith-based social action. In *political spaces* we see the churches involved in mediation, especially in dialogue with paramilitaries, acting as back channels of communication, facilitating engagement between the various factions, including paramilitaries and governments.

When analysed in these terms, it is important to note the opportunities and constraints that operate on churches and para-church organizations in entering socially strategic spaces. The minority/majority status of the churches significantly affects the level of engagement and its forms, since majority religions tend to be established churches linked to the state or the religion of the dominant group culture. This majority status can limit their role in peace processes; a majority church can also be constrained in the critical positions it can take, and, broadly speaking, may be fearful of offending sectors of their congregations. Minority churches can be more critical, but also very vulnerable. One way in which the majority churches managed the problems of engagement, should events go wrong, was to restrict the involvement to 'unofficial' activity, although church decision-making processes and governance structures also made it very difficult to arrive quickly at an "official" position. The distinction between "official" and "unofficial" peace work thus becomes another important element to the conceptualisation, for it helps churches manage the risks of public exposure both to themselves as institutions and to their members.

We contend that in order to understand the opportunities and constraints on religious peacework, the activities of the churches and religious bodies need to be located in a three-way relationship between themselves, civil society and the state, making the *church-civil society-state matrix* the chief conceptual tool for understanding religious peacemaking. Church-state relations shape the kinds of peacemaking done by majority and minority churches, restricting the majority churches in the extent to which they challenged majority community dominance and power relations, and making certain forms of minority church activity particularly vulnerable, whether these threats were real or imagined. The constraints imposed on majority and minority churches by church-state relations can be managed by different forms of official and unofficial intervention, allowing majority church peacemakers some autonomy when acting secretly and facilitating minority religious peacemakers, some of whom were in a double minority position and whose capacity for engagement required creativity in sidestepping official constraints. Churches mostly move officially quite late to develop policies for engagement, which is why unofficial forms of religious peacemaking can dominate as path-breaking activities during the worst of the violence.

190.5   Religious Peacebuilding in Northern Ireland

Not only does this matrix proffer methodological and conceptual leaps by facilitating cross-national comparisons, but it also provides the intellectual apparatus to understand the nature and forms of religious peacemaking in Northern Ireland. It moves us beyond the emphasis in the case study literature on personality, individual motivations and religious leadership in Northern Ireland, important as these are, by locating the churches' activities within the context of their wider relationship to global civil society and the British and Irish states.

Applying this model to Northern Ireland, it seems possible to delineate specifically religious factors when explaining the course of religious peacemaking there, as with the impact of church governance structures and forms of church leadership, theological relations and disputes that opened or closed opportunities for collaboration and networking, and the effect of different institutional forms, such as the organisational and bureaucratic differences between established and non-established churches, religious orders, para-church organisations, religious lobby groups, and others that mediated the capacity for religious peacemaking.

We contend, however, that these narrowly religious factors need to be located in a broader relationship between church, civil society and the state. After all, governance structures, organizational forms and styles of religious leadership, for example, can be managed and manipulated if the commitment to peace engagement is present, especially persuading some minorities to engage in imaginative types of mediation to circumvent the constraints. Religious peacemakers in the North of Ireland found ways around the restrictions imposed by insecure or frightened leaders or the restraints in established churches in moving synods or presbyteries toward a critical position against the state.

It remains the case, however, that as institutions, the churches contributed less than the individual members they comprise. This is why some of the most effective religious peace work was done in non-denominational organizations, such as the ecumenical communities, or in organizational settings outside the control of conservative church hierarchies, such as in monasteries, or it was done in secret. We can explain this kind of religious peacemaking best by locating it in the church-civil society-state relationship.

This conceptual apparatus is, we contend, a particularly enlightening way of explaining the paradoxical dynamics of religious peacebuilding in Northern Ireland. Our intellectual apparatus helps explain, for example, the individualization of religious peacemaking in Northern Ireland as the institutional church withdrew from prophetic leadership. Individualization is both the cause and effect of the churches' weaknesses and describes the *process* by which religious peacebuilding was conducted (done mostly by independents and mavericks rather than the institutional church) and its *outcome* (a focus on improving individual relations between Catholics and Protestants within a framework of "politics as usual" rather than dismantling collectivized religion). Individualization is the necessary consequence of a weak institutional church containing within it several highly committed mavericks and independents who sought to circumvent conservative and cautious church hierarchies.

Individualization, however, constitutes a problem. "The Troubles" were not located by these otherwise well-meaning and highly committed mavericks in the continued capacity of the churches to reproduce themselves as collective religions and in the social structural conditions that sustained two mutually exclusive ethnoreligious blocs. Symptom and cause were confused. The problem was perceived to be violence itself rather than a sectarian social structure of which the churches were themselves an integral part, so negative peace became the solution rather than positive peace, conflict transformation the emphasis rather than social transformation. This stance predicated even the efforts of mavericks and independents; it was about normalizing relations between people rather than attacking the system that distorted them in the first place. This sort of work was recognized by them (and others) as inherently political, pushing the institutional churches and the mavericks into political spaces where they felt uncomfortable (in varying degrees), but the aim was to demilitarize politics rather than change the society that created the conditions for abnormal politics.

This change occurred at the same time when religious peacemakers, of course, displayed very high levels of personal motivation to peacemaking (and often great bravery). The value of our conceptual approach, however, is precisely that it moves debate beyond personal motivation and moral commitments to peace work (which is the problem with the case studies of individual religious peacemakers) (see Little 2007) in order to focus on wider opportunities and constraints. It is unquestionable that some people in the churches lacked motivation while others had it aplenty, but personal motivation interacts with opportunity and constraint in such a way so as to disclose that despite the high levels of motivation in some religious peacemakers from the very beginning, the churches' peacemaking activities did not prove effective until the conditions were ready for it in the mid-1990s. And they *were* effective – when the time came.

For a quarter-of-a-century, faith-based peace activism in Northern Ireland was dominated by secret engagement between church figures and political and paramilitary leaders on the one hand and also by the question of improving relations between Catholics and Protestants on the other. The intellectual reflection exemplified by (a) the Irish School of Ecumenics' Moving Beyond Sectarianism project, (b) the encouragement of Protestants to rethink their identity, (c) the development of meaningful positive political relationships through individuals like Rev Ken Newell, Fr Gerry Reynolds, Rev John Dunlop and Rev Harold Good, (d) the activities of groups like Evangelical Contribution on Northern Ireland, and (e) the continued witness of inter-church communities, in the form of Corrymeela and others, stand as icons of what Christians have tried to do to challenge the terms and ameliorate the consequences, of the vicious conflict. To those atheists like Richard Dawkins who claim that religion inevitably kills, we agree that without religion there would have been no conflict in Ireland, but it is also certainly the case that without religion there would have been no ceasefires and thus no negative peace. The motivations of the religious peacebuilders never diminished over this period. What varied were the conditions that shaped their opportunities and constraints.

We can distinguish between internal and external opportunities and constraints. Among the external conditions, premier must be the spaces that were opened up for the churches as a result of developments in the political peace process that provided political opportunities, such as transitions towards a political strategy within the paramilitary organizations, the formation of a single military command in Loyalism with which to negotiate, the active interest of the Irish government in working with the British government in delivering their respective client groups, the good personal relations between John Major (British prime minister) and Albert Reynolds (Irish prime minister), as well as the involvement of other international third parties, especially President Bill Clinton. The deterioration in the level of violence that occurred in the lead up to the Hume-Adams talks and the Downing Street Declaration seemed so bad as to counteract the delaying effects of the 'latest atrocity syndrome' and to reflect the truism that violence in the North had to get worse before it got better. The back channel dialogue that various religious peacemakers had established over the years with Loyalist and Republican paramilitaries, which were so suited to sacred spaces as places of secrecy, confidentiality and anonymity, were able to be mobilized later to deliver support for the political peace process and ceasefires. The key religious figures included Fr Alec Reid, Rev Roy Magee and Archbishop Eames who orchestrated the maneuvers in combination with the respective governments.

The churches' long-standing contributions to the social peace process were not irrelevant to this. The extensive development of ecumenist contacts between clergy, congregations and denominations, the involvement of neighbourhood clergy in instances of local conflict mediation and dialogue, and the churches' participation in public peace initiatives and secular cross-community activities, which comprised the main activities by which societal healing and relationship building was attempted in the social peace process. This cooperation continued throughout "the Troubles," but there is no real evidence that they were effective in their own terms. However, the relationship between the social and political peace processes is recursive and spaces were opened up for the social peace process only by advances in the political peace process. Progress in political negotiations was facilitated by the social peace process, such as when church dialogue with protagonists, as a form of conflict mediation, developed later into back channel political communications that assisted the political peace process. Religious peacemakers, however, could not proactively initiate these back channel communications, perhaps with the exception of the Hume-Adams talks instigated by Fr Alec Reid. They had to wait until the external conditions made the paramilitaries and the governments *want* to utilize sacred spaces for the purpose. This requirement was outside the churches' control.

190.6   Systems of Clerical Political Engagement

There is a need for further research that places Northern Irish churches in these wider national and international political developments that explore how churches were manipulated by political actors for their own ends. Our interest, however, has been with the opportunities and constraints internal to the churches. These include the following: the institutional and leadership barriers to engagement with peace; the church systems of authority that constrained mavericks and squeezed the institutional spaces for prophetic leadership; the resort to unofficial interventions that made practitioners vulnerable, insecure and open to marginalisation by their own church leaders; the tendency toward both denominationalism within the churches and separatism from other civil society groups that prevented an umbrella alliance; their ambivalent attitude toward anti-sectarianism as a result of the realization that "politics as usual" formed part of their ethno-religious boundaries that sustained them as churches; and the dominance of "clergy manager" role expectations that limited the ambitions as well as the time of individual clergy.

It was easier for some religious peacemakers to evade these internal constraints and to exploit opportunities, notably non-parish clergy, those in religious orders, the independent para-church organisations, ecumenical communities, and the mavericks impervious to any censorship and control from church leaders. Clergy outside these categories desirous of involvement in the peace process chose highly ambivalent institutional locations outside the purview of church leaders. Only those with the tacit knowledge of the hierarchy, such as Methodists, could claim any authority for their peace work, further individualizing the process of religious peacebuilding in Northern Ireland. For the rest, it put them on the margins of developments, limited their availability for meetings, and encouraged them to voyeuristic involvement in the peace process, keen and enthusiastic, to be sure, but from a distance, urging on the mavericks well from behind. These sorts of church people were not against peace, but their institutional position made them vulnerable and afraid. The ambivalence of these institutional spaces, therefore, either pushed them towards a "fellow traveller" role or, as a way of supplying a source of social and moral support, garnered an attitude amongst them of "ourselves alone," fostering religious denominationalism and separatism from the rest of civil society.

190.7   Conclusion

It is clear that our model of religious peacebuilding has transgressed the theoretical/policy dichotomy, being both oriented to practice and theory. We suggest this model opens an interesting direction in the future for advocates of religious peacebuilding, for we need to better understand its nature so it can be better practiced. We also have tried to show that its practice in Northern Ireland helped pacify a violent conflict and turned a situation of religious intolerance into now one at best of religious indifference. This process is not the same as societal healing and reconciliation. Much work still needs to be accomplished in Northern Ireland for religious intolerance to transform into social reconciliation. We would like to close by drawing attention to the failure of religious peacebuilders to tackle the problems of post-violence in Northern Ireland.

The institutional church failed Northern Ireland during "the Troubles." Religious peacebuilding was individualized to lone peacemakers, independents and mavericks. And in the post-conflict stage, the peace process, as far as the institutional church is concerned, has come to an end. Taking no responsibility for the past, or their contribution to sectarianism, the institutional church does not accept it has any responsibility to the future, save ministering to the pastoral needs of their congregations on a piecemeal basis. Hesitant and uncomfortable in displaying prophetic leadership in the public sphere during the violence, the institutional church is at a loss to know what to do publicly after it. It is thus left again to individual religious peacebuilders to address the legacy of violence, equally piecemeal.

Victim groups feel neglected by the church as victimhood is reduced to a pastoral issue to be handled on a case by case basis inside private church space not in the public square. There is no public religious discourse on forgiveness, hope and compassion; still less on resentment and anger. Forgiveness as a process is often feared by liberal human rights activists and victims alike because it is assumed to mean amnesty, although it need not. But debates about what forgiveness means politically, and whether or not it first requires repentance, on which churches *ought* to take the lead after conflict, are not entering the public arena. And hope is not a word in the lexicon of the churches – at least not this-worldly hope. As part of the same neglect, the churches are silent on transitional justice issues. There is no religious discourse in Northern Ireland on human rights issues, on truth-recovery, or on other transitional justice themes like reparation, memory, restorative reintegration of ex-combatant prisoners and the like. Religious peacebuilders placed a very high priority on working with prisoners and their families, but not when released nor once they gave their imprimatur to the deal. Of course, one can cite a few examples where this is not so, of, for example, brave churchmen and women active in social witness, managing the risk of renewed outbreaks of violence, or dealing with the management of memory. But these are independent of the institutional church, done freelance by individual religious peacemakers (in some cases done after they retired from active ministry) and undertaken in conjunction with other civil society representatives in a way that they carry no strong religious stamp. There is no authoritative religious voice in public debate on post-violence issues. This is partly a skills issue, but also primarily a motivational one. Polarization has been left intact by the peace agreement and in this quiescent phase without overt violence to stir them, there thus is no motivation for the churches to get involved. The churches are uninterested in post-violence reconstruction.

By way of conclusion, therefore, we need to ask the key question. Are the churches capable of doing anything anymore, even if they wanted to? Their condition post-agreement is worse than before. Secularization and anti-clericalism diminishes their influence and respect. They lack moral legitimacy for having missed opportunities for prophetic leadership during "the Troubles." Individual peacemakers are aging, retiring from active ministry, or burning out, ill and moving out of Northern Ireland. Religious peacebuilding is in crisis as the individuals who bore its brunt depart and as the institutional church evades its responsibilities in the public square.

References

Barkley, J. M. (1967). *The antichrist*. Belfast: Presbyterian College.
Barnes, L. P. (2005). Was the Northern Irish conflict religious? *Journal of Contemporary Religion, 20*, 55–69.
Berger, P. (1967). *The sacred canopy: Elements of a sociological theory of religion*. Garden City: Doubleday.
Brewer, J. D. (2003a). Contesting Ulster. In R. Robin & B. Strath (Eds.), *Homelands* (pp. 283–304). Brussels: Peter Lang.
Brewer, J. D. (2003b). Are there any Christians in Northern Ireland? In A. M. Gray, K. Lloyd, P. Devine, G. Robinson, & D. Heenan (Eds.), *Social attitudes in Northern Ireland: The eighth report* (pp. 22–38). London: Pluto Press.
Brewer, J. D. (2010). *Peace processes: A sociological approach*. Cambridge: Polity Press.
Brewer, J. D., & Higgins, G. (1998). *Anti-Catholicism in Northern Ireland 1600–1998*. Basingstoke: Macmillan.
Brewer, J. D., Higgins, G., & Teeney, F. (2010). Religious peacemaking: A conceptualisation. *Sociology, 44*, 1019–1037.
Brewer, J. D., Higgins, G., & Teeney, F. (2011). *Religion, civil society and peace in Northern Ireland*. Oxford: Oxford University Press.
Bruce, S. (2011). *Secularisation*. Oxford: Oxford University Press.
Chambers, S., & Kopstein, J. (2001). Bad civil society. *Political Theory, 29*, 837–865.
Coward, H., & Smith, G. (2004). *Religion and peacebuilding*. New York: SUNY Press.
Creamer, C., Driscoll, J., Blair, N., & Bartley, B. (2011). Engineering healing and the Northern Ireland question. In S. D. Brunn (Ed.), *Engineering earth* (pp. 2089–2111). Dordrecht/New York: Springer.
Edwards, A., & McGrattan, C. (2010). *The Northern Ireland conflict: A beginner's guide*. London: Oneworld Publications.
Galtung, J. (1969). Violence, peace and peace research. *Journal of Peace Research, 6*, 167–196.
Hadley, M. (2001). *The spiritual roots of restorative justice*. Albany: State University of New York Press.

Higgins, G. (2000). *Great expectations: The myth of antichrist in Northern Ireland.* Unpublished PhD thesis, Queen's University of Belfast, Belfast.

Johnston, D. (2003). *Faith-based diplomacy.* Oxford: Oxford University Press.

Little, D. (2007). *Peacemakers in action.* Cambridge: Cambridge University Press.

McGarry, J., & O'Leary, B. (1995). *Explaining Northern Ireland.* Oxford: Blackwell.

Mitchell, C. (2006a). *Religion, identity and politics in Northern Ireland.* Aldershot: Ashgate.

Mitchell, C. (2006b). The religious content of ethnic identities. *Sociology, 40,* 1135–1152.

Mulholland, M. (2003). *Northern Ireland: A very short introduction.* Oxford: Oxford University Press.

Norris, P., & Inglehart, R. (2004). *Sacred and secular: Religion and politics worldwide.* Cambridge: Cambridge University Press.

Rafferty, O. P. (1994). *Catholicism in Ulster 1603–1983: An interpretative history.* Dublin: Gill and Macmillan.

Schlack, A. (2009). *The role of religion in peacebuilding and conflict transformation.* Saarbrucken: VDM Verlag.

Shore, M. (2009). *Religion and conflict resolution.* Farnham: Ashgate.

Smock, D. R. (2001). *Faith based NGOs and international peacebuilding: Special report.* Washington, DC: United States Institute of Peace.

Smock, D. R. (2002). *Interfaith dialogue and peacebuilding.* Washington, DC: United States Institute of Peace.

Smock, D. R. (2006). *Religious contributions to peacemaking.* Washington, DC: United States Institute for Peace.

Smock, D. R. (2008). *Religion in world affairs: Its role in conflict and peace.* Washington, DC: United States Institute for Peace.

Van Leeuwen, M. (2009). *Partners in peace: Discourses and practices of civil-society peacebuilding.* Farnham: Ashgate.

Walsham, A. (2006). *Charitable hatred: Tolerance and intolerance in England 1500–1700.* Manchester: Manchester University Press.

Chapter 191
Religion, Space and Peace in Sri Lanka: Transforming Spaces of Freedom Threatened by Violence into Islands of Civility

Shirley Lal Wijesinghe

191.1 Introduction: "Islands of Civility"

Peace has a spatial dimension because it is impacted by the political geography of the conflict. In some societies conflict is spatially restricted, leaving large parts free of violence. This can often have the effect of lessening the effort put into peacemaking if the violence is perceived to be somewhere else and thus someone else's problem. However, where there is a strong sense of territorial integrity and cultural integration, the conflict, even though geographically restricted, is never experienced as remote. In these cases, the non-violent zones become sites of peacemaking, where peace work is conducted in relative safety and security, even though intended for implementation elsewhere. There are many societies, however, where the conflict has no political geography, for it is universal and endemic, capable of breaking out anywhere, even everywhere. But even here, peace has a spatial dimension, for as Mary Kaldor (1999: 55*ff*) has pointed out, there are always "islands of civility" within the midst of the worst kinds of wars. They are often vulnerable, subject to attack, and always fragile, affected by the volatility of the violence and the narrowing space for peacemaking that war brings. Indeed, they can become sites of deliberate attack by parties who seek to keep the conflict going precisely because they proffer an alternative to war. They can be enduring, of course, and effective in the long term despite their short term nature, because they can symbolically come to represent what Brewer (2010: 138*ff*) calls "moments of transcendence," in which looking back the brief expression of civility made a difference, sometimes all the difference.

Islands of civility exist within all conflicts and their spatial dimensions are interesting. On occasions they can represent physical places, such as neighborhoods, buildings or sites where peacemaking is conducted or where examples of

S.L. Wijesinghe (✉)
Faculty of Humanities, University of Kelaniya, Kelaniya, Sri Lanka
e-mail: slwijesinghe@kln.ac.lk

© Springer Science+Business Media Dordrecht 2015
S.D. Brunn (ed.), *The Changing World Religion Map*,
DOI 10.1007/978-94-017-9376-6_191

co-operative, peaceful living survived; even camps where those fleeing the violence find refuge and security can become islands of civility, especially where they are protected by third parties, such as United Nations peacekeepers. Kaldor (1999: 110–11) cites many examples from the wars in the Balkans following the deconstruction of the former Yugoslavia. Brewer et al. (2011) use the example of Clonard Monastery to illustrate its role as a special place where religious peacebuilding was conceived and developed in Northern Ireland during "the Troubles," as that society's conflict is colloquially known.

Islands of civility can also be cultural spaces that transcend any particular physical space. Some attention has been devoted to the characteristics of religious or sacred spaces that make them suitable as islands of civility. For example, religious sites are often associated culturally with the idea of sanctuary, so that churches and temples become physical places where refuge is sought. Of course, where the conflict is religious or is between groups whose boundaries are marked by religion, religious sites offer no protection at all, as the examples of church burnings in Rwanda and Colombia illustrate. On the other hand, Brewer et al. (2011) have shown that despite the religious hue to conflict in Northern Ireland, "sacred spaces" as specific cultural sites have particular connotations suitable to peacemaking (2011: 102–7). Sacred spaces functioned effectively as back channels of communication between warring factions in Northern Ireland who could not be seen to be dialoguing in public because sacred space is associated with confidentiality, secrecy, trust and redemption. People were encouraged to enter into dialogue because of these connotations and persuaded to stick to it despite the travails whenever the dialogue proved difficult because of the very nature of the space in which it was taking place.

Islands of civility can also be constructed by people whose special legitimacy, noteworthiness or public acclaim accords them the personal charisma to attempt a "moment of transcendence." Politicians, celebrities, religious figures and the like can help construct spaces for peace where they have cultural legitimacy. More unusual is the case of converts to peace from amongst the former militant groups who denounce the violence and who stake a claim to peace, who do so often at great risk to themselves from their erstwhile militant colleagues who target them precisely to close the space for peacemaking. Nelson Mandela comes to mind as the quintessential example. More unusual still is the case of victims, who despite their injury, suffering and loss, speak out against continued violence, urging against revenge on their behalf and even, in some instances, calling for forgiveness towards the perpetrators and an end to further loss. The case of Gordon Wilson comes to mind in Northern Ireland (see Brewer 2010: 138), whose daughter was killed in a bomb by the Irish Republican Army but before the day was over appeared on television saying he forgave her killers, establishing for himself considerable public affection and acclaim. Atrocities, in other words, can, ironically, create spaces of peace as a reaction. But moments of transcendence like this are just that, momentary (see Brewer 2010: 202), and require considerable levels of personal charisma by the people who create them if they are to be sustained, making the island of civility short lived. Nonetheless, particular forms of cultural legitimacy, such as that held by religious figures and celebrities, can turn some people into peacemakers who make a difference.

It is also the case that islands of civility can be constructed around senses of shared community that survive the conflict. This sense of community can be based around shared interests, shared memories, shared occupational or leisure interests and the like, which give people across a divide a sense of togetherness and mutual identity that transcends the cleavage around which conflict adheres. Again these can be fragile, subject to collapse internally as the groups involved polarize as a result of the conflict itself, or to external attack as militants seek to break the bonds of community in order to sustain conflict. But these islands of civility can often be picked up again after the war is over to achieve a more rapid form of societal healing because the reconciliation is based on these earlier shared social networks. Significantly, of course, where these communal bonds do not exist, but there are still shared occupational experiences, or common class or material interests, work, land, or class position can be utilized as a resource to try to suggest that the people concerned ought to have a common communal bond. Religious peacemakers can be amongst the many different types of person involved in constructing this sense of community through highlighting people's shared experiences, especially in societies where religion retains its cultural legitimacy and importance in people's material lives. That communal conflict occurs mostly in poor countries where religion plays a significant role in distributing resources and constituting an alternative source of welfare only enhances the potential for religious actors to exploit poverty as a shared communal experience to construct an island of civility.

In practice, of course, islands of civility within specific conflicts have features of all these forms of physical and cultural space, and people of widely different sorts can utilize them for the purposes of peacemaking. The case discussed here from Sri Lanka illustrates the mix of religious and secular spaces and personnel in the creation of one particular island of civility, the pressures exerting on that space to make it fragile, and the potential for hope in the future through its resurrection. Caught between the Sri Lankan Security Forces and the militant group Liberation Tigers of Tamil Eelam (LTTE), the farmers of a border region in Sri Lanka suffered immensely without being able to cultivate their paddy lands. Many were shot and entire villages were victimized. With the help of an Anglican priest, a group of brave villagers were able to bring together the Sinhalese and Tamil farmers to cultivate their lands carving out an area out of the fists of the security forces and the LTTE through negotiations. The farmers of the ethnic divide worked together and cultivated their abandoned lands in spite of frequent threats by the armed parties of the conflict trying to reclaim dominance over the area. The carved out agricultural lands by the farmers carried the political adage "Peace to Paddy Fields." The case illustrates that conflict resolution calls for independent spaces free of the exercise of power by conflicting armed groups. Though fragmentary, these spaces are "landscapes of peace," or "islands of civility." This chapter explores the possibility of such landscapes of fragile peace being transformed into political landscapes of sustainable peace. Before we move to this however, it is necessary to introduce the Sri Lankan conflict.

191.2   The Sri Lankan Conflict

The history of the ethnic conflict in Sri Lanka goes back to British colonization. The minorities, Tamils and Christians, benefitted from missionary education which occasioned in greater possibilities for white color employment. This does not mean that all Tamils or all Christians had this privilege. For example, most of the Christians living in the western coastal belt and most of the Tamils living in Vanni area did not have access to English missionary education and as a result did not have access to privileged employment. The post-independence era offered greater access to government posts to the Sinhala Buddhist majority. This was due to "Sinhala Only" policy – Sinhala being declared as the official language – and restricted university entrance on district basis. The post-independence farmer colonization schemes were also considered to be detrimental to the Tamil population. The failure of the non-violent protests of the Tamils led to a deadly armed struggle and as a result thousands have been killed and many more mutilated in the three-decade civil war (Loganathan 1996: 1–117).

One of the major constraints for peace was the strong positions of the military and militants leaving little or no room for grassroots peace initiatives. Nevertheless, there were such initiatives, though often short lived, yet harnessing energies for positive peace. Vanni Peace Foundation (Vavuniya), Scouts of Vavuniya (Vavuniya), St. Anthony's Church (Vavuniya), Assembly of God (Vavuniya), Centre for Society and Religion (Colombo), Peace Committee of Nugelanda (Ampara), Peace Committee of Batticaloa (Batticaloa), Butterfly Garden (Batticaloa), *Sarasavi Diyaniyo* (an association of Tamil and Sinhalese university women), *Rukada* (a travelling puppet show for peace), Sunila Women's Development Movement (a group aiding women and children traumatized by war), *Ginikandulu* (a photo exhibition on peace), *Bingusara* (a street drama group), *Sadujanarava* (music and poster campaigns to raise awareness of the suffering caused by the war – a mobile team travelling from town to town), and *Kithusara* (a Roman Catholic group of Sinhalese youth dedicated to peacebuilding) are some of the groups engaged in significant grassroots initiatives for peace at the height of the violent struggle (Wijesinghe 2003: 174–176). Had these initiatives flourished and made a claim, the death, desertion and mutilation of thousands of Sri Lankans, especially youth, would have been spared.

Among these groups, the Peace Committee of Nugelanda is significant as it was able to claim a land of about 1,200 acres (4.86 ha) for cultivation purposes from the warring parties. This is a telling example of the importance of landscapes of peace in a violent struggle to achieve sustainable peace (Wijesinghe 2003: 178–180).

191.3   "Peace to Paddy Lands"

Sri Lanka is supposed to be the only country which departed from a river valley civilization to a "wewa" civilization. There is no English equivalent to "wewa." It is usually translated as "reservoir." The ancient Sri Lankans who could not depend on

Fig. 191.1 A reservoir in the vicinity provides water for paddy cultivation (Photo by S. L. Wijesinghe and Deepthi Silva)

river water, constructed the "wewa system." It is a complex system consisting of a forest, catchment area and even mud sluice gates and dates back to pre-Buddhist era. The "wewa" system was neglected during the colonial era, and post-independent Sri Lankan rulers began to revert back to river water for irrigation. Gal Oya scheme is one such project. Colonies were created in the Gal Oya irrigation scheme in Sri Lanka in 1950s (Fig. 191.1). These agricultural colonies were peopled by both Sinhalese and Tamil families. (One of the crucial issues of the conflict was the question of settling Sinhalese families in these colonies). Tamil colonies were on the western boundary of Batticaloa district and the Sinhalese colonies were situated on the northern and eastern boundaries of Amprara district (Fig. 191.2). The drama of peace takes place in a small village called Nugelanda, situated at a distance of about 500 m (1,640 ft) to the west of Maha Oya – Ampara trunk route close to Piyangala (Fig. 191.3).

The good relations among the Sinhalese and Tamil farmers begun in 1950s grew into mutual support and cooperation over the decades. The signs of trust that existed between the two communities were seen by Tamils deciding to study in Sinhalese schools with Sinhalese students and farmers of both communities collaborating among themselves.

The failure of the negotiations between the LTTE and the Sri Lankan government marked the disastrous massacre of over 900 police personnel of Batticaloa by the LTTE. The resultant offensive by the government against the LTTE in 1990 was a sweep up of militants. In this exercise, the military marched through the Tamil villages of the colonies. This decimated the long lived peace between the Sinhalese and the Tamils in the colonies. The involvement of the Sinhalese villagers in acts of looting the Tamil villagers on the run when the government forces were on the

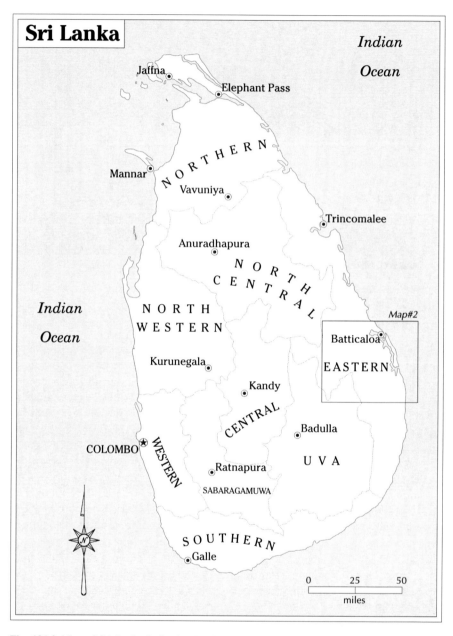

Fig. 191.2 Map of Sri Lanka indicating Batticaloa and Ampara areas (Map by Dick Gilbreath, University of Kentucky Gyula Pauer Center for Cartography and GIS; commissioned by the editor)

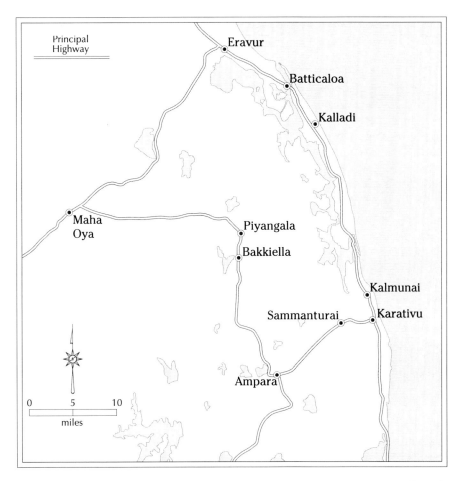

Fig. 191.3 Map indicating the road to Nugelanda from Maha Oya (Map by Dick Gilbreath, University of Kentucky Gyula Pauer Center for Cartography and GIS; commissioned by the editor)

offensive, brought an irreparable rift between the two communities. In the end the Tamil farmer families were confined to refugee camps.

These events brought disaster to Sinhalese families as well. They were unable to go to their paddy fields as they became targets of militant gunfire. The only solution available was to plough and work their fields under the security of the government forces. In spite of the assistance of the security forces providing cover during farming activities, some farmers were gunned down by the militants.

The situation of despair in the Sinhala villages prompted them to approach Rev. Nirmal Mendis the pastor of the Nugelanda Anglican Church asking him to assist them in their plight. This was in 1997 and at this juncture, Rev. Mendis commended that Sinhalese farmers identify themselves with the sufferings of their long time neighbors and collaborators, namely the Tamil farmers – now living in destitution in refugee camps – and workout a program which will allow both Sinhalese and Tamil farmers to cultivate their fields.

It took a while to change the mindset of the Sinhala villagers and the Buddhist religious leadership of the area to plan a meeting with the Tamil counterparts. The delay was due to instigation by some who thought the only feasible alternative to be the military protection. Suddenly everything changed with a brutal attack on the Sinhalese villagers by the LTTE in September 1999. (This has been interpreted as a retaliatory attack following air strikes on Tamil civilians in Mullaittivu district.) Many were killed and mutilated in this attack after which some fled the village and others spent nights in the church or the temples around. The incident created the urgency to search for a sustainable solution.

After a laborious preparation, a representation of the farmers of the Sinhalese villages and the Tamil villages (40 Sinhalese and 17 Tamils) and religious leadership including two leading Buddhist monks met at Wesley College in Kalmunai. Consultations with the religious, civil, military and rebel leadership included the hard part of preparations. At this meeting, convened by Rev. Mendis it was decided that both Sinhalese and Tamil farmers be allowed to cultivate their fields and their protection be assured. This involved the task of negotiating with the LTTE and the government security forces and the responsibility was vested upon the religious leadership present at the meeting. The representation also decided to call a second meeting with a larger representation.

In the interim Rev. Mendis planned a goodwill visit with the Sinhalese farmers to meet their Tamil counterparts who were living in dire conditions. They took presents and provisions as a gesture of reconciliation and met the Tamil farmers who were living in refugee camps. Furthermore, a greater effort was made in the interim to assure a larger representation at the second meeting. The second meeting took place at the Carmel Fatima National School at Kalmunai with a larger representation of the farmers and the clergy including the Bishop of Batticaloa. At this meeting the farmers established a Peace Committee and the Farmer Society. At a third meeting held in January 2000, attended by 75 Tamil and Sinhalese farmers, it was decided to meet at the disputed territory on the 11 February that year.

After continuous negotiations with the civil authorities, security forces, militia and the villagers themselves, the "Victory Day" dawned when about 1,000 Sinhala and Tamil farmers came together and established a "Peace Zone." The flags carrying the appellation "Peace to Paddy Fields" (Fig. 191.4) were positioned on the borders of the peace zone. The peace zone was a stretch of about 1,200 acre paddy (4.56 ha).

Having regained their lands, both Sinhalese and Tamil farmers cultivated them in peace, till one day a tractor was taken away by force by the LTTE militants. The fears began to grow and the lands were reclaimed by the warring parties leaving the farmers languishing once again. Cultivations continued with renewed fears, and peace was lost. It has been recorded that no farmer was killed or injured once the peace zone was established (though some security personnel were killed by militant attacks).

Fig. 191.4 Peace flag with "Peace to the Paddy Fields" written in Sinhala and Tamil. These flags were hoisted in the "peace zone" (Photo by S. L. Wijesinghe and Deepthi Silva)

191.4   "Space of Freedom"

This is an example of a "Space of Freedom" in the context of the deadly civil war in Sri Lanka. Unfortunately it was short lived only to the detriment of the poor farmer villagers. A space of freedom is a geographical entity carved out of the fists of warring parties as a safe haven for people threatened by violence. It is a free space, out of bounds of warring factions, where people could engage in their day-to-day activities freed from the fear of violent attacks. Such safe havens are assured by high levels of reconciliation realized within. The sustainability of such spaces demands the difficult task of negotiations with the extremists within, the personnel of the civil administration without, the support of the religious leadership – in situations where they are highly influential to do good as well as to instigate evil – and the deadly task of compromising with the military and the militia. These spaces are fragile as they depend on many factors for sustainability. They are vulnerable as warring parties could claim the territories at any given moment.

The multiplication of such efforts across the territory once called the no man's land which was basically arable land would have made a strong impact on peace-making in Sri Lanka. Unfortunately it was not to be and the conflict ended in a massacre of thousands, military, militants and civilians, both Sinhala and Tamil. This narrative is perhaps one example among thousands of such efforts in establishing free spaces of peace in countries affected by civil strife. Would there be a possibility signing a UN convention in the near future to safeguard such free spaces giving them a legal, political stature in contexts of conflicts for the purpose of safeguarding life and property?

191.5 Conclusion

Sri Lanka now has a victor's peace with the final military defeat of the LTTE, but there is considerable polarization of Tamils and Sinhalese. The extent of the violence squeezed the space for peacemaking and destroyed many of the islands of civility that existed, as the case here illustrates. There is a risk in Sri Lanka of what Brewer (2010: 22*ff*) calls cultural annihilation. Rather than introducing a shared culture which is a happy "moment" of reconciliation, the Tamil culture could be destroyed. It looks that there is no strategic interest by the West in Sri Lanka and there is also no visible third party intervention to persuade the authorities to expedite the work of sustainable peace; a few interested groups are keeping the conflict alive as a public issue. The future thus depends on the persuasion that true peace and security lies in recognizing Tamil culture as an important part of Sri Lankan society and respecting Tamil minority rights. For this to occur, the society needs to build on the experience and memory of shared moments of civility and commonality such as described here, and restore the co-operation that the violence destroyed. The spaces of freedom that the violence threatened need to be turned into spaces where genuine political freedom exists. The real job of peacemaking in Sri Lanka has only just begun.

Acknowledgement I wish to record a special word of appreciation and gratitude to Prof. John Brewer for encouraging me to bring to light this rather hidden heroic attempt narrated here and for the expertise offered in formulating the theoretical framework of the book chapter.

I am also grateful to Mr. Deepthi Silva, the research assistant who ventured to gather information on the Nugelanda saga in trying circumstances.

References

Brewer, J. (2010). *Peace processes: A sociological approach*. Cambridge: Polity Press.
Brewer, J., Higgins, I., & Teeney, F. (2011). *Religion, civil society and peace in Northern Ireland*. Oxford: Oxford University Press.
Kaldor, M. (1999). *New and old wars*. Cambridge: Polity Press.
Loganathan, K. (1996). *Sri Lanka. Lost opportunities. Past attempts at resolving ethnic conflict*. Colombo: Centre for Policy Research and Analysis.
Wijesinghe, S. L. (2003). Sri Lanka. Prophetic initiatives amidst deadly conflict. In M. A. Cejka & T. Bamat (Eds.), *Artisans of peace. Grassroots peacemaking among Christian communities* (pp. 166–195). Maryknoll: Orbis Books.

Chapter 192
Religion and the Social Reconstruction of Memory Amid Violence in Bojayá, Chocó (Colombia): Creating Transitional Justice from Below

Sandra Milena Rios Oyola

> *They have healed the wound of my people lightly, saying,*
> *'Peace, peace,' when there is no peace (Jeremiah 6: 14).*

192.1 Introduction

The confrontations between the multiple armed actors in Colombia have created a scenario of continuous threat against the human rights of civilians. The leftist guerrillas and the extreme right paramilitaries are not only armed actors fighting for a cause, but "they are extremely sophisticated and powerful mafia-like organizations, largely motivated by profit" (Human Rights Watch 2005: 3). In addition, the official armed forces have been involved in actions against civilians as a result of a strong policy of *body counting* and their integration into the paramilitary project.[1] The characteristics of the conflict vary according to each region; however, at a national level they have attempted to reach decisions in the direction of a negotiated solution to the conflict through the enactment of several failed attempts of peace negotiations in the last thirty years. The warlords' spoilers of peace, the continuous American funding of strong military "solutions" and the persistence of the socioeconomic conditions of inequality and poverty, have made of Colombia a country in ongoing conflict. At the same time, there are attempts to create paths for peace.

In 2005, the Law of Justice and Peace that made possible the reintegration of over 30,000 paramilitaries was followed by poor and controversial measures for the

---

[1] Paramilitaries have been "so fully integrated into the army's battle strategy, coordinated with its soldiers in the field, and linked to government units via intelligence, supplies, radios, weapons, cash, and common purpose that they effectively constitute a sixth division of the army" (Human Rights Watch 2001: 1).

S.M. Rios Oyola (✉)
Department of Sociology and the Compromise after Conflict Research Programme,
University of Aberdeen, Aberdeen AB24 3QY, Scotland, UK
e-mail: smroyola@abdn.ac.uk

© Springer Science+Business Media Dordrecht 2015
S.D. Brunn (ed.), *The Changing World Religion Map*,
DOI 10.1007/978-94-017-9376-6_192

protection and recognition of victims. There was a risk that the law would assure impunity towards the perpetrators and their alliances, which included influential political actors. As a response to this fear, civil society mobilized and demanded stronger policies of truth, justice and reparation. In this context, the *National Commission of Reparation and Reconciliation* created the autonomous group called *Historical Memory* (GMH). This group was in charge of finding out the historical circumstances that allowed the creation of the paramilitary project in Colombia. The group did not have the resources and mandates of a Truth Commission. Even before the official initiatives took place for the social reconstruction of memory as the GMH started in Colombia, there were multiple local grassroots efforts related to the social memory of past atrocities.[2] Many of those initiatives were led by the Catholic Church. To some extent, the GMH has integrated the local initiatives in its reports. Many of these reports focus on the responsibility of the state, even though there has been little official recognition of such responsibility. Consequently, the outcomes of the reconstruction of the historical memory of the conflict and past atrocities are in a gray area of truth, justice and reparation.

Colombian victims consider that their needs for truth, justice and reparation have not been met by the government despite the enactment of transitional justice laws such as the already mentioned Law of Justice and Peace (2005) and the Law of Victims and Restitution of Land (2010).[3] In absence of measures of transitional justice "from above" to respond to these needs, there have been important initiatives that address those needs from the grassroots. Again, many of those initiatives have been led by the Catholic Church particularly in areas of Colombia where the state does not have a strong presence and the Church has been a prevalent influential actor, such as in the Pacific coast of Colombia.

This chapter presents an analysis of the work of the Catholic Church and its initiatives for *transitional justice from below* (McEvoy and McGregor 2010; Nyamu-Musembi 2002) through the construction of social memory. It is based on a case study in Bojayá conducted in 2012 through fieldwork and documentary review.[4] Since the research has been based on a case study, it has limitations for generalization but at the same time it has the benefits of generating an analysis of the mechanisms of religious peacebuilding in societies in transition where the conflict is still ongoing. Since violence in transitional societies does not always stop after signing an agree-

---

[2] The GMH has produced reports on five emblematic cases of massacres in the country. "The methodological strategy employed by the GMH to analyze and illustrate the processes and dynamics of conflict in Colombian territory, is the reconstruction of emblematic cases. Emblematic cases that are identified by MH, condense multiple processes, modes and regional expressions of conflict and victimization that help define and make public the authorships, responsibilities, and the social and political impacts of contemporary violence" (Grupo de Memoria Historica 2012).

[3] Nearly twenty land rights activists have been murdered since President Santos took office, and threats against Afro-Colombian leaders have increased. This number could skyrocket if victims are returned to conflictive areas without effective protection from the government and if the root causes of the conflict are not addressed (WOLA – Washington Office on Latin America 2011).

[4] This chapter presents the preliminary results of the research and some of the aspects that are relevant to understand the case. More information and detailed analysis will be contained in the final thesis.

ment, religious peacebuilders need to develop particular strategies to work for peace in spite of the presence of armed actors as in the case of Bojayá.

I understand by "transitional justice from below" concept as the participatory processes of implementing of the discourses and practices of transitional justice by civil society. It also implies a broadening of the legalist view on transitional justice by including those socioeconomic injustices that caused the conflict (Lundy and McGovern 2008). Some scholars have commented that in the case of the transitional process in Colombia, the best outcome of an otherwise poorly executed transitional law has been the use of juridical tools by the victims at grassroots level. Some sectors of civil society and the Church see in these tools a framework to challenge a state-sponsored attempt to use transitional justice for impunity (Diaz 2010.

The Diocese of Quibdó[5] (DQ), in particular, has developed participatory processes since the 1970s in Chocó, a department in the Pacific Coast of Colombia (Fig. 192.1). The violence in Chocó was enhanced by the incursion of paramilitary actors and their confrontation with guerrilla armies in the late 1990s. On May 2, 2002 a massacre left 78 victims died in a Catholic Church in Bojayá, Chocó. The DQ has led initiatives for the social reconstruction of memory that can be regarded as participatory processes of transitional justice from Below.

192.2   Religion and Transitional Justice

The field of religious peacebuilding takes as a point of departure that religion can be both a source for peace and for violence. Even though, it may be "misleading to assume that religious actors have an inherent peacemaking capability" (Harpviken and Roislien 2008); there is evidence that supports the idea that religion can offer paths for peace (Appleby 2000; Philpott and Powers 2010; Shore 2009; Smock 2006). Appleby (2000) claims that both the religious peaceworker and the religious extremist are just different types of religious militants. They both feel they are in a deep contradiction with the injustice of the world and find in their faith a noble and sacred cause that calls them to change their reality. They both are committed to their cause to the point of sacrificing themselves and their loved ones. For instance, in Latin America, we find that the Theology of Liberation has fueled both leftist guerrillas, such as the ELN in Colombia, and the leadership of movements of peace and resistance like Mon. Romero in El Salvador.

Religion often matters for the construction of peace in transitional societies, but it is a powerful substance that only when used properly, is an instrument of healing and reconciliation (Amstutz 2005; Cejka and Bamat 2003; Gopin 2005). Religious

---

[5]This makes reference to the local Diocesan Church, but also to the several missionary religious orders such as the Hermanas Agustinas Misioneras, Misioneros del Verbo Divino, Los Misioneros Claretianos and the Hermanas Lauritas.

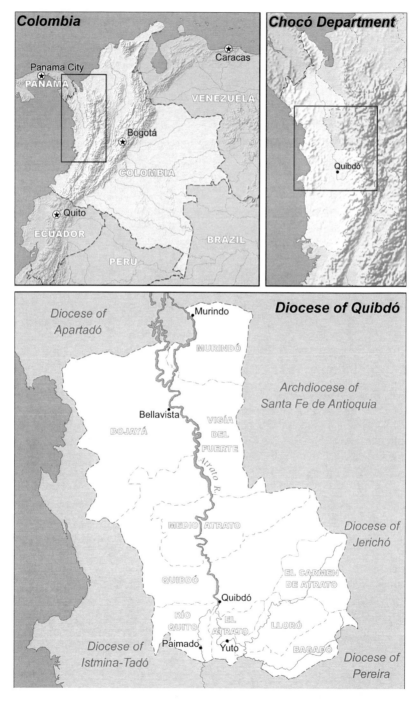

Fig. 192.1 Location of Bojayá, Chocó, Colombia (Map by Jeff Levy, University of Kentucky Gyula Pauer Center for Cartography and GIS; commissioned by the editor)

peacebuilding makes a difference compared to other kind of peacebuilding, not only because of the hermeneutics, trustworthiness and motivations that it provides, but also because its contribution is mediated by civil society and the state (Brewer et al. 2010). Amartya Sen (2006) claims that it is easy to overestimate the role of religion; therefore, religious authorities should not be established beforehand as fruitful conflict mediators. Douglas Johnston has identified some conditions under which faith-based intervention can be suitable for peacebuilding:

> 1) Religion is a significant factor in the identity of one or both parts to the conflict; 2) Religious leaders on both sides of the dispute can be mobilized to facilitate peace; 3) Protracted struggles between two major religious traditions transcend national borders, as has been the case over time with Islam and Christianity; and/or 4) forces of realpolitik have led to an extended paralysis of action. (quoted by Smock 2006: 2)

The literature on religious peacebuilding has been concerned mostly with the description of concrete cases in which religion has been part of the construction of peace in at least one of the roles mentioned above (Cejka and Bamat 2003; Shore 2009). Another important part of the literature focuses on the elements in the different religious traditions that can enhance positive values for peace, such as tolerance, reconciliation and forgiveness, among others (Gopin 2005; Biggar 2009). In the case of conflicts fueled by religious and ethnic identity, interreligious dialogue and religious literacy are valuable tools to build peace (Fox 2001; Hayward 2010). Religious peacebuilding has also contributed to the development of the reflection around Restorative Justice, for which the main goal is the reestablishment of the right relationship between antagonistic parties (Amstutz 2005; Hadley 2001). This statement has echoes of religious inspiration: "the confession and repentance of perpetrators, the forgiveness of victims, the empathetic acknowledgement of suffering on the part of other citizens, and the overcoming of enmity" (Philpott 2007: 17).

Societies that are confronted with atrocious pasts but are engaged at the same time in efforts to achieve peace often use transitional justice mechanisms. Philpott defines transitional justice "as the sum total of activities that states and citizens undertake to redress past political injustices in order to restore political orders in the present and in the future" (Philpott 2007: 4). The topics of peace, healing, reconciliation, redemption, forgiveness, hope, truth, restoration, love, compromise, memory, hate, sacrifice, martyrdom and resentment, among others, are usually considered part of the private realm, but become public when they are introduced into transitional justice (Brewer 2010). These elements are closely related to religious language and they, unavoidably bring the religious aspect back to the public arena.

192.3   The Conflict in Chocó and the Efforts of the Diocese of Quibdó

The Catholic Church has developed important work for peace in Colombia particularly since the 1980s. The Church has advocated the cause of peacebuilding through the facilitation of dialogues with the armed actors, attention to the displaced people and campaigns for the defense of human rights (Arias and González 2006; González 1997; Esquivia and Gerlasch 2009; Henao 2009). The DQ received the National Peace Award in 2005 for its continuous efforts to strengthen the social fabric through projects for social and economic development, and the denunciation of the violation of communities' human rights by the different armed actors in the Pacific coast. The DQ also has a very strong social pastoral team. This team has supported the creation of collective work in the peasant communities and the protection and defense of the collective territories through the support of their organizations.

However, since the incursion of the armed actors in the area the Church's *Option*[6] for the life of the communities has shifted from the defense of the territorial rights to the defense of their human rights. The conflict in the Pacific Coast is fueled by the multiple interests in the land. The majority of the population in this region is Afro-Colombian and indigenous. Thanks to Law 70 of 1993, the Afro-Colombian communities (that represent the 80 % of the population) have acquired collective territorial rights to over 5 million ha (12.3 mill acres) of the 10 mill ha (24.6 mill acres) of tropical forest that form this territory. Law 70 was a result of the intense participatory work of the grassroots organizations. Even though the law has brought territorial autonomy to the members of the communities, it also has had perverse consequences for them.

When the first processes of collective titling of the lands started in the region in 1997, paramilitary groups entered the zone, sometimes in collaboration with members of the Colombian army. In the Low Atrato they implemented military actions such as "Operación Génesis" that terrorized hundreds and left thousands of Afro-Colombian and native people displaced. The systematic attack on these ethnic groups, followed by the illegal appropriation of the land by transnational and national agribusiness and infrastructural megaprojects, have led some activists and scholars to speak of possible ethnocide in Colombia (Almario 2002; Wouters 2001).

In Bojayá, the leftist guerrillas of the FARC[7] have cohabited the territories for years and caused relatively little turmoil. Given that the presence of the State was almost nonexistent, confrontations with other armed actors were minimal. Violence increased dramatically when illegal extreme right self-defense paramilitaries arrived in the region. The confrontations between these groups and the techniques of terror used by the paramilitaries generated the displacement of thousands of local inhabitants.[8] They were victim of massacres, confrontations, sexual violence and the selective killing and disappearing of social leaders of the communities in order to spread terror. In addition, the communities of the Atrato suffered a shortage of food and supplies that provoked humanitarian crises due to the strict control of the river by the armed actors.

---

[6] The Options of the Church as a Diocese are established at a regional level by the Bishops, priests, religious missionaries and lay people. They represent the mandate of their work as communitarian organizations.

[7] The biggest irregular guerrilla army in Colombia (Fuerzas Armadas Revolucionarias de Colombia/ the Revolutionary Armed Forces of Colombia – *FARC*).

[8] The displacement that took place in the region was 5,230 people from 1997 to 1999 and 66,588 from 2000 to 2002 (CNRR 2010: 245).

The purpose of the de-territorialization in Middle Atrato directed by the armed groups has been to obtain better routes for the drug trade and arms smuggling, but also paving the way for projects such as gold mining and oil palm tree plantations (Franco and Restrepo 2011). It was during the years 1996–2005 that a context of direct violence was created and one of the cruelest massacres in the recent history of Colombia occurred. It happened in Bellavista (Head of the municipality of Bojayá), a village in the middle riverside of the Atrato on May 2, 2002. The massacre was the result of a fight between the AUC paramilitaries[9] and the FARC guerrillas; 79 civilians were murdered, 48 of them children.

In 2005 the paramilitaries were reintegrated into civil society. With their alleged reintegration, urban areas of Colombia experienced a shallow sense of peace. Nevertheless in regions like Bojayá the reality of the conflict is uncompromising. The new Criminal Bands[10] (known in Colombia as BACRIM) are coexisting with guerrilla members dressed in civilian clothes and blending in the town. This permanent confusion has fueled feelings of untruthfulness and has weakened solidarity among the peasant members of the rural communities. The BACRIM also have attempted to integrate the civilians into the conflict through networks of information and commerce.

192.4   Dos de Mayo: The Massacre of Bojayá

Bojayá is one of 31 municipalities of the department of Chocó, in the sub-region of the middle Atrato. The Atrato river is the main connection between 120 Afro-Colombian communities that form the COCOMACIA (Consejo Mayor Comunitario de la Asociación Campesina Integral del Atrato – Main Comunitarian Counsil of the Integral Peasant Organization of the Atrato). Floods are common in this region, where the population has learned to coexist with the humid forest and the river. The seat of the municipal government of Bojayá is Bellavista, which has 1,200 inhabitants.

In 1997, paramilitary troops entered Bojayá. They claimed that their goal was "cleansing"[11] the region from suspected guerrilla supporters. They had the support of some leaders of Bojayá (the mayor of Vigía del Fuerte, Wilson Chaverra, who was later publicly recognized as an important paramilitary leader in the region). During this period, the paramilitaries imposed strict control over the entry of food in the town. To prevent food scarcity, the missionary groups stored and distributed food for the villagers through communitarian shops (*tiendas comunitarias*). In an

---

[9] The AUC unified what had been a disparate array of paramilitary groups (Autodefensas Unidas de Colombia/United Self-Defense Units of Colombia – *AUC*).

[10] These are new criminal organizations that often are constituted by the ex-paramilitary soldiers who use their same criminal networks and modus operandi.

[11] For a vivid anthropological analysis of the practice of "cleansing" in Colombia see: M. Taussig, (2003) The Cleansing. "Law in a Lawless Land: Diary of a Limpieza in Colombia." Chicago: University of Chicago Press.

attempt to repress these activities, the paramilitaries killed some religious leaders, among them the influential Fr Jorge Luiz Mazo, priest of Bellavista and the Spanish missionary Inigo Eguiluz in 1999.

For over a decade, the expansion of the paramilitary project in the region spread. They established bases in the municipalities of Riosucio, Murindó, Beté and Quibdó and took control of the riverside populations of the Atrato River. During these years, the Diocese of Quibdó and other regional organizations denounced the open collaboration between the national army and the paramilitary groups (CNRR 2010: 42). This period corresponded with the development of the Pastoral Plan and the Options of the Diocese that encouraged the missionaries, lay people and priests to work for the poor communities and to defend their life in a common project. The process of *acompañamiento*[12] was oriented towards a continuous presence in vulnerable communities and the support of the victims. In 1993 *The Life, Justice and Peace Commission* (COVIJUPA) was created within the Socia Pastoral Office which specializes in monitoring human rights violations and breaches of international humanitarian law committed within the jurisdiction of the Diocese of Quibdó. COVIJUPA initially disguised its work by acting as praying groups in order to reach the most vulnerable and isolated communities. One of COVIJUPA's activities has been to create works of memorialization of the victims through regular meetings, demonstrations, festivals and by placing the pictures and details of the victims in some Catholic churches as *Chapels of Memory* (Fig. 192.2). They also have regular publications denouncing violations of the human rights in the communities.

Fig. 192.2 Capilla de la Memoria, Diocese of Quibdó, Quibdó (Photo by Sandra Rios Oyola)

---

[12]Acompañamiento is a term used to explain a range of activities that includes several forms of monitoring, promotion, and defence of the communities.

The guerrillas had a strong presence in Bojayá, Vigía del Fuerte y Murindó since 2000. During this period, the presence of missionaries and the DQ was crucial for the survival of the communities and the prevention of their displacement. They were working as strong teams with the support of the bishop. Even though they did not claim to be neutral, the missionaries consider themselves respected agents because of their continuous support of victims. This practice was not always a safe position, as a few members of these missionary teams were murdered in the region.

Violence continued in the area, affecting not only Bellavista, but multiple other towns. On 17th April 2002, troops commanded by the paramilitary leader known as "el Aleman" arrived to the municipality of Vigia del Fuerte and Bellavista. The Office of the High Commissioner for Human Rights of the United Nations (OHCHR), the Ombudsman, the Diocese of Quibdó, and other NGOs issued seven "early warnings"[13] to the national government considering the vulnerability of the population of 1,500 inhabitants in the area in regards to an imminent confrontation between the paramilitary bloc and the guerrillas (CNRR 2010: 13). However, there was not an immediate response by the government or the army. In despair over these events, a woman leader of the community read to the paramilitaries their "Declaration for Life and Peace," a document written by the community on 12th September 1999. It expressed the determined desire of the community to be left outside of the conflict. Again, the voice of the community was ignored.

During the confrontation between guerrillas and paramilitaries, the guerrillas fired four explosive gas cylinders. One of them reached the temple. It entered through the roof, hit the altar and exploded. The horror and the chaos were overwhelming (Gomez 2002). The evacuation of survivors through the crossfire was led by Fr Antun towards Vigia del Fuerte. The civilians in the boats were waving white flags and chanting: "Who are we? Civilians! What are we asking for? Respect our lives!":

> It was a mass of dead human flesh was all that was left. It was like a wall in the middle of the Church. I was in the priests' house and I was thrown against the door due to the impact. I started running, and pretty soon after that I picked up a little girl whose mother could not carry her because she had other kids with her. I put her in my shoulders while they were still shooting around us. The little girl stopped grabbing my shoulders once, and I thought they had shot her, then she grabbed my shirt again, she was still alive and I kept running (laughs). (Personal interview with a male survivor of the Massacre, 20th April 2012)

192.5   The Narratives of Social Memory of Dos de Mayo

There have been over 2,500 massacres in the last 20 years in Colombia and only the responsible of the 10 % of these massacres have been processed. Nevertheless the main narratives about the massacre of Bojayá that circulated in the media have been used to single out the massacre as an exceptional case. The horror of the experience

---

[13] The Early Warning System has been issued by the government in order to anticipate and prevent imminent human rights violations: M. Chernickl. (2004) Evaluation of Colombia's Early Warning System. Management Systems International. Retrieved on June 1, 2012 From: http://pdf.usaid.gov/pdf_docs/PDACR350.pdf

was expressed in rhetoric of terror that made it a unique event in which the forces of good and evil clashed, failing to notice that the violence and structural poverty of the region had produced a considerable numbers of victims in the previous years. For instance, only in Carmen del Darien (Chocó), have there been more victims than in Bellavista.

There is not a univocal narrative of the massacre of *Dos de Mayo*. The massacre was used by the Colombian government to expose the cruelty of FARC and have them be labeled as terrorists by the international community, including the Vatican (CNRR 2010: 265). The State expressed its "solidarity" with the victims, and it was represented by the media as another victim in the scenario. The responsibility was mainly directed towards FARC, leaving aside the accountability of the State and rendering the paramilitaries as a less important actor in the drama (United States Committee for Refugees and Immigrants 2003). As part of the project of finding out the historical reasons for the creation of the paramilitary project in Colombia, the GMH published a report labeled *"Bojayá: La Guerra sin Límites"* (Bojayá: The War without Limits) in 2010. The report exposed the responsibility of the state in the massacre. It explained how the state did not attend to any of the early warnings of the incursion of the paramilitaries and how despite the controls of the national army over the Atrato River, the state decided to ignore the paramilitary soldiers. Only a few of the victims of the massacre have received some monetary reparation but most of the survivors and displaced victims have still not seen any reparation, truth or justice.

For the community, the complicity of the government in the massacre was evident. They also see the massacre as the result of chronic marginalization and neglect by the government. For the victims the massacre was an predictable outcome of the chain of violence built in the Pacific coast. This chain has been developed by actions of violence such as the "Operacion Genesis" in the Low Atrato that led to the continuous displacement of the peasant population, the Naya massacre, and the incursion of paramilitary activity in the Middle Atrato. This is why, according to the inhabitants of Bellavista, this is a town self-identified as "de malas" (a town that is out of luck) due to its long history of oppression. The songs composed by the survivors served as an expression of their pain, but also as a way of "desahogarse" (venting out their suffering and frustration) to take the cold away from the body. These songs often mention that what happened in Bojayá is nothing compared to the long history of suffering (Millan 2009: 54). The songs follow rhythms that are recognized as traditional funerary songs known as "alabaos." In interviews conducted a few years after the massacre, some of the inhabitants expressed that "you live the pain, until you lose the rhythm of the pain" and "you have to get used to live through tears" (Millan 2009: 68–69)

The negative emotions of helplessness, anger and frustration come together in the idea of *mala-muerte* (bad death). The bad death occurs when the funerary rituals are poorly executed or incomplete. The connection between the community and their ancestors is strong and it is done through the funerary rituals. When this is broken, it has negative consequences for the lives of survivors. The funerals could not be conducted because people had to escape from the confrontation leaving the corpses behind without burial.

> The local custom "when a child under 12 dies is to hold a kind of party, to dance with the dead child; it's like a game" in which the child's godfather is the first to pick up the little body and dance with it, to the rhythm of drums and chants.
>
> The roots of the tradition go back to the days of slavery, when the slaves celebrated the "liberation" of the children from their bonds.
>
> "See how beautiful he looks/They're bringing him down/With bouquets of flowers/They are crowning him/" are the words of a traditional funeral song of the black community in the Chocó region. (Cariboni and Vieira 2007)

At the very least, one of the outcomes has been that after the massacre the State has finally been recognized in Bojayá. Even the president of the republic has witnessed the situation. This is why the national government has built a particular relationship of paternalism with Bellavista that, nevertheless, has not yet been translated into a concrete development plan for the town and the region. The Program for the reconstruction and development of the region (CONPES 3180, 2002) was issued a few weeks after the massacre, following the promise of relocation of Bellavista to a safer site further from the river. The efforts for integrating the voice of the community in the process of decision making for the relocation of the town were few and it affected the agency of the survivors who returned. The survivors of the massacre have received some humanitarian help and the reconstruction and relocation took place farther from the river and with less risk of flooding. However, this proposal is an incomplete development plan of development that does not provide for potable water, electricity or a sewer system. Moreover, the school and the hospital are in extremely poor conditions. The general feeling in the population is that the state has forgotten them, which also explains the persistent but failed efforts of bringing together members of the government for the tenth year commemoration of the massacre.

If the victims of Bojayá have not been satisfied in terms of reparation, they have felt despair in terms of truth. The declarations of the paramilitary leaders involved in the massacre have been not truthful, for instance they have blamed Fr Antun for locking the civilians in the Church. They have not assumed their responsibility for the events of the massacre itself (Semana 2008). The way in which the "reparation" was handled and people's self-perception as rightful victims are questioned as the administrative officials often believe that the victims are mere recipients of official charity, using tags such as "beneficiaries" or "users" rather than victims of the armed conflict (PIUPC 2005: 70).

192.5.1   Ten Years Later

Ten years after the massacre the frustration over the unfinished funerary ceremonials continues. The victims still remember with pain and helplessness. The displaced victims cannot visit the cemetery as often as they wish due to lack of resources. Those who stayed in the town feel that the anniversary of the massacre has been hijacked by journalists, politicians and the army. In 2012, the victims decided to hold separate political and religious events in order to protect the religious services.

The tenth commemoration of the massacre was called "the ten years of the ethnocide of Bojayá" by the communities, with the slogan "Bojayá: the path to dignity." For the social leaders that organized the event, the connection of the violence with their identity as ethnic minorities is crucial and it is part of a long story of their marginalization and exclusion. In Bellavista, the commemorative event was used to demand the unfulfilled promises of the State. The leaders of the villages surrounding Bellavista, also members of the municipality of Bojayá, worked together to create a list of demands that eventually were not heard since the members of the government did not attend. The event turned out to be highly political and there was some subtle hostility between the political and the religious leaders. Symbolic and religious actions were threatened to be cancelled when the social leaders declared a general assembly (strike) due to the absence of important members of the government.

Even though symbolic actions took place after all, the disagreement expresses a shift in the perception of the role of the Church in regards to the commemoration itself. Being a witness of the commemoration of ten years of the massacre means to observe the paradox behind the use of transitional justice mechanisms, even while the conflict has not yet stopped. The memorials such as the singing of the funerary songs introduced lyrics referring to their situation of poverty and abandonment (Fig. 192.3). There was also a theater play in which the young actors used the ruins of the old town as their backdrop (Fig. 192.4) and the children were carrying the names of their relatives who were victims of the massacre (Fig. 192.5). The Mass was led by the Bishop of Istmina and several other priests following a short pilgrimage. The Bishop used this space to address the demands of the survivors since the political event could not take place.

The claims for truth and justice, however, were stronger in Bogotá. The presence of the media and the armed actors might explain the shift in the priorities in the narratives of memorialization in both commemorations (Figs. 192.6 and 192.7). The Social Pastoral (DQ) organized an event in Bogotá that brought together activists, victims, academics, and media. They also invited members of the government, but only the representative of the Office of Victims from the government accepted the invitation. The Social Pastoral outlined the problems of the community of Bojayá:

> Each one of these communities requires and deserves the reestablishment of their rights and satisfaction of their claims of truth, justice and collective reparation. These claims should not be mistaken by the governmental development of focalized, desculturizadas and inconclusive investments, without the consultation of the locals in Bellavista. These actions correspond to protective measures in case of natural disasters and not to public policies that are structured and oriented towards the protection of the victims' rights. The victims need protection of the political violence according to the Colombian law. The Law regarding reparation, the International Humanitarian Right, and the Peoples Rights states it. (Diócesis de Quibdó 2012)

The conditions of the conflict are changing and they are influencing the way the work of *acompañamiento* by the religious peacebuilders is being developed. The lack of a visible structure of leadership among the new criminal bands has shifted the possibilities of an eventual informal bridging for the construction of peace. However, this shift opens up the possibility to encourage the *acompañamiento* of the communities of victims through the creation of mechanisms of social reconstruction of memory.

Fig. 192.3 Women of Pogó singing in front of the Patchwork Curtain sewed by the women of the community (Photo by Sandra Rios Oyola)

192.5.2   The Little Shoe and the Big Show

The massacre of Dos de Mayo had a high media coverage compared to many others which were overlooked by the media. The main discourse related to the cruelty of the guerrillas. For many years, a big banner in Bellavista read: "El 2 de mayo de 2002 aquí las Farc asesinaron a 119 personas. ¡Que no se nos olvide nunca!" (On 2nd May 2002 here the FARC assassinated 119 people. We shall not forget!). The banner was installed by the national army. It was a clear message with no room for other descriptions or voices. On a similar note, in the national news General

Fig. 192.4 Young actor in the play commemorating the massacre of Dos de Mayo (Photo by Sandra Rios Oyola)

Fig. 192.5 Child holding a flag with the name of his relative who was killed in Dos de Mayo (Photo by Sandra Rios Oyola)

Montoya appeared holding a little shoe and crying over the lost lives of children killed by the guerrillas. This same general appeared surrounded by paramilitaries in videos that were used by the GMH in their report. Despite evidence of the

Fig. 192.6 Head of the pilgrimage entering the Old Church San Pablo Apostol in the old town of Bellavista, Bojayá (Chocó, Colombia) (Photo by Sandra Rios Oyola)

Fig. 192.7 Old Church San Pablo Apostol on the day of the tenth commemoration of the massacre of Dos de Mayo (Photo by Sandra Rios Oyola)

complicity of the national army and the paramilitary groups, there have been no judicial actions against any of them. The guilt has been placed entirely on the guerrillas.

The memorable scene of the "crying general" was introduced in a play performed by the local theatre group led by a member of a Swedish NGO and sponsored by the DQ. In an interesting twist, one of the main characters appears dragging a little shoe as his pet. It represents the disparagement and manipulation of the emotions aroused by the massacre. It criticizes the political uses of *Dos de Mayo* and how it has had little positive impact for the locals.

Through these kinds of activities – songs, poetry, theatre, and other kinds of performances and artistic expressions – the community has been able to express what it could not say in other ways. They contested the official discourse on the history of the events of the massacre. It has been only recently that some of those initiatives were introduced in an official account through the work of the GMH.

192.5.3 The Role of the DQ in the Social Construction of Memory

The role of religion is important in the case of *Dos de Mayo*. In the first place, the massacre occurred inside the church, which was not only supposed to be the safest building, but also was a sacred religious space that represented a place of safety. The Church also represented an actor that was recognized as neutral by the armed actors. It was known that the DQ often had to be involved in the rescuing of civilians that were threatened by guerrillas or paramilitaries, or in helping to release kidnapped civilians. Therefore, the neighbors from Pueblo Nuevo in Bellavista did not doubt their decision to hide in the temple. After the explosion, it was Fr Antun Ramos who led the civilians to find refuge in the nearest town and the DQ which gave the immediate help to the survivors.

Shortly after the massacre, the team of the Augustine Sisters thought about the importance of collecting and constructing memory of the event. They knew that people would prefer to forget such an atrocity. They supported many activities that included a process of therapy for the women through sewing groups that they had already started a few years earlier with the group Guayacan. This was not an easy task because many of its members had died in the massacre. They created a patchwork curtain with the names and details of their loved ones (see Fig. 192.3). Also, the Mutilated Christ, the figure of Christ that used to be in front of the altar, became a symbol of the Dos de Mayo. There was masses and pilgrimages through the river up to Bellavista every *Dos de Mayo*. Also, an association called ADOM (Asociación Dos de Mayo) was created that represented the interests of the survivors of the massacre and that has its office in the DQ in Quibdó.

The DQ recognizes the importance of keeping and making publicly known the memory of the atrocities. The relevance of their work compared to other bottom-up initiatives is that they work with networks of social organizations and leaders that allow them to reach vulnerable and isolated areas in Chocó. This kind of work allows for the provision of a broader historical context for the crimes. They attempt to not isolate the case of *Dos de Mayo* as a unique massacre. This is why, in the Diocese's territory, they celebrate the 2nd May as the day of all victims.

The importance of the social construction of memory lies in its power to recognize the victims as rightful wronged civilians and to denounce the crimes committed against them. In other words, the claim that lies behind the initiatives of social memory is not only to warrant that the atrocities will not happen again, but that they will stop. However the violence and impunity in the region continue. Hence, the work of the religious peacebuilders becomes crucial for building the hope for justice in a country with high rates of impunity. According to Fr Jesus Florez[14] the purpose of remembrance is to move from the narrative of pain and resentment to the narrative of life and resurrection. He says that memory is about "a renewal of life and not of death, in order to reach the resurrection. Memorial works are done as an interpretation of reparation in the context of resurrection" (Personal interview, April 27, 2012).

The victims of Bojayá clearly understand the need for truth and justice, but the continuous presence of armed actors makes it difficult to be outspoken about this at a local level. Nevertheless, the Church has used its networks to commemorate and claim truth, justice and reparation not only in Bojayá, but also in other contexts such as Bogotá, where they held a memorial act at the National Museum of Colombia. In addition, the construction of social memory has been important in bringing together a community that seems weakened and fragmented due to their experience with violence, but also because some individuals have received any restitution.

The social construction of memory led by the religious peacebuilders such as COVIJUPA locates the struggle for justice beyond the solitary efforts of powerless civilians against powerful armed actors or the government. The religious peacebuilders place this struggle in a narrative that is larger than the event they are currently facing; they express it in biblical times of the struggle against the oppressor. This religious narrative allows them to interpret their history in terms of a chain of memory that offers them hope: "It is not for us, it is not even for our children, we do not know which generation will actually see the results of our work but we need to be constant" (Personal interview with Sister Elsa, member of COVIJUPA, April 14, 2012). They are situated in a struggle against the powerful that is larger than their own lives. Their memory work inscribes the story of the people in the light of the resurrection even if they do not see justice themselves.

---

[14] Fr Jesus Flórez is an important religious leader of the region. He is member of the Claretian Religious Community and a former director of the Claretian University Foundation in Quibdó (FUCLA).

192.6   Conclusions

The case of the *Massacre of Bojayá* not only represents the involvement of the Catholic Church in peacebuilding and the construction of memory in Colombia, but it also represents the challenges of the application of mechanisms of transitional justice in a context where the atrocities are too recent and the conflict is still happening. The response of the government and of some organizations has been confusing in terms of providing reparation. Their help is charity, rather than reparation. It is important to understand how the conflict and the massacre have been defined in order to understand the demands of the communities. It has been explained how the evasion of accountability by the government when all the guilt has been placed on the guerrillas, has led to a representation of a "solidarity" state rather than an accountable state.

This lack of accountability has been one of the aspects addressed by the initiatives of transitional justice from below. One of these initiatives led by the DQ has been the social construction of memory through participatory processes. The annual commemoration and the conservation of the old church, but also the songs, texts, and plays that have been transmitted are ways of remembering what happened, but also of bringing people together and helping them to be critical of their situation as recipients of help. An important outcome of this work has created subtle frictions when there are local leaders that attempt to exploit the political capital of the memorial.

The building of the new town corresponds to a mechanism of response in case of natural disasters, rather than a response to a situation of political violence. There is the general feeling that the report of the GMH has been useful to record their history, but it has been ineffective in terms of justice and truth. Following Lundy and McGovern (2008), the initiatives of transitional justice from bottom up have allowed for a broadening of the concept of justice. For instance, the local participatory processes are oriented towards the recognition of social and civil rights, including measures that could help them to achieve more equitable conditions of life and important cultural elements such as the notion of dignity. It is necessary to underline that the requirements for judicial truth and legal justice may be limited at the local level due to the continued presence of armed actors.

Since the situation of conflict is still present, the Catholic Church has an important role due to its privileged position not only as an actor on the side of the victims, but also because the DQ has built important networks with NGOs and other official agencies. This situation allows the DQ to denounce the crimes of the conflict through memorials to a broader audience. At the same time, they fight the unilateral narratives of memory that make the atrocity a singular and exceptional case and rather seek to place it in a chain of struggles and violence. In contrast to the history of the hierarchical Church in Colombia, the DQ due to its Theology of Liberation roots has built a path that is independent of the centers of power but not neutral. According to Philpott (2007), this is one of the requirements for a successful religious peacebuilding role. The DQ, through its Office of Pastoral Social and COVIJUPA, has encouraged victims in their demands for truth, justice and reparation, since the Church's networks can go beyond the local context where armed actors still have strong agency.

The DQ combines economic communitarian work inspired by their *option for the poor* with the denunciation of crimes against human rights and the defense of the memory of victims. The multiple roles covered by the DQ make them a strong actor in terms of transitional justice from below. It is clear that social memory on its own is not enough to build peace and can actually create feelings of despair in the victims, particularly in the face of the high rates of impunity within the country. This is where the feelings of hope and justice religiously inspired by the DQ make a difference in their participatory processes of construction of social memory.

References

Almario, O. (2002). Territorio, identidad, memoria colectiva y movimiento etnico de los grupos negros del Pacifico sur colombiano: Microhistoria y etnografía sobre el río Tapaje. *Journal of Latin American Anthropology, 7*(2), 118–139.

Amstutz, M. (2005). *The healing of nations: The promise and limits of political forgiveness.* Lanham: Rowman & Littlefield.

Appleby, R. S. (2000). *The ambivalence of the sacred: Religion, violence, and reconciliation.* Lanham, MD: Rowman & Littlefield.

Arias, R., & González, F. (2006). Búsqueda de la paz y defensa del "orden cristiano": el episcopado ante los grandes debates de Colombia (1998–2005). In F. Leal (Ed.), *En la Encrucijada: Colombia en el Siglo XXI* (pp. 173–206). Bogotá: Editorial Norma.

Biggar, N. (2009). The ethics of forgiveness and the doctrine of the just war: A religious view of righting atrocious wrongs. In T. Brudholm & T. Cushman (Eds.), *The religious in responses to mass atrocity interdisciplinary perspectives* (pp. 105–123). Cambridge: Cambridge University Press.

Brewer, J. (2010). *Peace processes: A sociological approach.* Cambridge: Polity Press.

Brewer, J., Gareth, H., & Francis, T. (2010). Religion and peacemaking: A Conceptualization. *Sociology, 44*, 1019–1037.

Cariboni, D., & Vieira, C. (2007). *A painful pilgrimage.* Retrieved January 16, 2012, from http://ipsnews.net/news.asp?idnews=39853

Cejka, M., & Bamat, T. (2003). *Artisans for peace.* Maryknoll: Orbis Books.

CNRR – Grupo Memoria Historica. (2010). *Bojayá: La guerra sin límites.* Bogota: Taurus.

Diaz, C. (2010). Challenging impunity from below: The contested ownership of transitional justice in Colombia. In K. McEvoy & L. McGregor (Eds.), *Transitional justice from below. Grassroots activism and the struggle for change* (pp. 189–215). Portland: Hart Publishing.

Diócesis de Quibdó. (2012). *Paper presented at the Commemoration of the Massacre of Bojayá at the National Museum of Colombia, Bogotá, 27 April 2012.* Retrieved May 08, 2012, from http://bojayaunadecada.files.wordpress.com/2012/04/diocesis-de-quibdo-bojaya-10-ac3b1os-despues.pdf

Esquivia Ballesta, R., & Gerlach, B. (2009). The local community as a creative space for transformation: The view from Montes de María. In V. Bouvier (Ed.), *Colombia: Building peace in a time of war* (pp. 295–310). Washington: U.S. Institute of Peace.

Fox, J. (2001). Religion as an overlooked element in international relation. *International Studies Review, 3*(3), 53–74.

Franco, V. L., & Restrepo, J. D. (2011). Empresarios palmeros, poderes de facto y despojo detierras en el Bajo Atrato. In M. Romero (Ed.), *La economía de los paramilitares: redes decorrupción, negocios y política* (pp. 269–410). Bogotá: CNAI – Random House Mondadori.

Gomez, P. (2002). *Los muertos no hablan.* Bogota: Aguilar.

González, F. (1997). *Poderes enfrentados, iglesia y estado en Colombia.* Bogotá: CINEP.

Gopin, M. (2005). *Holy war, holy peace.* Oxford: Oxford University Press.

Grupo de Memoria Historica. (2012). *Emblematic cases.* Retrieved May 10, 2012, from http://memoriahistorica-cnrr.org.co/archivos/arc_centro/emblematic_cases.pdf

Hadley, M. (2001). *The spiritual roots of restorative justice.* New York: State University of New York.

Harpviken, K., & Roislien, H. (2008). Faithful brokers? Potentials and pitfalls of religion in peacemaking. *Conflict Resolution Quarterly, 25*(3), 351–373.

Hayward, S. (2010). *Averting hell on earth religion and the prevention of genocide.* Special report. United States Institute of Peace. www.usip.org/files/resources/sr248.pdf

Henao, H. (2009). The Colombian church and peacebuilding. In V. Bouvier (Ed.), *Colombia: Building peace in a time of war* (pp. 173–191). Washington: U.S. Institute of Peace.

Human Rights Watch. (2001). *The "sixth division:" Military-paramilitary ties and U.S. policy in Colombia, 4 October* (2653). Retrieved June 10, 2010, from www.unhcr.org/refworld/docid/3c2b20470.html

Human Rights Watch. (2005). Smoke and mirrors: Colombia's demobilization of paramilitary groups, *17*(3B). Retrieved June 10, 2010, from www.hrw.org/reports/2005/07/31/smoke-and-mirrors-0

Lundy, P., & McGovern, M. (2008). Whose justice? Rethinking transitional justice from the bottom up. *Journal of Law and Society, 35*(2), 265–292.

McEvoy, K., & McGregor, L. (Eds.). (2010). *Transitional justice from below: Grassroots activism and the struggle for change.* Portland: Hart Publishing.

Millan, C. (2009). *"Ya no llega el Limbo porque la gente bailando está." Prácticas de memoria en Bojayá – Chocó.* Thesis in Social Anthropology. Bogotá: Universidad Nacional de Colombia.

Nyamu-Musembi, C. (2002). *Towards an actor-oriented perspective on human rights* (Working Paper 169). Sussex: Institute of Development Studies. Retrieved June 01, 2011, from www.drc-citizenship.org/system/assets/10.2734370/original/1052734370-nyamu-musembi.2002-towards.pdf

Philpott, D. (2007). What religion brings to the politics of transitional justice. *Journal of International Affairs, 61*(1), 93–110.

Philpott, D., & Powers, G. (Eds.). (2010). *Strategies of peace. Transforming conflict in a violent world.* New York: Oxford University Press.

PIUPC. (2005). *Bojaya, memoria y río. Violencia política, daño y reparación. Programa de iniciativas universitarias para la paz y la convivencia.* Bogotá: Universidad Nacional de Colombia.

Semana. (2008). *Rabia en Bojayá.* Retrieved June 01, 2011, from www.semana.com/nacion/rabia-bojaya/112284-3.aspx

Sen, A. (2006). *Identity and violence: The illusion of destiny (issues of our time).* New York: Norton and Company.

Shore, M. (2009). *Religion and conflict resolution. Christianity and South Africa's Truth and Reconciliation Commission.* Surrey: Ashgate.

Smock, D. (2006). Religious contributions to peacemaking when religion brings peace, not war (Peaceworks No. 55). Washington: U.S. Institute of Peace. Retrieved June 10, 2010, from www.usip.org/publications/religious-contributions-peacemaking-when-religion-brings-peace-not-war

United States Committee for Refugees and Immigrants. (2003). *U.S. Committee for refugees world refugee survey 2003 – Colombia.* Retrieved June 10, 2010, from www.unhcr.org/refworld/docid/3eddc48d18.html

WOLA, Washington Office on Latin America. (2011). *Dangerous precedent set by consultation process for Colombia's victims' law. Afro-Colombian rights undermined.* Retrieved May 10, 2012, from www.wola.org/news/dangerous_precedent_set_by_consultation_process_for_colombia_s_victims_law

Wouters, M. (2001). Ethnic rights under threat: The Black Peasant Movement against armed groups' pressure in the Chocó, Colombia. *Bulletin of Latin American Research, 20*(4), 498–519.

Chapter 193
From Nasser's Revolution to the Fall of the Muslim Brotherhood

Seif Da'Na

193.1 Historical Background

The geopolitical significance of Egypt's location and the political implications of its geography and history have been the focus of Egypt's great minds and distinguished scholars and historians for decades. They seem to have strongly informed the Nasserite local and regional policies from 1954 until 1970 (Nasser (1954) 1996). In his four-volume seminal work *The Personality of Egypt*, Jamal Hamdan highlights the role Egypt's geography played in its long history and details what he called "the regional personality," in which he perceives the country as a product of time and space (Hamdan 1967). In more than 4,000 pages, Hamdan eruditely explores the fascinating interrelationship between Egypt's geography and history and the total implication of Egypt's location. This is a work that not only employs a unique theoretical framework to rewrite Egypt's history but also a grand intellectual scheme that recalls Fernand Braudel's three-volume *La Méditerranée et le Monde Méditerranéen a l'époque de Philippe II* (Braudel 1972) that informed the writing of history for some time. (Braudel's work entails the recognition of the role of non-social limits and the inclusion and consideration of what might be characterized as non-social structures, for example, geomorphic, geophysical, and ecological forces – into the writing and making of a country's history.)

Hamdan's comprehensive study also analyzes the colonization era of the Middle East and North Africa in what he called "the grand epic of geography," highlighting the role non-social structures played in both colonization and resistance. He also analyzes the rise of radical political Islamic movements and considers them a sign of weakness of the Arab society. In short, Hamdan's fascinating work and critique of radical Islamism (unintentionally due to being a work of geography and history

S. Da'Na (✉)
Sociology and Anthropology Department, University of Wisconsin-Parkside,
Kenosha, WI 53144, USA
e-mail: dana@uwp.edu

© Springer Science+Business Media Dordrecht 2015
S.D. Brunn (ed.), *The Changing World Religion Map*,
DOI 10.1007/978-94-017-9376-6_193

in the first place) makes a very solid scholarly case for an Egyptian Arab nationalist ideology that informed the Nasserite era (a non-isolationist ideology that is contrary to the domestic Egyptian nationalism that preceded the 1952 era). Hamdan, however, was not the only Egyptian scholar who paid significant attention to this important issue.

Hussein Mu'nis' *Misr wa Risalatoha* (*Egypt and Its Mission*) is based on almost the same scholarly assumptions that informed Hamdan's work, but is written in a less sophisticated style and seems to have been intended for the public rather than for the scholarly community. This work was so influential that the introduction to its first edition was written by President Nasser himself (Mu'nis 1973). Mu'nis, again makes the case for Egypt's role based on existing realities shaped by geography, history, and location. Although this scholarly orientation dominated Egypt's geopolitical scholarship even before the 1940s, it primarily flourished during the Nasserite era as it seemed to both explain and justify the Nasserite domestic and regional policies. However, the Nasserite regional and domestic policies were not only informed by this geopolitical orientation only, but they did constitute the core of Nasserite revolution and its Arab nationalist brand.

Scores of other Egyptian scholars from various theoretical orientations and backgrounds also did not fail to recognize the implications of geography and history (for example, Ghurbal 1990; Wahida 1974; Fawzi 1960) although the specific historical moment of each historian seems to have informed their emphasis (for example, the rise of nationalism in the post Mohammad Ali period shaped the Egyptian historiography as did the eras of Nasser, Sadat, and Mubarak) (see Gorman 2003). The political significance and implications of this outlook lies primarily in two issues. First, it conceptualizes the interrelationship between local and regional realities as a basis for an Egyptian Arab nationalist ideology. Second, it lays the foundation for an anti-colonial and anti-imperial regional infrastructure under the Egyptian leadership.

193.2   Nasser's Revolution

The efficacy of history and geography on one hand, and conceptualizing the interrelationship between local Egyptian nationalism and Arab nationalism (or conceptualizing the Arab dimension and role of Egypt) are present in Egyptian historiography and politics regardless of the view or the orientation and conclusion. This lesson President Nasser seems to have understood very well as Egypt under his rule became the central regional and Arab country within the region as well as a recognized international power (Fig. 193.1). To Nasser, ignoring the significance of Egypt's geopolitics would not only result in the deterioration of Egypt's regional status and role, but it would undermine local stability as well (Sbahi 2013).

However both Sadat and Mubarak, and later the Muslim Brotherhood (MB) movement, failed to understand the political implications of Egypt's history and geography as well as its potential Arab and regional status. But unlike President Nasser their fate was sealed with popular discontent – the former was assassinated

Fig. 193.1 Nasser's 1952 revolution enjoyed widespread popular support (Photo by Hani Shetaea, used with permission)

and the latter toppled, while now the MB is facing its most significant challenge of survival in a century.

The core of Nasser's revolution was to restore Egypt's geopolitical role and status that were undermined by an isolationist ideology of Egyptian nationalism in the pre 1952 royal era. Nasser's view was simple: Egypt cannot be liberated from the influence of powerful foreign empires and by implication cannot achieve socioeconomic development alone. That Arab nationalist orientation of Nasser was not, therefore, a mere choice or ideological bias as much as a historical necessity (Nasser 1996). Based on this view, President Nasser managed to successfully transform the July 1952 military *coup d'etat* into a total revolution that radically transformed Egypt politically, economically, socially, and culturally.

While many historians credit President Nasser with policies that structured such transformation (nationalizing the Suez Canal, Agricultural Reform, and the construction of the High (Aswan) Dam), Nasser's revolution lies in his comprehension of the significance of Egypt's geopolitics, or what he called the "facts of time and space" that cannot be ignored (Nasser (1954) 1996). That is, the idea underlying Nasser's revolution was to question the pre-1952 "isolationist ideology of the domestic Egyptian nationalism" and to propose an Arab nationalist alternative to Egypt's liberation and development (Al Hafez 1978: 40). Nasser's revolution rested on the premise that Egypt cannot be liberated from the colonial and imperial influence and succeed in socioeconomic development without questioning the existing "isolationist ideology" that delimits Egyptian politics. His vision was to establish what he called "historical blocks" locally, regionally and internationally as requirements for liberation in the age of great empires (Nasser (1954) 1996). That is why he stressed in his *Philosophy of the Revolution* that "we should look beyond our borders" and that "Rafah (Egypt-Palestine border) is not the end of the border" (1996: 98). Such a view informed Nasser's policies on the Palestinian question (Arab-Israeli struggle) and the Arab World, Africa (and the countries of southern Africa), and the Islamic World.

On the question of Palestine, Nasser was clear: "Fighting in Palestine was a matter of self-defense" he wrote in his *Philosophy of the Revolution* (1996: 95). His policies in the Arab world were culminated with a unity with Syria that he considered Egypt's extension in Asia. As for Africa, Nasser was a founder of the "Organization of African Unity" and supported the African national liberation movements (for example, in the Congo). In the Islamic World, Nasser built an anti-colonial alliance and was central in the formation of the Non-Aligned Movement.

However, Nasser did not assume that success depends on ideology alone, but rather on considering the existing realities and creating new ones. Therefore, the success of President Nasser's idea was enforced with a series of policies that created what Nasser called an "historical block" locally, regionally, and internationally to counter the impacts of colonial and imperial powers. Such an idea explains much about the success of Nasser and the failure of later Egyptian rulers.

On all these issues, Sadat, Mubarak, and later the MB president Mursi would pursue different policies that resulted as Nasser anticipated domestic instability and also the loss of Egypt's regional and international status. For not only did Nasser's view entail a recognition of a dialectical relationship between the local and regional, but also the Nasserite view and the Egyptian historiography that informed it assume that Egypt's central regional role is informed by geography and history. Political leaders who ignore this role will fail to meet the domestic challenges. In the case of the MB, the failure recalls an ideological setback that befell this Islamic genre and was manifested most clearly in the performance of president Mursi as outlined below.

193.3   Re-imagining Islam

Similar to the rise of Arab nationalist ideology under Nasser, the hegemony of certain political Islamic orientations (including the MB and its offshoot Hamas in Palestine) that led to their electoral victory in both Egypt and Tunisia depended on their ability earlier to construct an anti-colonial and anti-imperial view of Islam. In addition, it depended on their ability to re-imagine Islam in the post "Sykes-Picot" regional context, that is, a series of post WW1 colonial regional arrangements that were agreed upon between Britain and France that resulted in dividing the region and created the present borders between the current Middle East states. The arrangement after 1980 depended on a revolutionary re-imaging Islam that took place in the context of the Arab-Israeli struggle (Hezbollah and Hamas being the best examples).

While the MB primary orientation might be characterized as reformist, rather than revolutionary, the century old movement evolved dramatically since the 1980s, especially after the rise of its Palestine branch, Hamas, as a resistance movement in Palestine as well as the influence of a new generation that increased over time (Rosefky Wickham 2013; Da'na 2009). In this sense, the popularity of this genre of political Islam did not rest on the so-called "Islamic" popular culture or an enormous infrastructure of a large number of mosques and religious institutions. Actually, these factors did not prevent the hegemony of the nationalist and leftist oriented political ideologies from developing, or even challenge a dominant negative popular view of the MB in the Arab world in the 1960s and 1970s.

Therefore, understanding the rise (and decline) of certain Islamic movements (including the MB), their politics and narrative must be seen within contemporary local, regional, and global rivalries, rather than primarily making reference to Quaranic texts. Religious oriented movements in the history of Arab-Islamic societies always expressed and signified social and political antagonisms that were social and political in essence. This is even true of the inception of Islam itself, as well as the subsequent Islamic movements (Bamyeh 1999). Intellectual and theological disputes between the various Islamic scholars and schools, therefore, always signified a political and social dispute (Da'na and Khory 2004).

While some Muslim scholars and especially those close to the MB Islamic orientation, spoke of a "new Islamic discourse" to signify the development of the Islamic socio-political thoughts (Elmessiri 1997), the origin of this Islamic political genre dates back to the nineteenth century reform movement initiated by Jamal Aldin Al-afghani. Al-afghani's subversive view of Islam was primarily shaped by resistance to European colonialism and his teachings influenced a generation of Islamic scholars among them the Mohammad Abdu. Abdu will later introduce his students at Al Azhar to some of the most subversive readings of Islam that he himself studied under Al-afghani (Rida 2006; Haj 2008). His students were introduced to many texts in theology that were not acknowledged or taught at al-Azhar, including *The Commentary of al-Taftazani on the Creed of al-Nasafi* (Haj 2008). Al-Nasafi's work is distinguished (and rejected by most fundamentalist Islamists) due to shared views with the Mu'tazila. That immediately put Abdu in conflict with the established "ulama of al-Azhar" since the Mu'tazilis, *Ahl al-Tawhid wa al-'Adl* ("People of Monotheism and Justice") sought to ground Islamic creedal system in reason.

But the most significant contribution to a revolutionary genre of political Islam came from outside the world of *Ulama* (Islamic scholars) and from outside the centers of Islamic scholarship in Egypt (Al Azhar), Iraq (Al Najaf) and Iran (Qom). Arab and Muslim Islamic academics and intellectuals such as Mohammad Iqbal, Ali Shariati and Malek Bennabi would contribute significantly to discovering the revolutionary potential of Islamic thoughts. Iqbal's view entailed a reconstruction of the religious Islamic thoughts (the title of his book) based on the philosophical views of his time (Iqbal (1934) 2013), Shariati's revolutionary Islamic scholarship was so influential that he is even considered the ideologue of the Iranian Revolution of 1979 (he was assassinated 2 years earlier) and Malek Bennabi, who became

famous for his theory "Susceptibility to Colonization" (1999, 2006a, b) even introduced an Islamic theory of imperialism. The central themes of the reform movement
were anti-colonialism, inter-denominational orientation, and the incorporation of
non-Islamic philosophy.

193.4   The Setback

It is possible to trace back the seeds of what might be called a setback in the Islamic
political thinking to the work of the most influential student of Al-afghani,
Mohammad Abdu. Abdu's views centered on an educational orientation (or teaching Islam to Muslims), thus failing to distinguish between what Bennabi called
"religion as a tool of self-puritanism (and preaching) and religion as a tool of
national revival" (Bennabi 2002: 54).

The setback was manifested most clearly in the performance of the MB in both
Tunis and Egypt immediately after the successful overthrowing of the heads of the
regimes in both countries. The movement's policies and agenda violated not only
the tenets of the Nasser's geo-politics and the reason underlying his successful
transformation of Egypt into a regional and international power, but also the principles of Islamic reform movement. In the former, the MB would maintain the same
political and economic policies of the deposed regimes while and in the latter it
would go against all principles of Islamic reform.

In the case of Palestine, President Mursi and the MB upheld the Egyptian-Israeli
agreement (Camp David agreement) contrary to the MB discourse and propaganda.
In addition, while Nasser formed unity with Syria, due to its geopolitical value,
Mursi's announcement on June 15, 2013 not only terminated the diplomatic ties
with Syria but even declared Jihad against the Syrian regime. As for Africa, the
crisis of the Ethiopian scheme to build a dam on the Nile was mishandled by Mursi's
administration and reached threatening military action against Ethiopia. In the case
of Islamic interdenominational orientation, the MB expressed sectarian tendencies
in their reactions to both the Bahraini and Syrian uprisings as well as restoring the
diplomatic relations with Iran. They supported the Syrian uprising on almost purely
sectarian grounds (arguing it is a Sunni uprising against an Alwite regime) while
dismissed the Bahraini uprising as a sectarian uprising (the Bahraini's majority
being Shia Muslims).

One of Egypt's leading journalists, Mohammed Hassanein Heikal, commented
on Mursi performance in the last three weeks of his rule and argued that "Mursi
committed three disasters in the past three weeks, each one capable of leading to the
downfall of the regime" (Dana 2013: 24). In this regard, Heikal specifies the president's handling of opposition plans to hold protests against his regime, in the same
way he dealt with Ethiopia's bid to build a major dam on the Nile, and his rally
appearance in which he cut off diplomatic relations with Syria.

Heikal's sentiments were echoed by Hassan Nafaa, a prominent professor of
political science at Cairo university, noting that the president has "lost his political

sense," particularly in his irresponsible appearance at an Islamist rally in support of the Syrian opposition, in which he declared cutting off diplomatic ties with Damascus.

The MB performance in Egypt led to a dramatic loss of their popular support, as it is evident in the gap between the votes the MB received in the Parliamentary election in 28 November–11 January 2011 (more than ten million votes or about 37 %) and the first round of presidential elections in which Mursi received almost half these votes (5,764,952). He managed to win the second round only with the support of other political parties.

193.5   Significance of July 3 *Coup d'etat*

After gaining power, the MB embarked on a comprehensive program to extend their reach into all the state's branches and institutions, in a process that became known as the "*Ikhwanization*" of the state (hijacking the state by MB or Ikhwan). They did not distinguish themselves politically and economically from the Mubarak regime, but actually maintained the same policies. Economically, Mursi did not only maintain the same neo-liberal economic policies, but his administration negotiated a very daring package with the IMF that was criticized by almost all political factions, and the economic conditions of Egypt deteriorated even more under his rule. Politically, the MB president upheld Mubarak's political principles. He expressed commitment to maintain the Camp David agreement and during his rule the Egyptian military destroyed more of Gaza's underground tunnels (used to bypass the severe siege imposed by Israel) than during Mubarak's rule. He also remained skeptical regarding Iran's diplomatic relations with Iran were never fully restored, signaling acceptance and support for concerns expressed by the US and Saudi regarding Iran's regional role.

However, while the MB's performance in Egypt (and Tunisia) might have been the result of a serious lack of political and economic experience, ignoring the implications of Egypt's history and geography might have been fatal and that might have contributed significantly to their demise.

The Egyptian MB was described as the "mother ship of political Islam in the region" (Al-Amin 2013) and given the regional centrality of Egypt, it is possible that the challenge facing the MB and political Islam in general, in the aftermath of ousting President Mursi on July 3, 2013 is unprecedented. The question goes beyond the mere loss of power by the MB or the future of political Islam. A weak MB might be more dangerous than a strong one. A weak MB means that the leadership will not be able to maintain discipline among dissatisfied and radical members (after all Mursi was democratically elected) who might resort to violence. Only a strong and intact organization might be able to control the radical elements and remain involved in politics rather than resorting to non-political options and violence. A weak MB also might mean that many dissatisfied members might join radical Salafist's ideological groups who continue to reject the political option.

Nonetheless, the July 3 *coup d'etat* (there is no other way to characterize overthrowing a democratically elected president followed by a military takeover) is a significant event and could steer not only Egypt but the whole region into various directions. As the experience of the Egyptian uprising have taught us, it is very difficult to foresee how events might eventually unfold. It is, however, possible to imagine many potential scenarios. Most of these scenarios depend on how the MB will react to the military coup, how they grasp the new situation, and what their priorities are.

While the bloody Algerian or Syrian scenarios would be difficult to duplicate in Egypt, due to Egypt's history and composition of the population (Egypt's Muslims are overwhelmingly Sunni, which prevents any sectarian conflicts and keeps the ongoing conflict in the political realm), it is doubtful that the MB would give up easily and surrender to the military coup. They realize that it is the future of their movement and the future of political Islam that are at stake. If the MB's priority at this point is survival and influence, not primarily restoring the presidency, as it seems until now, a bloody confrontation like Syria 2011–2013 or Algeria 1991 is unlikely. However, that does not eliminate smaller scale clashes altogether, especially in areas with weak governmental control (for example, the Sinai).

Regardless of which scenario might unfold, it is unlikely that the decade old MB movement will utterly disappear or become totally irrelevant. The coup will nonetheless weaken the movement, especially since the performance of the MB leadership to counter the effects of the coup reflect the same lack of political experience they MB displayed during Mursi's presidency.

As for the MB, it seems inevitable that the movement must reinvent itself should it seek to remain politically influential. The movement must revisit and reconsider its political and economic and social views that were central in its downfall. But most importantly the movement must reconsider its geopolitical outlook to devise a strategy to restore Egypt's regional and international status. Otherwise, the movement will continue to grow weaker and lose more of its influence.

193.6   Conclusion

The rise and fall of the Muslim Brotherhood movement has been discerned as a matter of politics, economics, theology, and culture; but, steeped in complexity; they were and are many, many things. This chapter focused only on the geopolitical dimension and the ongoing regional antagonisms and transformations that contributed to the downfall of the MB movement. It argued that similar to both Sadat and Mubarak, but unlike President Nasser, the MB failed to grasp the uniqueness of Egypt's geopolitics and the political implications of both Egypt's history and geography. Based on a long scholarly tradition of some of Egypt's distinguished scholars and historians, and based on a rereading of what I called "Nasser's Revolution," this synopsis both assessed and questioned examples of the political performance of president Mursi during his single year in power. It attempted to show that the Muslim Brotherhood movement's comprehension of the significance of Egypt's history and geography and by implications their failure to perceive the nature of the ongoing Arab popular uprisings have been one of the underlying factors that led to its political downfall. In addition, this synopsis highlighted the shortcomings of the MB theological doctrine and its failure to recognize the political implications of Egypt's unique location, geography and history.

However, given the regional centrality of Egypt on one hand and the Egyptian MB, the significance of toppling the MB president might affect the entire regional and domestic political dynamics in the Arab World. The coup, similar to previous events in the region, is re-steering events in the entire Arab World into previously unforeseen directions making it very difficult to imagine how events will eventually unfold. This is a lesson that previous revolutions also taught us.

References

Al-Amin, B. (2013, July 24). Political Islam's moment of reckoning. *Al-Akhbar English*. Retrieved May 8, 2013, http://english.al-akhbar.com/content/political-islam's-moment-reckoning

Al Hafez, Y. (1978). *Al Hazima wa al ideologia al mahzoma* [Defeat and the defeated ideology]. Beirut: Arab Development Institute.

Bamyeh, M. (1999). *The social origin of Islam: Mind, economy, discourse*. Minneapolis: University of Minnesota Press.

Bennabi, M. (1999). *Islam in history and society*. New Delhi: Kitab Bhavan.

Bennabi, M. (2002). *Wojhat al lam al islami* [The direction of the Islamic world]. Damascus: Dar Al-Fikr.

Bennabi, M. (2006a). *Shoroot al nahda* [The conditions of renaissance]. Damascus: Dar Al-Fikr.

Bennabi, M. (2006b). *Masa'alat al thaqafa* [The question of culture]. Damascus: Dar Al-Fikr.

Braudel, F. (1972). *The Mediterranean and the Mediterranean world in the age of Philip II*. New York: Harper and Row.

Da'na, S. (2009). Islamic resistance in Palestine: Hamas, the Gaza war and the future of political Islam. *Holy Land Studies, 8*(2), 211–228.

Da'na, S., & Khory, L. (2004). History and race consciousness in the Arab World: colonial capitalism and the construction of race in the Arab World. In C. Fluehr-Lobban (Ed.), *Race and identity in the Nile valley* (pp. 347–378). Trenton: Red Sea Press.

Dana, S. (2013, June 28). *Ma'sat al Sadat malhat Mursi: tashreeh mukhtasar li raees Ikhwani* [The tragedy of Sadat, the farce of Mursi: A brief anatomy of a Muslim Brotherhood president] (No. 204). Retrieved May 8, 2013, http://www.al-akhbar.com/node/185990

Elmessiri, A. (1997). *The West and Islam: Clash points and dialogues, features of the new Islamic discourse: Introductory remarks*. Retrieved June 20, 2008, http://www.muslimphilosophy.com/ip/21-cen.htm

Fawzi, H. (1960). *Sinbad misry* [Egyptian sinbad]. Cairo: Dar Al Ma'aref.

Ghurbal, M. S. (1990). *Takween misr ibra al osor* [The making of Egypt over time]. Cairo: General Egyptian Press Organization.

Gorman, A. (2003). *Historians, state and politics in twentieth century Egypt: Contesting the nation*. New York: Routledge.

Haj, S. (2008). *Reconfiguring Islamic traditions*. Palo Alto: Stanford University Press.

Hamdan, G. (1967). *Shaksiyat Misr: dirasa fi abkariyat al makan* [The personality of Egypt: A study in the genius of place]. Cairo: Dar Al Hilal.

Iqbal, M. (2013 [1934]). *The reconstruction of religious thought in Islam*. Palo Alto: Stanford University Press.

Mu'nis, H.n. (1973). *Misr wa Risalatoha* [Egypt and its Mission]. Cairo: Egypt's commission on authoring, translation and publishing.

Nasser, J. A. (1996 [1954]). *Falsafat al-thawra* [Philosophy of the revolution]. Cairo: Modern Arab Documentation publisher (Bait al Arab li al-tawtheek al-hadith).

Rida, R. (2006). *Tarikh al-Imam Muhammad Abduh* [Life and history of Imam Muhammad Abduh] (Vol. 1). Cairo: Dar Al Fadila.

Rosefky Wickham, C. (2013). *The Muslim brotherhood: Evolution of an Islamist movement*. Princeton: Princeton University Press.

Sbahi, H. (2013, June 24). Interview with al-hayat newspaper. *Al Hayat*. 2013. Retrieved August 5, 2013, http://daharchives.alhayat.com/issue_archive/Hayat%20INT/2008%20to%202013/Alhayat_2013/06-June-2013/06-General/2013-06-28/28p05-01.xml.html

Wahida, S. (1974). *Fi Osol al masa'ala al misriya* [The origins of the Egyptian question]. Cairo: Madboli Books.

Part XV
Virtual Worlds and the Visual Media

Chapter 194
A Breath of Narcissism: Hollywood as Proselytizer of Secular Religion

C.K. Robertson

194.1 Introduction

In the shifting religious landscape of the early twenty-first century, the role of film as both an art form and a multi-billion dollar industry is a fascinating and, at times, complex one. The following chapter begins in the erstwhile sleepy town of Hollywood, California, a microcosm of the still largely rural America. It ends amid the glitz and glamor of modern-day Tinsel-town. Along the way, voyeurism and narcissism transform the town—and the country—supporting a growing trend in the rest of American life away from traditional forms of faith and towards a more indefinable spirituality of sorts. Indeed, as will be seen here, it can be argued that the film industry in its first hundred years became a proselytizer of "secular religion," filling a role "that would be satisfied by a church or another religious authority" (Wikipedia 2012). As Burt Lancaster's likeable but corrupt revivalist tells a group of staid, out-of-touch ministers in the 1960 movie, *Elmer Gantry*, "Like it or not, we are in competition with the entertainment world." To understand this development and its ramifications, this chapter opens with a glance back to the advent of film, then moves to key themes and influences on culture and community, and concludes with an exploration of the interplay between faith and film in this new era.

C.K. Robertson (✉)
Presiding Bishop, The Episcopal Church, 815 2nd Ave, 4th Floor,
New York, NY 10017, USA
e-mail: crobertson@episcopalchurch.org

© Springer Science+Business Media Dordrecht 2015
S.D. Brunn (ed.), *The Changing World Religion Map*,
DOI 10.1007/978-94-017-9376-6_194

194.2   Precursors to Hollywood

Our story begins a little over 100 years ago, in a town of tree-lined boulevards,[1] where churchgoers of every variety bumped into one another on their way to Sunday services, and where anyone wanting a cold beer or a gin and tonic had to travel several miles away to purchase and drink it. Hollywood at the start of the twentieth century was, after all, a wholesome town, a God-fearing town, and very much a dry town.

All that soon changed. A new world dawned with the coming of motion pictures. Of course, film did not arrive in Hollywood *ex nihilo*. As early as 1888, Thomas Edison had filed a preliminary claim with the U.S. Patent Office for his idea of a "Kinetoscope," a device that would, as the famed inventor put it, do for the eye what the phonograph (another Edison invention) did for the ear. It was his assistant, William Kennedy Dickson, who actually made the Kinetoscope a reality, creating 35 mm celluloid film that could then be projected in front of an Edison light source for people to view. In 1894, Edison built the first movie production studio in his West Orange, New Jersey hometown. It was a dark and uncomfortable space that Edison nicknamed "the doghouse," but Dickson and other assistants dubbed it the "Black Maria," a term used for the police paddy-wagons of that time. Dickson eventually broke away from Edison, who was not initially impressed with the video projection system, unable to sync it the way he wished with his phonograph sound system. He preferred to see the application used by one person at a time via a kind of "peep show" device, and chose as the content everyday events such as the scene of Fred Ott sneezing, the first copyrighted film.

It might more accurately be said, however, that the true origins of Hollywood as we know it are to be found an ocean away on December 28, 1895, where, at the Grand Café in Paris, around three dozen Parisians paid a fee of one franc each to attend the first showing of brothers Auguste and Louis Lumiere's "flickers," so named because of the uneven exposure from frame to frame as a result of the minor variations in speed of those first hand-cranked cameras.[2] While understandably primitive by today's standards, those first moving pictures amazed and even startled viewers. The claim that audience members shrieked and ran for cover when a train came barreling towards them in the Lumieres' L'Arrivée d'un train en Gare de La Ciotat may not be wholly accurate, but it is certainly "not that far-fetched" (Parkinson 2012: 16).

Word got around, and in a short amount of time, the century of film was underway. Director Edwin S. Porter's 1903 movie, *The Great Train Robbery*, served up

---

[1] A longtime resident of the area described the bucolic nature of pre-movies Hollywood: "It was so lovely with the oak trees, holly bushes, greasewood, and poppies. Ferns grew under the trees and by the little stream beds." From Gregory Williams, "The Story of Hollywoodland," www.beachwoodcanyon.org/HISTORY.htm, accessed September 1, 2012.

[2] Although those frame-to-frame flickers were gone forever with the move to fully automated cameras at the coming of the "talkies," the designation stuck and to this day we often speak of movies as "flicks."

an action-packed tale that left audiences breathless, and the iconic ending scene of the bandit pointing his gun directly at the camera and shooting elicited a similar audience response as the earlier Lumiere train scene. Porter's twelve-minute film, distributed by the Edison Manufacturing Company, was an instant hit. While "no accurate statistics were kept" in those earliest years, it is estimated that "millions of people the world over saw [*The Great Train Robbery*], and it earned unexpected wealth for its exhibitors" (Spehr 1977: 40).[3] Soon movie studios were springing up all over, located largely in "converted houses, stores, and warehouses," (Spehr 1977: 45) wherein films were produced in an almost factory-like manner. In Edison's Black Maria alone, between 200 and 300 films, all very short pieces, were produced. New York became home to countless studios,[4] though it was not long before they found their way west instead.

194.3   Hollywood: Early Years

And so it was that the wholesome, God-fearing, dry town of Hollywood, California soon attracted what film mogul Jack Warner described as a fascinating assortment of people: "fast-buck characters and con artists, real estate chiselers, soda jerks, writers who never wrote, call girls with agents, hoodlums and gamblers."[5] They came to Hollywood largely because of its perennial sunshine, something lacking in New York. In practical terms, this equated to ongoing film production all year long.[6] And, as David Parkinson notes, Hollywood also offered "a low tax base, plentiful labor, limited union interference, cheap real estate, and a diverse landscape within a 50-mile radius that could stand in for anywhere in the world" (Parkinson 2012: 54). The studio heads who set up shop there were "street-smart men attuned to the dreams of the working-class Americans who made up their first audiences" (Britten and Mathless 1999: 28). These authors saw the medium's possibilities before any-one else died. These visionaries invested time, money, and a pool of talent to trans-form the former temperance colony into a bustling city with a single primary export: celluloid escapism, whether from the despair of economic depression, the dread of global war, or simply the doldrums of everyday life. Metro-Goldwyn-Mayer, or

---

[3] Neglect of statistics was not the only problem in the early years of film. Hollywood was also "woefully shortsighted" when it came to preserving their actual movies. "The studios saw little value in silent films following their initial theatrical runs." From John Bengston. 2011. Silent visions (Santa Monica: Santa Monica Press), p. 9.

[4] "Like New York, film is big. Like New York, it is larger than life." James Sanders, Celluloid Skyline: New York and the Movies (Alfred A. Knopf, 2001), p. 22.

[5] Jack Warner of Warner Brothers fame spoke these words, cited in Gregory Paul Williams, The Story of Hollywood (BL Press, 2011), p. 102.

[6] James Sanders notes that New York was by no means completely left behind, as "an invisible chain" continued throughout the twentieth century to connect "the industry's financial, administra-tive, and publicity headquarters in midtown Manhattan to its burgeoning production center in Southern California," Sanders, p. 5.

MGM, summed it up best with their studio motto: "Make it good ... make it big ... give it class!" While the first and last parts of that motto might be debated at times, no one would question that Hollywood would do things BIG. Their movies would go on to produce "a realm of shared material fantasy" (Hoberman 2003: xiv). Indeed, it wasn't long before many people came to know the town as the Dream Factory. Others would more cynically refer to it by another name: Babylon.

It was not simply the time, money, and talent that made such an unforeseen impact, positive or otherwise, on the public and on popular culture. It was, as Edison came to admit, the cinemas or movie houses themselves. The power of the medium was largely the result of people experiencing the movies *together*. With the creation of movie houses, mass dissemination and mass consumption, crucial features of the modern age, became possible. While the original "nickelodeons" Edison's and the Lumieres' time were often "rather humble ... with ceilings so low that the top of the picture sometimes was cut off" (Butsch 2000: 32) movie houses soon grew in size and also in extravagance. With the 1927 opening of the famed Roxy Theater, the so-called "Cathedral of the Motion Picture," cinemas came to be likened to the magnificent churches of old.

The analogy is an apt one.

194.4   The Movie Theater

However, unlike a religious gathering, there are few, if any, common denominators to the group that gathers in a movie house. The unique communal nature of the cinematic experience results from the fact that anyone who buys a ticket can enter the cinema and, thus, "the viewing happens among strangers, usually in seats of parallel rows" (Pedullà 2012: 25). There is, of course, an obvious similarity here to musical concerts or stage performances. Because film, however, is not live and thereby limited in its immediate scope, but rather mass-produced and shown simultaneously in countless venues, the cinema has been able to evolve into the true weekly place of gathering. "Week after week, we return almost religiously to the dim auditorium of a movie theater. We sit there in silence and awe, surrounded by strangers, as the little beam of flickering light lures us into a self-disbelief and possibilities without limits" (Kupŝe 1989: 34).

Perhaps it is the very anonymity of the movie house that draws so many different people through its doors. They make their offering and receive an entry ticket, "little old ladies from retirement homes [mingling] with 40-year-old musclemen whose beer bellies hand over their shorts" (Lumet 1996: 204). The disparate group enters an air-conditioned space, greeted by the all-pervasive aroma of hot dogs and popcorn covered in something called a butter-substitute. They sit in cushioned seats complete with cup-holders, watching commercials and previews of coming attractions. In the fullness of time, the lights dim further and there begins the feature film. And for roughly 2 h the people "laugh, cry, ponder [their] humanity, and escape from it entirely" (Elden 2011: vii) and all of this while sitting beside fellow patrons they do not know and likely will never see again. Theirs is a shared experience while simultaneously an utterly individualistic experience.

In this "church" the true ministers are the filmmakers themselves, offering their literally larger-than-life visions for all to see. These off-screen mediators of the "big picture" are often effective prophets of their times precisely because they are so much a part of it. Steven Spielberg, for example, is well known and well loved not only because of his vision and technical expertise, but because of his sensitivity to the constituency he serves. Decades ago, famed movie composer Irving Berlin, "one of the most important forces in 20th-century American music and theater," (Magee 2012: xi) once proclaimed, "The mob is always right." This philosophy explains not only his own success, but also the impact of great filmmakers from Chaplin to Spielberg, all of whom have found a way to tap into the desires and dreams of the people for whose enjoyment they have created their product. In the glory years of Hollywood, that influence equated to huge numbers of people standing in line to enter the movies houses: "During an average week in 1938, America's population of 120 million bought 80 million movie tickets" (Britten and Mathless 1999: 6). It was then, and remains today, a matter of giving "the mob" what they want, while also in the process shaping what they will want, and maybe even what they may not want … yet. Christian missionaries of times past understood this. By and large, the religious establishment as a whole has not, and has resented Hollywood's influence. By the time that movies had become the fourth-largest industry in America, "there were already rumblings about its often titillating product from….clergy" (Piazza and Kinn 2011: 11). Not surprisingly, some filmmakers responded in kind.

194.5   Depicting/Portraying the Religions

As early as 1915, Lois Webber's silent classic, *The Hypocrites*, provided a picture of the cruelty and self-indulgence of church members and monks alike, all exposed by a literally naked Truth.[7] Francis Ford Copolla's 1972 masterpiece, *The Godfather*, presents "pious" Mafiosi making baptismal vows to renounce evil while authorizing a series of gangland murders. In the 1998 film, *The Truman Show*, Jim Carrey's lovable title character is little more than a puppet in a world created and preserved by Ed Harris' godlike tyrant, Christoph. The *Matrix* trilogy, replete with theological sublimation, depicts the "messiah" of trapped humanity as a leather-clad, independent-minded rebel. And great is the number of filmmakers through the years who have critiqued traditional ministers, portraying them as lushes, lechers, and worse. Still others have opted instead to offer their own grand visions of God and existential meaning in creative ways.

---

[7] It is sadly ironic that, although Lois Webber was hailed by a 1916 film magazine as "Best Film Director in the World," even over D. W. Griffith, women were quickly ousted from directorial duties and relegated to "starlet" positions in Hollywood.

The name of Cecil B. DeMille, for instance, has become forever synonymous with religious spectacle. When he introduced his 1956 remake of his own earlier silent classic, *The Ten Commandments*, the director claimed that his depiction of the hidden years of Moses drew heavily from the annals of ancient historians like Philo and Josephus. Yet even a cursory glance at DeMille's 1932 silent film reveals not a scholarly tome, but rather a profitable vehicle for displaying shocking violence and sexual orgies, climaxing in the horrendous deaths of the immoral Egyptians in the Red Sea.[8] DeMille and filmmakers who have followed his lead have done with a camera lens what the on-screen Elmer Gantry does on a stage: offer a bold, bombastic, and sometimes gaudy presentation of religious "history" for the paying masses. Such presentations may be criticized, but they are not easily forgotten. To this day, when people think of the crossing of the Red Sea, it is not the reeds and tides described by modern biblical scholars that fill their imaginations, but Charlton Heston's outstretched hands before the massive parting waves. As one of Kevin Smith's characters in the 1999 dark comedy *Dogma* laments: "Mention something out of a Charlton Heston movie and suddenly everyone is a theology scholar."[9]

Hollywood has enjoyed an alternative to its own religious epics: if the spectacular does not work, there is always the sentimental. Directors of the many "angel movies" through the years may rightfully tip their caps to Frank Capra, whose holiday standard, *It's a Wonderful Life* (1946) set the standard for a very "human" angel in the character of tenderhearted and absentminded Clarence. The theological description of angels is pure fluff: "Teacher says that every time a bell rings an angel earns his wings." It is, however, more palatable to most audiences than the heavenly warriors of the book of Revelation .... and the larger message of the infinite importance of an individual life remains a profound one. Indeed, for most on-screen saints, seraphs, and spirits—from *Here Comes Mr. Jordan* (1941) to *City of Angels* (1999)—eternity in paradise may be okay, but a passionate night on earth usually sounds a lot better. Many people watching these movies, whether they attend a house of worship or not, would probably nod their heads and agree.

Movies are so effective in their messaging precisely because they act as "personal touchstones" for audiences, providing "soundtracks and wallpaper for our memories, and sometimes...part of our own histories" (Elder 2011: 41). Using the composer Irving Berlin again as an example, note how a Jewish immigrant gave moviegoers the

---

[8] DeMille's epics were "invariably spiced with romance and violence but often neatly wrapped in the cloak of a historical or biblical event." Britten and Mathless, p. 63. It was because of the lurid images of early directors like DeMille that the so-called Hays Code was created, to set industry standards of what would be considered appropriate. Another more subtle response from Hollywood itself was the creation of the Academy Awards, a way of lifting up the finest in their art, to set a standard not of morality as it were, but of excellence. The public enthusiastically endorsed the coming of the Oscars. See Piazza and Kinn.

[9] Smith's Dogma, one in a long line of films banned by the Roman Catholic Church, goes to extremes in its jabs at organized religion and its ministers, containing such outlandish characters as Rufus, the thirteenth apostle, left out of history because he was black; two inane, drug-taking prophets, Jay and Silent Bob; a sarcastic embodiment of the Voice of God, and public relations campaign by a radical Roman Catholic cardinal played by comedian George Carlin to replace the statues of Jesus as crucified with ones where he displays a toothy smile and offers a big "thumbs up."

memorable title song from the movie *Easter Parade*. He intuitively grasped what priests and preachers often missed: that the real highlight of that Christian holy day was not their sermons, but rather the carriage ride down the boulevard after church, as neighbors smiled and waved and showed off their Sunday finery. Likewise, as an immigrant who had been embraced by his adopted home, he understood the deep love of country that ran through his listeners and offered them the song "God Bless America." Given voice by popular singer Kate Smith in November, 1938, first on Armistice Day and then again two weeks later on Thanksgiving, the song was an immediate hit, so much so that Berlin actually had to deny that he was trying to replace "The Star Spangled Banner" as the national anthem with his own song.[10] This is why Berlin scholar Jeffrey Magee is able to assert with authority that the composer did more than create music; he gave people "a redemptive form of secular religion" (Magee 2012: xi). This was true not only of Berlin, but of the entire system of which he was a part. As veteran actor Lionel Barrymore said to rising star Jimmy Stewart when they were together in *It's a Wonderful Life*, "Jimmy, don't ever forget that … when you act you move millions of people, shape their lives, give them a sense of exaltation. No other profession has that power" (Hoopes 1994: xix).[11]

So what is at the heart of this powerful alternative system of which filmmakers, composers, actors, and, yes, audiences, are all integral parts? Miriam Hansen, a specialist on early cinema, notes that from film's earliest days there has been a narcissistic nature inherent in what people see and experience, something that was particularly evident in the silent films depicting scenes from everyday life, offering viewers "the possibility of seeing oneself or someone familiar on the screen" (Hansen 1991: 31). While Hansen argues that this changed the more that film as an art form evolved and matured, Steven Ross, historian and recipient of the Academy of Motion Picture Arts and Sciences' Film Scholars Award, asserts that Hollywood not only was, but remains, a town "where narcissism is only slightly less widespread than oxygen" (Ross 2011: 10). What matters is what matters *to me!*

We see this in movies like Robert Altman's *The Player* (1992), in which the main character, Mr. Mill, a studio executive who has been receiving a series of threatening notes from a script writer. In a fit of rage kills the person he believes to be the perpetrator, only then to enter into an affair with the dead man's lover. In a provocative conversation between Mill and June, we catch a glimpse, not only of their characters, but also of their world.

[Mill] "We should pay for our crimes, shouldn't we?"
[June] "I don't know. I think that knowing you committed a crime is suffering enough. If you don't suffer, maybe it wasn't a crime after all. Anyway, what difference does it make? It has nothing to do with how things really are."
[Mill] "You don't really believe that, do you?"

---

[10] The fact that the profits from "God Bless America" went to the Boy Scouts and Girl Scouts of America only added to Berlin's legend, making him something of a national "saint."

[11] Hoopes further notes, the government understood this influence. During World War II, some actors—including Jimmy Stewart, Clark Gable, and Henry Fonda—interrupted their film careers to serve their country as soldiers, sailors, and pilots. Others were called on to help with fund-raising efforts, to entertain troops, and, of course, "to continue making films that could rally the people back home," p. 334.

[June] "I don't know what I believe, Mr. Mill. It's just what I feel."
[Mill] "You know what you are, June whatever-your-name-is? You're a pragmatic anarchist."
[June] "Is that what I am? I never was sure."

That philosophy of "it's just what I feel" is far removed from a traditional Judeo-Christian belief that you should "do unto others as you would have them do unto you." But somehow the Golden Rule feels outdated and naïve in a world of temporary sets and facades.

In another telling scene, conscience-stricken Mill, while making love, tries to confess his dark deed to June, who only hushes him and tells him that nothing matters but that moment right then, that moment of pleasure and intimacy. The morning after, like a modern Lady Macbeth, her face hidden under a mud-bath shell, June's eyes reveal something…a sense of guilt or dread, perhaps. Mill receives a call and is whisked off to face a police line-up and the inevitable judgment for his actions. But through a fluke, he is acquitted. At the end of the movie, we see Mill literally get away with murder, rise up in the studio hierarchy, and go home to a now-pregnant June. And what is most disturbing is that he appears to be perfectly comfortable with it all. In the end, all that really matters, apparently, is what happens to him. He may call it pragmatic anarchism, but it is plain and simple self-interest all the same.

In the 2000s, on the cable channel HBO, an eight-season series produced by film actor Mark Wahlberg entitled *Entourage* has similarly focused on insider Hollywood, this time through the lives of a rising movie star, his agent, and his lifelong buddies who are along for the ride. No matter what is going on around them in the larger world, everything that matters to this star and his entourage is right there in the cocoon that is Hollywood, where the sun is always shining and the greatest worry is about the star's career or his manager's love life. For the viewer, the peek into that world week after week can almost seem surreal, the moving picture equivalent of reading a stack of celebrity tabloids. As one cast member admits, "There is just a fascination with Hollywood … the glamour and the spotlight. It's part of the scenery of Hollywood. You don't like it? I respect that … but there are other people who love it"[12] Indeed, the viewing public often finds itself in the role of voyeurs paying to see their favorite stars do what they themselves either dream of doing or fear doing.

194.6 Towards Secularism in Religion

But it is hardly all about self and selfishness. The greatest movies have tapped into, and adapted in new ways, timeless mythologies of good and evil, of heroism, of sacrifice for the greater good. From *The Great Train Robbery* on to today, westerns

---

[12] From an interview with Emmanuelle Chriqui in "Hollywood Sunset: A Farewell to Entourage," from Entourage: The Complete Eight Season (HBO, 2012).

gave us good guys and bad guys, often distinguished from one another by fairly superficial means such as the colors of their Stetson hats. War movies, once rallying points during actual wartime, continued on to show the extraordinary depths of heroism by otherwise ordinary "Joe's." During the end of the 1970s, a particularly cynical decade in American culture and, by extension, American film, George Lucas updated the old-fashioned good versus evil western by setting it "a long time ago in a galaxy far, far away," transforming movies forever. Besides the remarkable technological gifts that came from the creative force behind *Star Wars* (Industrial Light Magic, LucasSound, Pixar animation), Lucas also tapped into the concepts of Joseph Campbell and gave audiences a new mythology. Many books and articles have been produced exploring the *Star Wars* concept of the Force, but in the end it is less important what Lucas specifically intended by it than what it says about hope in the transcendent. Before *Star Wars* science fiction films or movies about the future were often more grim, and at certain times since then—notably at the turn of the millennium and following 9-11-01—apocalyptic films have still found an important place (Ostwalt 1998). Whether the future pictured on screen is bright or bleak, however, it is usually secular, in the sense that it does not promulgate a particular religious doctrine of salvation or apocalypse, but rather a more universal sense of something beyond ourselves.

Superhero movies, a genre now possible through many of the technological innovations of Lucas and others like him, have taken Hollywood by storm. From the moment that *Superman: The Movie* director Richard Donner helped us "believe that a man can fly," four-color heroes have leapt from the pages of comics to the big screen, often becoming box office hits. It is little wonder that comic books now "litter the offices of the studios" (Jones and Jacobs 1996: 212). They are modern day morality tales in brightly-colored garb, old mythologies in new guises (Robertson 2012: 46). Avi Arad and Kevin Feige, respectively the former and current heads of Marvel Studios, have helped us appreciate with Spider-Man how "with great power comes great responsibility," have reminded us through Captain America that old-fashioned values are still in style, and have taught us with the X-Men that difference and diversity are to be honored, not feared or taunted. Iron Man, Thor, the Hulk, Daredevil, the Fantastic Four, and of course the collection of heroes assembled as the Avengers all tap into archetypal dreams that lie within each of us. Stan Lee, the co-creator of most of these "marvelous" heroes confesses, "Somehow, though I really wasn't making an attempt to create any sort of mythology, one started to develop on its own" (Oropeza 2005: 41). It is little surprise, then, that masses of people beyond comic book aficionados have welcomed these mythical, archetypal characters, for in truth they inspire all who would still enter the doors of our secular churches.

Of course, it is impossible to speak of secular religion in film without recognizing a major theme, perhaps *the* primary theme in American movies. Whether it is a romantic comedy or an edge-of-your-seat thriller, the common denominator in most movies is the so-called Hollywood ending, which is by definition a happy ending. This is particularly interesting, given that life does not work this way. But this is because films are all about shooting a section of a life, not the whole thing, which often can be dull and uninteresting. As Alfred Hitchcock once famously remarked, "What is drama, after all, but life with the full bits cut out?" (Britten and Mathless 1999: 56). Like the cave painters of old, filmmakers are able to give us images that take hold of our imagination and often give hope. Frank Capra, who directed so many classic films in the 1930s and 1940s, made idealism the mark of his movies. "He evoked an America where the little guy can, by decency and strength of will, prevail over the powers of greed and venality" (Britten and Mathless 1999: 75). Few directors have matched Capra in this, but many more have offered a different, even more pervasive happy ending found as the main character somehow finds the love of his or her life. The notion that there is someone out there for me, and that I will somehow know it by my feelings, and that my soul mate and I will then live happily ever after is not only a fairy tale, but a basic Hollywood plot. The very notion in some religious traditions at one time or another of arranged marriages is anathema in movieland.

"Hollywood endings" have at times been challenged by filmmakers themselves. But the fact remains that for every film that ends in a depressing, "real-life" way, there are countless movies that leave audiences feeling good as the final credits roll. This is, after all, why many of them give their hard-earned money to see a movie in the first place. It is arguably another form of narcissism, but less distasteful than the extreme form found in *The Player*. But like June in that movie, it is still usually about "what I feel." And what most people want to feel is happy. Thus, again, at the heart of the secular religion that Hollywood continues to promote is a feelings-oriented, me-focused message, albeit in much more subtle garb. If my heroes on the big screen can in the end find love and enjoy a taste of happiness, then maybe, just maybe, I can as well. For far too many, guilt and the weight of responsibility come naturally. And for these, traditional religion may appear unattractive. Filmmakers go out of their way to make sure they do not come across to their audiences as "preachy," with the implicit message here being that preaching, a fundamental action in houses of worship, is a negative thing, something that makes people feel, well, guilty. So, many who will not darken the doors of a church or synagogue will continue to buy a ticket to the cinematic cathedrals not simply for entertainment, but even more to be comforted with the possibility of a happy ending for them.

194.7   Hollywood's Futures

But is there a happy ending for Hollywood itself? Once more, change is in the air in the film industry. Gabriele Pedullà argues that film, the mass art form of the twentieth century, is now past its prime. Much of this is again due to changes in technologies, as well as human social patterns in an increasingly globalized world. People today covet personal entertainment more than ever, but the means to obtain it have expanded dramatically. Beginning with the advent of television, movie houses have increasingly faced competition from other sources, so much so that "the age of cinema, it is commonly claimed, is now drawing to a close" (Pedullà 2012: 1).

Cable and satellite TV, tape and digital recording devices, internet streaming, video games, personal computers, iPods and iPads—all have transformed the entertainment world by allowing the viewer to choose when, where, and with whom she or he will watch, play, or otherwise experience entertainment. As Pedullà suggests, "we have become indifferent to whether we see a film on widescreen with Dolby Surround, on an eighteen- or forty-two inch TV, on a laptop, or on our mobile phones. And the new generation will not even feel deprived for never having seen the work of their beloved directors and divas on the big screen" (Pedullà 2012: 2). People still go to the movies, still enjoy the communal experience, but now that is simply one option among many. In many ways, the film industry itself is at least in part responsible for the shifts. As early as the 1970s, studios were noticing that the glory days, "the days when the dream factories were running full blast, peopled by almost mythical celebrities," were behind them (Eames 1975: 9). In this sense, movie houses face challenges similar to those of churches and synagogues: "The cathedral of cinema, like all sanctuaries… increasingly resembles a residual, albeit prestigious, space, destined to lose more ground" as it seeks to attract people who are increasingly distracted by alternatives of all kinds (Pedullà 2012: 132).

And yet, for Americans born in the latter half of the twentieth century, the movie theater remains a "primary arena for discovering and debating important moral, cultural, and religious issues" (Robertson 2002: 237). Famed director Martin Scorsese, who actually considered entering the priesthood when he was younger, rightly asserts that many movies "address themselves to the spiritual side of man's nature" (Robertson 2002: 237). Pastors and youth ministers still show movies to their groups and then lead discussions on what they saw and heard therein, though it may be through DVDs and web-streaming instead of actually going into movie houses. More than ever, actors are tapped for political and social causes, with some analysts suggesting that in the future "movie stars are more likely to bring greater sanity to political discourse than most news/entertainment show commentators" (Ross 2011: 417). And the messages of this secular religion—of patriotism and idealism, of hope and "true love," and most of all, of happy endings—continue to find their way onto the silver screen and resonate with audiences everywhere.

194.8   Retrospect and Prospect

And so we find our way back once more to that once quiet town in southern California, where, as the movie moguls arrived over a century ago, churchgoing residents of all types lined up along Hollywood Boulevard and prayed (Williams 2011: 30). Today, that same street looks quite different, and few people walking past the cafes and shops are praying or, if they are, it is most certainly for the lucky break to get out of waiting tables and into the part that will sail them to stardom. Now, a personal note, if I may. I myself have walked those streets many times, working with my friend and colleague Michael Rhodes, five-time Emmy Award-winning director and president of Film Clips, Inc. I have had the joy and frustration of seeing studios and guilds up close, and have found myself in moments that feel *surreal*. Perhaps that is the word for all of it. There is much good that comes from the proselytizers of secular dreams who fill those streets today. But it is all a bit surreal, like the haze that often hides the surrounding hills. Walking down Rodeo Drive and Sunset Boulevard, I know that I am "not in Kansas anymore." Neither am in the heaven of traditional religion. No, I am in Oz, where everything glistens and the sun is shining and people behind the curtains are making deals and producing spectacles. And even though I know that most of this is ethereal and fleeting, even though I admit that I am breathing in a bit of narcissism myself at that moment, I cannot help but look down at the stars on the street and make a wish that my own story will have a Hollywood ending.

References

Bengston, J. (2011). *Silent visions*. Santa Monica: Santa Monica Press.

Britten, L., & Mathless, P. (Eds.). (1999). *100 years of Hollywood* (Our American century). New York: Time-Life Books.

Butsch, R. (2000). *The making of American audiences: From stage to television, 1750–1990*. Cambridge: Cambridge University Press.

Eames, J. D. (1975). *The MGM story*. London: Octopus Books.

Elder, R. K. (2011). *The film that changed my life*. Chicago: Chicago Review Press.

Hansen, M. (1991). *Babel and Babylon: Spectatorship in American silent film*. Cambridge: Harvard University Press.

Hoberman, J. (2003). *The dream life: Movies, media and the mythology of the sixties*. New York: New Press.

Hoopes, R. (1994). *When the stars went to war*. New York: Random House.

Jones, G., & Jacobs, W. (1996). *The comic book heroes*. Grängesburg: Prima.

Kupŝe, J. (1989). *The history of cinema for beginners*. New York: Writers and Readers.

Lumet, S. (1996). *Making movies*. New York: Vintage Books.

Magee, J. (2012). *Irving Berlin's American musical theater*. New York: Oxford University Press.

Oropeza, B. J. (Ed.). (2005). *The gospel according to superheroes*. New York: Peter Lang.

Ostwalt, C. (1998). Visions of the end: Secular apocalypse in recent Hollywood film. *Journal of Religion and Film, 2*, 1.

Parkinson, D. (2012). *100 ideas that changed film*. London: Laurence King Publishing.

Pedullà, G. (2012). *In broad daylight: Movies and spectators after the cinema*. New York: Verso.

Piazza, J., & Kinn, G. (2011). *The academy awards*. New York: Black Dog and Leventhal.

Robertson, C. K. (2002). Sorcerers and supermen: Old mythologies in new guises. In J. McGrath (Ed.), *Religion and science fiction* (pp. 323–358). Eugene: Pickwick Publications.

Robertson, C. K. (Ed.). (2012). *Religion as entertainment*. New York: Peter Lang.

Ross, S. J. (2011). *Hollywood left and right*. New York: Oxford University Press.

Sanders, J. (2001). *Celluloid skyline: New York and the movies*. New York: Alfred Knopf.

Spehr, P. D. (1977). *The movies begin*. Dobbs Ferry: Morgan Press.

Wikipedia. (2012). *Entry on "secular religion"*, http://en.wikipedia.org/wiki/Secular_religion. Accessed 5 Sept 2012.

Williams, G. P. (2011). *The story of Hollywood*. Los Angeles: BL Press.

Chapter 195
Towards a Virtual Geography of Religion

Paul Emerson Teusner

195.1   Introduction

If geography is the study of people in space, then arguably virtual geography is the study of people in relation to virtual space, and relationships between them therein. This chapter will present an overview of theories, concepts and issues that lie in the study of religion online, and explore how the term "virtual" has been approached and apprehended in the use of Internet technology to create and disseminate religious experience and information. In order to do this, a distinction will be argued between two avenues of exploration: spatial geography and network geography. The former considers the shape and contours of aesthetic experience and communication online. The latter explores the connections between sites and users within the context of life offline.

195.2   Toward a Spatial Geography of Online Religion

In actuality, the Internet consists of one-dimensional data streams between computer servers and clients. The creation and development of user access technology, together with popular imagination, have given birth to the conceptualization of the Internet as a multi-dimensional "space". Arguably this began with the introduction of the Graphic User Interface (GUI) to Macintosh and Windows in the early 1980s GUI-based operating systems transformed the experience of computer information, from lines of worded code to pictorial symbols, allowing users to arrange

P. Emerson Teusner (✉)
School of Media and Communication, RMIT University, Melbourne, VIC 3000, Australia
e-mail: paulteusner@me.com

© Springer Science+Business Media Dordrecht 2015   3725
S.D. Brunn (ed.), *The Changing World Religion Map*,
DOI 10.1007/978-94-017-9376-6_195

information within a symbolic space (Campbell 2005: 12; Turkle 1996: 14). The use of a mouse to "click and drag" files of data added to the experience of "moving" information.

Intuitively, our language would construct the World Wide Web in the same way. Information stored online would be known as "web sites," that could be access by typing its "web address." Personal web sites would be known as "home pages", and audiences would be known as "visitors." As Søraker notes, information would be ordered according to "their topological location in relation to other virtual entities" (2011: 46).

Long before the Internet penetrated daily life, virtual worlds were part of popular imagination. William Gibson's 1984 novel, *Neuromancer*, introduced readers to "Cyberspace," a digital matrix connecting humans to a collective consciousness, releasing them from the limitations of embodiment to create new modes of being. So compelling was this Cyberspace an escape from real life that the novel's protagonist resorted to illicit drugs to replicate the experiences once disconnected from it.

Similarly the Internet has been socially constructed. For Wertheim, Cyberspace is an "immaterial space of mind" into which we transport ourselves, though physically we remain (1999: 41). Being a "place outside physical space," the Internet is communally produced, both being born of and bringing to birth new kinds of language and practice (1999: 303). Potentially, then, Cyberspace becomes like a "New Jerusalem," built by story and myth, a reflection of the world we desire, rather than the one we have (1999: 20). It is only natural that organized religions would aim to create a presence in Cyberspace. As Brasher notes, to do so "classifies cyberspace as a valued and value-producing home," and recognizing that its construction is spiritual as any human social endeavor (2001: 142).

A spatial geography of religion online considers how Internet users approach religious experience, identity and community in religious sites. Such sites include those who visually recreate offline religious spaces, such as the *Cyberhajj* Mecca site on Second Life, created by OnIslam.com, and the UK Methodist *Church of Fools* (www.churchoffools.co.uk). In both sites, traditional religious spaces (Haram Mosque and a cathedral, respectively) are represented, providing a visual frame for actions of, and interactions between, users, who are also visually represented by avatars. The aim of both sites is to create an aesthetic experience of being within a real-world religious space among other virtually co-located users. In the first instance, the aim is to educate about the experience of the real Hajj (Solayaman 2010). In the second, the aim is to produce an experience of virtual religious community.

This form of inquiry may not be limited to visually created spaces, but include email groups, group pages on social networking sites like Facebook, home pages which enable discussion or comments (such as web logs) and mobile device applications. While not necessarily providing a visual spatial aesthetic, they offer a setting for interaction between users that may be discursively constructed as a "space" or "place." In email groups, for example, while users do not share a visual online space, but receive messages through an email client program, communications between users allow for the interweaving of individual experiences to effect a common identity and sense of togetherness (Campbell 2005: 79).

The mobile device application developed for a First United Methodist Church in Shreveport, Louisiana, intends to recreate the experience of being at church through a mobile phone. This is done by making available video and audio recordings of choir performances and pastor sermons and listing the pastor's Twitter entries. While interaction in real time with others who access the application is not afforded by the software, much is done to produce an aesthetic experience of being part of the community.

In all these examples we see the endeavor to overcome physical distance by the creation of new space and the aesthetic experience of being connected to others within it. The apprehension of the term "virtual" in this sense is that Cyberspace is "another space," distinct from the "real world" and bound by symbol and simulation.

A spatial geography of religion online uncovers a number of issues in relation to religious practice among users, and the production and dissemination of religious information. One example is the question of whether simulated aesthetic experiences on a computer screen of religious practices may constitute authentic religious experience. This question is particular important in Christian and Muslim traditions, where material and embodied experiences are privileged (for example, sharing of food, facing in common direction during prayer). For other religions, where material practices are less privileged than visual or aural, simulated experiences appear to be questioned with less fervor. *Buddhist Space* (www.buddhistspace.com), for example, allows users to create virtual shrines for deceased loved ones that are built with text, pictures and video, and where visitors can contribute to shrines by lighting "virtual candles."

A similar issue stems from conceived theories of identity play that is allowed to users in online spaces where our embodied selves are hidden (see Lövheim 2005; Turkle 1996). In spaces such as email groups, the question of authentic experience of community is raised where users cannot verify the identity of others in the communicative space (see Larsson 2005). Recent studies have debated the severity of this issue, claiming that most users seek to present, as much as possible, authentic identities online, in the pursuit of relationships that flow between online and offline interactions (Hine 2000: 15; Kennedy 2006: 862).

How online spaces allow for new negotiations of traditional structures of religious authority is another issue in this field of inquiry. Papacharissi argues that all online spaces are structured to privilege certain uses, types of speech and forms of behavior (2009: 216). How the technology favors certain forms of behavior may present a challenge to traditional patterns of speech and behavior that determine in part, how authority is distributed among a group. When the *Church of Fools* first opened, for example, the opportunity to provide a sermon or speech was open to any user, not merely clergy present online. This presented issues regarding rules of behavior when speaking from "the virtual pulpit" which required negotiation from users and the site's moderators.

A spatial geography of religion online is arguably limited by its assumptions. The first of which is that online spaces are bounded in some respect, either by the design of the site or, in the case of email groups or membership to pages on social networking sites, the number of users who gather. The second is the limitation that arises when considering use of religious sites online in separation from the context in which they are used. That is, this exploration neglects how online religion influences and is influenced by offline religious practice and participation. The following approach to a geography of religion online will attempt to address these limitations.

195.3   Toward a Network Geography of Online Religion

Filoramo sees Cyberspace as "being at the opposite pole to the traditional sociological scenario of the great religions, composed of institutions, bureaucracy and tradition; it is non-stop chaos: anarchy, immediacy, fragmentation" (2003: 39). Dawson complements the assertion in his theorizing of the "disembodiment" of religious organization in online settings, resulting in the removal of material sacred objects and a greater dependence on the reflexivity of identity, community and authority (2005: 33). In these assertions the writers imagine the expression of religious ideas in words and practice completely free from the rule of the Church. This researcher suspects that they may be too extreme in their conceptions of religious organization both online and in the real world. Indeed, while religious institutions may be bound by bureaucratic systems, local, grounded communities are known to frequently negotiate official dogma, teachings and practices with personal and communal experience. The reverse may be true for religion online, where users negotiate new social conventions, language, identities, and the freedom thereof, with the knowledge, values and conventions of their offline religious experience.

This chapter has already noted that patterns of authority are being identified by users. Hyperlinks form the currency for various search engines that allocated authority to Internet places and voices. Research has noted that certain discourses attract this currency and, therefore, specific groups of users are awarded greater authority than others. Religious authority may be more or less "disembodied" from the offices of religious institutions in an online space. However, if we understand Cyberspace as a "space of flows," then we must take into account the discursive and symbolic patterns of this space that place constraints on interaction and expression. As Turner puts it, power "is not necessarily embodied; it is a switchpoint in the information flow" (2007: 123). If users of new online media engage in building of religious identity and community by navigating the information flows, they will recognize, accept, adopt, claim, reject, nurture and/or subvert voices appealing to authority, voices claiming authority and the systems by which that authority is legitimated. These systems may be imported from offline religious experience and knowledge, or may be found in Cyberspace.

A network approach to religion online acknowledges that users do not experience online religious content outside of an offline context, but negotiate that information within the flow between both online and offline contexts. This is particularly true in the modern West, where, Pang (cited in Thomas 2006: 389) argues, there is no longer a discrete step between being offline and being online.

> Designers and computer scientists are also trying hard to create a new generation of devices and interfaces that don't monopolize our attention, but ride on the edges of out awareness. We'll no longer have to choose between cyberspace and the world; we'll constantly access the first while being fully part of the second.

Moreover, the approach recognizes what Castells (2001) argues as the rise of personal relationships outside families and embedded communities (schools, churches, sporting groups, workplaces) as a dominant pattern of sociability, to the embodiment of "me-centred networks." Castells blames not the Internet on the rise of networked individualism, but sees it "provides an appropriate material support for the diffusion of networked individualism as the dominant form of sociability" (2001: 131).

Prayerbuddy (www.prayerbuddy.org) works within this understanding by providing information and helping establish networks between people who wish to develop personal spiritual disciplines in their own settings. As stated on the introductory page:

> Members of Prayerbuddy live ordinary lives in the world, but with a difference: they share rich wireless presence with each other day and night. Within this unceasing presence to one another they are interrupted together – six times each day – to pray an ancient monastic Divine Office as it is practiced at the Monastery of Christ in the Desert.

Rather than gathering people into a symbolic or discursive space for interaction, the main aim of *Prayerbuddy* is merely to introduce members to one another for the purposes of integrating spiritual practices in daily life.

Online web authoring applications, such as web logs (or blogs), also present a site for this form of inquiry. Rather than gathering a community of people, blogging is primarily an individual pursuit, to share personal thoughts and experiences in the public realm. According to Cheong et al., blogging provides an opportunity to share values with a known or imagined audience, and to integrate religious ideals with the experiences of everyday life (2008: 122). Blogging provides "an integrative experience for the faithful, not a "third place," but a melding of the personal and the communal, the sacred and profane" (2008: 107).

My own study of religious bloggers reveals that motivations to produce online content lie in a sense of dissatisfaction with traditional communal practices and doctrine and to seek like minded people with whom to evaluate their own religious participation and seek new forms of practice. In doing so they join a global conversation that spans a multitude of web sites, connected by hyperlinks and fuelled by discussions in posts and comment threads (Emerson Teusner 2011).

In these two examples it is shown that religious practice online, rather than simulating offline religious experiences, or creating new ones, to form distinct communities, is aimed at integrating with and enhancing offline practice. In a networked geography of religion online, virtual does not refer to "another space," but a system of information that sits at the periphery of all settings of social interaction.

Issues explored in this form of inquiry involve the formation and negotiation of common identity within a network of users who are tied to separate settings offline. These include the formation of common discursive practices and aesthetic values (see Emerson Teusner 2010), the assertion of authority among multiple web sites (see Howard 2011), or the formation of global tribes and diasporas (see Alonso and Oiazarbal 2010).

195.4   Conclusion

This chapter has introduced two modes of inquiry into religion online. The first recognizes the discursive construction of the Internet as virtual space, and seeks to map religious practice and participation with symbolically bounded spaces. The second recognizes the networked configuration of flows of information and interaction between online and offline places and maps connections between individual sites on the Internet, especially in regard to new online applications. Each mode highlights distinct questions and issues that pertain to religious life online. As new Internet technologies emerge and develop, so too we may see religious practice evolve online, with new questions calling for new models and approaches.

References

Alonso, A., & Oiazarbal, P. (Eds.). (2010). *Diasporas in the new media age: Identity, politics, and community*. Reno: University of Nevada Press.

Brasher, B. (2001). *Give me that online religion*. San Francisco: Jossey-Bass.

Campbell, H. (2005). *Exploring religious community online: We are one in the network*. New York: Peter Lang.

Castells, M. (2001). *The internet galaxy: Reflections on the internet, business and society*. New York: Oxford University Press.

Cheong, P. H., Halavais, A., et al. (2008). The chronicles of me: Understanding blogging as a religious practice. *Journal of Media and Religion, 7*(3), 107–131.

Dawson, L. L. (2005). The mediation of religious experience in cyberspace. In M. T. Højsgaard & M. Warburg (Eds.), *Religion and cyberspace* (pp. 15–37). London: Routledge.

Emerson Teusner, P. (2010). Imaging religious identity: Intertextual play among postmodern Christian bloggers. *Online – Heidelberg Journal of Religions on the Internet, 04.1,* Special Issue on Aesthetics and the dimensions of the senses.

Emerson Teusner, P. (2011). Networked individualism, discursive constructions of community and religious identity: The case of Australian Christian bloggers. In F. Comunello (Ed.), *Networked sociability and individualism: Technology for personal and professional relationships* (pp. 265–289). Hershey: IGI Global.

Filoramo, G. (2003). Religious pluralism and crises of identity. *Diogenes, 50*(3), 31–44.

Gibson, W. (1984). *Neuromancer*. New York: Ace.

Hine, C. (2000). *Virtual ethnography*. London: SAGE Publications.

Howard, R. (2011). *Digital Jesus: The making of a new Christian fundamentalist community on the Internet*. New York: New York University Press.

Kennedy, H. (2006). Beyond anonymity, or future directions for internet identity research. *New Media Society, 8*(6), 876.

Larsson, G. (2005). The death of a virtual Muslim discussion group: Issues and methods in analysing religion on the net. *Online – Heidelberg Journal of Religions on the Internet, 01.1*, Special Issue on Theory and methodology edited by Oliver Krüger. http://online.uni-hd.de/. Accessed 26 Apr 2006.

Lövheim, M. (2005). Young people and the use of the internet as transitional space. *Online – Heidelberg Journal of Religions on the Internet, 01.1*, Special Issue on Theory and methodology edited by Oliver Krüger. http://online.uni-hd.de/. Accessed 1 Apr 2006.

Papacharissi, Z. (2009). The virtual geographies of social networks: A comparative analysis of Facebook, LinkedIn and ASmallWorld. *New Media & Society, 11*(2), 199–220.

Solayaman, H. (2010). Cyber Hajj season begins in second life. *EMAJ Magazine*. Retrieved March 31, 2012, from http://emajmagazine.wordpress.com/2010/11/14/cyber-hajj-season-begins-in-second-life/

Søraker, J. H. (2011). Virtual entities, environments, worlds and reality: Suggested definitions and taxonomy. In C. Ess & M. Thorseth (Eds.), *Trust and virtual worlds: Contemporary perspectives* (pp. 44–72). New York: Peter Lang.

Thomas, S. (2006). The end of cyberspace and other surprises. *Convergence: The International Journal of Research into New Media Technologies, 12*(4), 383–391.

Turkle, S. (1996). *Life on the screen: Identity in the age of the internet*. London: Weidenfeld & Nicholson.

Turner, B. S. (2007). Religious authority and the new media. *Theory, Culture & Society, 24*(2), 117–134.

Wertheim, M. (1999). *The pearly gates of cyberspace: A history of space from Dante to the internet*. Sydney: Doubleday.

Chapter 196
The Creation of Secularist Space on the Internet

Christopher Smith and Richard Cimino

196.1   Introduction

The growing presence of secularist thought and discourse in the U.S. is directly related to the contemporary transformation of space brought about by the increased importance of electronic communication. In this chapter we examine the role of the Internet in creating a space for the formation of a more visible and assertive secularist identity that is informationally greater than the sum of its participants.

A significant expansion of websites and blogs dealing with atheism has appeared in the last decade, opening up active spaces for atheists to construct and share mutual concerns about their situation at a time when American public life is still largely functioning under a norm of religiosity in many contexts (Cimino and Smith 2010, 2011). These sites also form an important part of atheist activism, especially in the way they circulate information and encourage consciousness-raising among participants. We argue that it is only in understanding the inherently public and connected nature of the Internet that we can consider atheist interactions online to be both collective and encompassing activism.

Based on a model of narrowcasting against a backdrop of broadcasting, the Internet reveals and accents this plurality of smaller identity groupings within a larger general (and increasingly global) "secularist" collective. In this chapter we examine secularist websites and other forms of new media to better understand the nature of how secular space is created and sustained and its ramifications for atheist identity and activism. In discussing atheist online presence and activism it is impor-

C. Smith (✉)
Independent Scholar, Tecumseh, OK 78476, USA
e-mail: chris_schmitt4@yahoo.com

R. Cimino
Department of Anthropology and Sociology, University of Richmond,
Richmond, VA 23173, USA
e-mail: relwatch1@msn.com

© Springer Science+Business Media Dordrecht 2015
S.D. Brunn (ed.), *The Changing World Religion Map*,
DOI 10.1007/978-94-017-9376-6_196

tant to note that we are referring to diverse groups—ranging from "hard-core" atheists to agnostics, "Brights," a group that promotes a naturalistic worldview, and various kinds of humanists—which is the main reason we use the broad terms "secularists" and "secularism" to describe this phenomenon.

196.2 The Space for Atheism in Religious America

Atheism is not a new phenomenon, as it has had a consistent, if small, presence in the U.S. since it's founding. It, however, has also carried a significant amount of stigma. Part of this is due to the particular pattern of secularization in the U.S. As Jose Casanova, drawing on the comparative work of David Martin, has noted:

> [T]here is little historical evidence of tension between American Protestantism and capitalism. There [was] no manifest tension between science and religion in America prior to the Darwinian crisis at the end of the nineteenth century, and the secularization of the American university dates only from this period. The American Enlightenment had hardly any antireligious component. Even the separation of church and state that was constitutionally codified in the dual clause of the First Amendment had as much the purpose of protecting 'the free exercise' of religion from state interference as that of protecting the federal state from any religious entanglement. (Casanova 2006: 22)

In America, in contrast to much of continental Europe, "the triumph of the secular came aided by religion rather than at its expense" (Casanova 2006: 23). American religion has exerted influence in policy-making and the political sphere in ways that religion in other secular countries has not; religious symbolism has historically pervaded American public life; and personal and public identification with religion among public officials has been the rule instead of the exception. And if polls and surveys are any indication, atheism is still far from acceptable for many Americans. A 2007 survey conducted by Newsweek, for example, found that a substantial percentage of Americans, 62 % of those surveyed, would refuse to vote for an atheist candidate; a Gallup poll from 1999 found that 49 % of those polled would not vote for an atheist for president. The influence of conservative religion remains evident. Just 13 % of those polled in a 2004 CBS/New York Times survey thought that humans evolved; on the other hand, 64 %, nearly two-thirds, of those surveyed in a Pew Research poll from 2005 said they thought that creationism should be taught alongside evolution in American schools. Various Harris polls over the last 4–5 years continue to confirm that the majority of American adults (a steady 82 % from 2005 to 2009) continue to believe in God (Harris 2009). These findings provide an explanation as to why "[a]theists are at the top of the groups that Americans find problematic in public and private life, and the gap between acceptance of atheists and acceptance of other racial and religious minorities is large and persistent" (Edgell et al. 2006: 230). The atheist, viewed as a deviant "Other" that valorizes godlessness and enthusiastically endorses evolution and naturalism, acts as a symbolic scapegoat and an "enemy" that is used to draw a distinction between those who belong culturally and morally and those who do not (Girard 1972).

In this context, Stephen Mennell is right to ask if civil religion in American "has served as a functional equivalent to an established national church" (2007: 291). Although there is competition and differences between the various religious bodies in the U.S., they are united in their common belief in God. And "[u]nlike intolerant societies that find members of different faiths to be a threat, under civil religion, the person who poses the greatest threat to society is the person who has no faith: the godless individual" (Shaffer and Verrastro 2005: 75). In the case of civil religion, then, belief in God figures as the central mechanism that marks off the borders of cultural membership and constitutes the criteria of what it means to be American. This leaves the American atheist, much like Georg Simmel's "stranger," as someone who is near in terms of not being alien to the community, yet far socially/symbolically as someone who is not fully accepted in such a community.

On one hand, secularists identify with one another based on being excluded from the mainstream. The American mainstream—a single, hegemonic view of representing reality, which is at best an approximation of the many discourses in circulation—positions secularists as outsiders. The formation of a secularist "minority consciousness" can be seen as a consequence of heightened awareness of the increasing distance between their strongly held views and the views of the mainstream, which is in actuality a product of diminishing distance due to increasing access to the same experiential sphere. On the other hand, their collective identities are based on a positive affirmation and activist strategies and organizing centered on redefining the American situation in secularist terms. Such positive secularist affirmation and revitalization is related, at least in part, to the tension between a communication model based on "representation" and one based on "presentation" with the advent of electronic communication technologies and a mass Internet literary among secularists.

196.3   The Internet Beyond Representation

Rather than viewing the Internet as a tool mainly for communicating information and representing the world "out there," we argue that it should also be seen as a constituting a space or dimension where knowledge of the social world and the social world itself is actively shaped. While the creative aspects involved in the critical reception and interpretive readings of media events and products has long been stressed by theorists focusing on audiences (for example, Brooker and Jermyn 2003), our current media landscape of one-to-many and many-to-many communication highlights the agency involved not only in reception and interpretation, but also in response and creation.

The one-way medium of the press has played an important role in constituting "the public." The privileging of a shared printed language helped to undermine place bound relations and set the stage for the emergence of a "national culture" based on the sharing of texts and common cultural codes. The boundaries—both spatial and social—that once highlighted the limits of common experiences and community relations have become more akin to intervals across which various forms of communication and interaction at a distance take place. The Internet continues and alters this relation by encouraging new social and affective connections based not so much on representing the public or a mass spectatorship as blurring the border between the private and public and allowing "the public" to participate in presenting themselves *in public*.

The personal is increasingly public. From posting one's feelings on Facebook to tweeting one's thoughts on Twitter, from donating money to a charity or a political campaign to chatting via Skype, the Internet is not only a space for interpreting the world but actively producing it. Representation is no longer representation of something; rather, representation *is* something. In saying this, we are not arguing that there is no distinction to be made between lives online and off, only that the subjects using computer technology, the world represented through computer technology and computer technology itself are all aspects of the same reality.

With the advent and mass availability of the Internet, secularists have gained a new ability to create cultural messages, symbols, and codes relevant to their particular interests and needs. Today's secularists are not content to merely be represented or "reading with" one another; they desire to present themselves to the public. This disposition towards self exhibition is intensified with secularists appearing to themselves as themselves in a place they hold in common (cyberspace) that is not so much discrete images and bits of information, but rather the source of presentation itself insofar as being public today increasingly means being online. Part of the Internet's power and appeal for secularists is that it encourages forms of personal presentation and mobilizes them to action "in public, as themselves, unscripted and unrehearsed, as writers of their own texts and producers of their own public pronouncements and utterances …" (Carpignano 1999: 187).

196.4   Two Forms of Secularist Culture and Activism Online

Blogs and YouTube videos have become key nodes of secularist culture and activism. In our study of 13 secularist sites, we found they tend to promote a highly personalized mode of presentation (note 1). A no-holds barred style, such as that often found on P.Z. Myers' (2010) popular "Pharyngula" site, has appeal due to the author's uncompromising opinions and views. Readers—many of whom comment on posts—are not interested in getting an objective view of an issue or news story when they visit "Pharyngula;" they are interested in reading Myers' unique iconoclastic take on particular issues and news. As a site for Myers to promote his particular brand of atheism (often dubbed "militant" by himself and others), the blog functions as a secularist source of information set against not only the perceived deficiencies of the mainstream press, but also the perceived deficiencies of a more accommodating atheism. In a post from April of 2010 titled "Witless Wanker Peddles Pablum for CFI," for example, Myers castigates the secular humanist Center for Inquiry's Michael De Dora for his "willingness to accommodate *any* nonsense from religious

BS artists ...." In response, Massimo Pigliucci (2010) of the blog "Rationally Speaking" wrote a post called "PZ Myers is a witless wanker who peddles pablum." In the piece Pigliucci criticizes Myers' post for being "the latest example of an escalation (downwards in quality) in the tone and substance of the discourse on atheism," which he blames "broadly on the rhetoric of the new atheism (the only 'new' aspect of which is precisely the in-your-face approach to 'reason')." Pigliucci thinks Myers' (and other new atheists') *publicized* rhetoric gives atheism a bad name and contaminates more moderate secularists. Whereas the new atheists understand that well-publicized transgressions against norms can embolden the audience, Pigliucci opts for a more restrained approach. With Stephen Jay Gould, he maintains the idea that differences between academic fields should be respected and that properly philosophical and/or religious questions cannot be answered by way of science, whereas atheists like Myers and Dawkins promote the view that there are few limits to the critique of science. Of course, the boundary between science and non-science is becoming harder to maintain (outside of professional spheres and peer-reviewed publications) online, with experts and novices co-mingling and publishing in the same place—a situation no secularists likely see as progressive. Dissolving the public/private distinction also makes it more comfortable to just speak one's mind, going public in a less polished, professional, and civil manner and not being overly concerned with your audience (or, in contrast, knowing your audience and only speaking to them). In terms of generating press, given the disruptive publicity of transgressing a norm, the more outspoken approach and tactics of the new atheists clearly have the advantage over their more moderate colleagues.

Siding with Myers, Atheist Revolution's vjack and Austin Cline (who writes about atheism on about.com) wrote posts defending mockery (vjack 2011) and assertiveness (Cline 2010) as atheist strategies. As Cline, whom vjack quotes in his post, writes:

> The sad fact is, atheists were not getting positive press and love from the general public before the so-called "new atheists" and their more assertive tactics appeared. Being less assertive and more submissive is no way to promote change and there's absolutely no reason to think that it would make the situation for atheists in America any better.

Such blog posts (including all of the comments and commentaries they generate) reflect a more personalized mode of presentation as well as showing how messages can be easily refashioned to promote both similar (in the case of Myers, Cline, and vjack) and dissimilar agendas (in the case of Pigliucci). They also highlight the division that exist between those atheists content to attack religion and secular and other humanists' concern with promoting a positive system of secular ethics. While there are notable cases of different secularists joining ranks in a common cause, the divisions offline tend to be reflected online as well. In noting the links between secularist web sites we found significant divisions, with atheist sites and blogs regularly linking to each other but not necessarily to secular groups such as CFI and Ethical Culture, an older humanist movement, while other groups that disassociate themselves from the atheist label and seek to promote a new secularist identity, such as the Brights, tend not to link to other sites at all. There, however, are exceptions even among atheist groups. For example, Freedom from Religion, one of the oldest and largest activist atheist organizations in the U.S., provides no links to other atheist groups, while the American Atheists do provide such links. In all of this we can see acts of mobilization and countermobilization (on the part of Myers and Pigliucci) as well as internal boundary marking (Gamson 1997), as secular activists and secular organizations distinguish their particular brand of secular activism or organization from others within the same milieu. Such boundary marking—which is directly related to the influx of diverse participants sharing the same space—helps situate and distinguish individuals and groups even as a loose group consciousness unites them collectively. This is the aforementioned narrowcasting against a backdrop of broadcasting, which shows how virtual space heightens individuality, but always and only in relation to knowledge and information that is held in common. And as this common stockpile widens, the individuals and collectives that emerge become more diverse and varied.

Although such inflammatory pieces as Myers' would never be published in a mainstream media outlet, online it is broadcast. The slogan for YouTube, "Broadcast Yourself," may sound oxymoronic, since broadcasting has traditionally been associated with a privileged few, namely large media organizations. Today this is no longer the case. With the capabilities of the Internet network, atheists (as nodes in such a network) can broadcast their views and opinions, even as the audience is splintered. This ability to broadcast has increased the opportunity for geographically dispersed atheists to "come out" (often anonymously) and disseminate their ideas and views in no place in particular and to no one in particular (or to merely assertively "come out" by denying the Holy Spirit—without much further comment—as in the case of the Blasphemy Challenge (2006)). This, in turn, allows secularists to collectively collaborate in spreading secularist views and ideas without operating under any strong collective identity (in terms of a shared definition of what it means to be secular) or having similar goals (at least in the short-term) and priorities (regarding issues to emphasize), or even acknowledging that they are collaborating with one another. We can see this quite clearly when we look at YouTube.

The individual secularists that put up videos cannot be said to be *formally* collaborating with one another on YouTube; the relations among these various videos and individuals are too indirect. Instead of speaking to one another directly, the secularists posting videos more often than not are discussing and debating particular issues that are important to them personally from their particular point of view and background, not unlike Myers' and Pigliucci's blogs. Thunderf00t (2006) and The Amazing Atheist's videos (2006) are good examples. Thunderf00t's popular video series "Why People Laugh at Creationists," for example, seeks to show the foolishness of creationism and intelligent design by juxtaposing clips of creationists and other religious conservatives making controversial statements on science and creation followed by Thunderf00t's critique in voiceover. The substance and tone of Thunderf00t's videos suggest that of the professional, well-educated and articulate British academic expert exposing irrational—often American—beliefs and behavior. In contrast, theAmazingAtheist plays the role of the informed yet entertaining everyman. In producing videos that span a broad spectrum of subjects, many relating to current events and concerns that are not explicitly secularist (which, along with the comedic aspect, broadens the ability to reach beyond the deconverted), TheAmazingAtheist presents his views and arguments with a mix of humor and vitriol. In the course of debating, debunking, and critiquing other YouTubers, such as pro-creationist VenomFangX (2006) and libertarian HowTheWorldWorks (2008), Thunderf00t and TheAmazingAtheist have both taken advantage of the fair use clause in U.S. copyright law, which allows users to legally use segments of other's videos in their own. These exchanges also highlight "one of the current pitfalls of U.S. Copyright law: the application of the Digital Millennium Copyright Act (DMCA) Takedown Notices. This is a clause in the DMCA under which copyright owners who believe their work is being infringed upon (beyond fair use) via the Internet can merely notify the carrier to remove the material" (Farley 2009). Such a clause has allowed opponents, who in this case are mostly pro-creationists, to file false changes against secularists, the most infamous being the false claims issued by VenomFangX against some of Thunderf00t's videos. To combat such abuse, individual secularists filed counterclaims and spoke out against the abuse in their individual videos. Such countermobilization culminated in a group of secularist users, including thunderf00t, forming a "multinational alliance." This alliance set up a channel called DMCAabuse and created a video titled "Creationist DMCA abuse" (DMCAabuse 2008). They also drafted a statement that read in part:

> We all share an interest in science and we have respect for the advancements and benefits that science has brought us. The Internet is one example of this. We believe that the internet offers extraordinary and valuable opportunities for education, and through sites such as you tube, a forum for open discussion and exchange of views. In order for people to fully benefit from the Internet, freedom of speech, freedom of expression and a lack of censorship are essential. (DMCAabuse 2008)

This effort shows the potential of the Internet for outreach and collective action. As Farley (2009), discussing YouTube and the skeptical movement, confirms:

> YouTube is an excellent avenue for skeptical outreach on the Internet. Its ease of use and lack of fees lower the barrier of entry so almost any skeptic can participate. Fair use ensures a steady stream of source material to debunk. The high traffic of the site and its explorability make the skeptical message accessible to people who may not even be aware of organized skepticism. Any skeptic with minimal audiovisual editing skills should consider YouTube an outlet for their efforts. (Farley 2009)

It also shows how secular groupings are capable of springing up from below, in the moment, as the need/issue arises, without representational mechanisms that would seek to determine or steer such collective action from above. This presupposes a space of inhabitation where individual secularists can present themselves to one another and collaborate in advocacy and constructing cultural messages, in relation to each other as well as in relation to outsiders, opponents, allies, and the mainstream media. Through reuse in videos and reposting on blogs, we can also see how new messages and new videos can (re)emerge from the initial messages and videos that gave rise to them. Such reappropriation and resignifying practices are an important component of secularist online activity and mobilization, and a point around which political questions related to fair use and copyright tend to emerge.

196.5   The 'Institutionalized Individualism' of Secularists

The Internet allows individuals only weakly-tied through information distribution networks to inadvertently collaborate in the dissemination of ideas and advance certain issues without sacrificing their individual autonomy and identity for the greater good of an organization or movement. This is in contrast to the activist tradition of vanguardism that looked upon intellectuals (who were speaking on behalf of organizations or parties) to provide a correct analysis of the issues and historical situation in order to lead the masses. This does not mean that intellectuals such as Harris and Hitchens are not considered as leaders by many secularists. Clearly they are. It also isn't to deny the situationally advantageous position intellectuals like a Dawkins or a Myers occupy within the secularist milieu, one that is exceedingly "mind heavy" and intellectually oriented in posture and content. Clearly intellectual leaders are important for secularist activism due to the elite positions they hold and the cultural capital they are disproportionately able to expend.

In highlighting the difference with the tradition of vanguardism, we take particular note of the networked nature of practice online, which connects each actor to multiple intersecting networks. This "networked individualism" (Wellman 2001), which online secularist activity reflects and perpetuates, gives the singular secularist, as a node, the ability to directly influence such networks and "go public" (without formal institutional support) in a way and on a scale that didn't exist previously, changing both the relationship between intellectuals and institutions and leaders and "followers." What this shows is the fact that attractive cultural symbols or messages are as likely to come from the periphery as they are the center. In fact, one way to understand the success of new atheism is to say that it was an outsider position becoming popular, or a peripheral cultural production successfully moving towards the center. Our findings suggest that this success story is inextricably tied to the electronic communication, specifically cyberspace where "the center is everywhere."

Secularist activism *online* might be called a "soft" activism; for, however impassioned the debate and rhetoric, no matter how uncivil and contentious the flame wars and comments, and regardless of the fact that many secularists label themselves and others as militant, at the end of the day their social antagonism with each other and the broader world online is based on the expression of subjective opinions and intellectual "debate." Of course, having different ideas about the best way to proceed on various issues *without* having the ability to actually make collective decisions or reach a consensus regarding such issues while *online* doesn't necessarily mean that secularists are uninterested or incapable of doing so *offline*. In fact, secularists have engaged in coordinated, collective action in the form of legal cases, protests, rallies, and advertising campaigns. Secular alliances and strategies have shown some success against efforts by the religious right to restrict abortion and gay marriage and introduce intelligent design into public schools. For example, the Freedom from Religion Foundation, reporting a membership of 14,000, scored a major victory for secularism in 2010 when it won a lawsuit that declared the

National Day of Prayer to be a violation of the First Amendment (Hansen 2011). What this shows is that we as researchers need to be sensitive to the relation between what happens online and off, between the placelessness of virtual space and the particular place of actual events.

The interplay between online and offline presence, mobilization and activism was vividly on display at the 2012 Reason Rally, an event sponsored by 20 atheist and humanist organizations and drawing over 10,000 secularists to the National Mall in Washington, DC. The rally billed itself as a "coming out party" for atheists to publicly declare their unbelief and demand a place for themselves in politics and society. The self-conscious attention to the numbers of attendees at the event (with organizers estimating as many as 20,000 participants) was a way to encourage remaining secularists to come out of the closet as well as to demonstrate political clout. But the online presence at the rally was never very far away; secularist bloggers with large followings, such as Greta Christina and P.Z. Meyers, were given prominent billing at the event.

Master of ceremonies and comedian Paul Provenza punctuated his jokes and introductions of guests with frequent relays of greetings "tweeted" in by a scattered flock of secularists from Mississippi to South Africa and Iran. The large video screens positioned around the Mall allowed participants to view themselves as part of a significant gathering. The rally not only served to physically mobilize secularists in a particular location; it was also an event that acted to emotionally liberate and strengthen solidarity among participants in highlighting their common identity, allowing participants to come out and speak out as secularists. This degree of engagement among participants at a rally can to some extent be generalized insofar as place-based events are being "doubled" thorough the media today by being watched and heard elsewhere both during and after the event by means of broadcasting and electronic reproduction.

Although we focus on the pluralistic nature of secularist collective action online in this article, scholars have noted definite patterns regarding secularist identity and ideology (Kosmin and Keysar 2009; Pasquale 2010), as well as the process by which one becomes an atheist (Smith 2011). Yet focusing on cohesion and a unified identity misses the significant identity work taking place during "latency periods" (Melucci 1996), as well as the tensions and divisions related to identity formations within the secular milieu (Cimino and Smith 2011; LeDrew 2011). These tensions are exacerbated, at least partially, due to the fact that secularist collective identity is based on independence and freedom. As such, we can talk about an "institutionalized individualism" (Beck and Beck-Gernsheim 2002) among organized nonbelievers. Another factor has to do with the variety of collective histories and intellectual approaches in identifying with atheism (for example, LeDrew 2011), as well as the fact that, in many respects, this is a first generation *collective* or group identity, an identity achieved rather than given (Smith 2011). This helps explain some of the opposition to the notion of a secularist community among many secularists (who do not want to give the appearance of being even slightly imitatively or residually religious), even as they arguably constitute one.

196.6 Media and Secularist Collective Consciousness

Blogs and videos derive part of their significance in promoting secularist concerns and issues in public discourse. At a minimum, these media broaden secularist discussion, creating new forms of virtual association and new forums for outreach and advocacy based around both "political issues" concerned with inclusion and the separation of church and state, as well as "cultural issues" concerned with identities, norms and alternative values. In fact, part of the novelty of such media is that they are stretching the boundaries of the traditionally political, making the personal more political. It is against this backdrop that we can better understand the historical shift, noted by Nabors (2009) in his study of secularist organizations, moving from a predominantly institutional secularist activism focused on legal proceedings, building coalitions with religious organizations, and fighting for church-state separation, to a cultural secularist activism stressing the role of argument and debate and associating itself with science, actively trying to discredit religious belief outside the sphere of law, while advocating for change outside the boundaries of dominant secularist organizations. Such a distinction follows the division noted by social movement theorists that exists between political and cultural movements (Eder 1993). Defined in secularist terms, the former seeks to maintain the secular nature of government and challenge forms of domination at the level of the state, while the latter is primarily interested in cultural change and attempting to construct secularist, that is, science-based and science-informed alternatives at the various societal levels. In this respect, the latter is more "radical" in the sense that it is less inclined to see present social life as legitimate and less interested in maintaining the status quo (even within the secularist milieu). Of course, in practice these two modes may not be mutually exclusive, and often secularist groups may engage in both strategies (a point to which we return below).

Some challenges for understanding secular activism include the question of how to conceptually categorize such diverse subgroups in terms of a "collective identity" or "collective interest." Another issue is the relationship between secularist cyberactivism and secular organizations, on the one hand, and the relationship between secularist activism and politics (for example, public-policy making), on the other.

As one atheist author has noted,

> [A]theism is not itself an ideology; there is no such thing as an 'atheist mindset' or an 'atheist movement.' Atheism per se hasn't inspired and doesn't lead to anything in particular because it is an effect—not a cause—and there are countless reasons for a person to not believe in God, ranging from vicious to innocent to noble. The newborn baby lacks a belief in God, as does the Postmodern Nihilist, the Communist, and the Objectivist—but each for entirely different reasons having dramatically different implications. So lumping all of these together under the 'atheist' label as if that were a meaningful connection is profoundly confused. Yet this is exactly what the New Atheists do and encourage: they talk about how there are so many atheists out there, and advocate their banding together into an atheist community to seek fellowship, foster cultural change, build a political voice, and so on (Perkins 2008).

Perkins' is incorrect in stating that all the "New Atheist" authors uniformly see the need for a social movement. Harris and Hitchens (Humanist Network News 2007) have both stated that there is no need for one. Perkins does, however, point out why one needs to be careful and qualify the use of the term social movement when looking at secularists, especially online. The electronic secularist landscape is highly pluralistic, made up of multiple groups and minor subgroups, which are made up of individuals holding diverse ideas, philosophies, and political stances, even as they share *some* level of consciousness. As Hitchens (quoted in Cipolla 2007), referring to secularists has stated: "We're not a unified group. But we're of one mind on this: The only thing that counts is free inquiry, science, research, the testing of evidence, the uses of reason, irony, humor, and literature, things of this kind."

The common denominator among the wide spectrum of secularist groups online is minimal: a "no" to religion and irrationality and a "yes" to the authority of reason and science. Of course, such a minimal bond has its benefits, specifically with respect to the rapid spreading of information precisely due to the decentralized nature of secularists and secularist groups online—a decentralization which echoes the structure of the Internet. This spreading of information is related to secularists realizing they are not alone, and recognizing their commonality with their fellow secularists. For example, when the question "Would You Be an Atheist without the Internet?" was asked on the "Friendly Atheist" blog (2009), many respondents stated that while they would be an atheist without the Internet (many coming of age prior to the advent of the Internet) they would be less active, less informed, and less aware that they were not alone without it.

Secularists' recognition that there are like-minded individuals in society goes hand in hand with the expansion of the communicative conditions of contemporary media. Consciousness-raising from this angle "would not be limited to a set of assumptions derived from life experiences that are used to confront, challenge, or resist, from the outside, the dominant ideology" of theism but "could also be conceived as a product of an electronically defined common place that, by virtue of being electronically reproduced, can be considered a public space" (Carpignano et al. 1993: 113–114).

196.7 The Relationship Between Virtual Secularism and Organized Secularism

One of the more noteworthy aspects of secular activism online is the fact that much of it is coming from the bottom up, from individual users not formally affiliated with any organization. Traditionally secularists have had their protest represented and mediated through secular organizations, including their publications. Secularist organizations were the only channel for activism. At the time, such organizations offered a smooth transition between the "private" urge to freely speak one's mind and the "public" need for civility and building coalitions. In viewing the Internet as a form of social communication that simultaneously changes the nature of the experience secularists have with each other as well as the experience others have of secularism, we can see how the role of secular organizations has changed. Daniel Loxton is discussing the impact of digital media on skeptic organizations (which seek to debunk the paranormal and supernatural), but it may also apply to the case of secularists. He states:

> [The Internet] changes everything. It's true that digital outreach may bring new grassroots support to traditional skeptical organizations, but realizing that potential requires facing up to a more fundamental shift: traditional skeptical organizations are no longer the default leaders of the popular movement. Indeed, new skeptics may not even realize the traditional skeptical groups exist. (2009: 24)

As we saw when looking at YouTube and blogs, secularists online are advocating for change outside the avenues created for this purpose by the dominant secularist organizations, even as they build and draw on the work of such organizations. Such advocacy often emerges as a direct response by grassroot groupings that do not necessarily have tight links to secular organizations. These countermobilization efforts more often emerge out of immediate social and cultural needs than any pre-defined, long-term agenda or goal. Such online advocacy and activism has the potential to both strengthen and undermine secularist organizations' legitimacy simultaneously. On the one hand, it can strengthen such organizations by pursuing actions parallel to their activities. On the other hand, it can undermine such organizations insofar as the demands and desires of the activists cannot be adequately integrated into such organizations. To the extent this happens such organizations have two options: face a crisis of legitimacy or adapt and change to meet such demands. The latter option is arguably what the CFI opted for when Paul Kurtz was voted out as the center's chairman, later resigning altogether in the face of critiques from those he labeled "angry atheists." Of course, as Massimo Pigliucci noted in a blog post responding to an earlier draft of this chapter we presented at a conference: "We also have to remember that people like P.Z. [Myers] see CFI as accommodationist, and in their view not in sync with the grassroots." Irrespective of Myers' understanding of CFI, there can be no doubt of the massive influence of new atheism as both a phenomenon of cultural self-reference and (sub)cultural, oppositional practice within and beyond the secularist milieu. Future research will need to empirically clarify how such organizations simultaneously adapt to and influence such changes.

196.8   Secular Activism Online and Off

The cyberactivism of secularists has had an impact in cultural and organizational terms, with secularist organizations that had once restricted their activity to more accepted and standard strategies adopting the methods of protest initiated by those more radical and vocal individuals and groups. Again, CFI serves as an example of this, with Ronald Lindsey, CFI's current president and CEO, promoting projects that Kurtz objected to, such as International Blasphemy Day and a contest soliciting cartoons critical of religion. However, the question of whether such activity has had the same success in political terms as far as policy-making and agenda-setting is a more complicated question, partly because while we can make an analytical distinction between the institutional and the cultural logic of secularist activism, empirically they are different sides of the same coin.

It is thus inaccurate to paint the shift from instrumental to expressive or institutional to cultural, as a linear and all-or-nothing occurrence since information and influence has flowed in both directions and on both sides; organizations have adopted postures and methods of the cyberactivists and many such "independents" have supported or joined various organizations (Cimino and Smith 2011). Thus the idea that the primary target of secularist activists has strongly shifted from the political and legal spheres to the cultural and academic spheres, insofar as their *methods* have moved in such a direction, is misleading to some extent. Although there has been a greater emphasis placed on science and reason and discrediting religion along such lines, this has not completely displaced legal concerns, such as the struggle over the separation of church and state, a form of activism supported by secularists of all types. There are also secularist lobbying groups targeting the political sphere directly, which many independents support through financial contributions. This dual strategy confirms that when speaking of opportunity for secularist mobilization, such opportunity should not be framed in terms of the movement as a whole. Opportunity may shift in favor of a particular strategy and/or a particular segment—the new atheists, for example—and may influence and modify the rest of the relevant actors and organizations (Gamson and Meyer 1996; McAdam et al. 2001). All of this relates more to the relationship between secularist activism and secular institutions, and does not tell us what influence, if any, secularist cyberactivism has had on politics more generally.

There has been a fair amount of commentary on the link between new information technologies and politics. Goldfarb (2006) looked at the role the Internet played in Howard Dean's campaign, specifically in relation to grass-roots organizing and initiatives. Of course, secularists in the U.S. have a unique problem that Dean did not share: they don't have a channel to influence political parties in any direct way. In the U.S., few politicians can afford to align themselves explicitly with secularist issues or secularist personalities since the consequences to their careers may be too severe. And although secularists do arguably have the numbers, as well as growing influence in certain elite sectors, such as academia and entertainment, to exert substantial influence in the political arena, they do not have historically well established umbrella organizations or powerfully connected think tanks in place to do so (in contrast to their religious counterparts). There are, however, organizations such as the United Coalition of Reason and National Atheist Party that aim to unite secularists at the national level. Of course, political opportunities, resources, and the support of powerful allies are not the only issues a movement has to consider. In fact, without a mass constituency willing to support secularist issues and causes no institutional or structural change will suffice for success. A mass constituency, in fact, is a prerequisite for *any type* of social collective to emerge. And this precisely, in terms of developing a group consciousness rooted in an identity based around broadly similar agendas and ideas as expressed in collective action online and off, is where the positive attributes of secular cyber presence can be seen to be having the most influence.

196.9   Conclusion

Georg Simmel introduced his concept of the stranger to explain the contradictory experience of what it means to be both near and far simultaneously. As Simmel explains:

> The unity of nearness and remoteness involved in every human relation is organized in the phenomena of the stranger, in a way which may be most briefly formulated by saying that in the relationship to him, distance means that he, who is close by, is far, and strangeness means that he, who is also far, is actually near. (Simmel 1950: 402–3)

Secularists can be considered strangers in two ways. First, in a strictly spatial sense they can be considered strangers in crossing a boundary and entering a space where they were previously not visible to the degree they are now. Second, in a symbolic sense, as Edgell et al. (2006) show, they are positioned as distant from the norm of American religiosity, while at the same time they actively and symbolically position themselves apart and antagonistically against such a norm. Of course, Simmel's understanding of "strangeness" involved being co-present in a particular geographic locale. Yet, with the media to be close socially does not necessarily require geographic proximity. In a world of electronic mediation, the co-presence of subjects is no longer the necessary basis of social and communal relations.

This transition from place-to-space based forms of interaction and communication with the Internet does not negate Simmel's insights regarding the tension between near and far. And if we consider cyberspace to constitute, at least theoretically, an all-inclusive arrangement where all individuals and groups may be considered strangers to one another, we can see how there is no mainstream culture or host groups in such a "place": The center is everywhere and everyone is included. This, however, does not mean that tension or distance has been eased or erased. In fact, granting individuals and groups once firmly separated (by geography as well as social status and roles) access to the same experiential space is one of the principal agents for increasing "awareness of physical, social, and legal segregation," (Meyrowitz and Maguire 1993: 43) as well as awareness of social and symbolic similarities.

In the case of secularists, the initially limited conflicts (with antagonists) and solidarity (with allies) based around this shared presence rooted in local settings is generalized and broadened on the basis of the expansion of communicative networks. The process of formation of secularist consciousness, as an awareness of being with like-minded others as well as being an outsider in a largely religious society, coincides with the rise of secularist group organizations, publicity (for example, new atheism), and entering the public arena (both online and off) as a collective engaged in particular strategies and issues. Such a collective identity is based around broadly similar themes and ideas, even when individual secularists and particular groupings diverge on some of the particulars. The greater visibility online has thus opened up an opportunistic space for secularists, whose politics, philosophies, life experiences, and social statuses differ widely, to acknowledge one another, socialize, and construct cultural codes solely on the basis of their secularist identity. This allows secularists to temporarily suspend their other identities related to such spheres as work and family and enter a space—which can be considered public by virtue of being electronically reproduced—with their secularist identity at the forefront. In this, it is an effective way of getting around the lack of numbers (in rural areas especially) and the problem of trust and the stigma atheists often feel in their local settings. Highlighting their secularist identity online is something just for them, something personal and "private."

Electronic media, as it challenges the private/public distinction, gives these personal, intimate, and issue-based concerns a more public airing, allowing secularists to expand their capacity for communal relations beyond their local communities. The new media constitutes a space for secularists to imagine themselves as part of a community (with no sense of place in particular) even though they will likely never meet a lot of their fellow secularists. This sense of "being with" like-minded strangers in a common place helps secularist individuals and groups become conscious of themselves as a unified body of individualists and organize and engage in particular strategies both online and off as a self-conscious collective in a society and world that still largely self-identifies as religious. In short, the formation of a more assertive and active secularist collective identity serves as counterpart to the formation of a secularist group consciousness in the imagination. They mutually support and co-determine each other. In this respect, the Internet as a space for self-presentation and representation is a crucial component of secularist identity and practice today.

196.10   Note

For our study we focused on PZ Myers' blog and Thunderf00t's YouTube videos since they are among the two most popular secularist hubs online. We also analyzed other sites we felt were relevant, such as vjack's "Atheist Revolution" blog and the Rational Response Squad's "Blasphemy Challenge" videos. Our intention is to look at how secular discourses are shaped and reproduced through the use of technology; we are not striving to give a representative picture of all secularists or an overview of public attitudes among secularists online. The blogs, videos and sites identified and referenced all may be found via an Internet search, unless they are now inactive. All text from these blog, social media, and web sites were collected between September 2010 and October 2011 for this particular study.

References

Beck, U., & Beck-Gernsheim, E. (2002). *Individualization: Institutionalized individualism and its social and political consequences*. London: Sage.
Brooker, W., & Jermyn, D. (2003). *The audience studies reader*. Abingdon: Routledge.
Carpignano, P. (1999). The shape of the sphere: The public square and the materiality of communication. *Constellations, 6*, 177–189.
Carpignano, P., Andersen, R., Aronowitz, S., & DiFazio, W. (1993). Chatter in the age of electronic reproduction: Talk television and the "public mind". In B. Robbins (Ed.), *The phantom public sphere* (pp. 93–120). Minneapolis: University of Minnesota Press.
Casanova, J. (2006). Secularism revisited: A reply to Talal Asad. In D. Scott & C. Hirschkind (Eds.), *Powers of the secular modern: Talal Asad and his interlocutors* (pp. 12–30). Stanford: Stanford University Press.
Cimino, R., & Smith, C. (2010). The new atheism and the empowerment of American freethinkers. In A. Amarasingam (Ed.), *Religion and the new atheism: A critical appraisal* (pp. 139–156). Boston: Brill.
Cimino, R., & Smith, C. (2011). The new atheism and the formation of the imagined secularist community. *Journal of Media and Religion, 10*, 24–38.
Cipolla, B. (2007, May). Is atheism just a rant against religion? *The Washington Post, 26*, B09.
Cline, A. (2010). *Weekly poll: Are so-called "New Atheists" too assertive or not assertive enough? About.com, Agnosticism/Atheism*. Retrieved April 2, 2011, http://atheism.about.com/b/2010/07/22/weekly-poll-are-so-called-new-atheists-too-assertive-or-not-assertive-enough.htm
DMCAabuse. (2008). *Creationist DMCA abuse. Found on DMCA abuse's YouTube channel.* Retrieved October 22, 2011, http://www.youtube.com/DMCAabuse#p/u/11/aWZ9XcdbO0w
Eder, K. (1993). *The new politics of class: Social movements and cultural dynamics in advanced societies*. London: Sage.
Edgell, P., Gerteis, J., & Hartmann, D. (2006). Atheists as "other:" Moral boundaries and cultural membership in American society. *American Sociological Review, 71*, 674–687.
Farley, T. (2009). *Skepticism via YouTube*. The Committee for Skeptical Inquiry. Retrieved October 22, 2011, http://www.csicop.org/si/show/skepticism_via_youtube/
Gamson, J. (1997). Messages of exclusion: Gender, movements, and symbolic boundaries. *Gender and Society, 11*, 178–199.
Gamson, W. A., & Meyer, D. S. (1996). Framing political opportunity. In D. McAdam, J. D. McCarthy, & M. N. Zald (Eds.), *Comparative perspectives on social movements: Political opportunities, mobilizing structures and cultural framings* (pp. 275–290). Cambridge: Cambridge University Press.
Girard, R. (1972). *Violence and the sacred*. Baltimore: John Hopkins University Press.
Goldfarb, J. C. (2006). *The politics of small things: The power of the powerless in dark times*. Chicago: University of Chicago Press.
Hansen, S. B. (2011). *Religion and reaction*. Lanham: Rowman and Littlefield Publishers, Inc.
Harris Poll. (2009). *What people do and do not believe in*. Retrieved April 24, 2011, www.harrisinteractive.com/vault/Harris_Poll_2009_12_15.pdf
HowTheWorldWorks. (2008). *HowTheWorldWork's YouTube channel*. Retrieved April 2, 2011, http://www.youtube.com/user/HowTheWorldWorks
Humanist Network News. (2007, October 24). The new atheists on organized freethought. *HNN Podcast*. Retrieved July 6, 2012, http://www.americanhumanist.org/hnn/archives/?id=320&article=0
Kosmin, B. A., & Keysar, A. (2009). *American names: The profile of the non religious population*. Hartford: Institute for the Study of Secularism in Society and Culture.
LeDrew, S. (2011). *Scientism and the new atheism*. Unpublished paper presented at the meeting of the Association for the Sociology of Religion, August 20, Las Vegas, Nevada.
Loxton, D. (2009). The paradoxical future of skepticism. *Skeptical Inquirer*, November/December, pp. 24–27.

McAdam, D., Tarrow, S., & Tilly, C. (2001). *Dynamics of contention*. Cambridge: Cambridge University Press.

Mehta, H. (2009). Would you be an atheist without the internet? *Friendly Atheist*. Retrieved October 4, 2011, http://www.patheos.com/blogs/friendlyatheist/2009/01/15/would-you-be-an-atheist-without-the-internet/

Melucci, A. (1996). *Challenging codes: Collective action in the information age*. Cambridge: Cambridge University Press.

Mennell, S. (2007). *The American civilizing process*. Cambridge: Polity Press.

Meyrowitz, J., & Maguire, J. (1993). Media, place and multiculturalism. *Society, 30*(5), 41–48.

Myers, P. Z. (2010). Witless Wanker Peddles Pablum for CFI. *Pharyngula*. Retrieved April 2, 2011, http://scienceblogs.com/pharyngula/2010/04/witless_wanker_peddles_pablum.php

Nabors, B. (2009). *The changing forms of organized nonbelief: The case of the Pacific City atheists*. Unpublished paper presented at the meeting of the Association for the Sociology of Religion, San Francisco, August 2009.

Pasquale, F. (2010). A portrait of secular group affiliates. In P. Zuckerman (Ed.), *Atheism and secularity* (Vol. 1, pp. 43–87). Santa Barbara: Praeger.

Perkins, G. (2008). Why the new atheists can't even beat d'Souza: The best and worst in human history. *Diana Hsieh: NoodleFood*. Retrieved November 22, 2009, http://www.dianahsieh.com/blog/2008/05/why-new-atheists-cant-even-beat-dsouza.shtml?nc

Pigliucci, M. (2010). PZ Myers is a Witless Wanker Who Peddles Pablum. *Rationally Speaking*. Retrieved April 2, 2011, http://rationallyspeaking.blogspot.com/2010/04/pz-myers-is-witless-wanker-who-peddles.html

Shaffer, L. S., & Verrastro, S. (2005). The quasi-theory of godlessness in America: Implications of opposition to evolution for sociological theory. *Sociological Viewpoints, 21*, 73–85.

Simmel, G. (1950). The stranger. In K. H. Wolff (Ed.), *The sociology of Georg Simmel* (pp. 402–408). New York: The Free Press.

Smith, J. (2011). Becoming an atheist in America: Constructing identity and meaning from the rejection of theism. *Sociology of Religion, 72*(2), 215–237.

TheAmazingAtheist. (2006). *TheAmazingAtheist's YouTube channel*. Retrieved April 2, 2011, http://www.youtube.com/user/TheAmazingAtheist

TheBlasphemyChallenge. (2006). *TheBlasphemyChallenge's YouTube channel*. Retrieved October 22, 2011, http://www.youtube.com/user/BlasphemyChallenge

Thunderf00t. (2006). *Thunderf00t's YouTube channel*. Retrieved April 2, 2011, http://www.youtube.com/user/thunderf00t

VenomFangX. (2006). *VenomFangX's YouTube channel*, Retrieved April 2, 2011, http://www.youtube.com/user/HowTheWorldWorks

Vjack. (2011). When mockery is effective. *Atheist Revolution*. Retrieved April 2, 2011, http://www.atheistrev.com/2011/03/when-mockery-is-effective.html

Wellman, B. (2001). The rise of networked individualism. In L. Keeble & B. Loader (Eds.), *Community informatics* (pp. 17–42). London: Routledge.

Chapter 197
Technology and the Changing Geography of Religious Media

Thomas A. Wikle

197.1 Introduction

From radio and television programming to text messaging, email, and Internet websites, religious organizations are increasingly dependent on electronic media for sharing and exchanging information, ideas and beliefs. Religious broadcasting now arrives as a 24/7 stream of talk shows, news, entertainment, and other types of programming. Increasingly, religious messages are also posted on Internet websites, exchanged through computer chat rooms, or sent through email, text messages and blogs. Whereas the movement of religious information and messages was once limited by transportation systems, electronic media can be instantaneously transmitted to nearly every part of the world. In this chapter we explore how religions and religious groups have used electronic media to communicate and expand their influence within the United States and at the international level.

As with other areas of communication, religious media has evolved with changes in technology. For example, Johannes Gutenberg's invention of moveable type during the mid-1400s enabled the Catholic Church to print and distribute the Bible in large quantities. Technology has also facilitated the movement of religious ideas over long distances. For example, Christianity was introduced to the Americas in part through innovations in ship design and navigation while improvements in road and railroad construction opened isolated areas of North and South America to new religious beliefs. During the late 1800s developments in electronic technology such as sound recordings enabled religious music and sermons to be stored and transported.

Until the development of electronic communication, outdoor events or meetings in buildings such as mosques and cathedrals were the most practical venues for

T.A. Wikle (✉)
Department of Geography, Oklahoma State University, Stillwater, OK 74078, USA
e-mail: t.wikle@okstate.edu

© Springer Science+Business Media Dordrecht 2015
S.D. Brunn (ed.), *The Changing World Religion Map*,
DOI 10.1007/978-94-017-9376-6_197

bringing people together for large religious events. In the late 1800s the development of wired communications technologies played an important role in connecting distant places. However, while suitable for the synchronous exchange of information, the telegraph and telephone were not effective for reaching many people simultaneously. After World War I radio broadcasting offered an efficient way to reach people over large geographic areas. First used by religious groups for live broadcasts of Sunday services, religious content in the form of music, hymns and Bible readings was soon shared within programming broadcasted on newly established commercial stations such as New Jersey's Station WJZ and Pittsburgh's KDKA (Erickson 1992). With much of the U.S. population living in isolated rural areas during the 1920s, "clear channel" AM radio broadcasts from large cities such as St. Louis became important sources of entertainment and information (Craig 2004). Adding to the popularity of local religious programming was the development of syndicated radio shows that enabled commercial stations to offer live or prerecorded broadcasts of nationally known religious personalities such as Charles Fuller ("Old Fashion Revival Hour") and Walter Maier ("The Lutheran Hour"). Weekly radio shows also featured controversial religious figures such as Aimee Semple McPherson and Charles E. Coughlin.

Outside the U.S. religious broadcasting expanded rapidly during the interwar years through the establishment of commercial stations in Europe, Latin America and Asia. By 1922 broadcast stations carrying religious programming had been established in several cities in Great Britain including London, Manchester, and Birmingham (Dinwiddie 1968). With help from Guglielmo Marconi, the Catholic Church created Vatican Radio in 1931 as a central location for distributing papal doctrine (Matelski 1995). Religious stations were also established in some European colonies located in Africa and Asia. Patterned after the British Broadcasting Corporation (BBC), religious programming in India featured live broadcasts of singing and prayer recorded at mosques, Hindu temples, and Sikh Gurudwara. A few religious radio stations also operated in the absence of government approval. For example, Jewish residents of Tel Aviv operated Haganah Radio in the 1940s as a clandestine station that was moved periodically to avoid detection by British authorities.[1]

Within the U.S. an important change took place in the early years of radio that profoundly impacted religious broadcasting. To address increasing interference and congestion caused by thousands of new commercial and hobby stations Congress passed legislation to license and regulate radio transmissions. Among its provisions, the Radio Act of 1927 required stations operated by non-commercial groups, including religious organizations, to share frequencies with commercial broadcasters. Unable to afford new equipment many churches were forced to close their stations. Although a major setback to small broadcasters, the Radio Act included a provision aimed at sustaining non-commercial radio programming through a stipulation that commercial stations donate airtime for content considered to be in the public

---

[1] Clandestine religious radio stations continued to operate from ships after formation of the Jewish state.

interest. To address what became known as the "public interest requirement," many commercial broadcasters offered blocks of free airtime to the three largest religious organizations in the U.S.: the National Council of Catholic Bishops, the Federal Council of Churches (Protestant), and several Jewish organizations (Kay 2009).[2] Conspicuously absent were evangelical Protestant groups whose content was considered offensive to most listeners (Erickson 1992). In fact, network radio executives refused to broadcast evangelical programming even when approached to sell airtime. Undeterred, evangelical groups began working with smaller, unaffiliated radio stations that were eager to sell airtime.

197.2   Evangelical Broadcasting and Mission Radio

American evangelicals have searched for ways to "spread the gospel" since early colonization of North America. Denied permission to address church congregations while traveling through the American colonies in the 1740s, Preacher George Whitefield spoke instead in public squares and barns (Belcher 1857). During the 1800s evangelical Bible societies printed and distributed religious material while Methodist circuit riders brought "God's word" to isolated frontier settlements. When commercial broadcasting was introduced in the 1920s, evangelists recognized its potential for reaching people. With an emphasis on sharing what evangelical leaders described as a "personal experience of conversion," radio was used to promote a conservative Christian worldview, sometimes through fiery sermons carrying themes of fear or urgency. As Hangen observes, radio helped evangelists achieve, "…a toehold back into their preferred role as guardian of the nation's values" (2002: 19). In addition to communicating and disseminating ideas, evangelists used radio as a means of building financial support for their initiatives through listener donations.

At the same time they sought to establish a broadcasting presence in the U.S., evangelicals also took steps to extend their influence within other countries (Stump 1991). Using powerful "mission radio" stations, evangelicals broadcasted a blend of news, educational, and cultural programming. To reduce confusion that might take place with conflicting church doctrine their messages were nondenominational. Political views were also avoided. As noted by Hadden (1990), Christian evangelism focused on three populations: (1) persons living in "forbidden countries" including those under Communist control or located within nations with Islamic majorities, (2) non-Christians, including animists living in developing countries of Africa, and (3) Catholic populations of Latin America.

Among the earliest mission stations was Radio HCJB (Heralding Christ Jesus' Blessing), founded in 1931 using a 250 W transmitter located in Quito, Ecuador. Many mission radio stations were established using surplus or abandoned WWII

---

[2] Southern Baptists and Mormons were also provided with modest amounts of airtime. Broadcasters provided about 4 % of their airtime for religious programming.

shortwave radio equipment for reaching parts of Europe, North Africa, the Middle East and eastern Asia. For example, Trans World Radio began broadcasting across North Africa and southern Europe with a station in Tangiers, Morocco that had once been used for Nazi propaganda (Hadden 1990). As noted by Bruce (1990), the expansion of mission radio corresponded with a postcolonial decline in the number of western missionaries operating in newly independent countries.

The success of mission radio during the 1940s provided a source of encouragement to large commercial broadcasters such as the National Broadcasting Corporation (NBC) and Columbia Broadcasting Corporation (CBS) to establish radio networks in developing countries. However, by 1947 both NBC and CBS withdrew from international broadcasting when it was perceived to be unprofitable (Stoneman 2006). However, evangelical stations continued their expansion in Latin America, Asia and Africa, establishing stations in isolated or underserved areas whenever possible. In an effort to reach indigenous populations, programming was broadcasted using local languages. In some poor areas broadcasters distributed free transistor radios.[3] During the 1960s and 1970s evangelical groups began broadcasts directed at communist countries. For example, the Far Eastern Broadcasting Company (FEBC) developed programming aimed at Communist China from stations in the Philippines and U.S. Marianas Islands. Other evangelical organizations produced programming in English and Russian in an effort to reach populations located in the Soviet Union (now Russia and the Commonwealth of Independent States).

In recent years international efforts to liberalize state-controlled media have resulted in both opportunities and setbacks for religious broadcasters. For example, to address requirements of the International Monetary Fund for greater media freedom, some developing countries have privatized state-controlled radio and television stations. In Kenya religious groups that previously received free airtime on state radio and television must now pay for broadcasting time.

197.3   Translators and Low Power FM Radio

To attract listeners, stations operated by religious organizations frequently adapt secular programming formats that offer a blend of music, news and entertainment. However, the high cost of purchasing and operating a full power radio station in the U.S. has remained a substantial barrier for religious broadcasters. During the late 1970s religious broadcasters began using low power translator stations as alternatives to full power stations. Translators were originally designed to rebroadcast signals from full power "parent" stations on adjacent frequencies to reach places blocked by obstructions such as high terrain. Evangelical broadcasters such as American Family Radio (AMF) soon developed chains of translator stations that carried programming from a single parent transmitter. During the 1980s a new type

---

[3] In 1954 Sony marketed the first transistor radio for mass consumption.

of satellite-fed translator emerged that was capable of unattended operation anywhere a source of power could be located. From a single studio, broadcasters such as the Calvary Satellite Network (CSN) began operating hundreds of auto-mated "satellator" stations, each broadcasting an identical stream of syndicated reli-gious programming (Wikle and Comer 2010). In addition to bringing FM radio reception to many small towns and rural areas in the U.S., some evangelical groups established satellators in other countries.

During the mid-1990s U.S. telecommunications deregulation led to the consolidation of independent radio stations within mega networks such as Clear Channel Communications. Recognizing the potential loss of local programming, the U.S. Federal Communications Commission (FCC) began licensing low power FM radio (LPFM) stations. Like translators, the signals from LPFM stations are limited in range to a few miles. FCC requirements limited LPFM licenses to local, non-commercial organizations willing to broadcast at least eight hours of local pro-gramming content each day. With assistance from religious radio networks such as CSN, the largest proportion of LPFM licenses awarded by the FCC went to churches, church affiliated schools and colleges, and other religious organizations. However, because they may not interfere with existing FM radio stations, relatively few LPFM stations were established by religious organizations or other groups within large metropolitan areas (Wikle and Comer 2009). The overall pattern of LPFM and translator/satellator has been to bring evangelical radio programming to small towns and rural areas (Figs. 197.1 and 197.2).

Fig. 197.1 Religious translator stations in the contiguous U.S. (Map by T. Wikle)

Fig. 197.2 LPFM stations in the contiguous U.S. (Map by T. Wikle)

197.4 Televangelism

Because of the dominance of U.S.-based conservative Protestant groups within religious broadcasting, televangelism is perceived as a mostly America phenomenon. Early televangelism featured personalities such as Oral Roberts and Pat Robertson whose charismatic styles blended prayer, viewer call-ins and drama (Harrell 1985; Kay 2009). Televangelists often emphasize personal connections to viewers through interaction with studio audiences and viewer call-ins. Televangelism has been credited with fulfilling social needs for "ritual performances" (Alexander 1994) and with helping validate the faith and religious identity of viewers within secular society (Soukup 2002). Notably, televangelism has also impacted the location where people practice or experience religion. For some the experience of watching serves as a substitute for attending religious services (Gaddy and Pritchard 1985). Critical assessments of televangelism abound. For example, because it is not geographically based, televangelism has been blamed for diverting financial resources that previously supported local parishes (Buckser 1989). Since viewers do not need to leave their homes to participate in religious television, the overall effect of "teleministries" has been to weaken and sometime replace involvement within local communities.

During the 1960s programming featuring Billy Graham began airing on stations in Latin America, Africa, and Australia. Capitalizing on the popularity of Graham's broadcasts and previous success with mission radio, U.S.-based evangelical groups began exporting syndicated programming. By the early 1970s television and radio programming was offered through stations affiliated with large groups of stations such as the Trinity Broadcasting Network (TBN). Established as a single UHF television station in southern California, TBN's network now includes 33 full power stations and 6,000 affiliates in the U.S., Latin America, Asia, and Africa (Wilford 2009). The Christian Broadcasting Network (CBN) also operates a global network of stations. CBN's fulfills its mission to "Spread the Gospel," by distributing international editions of its popular "700 Club" and other programming such as "CBN News." Table 197.1 provides estimates of potential CBN viewers within selected regions and countries. Other religious broadcasters operating at the international level include the Catholic Church's Eternal Word Television Network and the Three Angeles Broadcasting Network (3ABN) operated by the Seventh Day Adventist Church.

Neuendorf (1987) suggests that three major developments have contributed to the growth of religious television broadcasting in the U.S.: (1) the expansion of charismatic and entertaining religious television programming, (2) new broadcast and cable television infrastructure, and (3) the adaption of secular programming formats to serving religious objectives. U.S. televangelism reached its greatest popularity in 1986 when 15 million U.S. households regularly viewed one or more of the ten most popular evangelical programs. Although suffering from a decline following the Jim Swaggart and James Bakker scandals, religious television has retained its importance. In recent years developments in cable television, satellite systems, and digital terrestrial broadcasting significantly increased the availability of network and local channels to locally-based religious broadcasters. This has enabled small religious broadcasters to maintain

Table 197.1 Potential television viewers of CBN network stations

CBN center	Region	Population (in millions)	Potential viewers (in millions)	Percentage of population
Africa	East and South	312	15	4.8
	West	158	50	32.6
	French	315	99	31.4
Asia	China	1,320	n/a	n/a
	Hong Kong	6.9	50	100
	India	1,130	131.6	11.6
	Cambodia	14	4.5	32.6
	Indonesia	178	178	100
	Laos	6.5	n/a	n/a
	Philippines	91.1	66.7	73.2
	Thailand	65.1	2.5	3.8
Canada		33	7.8	23.6
Europe		554	102	18.4
Latin America		500	54.4	10.8
Middle East and North Africa		371	320	86.2
Russia/CIS		404	102	25.2

Data source: www.cbn.com

local cable channels or operate low power television stations. Religious content also makes up a large portion of programming carried by non-U.S. broadcasters. Owned by the conservative Islamic daily newspaper Tükiye, Turkey's first nationwide Islamic television channel (TGRT) began broadcasting in 1993. With more than 2,000 congregations, Brazil's Universal Church of the Kingdom of God operates 30 stations making it the country's third largest television network (Kay 2009). Religious programming is also popular in Middle Eastern countries where television is the principal source of news and entertainment and a major factor shaping public attitudes.

197.5 Satellite Television

Among the most significant developments affecting religious broadcasting has been the expansion of satellite television. Satellite systems offer greater channel capacity and enable religious programming to reach areas where terrestrial television broadcasts cannot be received. Religious programming is currently carried on a wide range of government and privately owned satellites in geostationary orbit. These include Galaxy-V (U.S.), Intelsat-804 and Hotbird-5 (Europe), Brasilsat-B2 (Brazil) and JCSat-3 (Asia). In addition, the Catholic Church's Spanish language broadcasts can reach Latin America countries using Galaxy-IR and PanSat-3 satellites. Along with large religious broadcasters such as CBN, satellites are also used by small religious groups. For example, since 2005 Coptic Christians who make up only 10 % of Egypt's population, have operated a satellite channel called Aghapy Television. On a global basis American evangelical groups such as TBN operate the largest number of religious channels carried on satellite (Table 197.2).

In many countries terrestrial radio and television are tightly controlled by the government, making it difficult for non-state or minority religions or religious views to be represented in broadcasting.[4] Operating over large geographic regions, satellite television has become an alternative to state-controlled terrestrial broadcasting. For example, uncensored satellite broadcasts carry programming that features popular Islamic activist Amr Khaled (Anderson 2003; Moll 2010).[5] Islamic fundamentalist organizations also utilize satellite broadcasts. Hamas operates Al-Aqsa Television using both Arab-Muslim and western satellites while Hezbollah's Al-Manar Television can reach locations throughout the Middle East using the PalapaD satellite, operated by an Indonesian company.

---

[4] It is notable that the absence of a central authority during Lebanon's 1970 Civil War led to an open market for radio stations.

[5] In 2011 Time Magazine identified Khaled as among "The 100 most influential people in the World."

Table 197.2 Satellites used for the Trinity Broadcasting Network

Satellite	Region	Longitude
Galaxy-V	United States, Canada, Mexico	125°W
Intelsat-804	Europe, Middle East	64°E
Hotbird-5	Europe	13°E
Hotbird-3	Italy	13°E
ICSAT-3	Asia	128°E
Intelsat-701	Australia, New Zealand, S. Pacific	180°
PamAmSat-9	Central and South America	58°W
BrasilSAT-B2	Brazil	65°W
JCSAT-3	S. Asia	128°E

Data source: www.donaldfrobertson.com/religion_via_satellite.html

197.6 The Internet and Religious Media

Computers were used to support religious communication prior to development of the Internet. For example, linked through a computer bulletin board, thousands of people participated in an online memorial service for crew members of the Space Shuttle Challenger in 1986 (Campbell 2005a). The Internet has touched almost every aspect of modern life, from learning and entertainment to corporate advertising and shopping. For religious organizations the Internet has emerged as a tool of unprecedented utility, serving the needs of large or small organizations without time and distance limits associated with other types of media such as radio or television. As a device for information dissemination the Internet can be used to share written material, photographs, sound, or video recordings. Unlike print media it also functions as a venue for interactive communication, enabling the exchange of information among congregation members or with the outside world through chat rooms, blogs, video conferencing, or social networking. However, among conservative religious organizations the Internet remains a paradox. At the same time it provides religious organizations with a tool of unprecedented power, it also facilitates activities seen by conservatives as inappropriate such as pornography. The Internet may also influence power structures within congregations. As noted by Campbell (2005a) it has empowered "techies" responsible for managing website content, decentralizing authority over external messages.

197.7 Religion-Online and Online-Religion

Helland (2000) draws a distinction between religion-online, defined as importing traditional forms of religion to an online format and online-religion where the Internet offers new ways to worship or express personal beliefs. An example of religion-online can be seen in LifeChurch (Edmond, Oklahoma). Through its website (www.LifeChurch.tv) church members in remote locations can participate in real-time worship along with persons attending regular services. In this way LifeChurch connects online members with traditional nodes of religious activity.

Fig. 197.3 Outdoor advertisement for a religious website (Photo from www.truthmagazine.com/, used with permission)

Religion-online is also represented by websites that provide access to religious materials (Fig. 197.3). These range from the nondenominational "thought of the day" (www.beliefnet.org) to sites that post educational materials or religious doctrine specific to a faith such as "Buddhanet" (www.buddhanet.net) or "Renaissance: A Monthly Islamic Journal" (www.renaissance.co.pk). Radical Islamic insurgents in Iraq and fundamentalist Christian groups in the U.S. have also used the Internet to communicate or distribute religious propaganda. Messages carried through email or posted on websites have tremendous power in influencing public opinion and sometimes foreign affairs. For example, threats by a Florida pastor to burn a Qu'ran in 2011 resulted in widespread outrage and street demonstrations within several Muslim counties.[6]

A unique attribute of the Internet is its role as a meeting place beyond social control or established hierarchies (Kong 2001). The ability to meet and share ideas in the absence of church dogma is consistent with what Meyrowitz (1985) describes as "situational geography" where different types of people are brought together in a manner uninfluenced by formally distinct social roles. Often the "binding" force linking Internet users is a common interest in aspects of religion rather than geographic proximity (Hayles 1993; Baker and Ward 2002). As a gathering place for religious activities the Internet is increasingly a venue in its own right. As noted by Strate (1999), cyberspace has become sacred space where users interact as they participate in religious activities.[7] An example is illustrated by "Second Life" (www.secondlife.com), a virtual world where visitors participate in activities ranging from classroom lectures or discussions to worship inside three-dimensional representations of churches, Buddhist temples, or synagogues (Radde-Antweiler 2008). Second Life also supports pilgrimages, enabling Muslims to experience a

---

[6] The Koran burning was streamed live on the Internet with Arabic subtitles.

[7] The importance of the Internet as religious space is highlighted be a 1996 ceremony where Tibetan Monks in New York City blessed cyberspace for use in Buddhist religious practices (Campbell 2005b).

virtual Hajj as they travel to Mecca. Other groups have developed entire online congregations such as the Anglican's I-Church (Kellner 1996; see www.i-church. org/gatehouse/). Interest in online religion has been a catalyst for international conferences such as "Crossroads of Civilizations: Media, Religion and Culture" held in July 2012 in Eskişehir, Turkey and the "International Conference on Digital Religion" organized through the Center for Media, Religion an Culture at the University of Colorado in Boulder, Colorado in January 2012.

197.8   New Geographies of Religious Participation

Social mobility and other demographic changes over the last 50 years have contributed to a U.S. religious landscape that is extremely diverse. With competing faiths and denominations America is increasingly a nation of "religious consumers" (Kohut and Stokes 2006). As suggested by Warf and Winsberg (2010) this diversity has been driven, in part, by the expansion of large nondenominational (mega) churches that draw parishioners away from established denominations. Within an "open market" less affected by church doctrine and authority, the Internet has become an information clearinghouse for exploring religions and religious beliefs. The free exchange of religious ideas has also been supported by related communications technologies such as the cellular telephone. Cell phones have assisted in the organization of spontaneous events not sanctioned by government authorities. They have also been used by established religious groups to connect and communicate about religious activities. For example, in 2004 Pope John Paul II's midnight mass and Christmas day messages were broadcasted throughout Italy with help from two Italian cellular companies, Vodaphone and Elecom Italia Mobile (Roman 2006). In Great Britain an Islamic prayer alert sends quotes from the Qu'ran and other messages as reminders about prayer times.

197.9   The Cultural Imperialists

As suggested by Hunt (2000) electronic media has enabled religion and religious messages to be exported in a more systematic fashion than ever before. However, electronic media directed across international boundaries by evangelical groups is increasingly treated as hostile. In Islamic countries Christian evangelism is considered a form of "cultural imperialism" where western countries attempt to export their cultural influence (Brouwer et al. 1996). Iran and North Korea have taken action in response to evangelical broadcasting by jamming western radio and television frequencies. In comparison to terrestrial and satellite broadcasting, short wave signals are less susceptible to interference. As a result, evangelical organizations continue using shortwave radio to reach the populations of Communist and

Table 197.3 Largest number of religious shortwave stations, by country

Country	Stations	Percent
United States	41	32.6
Germany	10	7.9
Brazil	8	6.3
South Africa	5	3.9
United Kingdom	5	3.9
Australia	3	2.4
Chile	3	2.4
Cyprus	3	2.4
Ecuador	3	2.4
Italy	3	2.4
Russia	3	2.4
Sweden	3	2.4
Others	36	28.6
TOTAL (all counties)	126	100

Data source: www.shortwave.be/rel.html

Muslim countries. Nearly a third of these stations are operated by U.S.-based evangelical organizations (Table 197.3). Increasingly, evangelical groups also use the Internet to reach their targets prompting Iran and Saudi Arabia to block Internet websites considered to be inconsistent with Islamic teachings (McCormick Tribune Foundation 2007). Other countries such as Nigeria use legal mechanism to prevent broadcasters from airing unsubstantiated or sensational claims such "miracles." Despite vigorous efforts it has been difficult for government authorities to block all forms of religious media.

197.10   Conclusion

Electronic forms of media used for religious communication have evolved considerably over the last 100 years. Despite the introduction of newer technologies, radio and television have remained extremely important both within and outside of the U.S. A 2005 study by the Barna Group found that 46 % of Americans listen to Christian radio broadcasts each month while more than 15 % of U.S. adults listen to religious radio on a daily basis. Radio is especially important in regions of the developing world with low rates of literacy. In recent years increased channel capacity from the analog to digital conversion and the growth of cable and satellite television systems have greatly expanded opportunities for smaller organizations to participate in religious broadcasting. In addition, satellite broadcasting has allowed religious broadcasts to reach remote areas of the world and countries previously dominated by state-controlled terrestrial broadcasting. In recent years computer networking has enabled religious information to be shared with a global audience. Evolving technologies such as the Internet offer possibilities for new ways to participate in or experience religion in the absence of social control. The rapid evolution of electronic media has also driven adaptation in the ways religious organizations communicate and connect with people. This can be illustrated by the growing number of social networking sites devoted to religious themes and interests. As a communication tool for religious organizations, the Internet's full potential may yet to be realized.

As shown in this chapter, the geography of religious media has been significantly affected by American evangelical organizations. Using secular formats and entertainment such as "Christian Rock" and by expanding into underserved rural areas, evangelicals have greatly extended their geographic reach and influence within the U.S. In pursuing missionary objectives, evangelicals have also used radio, television, and the Internet as mechanisms for reaching millions outside the U.S. including persons living in isolated areas and places closed to Christian missionaries.

References

Alexander, B. C. (1994). *Televangelism reconsidered: Ritual and the search for human community*. Atlanta: Scholars Press.

Anderson, J. (2003). The internet and Islam's new interpreters. In E. Eickelman & J. Anderson (Eds.), *New media in the Muslim world: The emerging public sphere* (pp. 45–60). Bloomington: Indiana University Press.

Baker, P. M. A., & Ward, A. C. (2002). Bridging temporal and spatial 'gaps:' The role of information and communication technologies in defining communities. *Information, Communication and Society, 5*(2), 207–224.

Barna Group. (2005). *More people use Christian media than attend church*. www.barna.org/barna-update/article/5-barna-update/183-more-people-use-christian-media-than-attend-church. Accessed 26 July 2011.

Belcher, J. (1857). *George Whitefield: A biography with special reference to his labors in America*. New York: American Tract Society.

Brouwer, S., Gifford, P., & Rose, S. (1996). *Exporting the American gospel*. London: Routledge.

Bruce, S. (1990). *Pray TV: Televangelism in America*. London: Routledge.

Buckser, A. S. (1989). Sacred airtime: American church structure and the rise of televangelism. *Human Organization, 48*(4), 370–376.

Campbell, H. (2005a). Making space for religion in internet studies. *The Information Society, 21*(4), 309–315.

Campbell, H. (2005b). Spiritualizing the internet: Uncovering discourses and narrative of religious internet usage. *Online-Heidelberg Journal of Religions of the Internet, 1*(1), 1–26.

Craig, S. (2004). How America adopted radio: Demographic differences in set ownership reported in the 1930 1950 censuses. *Journal of Broadcasting and Electronic Media, 49*(June), 179–195.

Dinwiddie, M. (1968). *Religion by radio: Its place in British broadcasting*. London: Allen and Unwin.

Erickson, H. (1992). *Religious radio and television in the United States, 1921–1991: The programs and personalities*. Jefferson: McFarland.

Gaddy, G. D., & Pritchard, D. (1985). When watching religious TV is like attending church. *Journal of Communication, 35*(1), 123–131.

Hadden, J. K. (1990). The globalization of American televangelism. *International Journal of Frontier Missions, 7*(1), 1–10.

Hangen, T. J. (2002). *Redeeming the dial: Radio, religion and popular culture in America*. Chapel Hill: University of North Carolina Press.

Harrell, D. E. (1985). *Oral Roberts: An American life*. Bloomington: Indiana University Press.

Hayles, K. (1993, October). Virtual bodies and flickering signifiers. *Magazine, 66*, 69–91.

Helland, C. (2000). Online religion/religion online and virtual communities. In J. K. Hadden & D. E. Cowan (Eds.), *Religion on the internet: Research prospects and promises* (pp. 205–223). New York: JAI Press.

Hunt, S. (2000). 'Winning ways:' Globalism and the impact of the health and wealth gospel. *Journal of Contemporary Religion, 15*(3), 331–347.

Kay, W. K. (2009). Pentecostalism and religion broadcasting. *Journal of Beliefs and Values, 30*(3), 245–254.

Kellner, M. A. (1996). *God on the internet*. Foster City: IDG Books Worldwide.

Kohut, A., & Stokes, B. (2006). *America against the world: How we are different and why we are disliked*. New York: Times Books.

Kong, L. (2001). Religion and technology: Refiguring place, space, identity and community. *Area, 33*(4), 404–413.

Matelski, M. J. (1995). *Vatican radio: Propagation by the airwaves*. Westport: Praeger.

McCormick Tribune Foundation. (2007, April 26–27). *Understanding the mission of the U.S. International Broadcasting*. Wheaton: Cantigny Park.

Meyrowitz, J. (1985). *No sense of place: The impact of electronic media on social behavior*. New York: Oxford University Press.

Moll, Y. (2010). Islamic televangelism: Religion, media and visuality in contemporary Egypt. *Arab Media and Society, 10*, 1–16.

Neuendorf, K. A. (1987). *The history of social impacts of religious broadcasting*. Paper presented at the annual meeting of the Association for Education in Journalism and Mass Communication, San Antonio, August 1–4.

Price, M. (2007). *Religious communication and its relation to the state: Comparative perspectives*. Departmental papers, Annenberg School of Communications, University of Pennsylvania, January 1.

Radde-Antweiler, K. (2008). Virtual religion: An approach to a religious and ritual topography of second life. *Journal of Religions on the Internet, 3*(1), 54–72.

Roman. (2006). *Texting God: SMS and religion in the Philippines*. 5th international conference on media, religion and culture, Stockholm, July.

Soukup, P. A. (2002). Media and religion. *Communication Research Trends, 21*(2), 2–43.

Stoneman, T. H. B. (2006). *Capturing believers: American International Radio, Religion, and Reception, 1931–1970*. Ph.D. dissertation. Atlanta: Georgia Institute of Technology, School of History, Technology and Society.

Strate, L. (1999). The varieties of cyberspace: Problems in definition. *Western Journal of Communication, 63*(3), 382–412.

Stump, R. W. (1991). Spatial implications of religious broadcasting: Stability and change in patterns of belief. In S. D. Brunn & T. R. Leinbach (Eds.), *Collapsing space and time: Geographic aspects of communications and information* (pp. 354–375). London: Harper Collins Academic.

Warf, B., & Winsberg, M. (2010). Geographies of megachurches in the United States. *Journal of Cultural Geography, 27*(1), 33–51.

Wikle, T. A., & Comer, J. C. (2009). Barrier to establishing low-power FM radio stations in the United States. *The Professional Geographer, 61*(3), 366–381.

Wikle, T. A., & Comer, J. C. (2010). Translator networks and the new geography of religious radio. *Journal of Radio and Audio Media, 17*(1), 48–62.

Wilford, J. (2009). Televangelical publics: Secularized publicity and privacy in the Trinity Broadcasting Network. *Cultural Geographies, 16*, 505–524.

Chapter 198
Introducing the Study of Religions at The Open University: The Scope and Limitations of a Distance Learning Approach to the Study of Religions

Gwilym Beckerlegge

198.1   Introduction: Some Preliminaries About Studying at Open University

The Open University (OU) in Britain was founded in 1969 and took in its first batch of some 25,000 students in 1971. Since 1969, more than 1.6 million people have studied with the OU, and its student body at the time of writing (2011) numbers more than 260,000 students, ranging from access level to research students. OU students beyond the UK study its programs directly (in English), as in the Republic of Ireland (3,500 students) and mainland Europe, and through partnerships with other institutions, as in Eastern Europe and Russia (9,000 students). Students also study modules in institutions whose degrees have been validated by the OU (46,000 students). Since the 1970s, the OU in Britain has been involved to varying degrees in setting up and advising on the development of open universities in a number of regions, including the Arab Open University and Indira Gandhi National Open University.[1]

The OU in Britain has been characterised by its open entry, without prerequisite qualifications, to almost all undergraduate modules and its reliance upon part-time distance learning – what the OU now refers to as "supported open learning" (distance learning supported by a blend of online, telephone and face-to-face tuition) to maximise access to its module.[2] Although many courses do offer some face-to-face tuition, attendance at such events is optional and not a requirement upon which successful completion of a module depends. Whether a student passes a module

---

[1] For more information about the Open University and its faculties and centres, visit www.open.ac.uk/

[2] For a discussion of "blended" learning and teaching with reference to the department's work, see Sinclair (2011).

G. Beckerlegge (✉)
Department of Religious Studies, The Open University, Milton Keynes, UK
e-mail: g.beckerlegge@open.ac.uk

© Springer Science+Business Media Dordrecht 2015   3765
S.D. Brunn (ed.), *The Changing World Religion Map*,
DOI 10.1007/978-94-017-9376-6_198

depends entirely on the grades s/he receives for prescribed assessment tasks, each successfully completed module earning points towards the total number required for the award of a degree.

In Britain, OU modules are normally mediated by part-time tutors (Associate Lecturers) who offer academic support to students, taking a limited number of face-to-face tutorials or running an online forum (depending on the teaching strategy of the module). The allocation of students is still largely managed on a regional basis, England being broken into 10 regions, while Ireland, Scotland and Wales are administered as separate national regions. Thus, for example, in Wales, although teaching materials are generated in English, students may declare a preference for dealing with administrative matters in Welsh and may apply to be assessed in the medium of Welsh. Associate Lecturers grade assignments contributing to continuous assessment and offer feedback to the student, which is expected to be extensive as it fulfills the function of the student's individual tutorial. It is not uncommon to find within the same tutorial group students with little or no prior experience of post-16 education, let alone higher education, sitting alongside highly qualified graduates who have returned to study out of personal interest or in the hope of changing career. For example, many teachers have taken modules in Religious Studies in order to assist them in teaching Religious Education in schools, perhaps as a second subject. The majority of students in the Arts Faculty fall within the 25–55 age-band, with an average age at the time of writing of 32.

Students' learning is typically structured by the OU's distinctive published study packs (sometimes including items co-published with external publishing houses), although increasingly much of this material is shifting from print textbooks onto the OU's Virtual Learning Environment (VLE).[3] These packs were originally designed as part of the OU's mission to widen access so that, for example, the disabled or remote student would not be disadvantaged.[4] Within the OU, we now distinguish informally between 'OU classic' and 'OU lite' packs – the former containing, in addition to study blocks, specially produced films, DVDs, CDs and podcasts, illustration booklets etc., while the latter might be a more basic wrap-around pack of a study guide built on existing, published texts produced outside the OU.

Academic departments are responsible for the content of study packs and over-see the presentation of their modules, including setting assignments and managing the final assessment of students, whether by examination or the submission of extended essays. The production of a module, which is a highly time-consuming

---

[3] The speed at which the OU has adopted changing forms of technology to assist in the delivery of its modules, for example, from live transmission on the BBC to video and then DVD, and now increasingly from print to online digital formats, has always been governed by its access policy rather than exploiting its technological capacity to the full.

[4] Leaving aside questions about deficiencies in the OU's curriculum and manner of delivery, its very scale and the success that the OU has enjoyed might suggest that flexible forms of study and open entry have not proved sufficient in themselves to address the intransigent problems of access to higher education in Britain. The Religious Studies level 2 module *Introducing Religions*, however, recruits a significantly higher proportion of students from ethnic minorities, relative to other Arts modules.

team effort, involving academics, the module's administrator, editors, specialist librarians, IT support, the publishing department, and a commissioned audio-visual production company etc., typically takes about 3 years. There are rarely modules produced by one individual in the Arts Faculty's curriculum, unlike the typical provision of specialist courses for senior undergraduates in other universities. This is largely because the complexity and costs of the production process would make it impracticable to attempt to accommodate in this way the interests of individual academics within the curriculum. This significant commitment of financial and human resources explains why we have relatively few modules on offer at any one time and why they have to run for several years, with supplementary updating, to pay for their production. Although not every member of a department would participate in the production of every one of its modules or to an equal degree, the ambitious nature of these multi-media projects would make most OU academics keen to play a full role in module production – our "teaching."

This chapter will be devoted specifically to the teaching of Religious Studies within Britain's OU. The reason for this, apart from the writer's familiarity with this institution, is simply that Religious Studies is taught in the British OU by a well-established department. This is not the case generally in other open universities, either because many of these have a narrower, in some cases exclusively, vocationally-linked curriculum, or because for historical or cultural reasons Religious Studies as such is not taught in the public education system of that region. This chapter will offer a reflection on the cumulative experience of delivering the study of religions through the medium of distance learning within the OU for the better part of four decades.

198.2   Religious Studies Within the Context of the OU

198.2.1   *The Department and Its Pathway*

The Open University (OU) in Britain launched its first specialist course in the study of religions in 1978. Since then, what was formerly the Religious Studies "section" has achieved full departmental status (1993) within the Arts Faculty[5] and has developed a pathway of undergraduate study that contributes to a degree in Humanities or enables students to weave those modules that interest them into their route towards a different degree (subject to its requirements).[6] In this chapter I shall

---

[5] For a fairly recent overview of the Religious Studies department at the OU, see Beckerlegge (2005). Details of the department's current staffing and modules are available on the departmental website, www.open.ac.uk/Arts/religious-studies/index.shtml

[6] Sweeping changes in the funding of British higher education are designed to encourage a more program-based registration for degrees, which regrettably is likely to limit the flexibility that OU modular programs have offered students until now. The effects of some of these changes have already had a dramatic effect since this chapter was written.

explore some of the advantages that have accrued to Religious Studies through being offered in a university that is committed to the principle of open access at undergraduate level. I shall also examine some of the pedagogic challenges we face in attempting to introduce the study of religions through relatively "remote" methods to a large and highly diverse student body.

In his global survey of Religious Studies, Gregory Alles (2007: 25) referred to Religious Studies at the OU as one of "two prominent pioneering departments" established during the 1970s. The OU department is solely a department of Religious Studies, and not a department of Theology and Religious Studies, as is the case in the majority of British universities where the subject is taught. Admission policies, including greater acceptance of non-standard entry qualifications, and attitudes to part-time and mature students have changed in many universities over the decades since the OU was founded. But certainly within British higher education, which will provide the context for this discussion, the OU's social mission, which is stated in its charter, and style of teaching remain distinctive, as do the numbers and diversity of students it serves, its offer of open entry (that is, without prerequisite qualifications) to undergraduate study, and the fact that its style of teaching and student support is designed primarily for part-time students. OU students can take on the workload of full-time students simply by registering for more modules in a year, but part-time study is primarily what we do, rather than adjusting provision originally intended and designed for full-time students. If volume remains a distinctive feature of the OU's work, then handling it constitutes a major pedagogical and pastoral challenge.

Unlike the pattern commonly found in other British universities, all current Religious Studies undergraduate modules at the OU carry 60 credit points, and are equivalent to an entire year's workload of a part-time student – a recommended minimum of 16 h study time per week.[7] Students are required to complete a sufficient number of modules to acquire the requisite number of points for their chosen degree and a stipulated number of modules at levels 2 (intermediate, progression to honours) and 3 (honours level) for honours classification. Thus, unlike "years" of study, students may move between "levels" flexibly and, unless registered for a single honours degree, are generally permitted to take higher level modules without first completing a lower level module in the same subject, although the latter is recommended.

I shall now outline briefly the historical development of the Religious Studies curriculum, delivered through our various modules, before offering an assessment of its design and delivery. I realize that this section runs the risk of reading like a literature review or, even worse, a promotional guide, but it is through the OU's

---

[7] Other faculties in the OU do offer modules with lower credit ratings. Although student preferences change, experience in the Arts Faculty suggests that our students have found combining 60 credits modules an efficient way to achieve the 360 credits necessary for a degree. This is also more cost effective than producing and maintaining a plethora of 10 and 15 credits modules.

published study packs that students experience their university and it is the published and multi-media nature of OU material that makes its teaching so distinctive.[8]

198.2.2   Interdisciplinary Studies

In the early years of the OU, appointments to the section that would become the Religious Studies department were largely in religious history, and the primary role of the section was to contribute to the Faculty's interdisciplinary interest as reflected in the level 1 introductory module (then called the Arts "foundation course") at that time. In later versions (1970s/early 1980s) of this introductory module, there were units on the study of religions, contributed by Ninian Smart, and on West African religions. The "foundation course" of the late 1980s and early 1990s centered on the study of the mid Victorian period to which Religious Studies was able to make a substantial contribution on Victorian Christianity and controversies of that period. This course, however, did not permit a wider introduction to the study of religions or the discipline of Religious Studies.

In 1998, the Faculty launched a new level 1 introductory module (*Introduction to the Humanities*), which, unlike earlier Arts "foundation courses" was organized as a show-case of all the disciplines represented in the Arts Faculty. This included an introduction to the study of religions, and a linked study "Looking for Hinduism in Calcutta" (Beckerlegge 1998). These was supported by two films, "What is religion?" shot in Liverpool, and "Looking for Hinduism in Calcutta." which was filmed around the celebration of Durga Puja in that city.[9] This went down well with tutors who had some background and interest in Religious Studies and with students wishing to continue in the Religious Studies line, but less so with tutors and students whose interests were centred on European history and literature. The current, major Arts level 1 introductory module has re-adopted a more thematic, and less discipline-based, framework. Religious Studies makes several contributions to its different blocks, including a study of the Dalai Lama in an initial block on "Reputations" (Waterhouse 2008). The department continues its involvement in inter-disciplinary studies, to which the Arts Faculty remains committed at both level 1[10] and higher levels.[11]

---

[8] Subsequent references to OU module materials refer to illustrative, core texts rather than providing a comprehensive list of all the materials relating to a particular module.

[9] We are hoping to be able to provide more open access to archived film and audio material relating to the study of South Asian religious traditions, although plans for this are still at an early stage.

[10] For recent examples, see contributions to the modules The Arts Past and Present (Graham Harvey and Marion Bowman 2008), Voices and Texts (Helen Waterhouse with Fiona Richards 2010), and Making sense of things: an introduction to material culture (Bowman 2012; Paul-François Tremlett 2012a, b).

[11] See, for example, John Wolffe's (2004) study of William Wilberforce and slavery in the module From Enlightenment to Romanticism, c.1780-1830.

A huge advantage of the scale of the OU student body for subjects like Religious Studies is that they have a chance of viability. The current level 1 introductory Arts and Humanities module, *Arts Past and Present*, continues to attract in excess of 7,000 students per annum.[12] Like its predecessors, it provides an opportunity to present something of Religious Studies to a large audience, a platform for recruitment to specialized modules in Religious Studies. Each of the specialist modules offered by Religious Studies has typically recruited between 350 and 500 students per annum. Introducing Religions (level 2) has recruited a little under 500 students per annum. Its predecessors recruited far better than this, and we are uncertain whether this dip mirrors current levels of interest in the study of religions, or more varied options within the OU Arts curriculum and the British higher education sector more generally.[13]

198.2.3   The Religious Studies Pathway

A significant turning point in the development of the Religious Studies curriculum, which would have major implications for its future evolution and direction, came in 1978 with the launch of *Man's Religious Quest* (which soon became *The Religious Quest*!). This was a level 2 module that ranged widely over ten religious and secular traditions past and present, from the ancient Graeco-Roman world to contemporary West Africa, and embracing Abrahamic and Indic traditions. The module also explored some topics across traditions and opened with a substantial treatment of method in the study of religions. It was produced largely by consultant academics. Religious Studies staffing at that time was limited to three (not yet a formal department) and, during the production of the module, a Visiting Fellow (John Hinnells). It followed the by then increasingly familiar OU format of blocks of study units, a substantial reader, linked TV and radio broadcasts (later put on tape), and set books. The production of *Man's Religious Quest* was regarded as sufficiently significant because of its potential outreach for the journal *Religion* to review its materials (Mews et al. 1979). The reviewers' judgement suggested materials of an uneven quality and the reviewers, although supportive, drew attention to the module's lack of coherence, to which I shall return later in the chapter. *Man's Religious Quest*, which recruited

---

[12] The downside of the OU's catering for large numbers of students is that the university judges the viability of modules and programmes against expectations of substantial recruitment. The department's MA programme in Religious Studies has been earmarked for closure after the 2012 entry. Since its inception, it has attracted in the region of 40 students per intake, a figure that would delight most universities in the UK offering MA routes in Religious Studies, but which is regarded as unsustainable within the OU where production and assessment costs of all modules have to be recouped through student fees.

[13] The specialist Religious Studies modules referred to in this chapter have achieved very high ratings, in the region of 93 % on average, in student satisfaction ratings. The relevant data are regularly and independently gathered and analyzed by the university's Institute of Educational Technology.

very strongly and sometimes in excess of 1,000 students in some years, was subsequently turned into a shortened version (*The Religious Quest*), which ran until 1992. It was subsequently revised, expanded, and presented again (*World Religions*) from 1998 with additional new materials, films and books, when it recruited in the region of 650 students per annum. We have continued to incorporate in our most recent Religious Studies level 2 module two features of these preceding modules, which were less commonly found in comparable courses in other universities at that time; namely, a section on village practice and belief in the coverage of Hindu tradition, and an extended treatment of the Sikh tradition (originally made possible by Terry Thomas, a founder-member of the department).

In 1993, the by now department launched *The Growth of Religious Diversity: Britain from 1945*, a 30 credits module, written largely in-house with more limited use of consultants. The materials produced for this module have been used extensively in other universities (Parsons 1993, 1994; Wolffe 1993). At much the same time, we also converted an existing, highly successful module on *Religion in Victorian* Britain, first launched in 1988 (for example, Parsons 1988), into a full 60 points module. This expansion centered on the theme of Culture and Empire (see Wolffe 1997). It included a study of the presence of religions of South Asian origin and Islam in Victorian Britain (Beckerlegge 1997), thus enhancing its existing coverage of Christian denominations, British Judaism, and secular thought. Both these modules on British religious history reflected the expertise within the department at that time. This has widened with additional appointments to the department since the mid 1990s, as is evident from the expansion of the department's curriculum from that period to embrace, among other areas, folk and popular religion and contemporary Paganisms. We continue to face the familiar problem of a relatively narrow staffing base imposing limits on the development of our curriculum. Having said that, we are all expected to leave our comfort zones and contribute to widening the curriculum where this is feasible. When it comes to module production, we have been able to augment our resources through the use of academic consultants (for example, the authors of text books we adopt as set books, particularly in the production of audio-visual resources).[14]

The experience gained from the production and maintenance of its earlier modules has led the department to adopt certain broad principles that have shaped the modules that followed. To avoid overwhelming students, we have largely focused on the six religious traditions – Buddhism, Christianity, Hinduism, Islam, Judaism and Sikhism – that are studied in British schools through Religious Education, although these are examined in a global context and not narrowly with reference to British society. In keeping with the department's research interests, our modules typically address forms of religions from the nineteenth century, employing historical and social scientific methods – contemporary religions in recent historical perspective.

---

[14] Examples of this are audio discussions with Damien Keown, Kim Knott, and Eleanor Nesbitt, authors respectively of the volumes on Buddhism, Hinduism, and Sikhism in the OUP Very Short Introduction series, which were adopted as set books in 2006.

The design of our most recent level 2 module, *Introducing Religions*, has also been very much informed by lessons learnt from teaching *Man's Religious Quest*, and its accumulated revised versions. The lack of coherence of that earlier module, identified by *Religion*'s reviewers, was in part a consequence of it having been produced largely by external consultant academics who were not wholly immersed in the internal course production process. For example, attention to addressing specified themes is more likely to waver when the authors are not working constantly as a team. Similarly, when selecting set books we have found that individual volumes in the same series often address the declared themes of the series to different degrees and can vary quite dramatically in style and format. This preoccupation with uniformity might sound almost as though it runs against the individuality of academic insight and argument, but for students working remotely and in a more isolated manner, any idiosyncrasy in the teaching materials can be potentially disruptive. It can cost time as students wrestle with the unfamiliar. It can prove highly frustrating when a text fails to address a stated theme or provides such a sketchy treatment of it as to make the text of limited use to students. Also, it can lead to a lack of symmetry in the devising of assessment tasks in that it may prove impossible to set realistic tasks/options based on certain sections of the teaching materials. When planning *Introducing Religions*, we adopted a "template" design in order to maximise consistency of coverage throughout, and sought a series of set books that combined authoritative scholarship with a reliable series format.

Introducing Religions is a wrap-round study pack, a mixture of "OU classic" and "OU lite," which uses the Oxford University Press Very Short Introduction (VSI) series, but with substantial, illustrated study guides and audio and DVD material produced by the department. Five of the six traditions were covered in existing volumes in the VSI series, and we were fortunate that Eleanor Nesbitt (and Oxford University Press) agreed to produce an additional VSI volume to cover the Sikh tradition, for which we produced a study guide (Beattie 2006). *Introducing Religions* also makes some connections to religious communities specifically in Britain (as has our MA foundation module and *Islam in the West: The politics of Co-existence*, which was devised by David Herbert (2007) and ran from 2007 until 2010). In concentrating our resources on enriching the set books particularly through the production of audio-visual resources, we have been able to create a DVD that includes extensive visual material on all the religions covered in the module, both in Britain and their historic locations. To date, we have not routinely assessed students' oral skills through presentations. Attendance at face-to-face study sessions with tutors is optional in the OU because a mandatory requirement could constitute a barrier to study for the disabled, shift-workers, and others. We are able, however, via the DVD to set assignments that require students to examine iconography and other aspects of material culture. Web site analysis now provides a further alternative basis for assessment.[15]

---

[15] In general, the forms of assessment we use in Religious Studies remain heavily dependent on the essay format. This is partly because the outcome of the assessment has to be something than can be submitted (now electronically as a general rule) within a remote environment and does not

At the third level of the curriculum, the Religious Studies pathway until recently reflected and held in balance the two dominant strands of interest within the department; namely, the multi-disciplinary exploration of recent and contemporary manifestations of religion and spirituality, and religious history. The module *Religion Today: Tradition, Modernity and Change* (produced in 2002) explored the themes from sacred text to internet, religion and social transformations, civil religion, global religious movements and belief beyond boundaries. *Religion in History: Conflict, Conversion and Co-existence* (produced in 2005) began its examination of its stated themes in late antiquity and the period of the crusades. Its main focus was the Christian tradition, but its themes also led into studies of Christian-Jewish relations and the Holocaust, the Hindu Renaissance, and conflict and co-existence in Palestine/Israel. At the time of writing, these modules were approaching the end of their lives and have now just been replaced by one new, multi-media module, *Why Is Religion Controversial?*

The department's MA programme has been delivered in a similar way to its undergraduate programme (see fn. 12). Here too, the costs of our style of delivery preclude offering a wide range of taught modular options, and the structure of the degree is based on one extended foundation module and a dissertation module. The foundation module is designed to focus on research skills, incorporating examples of scholars' work, rather than work on specific religious traditions. The nature of the programme is made plain in promotional material. Some students, nevertheless, have found the reality of working on methodology and research skills throughout the foundation module not what they claim to have expected. There are entry qualifications to the Masters programme, but, in keeping with the ethos and aspirations of the OU, these are reasonably generous, requiring simply successful completion of an Honours degree and not necessarily in the subject the student wishes to study at Masters level. This perhaps highlights the kinds of problems that can surface in our environment where recruitment is dealt with remotely, rather than via direct interface with the department responsible for the programme, and which are compounded by the extent to which students are left to decide their level of preparedness for the programme of study they wish to undertake. This would seem the appropriate at which to proceed to offer some measure of assessment of our system.[16]

---

assume attendance at group face-to-face or online study sessions. These considerations do limit our options, although the Faculty has introduced a module in 2011 in which online group-work will be assessed, and this is likely to become the norm.

[16] Research students are admitted to the department, as in other universities, on the basis of selective, rather than open, entry. They are supervised by combinations of internal/external and central/regional supervisors, depending on the status of the student (full or part-time) and her/his location. The majority of these students have been from within the United Kingdom. An example of an externally-funded PhD project (under the AHRC/ESRC Religion and Society programme) is Naomi Stanton's work on the decline of Sunday Schools in the twentieth century. For information about the department's research degrees, visit www3.open.ac.uk/study/research-degrees/religious_studies.htm

198.3   Achievements and Pitfalls

In the final part of this chapter I shall highlight what I consider to be some of the successes/advantages of our approach and some of its weaknesses specifically in relation to teaching the study of religions. Several of these factors stem from the distinctive features and strategic advantages that the OU is able to exploit because of its size and reliance on flexible, blended learning.

198.3.1   Size of Student Body

We are able to introduce students to the study of religions in relatively large numbers, and this must surely be beneficial to society as a whole, fostering as it does a more informed, wider interest in religion in general. Our students' wide range of ages, backgrounds, and expertise makes them particularly rewarding to teach, and those students who participate in tutorials similarly benefit from this richness of experience.

198.3.2   Openness

The size of our student body is directly linked to our "open" entry policy. Significantly, this "openness" also applies to entry to modules at levels 1–3 (as explained previously, we do not deal in "years" as students may move up and down between levels as they wish, as long as they meet the requirements of their chosen degree). In fact, Arts, and many other OU, students have tended to date to favor breadth of study (but see fn. 7), and while some have gravitated towards a single honours route, where this has been available, many have preferred to balance two or more subjects during their progress towards a degree, or simply go for "pick and mix." Thus, many, conceivably even the majority of, students taking a particular module might not have followed the subject strand in which the module is located. Consequently, many will not have followed a route of progression from Religious Studies level 2 into Religious Studies level 3. This does create problems for the consolidation of knowledge and the internalization of the technical terms associated with the study of specific religions, especially when these are drawn from a range of languages. This problem, is exacerbated by the structure of our modules, which requires students to move between diverse regional and historical studies. In the past we have produced pronunciation audio guides, but students have not found these very user-friendly. They do value aids such as glossaries and timelines. Based on our experience with a large and diverse student intake, we no longer use diacritical marks and tend to use anglicized forms wherever possible. As texts books follow different conventions in this respect, and these differences might be evident in the set books adopted for a study pack, even providing a glossary for less confident students can prove difficult.

Many students will dip into our offerings, rather than working progressively through the levels, because following the Religious Studies strand is not their main or sole interest. Another problem relating to progression, which I shall deal with here, is that a student's year of entry (and the number of years taken to complete the degree) will govern what modules are on offer at any one time. As noted above, costs of production limit the range of modules on offer at any one time. Modules are approved for a specified length of time and are routinely withdrawn at the end of this term as part of our quality assurance process. OU departments do not automatically replace modules simply with an updated version in the same area/on the same topic. The extended nature of curriculum development with its periodic design of new modules leads us to balance some continuity with change.

198.3.3   The 60 Credits Module

The substantial module has huge advantages in that a study contained within it can be presented in some detail and at some depth. This does mean, however, that the typical part-time OU student will only take one such module in any given academic year. This makes for a rather rigid pathway, offering students limited choice between modules, and thus any flexibility and options have to be built into the modules and reflected in their assessment. Our reliance upon block/chapter length studies of particular religions in a limited number of substantial modules, in which these studies are typically embedded in thematic or period-based frameworks, may not coincide with a student's main interest. Students, consequently, might have to decide whether they wish to take the whole module for the sake of certain parts of it. The location of these studies of particular traditions and regions in these broader settings also suggests that we shall be unlikely to satisfy, and thus to attract, students who wish to pursue a specialist route and to be members of an academic community (department or institution) bound together by this common interest, for example, in South Asian studies. In trying to ensure that students do have options within the module, we have used forms of assessment that offer a choice of tasks on different traditions, or tasks that leave the student to provide relevant examples. This can allow students simply to seek out what is most familiar rather guiding them to widen their exposure to a range of religious traditions, including the less familiar (to them).

We have also faced problems when attempting to cover what some students might regard as the "drier" aspects of method, theory, and historiographical debates, and I am not convinced that we have succeeded as yet in overcoming this problem. An OU student studying alone and wishing to get to grips, say, with the study of Judaism, might well have limited patience, if confronted with an extended introductory block on problems of method and theory, or indeed the history of Religious Studies. This problem is not unique to the OU, but in other universities, where the student's year is typically broken into modules of 15–20 credits, a course on method can be taken alongside a course on a specific tradition in a mutually illuminating fashion. Our response to this problem is to drip-feed such problems into theme or tradition-based studies in order to highlight specific problems of method, but I would be surprised if our students emerged with a secure appreciation of the historical contours and methodological concerns of Religious Studies as a discipline.

198.3.4   The Published Course Materials

Leaving aside the merits of individual academic's contributions, the OU published study pack brings benefits and scope for a structured integration of teaching material. The use of "activities," for example, can direct students to consult readings from primary and secondary sources, possibly linked to interview material or film material on a related festival, ritual, or a building such as a *gurdwara*, and contextualize these within a given locality. Our teaching is immeasurably enriched by the integration of high quality film and audio material, in almost all cases commissioned and produced for that purpose, thus giving us a considerable amount of control over locations (within the constraints of funding). For those without first-hand experience of a religion they are studying, this visual material may constitute something of a "virtual field trip." One example of this is the 1 h film made in 1978 in the village of Soyepur near Varanasi. We revisited this village in 2005 and have made a film for our current *Introducing Religions* module, which includes a "then and now" segment on this one village, catching up on the histories of some of those interviewed back in the 1970s, changes in village practices and values, and changes in employment and the local economy affected by the village's proximity to Varanasi. Both films were made with the assistance of Joseph Elder (University of Wisconsin) and his research team. Apart from the programmatic, integrated nature of our module materials, we are able to support these, as mentioned above, with chronologies, glossaries, links to relevant internet sites, and access to e-books, online reference works and data bases, and e-journals. Now that the tipping point has been reached in ownership of, or at least access to, personal computers, we are increasingly able to lead out from the core study pack through our electronic library and into the resources on the web.

I no longer feel apologetic when faced with the criticism that the published study pack and linked reader act as a straight-jacket. In my experience, most undergraduate students in other universities follow core reading lists or increasingly rely, as library resources are squeezed, on readers produced by module tutors. Our students in the past have had less experience of independently locating and evaluating primary and secondary source material, but have not been confronted by empty library shelves when preparing assignments! The earlier access-based argument for reliance on the study pack has now been overtaken by the popular use of the internet, and this will require a sea-change in the attitudes of some of our students to their studies. Our current and future students will be expected to gather and evaluate material more autonomously as we now take on the task of encouraging them to make full use routinely of our electronic library.

198.3.5 Nuance and Gloss

With cohorts of students assigned to tutors throughout the country and beyond, it is quite difficult to recall and refine statements made in *published* material or to address unanticipated reactions or readings of the material. In campus-based study this can be done by the tutor responsible for a module airing the matter in class or in conversations with students. Our Associate Lecturers, of course, do this routinely but with their own tutorial group and to some extent on national online forums. But a student who wants to raise something about the very nature of the module is confronted with the prospect of communicating with "the module chair," the "curriculum manager" (module administrator) or the even more amorphous "module team," rather than a person with whom they are acquainted and know to be responsible for the generation and day-to-day management of their module. It is simply not possible to explain in text the extensive process of decision-making that has shaped a particular module.

Our students enter the OU through very different routes and at the outset a significant number have little or no experience of academic language(s) and some clearly do not take its full force. We declare repeatedly and prominently that we offer an academic study of religion and that we are not there to offer religious nurture or formation. Yet, a significant minority of students appear to struggle to recognize this as a methodological principle and to act upon it when preparing their assessed work. In a campus-based environment, the ethos of the department can help in communicating and reinforcing its ideals. We have agonised over questions about what to include and what to omit in modules (particularly in our level 2 introductory module), and matters such as the order in which we introduce particular traditions and how we assess them. For example, what is the best way to introduce Judaism and Christianity to ensure that both are understood on their own terms (Harvey 2005; Sinclair with Bowman 2005)? We declare that we shall treat and value all the traditions we study in the same way; does this mean that the work on each tradition should be assessed in exactly the same way – essays of the same length? Do we privilege the religion we study first and appear to devalue the one we come to last? I do not mean to imply that our students endlessly raise such questions with us, but rather to illustrate that teaching committed to books and DVDs takes on a status and authority different to lecture notes that are fluid and capable of carrying constant glosses and second-thoughts, in addition to outright corrections!

198.4 Conclusion

Prevailing economic conditions may well make students view registering for degree modules and programmes in the Arts/Humanities and Social Science in a new light, as they reflect on vocational relevance and renewed encouragement from schools and employers to seek scientific and technical qualifications.

Students in England (this is not yet true for all parts of the United Kingdom) have faced significantly increased costs from 2012, following a shift in government policy that has reduced government funding for undergraduate teaching by transferring the costs onto new student loan arrangements. What the impact of these external factors will be on recruitment to, and thus the future of, particular subjects remains to be seen. As an institution, the Open University is well-placed to respond to the needs of prospective students who wish to study while remaining in employment, but it too will depend on reliable streams of recruitment and across a sufficient range and diversity of programmes.

The Department of Religious Studies continues to offer a congenial home for students who wish to study religions as part of a more flexible degree route, rather than specialising in one area, and for those who prefer to study religions in a secular institution. To date, it is likely that fluctuations in the numbers of our students could be explained by the expanding range of modules offered by the Faculty of Arts, which has the effect of spreading students more thinly. Current attitudes to the prominence of reports about religion in the news might be another factor. The number of pupils in schools taking Religious Education to examination levels increased after 9/11, and we detected an even greater interest among our students in Islam in the same period. Now, however, we have anecdotal evidence that constant reports about Islamism in the media has had the effect of making some students less inclined to give their time to its study. There is no evidence to suggest, given our relatively stable levels of recruitment to date, that any dips could be accounted for in terms of secularising tendencies within British society more generally. In fact, my impression is that wider travel, the day-to day experience of students as members of a multi-cultural society, and the coverage of religion more generally on television continues to encourage adult learners to want to come back to an area of the curriculum that has not been served well in many British schools.

In this chapter, I have tried to show that the opportunities we have to promote the scholarly study of religions in Britain through the OU bring with them significant pedagogical challenges of which the power of our outreach is arguably one of the most important.[17]

The very investment made by the OU in the quality of its study packs limits the range and flexibility of the curriculum. Many of the challenges built into the OU form of delivery, which I have identified above, arguably present more difficult hurdles in the teaching about religions that tend to be less familiar to the majority of British students when they embark on their first module in the study of religions. This is partly because the majority of our students have little, if any, prior knowledge of this area, and partly because the very nature of our open, supported learning approach does not require students to remain within one line of study and thus to consolidate on a year-by-year basis.

---

[17] Until recently, the OU also had a considerable presence in Continental Western Europe. As the OU has recently taken a strategic decision to withdraw from this region, this former dimension of the department's outreach has not been discussed.

As the reviewers of our first substantial venture into the study of religions noted in *Religion,* "It is not usual for university teaching courses to be subject to public review…" (Mews et al. 1979: 116). The "routine" teaching of the OU is very much a public matter through virtue of its publication and relative accessibility. As a public record, its spans a period during which practitioners of Religious Studies abandoned the language of "man's religious quest" (both the gender-laden language and the notion of "questing") and "world religions," and became more sensitive to the need to study religions as something with fluid boundaries and "lived," rather than located in and defined by authoritative texts.

As this chapter has illustrated, some of the lessons we have learned through this experience have been quite prosaic, but of considerable value in trying to improve our teaching of a range of religious traditions for a large and very diverse body of students.[18]

Acknowledgements   This chapter is based on a paper presented at the Philosophical and Religious Studies Subject Centre's conference on Teaching Religions of South Asian Origin, 13 January 2011, which was subsequently published in the Subject Centre's journal (Beckerlegge 2011). In this chapter, however, I have widened my focus to consider the study of religions more generally, and not narrowly those of South Asian origin.

References

Alles, G. (2007). *Religious studies – A global view.* London/New York: Routledge.
Beattie, H. (2006). *Sikhism study guide* (A217 *Introducing religions*). Milton Keynes: The Open University.
Beckerlegge, G. (1997). Followers of Mohammed, Kalee and Dada Nanuk: The presence of Islam and South Asian religious in Victorian Britain. In J. Wolffe (Ed.), *Religion in Victorian Britain* (Culture and Empire, Vol. 5, pp. 222–267). Manchester: Manchester University Press in association with The Open University.
Beckerlegge, G. (1998). Studying religion. In *Religion and science in context* (Units 14 and 15 of Block 4 A103 *Introduction to the Humanities*, pp. 9–82). Milton Keynes: The Open University.
Beckerlegge, G. (2005, November). Religious studies in the UK: Studying religions at the Open University. *Bulletin of the British Association for the Study of Religions, 106*, 43–45.
Beckerlegge, G. (2011). Teaching about religions of South Asian origin at the Open University. *Discourse: Learning and Teaching in Philosophical and Religious Studies, 10*(2), 61–77.
Bowman, M. (2012). From sanctity to celebrity? Relics in contemporary contexts. In P.-F. Tremlett (Ed.), *Afterlives* (pp. 1–54). Milton Keynes: The Open University.
Harvey, G. (2005). *Judaism study guide* (A217 *Introducing religions*). Milton Keynes: The Open University.
Harvey, G., & Bowman, J. (2008). Sacred space and landscape. In D. Brunton (Ed.), *Place and Leisure* (AA100 *The arts past and present*, pp. 35–62). Milton Keynes: The Open University.
Herbert, D. (2007). *Study Guide* (AD252 *Islam in the West: Politics of co-existence*). Milton Keynes: The Open University.

---

[18] The department was awarded the 2002 Shap Working Party's annual prize for its contribution to the teaching and study of World Religions in Britain.

Mews, S., Pye, M., King, U., Cousins, L. S., Alexander, P. S., Hanson, R. P. C., Bosworth, C. E., & McKenzie, P. R. (1979). Man's religious quest: A review of Open University Materials. *Religion, 9*(1), 116–139.

Parsons, G. (Ed.). (1988). *Religion in Victorian Britain, I Traditions*. Manchester: Manchester University Press in association with The Open University.

Parsons, G. (Ed.). (1993). *The growth of religious diversity: Britain from 1945* (Traditions, Vol. 1). London: Routledge in association with The Open University.

Parsons, G. (Ed.). (1994). *The growth of religious diversity: Britain from 1945* (Issues, Vol. 2, p. 272). London: Routledge in association with The Open University.

Sinclair, S. (2011). Blended learning and tuition in religious studies: An Open University perspective. *Discourse: Learning and Teaching in Philosophical and Religious Studies, 10*(3). Available at http://prs.heacademy.ac.uk/view.html/PrsDiscourseArticles/228

Sinclair, S. with Bowman, M. (2005). *Christianity study guide* (A217 *Introducing religions*). Milton Keynes: The Open University.

Tremlett, P.-F. (Ed.). (2012a). *Afterlives* (A151 *Making sense of things: An introduction to material culture*). Milton Keynes: The Open University.

Tremlett, P.-F. (2012b). The tourist gaze and religious objects. In P.-F. Tremlett (Ed.), *Afterlives* (pp. 55–93). Milton Keynes: The Open University.

Waterhouse, H. (2008). The Dalai Lama. In E. Moohan (Ed.), *A100 The arts past and present* (pp. 197–229). Milton Keynes: The Open University.

Waterhouse, H. with Richards, F. (2010). Voices of authority and truth: The *Metta Sutta*. In L. Prescott (Ed.), *The voices and texts of authority* (A150 *Voices and texts*) (pp. 1–35). Milton Keynes: The Open University.

Wolffe, J. (Ed.). (1993). *The growth of religious diversity: Britain from 1945* (Issues, Vol. 2). London: Routledge in association with The Open University.

Wolffe, J. (Ed.). (1997). *Religion in Victorian Britain* (Culture and empire, Vol. 5). Manchester: Manchester University Press in association with The Open University.

Wolffe, J. (2004). William Wilberforce. In D. Johnson (Ed.), *Religion, exploration and Slavery* (A207 *From Enlightenment to Romanticism c.1780-1830*, pp. 51–86). Milton Keynes: The Open University.

Chapter 199
Facebook Gets Religion: Fund-Raising by Religious Organizations on Social Networks

Mark D. Johns

199.1 Introduction

Despite an ongoing recession, charitable and philanthropic giving in the U.S.[1] remained in excess of $300 billion in 2010. Contributions to religious congregations and organizations was the largest segment of this giving, consistently representing more than one-third of all giving and approximately one-half of all individual (that it, non-corporate) donations (Giving USA 2011). Because giving at this level represents a significant sector of the U.S. economy and involves a very large portion of the U.S. population, it is not surprising that non-profit organizations have utilized all conceivable means of reaching potential donors. Consequently, many of these organizations have maintained a long-established presence on the internet (Chen et al. 2006). Online social networking sites, such as facebook.com, are currently popular attractions for large numbers of users, and intentionally hold out the promise of reaching a large base of audiences attractive to businesses and non-profits alike.

Social networking websites such as facebook.com have become a primary means of communication among persons in a constantly widening range of demographic categories (Gordon 2009). Facebook has more than 900 million members (Facebook 2012b) and is expected to surpass one billion users worldwide sometime in 2012 (Lyons 2012). Further, online social networking sites such as Facebook are no longer appealing only to teens and young adults, to which they were originally targeted, but are now attracting an increasing number of mature adult users (Lenhart 2009).

---

[1] Many who write and research in the field of fund-raising (Friedman 2003; Ostrower 1995) distinguish between charity, which is giving to alleviate human suffering such as homelessness, hunger or disease, and philanthropy, which is giving to the betterment of all society, such as artistic organizations, museums, libraries, or schools. The present study focuses primarily on charitable giving.

M.D. Johns (✉)
Department of Communication, Luther College, Decorah, IA 52101, USA
e-mail: mjohns@luther.edu

© Springer Science+Business Media Dordrecht 2015
S.D. Brunn (ed.), *The Changing World Religion Map*,
DOI 10.1007/978-94-017-9376-6_199

As more and more individuals use Facebook to keep in touch with one another, more and more businesses, non-profits, and other organizations seek to utilize this site as a tool to establish and maintain relationships with customers, clients, and constituents (Li et al. 2007; York 2009). Facebook is supported by advertising and offers ads targeted according to the interests users themselves have expressed. A page for potential advertisers boasts that, "Advertisers can ask Facebook to show ads to an audience based on demographic factors such as location, age, gender, education, work history and the interests people have chosen to share on Facebook" (Facebook 2012c).

Facebook also offers local businesses, brands, products, organizations, artists, bands, or public figures the option to establish "pages" – a unique type of "fan" profile, through which they may relate directly with users who elect to "like" these particular entities. Some businesses and organizations put a great deal of effort and a great many resources into maintaining connections with Facebook users through their profiles – presidential candidate, and now President Barak Obama, is one prominent example (Westling 2007). There are even firms offering professional management of Facebook "fan sites" for businesses. However, these "fan" profiles are sometimes created by users loyal to a product or organization. Apple Computer, for example, apparently did not establish an officially sanctioned corporate Facebook profile until December 2011. Yet individual lovers of Apple products have created dozens of "fan" profiles for the company and its products, some with as few as four users who have clicked the "like" link for the particular profile, and one with more than 180,000 fans as of this writing. Recently, in 2012, Facebook has begun an effort to identify pages that actually belong to a company or business, distinguishing those from pages created by others simply because of love for a brand (Roberts 2005).

Since 2007, Facebook users have been allowed to create "causes" through an application initially created by Sean Parker, Facebook's first corporate president (Wortham 2010). The Causes application is presently maintained by a separate corporation in which Parker is still involved. Causes, may range from a sick friend at a local high school to an international organization raising funds and awareness about conquering a dread disease, from a college student taking an individual pilgrimage to a holy site to a global mission organization. Cause sites differ from business or political fan sites in that Facebook users who choose to link with these are termed not as "fans" but as "members." Further, Causes encourages members to recruit others to join, and tracks recruitment efforts. Causes also encourages members to raise funds through the application, which facilitates credit card transactions online. There appears to be little effort by Causes to vet Cause entries for legitimacy or to see to it that a cause fits into the categories (Animal Rights, Disaster Relief, Education, Environmental Conservation, Human Rights, etc.)[2] that the Cause creators have selected for it. The company does not wish to become arbiter of what is,

---

[2] The "Causes" application currently hides category for Religious organizations from its "Discover" tab, though many such organizations are registered in this category. For example, a search on the word "Christian" in the Causes application at this writing yielded nearly 10,000 titles.

or is not, a legitimate cause. However, it has recently made efforts to identify Cause pages that are actually maintained by the organization named so that they may be identified as distinct from those created by other interested parties or fans, and the application now marks organizations that prove they have non-profit, tax exempt status with the U.S. Internal Revenue Service.

Implied in the pitch to advertisers, in the encouragement to recruit to Causes, and indeed, in Facebook's stated mission to "make the world more open and connected" (Facebook 2012a: 3) is the notion that, through facebook.com, organizations will be able to reach and establish relationships with new customers, new clients, or new supporters. Indeed, just as Facebook continues to suggest new potential friends to users – encouraging "friend hunting" (Rice 2009: 99) – there is at least the sug- gestion that users will similarly engage in "cause hunting" or "fan site hunting." But much of what is known from research about how users view and make use of Facebook, and other online social networking sites, suggests that this implicit prom- ise cannot be fulfilled because users tend to connect with known others rather than to seek connections with strangers (Joinson 2008; Lampe et al. 2006, 2008). Indeed, a major corporation, General Motors, recently withdrew advertising from Facebook, just as other major studies have suggested difficulties with the advertising model (Barnett 2012). The same factors may well apply to religious non-profit organiza- tions – including international relief agencies, youth ministry organizations, camping ministries, missionary organizations and others – using Facebook tools to build support and raise funds. The problem – for non-profit religious organizations as for business enterprises – is that the effectiveness of Facebook as a tool for raising awareness and interest is extremely difficult to measure.

199.2 Facebook, Religion, Brand Communities, and Social Ties

Facebook has been identified as a site where religious identities are formed and displayed, and where religious groups have created a presence (Bobkowski 2008; Campbell 2010; Johns 2012; Rice 2009). These observations are consistent with those who have noted religious activity and interaction taking place in mediated form on the internet generally (Campbell 2005, 2010; Lövheim 2004). From the beginning, Facebook has offered users the option of indicating their personal religious beliefs or affiliations on their profile pages. Observers have also noted the formation of many hundreds of interest groups on Facebook focused around religious figures, groups, or themes. Through these profile options and affiliations Facebook facilitates religious categories as significant markers in identity construc- tion online (Johns 2012). Conversely, many religious organizations have sought to use the internet, and Facebook in particular, as tools for conveying religious messages, engaging in religious outreach, and as a means of providing ongoing spiritual support to adherents (Campbell 2010).

Traditional fund-raising theories affirm that the motivations for giving are complex, and that these vary based on a number of factors, including social class, religious background, gender, education, legal and tax environments, and other less quantifiable variables (Andreoni 1998; Ostrower 1995). However, fundamental to almost all fund-raising strategies is establishing and maintaining a solid and trusting relationship between the individual donor and the fund-raising entity. Oates (2003) has pointed out that religious charities, in particular, have been historically more successful when supporters have been directly involved in the operation and governance of the institutions providing services to those in need. This engagement has taken a variety for forms, but all involve direct communication between charitable institutions and their constituents. Dozier et al. (1995) have demonstrated that individuals enter into social relationships with organizations, developing emotive affinity, feelings of loyalty, trust, or commitment to companies, brands, non-profits, and other organizational entities by way of ongoing communications carried out in interpersonal events and/or through various media. This tendency of humans to anthropomorphize or humanize their relationships with corporate entities is frequently a foundation of marketing, public relations, or fund-raising campaigns.

199.2.1 Brand Communities

Muniz and O'Guinn (2001) have developed the concept of "brand communities" which are communities formed around loyalty to a particular product or brand. They define brand communities as, "social entities that reflect the situated embeddedness of brands in the day-to-day lives of consumers and the ways in which brands connect consumer to brand, and consumer to consumer" (418). The concept is related to brand love, as described by Roberts (2005). These are not mere groups of consumers, but devoted loyalists, such as those who create passionate social gatherings around Apple Computers, Saturn automobiles, or Harley-Davidson motorcycles. Muniz and O'Guinn (2001) note that these groups possess the three markers that sociologists regard as essential to community: consciousness of kind (that is, awareness of their own unique identity as a group), rituals and traditions, and communal codes of moral responsibility. However, basic social relationships within these communities are triangular rather than dyadic – that is, the brand is a party to every interpersonal relationship. Individuals always relate to the brand and relate to others in the community in terms of the other's relationship to the brand.

Hassay and Peloz (2009) have noted the intentional use of brand community by charity organizations as a means to increased identification and response among donors. They observe that, "the religious organization might be viewed as the archetypal consumption (that is, brand) community" (p. 38), and suggest that "supporters with higher levels of consciousness of kind (identification) with a charity [tend to] exhibit higher levels of participation in shared rituals (behavioral involvement) with that charity" (p. 39).

Further, Ulusu (2010) has expressly observed the use of Facebook by advertisers to promote and cultivate brand community. She notes that, although users typically ignore sidebar advertising on Facebook, they willingly spend time receiving the news feeds of brand sites and engage in interaction with other fans on brand pages. Thus, it would appear that Facebook, with tools such as Pages and applications such as Causes, offers the potential to create exactly the sort of triangular relationships necessary to establish brand community for charity organizations.

199.2.2   Social Ties

Communication theorists continue to debate the impact of various media technologies on the messages they carry and on the individuals who receive them. Haythornthwaite has argued "that the use and impacts of media are dependent on the type of tie connecting communicators" (2002: 385). Drawing on earlier research by, and in association with, Barry Wellman (Wellman et al. 1996, 2001), Haythornthwaite suggests that technological networks, such as Facebook, create sets of "latent" social ties. That is, the network establishes the possibility that two individuals – or an individual and a corporate entity, or a triangle of individual fans and a brand – may form a connection. However, the network does not necessarily motivate such a connection. While the network may be a necessary cause, it is not a sufficient cause for establishing relationships. According to Haythornthwaite, the situation is analogous to assigning two persons to individual offices in the same building. Doing so creates the circumstances under which they might happen meet in the hallway, but it does not guarantee that they will, nor does it make them any more likely to speak should they pass one another. The technology is a means for communication, but the nature of the social relationship(s) between persons determines whether, and to what degree, communication will take place.

Latent ties develop into weak social ties only when socially constructed circumstances push individuals to actually meet and communicate with one another. Persons co-located in the same office building may find themselves summoned to the same meeting, assigned to the same project, or negotiating for use of the same copy machine. Persons in online social networks may discover they share common friends or common interests, but even so, they are unlikely to initiate communication with one another, that is, to create a direct social tie, without some motivation. Haythornthwaite states that, "a technical implementation needs to be matched with a social implementation in order to effect connection among as yet unconnected others" (2002: 393).

Individuals who share strong social ties are those who have had multiple interactions over time, so that levels of trust and/or intimacy have developed. These are persons who frequently share resources, information, or social support. Such persons have high levels of motivation to communicate with one another, and do so using multiple channels, such as face-to-face, land-line telephone, cell phone, email, or other media. As new media channels become available, or as older channels become unavailable, they continue to find multiple ways to communicate. However, those with weaker social ties may be in the habit of using only one communication channel to maintain that tie. If the communication channel is altered – that is, if the cafeteria schedule is changed so that hallway encounters no long occur, the email list shuts down, or the characteristics of an online social networking site are altered – the opportunity for contact is eliminated and there is little motivation to seek alternative means for communication. Likewise, if the social circumstances are altered – if the weekly employee meeting is no longer held, the class no longer meets, or a project reaches completion – even if the means for communication remain intact, motivation to continue contact is lacking, and weak ties may atrophy. Thus, weak ties require both the ready means as well as the social motivation in order to be perpetuated, whereas strong social ties perpetuate themselves and motivate participants to seek means for their continuation (Haythornthwaite 2002).

Applying Haythornthwaite's reasoning to brand communities suggests that such communities are based on latent to weak social ties. Even if an individual's brand loyalty is strong, the opportunity to seek and enter into relationship with others with similar brand loyalty is dependent on both technical opportunity social motivation. The question, for non-profits as well as for businesses, is whether Facebook is capable of providing both opportunities and motivation.

199.3 The Challenge of Studying Facebook

This pilot project is intended to test ways of studying potential charity brand community among religious charitable organizations on Facebook. Commercial enterprises may have means of scraping data, measuring click-through rates, monitoring point of sale activity, or otherwise quantifying customer behavior. But the assumption here is that religious organizations will not have access to such means. Thus, the best way to understand the actions and motives of those who support religious organizations is to ask them directly. By surveying Facebook users and observing activity on several Facebook Pages and Causes entries, I attempted to discover the best method for determining the extent to which facebook.com is providing a new venue for support to these religious or para-church charity organizations. More specifically, I wished to determine a means to examine whether evidence of brand community exists, and to learn how Facebook may be utilized to strengthen social ties in these communities.

The project focused on five basic questions: (1) How did you become familiar with this cause? Did you learn about it from Facebook, or did you know about this organization before connecting with their page on Facebook? (2) If you discovered this organization through Facebook, what was it about their page or their reputation that attracted your attention? (3) Have you ever participated in any off-line, "real life" meetings, events, or activities related to this organization? (4) Have you ever contributed money (in any amount) to this organization? If so, did you do so through their Facebook page? And (5) How frequently do you visit this organization's page on Facebook? For what reasons do you visit this Facebook page?

199.3.1 Written Surveys

For purposes of this pilot study, a convenience sample of current undergraduate students at a Midwestern university in the U.S. (N = 37) was surveyed in March of 2011 using a written questionnaire administered during class time concerning their involvement with charities through Facebook. Appropriate IRB approval was obtained and participants granted their informed consent. Although all reported being active on Facebook, half (19) reported not participating in any causes on Facebook, while a half-dozen others could not recall whether or not they had ever participated in a cause. Among the dozen who did indicate participation in a cause, there were varying levels of certainty about their activity. Most could not recall the exact name of the cause, but only the general area of concern. For example, one wrote, "Yes, New Orleans Hurricane thing." Another wrote, "Human rights, anti-cruelty to animals." Two respondents, however, belonged to more than one cause and provided the full names of each. These respondents also reported a higher level of activity with their charities than most.

None of the college undergraduates responding to the pen-and-paper survey indicated that they had done any "cause shopping" online. No one reported spontaneously discovering a cause on the social network that was previously unknown to them. Most encountered their charity "in real life" and later discovered it was also represented on Facebook. Few reported returning to the organization page with any regularity. One respondent wrote, "I visit only when it pops up on my news feed." Another responded, "Never. If they message me things I will skim it." And yet another wrote, "Very rarely. I don't use them very much at all."

199.3.2 Email Questionnaire

Also for purposes of this pilot study, a convenience sample of 100 current undergraduate students and recent graduates from a Midwestern university in the U.S. were emailed a questionnaire over their summer vacation period concerning their involvement with charities through Facebook. Appropriate IRB approval was obtained and participants granted their informed consent. However, despite a series of follow-up email messages encouraging participation, the response rate was quite poor at only N = 19. The email questions were identical to those of the written survey and the results were essentially the same.

Of these, six indicated that they had never "liked" a charity or joined a "cause" on Facebook. Three more specified that they had only done so for a single cause, either because of a personal involvement offline, or as a favor to a friend. Four indicated that they had come across a charity on Facebook, but usually through the action of a Facebook friend who had recommended the organization or issued an invitation to a cause. Most had affiliated with organizations with which they were already familiar through their offline religious groups or campus involvement. One of the most active in Causes wrote, "I knew about all of these causes through personal involvement or friends who were involved, so I knew about them before connecting on Facebook." Another who has affiliated with a number of causes online wrote, "I have either used them, met the people running them, or had some other personal contact with them…. I need to be aware of them beforehand to feel a need to 'like' them."

Again, there was little evidence among these respondents of the formation of any sort of online community surrounding the charity. None reported going back to an organization's page more than occasionally after initially liking or joining it. One wrote, "I visited both of them when I liked or joined them and haven't been back." Another responded, "I have visited this page a couple times, but not frequently. I mostly hear from my friend on the organization than through facebook." Some do look for news on pages, such as the respondent who wrote, "Every few months, often if an update shows up on my page, then I remember to look at it." And another who wrote, "I visit [organization] Facebook page when they have a post or a news article that catches my attention, That happens once every couple of months."

199.3.3 Facebook Messaging

A convenience sample of a half dozen religious non-profit organizations was selected from the "Religion" category in the list of "causes" on facebook.com. These included two Christian denominational hunger relief and development organizations of long standing and solid reputation, one denominationally affiliated Christian national youth ministry organization, two interdenominational disaster relief organizations with Christian roots, and one international, interfaith network of grief support groups. The number of members joined to these causes on Facebook ranged from just over 400 to well over 280,000.

The Causes application currently shows a random selection of Cause members on the organization's page, but no longer allows a link back to the Facebook profile of the member. However at the time of initial investigation the application presented a random selection of members on each Cause's page, and provided links to each member's Facebook profile. However, shortly after beginning the project, the design of Causes pages was changed, eliminating the random presentation of a small number of members, and instead displaying a full membership list in the inverse order of the date on which they joined the cause – that is, those who joined most recently were presented first on the list, progressing back to those who were the first to join. To adjust to this change, a random number table was used to determine how far to count between listed members, so a dozen Causes members were contacted with the same questions used in the written survey. Contacts were made through the user message email system built into Facebook. However, at one point the anti-spam feature associated with this email system blocked me for sending too many personal messages. Response rates were very poor because, just as the Facebook message system perceived these messages to random users to be possible "spam" email, many users apparently perceived these messages in the same way. Lengthy consent and IRB legal text required also tended to make the messages appear longer and more complicated, further reducing the willingness of users to spend time reading such a request from a stranger. Nevertheless, at least one or two members of each selected cause did agree to participate in a brief interview conducted by email – eight participants in all. Responses were essentially parallel to those in the written questionnaire. The complexity of this approach and the minimal response call into question the suitability of such a method for a more detailed study. This method of approaching Causes users comes perilously close to violating terms of service, and is likely viewed as spam by most recipients.

199.3.4   Online Observation

Finally, and most recently in the continuing evolution of the formats of Facebook pages and Causes entries, three religious charity organizations were selected for an analysis of their web sites, Facebook pages, and Causes sites. These charities were chosen because they are large, well-established and well-known religious organizations representing three major faith traditions. No effort was made to select in any sort of systematic manner for this pilot study. They are simply chosen at convenience.

Samaritan's Purse is such a well-known organization. It maintains a very extensive, professional-looking up-to-date website at www.samaritanspurse.org. On its Causes profile it describes itself as an:

> Evangelical Christian organization providing physical and spiritual aid to hurting people around the world since 1970. Samaritan's Purse is a Christian non-profit non-political humanitarian organization that works worldwide to meet the physical and spiritual needs of people suffering from war, poverty, disaster, disease, famine, and persecution. The organization's president is Franklin Graham, son of Christian evangelist Billy Graham. (Samaritan's Purse 2007)

The Causes profile was created in 2007 and boasts 15,546 members and a total of $21,031 raised through the Causes application. However, the last news item posted to Causes is dated April 21, 2009. There was another about 6 months previous. The Cause appears to be dormant. However, Samaritan's Purse has a very active Facebook page with 506,464 "likes" and official news items posted at the rate of one or two each day, including photos and very professionally produced videos. The "wall" section of the Facebook time line contains frequent posts from individuals. However, as is typical for religious organizational pages on Facebook (Johns 2012), most of the posts appear to be independent declarations of appreciation for the organization rather than any sort of ongoing interaction between fans. Nevertheless, those who have "Liked" this organization will see regular posts to their Facebook news feed about its work, and will also see there any activity by friends interacting with the page.

American Jewish World Service (AJWS) is a similar sort of organization to that previously discussed, but rooted in a different religious tradition. AJWS likewise has a very professional looking website at ajws.org, which states that it is involved in similar sorts of global relief and development projects around the world. There are two Causes profiles for this organization – the one which Causes designates as the official non-profit organization is clearly abandoned, and another one that appears to be more up to date, with 455 members. However, the most recent news item on this more recent Causes profile is December 29, 2008. That profile states that, "American Jewish World Service (AJWS) is an international development organization motivated by Judaism's imperative to pursue justice" (AJWS 2008).

AJWS maintains a Facebook page with moderate activity. As of this writing it had been "Liked" by 7,335 persons. Official posts on the time line were news items generally spaced 3–4 days apart, although in the previous month there had occasionally been two posts in one day. Wall posts by Facebook users other than the organization itself were at the rate of one to four per day in the week prior to this observation. However, several of the posts were unrelated to the organization itself.

Islamic Relief USA is yet another international aid organization with a professional looking website at www.irusa.org. There is a profile for the organization on Causes that describes the organization's mission as, "To alleviate suffering, hunger, illiteracy, and diseases worldwide regardless of color, race, religion, or creed, and to provide aid in a compassionate and dignified manner" (Islamic Relief USA 2007). The most recent news posting on the Causes profile was dated August 27, 2010. The cause had 3,747 members at the time of observation, and $26,619 had been raised through the application at that time – a much higher amount per member than the other profiles observed.

The Facebook profile for Islamic Relief USA had 66,516 "Likes" and the time line showed that the organization posted one news item about its work each day, with some days two items posted, roughly every four to five days. Unlike the pages of the other organizations observed, this charity had managed to place a link to a site enabling online donations quite prominently on its banner. This page has "wall" posts shut off, so no outside persons may join in any sort of general discussion. However, comments on individual news items are allowed, and a few news story links had more than 100 comments, including links to videos or photos, but very few reacting to previous comments or attempting to engage in any discussion among participants. For example, one news item about aid for Syria drew 104 comments, almost all of which were either links to cell phone videos of the fighting, or complaints that the international community was not doing enough. No comment was seen that was a response to any previous comment in the stream, each was an independent item related to the Syrian conflict. Most of these related to the fighting itself and not to the aid effort begin promoted by the organization. Clearly, the comment stream was being used as a way to vent frustrations or to "broadcast" information about the crises, and not to enter into conversation with others.

These observations of the online presence of three somewhat similar religious organizations reveal that Facebook's early promise of assisting charities through applications such as Causes may have spun off into oblivion. However, it is also clear that Facebook pages themselves are being used quite actively by organizations and their "fans" or supporters to regularly share news and information about the relief and development work in which the organizations are engaged.

199.4   Ties to the Cause

In terms employed by Wellman et al. (1996) and Haythornthwaite (2002, 2005), the presence of "Causes" and organizational fan pages on facebook.com are enabling religious charitable non-profit organizations to strengthen social ties with supporters by opening additional channels of communication in the relationship. If Haythornthwaite (2002) is correct that weak social ties are characterized by limited numbers of communication channels, while stronger social ties are characterized by expanded numbers of channels used to maintain the relationship, it stands to reason that establishing additional channels of communication would serve to strengthen ties. Of course, Haythornthwaite (2002) is also emphatic that online social networks such as facebook.com provide only a latent tie between network users. The tie remains latent until there is some social motivation to engage. Although Facebook suggests itself as a site on which friends are "found" and new relationships may be created, Joinson (2008) and Lampe et al. (2006, 2008) suggest that users seek out Facebook "friends" with whom they are already acquainted offline. The surveys and interviews conducted in this study – limited as they are – suggest that the formation of links to organizations on Facebook take place in similar ways. Going beyond the pilot study phase, it would be important to explore how causes are being selected by Facebook users, and the extent to which ties to the charity organization are being strengthened by this additional channel of communication over time.

Observation of the several Cause sites showed no evidence of the formation of any sort of interaction among Cause members, either in the form of "wall" posts or in any sort of discussion forums. Respondents to the written survey, and in the online interviews, likewise indicated that they rarely, if ever, visited a charity's page in order to enjoy interactions with other fans or Cause members. As one respondent noted, "I have visited the page only once or twice since liking it, but I do get the alerts and appreciate the [newsfeed] updates."[3] This evidence suggests that Causes are not being utilized as a locus for the formation or extension of brand community. Muniz and O'Guinn (2001) observed that brand community requires the formation of triangular social relationships in which individuals not only relate to the brand, but relate to one another in terms of the brand relationship. More specifically, brand community exhibits consciousness of kind – that is, an awareness of a unique group of "us" who share loyalty to the brand. Further, these brand communities share in rituals and traditions through their social interactions, and these interactions are bounded by communal codes of moral responsibility (Muniz and O'Guinn 2001). To take this study beyond the pilot phase it would be important to discover whether these conveniently selected Causes are aberrant in this lack of social interaction among members, or if this is the norm for Facebook causes.

---

[3] The function of the news feed is somewhat random, gathering as many status updates or other activities fit on the web page. Thus, a posting to a particular organization's page may not be shown in the news feed of all members, especially if the member has a great many "friends," fan links and Causes links. If there is too much activity, some items will be left out.

Dozier et al. (1995) have demonstrated that individuals enter into social relationships with not only with individuals, but also with corporate entities. Thus, the ability to create more frequent brief reminders of the organization and its work in the news feed, alongside the status updates of friends, establishes a closer relationship between an organization and its supporters. The limited data in the present study suggests that those organizations posting to their Facebook Cause sites on a regular basis would seem to be utilizing this communication channel optimally. Despite the fact that members may seldom actually visit the cause's actual page, the regular news feed items remind members of their relationship and increase motivation for further strengthening ties. The indication would be that organizations that establish pages that are static and infrequently updated will realize little benefit, while those organizations that post too frequently run the risk of being seen as 'spamming' supporters and being 'hidden' from the news feed – an option available to users with a simple mouse click. Fournier and Lee (2009) suggest a similar strategy to commercial entities, and going beyond the pilot phase, a study of how cause members use the news feed, and their attitudes toward these posts, would be of significant benefit to religious charities attempting to use this function. The information would also help draw a more complete picture of how members relate to causes generally.

199.5   Conclusion

The methodological challenges highlighted by this study call into question whether the effectiveness of social network efforts by religious charities can be measured without access to considerable resources. Observation of pages and Causes sites is relatively easy, but offers little information about social relationships among members that may be taking place in other venues invisible from the Causes application or fan page. The email questionnaire format allowed for follow-up and clarification questions in addition to the basic queries. This offers far richer data and the opportunity for greater depth through probing questions, but is severely hampered by low response rates and anti-spam restrictions, as the small number of email interviews successfully conducted for this study demonstrates. This was somewhat mitigated in the present study by the unanimity of views expressed by the respondents. However, a more extensive study examining a wider range of causes would be necessary to confirm these preliminary findings. The large percentage of responding Facebook users who did not participate (or could not recall participating) in a cause suggests that a rather large sample would be necessary to fully explore Causes, which are just one of myriad functions on facebook.com, and religious causes represent an even smaller sub-set of the whole. None of the methods explored here offer an unencumbered view of the use of social networks by religious charities. Thus, while the present study raises some interesting questions concerning the use of Facebook or other social network sites as tools for the support of religious charities, definitive answers are, at least for the moment, out of reach.

References

AJWS. (2008, January 25). *Causes profile for American Jewish World Service*. Retrieved May 27, 2012, from https://www.causes.com/causes/58793-american-jewish-world-service/about

Andreoni, J. (1998). Toward a theory of charitable fund-raising. *The Journal of Political Economy, 106*(6), 1186–1213.

Barnett, E. (2012, May 16). Majority of Facebook users do not click on adverts. *Telegraph UK*. Retrieved May 28, 2012, from www.telegraph.co.uk/technology/facebook/9269966/Majority-of-Facebook-users-do-not-click-on-adverts.html

Bobkowski, P. (2008). *An analysis of religious identity presentation on Facebook*. Conference papers – International Communication Association (pp. 1–24). Retrieved February 10, 2010, from Communication and Mass Media Complete database.

Campbell, H. A. (2005). *Exploring religious community online* (Digital formation series). New York: Peter Lang.

Campbell, H. A. (2010). *When religion meets new media* (Media, religion and culture series). New York: Routledge.

Chen, Y., Li, X., & MacKie-Mason, J. K. (2006). Online fund-raising mechanisms: A field experiment. *Journal of Economic Analysis & Policy, 5*(2), Article 4. Retrieved July 20, 2010, from www.bepress.com/bejeap/contributions/vol5/iss2/art4/

Dozier, D. M., Grunig, L. A., & Grunig, J. E. (1995). *Manager's guide to excellence in public relations and communication management*. Mahwah: Erlbaum.

Facebook. (2012a). *Getting started guide*. Retrieved May 27, 2012, from www.facebook.com/business/ads/

Facebook. (2012b). *Newsroom, company info, key facts*. Retrieved May 27, 2012, from http://newsroom.fb.com/content/default.aspx?NewsAreaId=22

Facebook. (2012c). *Newsroom, advertising*. Retrieved May 27, 2012, from http://newsroom.fb.com/content/default.aspx?NewsAreaId=139

Fournier, S., & Lee, L. (2009, April). Getting brand communities right. *Harvard Business Review*, 105–111.

Friedman, L. J. (2003). Philanthropy in America: Historicism and its discontents. In L. J. Friedman & M. D. McGarvie (Eds.), *Charity, philanthropy, and civility in American history* (pp. 1–21). New York: Cambridge University Press.

Giving USA Foundation. (2011). *Giving USA 2011: The annual report of philanthropy for the year 2010*. Chicago: Giving USA Foundation.

Gordon, R. (2009). Social media: The ground shifts. *Nieman Reports, 63*(3), 7–9.

Hassay, D. N., & Peloz, J. (2009). Building the charity brand community. *Journal of Nonprofit & Public Sector Marketing, 21*, 24–55.

Haythornthwaite, C. (2002). Strong, weak, and latent ties and the impact of new media. *The Information Society, 18*, 385–401.

Haythornthwaite, C. (2005). Social networks and Internet connectivity effects. *Information, Communication, & Society, 8*(2), 125–147.

Islamic Relief USA. (2007, June 4). *Causes profile*. Retrieved May 27, 2012, from https://www.causes.com/causes/1337-islamic-relief-usa/about

Johns, M. D. (2012). Voting 'present:' Religious organizational groups on Facebook. In S. C. Gelfgren, D. Ess, P. H. Cheong, & P. Fischer-Nielsen (Eds.), *Digital religion, social media and culture (digital formations)* (pp. 151–168). New York: Peter Lang.

Joinson, A. N. (2008). *'Looking at,' 'looking up' or 'keeping up with' people? Motives and uses of Facebook*. Proceedings of the ACM conference on human factors in Computing Systems (CHI), Florence, Italy, pp. 1027–1036.

Lampe, C., Ellison, N. B., & Steinfield, C. (2006). A Face(book) in the crowd: Social searching vs. social browsing. In *Proceedings of the 2006 ACM conference on computer supported cooperative work* (pp. 167–170). Banff: ACM.

Lampe, C., Ellison, N. B., & Steinfield, C. (2008). Changes in use and perception of Facebook. In *Proceedings of the 2008 ACM conference on computer supported cooperative work* (pp. 721–730). San Diego: ACM.

Lenhart, A. (2009). Adults and social network websites. *Pew Internet & American Life Project.* Retrieved August 10, 2010, from www.pewinternet.org/Reports/2009/Adults-and-Social-Network-Websites.aspx

Li, C., Bernoff, J., Pflaum, C., & Glass, S. (2007). *How consumers use social networks.* Forrester Research, Inc. Retrieved August 10, 2010, from www.eranium.at/blog/upload/consumers_socialmedia.pdf

Lövheim, M. (2004). Young people, religious identity, and the internet. In L. L. Dawson & D. E. Cowan (Eds.), *Religion online: Finding faith on the internet* (pp. 59–73). New York: Routledge.

Lyons, G. (2012, January 11). *Facebook to hit a billion users in the summer. iCrossing.* Retrieved May 25, 2012, from http://connect.icrossing.co.uk/facebook-hit-billion-users-summer_7709

Muniz, A. M., Jr., & O'Guinn, T. C. (2001). Brand community. *Journal of Consumer Research, 27,* 412–432.

Oates, M. J. (2003). Faith and good works: Catholic giving and taking. In L. J. Friedman & M. D. McGarvie (Eds.), *Charity, philanthropy, and civility in American history* (pp. 281–299). New York: Cambridge University Press.

Ostrower, F. (1995). *Why the wealthy give: The culture of elite philanthropy.* Princeton: Princeton University Press.

Rice, J. (2009). *The church of Facebook: How the hyperconnected are redefining community.* Colorado Springs: David C. Cook.

Roberts, K. (2005). *Lovemarks?: The future beyond brands (2/e).* New York: Power House Books.

Samaritan's Purse. (2007, June 2). *Causes profile.* Retrieved May 27, 2012, from www.causes.com/causes/1110-samaritan-s-purse/about

Ulusu, Y. (2010). Determinant factors of time spent on Facebook: Brand community engagement and usage types. *Journal of Yasar University, 18*(5), 2949–2957.

Wellman, B., Salaff, J., Dimitrova, D., Garton, L., Gulia, M., & Haythornthwaite, C. (1996). Computer networks as social networks: Collaborative work, telework, and virtual community. *Annual Review of Sociology, 22,* 213–238.

Wellman, B., Haase, A. Q., Witte, J., & Hampton, K. (2001). Does the Internet increase, decrease, or supplement social capital? Social networks, participation, and community commitment. *American Behavioral Scientist, 45*(3), 436.

Westling, M. (2007). *Expanding the public sphere: The impact of Facebook on political communication.* Retrieved August 11, 2010, from www.thenewvernacular.com/projects/facebook_and_political_communication.pdf

Wortham, J. (2010, November 30). Jumo, from a Facebook founder, to focus on charities - *NYTimes.com.* Retrieved May 27, 2012, from www.nytimes.com/2010/11/30/technology/30jumo.html?_r=1

York, E. (2009). Adoptive parents, agencies connecting on Facebook. *Advertising Age, 80*(37).

Chapter 200
My (Second) Life's Mission: Landscapes of Virtual Reality Proselytization

Andrew Boulton

200.1 Introduction

Place and faith, bodies and bandwidth, the material and the virtual constitute a vibrant landscape of Christian mission work within the virtual reality world of Second Life. Second Life is a Virtual Reality three-dimensional online environment owned by Linden Lab, as I describe later. Increasingly, geographers recognize and study from a variety of perspectives the intrinsic relationships between the digital and the material in constituting everyday places and lives – particularly under the auspices of ubiquitous web connectivity. Worship, ministry and missional work are increasingly taking on similarly hybrid forms as traditionally place-based (and locationally discrete) notions such as the church and fellowship are rethought in light of emerging technologies and remediated modes of engagement with religious practice, performance and ideas. Recognizing the huge, growing, and global nature of the online audience, churches and activists of various kinds are increasingly engaging with digital technologies as necessary and cost-effective means to engage believers, mobilize missionaries, and reach potential converts wherever they work and play – including online.

In this chapter, I examine a recent iteration of the online proselytization movement: the virtual reality mission. Drawing on ethnographic research with believers, ministers and missionaries within a virtual reality world, I discuss the complex ways in which participants navigate "first life" and Second Life notions of place, faith and the church in pursuit of souls to save (and likeminded individuals for social interaction). Geographical literatures around scale and place suggest a methodological framework in which "virtual" religious places, subjectivities and belief systems are

A. Boulton (✉)
Department of Geography, University of Kentucky, Lexington, KY 40506, USA

Humana, Inc., 500 W Main St, Louisville, KY 40202, USA
e-mail: andrew.boulton@gmail.com

© Springer Science+Business Media Dordrecht 2015
S.D. Brunn (ed.), *The Changing World Religion Map*,
DOI 10.1007/978-94-017-9376-6_200

commensurable with and intrinsic to their "real" world counterparts. Specifically, the (very practical) debates around the methods, efficacy and ethics of online proselytization are illuminated with reference to ongoing debates within geography over the utility and value of the scale concept.

I proceed in this chapter by situating the virtual reality mission movement within a broader online missional context. I then provide a critical overview of Second Life, describing its key features and capabilities as a missional tool and research object, as well as the method (and methodological challenges) of inworld data collection. Thereafter, I move iteratively between case study material and critical geographical scale literatures extant in order to conceptualize place and faith within a virtual reality world. Future work, I suggest, could fruitfully be directed towards explicating in further detail questions of digitally mediated places, identities, and landscapes as they relate to religious practice.

200.2   Screening Religious Practice

The present utilization by mission workers of online platforms of various kinds is consistent with the ongoing, longer run development of Christian outreach efforts in which emerging and mass-market technologies have been mobilized towards proselytizing activities. Televangelism (a portmanteau of "television" and "evangelism") and its associated household name preachers arguably represented the high water mark, in the American popular political imagination at least, of Christian media influence and relevance (Hadden 1993). Cast as key actors in the emergence and promotion in the 1980s and 1990s of the so-called "new Right" coalition of social and fiscal conservatives, celebrity preachers attracted acclaim and vitriol second in degree only to the special brand of hatred directed toward conservative talk radio hosts.

Of course, we must be careful about conflating such mass-market, heavyweight "celebrity" evangelists – many of whom attracted scorn, from ideological/religious allies and foes alike, for perceived largesse or sundry ethical violations – with more-or-less spontaneous, or at least more localized, grassroots and community level outreach and missionary efforts. Thus, although the "Christian missionary movement" is far from a doctrinally and politically coherent entity, it is heuristically useful both to consider virtual reality missional work in the context of a genealogy of Christian missionary movements – from radio to television to the web – and to acknowledge the commonalities between a loose coalition of organizations and individuals in mobilizing mass communication technology to advance a broadly Christian agenda.

More recently, with widespread web usage, particularly among younger Christians, various online platforms emerged to facilitate the mission process. From message boards (and other "community" functions) to the booming missionary blog genre, wherein individuals share stories of their trials and tribulations from the field, the Christian missionary movement has comprised a significant and growing online component for more than a decade. As a practical exercise in keeping congregations up-to-date with events, schedules and developments, churches large and small – in line with organizations in every sector – increasingly take to social media in addition to, or instead of, traditional noticeboard and newspaper announcements. The low (financial and technical) barrier to entry enables interested individuals and organization to connect with likeminded others and to broadcast messages more widely than ever before via online social networks.

While online missional work – and proselytization, formalized or not – is frequently a decentralized, diffuse venture of more-or-less interconnected individuals, highly organized efforts exist at a variety of spatial scales. For example, Silicon Valley-based Global Media Outreach purports to have reached over 360,000 individuals worldwide to date – 57,000 of whom indicated acceptance of the gospel – via its network of web sites and infrastructure of online volunteers (http://globalmediaoutreach.com/). Similar organizations deploy various, sometimes innovative, methods to reach potential audiences. These range from run-of-the-mill organized online marketing campaigns – placing Google Ad Words campaigns on targeted web sites or specific You Tube search terms, and purchasing banner advertisements – to more subtle social or "Astroturf" campaigns in various online social/community environments. Astroturf – generally referring to fake (that is, paid or otherwise orchestrated) grassroots activity – is placed in scare quotes to indicate the problematic nature of any speciously simple distinction between "authentic" (freely chosen, personally motivated) and "fake" (orchestrated, mercenary, otherwise planned) activism. Nonetheless, to the extent that volunteers and donors collaborate and/or pool resources towards consistent social messaging, religious activists – like their private sector and secular counterparts – recognize the value of "meeting people where they're at," in the words of a self-declared Second Life missionary, Jakeu Burgess, whether on the street or online.

200.3   The "Virtual" Missionary Movement

Second Life is a Virtual Reality three-dimensional online environment owned by Linden Lab, and operated for profit: "a world with infinite possibilities and live a life without boundaries, guided only by your imagination", says the Second Life web site (http://secondlife.com/whatis/). Content – virtual buildings, vehicles, games, simulations (etc.) – created in Second Life is shared under Open Source agreements, with certain rights reserved for content creators (the users).

Second Life is accessed using viewer software and an internet connection. Residents (players, participants) enter the Second Life main Grid as avatars. An avatar in this case is a three-dimensional model. Avatars can be customized, appearance-wise, to resemble various (mainly) human-like forms. Avatars may alternate between male and female (or other) appearances; their dress, hair, and so forth may be changed at will; only the name of the avatar is fixed. Each user of Second Life, known as Residents, may operate a single avatar under a fixed user name. In this way, possessions – known as an Inventory, which may include costumes, tools, objects, scripts, and other virtual belongings – can accrue to an individual avatar: so can land holdings, rent and presumably an entire history of movements, transactions, and interactions throughout the virtual space (Ball et al. 2006). Also attached to the identity of the avatar is a profile. The information presented to other users within the profile is largely optional, but includes details of affiliations to User Groups within Second Life and, optionally, information about the first life person/people responsible for managing/using the avatar. User Groups are means by which people express their affiliation with spaces and causes: frequently, they are used to police access to particular member-only locations inworld.

Central to the operation of Second Life is an active, real economy and a powerful spatial logic replete with geographical analogs. The basic commodity is land. Land is sold by the (virtual) acre, and varies in price with various locational factors: proximity to desirable amenities (waterfronts, clubs), and qualitative judgments about the desirability of surrounding land parcels/buildings, for example. The market operates inworld (that is, within Second Life) and outside where land and real estate – architectural simulations/models – find their price as buyers and sellers are brought together in the marketplace, through online staples such as Ebay and dedicated Second Life-specific platforms. Land is divided, by oceans, into Islands. There is a finite amount of land on the main Island, but corporations or individuals may purchase their own Islands for commercial/promotional or other activities (and Linden Lab may add more land as it sees fit). Educational institutions receive favorable rates on Island purchases, and numerous Higher Education institutions worldwide have purchased islands there – among them, the Ohio State University, University of Louisville, and even Princeton and the University of Kentucky. Advocates see Second Life as, variously, a more democratic learning environment, and an ideal platform for distance learning (among other claims), and examples of professors seeking to integrate Second Life into their (political science, geography, sociology, gender and women's studies, anthropology, etc.) curricula are legion.

An intricate/intimate knowledge of Second Life is not required in order to carry out research in this virtual world. The main software features utilized in tracing the missionary network in Second Life, within the present case study, were the following:

1. *Search and Teleport functions.* The search engine within Second Life allows one to search for avatars and places using keywords. Alternatively, locations can be discovered by performing searches in Google (a readily accessible Google Maps mashup includes the Second Life main Grid); churches, such as SecondChurch referenced here, post their so-called SLURLs (Second Life Uniform Resource Locators) on their web sites which allows Second Life users to teleport directly to their location. The teleport function allows instantaneous travel to any location on the Second Life grid, located by a co-ordinate system.

2. *Instant Messaging (IM) and public chats.* These are the main way in which I carried out interviews with the research subjects; email contact was also made subsequently. Within Second Life there are two ways of conversing textually; live audio chatting is also available. IM is the key means by which private

conversations can be carried out between individuals. One's avatar need not be "physically" proximate to another (that is, represented in the same part of the gird) in order to send and receive messages. The second way of communicating is to engage in public chats: within a parcel of land, all avatars can hear (or read) conversations taking place there.

200.4   Shepherdcafé

Jean Corfield is a 35 year-old mother who lives with her husband and their two children in a modest ranch-style home in small town Tennessee. Her husband works full time in a larger town an hour's drive north, and now that their children are both in elementary school, Jean works as a childcare provider, looking after the children of friends, neighbors and commuters to the nearby light industrial works. She is a familiar figure in the local community where she is known, in particular, as an active member at St Joseph's church.

A lifelong Christian, Jean attends services at St Joseph's – congregation c. 45 – as frequently as possible, but family commitments and childcare responsibilities occasionally prevent her from doing so. Between soccer practice, last minute baby-sitting promises and frequent visits to her elderly mother's home on the other side of town, Jean is occasionally unable to attend Sunday service, let alone join in with and organize as many social events and fellowship opportunities as she would like. Growing up in small towns in this rural area southeast of Knoxville, a church-centered program of events and activities had previously been central to her social life. Her best friend, Diane, finds herself in a similar position: mother to three young children, and working fulltime, the free time available to her to pursue an active social life in the church community that had previously been the hub of her recreational activity is limited.

But when Jean and Diane are unable to make it to St Joseph's for the Sunday morning service, they instead attend church together sometime between midnight and three in the morning. Or, at least, they send Shandae and Goldie on their behalf: not family pets, but, rather, the names of their avatars in the virtual world of Second Life. Both (avatars/people) are members of SecondChurch, a chain of churches with locations across the southern United States – and one, a megachurch, on its very own island in Second Life. Shandae (First Life, hereafter FL, Jean) and Goldie (FL: Diane) attend services on the SecondChurch island. Shandae (FL: Jean) explains:

> it is just as good as being in real life church; being with others; receiving His word – when I can not attend in real life (personal correspondence)

Each week, when the rest of the family is watching TV next door, Jean settles down in the kitchen and begins to converse with her friends at Shepherdcafé, a virtual coffee shop in which avatars are encouraged to discuss Christianity. Frequently participants engage in broader debates around controversial moral, ethical and political topics. Shepherdcafé is billed as a Christian coffee shop, where avatars gather for "fellowship and friendship." It is here that Shandae (FL: Jean) and Goldie (FL: Diane) talk to each other using the Instant Messaging (IM) facility built into the Second Life software.

It was here, in Shepherdcafé too, that Trystan Smythe (FL: Andrew Boulton) first met Shandae (RL: Jean). Over the following couple of weeks, Trystan (I) engaged in several conversations with Shandae (FL: Jean) both within Second Life, in which we chatted informally using the Instant Messaging (IM) facility, and outside of Second Life where we corresponded informally via e-mail about her First Life, the intersections between her First and Second Life, and about her Life as an avatar. By talking too with other members of the Church, other curious visitors to the island – including a First Life pastor who has adopted a Second Life preacher avatar; others were not so willing to disclose their First Life identities – I began to trace/follow the networks of people/avatars and places through which the cafe is connected to other places within and outside of Second Life. I observed/accompanied some of these avatars as they negotiated other places within the Second Life Grid, and listened to their stories about how their Second Lives have become intricately tied up with their first. Central to their linking of first and Second Lives, as the sketch above suggests, is the practice of Christian faith, and, in particular, the way in which a missionary network extends between virtual places in Second Life, connecting virtual places with actual places, people with avatars, and creating an extensive network of engagement: a network, I suggest, that takes in and draws strength from its articulation – its engagement (Cox 1998) – in different places, and at different scales.

200.5 "SecondChurch *Is* a Real Life Church :):" Space and Scale in a Virtual Reality World

Virtual places – the virtual café, the virtual SecondChurch – are real, in that they are experienced as intensive, social and intersubjective spaces (Boehner et al. 2007; Yee et al. 2007), and linked via complex networks to actual, material places – the home, the computer, the neighborhood, the server space at Linden Labs, etc.

Scale is a core and contested concept in human geography, the prescience of which is debated ever more hotly in light of continuing remediation of places and lifeworlds under the auspices of ubiquitous digital connectivity. Recent geographical literatures on scale are particularly useful for drawing attention to complex interrelationships between networks of places, people and ideas that often confound or complicate simplistic territorial/hierarchical scalar thinking. In particular, these debates are significant to our understanding of a "virtual reality" phenomenon such as Second Life's missionary movement because they draw attention to commonalities – analytical and experiential – between "traditional" and digitally-mediated religious practice. Marston et al. (2005) argument in favor of "a human geography without scale", and the various more-or-less passionate reactions to that intervention (Collinge 2006; Hoefle 2006; Jonas 2006; Leitner and Miller 2007; Moore 2008 in particular) comprise the "latest salvo in the scale debates" (Jones et al. 2007: 264). For Moore (2008: 207), the idea of jettisoning scale wholesale is "strikingly wrongheaded." For Hoefle the idea that scale should be expunged from the geographical lexicon is unthinkable: "nary a word", he cautions, "about the paper outside [of Geography]" (2006: 242). In Hoefle's formulation, scale *is* geography: or, at least, scale is geography's unique gift to social thought; an understanding of scale is the unique contribution of the geographer. Remove scale, remove geography's – and, more shockingly, geographers' – raison d'être. Here, I suggest scale – or at least a critical appreciation of scalar thinking put forth by Marston, Jones and Woodward (and their critics) – provides a valuable framework for understanding the strategic and emotional engagement with religion among Second Life's participants.

But when Marston et al. (2005) talk about a geography without scale, what are they seeking to remove? In other words, one way to understand what a (presumably improved – analytically, politically) geography without scale might look like is to characterize such a geography in contradistinction from a deficient geography – that is, a geography *with* scale. This, unsurprisingly, is Marston, Jones and Woodward's (2005) starting point as they argue from the outset of their paper that, for geographers, the term scale is used so broadly that the sign is largely meaningless: "there is no agreement on what is meant by the term or how it should be operationalized," they state (p. 416). I am arguing that noting that there is no single, commonsense or uncontroversial definition of a concept or term does not provide prima facie grounds to abandon that concept.

Sayer's (1981) definition of a "chaotic conception" (as opposed to a "rational abstraction") might be useful here in illustrating this claim that, even if there is "no agreement" on the meaning of scale, *even if* it is a "chaotic conception" (Sayer 1981: 9), it need not be abandoned. In Sayer's terms, chaotic conceptions are those that conflate unrelated concepts or divide indivisible concepts. Thus, we can ask, Is scale, uniquely chaotic amongst geography's bedrock concepts (place, space, culture, etc.)? That a conception is chaotic does not provide grounds to eradicate the sign. Although Sayer (in Jackson et al. 2006) cautions against using chaotic conceptions to assert "explanatory weight," they can be and are, he suggests, employed as descriptive categories, useful heuristics, to practical, analytical, political effect. The relatively uncontroversial claim that other bedrock concepts in geography – place, space, culture – are similarly chaotic, having no clear, singular, agreed-upon delimitation of their ontological/epistemological statuses, their meanings, or their utility, is not, of course, up for discussion in Marston, Jones and Woodward's (2005) piece. Whilst it would be unfair to claim that the sole argument put forward by Marston, Jones and Woodward in favor of abandoning scale is that "there is no agreement" about its meaning and its status (ontological or not), this is one of their main lines of argumentation, and indeed their opening rhetorical salvo.

Scale might provide significant analytical purchase on questions of hybrid (virtual/material) spaces such as Second Life. My contention here is that scale can profitably be retained in the lexicon even if, in many ways – like space, like place, like culture – it is "chaotic". Indeed, as Jackson et al. (2006) argue, chaotic conceptions – even if we concede, as Marston et al. (2005) do, that scale is such a conception – provide interesting, and politically useful objects of study in their own right. Acknowledging that the concept of scale is chaotic would, in Sayer's terms, lead not to the wholesale jettisoning of scale, but to asking different kinds of questions: how is scale discursively and materially constructed? How are different versions of scale deployed?

The political deployment of scale and scalar strategies (Tyner 2006) relies not on scale's ontological existence, but on discursive rationalities – a logic – around its practical utility (Moore 2008). Scale is an ambivalent concept. In psychological literature, ambiguity refers to a lack of "inherent meaningful structure," whilst ambivalence refers to an object/term that denotes "two or more distinct structures" (Lassiter 2005: 164). Scale, I am suggesting, does indeed denote distinct structures; the task in any given research context becomes to distinguish clearly why and how a particular meaning is deployed, either practically in/by the actor/network under consideration, or analytically by the researcher (Moore 2008). How do missionaries mobilize a power geometry connecting distinct spaces and discourses in propagating a persuasive religious rhetoric?

For me, in taking on Marston et al. (2005: 422) well-made criticism of a kind of scalar fetishism – body, house, community, nation-state, etc. – it is crucial is not to treat scale, or scalar categories, as a priori. Rather, empirical work, in the manner of tracing out connections and relationships in a flattened ontological framework (Ibid.), guides the emergent conception/categorizations of scale in an iterative/ inductive process. The question is not whether scale, and any reference to scalar properties, entities, forces, processes (etc.) should be abandoned wholesale. The idea of abandoning scale is meaningless without a clear, agreed-upon referent for the sign "scale" which, as Marston, Jones and Woodward illustrate convincingly, does not pertain. Rather, the question becomes: which aspect of scalar thinking – what "distinct structure" denoted by the term – might be useful (or unnecessary) in a given situation? Use may be defined broadly: analytical and theoretical, pragmatic, political and practical.

An alternative way of hinging the discussion is offered by Moore (2008: 218) who makes a distinction between scale as a category of practice – the ways in which political movements/causes "frame" problems as "scalar" – and scale as a category of analysis. Scale may be treated, as Hoefle (2006) arguable treats it, as an ontological category; scales (the nation-state, the city, for example), although socially and historically constructed, nevertheless have a material reality. In this ontological appeal to scale, scales actually (qua materially) exist. The household has a material reality; households are nested within towns, nested within the nation-state, etc. Treating scale as an ambivalent concept, asking questions about how it is deployed in *practice*, and using scalar categories *analytically* in an iterative/inductive way avoids the either/or of scale as solely ideological or unproblematically ontological, while recognizing that in the case of a Christian missionary movement even the most "localized" efforts defer to infinitely transcendent articles of faith.

Where I concur with Marston, Jones and Woodward's critique is in their claim that a hierarchical conception of scale – as they characterize them it least (vertically arranged, pre-given scales-as-levels) – "(de)limits practical agency as a necessary outcome of its organization" (427). This is an argument about the ways in which hierarchical conceptualizations of scale relegate the social – social practice and the possibility for resistance and politics – to a scale, qua level, below (or separate from) the level at which real, underlying, causal forces are operating. Thus, whereas social practice takes *place* down here, on the ground (the "local"), explanation – process, "broader forces" (Ibid.) – occur up there (the "global"). Politically, such a crude vertical-hierarchical conception of scale is disabling. The example they give is of socially irresponsible management engaging in "macro-mystification" whereby the buck (appropriately) is passed upwards to scales above – to, for example, globalization or the global economy. The result is that political affectivity and engagement is not, in this scheme, encountered on the ground, and not attributable to specific actors; the materiality of processes, the agency of subjects, is elided; the social is displaced.

The vertical-hierarchical conception of scale outlined in the previous paragraph is rather crude, in the sense that those scholars who deploy a vertical conceptualization of scale tend not to deny the significance of place – or of site – in the way that Marston, Jones and Woodward suggest (see, for example, Brenner 1999). Rather, scaling up is precisely about accessing, and operating within, those *places* where (apparently natural, agentless) processes, such as globalization, actually play out, where the rules are made. Whether proponents of a vertical, hierarchical conception of scale adhere to a strictly ontological view of scale is, I think, irrelevant to the broader political project such a conception of scale can sustain. If scale is deployed strategically with reference to those sites and places in which social practice can make a difference, the ontological status of scale is irrelevant. "Scaling up" is precisely about recognizing that processes and struggles are grounded in place (Smith 1993), however these fit into a scaled imaginary. In these ways, scaling up or down, to engage in other spaces of engagement is, potentially, a politically/strategically significant movement (Tyner 2006).

Where the idea of a flat ontology becomes more persuasive, for me, is in its rejection of "prior, static conceptual categories" (Marston et al. 2005: 425). The keywords here are *prior* and *static*. I am arguing that a non-hierarchical conceptualization of scale should begin without a predetermined – prior – framework of scales into which the empirical data is made to fit. This is not to say that, in tracing out and identifying the key places and flows in a network, scales are not emergent: either actual, material scales (such as the home), or epistemologically useful heuristic scales that are less explicitly bounded in their materiality (such as what I identify as *scriptspace* in the following discussion).

200.6   Scaling Missionary Work: Towards Networked Landscapes of Virtual Worlds

i offer a lm [landmark] for church, and yes, i spread His word …try to anyway :) (Shandae, Shepherdcafé, April 14th, personal correspondence)

I take Shepherdcafé, where I met Shandae (FL: Jean), as my starting point; the quotation with which this section begins is taken from our first conversation there. My claim is not that there is a fixed, coherent, proselytizing network in Second Life, but by (virtually) hanging out at Christian places of worship, and other spaces of fellowship, like the Shepherdcafé, and communicating with the people/avatars in those places, it became clear that there is a substantial – in terms of numbers of people, spatial extent, and organization – missionary movement based around Second Life. The extent to which these efforts are "authentically" grassroots can not be gauged (and in some ways is beside the point).

In short, my claim here is that a modified version of Cox's (1998) spaces of engagement versus spaces of dependence model offers us a useful framework for apprehending the scalar dimension of mission work based around Second Life. Thinking through scale in terms of dependence and engagement offers a means of understanding the ways in which scale is deployed politically in ways that connect grounded, real, material, experienced networked places of dependence – actual and virtual – with networked places of engagement. This is not to argue for an ontological divide between the dependence and engagement moments as equivalent to, for example, local scale versus global scale in a hierarchical sense. It is rather to suggest – specifically in the case of this missionary movement – that it is useful to think in terms of a recognition that a cause might gain strength – depend – on its reference to and relationship with specific places, online and off. This suggests a horizontal conceptualization of scale whereby scale is redefined in terms of differential network extent/audience, and is deployed – analytically and practically (Moore 2008) – as it is encountered: on the ground, in place, through practice. The characterization of place on which this scheme depends refers both to virtual and actual places, the binary between which is broken down with reference to the interactive, co-construction of place in the human-computer interface (Dourish 2006), a discussion with which I conclude this chapter.

Shandae (FL: Jean) and others, direct a portion of their Second Life usage towards encouraging others to attend their particular place of worship, and to consider Christianity more generally. To this end, advertisements in explicitly Christian (and non-Christian) virtual places proclaim the existence of religious radio stations. Places such as Shepherdcafé offer to visitors free (virtual) gifts such as T-shirts and bibles. Individual church members do as Shandae (FL: Jean) noted: they provide "landmarks," or virtual directions to church, to avatars they encounter in places other than those explicitly coded (coded through content and performance rather than through computer code) as Christian. One individual with whom I spoke, Stevenson Eerie (FL: requested anonymity), claimed – and I was able to verify to my satisfaction – to be an ordained minister in First Life, practicing also as a minister in Second Life. He hosts services daily in Second Life, and plays host to the Women of Hope women's group – a forum for Christian women to meet in Second Life.

Although these missionary activities take place in virtual spaces between avatars – crudely, between cartoon characters controlled more-or-less skillfully by computer users – Jakeu and Stevenson begin to articulate a sense in which mission work in Second Life is linked, very directly, to aspirations of converting the people behind the avatars:

> I don't see a difference in spreading God's teachings in Second Life...it's all about thoughts and opinions anyway...it doesn't matter to Him if the person behind the computer looks like their avatar. (Jakeu Burgess, Shepherdcafé)
>
> My concern here in SL, is as God commanded, to win souls, to witness the gospel to those who've never heard it before. (Stevenson Eerie, personal correspondence)

Needless to say, the ideas and aspirations behind their avatars' missionary zeal extends – spatially and discursively – far beyond the confines of specific digital spaces. The earlier vignette – Jean/Shandae practicing and articulating her faith, interpersonal relationships, and performances that constitute Christian in First Life and Second Life – offers further empirical evidence that the divide between First Life and Second Life is not so clear as a binary between the real world and the virtual world. As recent work in geography and elsewhere shows, virtual worlds do not comprise a separate space, and, I am arguing, Second Life does not, in itself comprise a single scale (see Boulton and Zook 2013; Dourish and Bell 2007 on the increasingly diverse, embodied, material and sensual experience of technology). Rather, I am suggesting, practices within Second Life are intimately tied up with everyday material practices such that the virtual environment together with actual places comprise a series of emergent scales; rather than "creating a distinct sphere of practice", virtual worlds, "open up new forms of practice within the everyday world" (Dourish 2006: 305).

Part of this "opening up" new practice to which Dourish (2006) refers, is about recognizing the co-construction of place, and the ways in which experience of virtual/actual place is mediated in the human-computer interface. More prosaically, this collapsing of the binary between virtual and actual – crucial to my conception of scale – can be illustrated, very briefly, with reference to two sites/sights that are at once virtual and actual (first life and Second Life). The first of these I term *scriptspace*, the second a publication, *Things of Hope*. Scriptspace is the code, the program, the algorithms (etc.) to which Second Life is reducible. Scripts determine the range of motions for avatars, the potentiality for building, the laws of the economy, and so forth. These barriers and rules – from the physics of movement to the enforcement of virtual property rights – are not experienced, by the Second Life Resident, as physical, material barriers: rather, they are virtual barriers inherent to this world; they are the laws of physics/nature, as such. These coded rules of the game are equivalent to the concept of "code power", used by geographers to describe the structuring power of computer code in ordering and ranking digital representations of place (Graham et al. 2013). However, scriptspace is, in another sense, very much material/actual: servers at Linden Lab headquarters, web servers that relay the information, the hard discs of users' computers on which the Second Life software is stored, all have a material existence and all trace out part of a network that links this abstract, virtual scriptspace with material places and actors. Second, *Wings of Hope* – one of several virtual Christian publications – is available via subscription (deliverable as a PDF) through Second Life. Thus, flows of money connect the central places, virtual and actual, of the publication's production, to the virtual spaces of its distribution, and to the actual spaces of its consumption. A further twist: the magazine exists as a "real" print magazine, again broadening the network – in the sense of broadening the audience, the spaces of engagement – and enrolling both the actual and the virtual (places) in ways that break down any clear boundary between the two realms.

A recurring theme in conversations with people in Second Life, and not just with individuals prominently and publicly involved in the missionary movement is how, almost invariably in my experience, people tend not to distinguish between on the one hand the personality, soul or character of their avatar, and on the other, their real life self. Such terms as personality, soul, and character, not to mention self, are problematic – of course – but within this popular storyline, there is the sense in which the body of the avatar is no more a front, a shell, a performance than is the human body and the image one presents of it, in all its actual more-or-less scripted context-specific guises (Leidner 1993). In other words, in this formulation, behind, underneath, inside – the avatar *and* the person – lies something essential, and identical; the pronoun "I" is pervasive in talking about exploits within Second Life. For example, Alice Burgess (FL: Beth Kraemer, a librarian at the University of Kentucky and prominent advocate for Second Life use in education at UK) told me that she feels her professional reputation is tied both to her actions as Beth and to her actions as Alice. Alice (FL: Beth) might be something of an anomaly, given that her avatar identity is relatively well known (advertised in her Second Life profile, and noted in her email footer) and used, in general, as a part of her professional activities promoting the UK Island and Second Life use at UK. But even for avatars/individuals enrolled anonymously in the Second Life-based missionary project, this blurring of the real/virtual, the first life/Second Life divide is suggestive; as (first and Second Life) Pastor Stevenson Eerie argues:

> Are both worlds [First Life and Second Life] different to me? no, certainly not. God sees all of our actions and hears all of our thoughts. Because you may think this is but a game its another way to let satan in to take control of your life. My concern here in SL [Second Life], is as God commanded, to win souls, to witness the gospel to those whove never heard it before. To share in the love that Jesus has for everyone of all walks and ethnic backgrounds. (Personal correspondence)

There are important implications for a hybrid, non-hierarchical, networked conceptualization of scale – such as that which I am setting out – in this collapsing of the binary between real world and virtual world. In post-Lacanian psychoanalysis (Venn 2004), the idea of a "mirror stage" – that developmental moment in which a child comes to recognize himself/herself as an image/body presenting itself to the world – is potentially useful for theorizing the ways in which people come to see in the "body" of their avatar a projection-surface of self equivalent to the corporeal distance of the mirror stage. Complementarily, we could apply Haraway's (2000) cyborg concept: the avatar-human entity becoming "a condensed image of both imagination and material reality" (2000: 158). Theoretically and practically, the notion that virtual worlds might, in total, represent a discrete space or scale is thoroughly disrupted. As Jakeu Burgess told me during an informal chat at Sheperdcafe, [s]he chooses to engage with the conversation in Shepherdcafé via a laptop and wireless internet at Starbucks. By contrast, he told me, he is more likely to attend his virtual church from the privacy of his dorm room or, frequently, his bedroom at his parents' home. This mundane and anecdotal story resonates, however, with this co-construction/hybridity notion: that, rather than comprising a single, coherent space, the virtual environment comprises a set of places which are produced and experienced relationally with the material places from which they are accessed and with which they seek to engage. The experience of fellowship in the virtual Christian café is not independent of his First Life, embodied experience of the cof-fee shop, and vice-versa.

Simply enumerating those places that, I am arguing, comprise the networked places of engagement (borrowing from Cox 1998) and networked places of depen-dence is perhaps less useful than seeking to understand some of the complex and shifting ways in which these networks change through time and space. As I have begun to suggest, both the engagement and dependence moments comprise emer-gent networks of places (virtual and actual) rather than single spaces or coherent, a priori scales. One of Cox's most suggestive insights, for me, is that moving to a space of engagement is not necessarily to scale up towards something ever approxi-mating the global; likewise, spaces of dependence are not, simply, a grounded, authentic local. Thus, I want to suggest that the Christian missionary movement depends on networked places as diverse as the kitchen from which Jean Corfield (SL: Shandae) attends church, the dorm room from which Jakeu attends church services, and Shepherdcafé which serves as a hub, a meeting place, for Christians in Second Life. Jones' (1998) reading of Cox is useful here: the key point she empha-sizes is that spaces of engagement and spaces of dependence are not mutually exclu-sive; places can at once be places of dependence and places of engagement. Spaces of engagement and dependence are not necessarily coextensive, but neither does the space of engagement necessarily expand beyond, in the sense of extent, the spaces of dependence. Rather, in the case of the Christian missionary movement, net-worked places of dependence are, simultaneously, *part of* networked places of engagement (and vice-versa), although, I am arguing, they extend too beyond those places of engagement.

200.7   Discussion

What draws Jean Corfield to her computer each Monday night to engage in fellow-ship with other people/avatars? "Local" social conditions are, surely, part of this: her role within the home, the privacy afforded her in the space of the kitchen, for example. But, also, we can suggest, more broadly, that it is her (gendered, etc.) position within the labor market, her subject position in respect of her local, real world church, and myriad other dimensions of her positionality, that comprise the social/spatial conditions on which her participation in the mission movement depends. Significantly, Jean also finds (real) fulfillment and fellowship in these (virtual) encounters.

As work in cultural landscape studies is beginning to explicate, there is no necessary distinction between place and digitally mediated place (Graham et al. 2013; Boulton and Zook 2013). That is, when we talk about landscape we are talking about material places and ways of seeing that are (re)produced through computer screens, cell phones, traffic management, circulations of images and discourses, geospatial surveillance and myriad other more-or-less insidious and more-or-less taken-for-granted manifestations of code's work. Such imbrications of digital and material objects come to the fore in online missionary movements where the practice of missional zeal relies (and depends) on both physical locations and virtual spaces.

Thus, Second Life billboards that appear on land outside of explicitly Christian sites, the avatar who attempted to present me with a bible on a college "island" are ways in which other places, and other people, are drawn into the missional network. The *Wings of Hope* magazine speaks of real people, real faith and real issues, even as it engages virtual spaces and circulates among the avatars of Second Life. Much has been written about the possible liberatory potential of virtual reality worlds as spaces in which the constraints of "first life" are mitigated by, for example, removing physical markers of (or mobility limitations associated with) disability (for example. Jones 2006). More generally, as Gottschalk (2010: 511) suggests, when it comes to designing their virtual selves, many Second Life participants choose to "construct avatars [as] idealized versions of themselves." Typically, the "ideal" self is drawn (metaphorically and literally) rather narrowly as a normatively aesthetically appealing version of oneself; Second Life is replete with rather young, slender and well-dressed residents.

Aside from but related to these opportunities for transcendence of physical attributes, virtual worlds also entail the possibility for experimental behaviors and performances in and through the avatar. Ranging from the mundane to the risqué, research suggests that avatars' experimentation, roleplaying and relationships are not readily separable from real life concerns – frequently overlapping with, drawing on, and having positive (and negative) effects on "offline" lives and identities (Ibid.). With this understanding of the social and emotional investment of individuals within virtual worlds, Christian missionaries' engagement with Second Life is a significant and potentially fruitful activity which broadens the field of missional activity. Particularly, given the apparent openness of Second Life's participants to new ideas and experiences, "meeting people where they're at" logically includes meeting avatars within virtual worlds. In a world predicated on anonymity, fantasy and deceit, Christian missionals tread a delicate path between caricatured, virtual spectacle and sincere, transcendent belief in the ultimate reconciliation between (and saving of) Second and First Life selves.

Acknowledgment   Thank you to Stan Brunn and Andy Wood for their comments on earlier drafts of this chapter, and to the Second Life residents who shared their thoughts and experiences.

References

Ball, K., Lyon, D., Wood, D., Norris, C., & Raab, C. (2006). *A report on the surveillance society: For the information commissioner by the surveillance studies network*. London: Surveillance Studies Network.

Boehner, K., DePaula, R., Dourish, P., & Sengers, P. (2007). How emotion is made and measured. *International Journal of Human-Computer Studies, 65*(4), 275–291.

Boulton, A., & Zook, M. (2013). Chapter 36: Landscape, locative media and the duplicity of code. In N. Johnson, R. Schein, & J. Winders (Eds.), *The Wiley-Blackwell companion to cultural geography*. Oxford: Wiley-Blackwell.

Brenner, N. (1999). Globalisation as reterritorialisation: The re-scaling of urban governance in the European Union. *Urban Studies, 36*(3), 431–451.

Collinge, C. (2006). Flat ontology and the deconstruction of scale: A response to Marston, Jones and Woodward. *Transactions of the Institute of British Geographers, 31*(2), 244–251.

Cox, K. (1998). Spaces of dependence, spaces of engagement and the politics of scale, or: Looking for local politics. *Political Geography, 17*(1), 1–23.

Dourish, P. (2006). Re-space-ing place: 'Place' and 'space' ten years on. *Proceedings of the 2006 20th anniversary conference on Computer supported cooperative work* (pp. 299–308). Retrieved March, 6, 2013 from http://portal.acm.org/citation.cfm?id=1180875.1180921

Dourish, P., & Bell, G. (2007). The infrastructure of experience and the experience of infrastructure: Meaning and structure in everyday encounters with space. *Environment and Planning B: Planning and Design, 34*(3), 414–430.

Gottschalk, S. (2010). The presentation of avatars in Second Life: Self and interaction in social virtual spaces. *Symbolic Interaction, 33*(4), 501–525.

Graham, M., Zook, M., & Boulton, A. (2013). Augmented reality in urban places: Contested content and the duplicity of code. *Transactions of the Institute of British Geographers, 38*(3), 464–479.

Hadden, J. (1993). The rise and fall of American televangelism. *The Annals of the American Academy of Political and Social Science, 527*(1), 113–130.

Haraway, D. (2000). A cyborg manifesto. In D. Bell & B. Kennedy (Eds.), *The cybercultures reader* (pp. 291–324). London: Routledge.

Hoefle, S. (2006). Eliminating scale and killing the goose that laid the golden egg? *Transactions of the Institute of British Geographers, 31*(2), 238–243.

Jackson, P., Ward, N., & Russell, P. (2006). Mobilising the commodity chain concept in the politics of food and farming. *Journal of Rural Studies, 22*(2), 129–141.

Jonas, A. (2006). Pro scale: Further reflections on the 'scale debate' in human geography. *Transactions of the Institute of British Geographers, 31*(3), 399–406.

Jones, K. (1998). Scale as epistemology. *Political Geography, 17*(1), 25–28.

Jones, D. (2006). I, Avatar: Constructions of self and place in second life and the technological imagination. *Gnovis, 6*. Retrieved March 13, 2013 from http://gnovisjournal.org/files/Donald-E-Jones-I-Avatar.pdf

Jones, J. P., II, Woodward, K., & Marston, S. (2007). Situating flatness. *Transactions of the Institute of British Geographers, 32*(2), 264–276.

Lassiter, G. (2005). Are ambiguous figures actually ambivalent? *Individual Differences Research, 3*(3), 162–170.

Leidner, R. (1993). *Fast food, fast talk: Service work and the routinization of everyday life*. Berkeley: University of California Press.

Leitner, H., & Miller, B. (2007). Scale and the limitations of ontological debate: A commentary on Marston, Jones and Woodward. *Transactions of the Institute of British Geographers, 32*(1), 116–125.

Marston, S., Jones, J. P., III, & Woodward, K. (2005). Human geography without scale. *Transactions of the Institute of British Geographers, 30*(4), 416–432.

Moore, A. (2008). Rethinking scale as a geographical category: From analysis to practice. *Progress in Human Geography, 32*(2), 203–225.

Sayer, A. (1981). Abstraction: A realist interpretation. *Radical Philosophy* (Summer), 6–15.

Smith, N. (1993). Homeless/global: Scaling places. In J. Bird, B. Curtis, T. Putnam, G. Robertson, & L. Tickner (Eds.), *Mapping futures: Local cultures, global change* (pp. 87–120). London: Routledge.

Tyner, J. (2006). "Defend the ghetto:" Space and the urban politics of the Black Panther Party. *Annals of the Association of American Geographers, 96*(1), 105–118.

Venn, C. (2004). Post-Lacanian affective economy, being-in-the-world, and the critique of the present. *Theory, Culture & Society, 21*(1), 149–158.

Yee, N., Bailenson, J., Urbanek, M., Chang, F., & Merget, D. (2007). The unbearable likeness of being digital: The persistence of nonverbal social norms in online virtual environments. *CyberPsychology and Behavior, 10*(1), 115–121.

Chapter 201
Christianity and Digital Media

Tim Hutchings

201.1   Introduction

Early research on religion and the Internet assumed an inherent theological and sociological bias could be found within the medium, a tendency to favor certain religious traditions, demographics and patterns of interaction. Early commentator Jeff Zaleski suggested that "the Internet will favour those religions and spiritual teachings that tend toward anarchy and lack a complex hierarchy" (1997: 111), a point of view echoed more recently by Brenda Brasher (2001) and Christopher Helland (2005). According to Helland, "hierarchies and networks are two very different systems and the Internet was only really designed for one of them" (2005: 13).

As these observers noted, however, both hierarchical and network-oriented forms of Christianity have established a presence online. Conservative, moderate and liberal groups are all thriving. Different religious groups are bringing different values, goals, practices and resources to the Internet and finding different ways to engage with digital media. Scholars continue to struggle to work out exactly how media engagement is changing religion around the world, but academic attention has shifted away from arguments about the transformation of religion toward studies of the internet as an everyday part of religious culture (Campbell 2010a).

This chapter uses two theoretical approaches – the religious-social shaping of technology (Campbell 2010b) and the mediatisation of religion (Hjarvard 2008) – to explore some of the main areas of online Christian activity. By combining these two approaches, we can analyze how religious groups create media while also examining how groups are changed by their media practices. After introducing these concepts, I will illustrate them through a case study of the Anglican Cathedral of Second Life, an online church in a virtual world.

T. Hutchings (✉)
Post Doc, St. John's College, Durham University, Durham DH1 3RJ, UK
e-mail: t.r.b.hutchings@durham.ac.uk

© Springer Science+Business Media Dordrecht 2015   3811
S.D. Brunn (ed.), *The Changing World Religion Map*,
DOI 10.1007/978-94-017-9376-6_201

The remainder of this chapter is divided into three sections examining digital forms of leadership, community and proselytism. These issues are particularly important for the relationship between Christianity and digital media, although the third has received minimal academic attention. Proselytism has been the motivation for many of the most innovative Christian media projects, and I will focus here on the ways in which Western missionaries are redesigning and repurposing media devices to support their work in other regions of the world. Each section includes examples from my own fieldwork, combined with references to the work of other scholars covering as wide a range of Christian traditions and geographical regions as possible.

201.2 Theoretical Approaches: Shaping Technology, Shaping Faith

Religious groups actively evaluate technologies and decide which to adopt, negotiating with developers in some cases to build more appropriate devices and applications. Heidi Campbell describes this process as the "religious-social shaping of technology," comprising four distinct stages: religious communities reflect on their history of media use, consider their core values and practices, evaluate the technology and negotiate its redesign, and finally frame the technology through a group discourse that sets appropriate goals and boundaries for its use (2010b: 60–61).

Religious groups are constrained in their approach to media technologies by a range of factors, as this model acknowledges. Groups may decide that their history and values are incompatible with certain aspects of media culture or technology. Specific media applications may suit some religious traditions more than others. If religious communities decide to create new technologies to suit their needs, they must secure resources, technical skills and in some cases commercial support in order to do so effectively. Attempts to create a new discourse may be opposed by co-religionists or external audiences.

Stig Hjarvard (2008) has used the concept of "mediatisation" to describe the influence of media in contemporary society, and this approach can help to explore the impact of these pressures and limitations on religious communication. Building on David Altheide and Robert Snow's classic study of media culture (1979), Hjarvard proposes that media have their own "logic," their "institutional and technological modus operandi," including the "formal and informal rules" by which they operate (2008: 113). Anyone wishing to use a medium to communicate must adapt to its logic. Hjarvard sees mediatisation as a societal process akin to globalisation or secularisation, an argument that has been criticized (Lövheim 2011), but the basic idea of "media logic" can help balance Campbell's emphasis on the active agency of religious groups in their interactions with technology.

Mediatisation and the religious-social shaping of technology are complementary concepts, working together to help illuminate different aspects of the relationship between religion and media. Campbell's emphasis on history and values encourages us to expect a rich diversity of forms among Christian approaches to the internet, reflecting the variety of Christian ideas and practices across religious traditions and geographical regions. Hjarvard reminds us that every religious group attempting to use a digital technology must adapt its communication to suit that medium.

201.3   Study: The Anglican Cathedral of Second Life Case

In 2006, an Anglican businessman from New Zealand entered the virtual world of Second Life. Mark Brown quickly discovered that very few Christian organisations were building virtual churches, and decided to intervene personally. Working with a group of like-minded Anglican Second Life users, Brown commissioned a talented Christian designer to create an elaborate, visually-striking virtual church building in a traditional Gothic architectural style and set about generating publicity to attract a congregation (Fig. 201.1). Brown decided to name his new church "the Anglican Cathedral of Second Life," a move intended to maximise attention:

> I wanted to create buzz, and guess what, it has [laughs]. You know, when the media got hold of it I've been on TV, radio, gosh, I don't know how many, seriously, I don't know how many times, radio in the US, Australia, New Zealand, you know, newspapers, weblogs, a huge number of blogs, and a big part of it is a) Anglican, b) Cathedral?, and c) high technology, that creates a buzz. If I just build another, I don't know, just an open space with some pillows on the ground and a cross in the corner, I don't think it would have got the same. (Hutchings 2010a: 75–76)

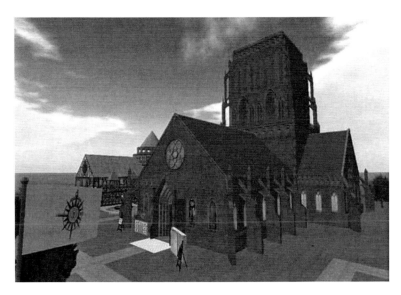

Fig. 201.1 The Anglican Cathedral of Second Life (Image created by Mark Brown 2008, used with permission)

The Cathedral's traditional style served other purposes as well. For Brown, this highly recognizable visual design is "a symbol of Christianity" (2010a: 75), a strategic attempt to appeal to post-modern taste for tradition (2010a: 78), and a way to help visitors "ground" the experience of online worship in a context they already find comfortable and easy to understand (2010a: 79).

The Cathedral was created as an independent community with no authorisation from the Anglican Communion, but quickly started seeking some kind of connection to its chosen denomination. At the time of my own research in 2009, Brown had already been ordained by his local bishop in New Zealand and had organised the first of a series of meetings with another Anglican bishop, a number of theologians and church lawyers in the UK. Through these conversations, the Cathedral sought to establish some form of official oversight for its activities and a theological basis for its digital ministry. The Cathedral's website now includes a Vision statement which lists a desire to be "recognised as an integral part of the Anglican Communion" among the church's ambitions and states that the community is "currently working under the authority" of both bishops (http://slangcath.wordpress.com/the-vision/).

The Cathedral is an interesting case study in religious technology. The Cathedral's development over time displays all four stages of Campbell's "religious-social shaping of technology:" reflection on tradition, identifying core values and practices, evaluation and design of media platforms, and attempts to shape Christian discourse. The Cathedral's builders have chosen to recreate traditional architecture, including an altar, but they have not created any way to use this altar for an avatar-based communion service, or to use the fountain outside for avatar baptisms. These are theological decisions, based on careful evaluation of Anglican theology. Several non-Anglican groups have reached different conclusions: online forms of communion and baptism have been offered since the 1990s (www.alphachurch.org) and are readily available for avatars elsewhere in Second Life.

The Cathedral has also undergone a process of mediatisation, adapting to the "media logic" of the virtual world. In order to create Epiphany Island, the Cathedral's leaders needed to raise money to rent virtual land from the company that owns Second Life, and find a designer with the skills required to construct their buildings. The creative negotiation and shaping of technology described by Campbell is possible only within the limits set by these basic media requirements.

"Media logic" includes a cultural dimension (Altheide and Snow 1979: 9). Media creators generate paradigms, genres and rules that audiences learn and come to expect. Audiences develop practices of media sharing, consumption and, particularly in digital spaces, participate in content creation and develop local norms for doing so. Over time, the Cathedral has encountered and evaluated many of the diverse cultures and subcultures of Second Life and chosen elements to embrace, accommodate or reject. Services of worship regularly end with displays of avatar dancing, but avatar weddings, another common Second Life practice, remain strictly forbidden. It is not uncommon for male Second Life users to create female avatars, and vice-versa, a practice that has been banned outright in many in-world churches. The Cathedral community has accepted this practice and includes a number of cross-gendered avatars, but online-offline gender correspondence is still required for church leaders. The Cathedral has been shaped by engagement with two very different cultures, Anglican and virtual, and has developed its own unique response.

201.4   Leadership in Digital Christianity

The Anglican Cathedral of Second Life demonstrates that communities and networks forming online can establish supportive relationships with religious institutions while also embracing new developments in practice and theology. The Cathedral is a good example of a religious community that has attempted to shape the internet in line with its core values, negotiating with the institutional, technological and cultural logics of the medium and generating new discourses to validate its activity.

The willingness of two Anglican bishops to engage in conversation with the Cathedral demonstrates a desire on their part to encourage, offer guidance to and potentially help regulate innovative online projects, rather than leading those projects personally. This approach has much in common with the attitude of the Vatican, which has hosted meetings of Catholic bloggers (Ivereigh 2011) and issued documents encouraging Catholics to use the internet appropriately (Pontifical Council for Social Communications 2002) while remaining somewhat cautious in its own online activity. The Pope has often called for Catholics to be active online in his annual address for World Communications Day (see, for example, Benedict XVI 2011) but official Vatican media platforms like vatican.va and youtube.com/user/vatican continue to block all public comments. Bishops of the Church have intervened to block a number of media projects that misrepresent Catholic theology, including attempts to offer confession by text message (Bell 2006). The much-publicised "Confession: A Roman Catholic App" received official approval in 2011 as an aid for Catholics wishing to prepare themselves for the sacrament (BBC 2011), only for the Vatican to issue a caution shortly afterwards when media reports began suggesting that the app could be used to receive absolution without talking to a priest at all (Wagner 2012: 149).

Early assumptions that the internet would favour free association and disadvantage established hierarchies continue to influence popular discourse in many areas, including religion, business and politics. This perspective overlooks multiple factors that can operate to consolidate power through digital media. Established leaders have the financial resources to hire talented designers and communicators. The digital technologies used by their opponents also empower their own grassroots supporters. Digital conversations are easy to monitor, and leaders can use them to track debate and identify opponents. Evgeny Morozov (2012) cites these strategies as just some of the ways in which authoritarian regimes – including the governments of China, Russia and Iran – are using digital media to retain power, but we can find close parallels among the digital activities of religious organizations.

The Anglican Cathedral's relationship with the Anglican Communion upholds traditional Anglican understandings of the role of bishops and the importance of authorized governance, but the Cathedral has also facilitated the emergence of a

T. Hutchings

new religious leader. Mark Brown was ordained by his local bishop, as we have seen, but quickly left the Cathedral to start a series of new online projects. To date, he has created a Christian consultancy company, a Bible study video blog and a number of popular Facebook pages. The most successful of these has been "The Bible," which attracted such intense activity that it ranked among the most active pages on the whole site for much of 2011 (Ward 2011). Brown's achievements clearly owe something to the connections he formed during his former career as CEO of the Bible Society New Zealand, but digital media have made it possible for him to make the resources he creates available to a significant global audience at minimal financial cost.

Leaders of established Christian organizations have also been quick to recognize the possibilities for global attention offered by communications media. An increasing number of churches are using video recording and broadcast technologies to enable one preacher to teach congregations at multiple locations, a form of "multi-site" ministry developed by North Coast Church in California in the late 1990s (North Coast, n.d.). Mediation here is integral to the church-going experience, which is constructed for the audience through recording equipment, screens, and – more recently – live satellite feeds.

Oklahoma-based megachurch LifeChurch.tv now invites visitors to attend one of eighteen physical sites across five states, all screening the same sermon via satellite. Each physical campus has its own worship band, local pastor and home study groups, but the power to preach is now centralized in just one individual, Pastor Craig Groeschel. In my interviews with church staff, the physical absence of the pastor was dismissed as a triviality that few visitors even noticed, while the use of screens was praised as an opportunity for immediate connection. Every congregation member can now see Groechel's face in close-up and feel his eyes gazing into theirs, an experience made possible only by mediation (Hutchings 2010b: 234). Groeschel's sermons are themselves highly mediated, sometimes incorporating pre-recorded location shoots and testimonies. One highlight of the annual sermon calendar, "At the Movies," uses clips from popular films to explain the Christian message and is particularly recommended for newcomers.

LifeChurch.tv uses the internet to share Groeschel's sermons and other materials with other churches worldwide through Open, a free resource site (open.lifechurch.tv). Supporter churches from Australia, Costa Rica and Germany have now joined the LifeChurch.tv Network (network.lifechurch.tv), showing a video sermon each week while remaining independent. The most committed can apply to join United and become a new LifeChurch.tv campus (united.lifechurch.tv).

LifeChurch.tv has also created an "online campus," Church Online, where visitors can watch the service on video, communicate in a volunteer-moderated chatroom and click a "Live Prayer" button to speak privately to a trained church representative (live.lifechurch.tv; Fig. 201.2). Visitors can join online study groups, and regular viewers are encouraged to invite friends to join them in their homes for "Watch Parties" (internet.lifechurch.tv/watchparty/). This online campus offers multiple advantages over traditional televangelism: churches can incorporate opportunities for moderated audience response into the broadcast to enhance engagement, avoid the costs and restrictions of buying airtime and make their programming accessible

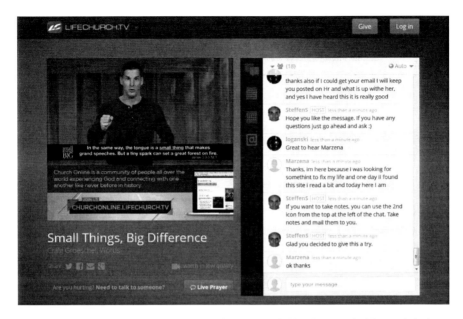

Fig. 201.2 Experience church online (Image from Live.LifeChurch.tv, used with permission)

Fig. 201.3 This real-time map displays the location of everyone participating in a specific Church Online event. In this case, 113 countries are represented (Image from www.LifeChurch.tv, used with permission)

worldwide. International appeal is boosted by a real-time translation tool created by church staff to help viewers communicate in more than 50 languages. Church Online now attracts more than 125,000 viewers each week, with up to 7,000 watching each event (staff member, private email communication, 2012), and more than 100 other online campuses have now been launched by large churches worldwide (Drinnon 2009).

During events, Church Online displays a map of the world marking the location of each viewer (Fig. 201.3). This simple visual bears silent witness to the international

Fig. 201.4 An inspirational tweet from American pastor Rick Warren, July 2012 (Image from Twitter.com/RickWarren)

appeal of LifeChurch.tv, validating the group narrative of divinely-powered growth. Martijn Oosterbaan has reported similar cartography on the websites of Brazilian Pentecostal churches. These maps reflect "dreams of a wholly Christian planet," inviting viewers to imagine themselves in a global community of believers (Oosterbaan 2011: 57).

Video preaching is only one of the ways in which Christian leaders are leveraging digital media to increase their impact. Social networking sites like Facebook and Twitter allow users to choose to follow individuals who produce content they appreciate, and religious leaders have risen to considerable prominence within these spheres. A recent *New York Times* article reports that certain Christian leaders were now attracted such intense engagement from their followers that Twitter has assigned staff members to analyse and replicate their success (O'Leary 2012). These leaders may attract far fewer followers than the most popular secular celebrities, but their followers retweet their messages – predominantly inspirational thoughts and Bible verses – with remarkable regularity (Fig. 201.4).

In her work on Christian pastors in Singapore, Pauline Cheong has proposed that religious authority is being reclaimed online through practices of "strategic arbitration" (Cheong et al. 2011). According to Cheong, the internet "proletarianizes, deprofessionalizes, and potentially delegitimizes leadership" by broadening access to knowledge, impacting the status of religious leaders as knowledge elites (2011: 944). In response, some pastors begin trying to relegitimize their authority by mastering online culture and training themselves in the skills of online communication. Tactics reported by Cheong include authoring online resources to mark their authority and compete with new rivals, restating the importance of distinctive skills such as biblical exposition, establishing principles of acceptable online behavior and claiming the authority to identify reliable online sources. Ministry patterns change to include online dialogue and preaching styles begin to incorporate digital presentation tools, responding to changing expectations (2011: 952). Pastors hope to use these adjustments to re-establish trust in their authority as sources of knowledge.

LifeChurch.tv's Craig Groeschel demonstrates one approach to contemporary leadership, using digital media to gather multiple online and offline audiences around one preacher and one church brand. We can understand this approach as an example of "strategic arbitration." LifeChurch.tv is creating deep libraries of online resources, including Craig Groeschel's sermons, to provide guidance on practical concerns for followers worldwide. A number of projects target online pornography, responding to an aspect of digital culture that evangelical Christians find particularly troubling and thereby reassuring supporters that Groeschel remains a relevant source of spiritual guidance (www.thepornevent.com). Media-rich sermons and sophisticated online projects also help to secure relevance, enabling the church to communicate effectively in contemporary culture and demonstrating technical skill. LifeChurch.tv's expertise in digital media enables the church to position itself as a leader in Christian innovation, offering advice, resources and software to other ministries, and expands the influence of the church and its pastor around the world.

201.5   Digital Networks and Online Communities

Church Online offers a moderated chatroom and encourages visitors to pursue deeper relationships elsewhere through LifeGroups, maintaining a tight focus while giving away digital resources for other groups and churches to use without supervision. In my own fieldwork, interviewees spoke highly of what they referred to as the "community" of Church Online, experienced through the chatroom, but many dismissed interpersonal friendships as unnecessary (Hutchings 2013). Their sense of community was based instead on the perception that the audience watching online was united in ideology and values. The chatroom was used during services primarily to perform ideological unity, as a space for praising God, the band and the preacher and affirming the work of the church. Any expressions of dissent, attempts at conversation or requests for prayer were quickly moved to a private chatroom by moderators. Persistent troublemakers are not expelled – they might, after all, benefit from the sermon – but they are muted from communicating in the chatroom.

This approach to community is quite different from that encountered at the Anglican Cathedral of Second Life. The Cathedral congregation is very much smaller than Church Online's audience, gathering ten to twenty individuals for each scheduled event, and the Cathedral's membership encompasses a considerably broader range of theological views. The Cathedral is a persistent virtual interaction space, available to anyone at any time of day, and during my period of fieldwork I encountered individuals outside the church leadership who spent many hours each week in conversation around the fountain in the church square. For some, regular social interaction with the community is far more important than attending services.

Much looser religious networks are also emerging online, marked not by shared access to a bounded space but by acts of directed communication. Twitter users can add hashtags to their messages, a series of letters beginning with a # sign, and other users can then locate every message including that specific hashtag through Twitter's search function. This is a simple way to create a conversation that can be watched and added to by any number of people, whether or not they have any pre-existing connection. The hashtag has been appropriated by Christian groups to create "synchronized prayer" networks (Cheong 2012: 198) like #VirtualAbbey and #Tworship (Heim and Birdsong 2010) and regular events like #chsocm, a weekly discussion for Christians interested in social media. Hashtags have also been used to create "faith memes" like #TOF, "Twitter of Faith," which challenges contributors to summarise their faith in a single tweet (Cheong 2012: 191).

LifeChurch.tv, the Cathedral and Christian Twitter hashtags have created quite different forms of online interaction, but all of the examples cited here affirm established religious authorities. Heidi Campbell (2010c) has found a similar tendency to support authority in the Christian blogosphere, reporting that her quantitative analysis of 100 randomly-selected Christian blogs recorded only one challenge for every 12 affirmations (2010c: 260). Campbell concludes that "Christian bloggers in this study were primarily focused on affirming their own beliefs rather than challenging other positions" (2010c: 270) and suggests that they might be attempting to build consensus among like-minded writers.

Campbell's research identified the exact types of authority each blog affirmed, and this fine-grained analysis uncovers some interesting counter-tendencies within her sample. References to God and the Bible were extremely common and almost exclusively positive, but human religious leaders received relatively frequent critique. Contemporary pastors, priests, church leaders and teachers received one challenge for every two affirmations, and references to televangelists were overwhelmingly negative. Relationships to religious authority, needless to say, can be complex and multi-layered.

Online networks have proven particularly valuable for individuals who feel marginalized, excluded or unable to find the religious resources they need in their local contexts. Campbell's *Exploring Religious Community Online* (2005) includes a case study of an email list called the Online Church, in which "the majority of members are visually or otherwise physically impaired, giving them a common bond, physically as well as spiritually" (2005: 89). "The freedom from their physical limitations" offered by specially-designed hardware and software was "an important factor for the growth of this community." The group is "a place of social acceptance" for members, where they are not judged on the basis of disability (2005: 92). Campbell reports that members frequently share personal stories of hardship, and identifies "validating each other's struggles and affirming each member's worth" as core group practices.

Experiences of personal revelation can also isolate an individual from their local religious context and prompt them to look online for more receptive audiences (Apolito 2005). Anna Rose Stewart (2011) reports that charismatic Christian women are finding new opportunities online to teach and share prophetic insights in ways that are denied to them as women offline. One of Stewart's interviewees, Patricia, feels that her attempts to communicate God's messages have been met with hostility in local churches, so she shares them in an online charismatic forum instead (2011: 1209).

Online networks may also serve individuals who feel ideologically marginalized in their local context. Koinonia Church, a Second Life group, promises to be "a safe place for those who have been hurt or rejected by their communities of faith"

Fig. 201.5 JenClothing, a leading LDS "modest dressing" company (Image from www.jenclothing. com/about-jen.html, used with permission)

(Koinonia, n.d.) for their sexual orientation. This emphasis on affirming diversity is built into Koinonia's architecture, which rejects traditional church designs in favor of an open central space without doors (Miczek 2008: 164; Fig. 201.5). Koinonia's website illustrates the faith position of the church by reposting a well-known progressive Christian document, the Phoenix Affirmations (Elnes 2006). Rev. Eric Elnes, author and editor of the Affirmations, has developed his own online network of progressive Christian supporters through a weekly web television program called Darkwood Brew (www.onfaithonline.tv/darkwoodbrew/).

Conservative internet users have also formed online networks to connect with like-minded others. Robert Howard (2011) has traced the emergence of an online movement devoted to charting the end of the world through meticulous and literal analysis of the Bible. Participants attempt to relate current world events to their shared understanding of this "End Times" narrative in order to make predictions about the future, a practice Howard refers to as "ritual deliberation." In the early 1990s, practitioners realised that their approach to the Bible was unacceptable to the majority of members of Christian Usenet groups, and they began using private email lists to support discussion groups that were only open to those who accepted their basic premises. These "tiny communities of the like-minded" (2011: 48) developed over time into a sizeable 'virtual ekklesia' (2011: 11) that now includes many hundreds of blogs and websites. One particularly valuable aspect of Howard's two-decade study is his focus on the relationship between technology and ideology, demonstrating that different media types favour different kinds of religious authority by placing more or less emphasis on the individual voice. Video blogs, he suggests, are used more by fundamentalists who claim charismatic authority, while text blogs and forums favour those who wish to provoke a greater degree of ritual deliberation.

Online networks can go beyond these ideological enclaves to mobilise support for local or societal transformations. In *The Church and New Media*, a recent edited volume of essays by Catholic media figures, blogger Thomas Peters writes about coordinating his readers to send emails in support of Catholic political and social causes (Peters 2011). Digital media can also be used to construct global movements, like Shawn Carney's anti-abortion prayer campaign 40 Days For Life. According to

Carney, "new media builds the movement:" "email is the way we communicate, invite, and train people to participate," bloggers help publicise the cause, and webcasts allow the leadership to communicate their vision to large national audiences (Carney 2011: 182). Other examples discussed include CatholicVote.org, a website that evaluates U.S. political candidates according to "the principles of Catholic social teaching" (Peters 2011: 167). The "true masculinity" website Fathers for Good uses podcasts, blogs and forums to help men "become selfless, devoted dads and spouses" and focuses particularly on the destructiveness of pornography (Carney 2011: 181), while the Ruth Institute uses new media to "reorient the views young people have toward marriage" (Carney 2011: 184). In each case, digital media are being used to advertise a message, engage audiences in discussion and promote personal, societal or political change through specific recommended actions.

The "modest dressing" movement is a particularly intriguing example of the use of digital media to connect like-minded individuals. Like many of the networks discussed above, "modest dressing" crosses regional boundaries without engaging with established religious leaders, but this movement also connects supporters across religious traditions. According to fashion researcher Reina Lewis, an increasing number of Christian, Muslim and Jewish women are seeking to dress in a way that meets their standards of modesty while still relating to global fashion trends (Lewis 2012). Fashion blogs provide valuable opportunities for these women to discuss their clothing ideas and encourage others to adopt their standards of dress. Companies dedicated to modest fashion are able to access this market niche by using websites to advertise and sell their creations (see Fig. 201.5). Most interestingly, Lewis and her research team report that many of these companies were founded with an explicit and distinctive religious ethos, but are now trying to rebrand themselves to reach out to new markets in different faith communities. This particular approach to modest dress has not yet been noticed by established religious authorities, Lewis argues, providing space for women to renegotiate their relationship to religious values and contemporary culture without official supervision or censure.

Modesty is not the only framework within which this renegotiation is taking place. As Karen Tice has shown (2010), some evangelical Christian women in the United States are also re-evaluating the beauty industry, using secular beauty pageants as opportunities to proselytise (McMichael 2009), establishing independent Christian beauty pageant contests (www.christianpageants.com) and setting up make-over consultancies (Griffith 2010), all aided by blogs and websites.

This section has covered a diverse range of groups, including loose networks, close-knit communities, like-minded enclaves and interfaith movements. Digital media are also increasingly important to our experiences of dying and mourning, playing a crucial role in maintaining the community of the living and the dead. News and information are shared online, blogs and forums provide opportunities to find support in suffering and grief, online communities arrange online funerals and memorial sites collect photographs and stories to commemorate the deceased (Hutchings 2012a).

Elizabeth Drescher has described one community of support and, eventually, mourning that emerged on a Christian Facebook user's wall at the end of her life. Friends 'prayed' on Facebook, "continuing to speak directly to Kirstin as she moved – their prayers expressing a shared theology – from this life to the next" (Drescher 2012: 215). These online communication practices were inspired by and framed within the theological worldview of participants, an excellent example of Campbell's religious-social shaping of technology.

Drescher's case study is full of Christian themes, but this kind of direct address to the dead has also been reported among non-religious online mourners. Researchers have repeatedly noticed the persistence of digital bonds between the living and the dead (Carroll and Landry 2010). Grieving friends continue to talk to the dead through the media they used in life, posting messages on their MySpace or Facebook wall, including expressions of loss, updates about everyday life activities, birthday greetings and requests for assistance. Funerals with no religious aspect have been organised in game environments, including World of Warcraft (Nagata 2010).

Online mourning encourages the public performance of practices that were formerly conducted in private, particularly the expression of intense emotion and the art of speaking to the dead. Participants renegotiate who can grieve in public, what they can say and who acts as gatekeeper and curator for the mourning process. There is an opportunity here for religious studies scholars to analyse cultural attitudes to death, prayer and the afterlife, an approach that has not yet been fully explored.

201.6   E-vangelism

The Anglican Cathedral and LifeChurch.tv's Church Online do share one common theme. Founders of both churches have cited proselytism as their key motivation, and at least in Church Online structures and policies have been put in place to attract non-Christian attention. Google Adwords link Church Online to pornography search terms, attempting to target individuals at a moment of moral decision (Hutchings 2010b: 240). "Online Missions" have been organized, gathering Church Online participants for short-term proselytising in MySpace and Facebook (Hutchings 2010b: 264). All Church Online events include a call to commit to Jesus, and participants are invited to respond by clicking a button. A free "What's Next Kit" including a Bible and a DVD of sermons is sent out anywhere in the world on request.

Online churches offer a number of advantages to would-be proselytisers. If an online missionary encounters a potential convert, they can invite them into an online community to attend online worship and listen to online preaching, continuing their relationship and avoiding the difficult task of finding an appropriate local church. Some individuals may be more willing to venture into a church online, where they can retain their anonymity and disconnect at any moment. Online church advocates have suggested that Christians in environments hostile to conversion might particularly

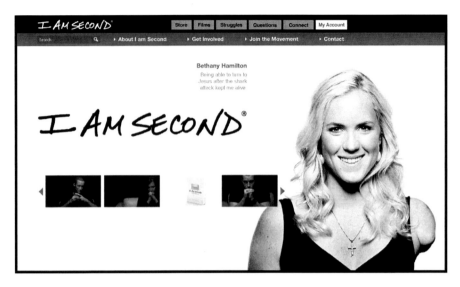

Fig. 201.6 "I Am Second" home page (Image from www.iamsecond.com, used with permission)

benefit from the privacy of digital communications. Others perceive a symbolic value to church-based evangelism: for Christian writer Douglas Estes, the virtual world is "by far the largest unreached people group on planet Earth" (Estes 2009: 29) and unless virtual churches are built there, "the cause of Christ is lost in that world" (2009: 38).

There are also disadvantages to the church-building approach. Only a minority of internet users participate in online communities, and individuals not yet interested in Christianity are unlikely to visit a site branded as an "online church." Online churches have tended to attract large numbers of already-committed Christians who generate a Christian-focused culture with a style of worship that closely reflects their offline tradition, and this can deter individuals whose experiences of offline churches have been less positive (Hutchings 2010a: 74).

One alternative model, the Bridge Strategy (Internet Evangelism Day, n.d.), focuses on creating spaces that engage with issues of non-religious interest in order to attract non-Christian attention. The site designer then includes their personal faith story or other spiritual content to encourage visitors to start engaging with Christian ideas. This is a well-known approach to proselytism offline as well as online, of course, and many evangelistic projects combine local and digital resources to maximize their impact. Texas-based I am Second (IaS) uses TV and radio adverts, physical merchandise and high-profile concerts to attract attention to its online video archive (www.iamsecond.com; Fig. 201.6). Each video is beautifully shot and edited and features a celebrity who speaks of overcoming struggles in their life by putting God first (Hutchings 2012b). Visitors are invited to contact an IaS volunteer in a private chatroom, by email or by telephone, and can then request an introduction to an IaS housegroup or associated church in their local area. Details of Christian doctrine are barely mentioned. IaS attracts attention through glossy design, celebrity glamour, real-life confessions and the promise of help in overcoming personal hardships, and visitors must start conversations with volunteers to find out what "being second" might actually involve.

We can understand the growing sector of Christian computer games as a form of "bridge" evangelism. Christian game designers are attempting to appropriate successful game styles and combine them with Christian messages and educational content to create a product that audiences will consider culturally acceptable, enjoy playing and find spiritually beneficial. Examples to date have met with mixed approval. Left Behind: Eternal Forces attempted to turn a successful series of books about the End Times into a strategy game and met strong criticism both for the awkwardness of its gameplay and for the supposed violence of its theology (Good 2011). Educational games are less ambitious, but less controversial. BIG Bible Town, for example, combines Bible stories with fairly conventional minigames designed to emphasise key themes (www.bigbibletown.com). In-game rewards can be used to develop the player's town or presented as gifts to the player's friends. Each town is focused on a church, and players are encouraged to donate a portion of their earnings to help their church grow. BIG Bible Town is designed to encourage young players to learn Bible stories, but also aims to socialize players into a certain understanding of Christian life: "kids… learn about missions and tithing and that when people work together as a church, they can do great things for their community" (BIG Bible Town, n.d.).

The rise of social networking sites presents new opportunities and challenges for evangelists. To attract friends and followers, the would-be evangelist must consistently produce content that those followers consider worth receiving, but once a network has been established material that promotes Christian faith can be easily distributed through it. IaS has developed an interesting, multi-layered presence on Facebook and Twitter, publishing quotes from videos and news about the movement every day (Hutchings 2012c). The IaS Facebook wall is a very active space where movement supporters share stories of their evangelistic activities, functioning as a semi-private zone for mutual support and encouragement, but IaS also publishes inspirational messages and advertises its videos and these updates can be shared thousands of times. The social media presence of IaS appears to have attracted an overwhelmingly Christian cohort of followers, but each follower has the opportunity to share IaS content with their own social networks as part of their everyday activity.

The emerging landscape of Bible technologies has also claimed a role in online proselytism. LifeChurch.tv's free mobile app, YouVersion, has now been installed more than 125 million times and used for almost 78 billion minutes of reading (http://now.youversion.com/). Users can access the Bible at any time in 471 languages worldwide, including many translated for YouVersion by teams of volunteers. Users can highlight and underline passages, access reading plans and commentaries, add private notes and share their comments with other users. YouVersion claims to be part of "a global shift where more and more people are engaging with God's Word," arguing that making the Bible constantly accessible is a way to make it relevant in people's lives and to increase the time they spend reading (YouVersion 2012).

The app also supports a more direct form of proselytism, encouraging users to share favorite verses through Facebook and Twitter as inspirational quotations. This public performance of devotion to the Bible serves multiple functions, encouraging fellow believers, allowing others to oversee the progress of the user's reading, and exposing non-Christian audiences to biblical texts.

Digital Bible technologies are also being used in international missionary work, helping Western organizations import and distribute Christian texts without attracting hostile attention. BibleLeague's Akses Digital Bible Library "securely provides digital Bibles to believers under the threat of persecution," packed into a MicroSD card that can be transferred and copied between devices, quickly concealed and "easily destroyed or erased if danger arises" (Bible League 2012). The Digital Bible Society has been distributing Christian texts in China on CD for 10 years, and is now designing SD cards to store libraries of biblical texts and interpretive resources in Arabic and Farsi. These cards will include Bible-reading software that does not require any installation or internet connection, a design decision intended to maximise security (Dyer 2012).

Other Christian missionary organisations are using digital technology to communicate the Bible through audio recordings. MegaVoice, based in Australia, has created a series of solar-powered MP3 players designed for missionary work in oral cultures (www.megavoice.com). The smallest model, the Envoy Micro (MegaVoice 2012), is advertised as simple to use, resistant to damage and "inconspicuous, making it safer to distribute to those living in places opposed to the Gospel." The Envoy Micro holds up to 400 h of content, uploaded by missionaries 'in the field' or preloaded to order by MegaVoice staff, and can be attached to external speakers for large-group listening. Recordings can only be uploaded with a special cable, ensuring that the message "cannot be erased or tampered with by the end user."

Mobile technologies have also been appropriated by missionaries. Organizations like Mobile Ministry Magazine (www.mobileministrymagazine.com) are trying to encourage Christian developers to move beyond Bible apps to create new ways to use cellphones in evangelism, focusing particularly on the opportunity to share short videos. According to Keith Williams, founder of Mobile Advance, "we now start out by sharing individual Scripture stories on our phone during a visit, and then transfer it by Bluetooth to those who are interested. If a person shows enough interest or has come to faith, we can then provide a mobile memory card" of additional Bible stories – all for a fraction of the cost of a MegaVoice player (Williams and Gray 2010: 141). Through his phone, Williams reminds his readers, "I have numerous spiritual poems and musical Psalms ready and waiting in my pocket, should the right opportunity arise" – and the nomads Williams works with are so comfortable with mobile technology that sharing video and music with them "feels completely natural."

Missionary projects like Akses, MegaVoice and MobileAdvance demonstrate a creative approach to digital media rooted in a specific religious context, tradition and set of values. Technology is being shaped in innovative and unusual ways, diverging from the standard paradigms of secular media design by focusing on security and reliability at the expense of connectivity and user-generated content.

Each organization relies on Western donors, however, and frames its activity for that audience through an evangelical Christian missionary discourse that dates back at least to the nineteenth century (Beidelman 1982). Non-Western peoples are starving in spiritual ignorance; only direct personal access to the Bible can save them; Western missionaries must use their advanced technologies to bring light into their darkness; converts are grateful for this work, even though they face persecution. Williams' report challenges this discourse in one interesting respect: the nomads he approaches are his equals in communication technology, with smartphones of their own.

201.7 Conclusions

A key theme throughout this chapter has been the renegotiation of boundaries. The internet connects global communication networks, overcoming restrictions once imposed by distance. Using digital media, Christian leaders can make their messages available worldwide, potentially multiplying their influence many times over. Believers can connect to networks of the like-minded, even if they are intellectually or physically isolated in their local contexts, and can sustain niche markets for books, clothing and other resources. Missionaries can find new ways to proselytize even in the most hostile environments. Access is restricted instead by quality of equipment, speed of connection, and in some cases by local blocks, filters and monitoring technologies.

Boundaries between leaders and followers are also renegotiated. The most effective online communicators are not necessarily those individuals who currently hold positions of authority within established religious organisations, and independent media producers can win considerable attention. To compete in this new media landscape, organisations are learning new skills, hiring new talent, encouraging supporters to engage online on their behalf and expending considerable capital to create high-quality online resources.

The boundaries between "church" and "life," public and private religiosity, and religious and non-religious activity are all blurred when religious leaders, organisations and individuals begin communicating with and monitoring one another through popular social network sites like Facebook. Church pastors can now direct daily messages to their congregation and invite continual engagement with their church community, all in the same spaces congregants use to keep in touch with friends and family. When Christians blog and tweet their prayers or use social networking sites to mourn, once-private acts become a public performance exposed to multiple audiences. This opportunity may be interpreted as a form of "witnessing" to outsiders, or as a way to encourage "accountability" among believers through mutual surveillance.

This chapter began by introducing two complementary approaches to religious media: the religious-social shaping of technology and the concept of mediatisation. The former highlights the active role of religious groups as evaluators and designers of technology, and emphasizes the importance of group history, culture and values in informing this process. The latter reminds us that every medium has its own logic, a constellation of institutional, technological and cultural norms that shape what can be communicated through that medium. Religious individuals and groups may shape their media, but they can only do so in negotiation with pre-existing media logics. In some cases, the communication that results from these negotiations may be quite different from anything those religious groups have created before, a cause for debate and concern among observers. In other cases, Christian communication genres have proven so well-suited to the logics of digital communication that their authors are significantly out-performing better-known secular rivals.

Much work remains to be done. The geographical focus of this edited volume draws attention to the scarcity of studies of digital religion in South America, Africa and Asia, compared to the relative wealth of studies focused on the English-speaking West. Minimal attention has been paid to mobile technologies so far, an issue that is becoming more and more pressing as smartphone devices and applications grow in sophistication and popularity. Proselytism is another understudied area, despite its significance as a motivator for the development of religious technologies. Further study is also needed to establish the ways in which religious communication is being shaped by new media logics. How, for example, does the use of digital video in evangelism projects like I am Second and MobileAdvance lead to changes in the message shared? How does storing the Bible on a digital device change the owner's relationship to sacred text?

The two theoretical approaches combined in this chapter provide useful orientations for scholars of religion and media. Each new digital technology or application raises new opportunities for religious groups and shapes their communication in new ways, and we must attend to both directions of influence in order to make sense of the changing geographies of contemporary religion.

References

Altheide, D., & Snow, R. (1979). *Media logic*. London: Sage.
Apolito, P. (2005). *The internet and the madonna: Religious visionary experience on the web* (A. Shugaar, Trans.). Chicago: University of Chicago Press..
BBC News. (2011, February 8). Catholic Church gives blessing to iPhone app. Retrieved June 01, 2012, from www.bbc.co.uk/news/technology-12391129
Beidelman, T. (1982). *Colonial evangelism: A socio-historical study of an East African mission at the grassroots*. Bloomington: Indiana University Press.
Bell, G. (2006). No more SMS from Jesus. In P. Dourish & A. Friday (Eds.), *Ubicomp 2006* (pp. 141–156). Berlin: Springer.
Benedict XVI. (2011, June 5). *Message of his Holiness Pope Benedict XVI for the 45th world communications day: Truth, proclamation and the authenticity of life in the digital age*. Retrieved June 1, 2012, from www.vatican.va/holy_father/benedict_xvi/messages/communications/index_en.htm
Bible League. (2012). *Akses Digital Bible Library can save lives and save souls!* Retrieved June 1, 2012, from www.bibleleague.org/akses-digital-bible-library?DESIG=13540
BIG Bible Town. (n.d.). *All about BIG Bible Town*. Retrieved June 1, 2012, from http://bigbible-town.com/whatsItAbout.asp?txt=1

Brasher, B. (2001). *Give me that online religion*. San Francisco: Jossey-Bass.

Campbell, H. (2005). *Exploring religions community online: We are one in the network*. New York: Peter Lang.

Campbell, H. (2010a). Internet and religion. In R. Burnett, M. Consalvo, & C. Ess (Eds.), *The handbook of internet studies* (pp. 232–250). Oxford: Wiley-Blackwell.

Campbell, H. (2010b). *When religion meets the new media*. Abingdon: Routledge.

Campbell, H. (2010c). Religious authority and the blogosphere. *Journal of Computer-Mediated Communication, 15*(2), 251–276.

Carney, S. (2011). Moving mountains: Building a digital movement. In B. Vogt (Ed.), *The church and the new media: Blogging converts, online activists and bishops who tweet* (pp. 177–189). Huntingdon: Our Sunday Visitor.

Carroll, B., & Landry, K. (2010). Logging on and letting out: Using online social networks to grieve and to mourn. *Bulletin of Science, Technology & Society, 30*(5), 341–349.

Cheong, P. (2012). Twitter of faith: Understanding social media networking and microblogging rituals and religious practices. In P. Cheong, P. Fisher-Nielsen, S. Gelfgren, & C. Ess (Eds.), *Digital religion, social media and culture* (pp. 191–206). New York: Peter Lang.

Cheong, P., Huang, S. H., & Poon, J. P. H. (2011). Religious communication and epistemic authority of leaders in wired faith organizations. *Journal of Communication, 61*(5), 938–958.

Drescher, E. (2012). Pixels perpetual shine: The mediation of illness, dying, and death in the digital age. *Cross Currents, 62*(2), 204–218.

Drinnon, D. (2009, October 8). *Directory of online churches and internet campuses (expanded)*. Retrieved June 1, 2012, from www.equipthem.info/2009/10/08/directory-of-online-churches-and-internet-campuses/

Dyer, J. (2012, February 28). *Powerful, secure Bible software for closed countries… and you!* Retrieved June 1, from http://donteatthefruit.com/2012/02/powerful-secure-bible-software-for-closed-countries-and-you/

Elnes, E. (2006). *The Phoenix affirmations: A new vision for the future of Christianity*. San Francisco: Jossey-Bass.

Estes, D. (2009). *SimChurch: Being the church of the virtual world*. Grand Rapids: Zondervan.

Good, O. (2011, July 14). *Christian game developers want to leave bad games behind*. Retrieved June 1, from http://kotaku.com/5821259/christian-game-developers-want-to-leave-bad-games-behind

Griffith, W. (2010, March 28). *Fearfully, wonderfully made: the one-day makeover*. Retrieved July 15, from www.cbn.com/cbnnews/us/2010/March/Fearfully-Made-The-One-Day-Makeover/

Heim, T., & Birdsong, T. (2010). *StickyJesus: How to live out your faith online*. Gallatin: Digital Scribe Press.

Helland, C. (2005). Online religion as lived religion: Methodological issues in the study of religious participation on the internet. *Online: Heidelberg Journal of Religions on the Internet, 1*(1), 1–16.

Hjarvard, S. (2008). The mediatisation of society: A theory of the media as agents of social and cultural change. *Nordicom Review, 29*(2), 105–134.

Howard, R. (2011). *Digital Jesus: The making of new Christian fundamentalist community on the internet*. New York: New York University Press.

Hutchings, T. (2010a). The politics of familiarity: Visual, organisational and liturgical conformity in the online church. *Online. The Heidelberg Journal of Religions on the Internet, 4*(1), 63–86.

Hutchings, T. (2010b). *Creating church online: An ethnographic study of five internet-based Christian communities*. Ph.D. thesis, Durham University, Department of Theology and Religion.

Hutchings, T. (2012a). Wiring death: Dying, grieving and remembering on the internet. In D. Davies & C. Park (Eds.), *Emotion, identity and death: Mortality across disciplines*. Farnham: Ashgate.

Hutchings, T. (2012b). I am second: Evangelicals and digital storytelling. *Australian Journal of Communication, 39*(1), 83–88.

Hutchings, T. (2012c). Are you ready to be second? Christian recruitment online. *Masaryk University Journal of Law and Technology, 6*(1), 1–13.

Hutchings, T. (2013). Considering religious community through online churches. In H. Campbell (Ed.), *Digital religion: Understanding religious practice in new media worlds*. Abingdon: Routledge.

Internet Evangelism Day. (n.d.). *The bridge strategy: A way to reach millions*. Retrieved June 1, 2012, from www.internetevangelismday.com/bridge-strategy.php

Ivereigh, A. (2011, May 3). Vatican bloggers meeting: Mission and responsibility. *America*. Retrieved June 1, 2012, from www.americamagazine.org/blog/entry.cfm?blog_id=2&entry_id=4180

Koinonia. (n.d). About. Retrieved June 1, 2012, from http://koinonia-church.org/about/

Lewis, R. (2012). *Modest dressing: Faith-based fashion and internet retail*. Retrieved June 1, 2012, from www.fashion.arts.ac.uk/research/projects-collaborations/modest-dressing/

Lövheim, M. (2011). Mediatisation of religion: A critical appraisal. *Culture and Religion, 12*(2), 153–166.

McMichael, M. (2009, June 17). *Pageant preachers*. Retrieved July 15, from www.christianitytoday.com/ct/2009/juneweb-only/124-32.0.html

MegaVoice. (2012). T*he Envoy micro solar-powered MegaVoice with onboard microSD memory car*. Retrieved June 1, 2012, from www.megavoice.com/envoy_micro.html

Miczek, N. (2008). Online rituals in virtual worlds: Christian online service between dynamics and stability. *Online: Heidelberg Journal of Religions on the Internet, 3*(1), 144–173.

Morozov, E. (2012). *The net delusion: How not to liberate the world*. London: Penguin.

Nagata, T. (2010, March 5). *The world of Warcraft raid – Four years later*. Retrieved July 15, 2012, from www.gamesradar.com/the-wow-funeral-raid-four-years-later/

North Coast. (n.d.). *Our history*. Retrieved June 1, 2012, from www.northcoastchurch.com/index.php?id=2633

O'Leary, A. (2012, June 2). Twitter dynamos, offering word of God's love. *New York Times*. Retrieved June 5, from www.nytimes.com/2012/06/02/technology/christian-leaders-are-powerhouses-on-twitter.html

Oosterbaan, M. (2011). Virtually global: Online evangelical cartography. *Social Anthropology, 19*(1), 56–73.

Peters, T. (2011). Changing the world: New media activism. In B. Vogt (Ed.), *The church and the new media: Blogging converts, online activists and bishops who tweet* (pp. 163–176). Huntingdon: Our Sunday Visitor.

Pontifical Council for Social Communications. (2002). *The Church and internet*. Retrieved June 1, 2012, from www.vatican.va/roman_curia/pontifical_councils/pccs/documents/rc_pc_pccs_doc_20020228_church-internet_en.html

Stewart, A. (2011). Text and response in the relationship between online and offline religion. *Information, Communication & Society., 14*(8), 1204–1218.

Tice, K. (2010). The after-life of born-again beauty queens. In M. Bailey & G. Redden (Eds.), *Mediating faiths: Religion and socio-cultural change in the twenty-first century* (pp. 105–118). Farnham: Ashgate.

Wagner, R. (2012). *Godwired: Religion, ritual and virtual reality*. Abingdon: Routledge.

Ward, B. (2011, August 15). *The Bible regains second place in page engagement*. Retrieved June 1, 2012, from http://allfacebook.com/the-bible-regains-second-place-in-page-engagement_b54875

Williams, K., & Gray, L. (2010). The little phone that could: Mobile-empowered ministry. *International Journal of Frontier Missiology, 27*(3), 139–145.

YouVersion. (2012, May 10). *The Bible App by YouVersion now installed on 50 million mobile devices*.RetrievedJune1,fromhttp://blog.youversion.com/2012/05/the-bible-app-by-youversion-now-installed-on-50-million-mobile-devices/

Zaleski, J. (1997). *The soul of cyberspace: How new technology is changing our spiritual lives*. San Francisco: HarperCollins.

Chapter 202
The People of the Nook: Jewish Use of the Internet

Ira M. Sheskin and Micah Liben

202.1 Introduction

Considered both an ethnic group and a religious group, there are about 13–14 million Jews in the world, constituting only about 0.2 % of the world population. The 6.7 million Jews in the U.S. constitute only about 2 % of the American population. The Jewish presence on the Internet is well out of proportion to these small percentages. This chapter will show that Internet usage by the American Jewish community is significant both as an educational resource and a communication tool. Thus, the "People of the Book" have embraced technology to also become the "People of the Nook."

First, we examine the percentages of Jews using the Internet both for general information about Jewish-related items and for information about their local Jewish communities. The extent to which various demographic and religious subgroups of American Jews use the Internet is also explored. Second, some of the uses of the Internet by the Jewish community are examined, including Jewish educational purposes, observing Jewish ritual obligations (*z'manim*, counting the *Omer*, *eruvim*, electronic *Yahrtzeit* boards), convening a *minyan*, and conducting research on the Jewish community.

I.M. Sheskin (✉)
Department of Geography and Regional Studies, University of Miami,
Coral Gables, FL 33124, USA
e-mail: isheskin@miami.edu

M. Liben
Judaic Studies, Kellman Brown Academy, Voorhees, NJ 08403, USA
e-mail: micah.liben@gmail.com

© Springer Science+Business Media Dordrecht 2015
S.D. Brunn (ed.), *The Changing World Religion Map*,
DOI 10.1007/978-94-017-9376-6_202

202.2   Use of the Internet by Jewish Households

It should not be surprising that Jews have taken to the Internet in large numbers. Levin (1996) had already published *The Guide to the Jewish Internet* two years before Google was incorporated as a company by Sergey Brin (a Jew from the former Soviet Union) and Larry Page (of part-Jewish origin).

In many ways, it is also not surprising that Jews have made extensive use of the Internet. Most Jewish households were probably early adopters of home computers, given that median household income for Jewish households in 2012 (based upon inflating the median income from the 2000–2001 National Jewish Population Survey (Kotler-Berkowitz et al. 2003)) was $64,000, compared to $56,000 for all American households. In addition, 60 % of adults age 25 and over in Jewish households have a 4-year college degree or higher compared to 28 % for all American adults age 25 and over (based upon the 2007–2011 American Community Survey at www.census.gov).

Table 202.1 shows that as early as 2000, the National Jewish Population Survey found that 40 % of Jewish adults had used the Internet for Jewish-related information in the past year (1999), a remarkable figure given that the Internet only really entered the public domain in a significant way in the mid-1990s. Since 2000, this same question has been asked in 25 local Jewish community studies with Internet usage varying from 29 % in Sarasota and South Palm Beach to 57 % in Portland (ME), 58 % in Washington, and 59 % in San Antonio. The data in this table suggest an increasing usage of the Internet for Jewish-related information: the correlation between Internet usage in a community and the date of the study is R = .557 (alpha = .002). In fact, controlling for percentage elderly in each community (recognizing that the elderly are laggards in this diffusion process and that the percentage elderly for these 25 communities varies from 10 % in Washington to 62 % in South Palm Beach) shows a partial correlation of .800 (alpha = .000).

Table 202.2 shows the results of a follow-up question to that which produced Table 202.1. Respondents who had responded in the affirmative to using the Internet for Jewish-related information in the past year were asked if they had used the Internet for *local* Jewish-related information in the past year. This is a significantly different usage than that in Table 202.1. Much of the usage in Table 202.1 may be for information about Jewish history, Jewish holidays, Jewish religious matters, Israel, and such. Usage reported in Table 202.2 would generally be usage that would connect the Internet user with their geographically local Jewish community, such as the time and place of local Jewish community events. Naturally, these percentages are lower than in Table 202.1, but nevertheless indicate impressive usage, varying from 9 % in Atlantic County and South Palm Beach to 36 % in Washington and 37 % in San Antonio, with a median of 21 %.

Table 202.3 shows the results from Table 202.1 crosstabulated by several variables. Recently, the first author compiled into one data set (named Decade 2000) all 22 local Jewish community studies that he was responsible for conducting between

Table 202.1 Jewish adults who used the Internet for Jewish-related information in the past year

Community	Year	Percent (%)
San Antonio*	2007	59
Washington*	2003	58
Portland, Maine*	2007	57
San Francisco	2004	55
Lehigh Valley*	2007	52
Detroit*	2005	50
St. Paul*	2004	50
Middlesex*	2008	48
New Haven*	2010	47
Minneapolis*	2004	46
Jacksonville*	2002	42
Tucson*	2002	41
Tidewater*	2001	41
Las Vegas*	2005	40
Rhode Island*	2002	40
Bergen*	2001	39
New York	2002	37
Miami*	2004	35
W Palm Beach*	2005	34
Westport*	2000	33
Rochester	1999	33
Atlantic County*	2004	32
Hartford*	2000	30
S Palm Beach*	2005	29
Sarasota*	2001	29
National Jewish Population Survey[+]	2000	40

Source: Ira Sheskin from the Decade 2000 data set
[+]NJPS 2000 data are for the *more Jewishly-connected sample*
*Communities that are part of the Decade 2000 data set

2000 and 2010 (asterisked communities in Table 202.1). Decade 2000, with a sample size of 19,800, is the largest random sample of American Jews ever compiled. Note that while the 22 community studies included do not form a national probability sample, the 19,800 interviews do randomly represent almost 547,000 Jewish households containing 1,247,000 persons, of whom about 1,081,000 are Jewish.

Overall, 42 % of respondents in the Decade 2000 communities used the Internet for Jewish-related information in the past year and 21 % used the Internet for *local* Jewish-related information. This discussion proceeds to examine variations in the 42 % figure.

Table 202.2 Jewish adults
who used the Internet for
local Jewish-related
information in the past year

Community	Year	Percent (%)
San Antonio	2007	37
Washington	2003	36
Lehigh Valley	2007	30
Detroit	2005	30
St. Paul	2004	29
Portland (ME)	2007	27
Minneapolis	2004	26
Las Vegas	2005	21
New Haven	2010	17
Middlesex	2008	17
Jacksonville	2002	16
Miami	2004	14
W Palm Beach	2005	11
S Palm Beach	2005	9
Atlantic County	2004	9

Source: Ira Sheskin from the Decade 2000 data set

The relationship with age is particularly interesting (although not surprising), with usage decreasing with age from 65 % of respondents under age 25 to only 17 % of respondents age 75 and over. Note that usage among respondents age 65–74 is about double the usage among respondents age 75 and over. All of this suggests that, as time passes, Internet usage for Jewish-related information will increase further. Consistent with this generational change, VERY few younger persons compared to older persons in the same surveys claim to always/usually read a local Jewish newspaper. That males are significantly more likely to use the Internet for Jewish-related information is almost certainly due to the prevalence of females in the age 75 and over category.

The Jewish community has a clear communal interest in educating its youth and encouraging in-marriage. Thus, the community is particularly interested in reaching households with children and non-elderly singles. Both groups make significant use of the Internet for Jewish-related information: 62 % for households with children and 51 % for non-elderly singles. Table 202.4 shows that significant percentages of non-elderly singles have used a Jewish Internet dating service, the most common of which is www.jdate.com. JDate has 750,000 annual users, about 2,500,000 of whom pay thirty dollars per month for the premium service. Of the 1.8 million single adult Jews in the United States, over 20 % are active on JDate (Schwarz 2013). Another such site, aimed at Orthodox Jews is www.SawYouAtSinai.com (Fig. 202.1). From a Jewish communal perspective, investing in websites is probably the best means of communication with these two household types.

The relationship with household income is also as expected with usage increasing from 20 % of households earning under $25,000 to about 55 % of households earning $100,000 and over.

Table 202.3 Jewish adults in the Decade 200 data set who used the Internet for Jewish-related information in the past year for various population subgroups

Characteristic	Population subgroup	For Jewish-related information (%)	For local Jewish-related information (%)
All	All	42	21
Age	Under 35	65	40
	35–49	59	35
	50–64	49	24
	65–74	33	11
	75 and over	17	5
	65 and over	24	7
Sex	Male	44	21
	Female	41	21
Household structure	Household with children	62	36
	Non-elderly couple	50	24
	Non-elderly single	51	31
	Elderly couple	33	10
	Elderly single	13	4
Household income	Under $25,000	20	9
	$25–$50,000	36	17
	$50–$100,000	48	26
	$100–$200,000	56	31
	$200,000 and over	54	27
Jewish identification	Orthodox	52	29
	Conservative	49	25
	Reform	44	22
	Just Jewish	31	13
Type of marriage	In-married	49	23
	Conversionary	62	36
	Intermarried	40	18
Synagogue membership	Member	55	29
	Non-member	35	16
Any adult visited Israel	On Jewish trip	54	30
	On general trip	44	19
	No	35	17
Donated to a Jewish charity in the past year	Yes	51	26
	No	32	14

Source: Ira Sheskin from the Decade 2000 data set

Respondents to these surveys are asked if they consider themselves Orthodox, Conservative, Reform, or "Just Jewish." Respondents who are "Just Jewish" (31 %) are much less likely to use the Internet for Jewish-related information than respondents in the other groups (44–52 %).

Table 202.4 Households
with single Jewish adults age
18–64 who ever used a
Jewish Internet dating service

Community	Year	Percent (%)
S Palm Beach	2005	26
St. Paul	2004	24
Middlesex	2008	21
San Antonio	2007	18
Miami	2004	18
Lehigh Valley	2007	17
Las Vegas	2005	17
New Haven	2010	16
Atlantic County	2004	15
Minneapolis	2004	15
W Palm Beach	2005	14

Source: Ira Sheskin from the Decade 2000 data set

Fig. 202.1 SawYouAtSinai.com, an online Jewish dating service (Image from www.SawYouAtSinai.com, used with permission)

While the surveys in Decade 2000 do not allow us to identify the "ultra-Orthodox," other evidence suggests varying patterns for this group. Among the "ultra-Orthodox," different reactions to the Internet may be noted depending on the individual Orthodox "sect." The *Chabad* sect, a *Hasidic* group that has outreach programs all over the world that attempt to involve less religious Jews in Orthodox Jewish practice, has made significant use of the internet (www.chabad.org) in their outreach efforts, maintaining an extensive website of educational material. *Aish HaTorah*, a more modern Orthodox group, does the same (www.aish.org). On the other hand, in May 2012, 40,000 ultra-Orthodox (*Haredi*) Jews (many from the *Satmar* sect) gathered at 42,000-seat Citi Field in New York, with an overflow crowd in the 20,000 seat Arthur Ashe tennis stadium watching via large screen TV, to discuss the risks of the Internet (Grynbaum 2012). In this case, the rabbis were concerned not just about exposure to such things as pornography on the Internet, but also simply exposure to the outside world: many ultra-Orthodox children are raised in a controlled environment in which they have no televisions and very limited access to many non-sectarian cultural events and popular culture. But even in this setting, a good portion of the attendees could be seen checking their email on their smart phones both during the presentations as well as on their way in and out of the stadia. Note that Orthodox Jews are less than 10 % of American Jewish households (Sheskin 2012) and the ultra-Orthodox are only a segment of this group.

Three types of marriages are identified in these surveys. An *in-marriage* is a marriage in which both spouses were born or raised Jewish and currently consider themselves Jewish. A *conversionary in-marriage* is a marriage in which one spouse was born or raised Jewish and currently considers himself/herself Jewish and the other spouse was not born or raised Jewish but currently considers himself/herself Jewish (a "Jew-by-Choice"). An *intermarriage* is a marriage in which one spouse currently considers himself/herself Jewish and the other spouse does not currently consider himself/herself Jewish.

Usage is highest (62 %) for conversionary in-married couples, probably because the Jew-by-Choice in the marriage may find himself/herself in situations in which he/she is lacking basic knowledge of Jewish matters and finds the Internet a non-threatening, easily accessible resource for finding information without asking questions of a human being who might question their level of commitment to and understanding of their new faith if they reveal a lack of certain basic knowledge. It is probably for a similar reason that usage is relatively high for intermarried couples (40 %), only 9 percentage points behind the in-married couples.

Table 202.3 also shows, not unexpectedly, that Internet usage is higher among Jews who are more involved in Jewish life, either by joining a synagogue (generally a religious connection), visiting Israel (generally an ethnic connection), and donating to a Jewish charity.

202.3   Uses of the Internet

From the proliferation of mobile applications and web-based communication tools to the ever-growing storehouse of information on the Internet, modern technology has made a significant and lasting imprint upon Jewish religious practice. As one op-ed writer put it, "I've become an improved Jew. I learn more, say more blessings, and didn't have a problem remembering a single day in the *Omer* [a period of ceremonial counting between the festivals of *Passover* and *Shavuot*]… So what caused this recent growth spurt in my Judaism? I got an iPod Touch."[1]

The information revolution triggered by the Internet, and its effects on Jewish life, are extensions of an equally significant revolution that preceded it, namely electronic storage media. With the advent of CD ROMs and USB hardware, massive databases could be made easily searchable and portable. For Jewish students and scholars alike, this meant being able to navigate the hitherto bottomless "sea of Talmud" with a touch of a button, via massive databases such as the Bar Ilan Responsa Project (www.biu.ac.il/jh/Responsa/). Today, the Internet provides widespread access to such databases on-line, searchable with the click of a mouse and often free of charge. Organizations such as the Spertus Institute for Jewish Learning and Leadership in Chicago (www.spertus.edu) enable members to access multiple databases in their e-library, including the Bar Ilan Responsa, Soncino Classics Library, and multiple periodical and journal collections. Dozens of mobile applications with digital, downloadable versions of the Hebrew Bible, Talmud, Rabbinic commentaries, Kabbalistic texts, Midrash and liturgical works have been created by a variety of developers. Simply put, Internet users today have unprecedented access to Jewish texts, resources and even ritual items such as digitized prayer books. Deuteronomy (30:11–14) states, "This commandment is not in Heaven, or far across the sea; rather the word is very close to you that you may do it…." Indeed, the Internet has rendered Jewish learning and even the performance of many Jewish rituals more accessible than ever.

The advent of the Internet has certainly been accompanied by concomitant difficulties, both *halachic* (pertaining to Jewish law) and sociological, for Jewish leaders and rabbinic authorities. Cohen (2012) discussed several significant *halachic* concerns engendered by use of the Internet, including sexual modesty and on-line relationships, e-commerce and "virtual acquisition" in Jewish law, and issues surrounding the prohibited use of electricity during the Sabbath. Additional examples of *halachic* questions abound; recent papers published by the Conservative Movement's Committee on Jewish Law and Standard (CJLS), including papers by Leff (2007), Dorff and Spitz (2001), and Dorff and Hearshen (2010), have dealt with the Internet in relation to such varied topics as intellectual property, workplace privacy, and violent video games. Typically, rabbinic authorities within the Conservative Movement (which is the religiously centrist Movement in Judaism) strive to integrate and benefit from modern technology, while still remaining within

---

[1] http://www.jewishpress.com/sections/scitech/how-apple-made-me-a-better-jew/2009/06/17/0/?print

a traditional *halachic* framework; this trend is evident in the right-of-center Orthodox Movement, as well. By contrast, in *Haredi* (or ultra-Orthodox) communities, use of the Internet has been highly circumscribed and even banned entirely by some rabbinic leaders. On the other end of the spectrum, liberal Movements, like the Reform Movement, do not view the traditional corpus of *halacha* as binding, and thus have embraced the use of the Internet for Jewish practice in ways which are innovative, but also beyond what some consider *halachically* acceptable.

Dilemmas regarding Internet usage will continue to emerge and be debated, with philosophical considerations about the synthesis of religion with modernity yielding different standards among various Jewish Movements (ultra-Orthodox, Orthodox, Conservative, Reform). As a practical matter, however, the ubiquity of the Internet in Jewish religious life today is irrefutable. We should not overstate the case; to be sure, Shandler (2009) has noted the tendency for creators of new Jewish media resources to describe their endeavors as "transformative undertakings, sometimes on a grand scale," whereas in fact, these same innovations often "rely upon older technologies, traditional media practices and canonical works." Notwithstanding this observation, the widespread use of the Internet and new media among Jews can hardly be denied, nor can the positive impact of this phenomenon upon Jewish practice. Indeed, among the mainstream Jewish community, employing mobile applications and other Internet tools for a range of Jewish practices is viewed not only as acceptable but also *de rigueur*.

The profusion of mobile applications with a wide array of Jewish content is evident on Jewish iPhone Community (www.jewishiphonecommunity.org), an on-line magazine which catalogues Jewish applications as they are released for Android, iPhone, and Blackberry devices. The site lists hundreds of available applications with Jewish content, which can be divided loosely into three categories—resources, games, and novelties. Many of the products in the latter two categories are "kitschy," such as the iShofar application which simulates the sound of a *shofar* (ram's horn) for Rosh Hashanah. Given that the obligation on Rosh Hashanah is to hear the sounding of an actual *shofar*, the application serves no functional purpose from a ritual perspective (though it certainly may be used for educational purposes). But applications such as these, which are intended more for fun than for actual ritual use, still have great significance to the extent that they generate excitement and even pride among Jewish users, and serve as entry points for continued Jewish engagement.

A variety of websites cater specifically toward outreach and "engagement." Shandler (2009) has chronicled organizations such as Chabad and Aish Ha-Torah, which were among the early leaders in harnessing the power of the Internet for strengthening Jewish involvement and affiliation. However, not all outreach and engagement sites focus solely on Torah study or religious matters. For example, the strengthening of Jewish identity was the impetus behind Jewish Rock Radio (JRR) (www.jewishrockradio.com). The flagship program of Judaism Alive (www.judasimalive.net), JRR launched a website in 2009 with the mission of "strengthening Jewish identity and connection in Jewish youth and young adults, by providing a mass communication channel utilizing the power of music to inspire and educate" (http://jewishrockradio.com/#/about/mission). Jewish Community Centers and other communal organizations list places of interest, kosher food, social programs and other services, and of course, there are countless Jewish news sources on-line, including the digital versions of Israeli newspapers, such as *Haaretz*, in both Hebrew and English.

Of particular interest here, however, is the content that falls specifically into the category of religious and educational resources. This section will briefly examine the role of the Internet in enhancing Jewish education as well as the performance of ritual obligations—including time-bound rituals (*z'manim*) in general, counting of the *Omer* period in particular, and use of enclosures (*eruvim*) for the Sabbath. We will then explore in greater detail one ritual in particular that has generated both popularity and controversy among Jewish leaders, namely convening *minyanim* (prayer quorums) for on-line prayer services.

202.3.1   Jewish Education[2]

As technology and on-line resources have increasingly become part of the landscape in secular education, Jewish institutions have followed suit, and in some cases even led the way. The Union for Reform Judaism (URJ) has developed an on-line platform for its entire network of congregational Hebrew schools; in August 2012 it rolled out its Mitkadem curriculum in digital format, allowing students to communicate virtually with each other and their teachers, work at their own pace in the classroom and work remotely with teachers outside of class.[3] Prager (2012) cites a study sponsored by the Avi Chai Foundation showing that 25 % of Jewish day schools are now offering some form of on-line learning, with a mix of secular and Judaic topics. Companies like Tomorrow's Genius and Bonim B'yachad, both centered in Israel, offer an array of on-line classes that Jewish schools utilize for enrichment, tutoring or electives. In September 2012, Yeshivat He'Atid (Hebrew for the *yeshiva* or "Talmudic academy" of the future) opened with blended learning in classes as young as pre-K; a model which they believe will both increase personalized learning and decrease tuition costs. Educational software programs such as Gemara Berura, which enable students to break down Talmudic texts and color-code them in on-screen flow charts, have developed web-based versions for students to use on iPads or other tablets. In higher education, on-line Jewish learning is commonplace through academic institutions like the Lookstein Center in Israel, which offers numerous courses for credit in a distance learning format. Thousands of Israeli educators and college students have completed on-line courses through the Lookstein Center since 2001 (http://www.

---

[2] The entire Autumn 2012 issue of *Contact* magazine, published by The Steinhardt Foundation for Jewish Life was devoted to "Technology and Jewish Education" as was the Spring 2006 issue "The Internet: Pathways and Possibilities."

[3] http://www.jta.org/news/article/2012/08/20/3100576/looking-for-a-jewish-education-theres-an-application-for-that

lookstein.org/distance_ed/). There is even an independent rabbinical college—the New York based Jewish Spiritual Leaders Institute (www.jsli.net)—that functions completely on-line via chat-rooms and webinars.

On the web itself, the educational resources available are virtually limitless. To cite a minute representative sample, www.JTSA.org provides scholarly *divrei Torah* (homiletic teachings) about the weekly Torah portion which can be emailed in digest form; www.Zieglertorah.org provides educational podcasts along with downloadable source sheets to accompany the lessons; and Kosher-Tube (www. koshertube.com) lists over 12,000 YouTube-style videos of classes, sermons and lectures, boasting over 400,000 video hits per month. www.AskMoses.com enables users to discuss their queries with experts via live chats on their website, and www. PartnersinTorah.org connects people who want to teach with people who want to learn. Cohen (2012) expounds upon the extent to which individuals from different Jewish Movements utilize "virtual rabbis" through sites such as Kipa (www.kipa. co.il) and Jewish Answers (www.jewishanswers.org) for on-line "rabbinic counseling," which is a growing trend particularly in the Modern Orthodox community.

A variety of Internet-based projects specifically designed for collaboration in the field of Jewish education are also extant. For example, YU2.0 (www.YU20.org) is a "community of practice" made up of educators who are invested in sharing resources and staying current on the latest technological innovations in education. Their website features blogs, webinars, discussion forums, videos, and research reviews. Likewise, blogs such as Tech Rav (techrav.blogspot.com), and JLearn2.0 (www.jlearn2.0.org) serve as hubs for research and data on technological innovations in Jewish education. Other sites such as Sefaria (www.sefaria.org) share open source codes for collaboration. Sefaria is a not-for-profit project that aims to bring digital versions of Jewish texts into the public domain, and allows scholars and developers to actively engage in the process of contributing to and building the site. In a similar vein, the Open Siddur Project (www.opensiddur.org) is a collaborative digital-to-print publishing application that enables users to create their own *siddurim* (prayer books) by uploading, transcribing, and sharing liturgical manuscripts of public domain and free-culture licensed texts.

202.3.2   Jewish Ritual Obligations

This section examines the use of the Internet to assist in the observance of *z'manim*, counting the *Omer*, utilizing *eruvim*, and the use of electronic Yahrtzeit boards.

Z'manim—Set Times. In Jewish law, many *mitzvot* (ritual commandments) must be performed at set times or within specific time frames, termed *z'manim* (Hebrew for times). If a worshipper misses the window of opportunity for a particular *mitzvah*, he generally cannot make it up; that is, he has lost his opportunity to fulfill the *mitzvah* and so commits an *aveirah* (transgression). The Internet effectively reduces the omission of time-bound *mitzvot* and the resultant commission of *aveirot* because it enables easy access to information, and provides efficient notification tools.

Take, for example, the verses of *Sh'ma* from Deuteronomy 6:7 (beginning "Hear O Israel … the Lord is one"), which serves as a sort of credo for the Jewish people. This passage must be recited twice daily—in the words of the Bible, when you "arise" and when you "lie down." Hillel, the great Sage of the Mishnah, explained the Bible's wording to mean that the *Sh'ma* should be recited in the morning and evening, namely, the times of day when people generally "arise" and "lie down." Yet even with Hillel's explanation, the Bible's instructions here are inherently subjective— if I typically arise after noon, can it still considered morning for me? Or have I lost my window of opportunity to perform the commandment? To avoid such ambiguity, the Talmudic Rabbis (Tractate *B'rachot*) established objective rules for determining the set time frames to recite the *Sh'ma* and other prayers. The precise rules that were developed for calculating *z'manim*, however, became increasingly complex and even arcane; fortunately, the Internet allows modern Jews to keep track of *z'manim* with ease.

Among some of the *z'manim* an observant Jew must track to recite daily prayers within their acceptable time-frames are *alot hashachar* (dawn), which is the earliest one can recite *Sh'ma* in the morning; *plag haminchah*, which is the earliest time one may light Sabbath candles on a Friday afternoon; and *shkiah* (sunset), which is the latest time for reciting the afternoon prayer. The calculation of these *z'manim* is based on various astronomical considerations, such as the number of daylight hours (which varies by season) and the angle of the sun in a given locale.

A layperson who wanted to calculate all these times would require both signifi- cant *halachic* expertise as well as mathematical skills, not to mention access to the astronomical data for a given location (as may be found in an almanac or some newspapers). Today, the Internet has simplified this. Websites such as www. Myzmanim.com allow users to provide a zip code or search millions of locations for precisely calculated *z'manim*, and many synagogues post calculations for their own communities on-line (for example, http://yisd.youngisrael.org/content/zmanim). Other sites such as www.Kaluach.org and www.Hebcal.com both offer *z'manim* search capabilities, as well as date conversions between the Jewish and secular cal- endars, and are compatible for use with cell phones, Windows, Google Calendar, Outlook and other programs. Among the many popular Jewish smart-phone applications developed by Rustybrick (www.Rustybrick.com/apps) is one called "Shabbat Shalom," a free application which enables users to easily check times for lighting Sabbath candles and for performing *Havdalah* (the ritual to end the Sabbath) in any location around the world.

202.3.2.1 Counting the Omer

Another example of a time-bound *mitzvah* is the daily counting of the *"Omer"* (Hebrew for sheaf), which marks the barley harvesting period in the weeks between the holidays of Passover and Shavuot. During this period, a blessing is recited each evening and the coming day is ceremoniously enumerated. This ritual is repeated until the day of Shavuot (the Feast of Weeks), which is the traditional date upon

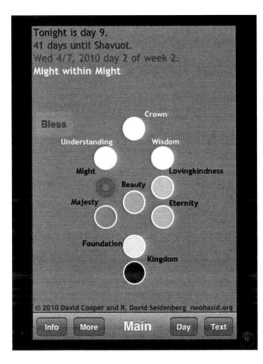

Fig. 202.2 Omer counters for mobile devices (*Left image* from www.neohasid.org, used with permission. *Right image* from www.rustybrick.com, used with permission; the app can be downloaded at rustybrick.com/apps)

which the Israelites received the Torah at Sinai. To help facilitate the performance of this *mitzvah*, Rustybrick offers a "Count the Omer" application (one of several *Omer* counting applications available) (Fig. 202.2). The application includes the special blessings that accompany the nightly counting, and sends reminders to your phone throughout the *Omer* period, helping users avoid missing days. This is significant, because according to most *halachic* sources, once an individual has missed even one day he/she is no longer permitted to say the daily blessings for the duration of the entire *Omer* period; effectively forfeiting their place in the counting. As indicated above, this would result in the distressing prospect of missing out on performing a *mitzvah*, and thus committing an *aveirah*. Moreover, the counting of the *Omer* has taken on added significance over time; beyond merely a rote ritual with prescribed texts to recite, the *Omer* period has become wrapped up in *Kabbalistic* (Jewish mysticism) imagery, whereby each day signifies one of the *sefirot*, the mystical emanations by which God's divine attributes are revealed in the world. (In light of this, the iPhone *Omer* application helpfully provides an on-screen digital diagram of the *sefirot* for each day.) All these factors heighten the experience of the counting of the *Omer*, and help explain both the significance and popularity of the ritual itself as well as that of the Internet tools which facilitate its full and proper performance.

202.3.2.2 Eruvim – Sabbath Enclosures

To safeguard the Sabbath as a day of rest set apart from the ordinary work week, Jewish law stipulates that various categories of "creative labor" ("*melacha*") are not to be performed. The specific types of labor from which religious Jews refrain on the Sabbath were derived from the building of the Tabernacle in the desert (Exodus 25–40), and include the act of transferring items from one domain to another. Jewish law, therefore, prohibits moving any objects from a "private domain" to a "public domain" and vice versa on the Sabbath.

In the past, cities and neighborhoods were commonly surrounded by walls. During the Middle Ages, European Jews typically lived within ghetto walls. In such cases, the whole area could be regarded as one "private domain," and thus carrying on the Sabbath was permissible. However, for the sake of Jewish communities in un-walled cities, a legal workaround was developed by the Rabbis in the Talmud (Tractates Shabbat and Eruvin) to enable the transferring of items on the Sabbath, namely the creation of a technical enclosure colloquially known as an *eruv* (pl. *eruvim*). The *eruv*—which may be comprised of poles, wires, fences, and natural boundaries—forms a periphery around multiple private and (previously) public domains, creating one large area wherein carrying is permissible.

More than 200 *eruvim* currently exist throughout the United States. Typically, wires are run between utility poles to indicate the geographic boundaries of the *eruv*. In some places, when Interstate highways are raised above city streets, the wall of the Interstate or the sound barriers along the Interstate may be used for parts of the *eruv* boundary. As wires may come down during a storm, *eruv* boundaries are checked on a regular basis. On a practical level, knowing the status of the *eruv* is a matter of convenience. If an *eruv* is "up" (in use), observant Jews are allowed to push a baby stroller to synagogue or carry a bottle of wine to a neighbor's Sabbath meal (Fig. 202.3).

On another level, however, even more is at stake in the status of the *eruv* than merely these conveniences; for if one were to transfer items on the Sabbath in the absence of an *eruv*, it would amount to a "*chilul Shabbat*" (or desecration of the Sabbath), for which the *halachic* and theological consequences are considered severe. Being certain of the *eruv's* status is therefore paramount, and to that end, the Internet has significantly enhanced an individual's ability to keep abreast of an *eruv's* condition in any given community. *Eruv* associations maintain websites showing detailed maps with *eruv* boundaries and messages that can be checked to make certain the *eruv* is intact. (See, for example, www.miamibeacheruv.com.) Many of these sites also enable users to register for "alerts" which are sent to an email address or mobile device (see http://www.denvereruv.org/subscribe.html). There are dozens of *eruv*-related handles on Twitter, as well as groups like "Metropolitan Eruvin" on LinkedIn, which provide users with real-time updates. Such websites and social networking platforms—which serve to disseminate information about *eruvim* widely, efficiently and accurately—reduce the potential for committing a *chilul Shabbat* and ultimately enhance Sabbath observance.

Electronic Yahrtzeit Boards. Jews are required to commemorate annually the deaths of parents, siblings, spouses, and children. Synagogues have traditionally

Fig. 202.3 Eruv alerts (Image from www.DenverEruv.org)

sold Yahrtzeit plaques which are placed on Yahrtzeit boards on the walls of synagogues. In some cases, these boards are being replaced with computers connected to the Internet and a large screen display (Fig. 202.4). Names and other information about the deceased are then continuously scrolled on the screen. The display is controlled over the Internet by a company which changes the information weekly to reflect the appropriate deaths in the appropriate weeks.

202.3.3   Convening a Minyan over the Internet? Problems, Possibilities and Practice

We have seen how the rituals above—counting the *Omer*, reciting the *Sh'ma*, preparing for the Sabbath, commemorating *Yahrzeits*—may be enhanced by the use of the Internet. However, they are not rituals that can be performed through or via the Internet, per se; the Internet is merely a tool to help facilitate the performance of

Fig. 202.4 Electronic
Yahrzeit board (Photo by Ira
Sheskin and Micah Liben)

these rituals and *mitzvot*. In addition, they are all, to a certain degree, "individual" *mitzvot* that one person can perform when alone. But Judaism emphasizes community, including communal prayer, and, of course, connecting with other people is one of the great benefits of using the Internet. Indeed, an increasingly widespread phenomenon on Facebook is the existence of pages for wishing "*refuah shlemah*" (get well soon) and for organizing the recitation of *Tehillim* (Psalms) on behalf of individuals who are ill. The organization KEY, which stands for *Kulanu Yachad Yehudim*, or "We are All Jews United Together," has likewise created international "*Tehillim* campaigns" via the Internet (http://www.shemayisrael.co.il/orgs/key/index.htm). Meanwhile, the Internet serves to bridge worshippers throughout the world with their spiritual homeland in Israel; through sites such as Kotel-Cam (english.thekotel.org/cameras.asp), worshippers can view real-time services and other activities taking place at the Kotel, the Western Wall in Jerusalem, and in lieu of placing messages in the cracks of the Wall (a long-time Jewish tradition) worshippers can submit virtual messages via the website (Fig. 202.5). Given the Internet's communication capabilities and the already widespread use of the Internet for various types of

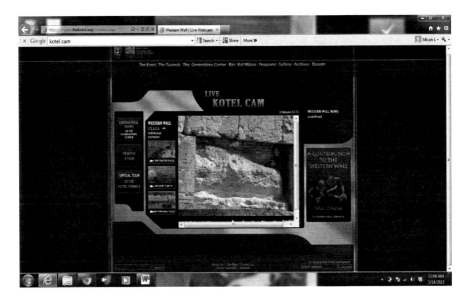

Fig. 202.5 The Kotel website provides a live view of the Western Wall in Jerusalem (Image from English.thekotel.org/cameras.asp, used with permission)

prayer-related activity, it would seem that convening a full communal prayer service over the Internet would be a natural and beneficial next step in the evolution of on-line ritual practices.

In 2006, Temple Beit Israel took that step. But the worshippers in this case are not real people, *per se*, and Beit Israel only exists on screen—it was the first synagogue constructed in the Internet's popular virtual world, "Second Life." The virtual synagogue boasts hundreds of members from across the globe who meet each week (in the form of avatars) to light Sabbath candles on screen. Strictly speaking, Second Life is essentially a game; the rituals performed therein do not fulfill real-world religious obligations, as made clear by the leaders of these virtual ceremonies (http://www.davening.net/jpost.html). Nevertheless, for many unaffiliated Jews, such rituals may be the only ones they perform, and some have reported that their on-line Jewish experiences have inspired real-life practice. The presence of synagogues in Second Life attests to the growing intersection between the Internet and Jewish religious observance. And it raises to the fore the complex question for contemporary decisors of Jewish law: While virtual candle-lighting via avatars may not exactly be kosher, what about real people connecting via the Internet to recite prescribed prayers? From the perspective of *halacha*, what requirements would need to be met to convene live Jewish prayer services over the Internet?

At the heart of this question are the parameters for communal prayer. Daily prayer services may be recited individually and indeed many observant Jews are punctilious in doing so. However, while individual recitation of most prayers is technically acceptable, *halacha* prizes the communal nature of prayer, and in fact mandates that certain key prayers be recited only in the presence of a *minyan,* or prayer quorum. Therefore, the act of praying individually thrice daily is considered meritorious, yet it is inferior to the act of joining together with others to hear and recite prayers communally in a *minyan.*

Literally translated as a "count," a *minyan* is defined as a quorum of ten worshippers who have reached the age of *bar/bat mitzvah*, that is, 13 years old for males and 12 for females. (Only Jewish males count toward a *minyan* in Orthodox synagogues, since, from a traditional *halachic* perspective, women are not obligated to perform time-bound *mitzvot* such as daily services, and therefore cannot discharge the obligation for others; however, in egalitarian synagogues such as in the Conservative and Reform Movements, Jewish women are considered obligated in the *mitzvot* to the same degree as men and thus count equally toward a *minyan.*)

One of the prayers requiring the presence of a *minyan* is the Mourner's *Kaddish,* a hymn recited by worshipers in mourning. For one who has experienced the loss of a close relative, it is a *halachic* obligation to recite the Mourner's *Kaddish* daily during the 11 months following the funeral, and every year thereafter on the anniversary of the death. Traditionally, this ritual's purposes are to aid the healing of the mourner as well as to provide a "spiritual boost" for the soul of the deceased. Significantly, the obligation to recite *Kaddish* for family members occupies a central spot in the Jewish consciousness, so much so that many otherwise unobservant Jews make a point of attending services to recite it. Thus when a mourner is rendered unable to recite *Kaddish* because an insufficient number of eligible persons are present to form a *minyan*, it is painful not only for the mourner but also to the community which could not adequately facilitate his (or her) performance of this ritual.

Both because communal prayer is paramount in *halacha*, and due to the special importance of enabling mourners to recite *Kaddish,* Jewish communities in general, and synagogues in particular, assign a high priority to convening a *minyan* for daily services as consistently as possible. It is within this broad context that rabbis and *halachic* authorities in recent years have begun exploring the feasibility of utilizing new technology to convene a *minyan* over the Internet. For Jews in far-flung communities, or those physically incapable of travelling to a synagogue or elsewhere, virtual assembly through the Internet could theoretically provide an accessible option for praying "communally" with a *minyan*. Indeed, virtual prayer services today have proliferated, although, as explained below, rabbinic authorities disagree as to the *halachic* acceptability of virtual *minyanim.*

Early *halachic* sources certainly knew nothing of modern technology; thus contemporary rabbinic authorities extrapolate from precedents set in earlier sources to the realities of modern life. In this case, the point of departure is the *Shulchan Aruch (Orah Hayim 55)*, a sixteenth century *halachic* code that still remains a definitive legal guide for religious Jews. The relevant text regarding communal prayer, states in part: "The ten [who constitute the *minyan*] must be in one area, with the leader among them… If a person is standing outside the synagogue and there is a window, if he/she shows his face [through the window] he/she may be counted. If a few worshipers are inside and a few are outside, and the leader is

positioned in the entranceway, then he/she connects them [to form one *minyan*]."
The *Shulchan Aruch* defines the parameters of a *minyan* in clearly physical terms—
even the doors and windows are considered. Given this emphasis on physical
proximity, it would seem unlikely that cyberspace could be considered an acceptable medium for convening a *minyan*. Indeed, Orthodox authorities have ruled
accordingly that an on-line minyan is not acceptable; see, for example, "Why a
Skype Minyan is Not Ok"[4] at www.Chabad.org.

The Conservative Movement examined the *halachic* permissibility of convening
a virtual *minyan* in a Responsum paper by the CJLS. The paper came to a similar
decision as the Orthodox ruling:

> A minyan may not be constituted over the Internet or any other medium of long distance
> communication. Only physical proximity, as defined, that is being in the same room with
> the prayer leader, allows a quorum to be constituted. (Reisner 2001)

Notwithstanding this stricture, the CJLS paper also included some leniencies,
which opened the door to use of the Internet for prayer under certain conditions.
Reisner (2001) cites a countervailing *halachic* passage: "If a person passes by a
synagogue, and hears the sound of the *shofar* being blown, if he was attentive then
he fulfilled his obligation" (*Mishnah Rosh Hashanah* 3:7). This source indicates
that hearing alone—along with proper concentration—is enough to fulfill one's
halachic obligation for certain rituals, despite the worshipper not being physically
present among the rest of the congregation. Using this text as a springboard, Reisner
(2001) goes on to conclude, "Once a quorum has been duly constituted, anyone
hearing the prayers being offered in that *minyan* may respond and fulfill his or her
obligations thereby, even over long distance communications of whatever sort." In
other words, while individuals on disparate computers could not connect to form a
minyan in cyberspace, they could use Skype or a similar tool to "join" a live *minyan*
that had already been convened. The Reisner paper was passed by the CJLS, effectively making it the standard of practice for U.S. Conservative congregations.

Since the CJLS paper was approved, the use of the Internet to include people in
their services has become increasingly common among Conservative synagogues.
Temple Emunah, in Lexington, MA, set up a two-way audio/visual connection in
their daily weekday *minyan* to enable a congregant to participate while she was
abroad. The rabbi commented that, despite initial concerns about on-line participation weakening the bonds of community, the effect was in fact the opposite: "It's
actually been a very powerful experience, as members of the *minyan* have gotten to
know Maxine, schmoozing with her after services over Skype."[5]

With regard to "Skyping into a minyan," the CJLS paper allows for a hierarchy
of potential courses of action, depending on various circumstances. At the top of the
hierarchy is physical attendance at an actual *minyan*, below which is the two-way
connection of the kind set up by Temple Emunah. "Only in rare or exigent circum-

---

[4] http://www.chabad.org/library/article_cdo/aid/1783077/jewish/Will-the-Synagogue-Ever-Go-
Virtual.htm#footnoteRef1a1783077

[5] www.thejewishadvocate.com/news/2011-08-12/Top_News/Kaddish_courtesy_of_Skype.html

stances," writes Reisner, "should one enact the third, and least desirable, method of fulfilling one's obligation to pray with a *minyan* by attaching oneself to that *minyan* through a one-way audio vehicle, essentially overhearing them as one standing outside the synagogue." Despite placing this one-way method at the bottom of the list, Reisner nevertheless does allow for it, and indeed it is commonly utilized. Thus for example Congregation Shirat Hayam in Swampscott, MA, webcasts its services in one direction, so that worshippers may join on-line and fulfill their individual obligations, yet they are neither seen nor heard, and thus do not participate in the communal service together with the *minyan* (see www.shirathayam.org/shulcast/).

A similar model is followed by www.ourjewishcommunity.org, a site created by an unaffiliated synagogue in Ohio, which live-streams its High Holiday services around the globe. Their well-developed virtual site—replete with video archives of rabbis' sermons, holiday services, and Passover *seders*—itself serves as a destination for on-line congregants, and in turn feeds into the activities of the actual brick and mortar congregation. Temple Emanuel, a Reform synagogue in New York, likewise began broadcasting Passover seders. Because they do not consider themselves bound by *halacha*, more leeway exists in the Reform and unaffiliated communities to broadcast participatory services on Passover and other *Yamim Tovim* (Festivals), leeway which Conservative and Orthodox synagogues do not have due to *halachic* prohibitions on the use of electricity on Festivals.

Even more different is www.Oneshul.org, an offshoot product of www.PunkTorah.org, which is a website dedicated to reaching unaffiliated Jews, and whose "multimedia network spreads a message of love, inclusion and hope to thousands of people around the world." Unlike www.ourjewishcommunity.org, which is connected to an actual congregation, Oneshul is an exclusively virtual synagogue— with no building or clergy, just an on-line independent *minyan*. Oneshul has come under fire from right-wing critics, who argue that they are violating *halacha* by convening a *minyan* via multiple computers, instead of allowing people to "attend" a *minyan* that has already been convened (indeed, even their liberal counterparts at www.ourjewishcommunity.org have actual *minyanim* in place when they broadcast their High Holiday or Passover services). But unlike their liberal counterparts, the Oneshul founders have sought to justify their creation with references to *halachic* sources, rather than pushing aside the authority of *halacha* altogether. The ongoing dialogue between Oneshul and its critics, found within the pages of reputable sources like *The Forward* (http://zeek.forward.com/articles/117182/), speaks to the high level of interest this issue has captured in the Jewish community.

The growing changes in social norms regarding on-line technology, and the extent to which Jews have embraced virtual communication and communities through Google Hangouts, Face Time and other tools, have seemingly had an effect on the Conservative Movement's position, as well. Eleven years after Reisner's initial CJLS ruling, a new *halachic* reference guide issued by the Conservative Movement indicates an openness to potential new uses of the Internet for virtual *minyanim*. In her chapter on Prayer, Reiss Medwed (2012: 12) writes, "An individual connected to the rest of the worshippers solely by electronic means cannot be counted as part of the basic quorum," but that "as our communities evolve into the future, questions about the precise relationship between prayer and technology will continue to be addressed."

202.3.4   Use of the Internet for Research on the Jewish Community

The previous sections showed some of the uses of the Internet for meeting religious obligations. Table 202.5 shows seven common terms that specifically relate to Jews and Judaism. Each term was searched using the Google search engine, resulting in a total of 623 million "hits." Obviously, not all of these websites are specifically "Jewish" websites. But the amount of information available on the Internet about Jewish topics is clearly voluminous.

Thus, academic researchers and others have made significant usage of the Internet for information about Jews and the Jewish Community. The six websites detailed below deserve special mention because of the nature of the material they provide or their high levels of usage.

1. www.jewishdatabank.org provides the data and reports from more than 200 Jewish community studies. More than 100,000 documents were downloaded from this website in 2011. A report entitled *US Jewish Population 2010* (Sheskin and Dashefsky 2011) was downloaded more than 20,000 times in 2012. When this report in previous years was published as part of the *American Jewish Year Book* about 400 hard copies were sold. This example illustrates both the power of the Internet to make information accessible to many people and the significant interest in Jewish-related topics that exists on the part of both Jews and non-Jews (Fig. 202.6).
2. www.bjpa.org is the Berman Jewish Policy archive at NYU Wagner is the central electronic address for Jewish communal policy. BJPA offers a vast collection of policy-relevant research and analysis on Jewish life to the public, free of charge, with holdings spanning from 1900 until today (Fig. 202.7).
3. www.jewishvirtuallibrary.org is the most comprehensive on-line Jewish encyclopedia in the world, covering everything from anti-Semitism to Zionism. There are more than 16,000 articles and 7,000 photographs and maps that have been integrated into the site. The Library has 13 wings: History, Women, The Holocaust, Travel, Israel & The States, Maps, Politics, Biography, Israel, Religion, Judaic Treasures of the Library of Congress, and Vital Statistics and Reference (Fig. 202.8).

Table 202.5 Frequency of Internet matches for Jewish-related words

Search word	Number of matches (in millions)
Hebrew	214
Jewish	184
Jew	63
Yiddish	62
Rabbi	44
Judaism	39
Synagogue	17
Total	623

Source: Google search in February 2012

Fig. 202.6 Berman Jewish DataBank website (Image from www.jewishdatabank.org, used with permission)

Fig. 202.7 Berman Jewish Policy Archive website (Image from www.bjpa.org, used with permission)

Fig. 202.8 The Jewish Virtual Library (Image from www.jewishvirtuallibrary.org, used with permission)

4. www.myjewishlearning.com is a website aimed at both adults and children providing basic information on Jewish history and practice.
5. www.jta.org is the definitive, trusted global source of breaking news, investigative reporting, in-depth analysis, opinion and features on current events and issues of interest to the Jewish people.
6. www.mosaicmagazine.com is a collator and originator of Jewish ideas on the web. Each day, an original feature piece focuses on an issue of contemporary interest and enduring relevance. The Editors' Picks draw on sources ranging from daily opinion pages to weekly and monthly magazines, academic journals, books, blogs, think-tanks, universities, and on-line learning sites.

202.4 Conclusion

Notwithstanding the rapidity with which the majority of the Jewish world has embraced technological advances such as the Internet, there remains a certain conservative strain among rabbinic authorities which is typical of religions in general. To be sure, there has always been a tension in Jewish sources about the use of technology, a subject expanded upon by Gerstenfeld and Wyler (2006). In biblical texts and Rabbinic commentaries, advances in technology have been subordinated in certain cases. For example, Deuteronomy (27:5–6) states that God's altar was to be built of hewn stones without the use of iron tools. Rabbi Yochanan ben Zakkai (Mekhilta) explained that iron was excluded due to its association with swords and warfare, an unfit symbol for God's altar which was meant to symbolize forgiveness. While it surely would have been easier to build the altar using iron, this technological tool of the day was not used. Put another way, the notion that, "just because we *can,* does not mean we *ought to*" is a theme that runs through *halachic* literature. In modern times, this notion has manifested itself in the realm of *kashrut* (dietary laws). The laws of keeping kosher specify that one may not digest a whole living creature (such as a bug). Modern technology enables Jews to be ever more precise in the fulfillment of such commandments—with microscopes one could check for living organisms on leaves or in vinegar, yet, authorities such as the Aruch Hashulchan (84:36) have ruled that this is not *halachically* required. Once again, despite having the technology to check for microscopic organisms, the rabbis suboptimized the use of technology due to a variety of other concerns.

Warshavski (2012) identified another "Luddite" strain regarding technology in Jewish history in the transition from scrolls to bound texts (codices). Centuries after peoples around them adopted the codex format, Jews continued to hand-write Torah scrolls on parchment. Indeed the "technological format" of the scroll has become sanctified, following a comprehensive *halachic* framework, which makes the Jewish culture unique in its contemporary active utilization of scrolls. As our modern society races toward digitization, it is interesting to consider the lifestyle of traditionally observant Jews who, for *halachic* reasons, do not use electricity on the Sabbath. Indeed, as those in secular society truly do become "People of the Nook," it is possible that the only people in the next generation to make regular use of printed books will be observant Jews on the Sabbath! The reactionary forces in religious life may be problematic to the extent that they hinder educational and cultural opportunities in communities like the ultra-Orthodox. However, a certain degree of "technological fixedness" does allow us to slow down and examine the implications of progress upon our culture and values.

The Internet will certainly continue to play an extremely significant and positive role in Jewish religious life, as both an educational medium and a tool for performing religious tasks. Many communities will continue to enhance their prayer services by using the Internet to include worshippers across cyberspace via Internet broadcasts. However, because the communal nature of the physical synagogue service is central in *halacha* and Jewish tradition, the fully virtual *minyan* is not likely to become widespread among centrist and traditional Jewish communities. With regard to the digitization of sacred texts, a growing body of *halachic* literature has already formed dealing with questions such as cutting and pasting or deleting of God's name on screen (according to Jewish law, texts with the Tetragrammaton name may not be defaced or discarded, but rather must be buried). Beyond such tangible concerns, however, ethereal questions remain; how will the shift in medium affect users' attitudes toward the content of the text? As sacred texts become increasingly accessible to Jewish users in digital format, especially via mobile devices, more research may be done regarding the changing attitudes toward digital texts for People of the Book (Fig. 202.9).

From a geographic viewpoint we note that Judaism, like other faiths, puts significant emphasis on community and physical proximity. The use of the Internet to

Fig. 202.9 The traditional Torah Scroll meets the tablet computer (Source: Ira Sheskin)

form community by overcoming geographic space at almost no cost is, on the one hand, an exciting opportunity, allowing people to participate who might otherwise be unable to do so because of time and cost constraints or due to physical limitations. On the other hand, by allowing such usage, does the community downplay the importance of the physical proximity that allows one to comfort a mourner by a hug or a pat on the back?

References

Cohen, Y. (2012). Jewish cyber-theology. *Communication Research Trends, 31*(1), 4–13.

Dorff, E., & Hearshen, J. (2010). *Violent and defamatory video games*. New York: Rabbinical Assembly [EH 21:1.2010].

Dorff, E., & Spitz, E. K. (2001). *Computer privacy and the modern workplace*. New York: Rabbinical Assembly [HM 331:1.2001].

Gerstenfeld, M., & Wyler, A. (2006). Technology and Jewish life. *Jewish Political Studies Review, 18*(1–2), 119–150.

Grynbaum, M. M. (2012, May 20). Ultra-Orthodox Jews rally to discuss the risks of the internet. *New York Times.*

Kotler-Berkowitz, L., Cohen, S. M., Ament, J., Klaff, V., Mott, F., & Peckerman-Neuman, D. (2003). *Strength, challenge and diversity in the American Jewish population*. New York: United Jewish Communities.

Leff, B. (2007). *Intellectual property: Can you steal it if you can't touch it?* New York: Rabbinical Assembly [HM 203:1.2007].

Levin, M. (1996). *The guide to the Jewish Internet*. San Francisco: No Starch Press.

Prager, Y. (2012). A first look at on-line and blended learning in Jewish day schools. *Contact, 15*(1), 11.

Reisner, A. I. (2001). *Wired to the Kaddosh Barukh Hu: Minyan via internet*. New York: Rabbinical Assembly [OH 55:15 2001].

Reiss Medwed, K. G. (2012). Prayer. In M. Cohen & M. Katz (Eds.), *The observant life: The wisdom of Conservative Judaism for contemporary Jews* (p. 12). New York: The Rabbinical Assembly.

Schwarz, S. (2013). *Jewish megatrends: Charting the course of the American Jewish future.* Woodstock: Jewish Lights Publishing.

Shandler, J. (2009). *Jews, God and videotape: Religion and media in America* (p. 277). New York: NYU Press.

Sheskin, I. M. (2012). *Comparisons of Jewish communities: A compendium of tables and bar charts.* Storrs: Mandell Berman Institute, North American Jewish Data Bank/the Jewish Federations of North America at www.jewishdatabank.org

Sheskin, I. M., & Dashefsky, A. (2011). Jewish population in the United States, 2011. In *Current Jewish Population Reports, Number 4 – 2011.* Storrs: Mandell Berman Institute, North American Jewish Data Bank, The Association for the Social Scientific Study of Jewry/The Jewish Federations of North America.

Warshavski, A. (2012). Islands of value in a rapidly changing world. *Contact, 15*(1), 3.

Chapter 203
Mapping Japanese Religions on the Internet

Danilo Giambra and Erica Baffelli

203.1 Mapping Japanese Religions Online: Understanding the Digital Space

203.1.1 The Japanese Internet: A Localized Variety

The Internet is becoming *de facto* heterogeneous and reflects the multifaceted complexity of the world. Users, through their various cultural identities and the use of different languages make the web complex and diverse. In particular, diversity online becomes paramount in online environments that aim at multicultural and multilingual groups of users, creating highly heterogeneous environments, where social interaction and discussion of general topics is per se a practice of intercultural communication. Web forums, online chat rooms and bulletin boards (BBS) at first, then, more recently, the Social Networks (SNS) are particularly important in this regard, since they often (but not always) go beyond geographical borders, offering their services to users regardless of their physical location.[1] However, highly localized and linguistically homogeneous varieties of the Internet within the digital space represent a reality nowadays, and the adoption of one language over another[2] is

---

[1] Some of the Social Networks and other similar social platforms only allow users residing in a specific country to access their services, or have policies that make it difficult to create accounts from outside a designed country. This is the case, for example, of Mixi.jp.

[2] Usually a language other than the various Englishes (McArthur 1998; Crystal 2001, 2012).

D. Giambra (✉)
Department of Theology and Religion, University of Otago-Te Whare Wānanga o Otāgo, Dunedin, New Zealand/Aotearoa
e-mail: danilo.giambra@otago.ac.nz

E. Baffelli
School of Arts, Languages and Cultures, University of Manchester, Oxford Road, Manchester M1# 9PL, UK
e-mail: erica.baffelli@manchester.ac.uk

© Springer Science+Business Media Dordrecht 2015
S.D. Brunn (ed.), *The Changing World Religion Map*,
DOI 10.1007/978-94-017-9376-6_203

generally the first and most obvious element characterizing these culturally defined environments (Herring 1996; Herring and Danet 2007). The Internet, then, is no longer seen as a shapeless or borderless homogeneous alternative space, where people lose their offline identities to gain the status of indefinite and standardized *users*, nor does it represent a utopian alternative to the offline realities. Users are real people actively involved in the process of *making* the Internet, using available online resources and tools, engaging in interactive online social activities, paying the bills through secured connections, and reading passages of their preferred religious texts on the monitors of their computers and mobile phones, and so on. Users access and bring to the Internet their backgrounds. Also their experiences, their histories, their understanding of the world, their religious beliefs are brought online, invading an ever growing multicultural space.[3] On the other hand, culturally defined spaces are also created along multicultural ones, in a space that doesn't follow the linearity of the physical world, rather adopting a more flexible and permeable structure.

Adopting the terminology from sociolinguistics, we can then describe each culturally defined piece of the Internet as a *variety*. Varieties of the Internet share common traits with each other, in which they all are subelements of a greater technologically defined whole, although they show significant differences at the cultural level. This chapter focuses on the Japanese variety of the Internet–which we call "Japanese Internet"–and more specifically on how Japanese New Religions communicate within this section of the digital space.

203.1.2   Religion and the Internet in Japan

Since the 1990s religious organizations in Japan have demonstrated their ability to adapt to the Internet and to adopt its technologies to their benefit, actively reshaping their identities and their image strategies as a result. For example, Shinto shrines started to build websites, at times allowing their devotees to post prayers and money offers to the shrine directly online, as in the case of Tamō Hachiman shrine (www.netwave.or.jp/~hachiman/), which supported active worship on the Internet, defined in Japanese as *intānetto sanpai* (Reader and Tanabe 1998: 220–222). Similarly, Buddhist temples and Christian churches created their websites and provided email addresses, online religious materials, forums and all sort of online services (Fukamizu 2000). Also the so-called New Religions (*shinshūkyō*) started to engage with the Internet in mid-1990s, opening official websites to promote their activities. However, during these first years of growth of religious internet use in Japan, a dramatic event happened that would influence the development of the relationship

---

[3] Different linguistically defined varieties of the Internet are significantly growing. For instance, the English speaking web represents the biggest portion of all Internet users in the world just by 3 % as of 2011 (27 %), where the Chinese speaking web is following with its 24 %. The Japanese variety of the Internet, coming in the third place, counts 8 % of all Internet users. Source: www.smartling.com/globalweb

between new media and religion (and, more generally, the perception and definition of religion in the Japanese context). On March 1995 a new religious group called Aum shinrikyō[4] released sarin gas in the Tokyo subway, killing thirteen people and injuring thousands. After this tragic event, religious organizations had to reconsider their communicative strategies and their relationship with media. In particular, the Internet was seen as a space difficult to control and many groups (especially new religions) became very cautious in exposing themselves publicly online.

One of the earliest mapping uses of the use of the Internet by religion in Japan was the ARI (Archive for Religious Information) (http://ari.shukyo-gaku.net/) project started in 1998 at the Institute for Japanese Culture and Classics at Kokugakuin University (Tokyo, Japan). The archive included around 6,000 websites from religious institutions and individuals. In 1999 a projected titled "Self-representation and self-understanding of religious communities on the Japanese Internet – the WWW as a source for Japanese Studies" was started in the Japanese Department of the University of Tübingen (Germany) under the supervision of Professor Klaus Antoni, Dr. Birgit Staemmler, and Dr. Petra Kienle. The website created by the research team included an extensive catalogue of "JapanCyberReligion" (www.uni-tuebingen.de/cyberreligion).

Regarding an early analysis on the topic, in 1998 Tamura Takanori[5] (1997) published an overview on the use of the Internet in Japan both by religious organizations and individuals. In his article Tamura focused on the influences of the Internet on religion, and examines different types of religious activities online, such as the use of words for religious practice, the presence and functions of religious symbols and places online, and online counselling. In 1999 Kurosaki Hiroyuki (1999) published a comparative study of Japanese religions on the Internet, focusing on the case of Shinto shrines online. In 2003, Petra Kienle and Birgit Staemmler (2003) published a chapter in the edited volume *Japanese Cyberculters* (Gottlieb and McLelland 2003), comparing Tenrikyō and Jeovah Witnesses' online self-representations (Kienle and Staemmler 2003). In 2007 a few more articles on Japanese religion and the Internet were published on a special issues of the *Journal of Computer Mediated Communication*. Among them Fukamizu Kenshin (Fukamizu 2007) provided an analysis of the presence of traditional Japanese Buddhism on the Internet within the Computer-Mediated Communication framework. Fukamizu's work refers to Shimazono Susumu's analysis of the modernization of religion (Shimazono 2004) and dwells on the idea of a *horizontal communication*, through which users of religious services online would have become more aware of and critical toward religious ideas and beliefs. More recently, Erica Baffelli, Ian Reader and Birgit Staemmler (2011) have published an edited volume investigating several issues related to Japanese religion and the Internet, including discussions related to religious authority. At the University of Tsukuba, Tsujinaka Yutaka and Leslie Ktach-Kawasaki (2011) published a monograph on Japan and the Internet, covering a wide number of topics, including politics, religion and education.

---

[4] On Aum Shinrikyō founded in 1984 by Asahara Shōkō, see Reader (2000).

[5] Japanese names in this chapter are given in the Japanese order, with the surname first.

203.2   Japanese New Religions: A Brief Introduction

The term Japanese New Religions is commonly employed to refer to religious movements founded in Japan from the end of the nineteenth century onwards. In Japanese multiple terms have been used to define these religious groups, including *shinshūkyō* (literally, "new religions"), *shinkōshūkyō* ("new arisen religion," the term has however a pejorative nuance), and *shinshinshūkyō* ("new, new religions"), the latter used to distinguish groups founded from the 1970s from the older ones.[6] In general the 'newness' of these religious movements refers more to revisited ideas, practices and rituals, rather than on the very year of their founding. In fact, some of these movements are more than a 100 years old, as in the case of Tenrikyō, founded in 1838 by Nakayama Miki (Inoue 1996: 216).

These religious movements usually focus on the figure of a charismatic leader, and cannot be considered sects, temples or affiliated branches of established religions (*kisei shūkyō*), namely Shinto shrines and Buddhism sects. Another characteristic shared by many Japanese New Religions is the focus on "this-worldly benefits," or *genze riyaku* (Reader and Tanabe 1998: 1–2), that is on achieving immediate and even material goals during one's lifetime (however, Japanese religion is traditionally concerned with *genze riyaku*, see Reader and Tanabe 1998: 14–15).

Japanese New Religions often combine religious beliefs and rituals from different religious traditions, and in some cases they can be defined syncretistic. Alternatively, some of these religious movements emphasize their special bound with a (self-claimed) pure and an original lost or decayed religious tradition, which presents their religion in more conservative terms.[7]

203.2.1   Presence of Japanese New Religions on the Japanese Internet

Many Japanese New Religions have an official website. However, most of them have only established a marginal presence on the web. Recalling Christopher Helland's dichotomy (2000) between "online religion" and "religion online," these movements remain at the level of 'religion online', creating official websites that reproduce information already available in other media, in particular printed material, and without providing interactive ways for users to join discussion or to interact with the leader.

On the other hand, users have contributed considerably to establishing official and unofficial groups that focus on religious related topics as well as in sharing their

---

[6] An useful discussion on the use of Japanese terminology can be found in Inoue (1994) (2–5) and Astley (2006).

[7] For a broader discussion on Japanese New Religions see Reader (1991, 2005), Mullins et al. (1993), Inoue (1996), Baffelli (2002), Hardacre (2004).

ideas, materials and experiences with other members online. Generally, Japanese New Religions' official websites have been static and they have demonstrated very little innovation over the years, while individuals have been more flexible and made ample use of potentialities offered by the so-called Web 2.0 (Baffelli 2002: 196). Many religious organizations in Japan have not yet decided to invest largely on the Internet. Some New Religions have started exploring this medium, spreading over existing social platforms, but also creating their own online communities, and sometimes their own platforms, software and social applications. Below we will briefly examine some examples of the presence of these religious movements on the Japanese Internet.

203.2.2 Web Pages, Email Services and Personal Blogs

The simplest form of presence for a religious organization online is the official web page. Usually, the website contains basic information on the group, including information on the legal status of the movement.[8]

A short biography of the founder is often included and presented in a separate section, including perhaps an index page. Since new religious movements are usually built around the figure of a charismatic leader, it is important to analyse how much space is devoted to the leader, and whether or not pictures or images are present. In the case of *Kōfuku no Kagaku* (Happy Science),[9] for instance, the complex theology of the movement occupies a whole section on their official website. Stylized images picture the sequence of the previous lives of the leader, Ōkawa Ryūhō, who is believed by his followers to be a living Buddha and the reincarnation of other prominent religious figures, including Gautama Siddhartha. Sometimes a private secured section for members' use might be available. A username and a password will be required to access the private space. Other religious websites chose a more open approach and openly publish on the web the prayers of their members, or share links to videos taken at rituals, digital copies of religious texts, audio files containing preachers or the reading of the sutras, and even religious *manga* and *anime*.

Agonshū,[10] for example, shares publicly online videos to present the movement and to advertise its well-known religious event, the Star Festival (*hoshi matsuri*), through links that reconnect directly to the YouTube video-centred social platform.[11]

---

[8] In accordance to the current Japanese law, religious organizations are legally classified in different ways, according to established criteria. For instance, some organizations are registered locally through their respective prefectures, others are registered directly through the Japanese Ministry of Education, Culture, Sports, Science and Technology.

[9] Kōfuku no Kagaku (lit. Science of Happiness) was founded by Ōkawa Ryūhō in 1986.

[10] Agonshū was founded in 1987 by Kiriyama Seiyū.

[11] www.agon.org/youtube/index_j.html. Agonshū has historically used the new media to maximize the diffusion of its teachings. As an example, it was among the first to use satellite broadcasting of rituals (Reader 1991).

Usually, the physical address of the headquarters is provided in the homepage, followed by a telephone number; for those religious groups who are more active online, an email address is often provided. The email service has been one of the first interactive services available to the religious communities on the Internet. In particular, religious online counselling made ample use of the email service. Tenrikyō, for instance, was one of the first new religious movements in Japan to use it, creating a website called Counselling Room (Kawabata and Tamura 2007: 1011).

Recently, interactive biographies online, commonly called blogs, have blossomed online. Religious charismatic leaders largely benefit from such platforms, since the focus is not just on the movement, but also on the individual. Jōyū Fumihiro's blog (www.joyus.jp/hikarinowa/) named after the founder of the religious movement Hikari no Wa (lit. Circle of the Rainbow), provides an interesting example. This religious organization has been founded in 2007 and emerged as a split movement from Aum Shinrikyō (now renamed Aleph). For Jōyū the Internet, and in particular his blogs and social networking services, has provided a vital tool to spread Hikari no Wa's message and to try to distance the new group from Aum Shinrikyō. Indeed, because of the former association with Aum, Hikari no Wa has very limited access to other media and the Internet has represented a unique opportunity for the group to rebuilt itself and create a new image.

203.2.3 Bulletin Boards, Public Forums and Anonymous Chat Rooms

Beyond static web pages of religious movements, interactive online platforms such as the bulletin boards (BBS) allowed religious users to share their opinions and to engage with group discussions. One of the main characteristics at the basis of this typology of services, however, is that users can stay anonymous and use a nickname. A famous example of BBS in Japan is *Ni-channeru*, a service started in 1999 and still active to present date (www.2ch.net/). This platform is extremely popular and there is virtually no one among the Japanese youth who does not know it (Inoue 2012: 137). This platform has been usually used by numerous *nanashisan* (lit. Mr/Ms nameless), often to criticize new religions and rarely to present a positive image of them.

Although over the last few years, these types of services have been replaced by Social Networks, BBS have represented an important advancement in the history of the Internet, providing highly populated and active virtual rooms where all topics, and especially those perceived as taboos, are freely discussed. Users' identities are in fact "safe" in such environments, reducing considerably the anxiety of revealing personal opinions, expressing harsh criticism or thoughts that might been considered unacceptable in some contexts. On one hand, BBS allowed users to discuss topics such as religion asynchronously. There is no need to be online at the same time, and users can think and create their posts without having to rush ideas.

Religious movements such as Konkōkyō[12] and Tenrikyō, for example, have used these platforms to attract new members seeking for religious answers on online communities (Kawabata and Tamura 2007).

On the other hand, chat rooms are based on synchronous communication. Users must be connected at the same time and run software or open a web page to access the service. Online chats brought online an accessible system to communicate with multiple people from different locations in real time. Chat rooms are very popular even today and the service is now often implemented within private websites, allowing users that visit a web page to communicate in real time with each other. To date, Japanese religious movements' use of chat rooms has been very limited. Nonetheless, it would be interesting to see if this kind of service will be used more in the future, or if it will be implemented directly by search engines such as Yahoo! Japan and Google. co.jp. This development would bring real time communication at the meta-level, opening new online channels for discussing topics such as religion.[13]

203.3   Social Media Platforms and Social Applications

Social Networks represent the ultimate tool for people to join communities online, make new friends, catch up with the old ones, as well as play games, share images and videos, share the music they are listening to etc. Within complex platforms such Mixi.jp or Facebook Japan, users can create personal or group profiles. Furthermore, users can choose the level of privacy they want for their pages and posts, choosing from restricting the access only to few people or to publish publicly without any sort of restriction.

Japanese New Religions started to create profiles on the Social Networks at a different pace, however, it appears that official pages of religious leaders and groups are increasing. Tenrikyō communities on Mixi.jp, for instance, have been created since 2005 (Mixi.jp was started in 2004). Seichō no Ie chose to use Facebook instead, and Taniguchi Masanobu's (the leader) English page was started in December 2010, followed by the Japanese page in September 2011.

Each Social Network presents specific characteristics. Mixi, for instance, is only available in Japanese and has a very strict policy for registering new members (www.mixi.jp, or mixi.jp). Social Networks produce mostly self-centered networks for users where the individual is central, and the acquaintances are all dispersed.

---

[12] Konkōkyō was founded by Kawate Bunjirō in 1859.

[13] Contents on the Internet travel easily over multiple social platforms, passing through official and unofficial digital spaces. Alongside this practice, web search engines are becoming more and more "social," providing registered users with a variety of new services, and might soon implement tools for synchronous communication among users connected to the same url, thus making interactive their indexed internet addresses.

That purposes is evident just by looking at the very meaning of Mixi, whose title was created by putting together the English words "mix" and "I".

In Mixi it is possible to browse and join multiple communities. Many focus on religious matters or are official communities linked to a religious movement. As an example, Mixi hosts many communities related to Tenrikyō,[14] including official ones (created by the group or its members), but also critical ones (that is, anti-Tenrikyō, created by non-members or ex-members). The number of users joining each community is relatively high, the largest group counting 1886 declared members.[15] This community (created in December 2005 and still operating) offers several topics for discussion and is open to a wider public, including both members and non-members of Tenrikyō. Communities on Mixi establish their own rules and some users will moderate discussion, mostly to avoid that ideas disapproved by the group might be shared.[16] The Tenrikyō community in Mixi really seems to work as a second home for current members, but also as an interactive way non-affiliated people, for example, users of the 'Social Space', can get information and share comments, even before they actually attend any lecture or ritual in person.

Also, religious leaders are creating official profiles to share thoughts, small autobiographies, pictures and videos with their followers. Religious events can also be organized on Facebook and shared publicly on the network. The official page of the president of the religious organization Seichō no Ie, Taniguchi Masanobu, is available both in Japanese[17] and in English.[18] It maintains a commenting feature activated on the "wall", which allows other users to interact.[19] Both the Japanese and the English pages are shared publicly on the network, so contents can be traced using a common web search engine. Other interesting pages built by the movement on Facebook are the "Community of Mutual Love" (*Seichō no Ie Sōaikai*)[20] and the community page for the relief after the 2011 Tohoku earthquake and tsunami (started on 9 April 2011).[21] This latter example is particularly interesting as it shows how religious and charitable activities continue, and possibly find a maximized audience, on the Social Networks. In fact, Seichō no Ie used this Facebook page to promote fundraising activities held offline, such as music concerts, and to organize and manage groups of volunteers willing to go in the field to provide help to the

---

[14] As of 15/06/2012, 12 communities counting more than 100 members.

[15] As of 28/11/2011. The number is indicative of joining members only, therefore it does not include lurkers, if any.

[16] When creating a community, it is possible to add a description. Many users also provide a policy for entering and behaving inside the community, sometimes limiting the access to specific typologies of users.

[17] www.facebook.com/pages/成長の家総裁/280856148591794. It is possible to retrieve this page also searching 成長の家総裁 on Google.

[18] http://pt-br.facebook.com/Seichonoie.President. Also this page is searchable on Google.

[19] It must be noticed that comments can be canceled at any time by the administrator(s) of the page.

[20] www.facebook.com/pages/生長の家相愛会父親教室/214545931912279

[21] In Japanese, Seichō no Ie Higashi Nihon Daishinsai Kyūen Komyuniti (Community for the Relief of Eastern Japan Great Tragedy). http://facebook.com/pages/生長の家-東日本大震災救援コミュニティ/181060105275384

victims. As the group founder writes in the descriptive section, users are welcome to post articles and share pictures about the state of safety around the affected area. Also, users can place demands for goods and tools that are needed as well as moral support. Volunteers are also asked to post updated information about their activities. Pictures showing the teams of religious help-providers at work are still publicly available on Facebook. The number of users that appreciated the page and its contents by clicking on the *like* button is not particularly great (200 likes, as of 1 July 2012), although it is impossible to estimate how many users have actually seen the page. The group has now changed the appearance of the page, adopting the new "Facebook timeline look," and changed its privacy settings, restricting public access to many of the contents previously present on the page. Seichō no Ie's initiative to use Facebook as an effective and fast tool for communicating with its members and other users all over the country is just one example of how new media can be exploited effectively by religious organizations.

Some religious movements might not be satisfied with creating profiles or pages on pre-existing SNSs. Seichō no Ie, for instance, has also created its own religious network called *Postingjoy* (http://postingjoy.com) to gather all the members in a closed, religious digital environment. Members can share and comment "joyful" articles and pictures, in so doing making their lives more "cheerful and abundant." Many will also share sentences from the books written by Seichō no Ie reverends, update the community with their achievements in life, as well as providing their understanding of religious readings and asking for the community approval. *Postingjoy* is available in three languages (Japanese, English, and Portuguese), providing linguistically defined spaces where users from the different world regions are more likely to interact and share thoughts and experiences. Seichō no Ie has also developed an Android and i-Phone free social application, making *Postingjoy* directly available through mobile devices.

203.4 Conclusions

This chapter has provided an overview of the presence of Japanese religion on the Internet, citing examples of how religious organizations, and in particular new religious movements, are making use of this medium of communication to:

(a) Disseminate basic information about the movement and its leader (representational function).
(b) Spread religious teachings and materials through links, hypertexts, streaming and downloading of audio and video contents, anime and manga.
(c) Proselytize.
(d) Reinforce the unity and cohesion of the group, by creating new and attractive official digital spaces for the members, where communication is controlled and monitored to a certain extent.
(e) Communicate effectively and in a faster way with followers and sympathizers, serving both internal and external communication.

(f) Co-ordinate activities and enterprises that go beyond the impasse of the online/ offline dichotomy, using the Internet medium as a powerful tool to even manage emergencies at the national level.

(g) Build new presentational personal pages, where the leader's image is reinforced and re-shaped through the direct sharing of thoughts, links, materials, personal activities, but also through the direct interaction with other users in multiple social platforms (comments and direct messages).

Religious organizations in Japan have adopted different policies about establishing an active presence online, and many have chosen not to make use of the more interactive features available through the Internet. As we have shown, however, some groups are expanding their presence over more interactive online environments, such as the Social Networks, at times creating their own platforms and social applications to host their online religious communities.

Establishing a presence on the Internet requires time and money investments. Religious organizations will often employ external companies to develop their websites and their social platforms, and will establish committees to supervise and implement the use of media materials and online services. Multiple platforms are usually used to provide the religious organization with the maximum visibility. However, at present it seems that the materials and the messages posted online are often merely copied and pasted through the different platforms. As an example, it is not uncommon to see Twitter contents reproduced either on Mixi.jp or Facebook Japan. Even internationalized religious organizations tend to replicate their contents and material on the different platforms. However, internationalized religious organizations' localized platforms often show a greater level of interaction, and it is more common to see answers and further comments behind the original posts and materials shared by the organization or its individuals.

As we have seen for Seichō no Ie, it is interesting to note that the movement has built its English Facebook page before the Japanese one. Although Facebook Japan is only recently gaining popularity in the country, Seichō no Ie doesn't own a page on Mixi.jp, and used mainly Twitter (Twitter.com/SEICHO_NO_IE) for communicating updates to its members.

This might be an indicator of a change in the policy of some new religions on the matter of Internet usage, in favor of using more interactive social platforms (SNS) that will work alongside the personal blogs and pages centred around the figure of the religious leader, more sophisticated websites, and the sharing of multimedia resources.

References

Astley, T. (2006). New religions. In P. Swanson & C. Chilson (Eds.), *Nanzan guide to Japanese religions* (pp. 91–114). Honolulu: University of Hawai'i Press.

Baffelli, E. (2002). Il Sacro in Internet. L'esempio Delle Nuove Religioni Giapponesi. *Annali Di Ca' Foscari, 33*, 239–264.

Baffelli, E., Reader, I., & Staemmler, B. (2011). *Japanese religions on the internet: Innovation, representation and authority*. New York: Routledge.

Crystal, D. (2001). *Language and the internet*. Cambridge/New York: Cambridge University Press.

Crystal, D. (2012). *English as a global language*. Cambridge/New York: Cambridge University Press.

Fukamizu, K. (2000). The ability of the website of religion: The case of the temples and churches in Hiroshima. *Religion & Society, 6*, 47–60.

Fukamizu, K. (2007). Internet use among religious followers: Religious postmodernism in Japanese Buddhism. *Journal of Computer-Mediated Communication, 12*(3). Retrieved May 16, 2012, from http://jcmc.indiana.edu/vol12/issue3/fukamizu.html

Gottlieb, N., & McLelland, M. J. (2003). *Japanese cybercultures*. London: Routledge.

Hardacre, H. (2004). Religion and civil society in contemporary Japan. *Japanese Journal of Religious Studies, 31*(2), 389–415.

Helland, C. (2000). Online-religion/religion-online and virtual communitas. *Religion and the Social Order, 8*, 205–224.

Herring, S. C. (1996). *Computer-mediated communication: Linguistic, social and cross-cultural perspectives*. Amsterdam/Philadelphia: J. Benjamins.

Herring, S. C., & Danet, B. (2007). *The Multilingual internet: Language, culture, and communication online*. New York: Oxford University Press.

Inoue, N. (1994). *Shin Shūkyō Jiten: Honbun Hen. Shukusatsuban shohan*. Tōkyō: Kōbundō.

Inoue, N. (1996). *Shinshūkyō kyodan, jinbutsu jiten*. Tokyo: Kobundo.

Inoue, N. (2012). Media and new religious movements in Japan. *Journal of Religion in Japan, 1*(2), 121–141.

Kawabata, A., & Tamura, T. (2007). Online-religion in Japan: Websites and religious counseling from a comparative cross-cultural perspective. *Journal of Computer-Mediated Communication, 12*(3), 999–1019.

Kienle, P., & Staemmler, B. (2003). Self-representation of two new religions on the Japanese internet. Jehovah's Witnesses and Seicho No Ie. In N. Gottlieb & M. J. McLelland (Eds.), *Japanese cybercultures* (pp. 222–234). London/New York: Routledge.

Kurosaki, H. (1999). Toward the comparative study of internet usage by Japanese religions – Shinto Shrine Websites. *Transaction of the Institute for Japanese Culture and Classics, 83*, 35–48.

McArthur, T. B. (1998). *The English languages*. Cambridge: Cambridge University Press.

Mullins, M., Shimazono, S., & Swanson, P. L. (1993). *Religion and society in modern Japan: Selected readings*. Berkeley: Asian Humanities Press.

Reader, I. (1991). *Religion in contemporary Japan*. Honolulu: University of Hawaii Press.

Reader, I. (2000). *Religious violence in contemporary Japan: The case of Aum Shinrikyō*. Honolulu: University of Hawai'i Press.

Reader, I. (2005). Perspective chronologies, commonalities and alternative status in Japanese New Religious Movements Defining NRMs Outside the Western Cul-de-sac. *Nova Religio: The Journal of Alternative and Emergent Religions, 9*(2), 84–96.

Reader, I., & Tanabe, G. J. (1998). *Practically religious: Worldly benefits and the common religion of Japan*. Honolulu: University of Hawai'i Press.

Shimazono, S. (2004). *From salvation to spirituality: Popular religions movements in modern Japan* (Japanese Society Series). Melbourne: Trans Pacific Press.

Tamura, T. (1997). Information of religion on the internet: possibilities and dangers. *Religion & Society, 3*, 119–136

Tsujinaka, Y., & Tkach-Kawasaki, L. (Eds.). (2011). *Japan and the internet: Perspectives and practices* (Vol. 1, CAJS monograph). Center for International, Comparative, and Advanced Japanese Studies: University of Tsukuba. http://cajs.tsukuba.ac.jp/monograph/

Chapter 204
Virtual Buddhism: Online Communities, Sacred Places and Objects

Louise Connelly

204.1 Introduction

Until recently, there has been a dearth of research examining Buddhism online, however, in the last few years a number of studies have emerged. This chapter provides an insight into Buddhism on the Internet and directly contributes to our understanding of the relationship between media, religion and culture. By examining Buddhism online it helps us to identify the position of Buddhism in society, the possible implications for the Buddhist religion both online and offline, as well as how people engage and communicate in a place (cyberspace) that is not constrained by geographic boundaries. The terms "cyberspace" and "internet" will be used interchangeably, with the understanding that the definition includes a space, a location, a place to be, as well as the actual technology and the platform in which information is shared, therefore, "despite its lack of physicality, cyberspace is a real place. I *am there* – whatever this statement may ultimately turn out to mean" (Wertheim 1999: 229). If cyberspace is a "real place," can we assume that the social and religious constructs of the offline world apply online or does technology alter these? By focusing on these kinds of problems we can begin to understand the relationship between media, religion and culture.

One of the earliest examinations of Buddhism online is by Charles Prebish (2004) in *The Cybersangha: Buddhism on the Internet* where he provides a descriptive overview of online journals, communities and discussion forums such as Buddha-L. Since then, research of virtual Buddhism has included an exploration of Buddhist communities and rituals within the online world of *Second Life* (Grieve 2010; Connelly 2010, 2012), Buddhist blogs (Lee 2009) and communities and identity online (Kim 2005; Ostrowski 2006; Busch 2010). In addition, there has been a

L. Connelly (✉)
Institute for Academic Development, University of Edinburgh,
7 Bristo Square, Edinburgh EH89 AL, Scotland, UK
e-mail: louise.connelly@ed.ac.uk

© Springer Science+Business Media Dordrecht 2015
S.D. Brunn (ed.), *The Changing World Religion Map*,
DOI 10.1007/978-94-017-9376-6_204

focus on Buddhism on the internet at the *American Academy of Religion* conference (2011), the *Digital Dharma* symposium in California (2011), and on the *Network for New Media, Religion and Digital Culture Studies* website (http://digitalreligion. tamu.edu/). Discussions have focused on a wide range of topics, including the use of new media; the need to develop methodologies and typologies for the examination of Buddhism online; virtual material culture and specific Buddhist communities and identity online. There has also been the suggestion that within the field of media, religion and culture, there is a subfield emerging – digital religion, which can be defined as the space where technology and culture meets and provides a bridge between the offline and online world (Campbell 2012). The study of Buddhism on the internet and the introduction of new theoretical frameworks, highlight that there is a growing trend of using the internet to engage with religion.

To examine the nuances of a complex online culture requires robust methodologies that draw from different disciplines (Dawson and Cowan 2004; Campbell 2010), including, but not limited to, religious studies (Wagner 2012), media studies (Campbell 2012), material religion (Morgan 2010) and psychology (Turkle 2011; Campbell and Connelly 2012). There are four prominent themes which emerge, these are authority, community, identity and ritual (Dawson and Cowan 2004; Campbell 2010, 2012; Wagner 2012). This chapter will primarily focus on the discussion of Buddhist communities and ritual online while acknowledging issues relating to Buddhist identity and authority. The term "virtual Buddhism" has been used as an umbrella term to incorporate the complexities related to these themes. What will be demonstrated is the transition of these themes from the offline to the online world and vice-versa, for example, the facilitation of Buddhist meditation ritual within *Second Life* which might simultaneously be practiced offline. The examination of "virtual Buddhism" will be illustrated by examples of virtual places, virtual ritual and religious objects found in the online world of *Second Life*, as well as how new media, such as blogging and mobile applications, are being used by Buddhists and non-Buddhists alike to engage with Buddhism both online and offline. This chapter will introduce the reader to possibilities and the impact that online Buddhism may have for traditional forms of Buddhism. It is not my intention to draw definitive conclusions; instead, this discourse provides a starting point for further discussion, as there remains "substantive insights" to be gleaned in relation to both Asian religions and "Buddhist communities' use of new media" (Campbell 2010: 190–191).

204.2   Buddhist Communities and Sacred Places

The Buddhist community online has been defined by Prebish as either webpages providing information for members or "virtual temples created by traditional *sanghas* as an *addition* to their programs" or purely online communities that do not have a presence offline (Prebish 2004: 145). The different types of communities explored in this chapter will highlight that online Buddhist communities are complex and may not easily fit into one of these categories.

It is common for an online Buddhist community to be referred to as a *sangha* or a cybersangha (Prebish 2004; Busch 2010) and this may include communities found on websites, blogs, micro-blogs (for example, Twitter), virtual worlds (for example, *Second Life*) and online forums. One of the earliest Buddhist communities online was found via the discussion list. Many of the online lists originated in the 1990s, including the academic lists of Buddha-L (created by Richard Hayes at McGill University), ZenBuddhism-L (founded August 1993 by Matthew Ciolek, Australian National University), and Tibet-L (founded 1994) (Prebish 2004: 137–138). Prebish maintains that many of the original online Buddhist communities, including websites and lists are no longer in existence as the personal face-to-face contact and "shared practice in real space" is in direct contrast to what the cybersangha can offer (Prebish 2004: 147). However, although this may be true for some communities, others maintain a purely online presence, such as those related to blogs.

204.2.1   Buddhist Communities – America

A number of recent studies provide descriptive accounts of online Buddhist communities and question why people choose to engage with a community online rather than offline. Ostrowski's examination of Buddhism in America focuses on the demographics of users, the type of information being accessed, and the perception of a Buddhist community online. She concludes that the people accessing Buddhist sources online are primarily "white (72 %), have been raised as Catholics (27 %) or Protestants (26.1 %), and are not members of a Buddhist temple or meditation centre (74.5 %)" (2006: 97). Moreover, the online Buddhist community has developed due to a lack of access to offline Buddhist communities as 32.6 % of surfers accessed the internet for information on Buddhism due to the absence of opportunities for engagement with offline temples or teachers. Ostrowski argues that the growth of online Buddhism may change as more people engage with the cybersangha and, consequently the online community may eventually dissipate the "ethnic and cultural differences" and challenge traditional forms of Buddhism (2006: 102).

204.2.2   Buddhist Communities – Korea

Other studies of online Buddhist communities include Kim's analysis in Korea which focuses on the Chollian Buddhist Community (C-BUD) between 1996 and 1997. C-BUD is a web space providing sermons, meditation, chat rooms, debates and information about offline Buddhist activities. Kim claims that the success of the site has resulted in the community "develop[ing] into an alternative religious organization that satisfies the multi-pattern needs of contemporary individuals." This appears especially true of city-dwellers, who may not be able to visit Buddhist temples, which are, in many cases, still situated in rural areas (2005: 147). Kim, like

Ostrowski highlights that people engage with online Buddhist communities due to a lack of access to offline communities, but again this may not be true of all communities. Certainly, the internet has made it easier to access information and become part of a community without geographic boundaries, especially where access to a community offline is not possible. However this has possible implications for maintaining traditional forms of Buddhism which has led to the emergence of new forms of Buddhism or "virtual Buddhism," such as those found in *Second Life*.

204.2.3   Buddhist Communities and Sacred Places in Second Life

These new communities on the internet raise the question as to whether or not the online activity provides an authentic form of Buddhism or something new and different. Busch states that the online Buddhist sangha "transcends boundaries between the variety of Buddhist traditions, sects and their distinctive cultural differences" (2010: 62). This understanding of Buddhism is affirmed by Delani Gabardini (in-world name), the co-founder of The Buddha Center in the virtual world of *Second Life* (http://maps. secondlife.com/secondlife/Buddhism/138/130/21/), as she describes the eclectic type of Buddhism found at the Center as "universal Buddhism" (Connelly 2010: 15). If we look more closely at *Second Life*, we might begin to understand the grounding for such claims.

The *Buddha Center* was founded in 2008 and can be said to be one of the largest Buddhist communities in *Second Life,* with more than 2,500 participants (Connelly 2010: 14). It is host to a large Buddhist temple, meeting places, a library, shops selling Buddhist objects and regular activities, such as dharma talks, meditation, and other events. This is not the only Buddhist community in *Second Life* and others include *Kannonji* (http://slurl.com/secondlife/SnowlionMountain/207/97/22) and the *Upaya Mountain* Zen communities studied by Grieve (2010). Some of these communities are clearly associated with a specific sect of Buddhism, whereas others, such as the *Buddhist* Center represent different Schools of Buddhism, such as Tibetan, Zen and Theravada. In addition, there are often key similarities between the different Buddhist communities in *Second Life*, such as the design of the environment which includes elaborate virtual temples and objects, as well as the type of religious practices offered, such as chanting or silent meditation (Fig. 204.1).

The Zen Buddhist community, which Grieve calls the *Upaya Mountain Zen Retreat,* practices silent online Zen meditation (*zazen*) by being immersed in the 3D world of *Second Life*. He defines "immersion" as "being there" or "having presence" and recognizes that the notion of virtual embodiment enables participation within a community and integration within the online culture. A noticeable feature of integration is through the creation of an avatar which can be dressed in Buddhist religious attire (Connelly 2010; Grieve 2010). This feature particularly applies to the facilitators at *The Buddha Centre* and *Upaya Mountain Zen Retreat*, as they will often wear monastic robes and mala meditation beads and on occasion have been

Fig. 204.1 An avatar meditating within the Buddha Center, Second Life (Image from http://secondlife.com, used with permission)

seen to carry a handheld prayer wheel. The depiction of cultural and religious dress and artefacts aim to recreate an intentionally "authentic" form of Buddhism and Buddhist identity to those who engage with this community. This results in the participant being presented with a facilitator who is perceived as authentic and therefore, the practice (in this case, Zen meditation) is thereby deemed approximate to what might be found offline. In addition to the facilitators dress, many of the avatars found at the different Buddhist locations will often have a label identifying the name of the Buddhist group showing in addition to their own name (above their head). This intentional identification and association with particular Buddhist communities in *Second* Life further highlights the complexity and nuances of the online Buddhist culture (Connelly 2012).

In addition to the communities in *Second Life*, there are a number of virtual sacred places, such as the various Buddhist temples and the Deer Park located in the *Buddha Center* in *Second Life* (http://slurl.com/secondlife/Buddhism/138/131/22/). The use of the term 'sacred' is often contested and for some this is even more so in the context of virtual worlds (Wagner 2012: 79). Wagner concludes that there is a "virtual sacred" which is different to the ordinary application of the "sacred" offline. The virtual sacred is defined as "here, now, rich, obvious, visual, and constantly in flux" (2012: 79). To illustrate how this might apply, we can consider the Deer Park at the *Buddha Center*. The intentional naming of the location deliberately aligns with the offline location, where the Buddha gave his first sermon. Although the design of the online Deer Park has changed on more than one occasion, the elaborate visual environment includes large stones for sitting on, which are placed around a fire, a large Buddha statue, trees and deer. The activities which take place include talks and discussion sessions. In fact, the founder of the *Buddha Center* (Zino March, in-world name) states that he "intentionally creates the environment to be

Fig. 204.2 Main temple, Buddha Center, Second Life (Image from http://secondlife.com, used with permission)

similar to that found in real life" (Connelly 2010: 21). Defining what makes something sacred is challenging, however, the intentionality in design to replicate offline sacred places indicates that there is an expectation that locations such as the Deer Park in *Second Life* should be categorised as sacred, or at the very least "virtual sacred" (Fig. 204.2).

204.2.4 Buddhist Communities in Blogs

Second Life provides a highly visual and interactive form of Buddhism, but communities also exist in other areas of cyberspace such as blogs, social media sites (such as Facebook) and online forums. Blogs have been extremely popular and have been maintained by both Buddhists and non-Buddhists alike. Within the blogosphere (community of blogs) there are a large number of Buddhist blogs which provide information, personal insights into different aspects of Buddhism, or even the bloggers personal belief structure. Rettberg argues that "Blogs are a social genre. Bloggers don't simply write their "Dear Diary," they write into the world with the clear expectation of having readers. That readership does not necessarily need to be very large" and so blogs are similar to a personal diary but are also performative and public (Rettberg 2008: 57). A particular blogs readership can be indicated by certain features, such as readers leaving comments or the presentation of a blogroll (list of other blogs) which can result in reciprocal links from other bloggers, such as that found on *The Buddhist Blog* (http://thebuddhistblog.blogspot.co.uk/). The blogroll is not unique to this blog, as "41 % of bloggers say that they have a blogroll or friends list on their blog" (Lenhart and Fox 2006: 7). It can also be argued that "Readership may also be suggested – though not necessarily guaranteed – by linking from one blog to another on a blogroll" (Lenhart and Fox 2006: 34), consequently creating a Buddhist community online through associated blogs and the use of features such as comments or the blogroll.

The Buddhist blog community also has a unique award which is known as the Blogisattva Award (www.blogisattva.org/). This acknowledgement is given to English language Buddhist blogs and includes 22 nomination categories. Examples of blogs who have won, include The *American Buddhist* blog (www.patheos.com/blogs/americanbuddhist/) in the category of "Best Philosophical Blog" (2006) and "Philosophy and Psychology" blog (2008); *The Buddhist Blog* (http://thebuddhist-blog.blogspot.co.uk/) which won in the category for 'Best Achievement for Use of Quotations within a Blog' (2007); and *BuddhistGeeks* (www.buddhistgeeks.com/) won in the category of "Best Achievements in Wide Ranging Topic Interests" (2010).

Notably, many of the blogs mentioned also have a Facebook page or Twitter account and so expand their community through the use of other online platforms. Arguably, this might be seen as one of the most prolific changes in computer mediated communication (CMC) as there now exists integrated communication between different platforms, potentially enabling a wider global community to develop. However, it is not only individuals who build and maintain communities, but traditional offline Buddhist groups such as the Kagyu Samye Ling Tibetan Buddhist Centre in Scotland (www.samyeling.org/) use Twitter, Facebook and a website to share information, events, scriptures and much more. In a similar way to individuals, traditional Buddhist groups are using new media to engage with and build communities online.

204.2.5   Blogs as a Means of Religious Practice

Blogging can be used to build communities, but they can also be used for religious practice. Lee, who was once a Won Buddhist monk, analyses how the Won Buddhist tradition uses new media (blogs) as a means of religious practice. He concludes that the use of blogs is a form of religious practice, as the practice of finding "the self" is partially undertaken by maintaining a blog; "practicing spirituality habituates spiritual or religious practices such as prayer, meditation, chanting or keeping a diary, timelessly and placelessly in a way to cultivate the self" (2009: 99). He claims that blogging enables the monks and nuns to effectively "deterritorialize the intensities of the self" and in doing so enabling them to gain insight into the Buddhist philosophical context of not-self (Lee 2009: 100). Not-self (*anatta*), impermanence (*anicca*) and unsatifactoriness or suffering (*dukkha*) are the three central principles of Buddhism. Therefore, using blogs for religious practice enables the Won Buddhist monks and nuns to gain a greater understanding of Buddhist philosophy as well as writing about their own spiritual journey. Blogging has also enabled the formation of a community which engages both Buddhists and non-Buddhists alike, as the

"Won Buddhist monks and nuns have successfully adopted this new form of communication to foster friendly relationships with existing and potential believers" (Lee 2009: 112). The use of blogs as a means of religious practice indicates how new media can be used and embraced by traditional forms of Buddhism for personal development and to create and maintain a Buddhist community.

204.2.6   Buddhist Communities and Twitter

While blogging remains a popular medium for spreading information about Buddhism and creating a Buddhist community within the wider blogosphere, there are also a number of Buddhists using micro-blogging, such as Twitter (launched in 2006) to tweet their news and views in 140 characters or less. There are examples of organisations, non-Buddhists and ordained Buddhists using Twitter. These include the Dalia Lama who has 4,023,376 followers (@DalaiLama on 14 April 2012), Arjia Rinpoche with 2,448 followers (@TMBCC on 20 April 2012) and the Vietnamese Buddhist monk Thich Nhat Hanh with 58,102 followers (@thichnhathanh on 20 April 2012). Arjia Rinpoche claims to be "preserving Mongolian and Tibetan Buddhist culture and traditions" (taken from his Twitter profile) while using Twitter as a means to engage others and promote his beliefs. On his Twitter profile he provides a link to the offline Buddhist community of the Tibetan Mongolian Buddhist Cultural Center which is located in Bloomington, Indiana (http://tmbcc. net), thus creating a bridge between the offline and online world and highlighting how traditional forms of Buddhism are engaging with new media.

Buddhist communities online are created and maintained by individuals or organizations who blog, tweet or facilitate in online worlds. Their reasons for doing so may include curiosity, to participate in religious practice or to promote traditional forms of Buddhism commonly found offline. Individuals may join Buddhist communities online for different reasons, some of which have been outlined by Kim (2005) and Ostrowski (2006). The nuances found in online communities are complex and require further exploration to understand how technology is being used to create, develop and maintain communities.

204.3   Ritual and Virtual Objects

Ritual and the use of objects are important components of Buddhism since they provide the follower with an aid to spiritual awareness in order to ultimately obtain liberation from *samsara* (the cycle of rebirth). Wagner argues that "both religion and virtual reality can be said to be deeply concerned with some sense of the "other." That is, both are concerned with a mode of being that lies beyond our ordinary day-to-day experience" (2012: 2). Rituals can assist with realizing and engaging with this "other." Online, there are a number of Buddhist rituals which include the use of virtual objects which, for some, may be used to point to a sense of "other." This section will provide examples to illustrate how ritual can be negotiated between online and offline spaces and how virtual objects are engaged with online in order to provide a more complete experience and to replicate the components of offline rituals.

204.3.1   The Prayer Wheel

Blogs and social media sites often depict recognizable Buddhist artefacts, such as the Tibetan prayer wheel, images of Buddha statues or the eight spoke wheel of Buddhism (also known as the wheel of dharma). The relationship between virtual objects and ritual is important, as "materiality *mediates* belief…material objects and practices both enable it and enact it" (Morgan 2010: 12). Consequently this enables people to make sense of the world in which they live and the virtual space in which they are connecting with. If we consider the Tibetan prayer wheel in different contexts, such as blogs, mobile applications and in *Second Life*, it can help us to distinguish whether they are similar or dissimilar to traditional offline prayer wheels and consequently, identifying if they have the same purpose online as offline.

The prayer wheel, found offline, is usually touched and spun by hand. The intended purpose is to spin the prayer wheel and release the prayer into the environment which in turn results in merit for the person who has spun the prayer wheel. Merit is central to the Buddhist understanding of salvation and liberation, as the accumulation of merit "is connected to a persons' ethical and moral living and the aspiration for a better rebirth and the eventual release from the cyclic existence (*samsara*)" (Connelly 2010: 17). It is questionable whether the same purpose and outcome can be associated with a virtual artefact.

The myriad of digital prayer wheels for both the mobile phone and the internet highlights the significance of this image as a culturally recognizable symbol. The online prayer wheel often continually spins, without interaction. Examples of these can be found on the webpages of the *World Oneness Center* (www.worldoneness-center.com/index.html) and the *Digital Tibetan Buddhist Alter* blog (http://tibetan-altar.blogspot.co.uk/2009/03/prayer-wheel-design-and-engineering.html).

Prayer wheels are also found contained inside virtual *stupas* or as handheld virtual artefacts within *Second Life*. Like the offline world, many of the prayer wheels require the avatar to touch it, in order for it to spin. For some, such as Zino March, the founder of *The Buddha Center* and the creator of many of the objects, the virtual prayer wheel results in the same meritorious action as those engaged with offline "as long as it is done with the same intention" (Connelly 2010: 18). Therefore, the use of virtual artefacts or participation in ritual online can be said to be similar to that found offline, as long as there is the same intention (Fig. 204.3).

The prayer wheel can also be downloaded from the internet to a smart phone and these can be positioned within a number of categories from Wagner's typology of religious mobile apps, which include (i) prayer apps, (ii) ritual apps, (iii) sacred text apps, (iv) social media apps, (v) self-expression apps, and (vi) focusing/meditation

Fig. 204.3 Virtual prayer wheels in Second Life (Image from http://secondlife.com, used with permission)

apps (2012: 102–105). Examples of the prayer wheel mobile app can be found at *Prayer Wheel* (http://prayerwheel.net/), or Google Play (https://play.google.com/store/apps/details?id=com.lunesu.prayerwheel&hl=en), or on iTunes, *My Prayer Wheel* (http://itunes.apple.com/us/app/my-prayer-wheel/id468643041?mt=8). The use of mobile apps, also indicates the growing trend in mobile technology which is likely to increase in the future, as 35 % of American adults own a smartphone and 87 % use this to access the internet or email (Smith 2011: 1). The use of mobile technology to engage with Buddhism highlights a more individualistic approach to religion than the communities found in *Second Life* or blogs. This distinction is important when considering the relationship between technology and religion, as well as the possible future for the Buddhist religion worldwide.

204.3.2   *Meditation in Second life*

In *Second Life* there are a number of rituals and practices some of which are similar to those found offline, such as the act of prostrating performed when entering a temple or participating in silent meditation. The act of prostrating is carried out by the avatar and "is done three times and so reflects the devotion to the "three refuges" of the *Dhamma* (truth), the Buddha and the *sangha* (community) as well as being a mark of respect (Connelly 2010: 24). Another example of ritual is the meditation ritual at *the Buddha Center* which can be simultaneously practiced offline. The meditation session usually begins with the avatar sitting cross-legged on 1 of the 22 meditation cushions which are positioned in front of a large Buddha statue and which surround a pond. Similar to real-world temples, there is the presence of (virtual) incense and flowers or plants. Many of the virtual objects compensate for senses which are absent, such as touch, smell and taste and, therefore, imply the associated sensory experience (Connelly 2010). It can be argued that the visual component is heightened and emphasised, as "images get our attention and maintain a larger portion of it because our memories and feelings are intermingled with the brain's sensation of sensory stimuli" (Morgan 2005: 39). The virtual objects are used by the facilitator who sits beside a singing bowl or gong and this is used to begin and end the meditation session, thus providing continuity between offline and online styles of meditation (Connelly 2012). The authenticity of the meditation is further maintained as some of the facilitators at the *Buddha Center* are ordained Zen priests, monks or Tibetan lamas in the offline world.

The meditation session at *The Buddha Center* may be delivered in voice (by the facilitator) or through text. Participants are often issued with virtual note cards that contain scriptures, Zen koans or other information to enable them to participate fully and gain a richer experience. The participants are led in the session by the facilitator and while some may only participate in virtual meditation, many participants will simultaneously meditate offline. The reason that individuals engage with virtual meditation varies, but some have stated that because they do not have access to Buddhist meditation offline the facilitation via a virtual environment, enables them to be part of a community and to easily practice Buddhist meditation (Connelly 2012), thus corroborating some of the reasons put forth by Kim (2005) and Ostrowski (2006).

Furthermore, this particular virtual environment has many features associated with the offline environment, such as offline temples and artefacts. This environment could be said to provide a sense of authenticity as "Religious seeing is set within a series of ritual rules that *frame* the experience for the viewer providing boundaries for *what* is seen and *how* objects are seen" (Plate 2002: 162). Therefore, the use of objects, whether in blogs, websites, mobile apps or *Second Life*, provide culturally recognizable objects that are, for some, perceived to have the same meritorious outcome and purpose as those found offline. In addition, some online rituals are not purely enacted online, but extend into the offline world and enable individuals who cannot access Buddhist practices offline to engage with "virtual Buddhism."

204.4   Impact of New Media on Buddhism

When discussing the impact of new media on Buddhism, there are three areas which require further consideration. These include the way in which religion on the internet might reflect larger societal trends; the use of technology for spiritual practices; and the complexity of authority and authoritativeness in relation to online-offline religious practices. Helland (2005) argues that there is no longer a clear demarcation between online religion (informational sites) and religion online (participatory sites), since the "Internet constitutes the space where individuals and groups live out their social and spiritual lives, and offline boundaries and relations often inform the online sphere" (Campbell and Lovheim 2011: 1).

The blurring of boundaries between offline and online environments has highlighted issues of authenticity and authority for religious groups and individuals. An example of this is found in the E-Sangha community. Busch argues that "the content and physical structures of E-sangha, which are controlled by the forum's own selected authorities (the forum's creator and moderators), dictate to the user who is considered a true Buddhist." This implies that "the Web actors that manage these spiritualizing processes [are] able to establish themselves as ideological authorities outside traditional institutional structures that prescribe religious legitimacy to specific individuals" (Busch 2010: 59). Therefore, such online communities are controlled by individuals who may deviate from the "traditional" forms of Buddhism and ideologies. On the other hand, there are online communities who try to maintain authenticity and authority in cyberspace by ensuring that facilitators or teachers are recognized as such offline. An example of this is the ordained Zen Buddhist priests who facilitate Buddhist practice and teachings in *Second Life* (Connelly 2010, 2012). The issue of authenticity has also been highlighted in relation to mobile apps by Wagner, who maintains "that we can learn something about our streaming selves by looking at how we use, display, and interact with religion via our mobile devices" (2012: 105). The religious experiences and rituals are no longer controlled and facilitated by "recognised religious authorities" but by anyone who has access to the internet or a smart mobile phone (2012: 105).

The use of new media may have an impact on behaviors, attitudes and traditional forms of Buddhism, as more individuals and organisations across the world use the internet for practicing ritual, building communities or creating sacred spaces online. While some may argue that religion is on the decline, the use of new media and technology enables experiences, which for some at least, may be deemed "religious." Platforms such as virtual worlds may offer a type of "proxy religion" for those who engage with Buddhism on the internet (Wagner 2012: 7) and mobile apps such as *iShrine Virtual Buddhist Shrine* (http://download.cnet.com/iShrine-Virtual-Buddhist-Shrine/3000-2135_4-10972367.html) enable a person to easily engage with Buddhism through individual practice. Both individuals and organisations are using the internet to engage with Buddhism simultaneously online and offline. The possible change in religious practice, sacred places and community engagement highlights how technology is impacting on different aspects of our day to day lives and religious experiences, as more and more of us become part of the "always on" society (Turkle 2011).

204.5   Conclusion

Online Buddhist communities, sacred places and virtual objects are found on websites, discussion lists, social media (e.g. blogs, Twitter, Facebook) and in virtual worlds (e.g. *Second Life*). For some, the online Buddhist community and engagement with religious objects and practice is contained to purely online, for others it negotiates between the offline and online world. In addition, the changes in technology have also resulted in a rise in Buddhist apps for mobile technologies (e.g. *iShrine*) and this has resulted in a type of Buddhism which is individualistic and available at any time of the day.

The changes in how we use the internet and what we use it for has raised concerns relating to the type of Buddhist communities which are developing, the authenticity of the information being presented and consequently, whether traditional Buddhism is being threatened. Is online Buddhism spawning a new type of Buddhism, a "Virtual Buddhism?" Understanding digital religion requires us, "to consider not only the tradition a religious community comes from but also the particular characteristics and lived practice" of a group in relation to how they use technology (Campbell 2010: 20). Moreover, the exponential growth of the internet impacts not only on the Buddhist religion, but also society at large and this has been highlighted by scholars in a number of disciplines, including media and communication studies (Busch 2010; Campbell 2012), religious studies (Wagner 2012) and psychology (Turkle 2011; Campbell and Connelly 2012).

Furthermore, the increase in using the internet to engage with religion is examined by Hoover, Clark and Rainie who claim that 64 % of Americans use the internet for religious purposes (2004: i). For others such as Turkle, she presents a cautiously optimistic view of the internet, emphasising that online there are both "places to be yourself," such as blogs, as well as "places where one constructs an avatar – from games to virtual communities – where people go to find themselves, or to lose themselves, or to explore aspects of themselves" (Turkle 2011: 209), such as the *Buddha Center* in *Second Life*.

In order to monitor changes to the Buddhist religion both online and offline, it is necessary for researchers to continually adapt methodologies and create typologies to map the Buddhist cyberspace (Dawson and Cowan 2004; Grieve and Veidlinger, forthcoming). Consequently, this will help to identify how the internet is being used, for what and by whom. Further exploration is needed in order to fully appreciate what "virtual Buddhism" is and the implications this has for behaviors, attitudes, the Buddhist religion and the wider society.

References

Busch, L. (2010). To come to a correct understanding of Buddhism: A case study on spiritualizing technology, religious authority, and the boundaries of orthodoxy and identity in a Buddhist web forum. *New Media and Society, 13*(1), 58–74.

Campbell, H. (2010). *When religion meets new media.* Oxon: Routledge.

Campbell, H. (2012). *Digital religion: Understanding religious practice in new media worlds.* New York/London: Routledge.

Campbell, H., & Connelly, L. (2012). Cyber behavior and religious practice on the internet. In Z. Yeng (Ed.), *Encyclopedia of cyber behavior* (pp. 434–445). Hershey: IGI Global.

Campbell, H., & Lovheim, M. (2011). Rethinking the online-offline connection in religion online. *Information Communication & Society, 18*(4), 1083–1096.

Connelly, L. (2010). Virtual Buddhism: An analysis of aesthetics in relation to religious practice within Second Life. *Heidelberg Journal of Religions on the Internet, 4*(1), 12–34.

Connelly, L. (2012). Virtual Buddhism: Buddhist ritual in Second Life. In H. Campbell (Ed.), *Digital religion: Understanding religious practice in new media worlds* (pp. 128–135). New York/London: Routledge.

Dawson, L. L., & Cowan, D. (Eds.). (2004). *Religion Online: Finding Faith on the Internet*. New York/London: Routledge.

Grieve, G. (2010). Virtually embodying the field: Silent online meditation, immersion, and the Cardean ethnographic method. *Heidelberg Journal of Religions on the Internet, 4*(1), 35–62.

Grieve, G., & Veidlinger, D. (Eds.). (forthcoming). *The pixel in the lotus: Buddhism, the internet and digital media*. New York: London, Routledge.

Helland, C. (2005). Online religion as lived religion: Methodological issues in the study of religious participation on the internet. *Heidelberg Journal of Religions on the Internet, 1*(1), 1–16.

Hoover, S., Clark, L., & Rainie, L. (2004). Faith online. *PEW Internet and American Life Project*. Retrieved March 2, 2012, from www.pewinternet.org/Reports/2004/Faith-Online.aspx

Kim, M.-C. (2005). Online Buddhist community: An alternative religious organization in the information age. In M. T. Hojsgaard & M. Warburg (Eds.), *Religion and cyberspace* (pp. 138–148). Oxon: Routledge.

Lee, J. (2009). Cultivating the self in cyberspace: The use of personal blogs among Buddhist priests. *Journal of Media and Religion, 8*, 97–114.

Lenhart, A., & Fox, S. (2006). *Bloggers*. Washington, DC: PEW Internet & American Life Project.

Morgan, D. (2005). *The sacred gaze: Religious visual culture in theory and practice*. Berkeley/London: University of California Press.

Morgan, D. (Ed.). (2010). *Religion and material culture: A matter of belief*. Oxon: Routledge.

Ostrowski, A. (2006). Buddha browsing: American Buddhism and the internet. *Contemporary Buddhism, 7*, 91–103.

Plate, S. B. (2002). *Religion, art, and visual culture: A cross-cultural reader*. New York/Basingstoke: Palgrave.

Prebish, C. S. (2004). The Cybersangha: Buddhism on the internet. In L. L. Dawson & D. Cowan (Eds.), *Religion online: Finding faith on the internet* (pp. 135–147). New York/London: Routledge.

Rettberg, J. W. (2008). *Blogging*. Cambridge/Malden: Polity Press.

Smith, A. (2011). *35 % of American adults own a smartphone*. Retrieved March 2, 2012, from www.pewinternet.org/~/media/Files/Reports/2011/PIP_Smartphones.pdf

Turkle, S. (2011). *Alone together: Why we expect more from technology and less from each other*. New York: Basic Books.

Wagner, R. (2012). *Godwired: Religion, ritual and virtual reality*. New York: Routledge.

Wertheim, M. (1999). *The pearly gates of cyberspace: A history of space from Dante to the internet*. London: Virago.

Chapter 205
German-Based Cyber-*Daʿwah* 2.0: Back to the Roots with Forward Technology

Erik Munder

205.1 Introduction

The term *Daʿwah* (Wrogemann 2006: 19; Harms 2007: 38; Wiedl 2008: 21; Jones 2005: 225–26) is Arabic and translates as invitation or call. The theological concept for which it stands spans from the invitation, addressed to men and women by God and the prophets, to believe in the true religion, Islam, to the call for the dead to rise from the tomb on the Day of Judgment, on to a more profane understanding as an invitation to a meal (Bearman 2006). As a normative and directive source for *Daʿwah*, Muslims refer to the *Qurʾān* (Sura 16:125) and *Sunna* (Bearman 2006).

This article deals with the phenomenon of German-based Cyber-*Daʿwah* on the World Wide Web. The subjects of examination are websites that offer *Daʿwah*-related content in significant amounts or show an agenda towards *Daʿwah* which can be considered as a contemporary and state-of-the-art form of *Daʿwah* themselves.

The groundwork for this article is my research[1] in which I analyzed and interpreted such websites and the intentions behind them. As a theoretical background, I gratefully acknowledge the pioneer work of several scholars of religions who made the online-realm comprehensible and accessible more than a decade ago. My very own humble pioneering attempt was to introduce systematic content analysis as a methodical approach to the study of religions, borrowing contributions from the media and communication sciences. The purpose was to identify, analyze and compare the most salient websites regarding its contents and the techniques used to accomplish *Daʿwah* online. My main interest, aside from the technical framework, was to identify the particular subjective interpretations of Islam, which the authors

---

[1] Within the framework of my master thesis titled 'Cyberdaʿwa' im deutschsprachigen Internet – eine religionswissenschaftliche Analyse.

E. Munder (✉)
Institut für Vergleichende Kulturforschung - Kultur- u. Sozialanthropologie
und Religionswissenschaft, Universität Marburg, 35032 Marburg, Germany
e-mail: erik.munder@gmx.de

© Springer Science+Business Media Dordrecht 2015   3883
S.D. Brunn (ed.), *The Changing World Religion Map*,
DOI 10.1007/978-94-017-9376-6_205

and operators provided on their websites, whether they are liberal or restrictive. The results were intriguing: Muslims muster a great deal of flexibility to combine traditional content and customs with the high-tech possibilities of today's Information Age. They use nearly every technical feature the World Wide Web and other parts of the Internet have to offer. My results also indicated that Sunni fundamentalist groups like *Salafīya* or *Wahhābīya* are the prevailing force behind German-based Cyber-*Daʿwah*, offering corresponding content that is presented as mainstream-Islam.

In this article I will introduce briefly the concept of *Daʿwah* and Cyber-*Daʿwah* in its contemporary form, give an overview about the theoretical and methodical background of my research, summarize my findings and interpretations and revisit the websites to examine the changes that have taken place since my initial analysis in December 2010.

205.2   From Daʿwah to Cyber-Daʿwah

As the terminus technicus in this article, I understand Daʿwah as any Muslim activity and/or content in order to wittingly influence other Muslims and/or Non-Muslims to change their understanding of Islam as well as their everyday life. These are, of course, always based on the particular interpretation of Islam of the Daʿwah -practicing Muslims, the Dāʿī (Bearman 2006). Thus, Daʿwah could be understood as the concept of missionary work within Islam, but since there is no uniform, inter-religious term to work with, the Christian roots and connotation of missionary work would greatly obfuscate the use of it in the present context (Wrogemann 2006: 3; Harms 2007: 1; Wiedl 2008: 21).

The concept of Daʿwah is closely connected to another essential theological concept of Islam, the Ğihād (Bearman 2006). In its early centuries when Islam was spreading across the Arabian Peninsula and gaining more and more influence, Muslims customarily offered their enemies the conversion to Islam before a battle to become brothers and sisters in faith and omit fighting. After battle, the defeated again had the chance to convert to Islam. If they belonged to faiths possessing divine scriptures, as Christians, Jews, Zoroastrians, Sabaeans and – according to some Muslim scholars – even Hindus, they were part of the ʾAhl al-Kitāb (Bearman 2006), which can be translated as People of the Book. A devotee of these faiths had the chance to become a Dimmī (Bearman 2006) and thus was free to practice his/her religion protected by Muslim authority, but also obliged to pay a tax named Ğizya (Bearman 2006). Eventually, many converted to Islam due to social pressure and financial as well as occupational advantages (Wrogemann 2006: 32–40; Harms 2007: 45).

Reasons for this rather liberal permissiveness are found in the Qurʾān (Sura 2:256). Although there are also passages with very divergent intentions (Sura 9:5 and 9:29), the Ğizya became a lucrative income. Especially in the prime of Islamic propagation, around the eighth century, Muslims in occupied areas were often outnumbered by devotees of different faiths, like Hindus for example (Wrogemann 2006: 32–40; Harms 2007: 45). Leaving existing social structures intact while profiting thereby was easier, less risky and more sensible. Still, the ultimate goal was a conversion to Islam of every human being. In fact, in the context of Islam, every human being is born Muslim and thus, via conversion, only returning to the actual and original faith (Bearman 2006).

It is a holy duty for every Muslim to perform Daʿwah in his/her life. Even if he/she never has the chance to practice outer Daʿwah due to a lack of contact with non-Muslims, there is always the opportunity to practice inner Daʿwah among fellow believers.

The multitude of ways and means to invite other human beings to Islam, or a better understanding of it, is very extensive: every form of information and communication concerning Islam, addressed at non-Muslims or Muslims, considered off the right track by the Dāʿī, can be considered as Daʿwah. Even the silent effort of persuasion by living an exemplary, godly Muslim life could be meant as a serious act of Daʿwah (Poston 1992: 117–18; Jones 2005: 225–26; Harms 2007: 39–40; Wiedl 2008: 58).

Similarly, Islam-related content within the World Wide Web which shows similar intentions, is Cyber-Daʿwah in a broader sense. The focus of my research, however, is on the more explicit manifestations of Cyber-Daʿwah.

205.3   Islam 2.0

Islam has always been very flexible and innovative regarding new technologies and their implied potentials. Of course, the Internet is no exception with its new possibilities of communication and the methods to produce, offer, consume and share information, that is, accessible to nearly everybody, at least in most countries. In this way Islam is spread into the virtual world.

In his book iMuslims, Gary Bunt presents the above mentioned processes in detail and provides a comprehensive overview of Islamic websites, the Blogosphere and Online-Ǧihād. Despite his rather phenomenological approach, he enriches the methodological and theoretical framework for Islam-related research in Cyberspace with two conceptual terms: the cyber-Islamic environment (CIE) and iMuslims (Bunt 2009: 1).

According to Bunt, virtual places, structures and communities referring to Islam in general and/or providing activities and functions which are somehow linked to or invoke the Muslim identity and/or personal conceptualization of the user; these are considered as CIEs. Thus, iMuslims are Muslims who are linked to and use these CIEs to substantiate their Muslim identity and/or share their personal conceptualization of Islam. Bunt points out the massive impact these CIEs and their ongoing development have on traditional structures within the Islam world. There is an ongoing, expanding reconfiguration of networks, communication channels and religious aspects like authority, identity, holy knowledge and space. Concurrently, these modern CIEs are interweaved extensively with Islamic tradition and history, invoking and referring to them constantly (Bunt 2009: 37–39; 118–119).

This technological development brought along the novelties of the Online-Šahāda (Bearman 2006) and the Online-Fatwā (Bearman 2006), the former to facilitate Cyber-Daʿwah on a massive scale and the latter to provide authority based on an official, traditional religious foundation. The Šahāda is the first of the five pillars of Islam faith and the rite of passage any convert has to perform to become a Muslim. In regards to the confession of faith, it is the paramount goal of every act of outer Daʿwah, but the traditional prerequisites include witnesses, which of course complicates Daʿwah-activities in Cyberspace. This problem was remedied by another fusion of Islamic tradition and state-of-the-art technology, the Online-Fatwā. Sheikh Hamad Al-ʿAlī and Sheikh Ahmad Kutty, both well known scholars and active contributors to the popular Sunni website www.islamonline.net, endorsed the Online-Šahāda in general, if certain requirements are met (Bunt 2009: 88).

Today, more and more Muslims, especially in the western world, identify rather with a certain website than a local community or mosque (Bunt 2009: 10). For Muslims who are not accepted as religious authorities in traditional Islamic hierarchies, Cyberspace provides a platform to make one's mark on a global scale while being independent from conventional Islamic structures. This novel form of Online-Authority constitutes a problematic challenge to traditional Islamic authorities, especially to those who oppose the Internet or plainly ignore it (Bunt 2009: 37).

Another related issue is the difficulty distinguishing between the various interpretations of Islam. Most Islam-related websites present themselves as the one and only true interpretation of Islam, even if they have a very distinctive ideological agenda that differs significantly from the majority of their fellow Muslims worldwide. Particularly fundamentalist movements such as Wahhābīya (Bearman 2006) and Salafīya (Bearman 2006) seem to gain more and more ground in Cyberspace (Bunt 2009: 10; 33). My previous research indicates that these groups, at least within the German-based World Wide Web, tend to disguise their conceptualization of Islam as moderate and open. Defining and differentiating fundamentalism and Islamism is a delicate matter. As a scholar it is essential to avoid any swift judgments or false suspicions. Then again, a demand for literal utilization of Šarīʿa (Bearman 2006) punishments or the traditional role of women leads to an inconsistency with the Universal Declaration of Human Rights and is problematic even linked to an otherwise rather moderate interpretation of Islam. For my research, I follow the definition of Bowker and Golzio (2003) who define the fundamentalist as those who return to what they believe are the essential truths and customs of their faith. Thus, fundamentalist Muslims are those Muslims who claim the literal truth of the Qurʾān and the validity of its lawful and ritual commandments for present-day humanity. They do not incite changes, but insist on the changes being in line with the values of the Qurʾān, for example, in concordance to the Šarīʿa (Bowker and Golzio 2003: 328).

205.4   Religion 2.0 and Systematic Content Analysis

The majority of the scholars of religions avoided Cyberspace as a field of research for a very long time and some still do. Whether this is due to a lack of technical know-how or the necessary recognition that the Internet as an indispensable sphere of contemporary culture, I cannot say. But there are also those who already conducted research online when the Internet was still a rather minimalistic arrangement of bulletin and message boards (Helland 2004: 24). Over the years this mostly scripture-bound communication platform has evolved into the multimedia compound structure we know today. The number of users grew along with the technical distribution, simplification and, of course, the economic advantages. Along with the depth and complexity came the problems of methodology and theory construction regarding cultural phenomena online (Meier 2008: 149; Krüger 2005b: 1; Bunt 2009: 19; Rössler and Wirth 2001: 281).

Today, thanks to the developments generally labeled as social media and web 2.0, every user is an author at the same time. Thus, the content of websites can change dramatically or even vanish within minutes, especially in social networks or weblogs (also known as blogs). In addition to this ephemeral nature, scholars who do research within and on this media compound face other far-reaching problems. Within the German research landscape, Oliver Krüger and Gernot Meyer have contributed important groundwork for future research on religions in Cyberspace and helpful insights in the problematic nature of this environment. Both emphasize a comprehensive review of the applied methods to identify blind spots or personal dispositions and cope with the discrepancy between virtual and concrete realities. The perceptibility of communication processes and social structures within online communities are limited – for normal users, it is mostly hidden who is linked and communicates how and what with whom. Customary concepts and terms like membership do not apply in Cyberspace and the influence of online content to the non-virtual life varies from user to user (Krüger 2005a; Meier 2008: 150).

Among the anglophone pioneers of the study of religions in Cyberspace is Christopher Helland, whose contribution to theory and methodology for investigating religion in Cyberspace is crucial. His distinction between *online religion* and *religion online* derives from an earlier stage of the World Wide Web when most websites were simple documents consisting of text and pictures.

At the beginning of the twenty-first century, websites with religion-related content could be subdivided in two opposite types. Websites that offer information about religions, its structures and communities are classified as *religion online*. These websites are merely online representations of existing religion-related offline presences which they refer to as offering only restricted and controlled top-down and one-to-many communication (Helland 2000: 207, 219).

Websites classified as *online religion* constitute new forms of religious manifestation. They use the Internet far beyond its role as a communication medium for traditional content but as a platform for religious content and activities that have no counterpart or reference in real life up to genuine Online-Religions that entirely exist within the Cyberspace. As a prerequisite, these platforms offer or consist of open, unstructured and non-hierarchic, many-to-many communication forms (Helland 2000: 207).

Due to the rapid development of online technology and its spread around the globe, this distinction became more and more unacceptable or improper and subject to criticism (Young 2004: 94). Websites became more sophisticated and users became authors at the same time and with it; using the internet became a complex activity rather than a simple way providing information or offering information for others to absorb. Thus, most contemporary religion-related websites offer both possibilities, *religion online* and *online religion*. Helland took up the critique (Helland 2005: 13) and refined his theory to make it suitable with the more complex status quo. Additional distinctions indicating the extent of religious participation or simply information provision and especially references to online or offline activities acknowledged today's innovations in the World Wide Web (Helland 2005: 6). But allocating websites inside this theoretical coordinate system was indefinite and invoked further problems (Helland 2005). Finding an accurate definition of what constitutes religious interaction and how it is different from usual website-interaction on religion-related websites is very tricky to determine, for whatever reasons, why users visit such websites or to what extent authors and operators control or censor the available interactive functions. Still, the dichotomy *online religion/religion online* and the associated questions of user interactivity within religious Cyberspace are essential contributions to the study of religions online (Helland 2005: 206).

The intentions of users and authors using the World Wide Web are only accessible through methods of empirical social research which would take huge amounts of time. My research examined questions concerning how contemporary Cyber-Daʿwah websites work rather than why they exist and with what were intentions visitors and authors who use them. The ideal research method for my purpose is a systematic content analysis, which I adapted from the media and communication sciences. It is defined as an empirical method for both systematic and intersubjective comprehensible descriptions of content and formal attributes of messages (Früh 2007: 27). In their book *Inhaltanalysen im World Wide Web*, Rössler and Wirth divide the systematic content analyses of websites into five categories: (1) research questions and hypotheses; (2) random samples and analysis units; (3) instruments; (4) realization and quality assurance; and (5) evaluation/interpretation. They also provide helpful information concerning specific issues important in the research field. The massive dynamics associated with the development of online technology results in incremental complexity, volatility and reactivity of the content, rendering a systematic analysis very difficult, but not impossible. In order to create comparability between websites and contents, Rössler and Wirth suggest five dimensions of coding units: (1) structural parameters like the number of sub-pages; (2) screendesign; (3) interactivity; (4) specific content (like *Daʿwah*-related content in my case); and (5) usability (Rössler and Wirth 2001: 281).

For my research, I used these as a guideline and focused on the aspects most relevant for a subsequent interpretation that would be useful in the study of religions.

205.5   German-Based Cyber-Daʿwah

Using Google is not the only way to find specific content about a subject on the World Wide Web. It is, however, the easiest and most effective vehicle used today. There remain different search engines available, but the alternatives are, at the most, equivalent to Google in indexing and processing the vast amount of data that are available and rapidly increasing. The results of a search query may vary significantly depending on the algorithm and other parameters, such as the location of the used computer or previously visited websites which the scholar is not able to see or control thoroughly. Link catalogues and collections were handy devices in the early days of the World Wide Web, but they have associated with them similar and worse disadvantages than search engines do, as the content is subjectively selected by its author of the catalogue.

For my research, another good reason to choose Google as the source of my data was the intersubjectivity of the findings. To be able to find what other users have found and also will find too, scholars need to use the same tools within the same technical framework combined with equal settings insofar as possible. The majority of users, not only in Germany, *Google* their way through the World Wide Web while using the Firefox browser ("StatCounter Global Stats", 2012).[2] To obtain broad but also proper results, I analyzed the combined first hundred, exclusively German results for the keywords *Daʿwah* and *Islam*, using Google via the Firefox Browser, without the safe search function to avoid possible censorship of questionable websites.

After sorting out the irrelevant and duplicate results, thirteen *Daʿwah*-related websites remained for further analysis:

1. www.pierrevogel.de
2. http://dawah.de
3. www.diewahrereligion.de
4. www.salaf.de
5. www.dawa-shop.de
6. www.dawa-zu-islam.de.tl
7. www.islampuls.com1
8. www.dawaffm.de
9. http://deutsch-muslime.de
10. www.ezpmuslimportal.de
11. http://diewahrheitimherzen.net
12. www.wegzumislam.com
13. www.way-to-allah.com

Due to the many similarities in content, it seemed useful to arrange the identified websites regarding their approximate size, beginning with the smaller ones ana-

---

[2] http://gs.statcounter.com/#all-search_engine-DE-monthly-200812-201204-bar; http://gs.statcounter.com/#all-browser-DE-monthly-200812-201201-bar; accessed April 21st, 2012.

Fig. 205.1 HTML graph of
www.pierrevogel.de. The
HTML Graph Applet is a tool
that visually represents the
link-depth of websites (Image
created by Erik Munder using
www.aharef.info/2006/05/
websites_as_graphs.htm)

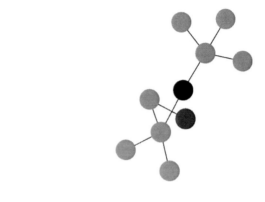

Fig. 205.2 HTML graph of
www.diewahrereligion.de
(Image created by Erik
Munder using www.aharef.
info/2006/05/websites_as_
graphs.htm)

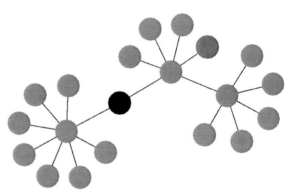

lyzed most thoroughly and then continue while focusing incrementally on salient
differences. Unfortunately, only one of the thirteen websites offered a sitemap at all.
In search of a solution to that problem, I discovered the *HTML Graph Applet,*
("Webpages as Graphs", 2012)[3] a tool able to visualize the link-depth of websites
(Figs. 205.1, 205.2, 205.3, and 205.4).

205.6   Results and Interpretation in 2010

The identified websites show both obvious resemblances and striking differences
regarding content, structure, options of interactivity and targeted audience. This heterogeneity derives from differing approaches to and the underlying intentions of each
venture. Some of the websites are merely private projects, only attended to in the
author's spare time, other authors or *Dāʿī* have a rather professional approach with their

---

[3] http://www.aharef.info/static/htmlgraph/; accessed April 21st, 2012.

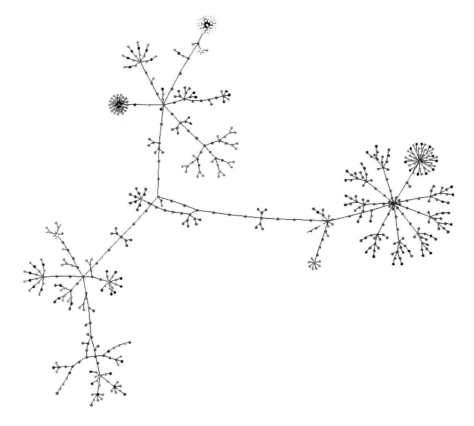

Fig. 205.3 HTML graph of www.ezpmuslimportal.de (Image created by Erik Munder using www.aharef.info/2006/05/websites_as_graphs.htm)

sense of mission standing out starkly. Nevertheless, every Daʿwah-related website refers eventually to the same sources, those being, Islam in general, Qur'ān and Sunna.

After analyzing and comparing the collected data, I developed a typology based on the results. Thus, an imagined prototype-Daʿwah-website that combines all basic or conventional elements found not on all but on most of the prospected websites, would look as follows:

Screendesign

- Layout and decoration mix tradition with state-of-the-art-design, that is, colorful graffiti-like lettering placed next to traditional Islamic calligraphy and suras as well as up-to-date animations in the video content which hint at the targeted youth audience.
- The packaged contents are structured and organized to suit the various visitor groups; highly visible navigation elements lead to sub-pages that offer information, content and topics for non-Muslims, new Muslims, Muslims in general and Muslims eager to do *Daʿwah*.

Fig. 205.4 HTML graph of www.way-to-allah.com (Image created by Erik Munder using www.aharef.info/2006/05/websites_as_graphs.htm)

Options of Interactivity

- Visitors can convert instantly by means of the online-Šahāda via video-chat or a telephone call with one of the website operators. An introduction (text, audio or video format) gives detailed instructions about the procedures for conversion.
- It is possible to donate money in one way or another, to support the website itself or to other related *Daʿwah*-activities.
- A search function or topic index is available, as is the possibility of downloading all or parts of the offered content (for example, as doc, pdf, mp3, mp4, flv, avi, mpg files).
- There are functions to connect the *Daʿwah*-website with social network profiles of the visitors like Facebook, Myspace etc. These may act as disseminators and advertise the project.

- Visitors may keep track up to date on the contents via email-newsletter, dynamic bookmarks, web feeds or micro-blogging services like Twitter.

Specific Da'wah-content

- *Da'wah*-activities are directed towards those inside and outside of the Muslim community which aim to convert non-Muslims and win Muslims over to the underlying interpretation of Islam.
- The offered content is generated or processed and provided by a small group of active members in which converts play a significant role as key figures, initiators and/or solitary operators of the website.
- Tutorials for recent converts concerning the right understanding and proper practice of the new faith are available as well as guidance for improving the *Da'wah* of already active Muslims.
- Interviews with converts about their path to Islam, personal experiences and the implied positive changes in their lives have a prominent placement on the website
- Among the authors of provided books and documents are the names of renowned, even famous representatives of Islamic Fundamentalism, e.g. *Wahhābīya* and *Salafīya* groups.
- Video contents often contain Arabic greetings and blessings. Apart from that they feature informal and simplistic language, youth language or even slang and idioms.

Although, the websites utilize for the most part every available form of media and technology for information supply and online-communication, online-videos seem to be the favored content offered. They provide advantages for both the authors and the audience. Being very easy and inexpensive to produce and supply, they are also extremely comfortable to consume compared to text content. If the websites offer no self-produced video content, they embed third-party videos, which are hosted on Youtube and other sources or link to websites with relevant content. Most of the *Da'wah*-websites, however, rather appear as a kind of online-archive for self-produced videos. The amount of data grows with every statement regarding current events or topics, every lecture and other offline-activity being visually documented and offered on the operated website or external video portals. With this growth is the propagation of the material as well as the visibility in public.

On every one of the websites analyzed, the authors strive to proclaim their liberal image, to dissociate themselves from terrorism and to call for a respectful dialogue between the different faiths.[4] Notwithstanding, on almost every website, content can be found which is clearly of Sunni-fundamentalist origin or propagates a corre-

---

[4] e.g.,www.pierrevogel.de/index.php/ueber-uns/anschlag-toulouse.html; http://dawah.de/index_main/wir_ueber_uns.htm; www.salaf.de/service/service_ueber_uns.html; http://deutsch-muslime.de/; http://diewahrheitimherzen.net/faqs; www.way-to-allah.com/wsw.html; accessed April 21st, 2012.

sponding point of view.[5] Some websites have obvious ideological connections to the *Wahhābīya* and *Salafīya* groups or the Muslim Brotherhood.[6] Other websites link only to liberal content of otherwise straightforward fundamentalist sources, which obfuscates their classification. Whether they disregard, ignore or lack the knowledge of the negative connotations in favor of the content is clearly impossible to certify without specific methods of empirical social research.

Pierre Vogel, former professional boxer who converted to Islam and spent 2 years in Mecca studying Arabic and the Qur'ān, is one of the most prominent representatives of the German Cyber-*Daʿwah* scene. His name and video lectures appear on almost every one of the *Daʿwah*-websites.[7] Simultaneously, he dissociates himself from violence and terrorism while still demanding martial punishments according to the *Šarīʿa* for German courts.[8]

Converts in general play a very important role within the German Cyber-Daʿwah scene. Aside from Pierre Vogel (Abu Hamza), Sven Lau (Abu Adam Hussein) (http://SvenLau.de/), Kai Lühr,[9] Mustafa Isa Seelmann http://deutsch-muslime.de/index.php/impressum) others hold key positions within the German Cyber-Daʿwah-scene. They bridge the cultural gap between German society and the Muslim Daʿwah-activists, diluting the fear of contact thanks to their impersonations of both worlds.

205.7   Revisited in 2012

Much has changed on the websites' surfaces and with the people behind them since my last visited in 2010. Many links within my thesis that pointed to a content which substantiated my findings are not available anymore. Two websites are offline

---

[5] e.g.www.pierrevogel.de/index.php/home/viewvideo/252/scharia/sind-steinigungen-zeitge-maess.html; http://dawah.de/index_left/de_filme.htm; www.diewahrereligion.de/jwplayer/player.html?skin=darksunset.zip&plugins=viral-2,fbit-1&autostart=true&file=www.diewahrereligion.de/jwplayer/videos/Die%20Strafe%20fuer%20Homosexualitaet.flv;   www.salaf.de/sira&tarih/PDF/sir0022_Biographie%20von%20Imam%20Muhammad%20Ibn%20Abdilwahhab.pdf; http://deutsch-muslime.de/eBook/Zeichen-auf-dem-Weg.pdf; www.wegzumislam.com/die-frau-im-islam/stellung-der-frau-im-islam/291-ruf-nach-gleichen-rechten-von-frauen-und-maennern; accessed April 21st, 2012.

[6] e.g.www.ufuq.de/newsblog/77-wwwsalafde-ersetzungen-von-schriften-und-predigten-saudischer-gelehrter;www.ufuq.de/newsblog/1120-deutsch-undoder-muslim-muslimische-jugendliche-in-deutschland;www.ufuq.de/newsblog/1118-die-dawa-arbeit-islamischer-vereine-zwischen-religioeser-bildung-und-missionarbeit-am-telefon; accessed April 21st, 2012.

[7] www.pierrevogel.de/index.php/home/viewvideo/180/wer-ist-pierre-vogel/pierre-vogel-mein-weg-zum-islam.html;   www.zeit.de/2007/41/Islam-Prediger;   www.faz.net/aktuell/politik/inland/islam-vom-boxer-pierre-vogel-zum-prediger-abu-hamsa-1595519.html#00MD58; www.youtube.com/watch?v=qJ_6VUYqdGk; accessed April 21st, 2012.

[8] www.youtube.com/watch?v=cMhXGswLvdg;www.pierrevogel.de/index.php/home/viewvideo/225/scharia/handabhacken-bei-diebstahl.html; accessed April 21st, 2012.

[9] www.spiegel.de/international/spiegel/0,1518,460364,00.html; www.youtube.com/watch?v=NbdhCAdeOIU; accessed April 21st, 2012.

(www.dawa-shop.de/ and www.Islampuls.com) and a third is a private project which appears to be unmaintained and unchanged since 2009 (www.dawa-zu-islam. de.tl). Due to personal and contextual overlapping, the ten remaining websites can be assorted into five groups, divided into the amount of content, public impact and update frequency:

1. www.pierrevogel.de/www.ezpmuslimportal.de (now http://muslimtube.de)/http://
   diewahrheitimherzen.net/www.wegzumislam.com/www.salaf.de
2. www.diewahrereligion.de/www.dawaffm.de
3. www.way-to-allah.com
4. http://dawah.de
5. http://deutsch-muslime.de

The websites of group one are connected to Pierre Vogel, Sven Lau, the former registered society *Einladung zum Paradies e.V.* (*EZP*) and their associates. The EZP society, an important pillar of the German Cyber-*Da'wah*-scene, was disbanded in 2011 after there were local public commotions against their daily prayers in the streets and interference by government bodies.[10] After that incident, the existing members joined other groups to continue their work or started their own projects (http://SvenLau.de).

Group two, led by Ibrahim Abu Nagie, filled the gap with their own public-provoking offline-*Da'wah*-activities. This second important pillar of the scene dominates recent media and the agenda of the *German Islam Conference*[11] with their distribution of free copies of the Qur'ān.[12] The third and fourth websites are also operated by registered societies: *Way to Allah e.V.* and *Islamische DAWAH Gemeinschaft Deutschland e.V.* The fifth website was initiated by the German convert Mustafa Isa Seelmann, but has not been updated since April 2011.

There have been not many notable changes according to the ideological classification of the offered content discussed above or the used online-techniques. Furthermore, they offer general information about Islam, the possibilities of online-conversion, statements about recent events, a mix of moderate and fundamentalistic content offered as mainstream Islam, extensive archives of downloadable video, audio and/or textual media, as well as links to social networks and update-services. Table 205.1 shows my summarized findings of 2010 and 2012.

---

[10] www.rp-online.de/niederrhein-sued/moenchengladbach/nachrichten/sven-lau-im-polizei-ver-hoer-1.1311164;   www.rp-online.de/niederrhein-sued/moenchengladbach/nachrichten/sven-lau-verkuendet-ezp-aufloesung-1.1315428; accessed April 21st, 2012.

[11] www.deutsche-islam-konferenz.de/SubSites/DIK/EN/Startseite/home-node.html?__nnn=true; accessed April 21st, 2012.

[12] www.economist.com/node/21553078;   www.dw.de/dw/article/0,,15895927,00.html;   www. dawaffm.com/neu/index.php?option=com_hdflvplayer&id=37; accessed April 21st, 2012.

Table 205.1 Summarized findings

Website	Associated names	Daʿwah content	Fundamentalistic content
http://www.pierrevogel.de	Pierre Vogel, Ibrahim Al Almani, Muhamed Ciftci (Abu Anes), Efstathios Tsiounis (Hamza Abu Alia), Sven Lau (Abu Adam Hussein), Einladung zum Paradies e.V., Metwally Galal	Video-lectures, Online-Šahāda, Chat, Telephone, Daʿwah-Tutorials,	Video-lectures by Pierre Vogel, Ibrahim Abu Nagie and Abu Ameenah Bilal Philips;
http://muslimtube.de		Translations of Islamic Scriptures, Social Networks, *Dawa-TV* Live-Stream	Books by Muhammad b. Abd al-Wahhāb, Abd al-Azīz ibn Abd Allāh b. Bāz, Ibn Taimīya, Abu Ameenah Bilal Philips
http://diewahrheitimherzen.net			
http://www.wegzumislam.com			
http://www.salaf.de			
http://www.diewahrereligion.de	Ibrahim Abu Nagie, Abu Dujana, Abu Abdullah, Sheikh Abdellatif, Abu Bilal und Abu Ubayda	Video-lectures, Contact Form, Telephone, Live Conversions,	Video-lectures by Abu Nagie
http://www.dawaffm.de		Donation, Social Networks	
http://www.way-to-allah.com	Adel Zaghdoud, Way to Allah e.V.	Video-lectures, Book renting, E-Books, Al Faysal TV Live-Stream, public information booth, cure of souls in prisons	Video-lectures by Pierre Vogel and Ferid Heider
			Books by Ibrahim Abu Nagie, Abu Ameenah Bilal Philips and Abd al-Azīz ibn Abd Allāh b. Bāz
http://dawah.de	Islamische Dawah Gemeinschaft Deutschland e.V.	Texts about Christianity and Islam, Links to other Daʿwah-websites	Video-lectures by Khalid Yasin
http://deutsch-muslime.de	Mustafa Isa Seelmann, Abdel Aziz	Texts, Audio, E-Books	Books by Muhammad Rassoul, Abu Ameenah Bilal Philips, Saiyid Qutb, Harun Yahya

Source: Erik Munder

205.8   Conclusion

Cyber-*Daʿwah* unites an essential part of the Islamic faith, the religious act of invitation to Islam, and the possibilities of modern communication technology. Tradition is simultaneously passed on as well as surmounted by this contemporary form of *Daʿwah*. The theoretical and theological foundations of *Daʿwah* have been adjusted over the centuries regarding the external circumstances and ideological intentions of the different groups. In this way, *Daʿwah* became a highly dynamic phenomenon of Islam, a combination consisting of brand-new elements and ones as old as the faith itself, which will develop even further, adapting possibilities we cannot think of yet. This process is exceedingly apparent when we look at Cyber-Daʿwah. The Šahāda – the first pillar of the Islam faith and core of Muslim identity – was set on-line and into new contexts by Islamic scholars to reach into a new space that was scarcely affected by Daʿwah before. This space became a venue for Daʿwah-activity by Muslims of all the different Islamic convictions, but German-based Cyber-Daʿwah is evidently shaped to a great extent by Sunni-fundamentalist groups, propagating their interpretation of Islam as the Muslim mainstream.

References

Bearman, Bianquis, Bosworth, van Donzel, & Heinrichs (Eds.). (2006). *Encyclopaedia of Islam* (Vol. II, p. 168. Online Edition. Second Edition). Leiden: Brill.

Bowker, J., & Golzio, K.-H. (Eds.). (2003). *Das Oxford-Lexikon der Weltreligionen*. Frankfurt am Main: Fischer-Taschenbuch-Verlag.

Bunt, G. R. (2009). *iMuslims. Rewiring the house of Islam*. London: Hurst & Company.

Früh, W. (2007). *Inhaltsanalyse. Theorie und Praxis*. Konstanz: UVK Verl.-Ges.

Harms, F. (2007). *Cyberdaʿwa – Islamische Mission im Internet. Vorausetzungen, Analyse und Vergleich von daʿwa-Sites im World Wide Web*. Aachen: Shaker.

Helland, C. (2000). Online-religion/Religion-online and virtual communitas. In J. K. Hadden & D. E. Cowan (Eds.), *Religion on the internet. Research prospects and promises* (pp. 205–223). London: JAI.

Helland, C. (2004). Popular religion and the World Wide Web. A match made in (cyber) heaven. In L. L. Dawson & D. E. Cowan (Eds.), *Religion online. Finding faith on the Internet* (pp. 23–35). New York: Routledge.

Helland, C. (2005). Online religion as lived religion. Methodological issues in the study of religious participation on the internet. *Online – Heidelberg Journal of Religions on the Internet, 01.1.* via http://archiv.ub.uni-heidelberg.de/volltextserver/frontdoor.php?source_opus=5823&la=de

Jones, L. (Ed.). (2005). *Encyclopedia of religion*. Farmington Hills: Thompson Gale (Encyclopedia of religion, 4). 2nd ed)

Krüger, O. (2005a). Discovering the Invisible Internet. Methodological Aspects of Searching Religion on the Internet. *Online – Heidelberg Journal of Religions on the Internet, 01.1.* via http://archiv.ub.uni-heidelberg.de/volltextserver/frontdoor.php?source_opus=5828&la=de

Krüger, O. (2005b). Methods and theory for studying religion on the internet. Introduction to the Special Issue on Theory and Methodology. *Online – Heidelberg Journal of Religions on the Internet, 01.1.* via http://archiv.ub.uni-heidelberg.de/volltextserver/frontdoor.php?source_opus=5822&la=de

Meier, G. (2008). Internet. In M. Klöcker & U. Tworuschka (Eds.), *Praktische Religionswissenschaft. Ein Handbuch für Studium und Beruf* (pp. 149–161). Köln: Böhlau.

Poston, L. (1992). *Islamic Dawah in the west. Muslim missionary activity and the dynamics of conversion to Islam.* New York: Oxford University Press.

Rössler, P., & Wirth, W. (2001). Inhaltsanalysen im World Wide Web. In W. Wirth & E. Lauf (Eds.), *Inhaltsanalyse. Perspektiven, Probleme, Potentiale* (pp. 280–302). Köln: Halem.

Wiedl, N. (2008). *Da'wa – der Ruf zum Islam in Europa.* Berlin: Schiler.

Wrogemann, H. (2006). *Missionarischer Islam und gesellschaftlicher Dialog. Eine Studie zu Begründung und Praxis des Aufrufes zum Islam (da'wa) im internationalen sunnitischen Diskurs.* Frankfurt am Main: Lembeck.

Young, G. (2004). Reading and praying online. The continuity of religion online and online religion in internet Christianity. In L. Dawson & D. Cowan (Eds.), *Religion online. Finding faith on the Internet* (pp. 86–97). New York: Routledge.

Chapter 206
The "Almost" Territories of the Charismatic Christian Internet

Anna Rose Stewart

206.1 Introduction

Discussion of the various technologies of the internet often draws upon the semantic field of geographical space (Davies 2006: 60). Internet users *search* for information and *visit* different *sites* on the world wide web, using browsers that bear names such as *Navigator*, *Safari*, and *Explorer* and that very often incorporate images of the globe into their logos. The linking of these technologies with experiences of space is indexed explicitly in the virtual environments described by researchers studying contemporary religion. "Virtual pilgrimages" to Mecca or Israel are constructed through series of interlinked websites (MacWilliams 2002), and online churches such as St. Pixels provide a simulated environment for prayer and contemplation (Jenkins 2008). Elsewhere, religious practitioners consider the possibility of making sacred space out of pre-existing forms of interactive platform. The Christian users of discussion forums or visitors to particular blogs may come to think of their online community as a "church" (Campbell 2005; Hutchings 2011), and Wiccans can come to an experience of the ritual circle through engaging with others in prayer and incantation over instant relay chat (O'Leary 1996: 797).

As online technologies have grown in their ubiquity and become increasingly interwoven into the everyday experience of many users, it seems appropriate to consider whether and how these concepts of space extend beyond the browser. In his discussion of the social experience of geography, Thrift draws our attention to the changes that modern technologies bring to our experience of "the local" (Thrift 2008). We are able, for example, to use email and social networking platforms to keep up with global news stories, to peer into the daily lives of strangers, to donate funds to the victims of distant earthquakes and to forge acquaintance or maintain

A.R. Stewart (✉)
Department of Religious Studies, University of Kent,
Canterbury CT2 7NF, UK
e-mail: stewart.anna84@gmail.com

© Springer Science+Business Media Dordrecht 2015
S.D. Brunn (ed.), *The Changing World Religion Map*,
DOI 10.1007/978-94-017-9376-6_206

social contact with family and friends on the other side of the world. The remote interventions enabled by the internet have interesting analogues in the existing grammars of religious practice. Processes of cultivating and maintaining a kind of proximity, even where physical presence is lacking, form an important part of the ritual life of many religious traditions (Engelke 2007). Spoken genres such as prayer or incantation may project the speaker into interaction with distant spaces and spiritual agents, or the listener into encounters with Gods and ancestors (Keane 1997, 2004). Religious practitioners are presented with a world of interconnection in which personal participation in a shared narrative and conceptual framework is possible (see Lurhmann 2006). In this chapter I present findings from my own research with Charismatic Christians in the U.K., a religious culture that is greatly preoccupied with the spiritual potentials of place-making, and for whom the internet presents a particularly compelling prospect.

206.2   Immanent Territories

In his 2008 ethnography of *Second Life*, Tom Boellstorff describes a simulated landscape whose users, in the form of avatars, are able to open businesses, beautify their surrounding and themselves, make friends, and explore. He settles on the appellation "virtual" in his description of this world, attracted to its vernacular interchangeability with the word "almost." Virtuality, for Boellstorff (2008: 19), "approaches the actual *without arriving there.*" The term "approach," like "almost" provides a sense of movement *towards* that seems particularly apt as we seek to chart the ever-unfolding social settings of the internet. Discussing the use of internet technologies in Trinidad, Miller and Slater (2000) frame much of their discussion of the effects of these technologies in terms of expansion. Personal websites, for example, provide a setting in which their Trini informants are able to act within an expanded social frame, projecting their own presence and their representations of their nation into the world. Critically, this expansion takes place along lines created in cultural aspiration. Observing the modern, cosmopolitan face that so many of their informants choose to present to the world through their websites, Miller and Slater comment that these new technologies seem to provide a means through which they are able to "deliver on the pledges that they have already made to themselves about themselves" (2000: 11).

The sense of reaching *towards* fits well with the cultural practices and preoccupations of Charismatic Christians. The term "Charismatic" comes from the Greek for "gifts." Charismatic or "Spirit Filled" Christians believe in and practice the "gifts" of the Holy Spirit. Church services are punctuated by the practices associated with these gifts, such as speaking in tongues, praying through the laying on of hands, and delivering prophecy. In keeping with the practices of the wider Evangelical movement, within which Charismatic faith is often included, Believers tend to frame their own personal faith in narrative form. Charismatic speakers will tend to possess an account, or *testimony*, of a time during which they perceived the presence of the Holy Spirit of God clearly for the first time, and they became "Saved" or "born again" (Stromberg 1993; see also Harding 1987). Crucially, this emphasis on personal conversion is maintained even after the point of initial salvation. Charismatic Christians describe their faith as a journey or "walk" *with* God that is simultaneously a walk *towards* God. This walk involves "coming to God" continually. Through practices such as listening to preaching, reading the Bible, and participating in Spirit-Filled worship, Charismatic Christians both perform and reaffirm their own conversion continually (Lurhmann 2004). If the experience of becoming 'born again' is constant, so is the experience of standing outside of conversion; this is a faith that is always almost achieved (Stromberg 1993; Lurhmann 2004).

Spirit-Filled Christian movements have been found in a wide range of global manifestations. Churches range from the traditions of asceticism and Biblical fundamentalism seen in American Apostolic Pentecostalism (Scott 1994) to the "health and wealth" Gospel of the Global Prosperity Movement (Coleman 2000) to the Masowe WeChishanu Christians of Zimbabwe, known colloquially as the "Christians who don't read the Bible" (Engelke 2007). For their own part, many Christians within these diverse streams are often found to demonstrate intense preoccupation with geography. In their readings of Scripture Spirit-Filled Christians often dwell particularly on the account of the workings of the Holy Spirit at Pentecost, during which members of the early church were gifted with the tongues of other languages so they might be able to proselytize to members of every nation on earth. The Holy Spirit remains, for many of these Christians, a resource for "empowered global witness" (Macchia 1999: 19). These groups have historically placed considerable emphasis on the importance of mission to distant nations (Robbins 2004: 164), and have been found in many contexts to be at the forefront of media outreach through radio broadcasts, televangelism, and increasingly the internet as they pursue a vision of a "faith which conquers the world" (Coleman 2000: 49).

Narratives of territorial expansion are particularly salient in the British Charismatic churches in which I have conducted fieldwork. A 2005 survey of British Churches found that around 6.3 % of the population of the U.K. attend church on a Sunday, a figure in keeping with the steady decline of British Christianity in recent decades (Brierly 2006). The majority of Charismatic Christians are intimately familiar with these figures. The term that is used most frequently to describe this state of national spiritual dissolution is "Lost." The condition of being lost is attributed to both the nation itself and the individuals within it. When Christians describe this loss they touch upon a private sense of disorientation and confusion that is understood to typify the lives and experience of the Unsaved, and the moral relativity and meaninglessness that they see reflected in the public spheres of media and policy making. It is important to note, however, that this concept also implies the possibility of future gains; those who are lost can be found, territories that are lost can be regained. Like the Believers themselves, the world in which they live, won through the work of Jesus Christ on the cross, must be constantly won again.

Christian discussion of this struggle is laden with the semantic field of armed conflict (see also Coleman 2000: 178). Christians are frequently advised by teachers and leaders to "develop a wartime mentality" in their dealings with the wider culture (see, for example, Piper 1990: 78), and I have heard evangelism in the local context described as an "air war," as materials are distributed from church buildings and members of the congregation go out to knock on doors and offer prayer to strangers. This perpetual work of evangelism is often conceptualized, in keeping with military tropes, as an advance through the taking of territory. At one Christian conference I attended the preacher paused in his speech about the ongoing growth of their church as a map of Western Europe was projected on the massive screens behind him. The map had been annotated with large arrows which extended from the U.K. across the Channel into the mainland. The preacher explained that the group had built a strong base in the U.K. "like an arrow being pulled back" the better to be shot out into the rest of the world. This explanation of the image was greeted with enthusiastic clapping and cheering by the gathered audience.

Space is not only mapped and described in the verbal habits that make up Charismatic life, it is made. In his works on religious speech Keane (1997, 2004), describes the "scenes of encounter" generated in the standard forms of religious speech that we see in ritual genres. As speakers address unseen spiritual agents, such as God or Jesus Christ, these agents are both rendered present and endowed with a particular kind of character. Those praying on their own, for instance, are drawn into an experience of God as omnipresent, as intimately aware of the everyday detail of one's life, and as possessing the power to directly affect these circumstances. This speech, in essence, creates the kind of encounter that it describes. This is brought into sharp focus by my Charismatic informants, who will literally invite God into a meeting, through the use of the phrase "Come, Holy Spirit" or "Come, Lord Jesus." Prayer extends not only the presence of God but the presence of the speaker, allowing petitioners praying in a church hall on the south coast of England the ability to extend some form of influence on events taking place half a world away.

The habits of personal extension that we see in prayer and other forms of ritual speech are carried more broadly into daily life, particularly in Charismatic engagements with Holy Scripture. The Bible, or "The Word," is often described not only as a book to read but as a place to *go*. Christians dealing with difficult situations and daily uncertainties are advised by teachers to "go to Scripture" or "get into The Word." Practices of daily Bible study and weekly preaching render the text of the Bible a living reality that can be *entered into*. My informants constantly relate their own circumstances to the trials and tribulations of Biblical figures and believe that Scripture contains a blueprint of unfailing relevance to their own lives. These conventions extend to other habits of religious speech. My informants often talk about the need for Christians to "speak Truth into" discourses of Atheism or to "bring the Word" to a difficult situation. Through the ongoing work of Bible study, prayer, prophecy and preaching, the spiritual life of the individual takes on great stature and significance as part of a wider story of the redemption of the world (see also Coleman 2011).

206.3   In Arm's Reach

These elisions of word and space bring us back to the internet. One of the most significant benefits brought by online technologies for members of these global and globalizing religious cultures is their ability to foster a sense of viable religious community across disparate contexts. Campbell (2007: 1050) notes in her own discussion of Christian internet use that engagement in online communities engendered in her informants a sense of a participation in a "global Body of Christ" beyond their own congregation. For Charismatic Christians in the U.K. this seemed to be a particularly valuable experience, as many chose to identify themselves strongly with North American ministries and teachers whose work is readily accessible online. The prominent conservative Christian leader and writer, Jon Piper, for example, maintained a strong presence in my research despite his physical distance from the meeting halls and living rooms that I worked in, as Christians downloaded recordings of his preaching from his Minneapolis congregation, and subscribed to his Twitter feed. During my fieldwork I met a school teacher from Zimbabwe who lamented what he saw as the lack of exciting and innovative Christian leaders in his own area, but remarked to me that through the internet prominent churches in the U.S. were "just around the corner" from his classroom in Harare.

The particular kinds of online sources consulted by my informants are often organized at the level of the congregation. Preachers recommend internet-based resources during weekly sermons and the names of websites or Twitter feeds considered to be spiritually edifying are shared among church members meeting in living rooms and coffee shops. These streams of recommendation lead to a fairly even consensus among members of any given church about the kinds of resources consulted. A preacher whose work is universally admired and widely referenced among members of one congregation, for example, might be all but unknown in a church down the road, whose members draw upon very different constellations of teachers and ministries that suit their own particular doctrines and styles of worship. This patterning of internet use provides a sense of a shared backdrop lying behind the congregation, that extends outside of the church context into the settings of everyday life. Christians listen to preaching or watch live videos of revivals taking place thousands of miles away while they do housework or breastfeed or travel to work. During my fieldwork I was often informed that the national media in the U.K. misrepresents the state of Christianity. One man in his 60s remarked that carrying out research must have been an enlightening experience for me as a non-Christian, "Because you'd think from the newspapers and television that Christianity in this country is dead. But when you scratch the surface, the church is everywhere." In the context of a secular state, these technologies helped to foster this sense of encompassing presence.

Internet technologies not only create a world in which the "church is everywhere," they also provides a means through which this revealed landscape can be navigated. Although the UK has a large number of church buildings and a great many churches, within Charismatic Christian streams it is unusual to find people who worship at the church that is located nearest to their home; many of the Christians I met would travel considerable distance to church on a Sunday. Charismatic Christianity in the U.K. comes in many forms, with very liberal, progressive churches sharing patterns of 'Spirit-Filled' practice with much more socially conservative congregations. In addition, "worship," the time of music, prayer, and active engagement with God that forms a key part of the Charismatic church meeting, is considered by many of the people that I speak to be as if not more important than the preaching and leadership. Many of my informants reported settling on their current congregations only after spending some time researching their beliefs, leadership structure, or their style of worship through the internet.

206.4   The New Media Frontier

Territories of the sacred are balanced, in wider Christian discussion of the internet, against territories of the profane. The general godlessness of the world and particularly the U.K. is understood, at its most fundamental level, as a reflection and result of the ongoing battle between good and evil. Discussion of the internet in preaching and church meetings frequently focuses on the spiritually damaging potentials of online technologies. For my informants, the most pressing concern is what is widely viewed as an epidemic of online pornography consumption among men. A variety of solutions have been proposed, many of which seek to add a layer of "accountability" to private web use; either through special programs that send ones browsing history to other church members or through discussing net use in church groups. Other concerns relate to the possibility that time that might be spent in spiritually useful pursuits becomes leached away through online distraction, and that social networking platforms such as Facebook are exchanging depth and meaning in human relationships for a superficial and illusory breadth.

The varied terrains of the internet not only reflect the ongoing battle between spiritual degradation and renewal that consumes the entire world, but also comprise an important battleground in itself. In 2004 a group of Christian writers and leaders came together at Biola University, a private Evangelical university situated on the outskirts of Los Angeles to discuss the growing imperative of digital evangelism. The title of the publication that arose from these meetings, *The New Media Fronter* (Reynolds et al. 2008), neatly expresses the Christian understanding of the internet as both a setting and a tool for the expansion of the faith. The contours of this new frontier are experienced and illustrated clearly in the manner in which Spiritual Warfare has become incorporated into online practice. "Spiritual Warfare" is the name given to practices of prayer and prophecy particularly associated with Spirit-Filled forms of Christianity that has been observed in groups as diverse as Haitian Pentecostalism (McAlister 2005: 252–254), the African Aladura Church (Adogame 2004), and Charismatic Catholics in North America (Csordas 1994: 41). The practices of Spiritual Warfare bring Believers into direct conflict with the forces of Satan that occupy the world; practitioners may pray in tongues to drive out demons or receive prophecies from God that describe the form of present and future battles for Spiritual dominion over Earth and its inhabitants. Explanations of this discipline often emphasize physical presence and geographical space (Griffith 1997: 191–196; McAlister 2005: 253). My informants have related accounts of being urged by God to pray as they looked out of windows on to views of the city, or as they walked down particular streets, specifically with the intention of bringing the influence of God into that area and pressing against the presence of Satan.

Of the Christians that I spoke to about Spiritual Warfare, Patricia was the most enthusiastic. Although she attended a church whose leaders and laity were generally wary of Charismatic forms of worship, she had in the past attended more Spirit-Filled congregations, and remained fascinated with the possibility of taking part in Spiritual Warfare herself. In the years since leaving her previous congregation, she had been able to find a massive amount of information and discussion related to Spiritual Warfare through web searches, and made contact with other Christians who shared her interests. Patricia maintained an active Youtube account that she used to connect with these new contacts, most of whom were based in the United States, and learn more about these practices. However, she soon discovered that these videos, and the comments lists below them, provided a platform for more combative forms of spiritual contact. The ubiquity of Youtube as a video hosting service means that Christians uploading videos about their own faith are likely to draw the attention of cynical and sometimes hostile observers. Patricia had found in these meetings a forum into which her offline practices of Spiritual Warfare could be extended. In an email to me, Patricia made the following comments on this practice:

> I mentioned Spiritual warfare and I've seen this to be more visible and pronounced on youtube.
> I've seen much zeal from Christians in their dedication to the Lord and in their ministry. I've even seen others attacking Christianity and declaring the devil, and haven't seen this offline and even when they are attacking the bible and Christianity.
> I can see a clearer battle between good and evil on youtube. I see this as existing in the world, but online and particularly on youtube this spiritual battle seems to be more ahead and progressed in comparison to offline. Maybe because of the collectiveness the internet provides and the anonymity there. It suggests to me to be in the forefront and showing what the world is expecting in the future, with the internet being that bit ahead of the rest of the world.

Patricia's comments indicate that the power and salience of online Spiritual Warfare is bound up in the visibility of the unfolding conflict. In the comments lists underneath youtube videos, Patricia is able to "see" more clearly the spiritual conflicts that she believes lie beneath the surface of everyday reality. The criticisms that dissenting commenters leave on Christian videos, and the similarly impassioned responses from Christians are perceived not primarily as a verbal debate, but as a materialization of the battle between the Unsaved and redeemed worlds that lies at the heart of her faith. While the internet is often described in terms of the anonymity that can be afforded through its use, here we see the web approached as a vehicle through which that which might otherwise be obscured is rendered visible. Patricia once told me that she would be worried to discover that any of the people that lived near her were Atheists, a group that she understood to be excessively hostile and aggressive to Believers, but was aware that she had no way of knowing. Online, however, the heart of the user that might be veiled in their everyday interactions is laid bare.

The spiritual significance of this kind of materialization can be seen in other contexts. The idea of persecution is an important point of identity within conservative Christianity in the U.K., and the decline of national Christianity is often described as a result of a hostile rejection of the Gospel message rather than mere apathy toward it. Speakers seeking evidence of this hostility find ample material online. During one church meeting the leader described a recent article that had been in the local newspaper about the possibility of a new church opening nearby. The leader spent little time remarking on the article itself or its content, instead stressing the negative comments that the online version of the article had attracted. These comments, he argued, showed the "hatred" that the city holds for the Christian message. Similar discussion of the internet as a kind of stage in which the constant attacks faced by Christianity and Christians are rendered visible could be heard throughout the churches in which I conducted research. The Pastor of one church incorporated a warning into a presentation about plans for expanding the congregation into new sites. Although this growth would act like a beacon to draw in more of the unconverted of the city, they would be sure to meet resistance. He predicted to his assembled congregation that in the coming months "things will be said in blogs and through Twitter" that would seek to undermine this growth.

Despite these difficulties Christians continue to approach the internet as a tool through which to evangelize and a setting in which the Lost can become Saved. Some churches incorporate the web presence of their own members into their strategy. One large congregation holds regular promotional drives during which members are encouraged to "donate" their Facebook status updates and Twitter messages to advertising the church. Many of the Christians that I have met will go beyond this, and regularly post their own affirmations of God, prayer requests, and reflections on spiritual life to their Facebook wall, or incorporate passages from Scripture into their email signatures. The aim of these practices, as with religious language practice in general (Keane 2004), is to shift the boundaries of "the local" to include a more transcendent aspect. "A Tweet about the power of God," notes John Piper in his own advice for online evangelism, "may distract someone from online pornography and cause him to look up."

In these statements we see the hope that animates Charismatic discussion of the internet; that words deployed correctly can reach into the lives of those who most require salvation. The widely-used concept of "web presence" takes on an urgent spiritual dimension for Believers who often referred to themselves as "ambassadors for Christ." Even the smallest church that I studied, a congregation that at the time had a regular attendance of around 12 people, had an extensive and visually appealing website, that, due to the technical expertise of one member, fared particularly well in Google search rankings. The tallies provided by hit-counters on church websites and the numbers of downloads that a particular sermon has attracted are related in church settings and Christian media with great enthusiasm. Each number, according to the logic of Charismatic practice, is another witness to the message, a figurative expansion of the Kingdom of God. This excitement stems from a specific understanding of the power of the Gospel message and the nature of its reception. Charismatic Christians often describe the inescapable pull of the Gospel message, a message which, correctly translated, can be readily apprehended regardless of personal circumstance or global context (see Coleman 2000: 127; Tomlinson and Engelke 2007: 21). To be exposed to the message correctly, according to these Christians, is to be as good as saved. In practice, however fields of reception may be less straightforward than these visions suggest.

206.5   An Audience of One

For all that the Christians I work with stress their commitment to maintaining a Christian presence in digital territories that would be otherwise devoid of spiritual value, discussion of the internet is peppered with concern about the hazards of this position. During my fieldwork I heard many different accounts of disagreements sparked by the incautious or improper use of the internet by other Christians. One church member in his twenties, for example told me that a few months before my research began there had been some controversy after another member of his congregation mentioned on Facebook that he was planning on moving in with his girl-friend without marrying, Other members of the church began to post comments criticizing his decision and eventually the discussion became so heated that one of the leaders of the church had to "step in." The leader urged everyone concerned to move their discussion from Facebook walls to face-to-face contexts, reminding them that their behavior, in a public forum, was bringing their church and by extension Christianity into disrepute *in front of* non-Christians. Despite the rather combative tone in which the use of the internet is often described, actual open conflict is generally avoided. Talking to Christians who often read blogs hosting debate between Atheists and Believers, I was struck by how frequently my informants would describe thinking and praying before posting a comment, in the fear that they might unwittingly present a negative face to Christianity, and thereby prevent readers from becoming saved.

This intense preoccupation among Christians regarding how their online activities might look to the Unsaved may in many cases far outstrip the actual presence of an Unsaved audience. An unusual thing about Evangelical media in general is that it is largely consumed by those who already believe. Christian preaching, books, courses and television and radio broadcasts deliver appeals to salvation to audiences that are, for the most part, already converted (Hunt 2003; Schultze 1996: 64). Although the nature of the online medium makes the religious affiliations of those exposed to these messages difficult to assess (see Livingstone 2004), the patterns of engagement that we see online suggest that the participation of non-Believers with these sites is minimal. Sites like Youtube and Facebook form online spaces whose ubiquity demands shared presence by many different people. Outside of these sites, however, and the rather more specialist blogs on which Atheists butted heads with Believers over complex theological, historical, and doctrinal questions, I found few Christian websites in which non-Christian engagement was evident in any way. In some cases the absence of the Unsaved seemed to be by design. One blogger told me that while she usually advertised new posts to her blog to the wide circle of his acquaintance via Facebook, pieces of writing on particularly controversial issues were not promoted outside of her own church community. That way, she felt, those who were interested to read the "Christian perspective" on the nature of Hell or physical disability, for instance, could look it up, rather than having it "pushed in their faces."

The restricted frame of many ostensibly Evangelizing texts indicates some value in this online activity that goes beyond the ability to engage with those outside of Christianity, and suggests the importance of an audience that is rather more immediate. Mission to uncertain and dangerous territories brings the force of salvation to those on both sides of the mission encounter (Ustorf 1998: 595); the process of asserting faith has spiritual significance for speakers as well as listeners (Stromberg 1993). Reflecting on the principles that ought to be applied by Christian bloggers, one prominent writer advised others to avoid excessive preoccupation with the judgments that other members of the Christian 'blogosphere' might make of their work and instead attempt, as the popular Scriptural quote goes, to keep their eyes fixed on Jesus. "We write always," the blogger reminded his readers, "with reference to an audience of One." For this writer, the audience is the Christian God who, as I was informed by many of the bloggers I spoke to, reads every word published online. Looking at these transmissions and the manner in which their authors speak about their significance to their own lives, we see that a particularly critical member of the audience may be the author himself/herself.

Kong (2001) suggests that one of the more interesting prospects of new media technologies is their ability to confuse the boundaries between classifications of inner and outer, personal and public. For my informants, materializing their affirmation of a particular sermon on their Facebook wall, or sharing a link with friends allows an experience of the broader significance of their private religious practice and personal engagements with God. Several of the Christians that I spoke to commented that their habitual sharing of daily life in the form of blogging or posting on message boards had made them more sensitive to the spiritual dimensions of events taking place in the world around them. One female blogger told me that her journey to the internet cafes in which she composed her entries had become a "Holy Walk." She explained that she had come to pay extra attention to the events taking place around her that might be incorporated into her writing, and believed that God himself had a hand in steering these events, so that she might have something to write about. Similar habits of everyday inspiration were reported by many women that I spoke to. While these might in the past have inspired particular kinds of Bible reading or prayer requests, the sense that their own experiences could be valuable for others, and the experience of publishing these on a global stage, was described as a significant source of self-esteem and an increasing awareness of their own value as spiritual agents. Several remarked to me that the knowledge that their normal lives were so laden with spiritual potential had led to a noticeable "growth" in their own faith.

This "growth" was always framed by Christians as an ongoing process. The spiritual meaning of these everyday conversions of experience to text is realized in time. Thrift (2008: 164) notes that one important way that evolving media technologies change what counts as the space around us is by giving us new ways to archive and remember. The ready availability of photographs, audio and visual recordings and written materials allows the past to become part of a living present. It is common practice among bloggers, Christian or otherwise, to compose retrospective commentaries on their previous entries after some significant period of time is past (Reed 2005: 231). Many of the Christian bloggers that I have spoken to emphasized the impact that engaging with their blogs in this way has had upon *them* as writers. Reading earlier posts and considering the changes that they had experienced since writing them made the central Christian narrative of the "walk" towards God real and apparent, as they saw their own progress charted along the time-line of their posts.

The temporal dimensions of this online journeying is further illustrated through the prominence of Israel in the internet activities of many Christians. Israel is particularly marked in many Evangelical and Charismatic streams of Christianity as a critically important site in both the past and the future of God's relationship with humanity. During my research I encountered several people who would use the internet to maintain some kind of regular connection with this sacred territory. Some downloaded videos of Christian teaching and prayer events that were delivered from locations mentioned in the Bible, others, mindful of prophecies that Jesus' return will be presaged by upheavals in Jerusalem, maintained an eye on current events and the national political situation. One woman used the internet as a platform through which to encourage other Christians to pray for the salvation of the Israeli people (See Stewart 2011: 1209). As is the case with the unfolding narratives of blogs, these practices allow Christians to experience for themselves direct connections between their own lives and the spatial and narrative centers of spiritual events.

206.6   Discussion and Conclusion

...we need to look squarely at the problem, not to deny it, not to pretend that it isn't there, not to live in our own ghetto, not to just live in our own comfortable life. There is a battle, we need to look at those horses and chariots. But then we mustn't stay there intimidated by what we see. God gives us a bigger picture. God shows us his perspective and he shows us that the army, the heavenly armies are there in order to fight on our behalf. And when you get that vision it completely changes your perspective... (Taken from a sermon entitled 'Spiritual Warfare', recorded July 8, 2010)

The quote above was part of a speech delivered at a large Christian conference that I attended in the summer of 2010. The final night of this gathering of thousands of Believers was given over to an extended period of prayer and singing. Christian friends, telling me what I could expect from the evening, mentioned the key hallmarks of Spirit-Filled worship, such as exorcisms, worshipers struck to the ground by spiritual force, and speaking in tongues. As the session began, I noted that although most of those in attendance chose to stand or sit to the front of the stage many seemed to be observing from the edges. An informant later explained that the Holy Spirit would allow some of those watching a view of the otherworldly forces and agents that sent ripples of spiritual power and activity through the assembly. He told me that he had attended large meetings before during which he had sat on the sidelines and seen with his own eyes the forms of angels and demons as the crowd worshiped, prayed, and sang. An important part of the appeal of the internet for Charismatic Christians is its ability to provide its own kind of elevated viewpoint from which to take in the scale of revival. The browser window opens up vistas from which large scale movements of truth and lies can be seen.

I noted in the introduction to this chapter that religious internet users often approach these technologies in terms of their potential to comprise ritual spaces in themselves. In this chapter, I have focused on the use of the internet as a way of conceptualizing and interacting with a spiritual space that extends outside of the browser and hopes, eventually, to encompass the entire world. In 2004 group of Christians came together to discuss new technologies and described *The New Media Frontier* as if it were a boundary brought into being by their very existence. But when we look to Charismatic practice we see that this frontier, and the territories that lie on either side of it, are actively constructed in practice. As Charismatic Christians download messages, share them with others, pray over divided and hostile comments lists and offer their own testimonies of faith, they come to experiences of a territory of faith that is constantly expanding and a fallen world that is on the attack.

This battle extends into the hearts of Believers themselves. For all that my informants stress about the need to bring about a change in the hearts and minds of the Unsaved, the most salient boundary between Saved and Unsaved lies within the Christian self (see Stromberg 1993). The labor of the online evangelist creates a space of ongoing redemption in the speaker as well as in the world that she speaks to. These Christians place themselves at a nexus of words that come in from the redeemed landscapes of global Christianity and those that are transmitted to the Unsaved world (see also Engelke 2007: 172). The internet lends a further layer of everyday salience to this positioning. Blogging and posting links and thoughts on Christianity on Facebook or through Twitter allows Christians to map the world of interconnection in which they place themselves; as they stumble across the right church, realize that greater plan behind chance encounters on the street, and extend prayer to the spiritually critical territories of Israel. All of this is activity is understood by my informants to be part of the more significant journey, the endless movement *towards* God.

The sense of immanent achievement bound up in the Charismatic use of the internet is neatly expressed by Patricia, who describes these technologies as "that bit ahead of the rest of the world." As an adherent of a prophetic faith, her comments are telling. For members of a faith in which the future is already outlined in narrative terms, there is a clear sense of a divine trajectory behind the human use of these technologies. Like the texts of Scripture, which come to be "inhabited" by Christians who understand their daily lives in terms of its stories and proscriptions (Lurhmann 2006), the internet allows them to step into a wider narrative. The "pledge" that these Christians "have made to themselves about themselves" (Miller and Slater 2000: 11) is that through their own efforts the Kingdom of God will be larger tomorrow than it is today. There is, of course, no end to this pursuit; to move forward only shifts the horizon. In the lives of these Christians we see emerging technologies deployed as a tool for the making of spaces that are in constant transition. Christians position themselves in relation to God, the Devil, and unbelievers in the continual making and remaking of these shifting geographies. In the Charismatic internet, landscapes of divine encounter with God are constantly reachable and yet always unreached.

References

Adogame, A. (2004). Engaging the rhetoric of spiritual warfare: The public face of Aladura in diaspora. *Journal of Religion in Africa, 34*(4), 493–522.

Boellstorff, T. (2008). *Coming of age in second life*. Princeton: Princeton University Press.

Brierly, P. (2006). *UK Christian handbook: Religious trends 6*. London: Christian Research.

Campbell, H. (2005). *Exploring religious community online: We are one in the network*. Oxford: Peter Lang.

Campbell, H. (2007). Who's got the power? Religious authority and the internet. *Journal of Computer-Mediated Communication, 12*(3), 1043–1062.

Coleman, S. (2000). *The globalization of charismatic Christianity*. Cambridge: Cambridge University Press.

Coleman, S. (2011). 'Right now!': Historiopraxy and the embodiment of Charismatic temporalities. *Ethnos: Journal of Anthropology, 76*(4), 426–447.

Csordas, T. (1994). *The sacred self: A cultural phenomenology of charismatic healing*. Berkley: University of California Press.

Davies, J. (2006). Escaping to the borderlands: An exploration of the internet as a cultural space for teenaged Wiccan girls. In K. Pahl & J. Rowell (Eds.), *Travel notes from the new literacy studies: Instances of practice* (pp. 57–71). Clevedon: Multilingual Matters.

Engelke, M. (2007). *A problem of presence: Beyond scripture in an African church*. Berkeley: University of California Press.

Griffith, R. (1997). *God's daughters: Evangelical women and the power of submission*. London: University of California Press.

Harding, S. (1987). Convicted by the Holy Spirit: The rhetoric of fundamental Baptist conversion. *American Ethnologist, 14*(1), 167–181.

Hunt, S. (2003). The Alpha program: Some tentative observations of state of the art evangelism in the UK. *Journal of Contemporary Religion, 18*(1), 77–93.

Hutchings, T. (2011). Contemporary religious community and the online church. *Information, Communication, and Society, 14*(8), 1118–1135.

Jenkins, S. (2008). Rituals and pixels: Experiments in online church. *Online – Heidelberg Journal of Religions on the Internet, 3*(1). Retrieved January 28, 2008, from www.ub.uni-heidelberg.de/archiv/8291

Keane, W. (1997). *Signs of recognition: Powers and hazards of representation in an Indonesian society*. London: University of California Press.

Keane, W. (2004). Language and religion. In A. Duranti (Ed.), *A companion to linguistic anthropology* (pp. 431–448). Oxford: Blackwell.

Kong, L. (2001). Religion and technology: Refiguring place, space, identity and community. *Area, 33*(4), 404–413.

Livingstone, S. (2004). The challenge of changing audiences: Or, what is the audience researcher to do in the age of the internet? *European Journal of Communication, 19*, 75–86.

Lurhmann, T. (2004). Metakinesis: How God becomes intimate in contemporary U.S. Christianity. *American Anthropologist, 106*(3), 518–528.

Lurhmann, T. (2006). Learning religion at the Vineyard: Prayer, discernment and participation in the Divine. Retrieved December 11, 2009, from http://martycenter.uchicago.edu/webforum/index.shtml

Macchia, F. (1999). The struggle for global witness: Shifting paradigms in Pentecostal theology. In M. Dempster, B. Klaus, & D. Petersen (Eds.), *The globalization of Pentecostalism* (pp. 8–29). Carlisle: Regnum Books.

MacWilliams, M. (2002). Virtual pilgrimage on the internet. *Religion, 32*(4), 315–335.

McAlister, E. (2005). Globalization and the religious production of space. *Journal for the Scientific Study of Religion, 44*(3), 249–255.

Miller, D., & Slater, D. (2000). *The internet: An ethnographic approach*. London: Berg.

O'Leary, S. (1996). Cyberspace as sacred space: Communicating religion on computer networks. *Journal of the American Academy of Religion, 64*(4), 781–808.

Piper, J. (1990). *What's the difference? Manhood and womanhood defined according to the Bible*. Wheaton: Crossway Books.

Reed, A. (2005). My blog is me: Texts and persons in UK online journal culture (and anthropology). *Ethnos, 70*(2), 220–242.

Reynolds, J., Overton, R., & Hewitt, H. (2008). *The new media frontier: Blogging, vlogging, and podcasting for Christ*. Wheaton: Crossway.

Robbins, J. (2004). The globalization of Pentecostal and charismatic Christianity. *Annual Review of Anthropology, 33*, 117–143.

Schultze, Q. (1996). Evangelicals' uneasy alliance with the media. In D. Stout & J. Buddenbaum (Eds.), *Religion and mass media: Audiences and adaptations* (pp. 61–73). Thousand Oaks: Sage.

Scott, S. (1994). "They don't have to live by the old traditions:" Saintly men, sinner women and an Appalachian Pentecostal revival. *American Ethnologist, 21*(2), 227–244.

Stewart, A. (2011). Text and response in the relationship between online and offline religion. *Information, Communication, and Society, 14*(8), 1204–1218.

Stromberg, P. (1993). *Language and self-transformation: A study of the Christian conversion narrative*. Cambridge: Cambridge University Press.

Thrift, N. (2008). *Non-representational theory: Space, politics, affect*. Abingdon: Routledge.

Tomlinson, M., & Engelke, M. (2007). Meaning, anthropology, Christianity. In M. Engelke & M. Tomlinson (Eds.), *The limits of meaning: Case studies in the anthropology of Christianity* (pp. 1–38). New York/Oxford: Berghahn Books.

Ustorf, W. (1998). Global topographies: The spiritual, the social, and the geographical in the missionary movement from the West. *Social Policy and Administration, 32*(5), 591–604.

Chapter 207
Christian-Atheist Billboard Wars in the United States

Daniel H. Olsen

207.1 Introduction

Culture wars, according to Don Mitchell (1995, 2000), occur when cultural groups attempt to create and solidify their group identity by differentiating themselves from other cultural groups. This is done through constructing discourses that inform group members who is considered an "insider" and who is an "outsider" to that group; that is, who belongs and who does not (see Cresswell 1996). This construction of various "Others" is not just a metaphorical and ideological exercise; it is also geographical in nature as "arguments over "culture" are arguments over real spaces, over landscapes, over the social relations that define the places in which we and others live" (Mitchell 2000: 5–6). Cultural groups also define certain places and landscapes as being in opposition to the ideologies of other cultural groups (Massey 1995; Sibley 1995: 101–110, 185).

Today there is a culture war taking place on the billboards along major highways in the United States. While religious groups have long used billboards and signs to promote evangelical messages to travelers as well as to Christianize the American landscape, recently billboards have been used by atheist and humanist groups to contest the idea of a Christianized landscape and to create a space where their non-religious views and discourses can be promoted and accepted. The uses of highway billboards by these latter groups have in turn been contested by a number of Christian groups with the result that in some cases both groups have erected competing billboards across the road from each other in an attempt to counteract what they see as subversive messaging.

D.H. Olsen (✉)
Department of Geography, Brigham Young University, Provo, UT 84660, USA
e-mail: dholsen@byu.edu

© Springer Science+Business Media Dordrecht 2015   3913
S.D. Brunn (ed.), *The Changing World Religion Map*,
DOI 10.1007/978-94-017-9376-6_207

The purpose of this chapter is to examine how the highway billboard acts as a contested space between evangelical and atheist/humanist groups, both who seek to influence people who are on the move. After examining the different types of religious billboards and signs one might experience along American's highways and the possible goals of those who erect them, the contested nature of these billboards will be discussed. Finally, ideas for future research are noted with a specific focus on other ways that Christian groups attempt to Christianize the American landscape.

207.2   Religious Billboards and Signs

In the United States there is a long history of utilizing highways and roadways for advertising purposes. Billboards and signs in particular have become the popular medium through which advertisers have attempted to seduce people as they move through the American landscape to purchase everything from war bonds to rock shows (Gudis 2004; Jakle and Sculle 2004; Basten 2007). Some enterprising companies and organizations have used religion as a theme in their billboard marketing to attract tourists and other passers-by to visit their businesses or events. For example, along Interstate 90/94 in Wisconsin one can see billboards advertising Noah's Water Park in Wisconsin Dells complete with cartoon characterizations of Noah, his ark, and assorted animals. Along the same stretch of highway, homemade signs advertising The House on the Rock Resort, also located in Wisconsin Dells, can been seen; the name of the resort is a possible allusion to the biblical passage in Matthew 7:24–27 regarding the story of the wise man and the foolish man. Religion can also be seen as a witticism in order to gain people's attention. One example is another sign again along Interstate 90/94 advertising Reverend Jim's Road House with the phrase "Food and Spirits" in large letters at the top of the sign. However, anyone who looks up Reverend Jim's Road House on the Internet (www.revjims-roadhouse.com) will note that there is nothing religious about this establishment![1] Another example is seen in Fig. 207.1 where attendance at college football games is advertised through the use of religious imagery and language. In each of these cases a religious theme was used to attract visitors to visit and consume tourist services.

At the same time, religious groups and individuals have long used highway billboards and signs to promote evangelical messages to travelers as well as to Christianize the American landscape.[2] In certain parts of the Southeastern and

---

[1] These three examples are based on the author's experiences traveling through Wisconsin in the summer of 2011.

[2] For the purposes of this paper, billboards are defined as very large outdoor panels that display advertisements in highly visible public places, such as along highways or on the sides of buildings, which are owned by private businesses and rented to corporations or groups. Signs, on the other hand, tend to be either hand-made by the person or organization who creates and places the signs and are typically small in size as compared to billboards. Billboards also tend to be higher off the ground to attract attention from a long distance, whereas signs are closer to the ground and can only be seen by passers-by from a short distance.

Fig. 207.1 Sign outside of Provo, Utah, suggesting that religious experiences can take place at Brigham Young University football games (Photo by Chad Emmett, used with permission)

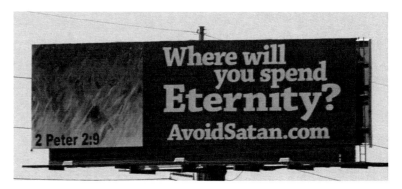

Fig. 207.2 Sign in Autauga County, Alabama asking travelers where they will spend their eternity (Photo by Daniel McGowin, used with permission)

Midwestern United States, it is difficult to travel very far without seeing a billboard or roadside sign with some sort of religious theme (York 2007). There are many different types of religiously themed billboards or signs one might see on their travels. For instance, many religiously-themed billboards and signs erected by church groups or organizations advertise church worship service times or the times and locations of Bible camps and evangelical revivals. In these cases, the affiliation of the particular church or religious organization is noted, with either a contact website, e-mail, telephone number, or a physical address affixed to the billboard or sign for those interested in contacting the Church or organization for religious information, literature or spiritual guidance. In other cases, a catchy statement or a scriptural quote is placed on a billboard or sign with a website for people to visit when they return home to search for more information. For example, in Fig. 207.2 a billboard asks the pointed question, "Where will you spent eternity?" with a reference to the biblical

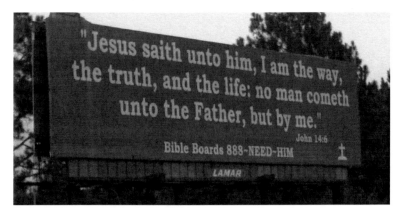

Fig. 207.3 A billboard just North of Panama City, Florida, promoting the idea that Jesus saves (Photo by Elizabeth Steiner, used with permission)

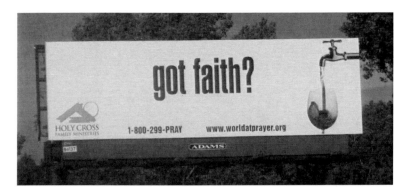

Fig. 207.4 A Holy Cross Family Ministries billboard asking people if they have faith in Jesus Christ (Photo by Daniel Olsen)

passage 2 Peter 2:9 listed in the bottom left corner of the billboard.[3] Supposedly this billboard is there to inspire people who see the sign to read the scripture and visit the website in order to be saved from an eternity in hell. Another example can be seen in Fig. 207.3, where a phone number under a scriptural quotation from John 14:6 leads one to contact *Need Him Ministries*, an organization that, according to their website (www.needhim.org), helps people to know where they are on their journey of faith and how they can move their lives closer to Jesus. Another example comes from I-94 in Michigan where a billboard simply reads "Got Faith"? with a picture of water turning into wine, reminiscent of the story from John 2:1–11 where Jesus miraculously turns the water into wine (Fig. 207.4). The listed phone number and website

---

[3] 2 Peter 2:9. "If this is so, then the Lord knows how to rescue the godly from trials and to hold the unrighteous for punishment on the day of judgment" (New International Version).

encourage interested persons to contact the Holy Cross Family Ministries whose mission is to strengthen families through prayer with a particular emphasis on using the rosary or Catholic prayer beads (www.hcfm.org).

There are other religiously-themed billboards and signs, however, which do not claim any affiliation with a church or religious organization. Rather, these anonymous billboards and signs are designed to promote a more general Christian theme to the passer-by. For example, in Fig. 207.5 a sign in Autauga County, Alabama encourages people to "Go to Church" or else "the Devil will get you!" Another example can be seen in Fig. 207.6, where the GodSpeaks campaign posts billboards

Fig. 207.5 Sign in Autauga County, Alabama encouraging passers-by to go to church (Photo by Daniel McGowin, used with permission)

Fig. 207.6 A photo of a GodSpeaks billboard as a part of a larger campaign to reach out to people on the move (Photo by Bethann Wong at Shamar.org, used with permission)

Fig. 207.7 A billboard in Minneapolis, Minnesota asking people to reflect on their origins and suggests that a belief in evolution is a de-evolution of humanity (Photo from www.whoisyourcreator.com, used with permission)

that contain one-line quotes attributed to God such as "Feeling lost? My book is your map" or "I love you" or "You think it is hot here?" or "Big bang theory, you've got to be kidding." Not only are these billboards meant to be humorous at times and generate discussion, but the lack of attributing these billboards to any one particular organization allows God's identity, not the anonymous sponsors, to become the focus on this advertising campaign (Tse 2005; McIntosh 2012).

These Christian-themed billboards and signs are, of course, a part of Christian evangelical efforts, where billboards and signs are used to relay information about a particular set of beliefs to others who do not hold their Christian beliefs. This is in part attributed to Christ's admonition to Christians, that is, to "…go and make disciples of all nations, baptizing them in the name of the Father and of the Son and of the Holy Spirit, and teaching them to obey everything I have commanded you" (Matthew 28:19–20). At the same time, the content of many religious billboards and signs that are sponsored by religious groups or organizations are specifically aimed at fighting against what they perceive as the decaying morals that are threatening America, such as homosexuality, evolution (Fig. 207.7), abortion, and atheism, with some billboards presenting the view that being anti-God is the equivalent of being anti-American (Fig. 207.8).

In addition to religious groups and organizations raising billboards and signs, some individuals also feel the call to create religious signs to evangelize to those passing by. For example, a newspaper article reported that a man in Kentucky felt that God wanted him to place a religiously-themed billboard on his properly adjacent to a major highway because he felt God was calling for him to do so. According to him, "I put these up for the people who want them on their property. They want the signs on their property… The Lord put this on me … to do, and it's not easy to do … and I don't think I put up as many as I should" (Jordan 2010). Another reason why these religious billboards are placed along America's highways may be because

Fig. 207.8 A billboard in Monongah, West Virginia suggesting that to be an atheist is the not only anti-American, but also treasonous (Photo publicly posted at http://nonprophetstatus.files.wordpress. com/2010/09/billboard.jpg)

erecting them is considered an expression of religious identity and belief. In essence, these billboards identify the owners of the land as being Christian believers; not only marking their location as consecrated, but also the region in which they live. As such, Christian-themed billboards and signs act to sacralize the American landscape in some small way (Beal 2005).

207.3   The Contested Nature of Religious Billboards

While signs and billboards along highways has become the norm, their content and placement of have been contested as early as 1924, with judges, highway authorities and civic organizations expressing concerns about the proliferation of certain bill-boards in regards to their placement, their effect on landscape aesthetics and their effect on driver safety (McFarland 1924). With regards to religious billboards and signs, their content has also recently been contested by atheist and humanist groups who are using the same medium to promote their non-religious discourses and views. A few of the many examples of atheist billboards are noted here. For example, in Austin, Texas, the Austin Coalition of Reason sponsored a billboard that asks passers-by, "Don't Believe in God? Join the Club" (Flynn 2010). The Freedom From Religion Foundation began a billboard blitz in Atlanta, Georgia; Louisville, Kentucky; New Orleans, Louisiana and Tulsa, Oklahoma in September of 2010 that advocated the separation of church and state. The billboards used phrases such as "Heathen's Greetings" and "Praise Darwin, Evolve Beyond Belief" to capture the attention of passers-by (Poole 2010). In March of 2011 the Center For Inquiry

Fig. 207.9 A billboard in Orange Country, California, sponsored by Backyard Skeptics asking people to reject religious superstition (Photo by Bruce Gleason, used with permission)

placed four "Living Without Religion" billboards in Indianapolis, Indiana that stated "You don't need God to hope, to care, to love, to live" (Matthews 2011). In Orange Country, California, a group called *Backyard Skeptics* recently unveiled new billboards reading "Make This A Better World—Reject All Religious Superstitions." Under this sentence is a list of 35 religious superstitions, including talking snakes, resurrection, Satan, Jihad, Voodoo, world floods, halal and salvation (Kramer 2011b) (Fig. 207.9).

The purposes of these billboards, broadly speaking, are both to contest the Christianization of America as well as to raise the awareness of those who do not believe in the existence of a higher power. As Fred Edwords, the national director of the *United Coalition of Reason* has stated: "The point of our ongoing nation-wide awareness campaign is to reach out to the millions of atheists and agnostics living in the United States…. Such non-theists sometimes don't realize there's a community for them because they're inundated with religious messages at every turn. We hope our effort will serve as a beacon and let them know they aren't alone" (Old 2011). As such, these billboards act as an outreach tool to bring atheists into communication with each other and to recognize that they are not alone in their skepticism about religion, as exemplified in the billboard in Columbia, South Carolina which reads "Don't believe in God? You are not alone" (Olds 2011) (Fig. 207.10). At the same time, the messages are also designed to help integrate atheist views into the mainstream. As the *Center For Inquiry* (2010) notes, the purpose of their billboards are to declare that "people who are [not] religious can live a meaningful life without God" and fight against the stigma of being a non-believer in American society who describes himself/herself as an atheist, agnostic or a humanist.

However, these billboards have raised the ire of some Christian individuals and groups, who accuse these humanist groups of doing what many non-religious groups have accused Christians of doing—proselytizing (Poole 2010). In some

Fig. 207.10 A billboard in that is part of an outreach to other atheists telling them that it is okay to be an atheist and that they are not alone (Photo from Wichita Coalition of Reason, used with permission)

instances the growth of humanist billboard sponsorship has been viewed as an attack on the Christian faith, leading to a series of "billboard wars" in some American urban centers. For example, there is a growing feud in Tampa Bay, Florida, between the local Christians and Atheists. When a retired business leader sponsored 10 billboards in the area condemning the idea of the separation of church and state, the *United Coalition of Reason* erected billboards in "retaliation," reading: "Are You Good Without God? Millions Are" (Clabough 2010). Another example comes from the Lincoln Tunnel that connects New York to New Jersey, where the *American Atheists* sponsored a billboard implying that the Christmas story was a myth ("You know it's a myth. This season celebrate reason") on the New Jersey side of the tunnel while the *Catholic League* sponsored a pro-Christmas billboard on the New York side ("You Know it's Real. This Season, Celebrate Jesus") (Dolan 2010; Kramer 2010).

As a part of this ongoing discourse battle over religion/non-religion via billboards, there has been a recent spate of cases where these "godless" billboards have been subjected to defacing. One such example is the vandalized billboard in Charlotte, North Carolina that was sponsored by the *North Carolina Secular Association*. The billboard contained the phrase "One Nation Indivisible," a purposeful play on the Pledge of Allegiance phrase "One Nation Under God." Someone or some group added the phrase "Under God" via spray paint to the bottom of the billboard. Ironically, the vandalized billboard just happened to be located along the Billy Graham Parkway! (Rivenbark 2010). Another example comes from Watertown, Wisconsin, where yellow paint was thrown onto a billboard sponsored by the *Freedom From Religion Foundation* that stated: "Enjoy Life Now, There Is No Afterlife" (Mehta 2010). Interestingly enough, however, the American Atheists billboard mentioned above was replaced recently by a billboard

Fig. 207.11 A billboard sponsored by American Atheists prior to being vandalized. Not only was this billboard criticized for its anti-bible stance, but also because it was viewed by some as racist (Photo from PA Nonbelievers, panonbelievers.org, used with permission)

sponsored by the Manhattan-based Times Square Church with the message "God is…" in large bold letters with more lightly-shaded sentence endings such as "alive," "good," and "incredible," and "aware of your struggle." Soon after this billboard was pasted someone spray painted the word "DEAD" at the bottom of the billboard, following the "God is dead" phrase made famous by German philosopher Friedrich Neitzsche (Kramer 2011a). In Harrisburg, Pennsylvania, a billboard sponsored by *American Atheists, Inc.* and *Pennsylvania Nonbelievers* was vandalized after having been up for only one day. The billboard in question depicted a slave wearing a spiked collar with a quote from Colossians 3:22, in which the apostle Paul wrote, "Slaves, obey your masters," and was a response to Pennsylvania lawmakers who had recently passed a bill making 2012 "the Year of the Bible" (Kramer 2012) (Fig. 207.11). These examples of competing billboards and the acts of vandalism to atheist billboards suggests that atheist billboards are seen by a segment of the American population as being subversive to the Christianizing of the American landscape and the history of America as a Christian nation and are not appropriate to be displayed in such a prominent location like highway billboards.

207.4   Conclusion

This chapter has briefly focused on religious billboards and signs along America's highways, looking at different types of billboards and signs and the reasons why they are found along America's highways (granted the reasons were in many ways speculative because of the lack of ethnographic research in this area). As noted above, highway billboards have become a space of competing discourses between evangelical Christians and atheist and humanist groups who see the billboard as a medium to promote their respective messages to the American public. Attention was also paid to the nature of atheist and humanist billboards as it relates to their anti-religious messages, which in turn have been contested by Christian organizations who feel that atheist/humanist billboards are at odds with the belief that the United States is a Christian nation. One must note, however, that the content of Christian highway billboards and signs can also be found on posters on buses and in subways, and are not only found in the United States but also in other countries. But even in these other countries these religiously-themed signs and billboards are being contested by atheist groups, other non-Christian organizations and in some cases governments. For example, in 2009 the British Humanist Association placed advertisements on 200 buses reading: "There's probably no God. Now stop worrying and enjoy your life" in an attempt to counteract posters put up by religious groups which threatened non-believers with "eternal damnation" (Beckford 2009). In Australia, a group called Aussie Christians erected billboards that read "Dear Aussie Muslims, glad you want to talk about Jesus. Love to chat more" in response to a series of billboards which stated that Jesus was "a prophet of Islam" and contained a website and a toll free number that people could access to find out more about Islam (Adams 2011). And in South Africa, the government's Advertising Standards Authority recently made a controversial ruling to remove a church billboard because it was seen to be anti-atheist (Gryboski 2012; MacLean 2012).

However, this chapter has only begun to touch upon the topic of the contested nature of billboards and signs along America's highways as it relates to the Christian and atheist discourses that grace America's highways. Much work remains to be done on this topic, particularly in light of the recent discussions about the future composition of American society regarding pluralism, secularism and fundamentalism (Yinger 1967; Skirbekk et al. 2010), where secular and humanist groups are attempt to carve out a niche within a resurgent evangelical discourse relating to a lack of cultural accommodation regarding atheist and non-Christian groups (Shibley 1996). With the growing political divide in the United States between evangelicals on the right and humanist groups on the left (Diamond 1995; Wald and Calhoun-Brown 2011), it will be interesting to see whether or not the content of these signs will focus more on political discourses rather than promoting Christian or atheist ideals. A mapping of the number and location of Christian and atheist billboards over time may prove fruitful in terms of where Evangelical and atheist groups have greater influence within the United States. As well, with future demographic shifts in the United States favoring Muslim, Hindus, Buddhists and other non-Christian faiths (Skirbekk et al. 2010), it will be interesting to see if these groups also promote their religious ideals through the medium of highway billboards and signs.

Along with religious and atheist-themed billboards and signs, there are many other aspects of the religious landscape in North America that geographers of religion and others could investigate. For example, mapping the locations of religious and atheistic billboards and signs throughout the United States, as noted above, would give one a better understanding of regional variations. Another topic would be the motivational reasons behind the creation and operation of these billboards and signs. Another example of research that could be done on the visible religious elements that potentially could impact people who observe them when traveling could be the examination of religious "big things" like the building of a life-sized replica of Noah's Ark along Interstate 68 in western Maryland by God's Ark of Safety Ministry or Field of the Woods in Murphy, North Carolina, which advertises the world's largest Ten Commandments (Watts 2004). How about religious theme parks such as the *Holy Land Experience* (www.theholylandexperience.com), a living biblical museum and a theme park in Florida that recreates various landscapes and events related to the Bible and the life of Jesus Christ (Rowan 2004; Beal 2005; Fafard and Lukens-Bull 2006; Lukens-Bull and Fafard 2007), or the *Creation Museum* (Chapman 2009; Maynard 2011) and its counterpart, *Ark Encounter*, a \$150 million biblical theme park (with some state funding) currently being built that will include another full-sized Noah's Ark (Prothero 2010); both are located in northern Kentucky. One could also expand the scope of inquiry to include the use of religious bumper stickers, religious graffiti, license plates and painted vehicles, or elements that create worship environments for travelers, such as wayside shrines, grottoes and chapels, drive-in chapels, or mobile chapels (MacDowell 1982). These mediums could also be examined from an atheist-humanist perspective as well. All of these topics, and probably more, should be under the purview of not only academics who have an interest on the intersections of religion and travel (for example, Badone and Roseman 2004; Timothy and Olsen 2006; Stausberg 2011), but also religious leaders and theologians who are concerned about how to effectively reach people who are "on the move" (Olsen 2011).

References

Adams, D. (2011). Christian group launches billboard campaign in response to poster linking Jesus and Islam. *Sight Magazine*. Accessed May 17, 2011, www.sightmagazine.com.au/stories/feature/billboard16.6.11.php

Badone, E., & Roseman, S. R. (2004). *Intersecting journeys: The anthropology of pilgrimage and tourism*. Chicago: University of Illinois Press.

Basten, F. E. (2007). *Great American billboards: 100 years of history by the side of the road*. New York: Ten Speed Press.

Beal, T. K. (2005). *Roadside religion: In search of the sacred, the strange, and the substance of faith*. Boston: Beacon Press.

Beckford, M. (2009). Atheist buses denying God's existence take to streets. *The Telegraph*. Accessed May 16, 2011, www.telegraph.co.uk/news/newstopics/howaboutthat/4141765/Atheist-buses-denying-Gods-existence-take-to-streets.html

Center for Inquiry. (2010). *About the Center for Inquiry*. Accessed September 14, 2011, www.centerforinquiry.net/about

Chapman, B. (2009). Kentucky's Creation Museum brings Genesis to life. *Washington Post*. Accessed September 14, 2011, www.washingtonpost.com/wpdyn/content/article/2009/06/12/AR2009061201547.html

Clabough, R. (2010). Atheists retaliate against Christian billboards. *The New American*. Accessed September 14, 2011, http://thenewamerican.com/culture/family/2941-heathens-retaliate-against-christian-billboards

Cresswell, T. (1996). *In place/out of place: Geography, ideology, and transgression*. Minneapolis: University of Minnesota Press.

Diamond, S. (1995). *Roads to dominion: Right-wing movements and political power in the United States*. New York: Guilford Press.

Dolan, L. (2010). Dueling billboards face off in Christmas controversy. *CNN*. Accessed May 17, 2012, http://religion.blogs.cnn.com/2010/11/30/dueling-billboards-face-off-in-christmas-controversy

Fafard, M., & Lukens-Bull, R. (2006). Next year in Jerusalem: A 21st century, medieval pilgrimage site. *Osprey Journal of Ideas and Inquiry*. 5. www.unf.edu/dept/scholarprograms/osprey-journal2006/FafardLukensBull.doc

Flynn, E. (2010). *An atheist billboard goes up in Austin*. Accessed September 14, 2011, http://eileenflynn.wordpress.com/2010/06/28/an-atheist-billboard-goes-up-in-austin/

Gryboski, M. (2012). Church billboard ruled 'anti-Atheist,' but even skeptics disagree. *The Christian Post*. Accessed May 16, 2011, www.christianpost.com/news/church-billboard-ruled-anti-atheist-but-even-skeptics-disagree-67372

Gudis, C. (2004). *Buyways: Billboards, automobiles, and the American landscape*. London/New York: Routledge.

Jakle, J. A., & Sculle, K. A. (2004). *Signs in America's auto age: Signatures of landscape and place*. Iowa City: University of Iowa Press.

Jordan, N. (2010). A sign from God. *BG Daily News*. Accessed September 14, 2011, http://bgdailynews.com/articles/2010/06/ 13/news/news1.txt

Kramer, H. (2010). Billboard wars: American Atheists vs. the Catholic League. *LA Atheism Examiner*. Accessed September 14, 2011, www.examiner.com/atheism-in-los-angeles/billboard-wars-american-atheists-vs-the-catholic-league

Kramer, H. (2011a). The "God Is" billboard that replaced atheist billboard in NJ has been vandalized. *LA Atheism Examiner*. Accessed September 14, 2011, www.examiner.com/atheism-in-los-angeles/the-god-is-billboard-that-replaced-atheist-billboard-nj-has-been-vandalized

Kramer, H. (2011b). New atheist billboard: "Reject All Religious Superstitions." *LA Atheism Examiner*. Accessed May 17, 2012, www.examiner.com/article/new-atheist-billboard-reject-all-religious-superstitions

Kramer, H. (2012). Atheist billboard with Bible quote on slavery vandalized on 1st Day Up. *LA Atheism Examiner*. Accessed May 17, 2012, www.examiner.com/article/atheist-billboard-with-bible-quote-on-slavery-vandalized-on-1st-day-up

Lukens-Bull, R., & Fafard, M. (2007). Next year in Orlando: (Re)creating Israel in Christian Zionism. *Journal of Religion & Society*, 9. http://moses.creighton.edu/JRS/pdf/2007-16.pdf

MacDowell, B. (1982). Religion on the road: Highway evangelism and worship environments for the traveler in America. *Journal of American Culture*, 5(4), 63–73.

MacLean, S. (2012). Church's "Offensive" anti-atheist billboard is banned by advertising watchdog. *Daily Mail*. Accessed May 17, 2012, www.dailymail.co.uk/news/article-2087778/Churchs-anti-atheist-billboard-banned-advertising-watchdog.html

Massey, D. (1995). The conceptualization of place. In D. Massey & P. Jess (Eds.), *A place in the world? Place, cultures and globalization* (pp. 45–86). Oxford: Open University/Oxford University Press.

Matthews, S. (2011). 'You don't need God' billboards sure to stir controversy. *The INDYChannel.com*. Accessed September 14, 2011, www.theindychannel.com/news/27062930/detail.html

Maynard, T. (2011). And on the eight day, God created rhetoricians: A case study of the Creation Museum. *Young Scholars in Writing*, 8. http://cas.umkc.edu/english/publications/youngscholarsinwriting/documents/8/8%20-%20And%20on%20the%20Eighth%20Day,%20God%20Created%20Rhetoricians.pdf

McFarland, J. H. (1924). The billboard and the public highways. *Annals of the American Academy of Political and Social Science, 116*, 95–101.

McIntosh Jr., W. F. (2012). Come, let us reason together. Together things look foolish to wise. *Morganton News Herald*. Accessed May 18, 2012, www2.morganton.com/lifestyles/2012/may/03/come-let-us-reason-together-things-look-foolish-wi-ar-2239799/

Mehta, H. (2010). Another atheist billboard vandalized. Accessed September 14, 2011, www.patheos.com/blogs/friendlyatheist/2010/08/31/another-atheist-billboard-vandalized/

Mitchell, D. (1995). There's no such thing as culture: Towards a reconceptualization of the idea of culture in geography. *Transactions of the British Institute of Geographers, 20*, 102–116.

Mitchell, D. (2000). *Cultural geography: A critical introduction*. Oxford: Blackwell.

Old, J. (2011). Atheist group places "Godless" billboards just before holidays. *WISTV.com*. Accessed May 17, 2011, www.wistv.com/story/16103513/athiest-group-places-godless-billboards-just-before-holidays?clienttype=printable

Olsen, D. H. (2011). Towards a religious view of tourism: Negotiating faith perspectives on tourism. *Journal of Tourism, Culture and Communication, 11*(1), 17–30.

Poole, S. M. (2010). Billboards shun religion, promote separation of church and state. *The Atlanta Journal-Constitution*. Accessed September 14, 2011, www.ajc.com/news/cobb/billboards-shun-religion-promote-608792.html

Prothero, S. (2010). My take: Where's the outrage over Noah's Ark Park? *CNN.com*. Accessed September 14, 2011, http://religion.blogs.cnn.com/2010/12/02/my-take-wheres-the-outrage-over-the-noah-ark-park/

Rivenbark, J. (2010). 'Godless' billboard in Charlotte defaced with "Under God'. *WISTV.com*. Accessed September 14, 2011, www.wistv.com/Global/story.asp?S=12722037

Rowan, Y. (2004). Repackaging the pilgrimage: Visiting the Holy Land in Orlando. In I. M. Yorke (Ed.), *Marketing heritage: Archaeology and the consumption of the past* (pp. 249–266). Walnut Creek: AltaMira Press.

Shibley, M. A. (1996). *Resurgent Evangelicalism in the United States: Mapping cultural change since 1970*. Columbia: University of South Carolina Press.

Sibley, D. (1995). *Geographies of exclusion: Society and difference in the West*. London: Routledge.

Skirbekk, V., Kaufmann, E., & Goujon, A. (2010). Secularism, fundamentalism or Catholicism: The religious composition of the United States to 2043. *Journal for the Scientific Study of Religion, 49*(2), 293–310.

Stausberg, M. (2011). *Religion and tourism: Crossroads, destinations and encounters*. London/New York: Routledge.

Timothy, D. J., & Olsen, D. H. (Eds.). (2006). *Tourism, religion and spiritual journeys*. London/New York: Routledge.

Tse, R. (2005). Billboards showcase messages 'from God' nationwide. *The Christian Post*. Accessed May 16, 2011, www.christianpost.com/news/billboards-showcase-messages-from-god-nationwide-13458

Wald, K. D., & Calhoun-Brown, A. (2011). *Religion and politics in the United States*. Lanham: Rowman & Littlefield Publishers, Inc.

Watts, J. W. (2004). Ten Commandments monuments and the rivalry of iconic texts. *Journal of Religion & Society, 6*, 1–12.

Yinger, J. M. (1967). Pluralism, religion and secularism. *Journal for the Scientific Study of Religion, 6*(1), 17–28.

York, J. (2007). *With signs following: Photographs from the southern religious roadside*. Jackson: The University Press of Mississippi.

Printed by Books on Demand, Germany